Advances in Intelligent and Soft Computing 79

Editor-in-Chief: J. Kacprzyk

Advances in Intelligent and Soft Computing

Further volumes of this series can be found on our homepage: springer.com

Vol. 64. E. Tkacz, A. Kapczynski (Eds.)
Internet – Technical Development and Applications, 2009
ISBN 978-3-642-05018-3

Vol. 65. E. Kącki, M. Rudnicki, J. Stempczyńska (Eds.)
Computers in Medical Activity, 2009
ISBN 978-3-642-04461-8

Vol. 66. G.Q. Huang, K.L. Mak, P.G. Maropoulos (Eds.)
Proceedings of the 6th CIRP-Sponsored International Conference on Digital Enterprise Technology, 2009
ISBN 978-3-642-10429-9

Vol. 67. V. Snášel, P.S. Szczepaniak, A. Abraham, J. Kacprzyk (Eds.)
Advances in Intelligent Web Mastering - 2, 2010
ISBN 978-3-642-10686-6

Vol. 68. V.-N. Huynh, Y. Nakamori, J. Lawry, M. Inuiguchi (Eds.)
Integrated Uncertainty Management and Applications, 2010
ISBN 978-3-642-11959-0

Vol. 69. E. Piętka and J. Kawa (Eds.)
Information Technologies in Biomedicine, 2010
ISBN 978-3-642-13104-2

Vol. 70. Y. Demazeau, F. Dignum, J.M. Corchado, J. Bajo Pérez (Eds.)
Advances in Practical Applications of Agents and Multiagent Systems, 2010
ISBN 978-3-642-12383-2

Vol. 71. Y. Demazeau, F. Dignum, J.M. Corchado, J. Bajo, R. Corchuelo, E. Corchado, F. Fernández-Riverola, V.J. Julián, P. Pawlewski, A. Campbell (Eds.)
Trends in Practical Applications of Agents and Multiagent Systems, 2010
ISBN 978-3-642-12432-7

Vol. 72. J.C. Augusto, J.M. Corchado, P. Novais, C. Analide (Eds.)
Ambient Intelligence and Future Trends, 2010
ISBN 978-3-642-13267-4

Vol. 73. J.M. Corchado, P. Novais, C. Analide, J. Sedano (Eds.)
Soft Computing Models in Industrial and Environmental Applications, 5th International Workshop (SOCO 2010), 2010
ISBN 978-3-642-13160-8

Vol. 74. M.P. Rocha, F.F. Riverola, H. Shatkay, J.M. Corchado (Eds.)
Advances in Bioinformatics, 2010
ISBN 978-3-642-13213-1

Vol. 75. X.Z. Gao, A. Gaspar-Cunha, M. Köppen, G. Schaefer, and J. Wang (Eds.)
Soft Computing in Industrial Applications, 2010
ISBN 978-3-642-11281-2

Vol. 76. T. Bastiaens, U. Baumöl, and B.J. Krämer (Eds.)
On Collective Intelligence, 2010
ISBN 978-3-642-14480-6

Vol. 77. C. Borgelt, G. González-Rodríguez, W. Trutschnig, M.A. Lubiano, M.Á. Gil, P. Grzegorzewski, and O. Hryniewicz (Eds.)
Combining Soft Computing and Statistical Methods in Data Analysis, 2010
ISBN 978-3-642-14745-6

Vol. 78. B.-Y. Cao, G. Wan, S. Chen, and S. Guo (Eds.)
Fuzzy Information and Engineering 2010, 2010
ISBN 978-3-642-14879-8

Vol. 79. A.P. de Leon F. de Carvalho, S. Rodríguez-González, J.F. De Paz Santana, and J.M. Corchado Rodríguez (Eds.)
Distributed Computing and Artificial Intelligence, 2010
ISBN 978-3-642-14882-8

Andre Ponce de Leon F. de Carvalho,
Sara Rodríguez-González,
Juan F. De Paz Santana, and
Juan M. Corchado Rodríguez (Eds.)

Distributed Computing and Artificial Intelligence

7th International Symposium

Springer

Editors

Andre Ponce de Leon F. de Carvalho
Department of Computer Science
University of Sao Paulo at Sao Carlos
Sao Carlos, SP, Brazil

Sara Rodríguez González
Department of Computing
Science and Control
Faculty of Science
University of Salamanca
Plaza de la Merced S/N
37008 Salamanca
E-mail: srg@usal.es

Juan F. De Paz Santana
Department of Computing
Science and Control
Faculty of Science
University of Salamanca
Plaza de la Merced S/N
37008 Salamanca
Spain
E-mail: fcofds@usal.es

Juan M. Corchado Rodríguez
Department of Computing
Science and Control
Faculty of Science
University of Salamanca
Plaza de la Merced S/N
37008 Sa
E-mail: corchado@usal.es

ISBN 978-3-642-14882-8 e-ISBN 978-3-642-14883-5

DOI 10.1007/978-3-642-14883-5

Advances in Intelligent and Soft Computing ISSN 1867-5662

Library of Congress Control Number: 2010932153

Typeset & Cover Design: Scientific Publishing Services Pvt. Ltd., Chennai, India.

Printed on acid-free paper
5 4 3 2 1 0
springer.com

Preface

The International Symposium on Distributed Computing and Artificial Intelligence (DCAI´10) is an annual forum that brings together past experience, current work and promising future trends associated with distributed computing, artificial intelligence and their application to provide efficient solutions to real problems. This symposium is organized by the Biomedicine, Intelligent System and Educational Technology Research Group (http://bisite.usal.es/) of the University of Salamanca. The present edition has been held at the Polytechnic University of Valencia, from 7 to 10 September 2010, within the Congreso Español de Informática (CEDI 2010). Technology transfer in this field is still a challenge, with a large gap between academic research and industrial products. This edition of DCAI aims at contributing to reduce this gap, with a stimulating and productive forum where these communities can work towards future cooperation with social and economical benefits. This conference is the forum in which to present application of innovative techniques to complex problems. Artificial intelligence is changing our society. Its application in distributed environments, such as internet, electronic commerce, environment monitoring, mobile communications, wireless devices, distributed computing, to cite some, is continuously increasing, becoming an element of high added value with social and economic potential, both industry, life quality and research. These technologies are changing constantly as a result of the large research and technical effort being undertaken in universities, companies. The exchange of ideas between scientists and technicians from both academic and industry is essential to facilitate the development of systems that meet the demands of today's society.

This symposium continues to grow and prosper in its role as one of the premier conferences devoted to the quickly changing landscape of distributed computing, artificial intelligence and the application of AI to distributed systems. This year's technical program presented both high quality and diversity, with contribution in well established and evolving areas of research. This year, 123 papers were submitted from over 11 different countries, representing a truly "wide area network" of research activities. The DCAI technical program has 88 selected papers (74 long papers, 10 short papers and 4 doctoral consortium).

We thank the Local Organization members and the Program Committee members for their hard work, which was essential for the success of DCAI´10.

Salamanca
September 2010

Andre Ponce de Leon F. de Carvalho
Sara Rodríguez González
Juan F. De Paz Santana
Juan M. Corchado Rodríguez

Organization

General Chairs

Sigeru Omatu	Osaka Institute of Technology (Japan)
José M. Molina	Universidad Carlos III de Madrid (Spain)
James Llinas	State University of New York (USA)

Scientific Chair

Andre Ponce de Leon F. De Carvalho	University of Sao Paulo at Sao Carlos (Brazil)

Organizing Committee

Juan M. Corchado (**Chair**)	Universidad de Salamanca (Spain)
Sara Rodríguez-González (**Cochair**)	Universidad de Salamanca (Spain)
Juan F. De Paz (**Cochair**)	Universidad de Salamanca (Spain)
Javier Bajo	Universidad de Salamanca (Spain)
Dante I. Tapia	Universidad de Salamanca (Spain)
Cristian I. Pinzón	Universidad de Salamanca (Spain)
Rosa Cano	Universidad de Salamanca (Spain)
Belen Pérez-Lancho	Universidad de Salamanca (Spain)
Angélica González-Arrieta	Universidad de Salamanca (Spain)
Vivian F. López	Universidad de Salamanca (Spain)
Ana de Luis	Universidad de Salamanca (Spain)
Ana B. Gil	Universidad de Salamanca (Spain)
Fernando de la Prieta Pintado	Universidad de Salamanca (Spain)

Carolina Zato Domínguez	Universidad de Salamanca (Spain)
Mª Dolores Muñoz Vicente	Universidad de Salamanca
Jesús García Herrero	Universidad Carlos III de Madrid (Spain)

Scientific Committee

Adriana Giret	Universidad Politecnica de Valencia (Spain)
Agapito Ledesma	University Carlos III of Madrid (Spain)
Alberto Fernández	Universidad Rey Juan Carlos (Spain)
Alicia Troncoso Lora	Universidad Pablo de Olavide, Sevilla (Spain)
Álvaro Herrero	Universidad de Burgos (Spain)
Ana Carolina Lorena	Universidade Federal do ABC (Brazil)
Ana Cristina García Bicharra	Universidad Federal Fluminense (Brasil)
Andre Coelho	Universidade de Fortaleza (Brazil)
Ángel Alonso	Universidad de León (Spain)
Ângelo Costa	Universidade do Minho (Portugal)
Antonio Berlanga de Jesús	Universidad Carlos III, Madrid (Spain)
Antonio Moreno	Universidad Rovira y Virgili (Spain)
Araceli Sanchos	Universidad Carlos III de Madrid (Spain)
B. Cristina Pelayo García-Bustelo	Universidad de Oviedo (Spain)
Beatriz López	Universitat de Girona (Spain)
Bogdan Gabrys	Bournemouth University (UK)
Bruno Baruque	Universidad de Burgos (Spain)
Carina González	Universidad de la Laguna (Spain)
Carlos Carrascosa	Universidad Politécnica de Valencia (Spain)
Carlos Soares	Universidade do Porto (Portugal)
Carmen Benavides	Universidad de León (Spain)
Choong-Yeun LIONG	University Kebangsaan (Malaysia)
Daniel Gayo Avello	Universidad de Oviedo (Spain)
Daniel Glez-Peña	Universidad de Vigo (Spain)
Darryl Charles	Univerty of Ulster (North Irland)
David de Francisco	Telefónica I+D (Spain)
David Griol Barres	Universidad Carlos III de Madrid
Davide Carneiro	Universidade do Minho (Portugal)
Deris Safaai	University Teknologi Malaysia (Malaysia)
Eduardo Hruschka	Universidade de Sao Paulo (Brazil)
Eladio Sanz	Universidad de Salamanca (Spain)
Eleni Mangina	University College Dublin (Ireland)
Emilio Corchado	Universidad de Burgos (Spain)
Eugénio Oliveira	Universidade do Porto (Portugal)
Evelio J. González	Universidad de la Laguna (Spain)
Faraón Llorens Largo	Universidad de Alicante (Spain)
Fernando Díaz	Universidad de Valladolid (Spain)

Fidel Aznar Gregori	Universidad de Alicante (Spain)
Florentino Fdez-Riverola	Universidad de Vigo (Spain)
Francisco Pujol López	Universidad de Alicante (Spain)
Fumiaki Takeda	Kochi University of Technology (Japan)
Gary Grewal	University of Guelph (Canada)
Germán Gutiérrez	University Carlos III of Madrid (Spain)
Helder Coelho	Universidade de Lisboa (Portugal)
Hideki Tode	Osaka Prefecture University (Japan)
Ivan López Arévalo	Lab. of Information Technology Cinvestav-T.(Mexico)
Javier Carbó	Universidad Carlos III de Madrid (Spain)
Javier Martínez Elicegui	Telefónica I+D (Spain)
Jesús García Herrero	Universidad Carlos III, Madrid (Spain)
Joao Gama	Universidade do Porto (Portugal)
José M. Molina	Universidad Carlos III de Madrid (Spain)
José R. Méndez	Universidad de Vigo (Spain)
José R. Villar	Universidad de Oviedo (Spain)
José V. Álvarez-Bravo	Universidad de Valladolid (Spain)
Juan A. Botia	Universidad de Murcia (Spain)
Juan Manuel Cueva Lovelle	Universidad de Oviedo (Spain)
Juan Gómez Romero	Universidad Carlos III de Madrid
Juan Pavón	Universidad Complutense de Madrid (Spain)
Kazutoshi Fujikawa	Nara Institute of Science and Technology (Japan)
Lourdes Borrajo	Universidad de Vigo (Spain)
Luis Alonso	Universidad de Salamanca (Spain)
Luis Correia	Universidad de Lisboa (Portugal)
Luis F. Castillo	Universidad Autónoma de Manizales (Colombia)
Luís Lima	Polytechnic of Porto, (Portugal)
Manuel González-Bedia	Universidad de Zaragoza (Spain)
Manuel Resinas	Universidad de Sevilla (Spain)
Marcilio Souto	Universidade Federal do Rio Grande do Norte (Brazil)
Margarida Cardoso	ISCTE (Portugal)
Maria del Mar Pujol López	Universidad de Alicante (Spain)
Michifumi Yoshioka	Osaka Prefecture University (Japan)
Miguel Ángel Patricio	Universidad Carlos III, Madrid (Spain)
Miguel Rebollo	Universidad Politecnica de Valencia (Spain)
Naoki Mori	Osaka Prefecture University (Japan)
Nora Muda	National University of Malaysia (Malaysia)
Norihiko Ono	University of Tokushima (Japan)
Norihisa Komoda	Osaka University (Japan)
Oscar Sanjuan Martínez	Universidad de Oviedo (Spain)
Paulo Novais	Universidade do Minho (Portugal)
Pawel Pawlewski	Poznan University of Technology (Poland)
Rafael Corchuelo	Universidad de Sevilla (Spain)

Ramón Rizo	Universidad de Alicante (Spain)
Ricardo Campello	Universidade de Sao Paulo (Brazil)
Ricardo Costa	Polytechnic of Porto (Portugal)
Rodrigo Mello	Universidade de Sao Paulo (Brazil)
Rubén Fuentes	Universidad Complutense de Madrid (Spain)
Rui Camacho	Universidade do Porto (Portugal)
Teresa Ludermir	Universidade Federal de Pernambuco (Brazil)
Tzai-Der Wang	Cheng Shiu University (Taiwan)
Vicente Botti	Universidad Politécnica de Valencia (Spain)
Vicente Julián	Universidad Politécnica de Valencia (Spain)
Victor J. Sosa-Sosa	LTI Cinvestav (México)
Zbigniew Pasek	IMSE/University of Windsor (Canada)

Contents

Feature Selection Method for Classification of New and Used Bills*

Sigeru Omatu, Masao Fujimura, and Toshihisa Kosaka

Abstract. According to the progress of office automation, it becomes important to classify new and old bills automatically. In this paper, we adopt a new type of sub-band adaptive digital filters to extract the feature for classification of new and fatigued bills. First, we use wavelet transform to resolve the measurement signal into various frequency bands. For the data in each band, we construct an adaptive digital filter to cancel the noise included in the frequency band. Then we summarize the output of the filter output in each frequency band. The experimental results show the effectiveness of the proposed method to remove the noise.

Keywords: bill classification, adaptive digital filter, sub-band analysis, acoustic diagnosis system, neural networks.

1 Introduction

By the progress of the office automation, it is important to classify the bill into new one and fatigued one in ATM or a bank machine. Basically, two methods are useful to do it. One is to use images of bills and another is to use acoustic data of bills. Former one has been well developed but there are too many variations for classification and the reliability is not so high [1]. Thus, we consider the case to use the acoustic data measured from a bank machine. A bank machine makes some sounds when the bill has been passed through it. But various kinds of noises are included at the same time such as sounds of gears, a motor, and background noise.

Sigeru Omatu, Masao Fujimura, and Toshihisa Kosaka
Faculty of Engineering, Osaka Institute of Technology,
5-16-1, Omiya, Asahi-ku, Osaka, 535-8585,Osaka, Japan
e-mail: omatu@rsh.oit.ac.jp
http://www.oit.ac.jp

* This work was supported by Research Project-in-Aid for Scientific Research (2010) No. 20360178 in JSPS, Japan.

A.P. de Leon F. de Carvalho et al. (Eds.): Distrib. Computing & Artif. Intell., AISC 79, pp. 1–8.
springerlink.com

In order to classify a bill into usable or fatigued one, we must extract only bill sound from the noisy measurement data. In this paper we will remove the noises by using adaptive digital filters and a neural network. First trial of the approach was done by Kang et al. [2] where two-stage adaptive digital filters were introduced. In a real bank machine, various types of acoustic noises were included stated above. Thus, two-stage adaptive filters might not be sufficient to remove the noises. Furthermore, it is difficult to find a suitable order of the adaptive digital filters.

The proposed method has three steps. First, we decompose the measurement signals into low frequency part and high frequency part by using discrete wavelet transform. Then for each frequency band, we apply the adaptive digital filter. After that we compose the filtered signals obtained in each frequency band and then we make the nonlinear predictor by using neural network of layered type. The output of the layered neural network is assumed to be the acoustic data of the bill. Finally, we apply the procedure to real data obtained by a bank machine to show the effectiveness of the proposed approach.

2 Adaptive Digital Filter

The adaptive digital filter has been proposed by Widraw [3] and it was applied to real data processing to reduce noises. The basic form of the adaptive digital filter is shown in Fig. 1.

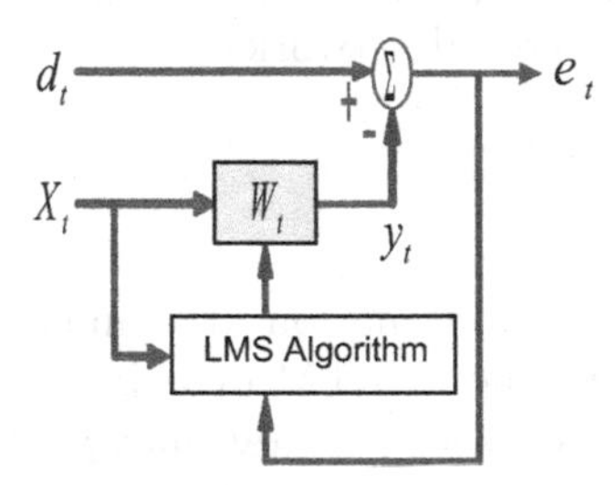

Fig. 1 General structure of adaptive digital filter.

In this figure d_t is a desired signal at time t and the input and the output of the filter are X_t and y_t, respectively which are given by

$$X_t = [x_t, x_{t-1}, \ldots, x_{t-N+1}]^T, \qquad y_t = \sum_{k=0}^{N-1} w_{k,t} x_{t-k}, \tag{1}$$

where $w_{k,t}$ is a weight of the filter and W_t is given by

$$W_t = [w_{0,t}, w_{1,t}, \ldots, w_{N-1,t}]^T. \tag{2}$$

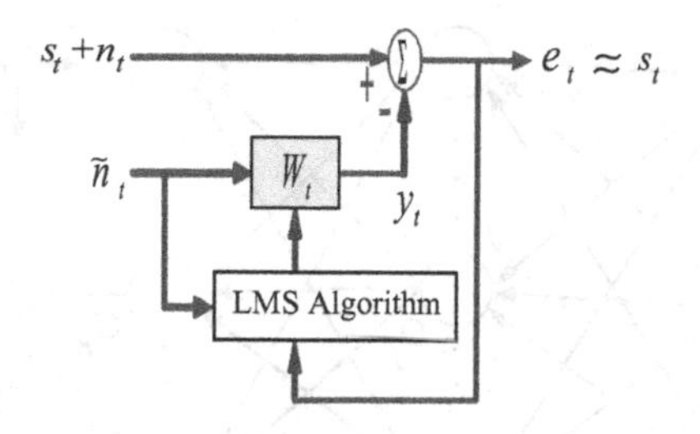

Fig. 2 Noise canceller by adaptive digital filter.

Furthermore, we define the error e_t of the filter by

$$e_t = d_t - y_t. \tag{3}$$

Using the least mean square(LMS) algorithm by [3], we can obtain the following recursive relation about W_t

$$W_{t+1} = W_t + 2ue_tX_t \tag{4}$$

where u is a constant.

If we set the desired signal d_t as

$$d_t = x_{t+1}, \tag{5}$$

then it means the one-time ahead predictor.

If we set the desired signal d_t and the input X_t as

$$d_t = s_t + n_t, \qquad X_t = [\tilde{n}_t, \tilde{n}_{t-1}, \ldots, \tilde{n}_{t-N+1}]^T, \tag{6}$$

then it is called as a noise canceller where $\tilde{n}$ means similar to n with high correlation. The noise canceller will cancel the noise n_t by using the correlation of $\tilde{n_t}$ and the output error e_t will approach to the sinal s_t as shown in Fig. 2. Noise cancelling techniques have been adopted in many application problems, for example, earphone, signal separation, acoustic diagnosis, *etc.* [3].

The noise property is not constant and changing along time in real environment as shown in the later. Thus, adaptive digital filtering is preferable although the order of the filter should be determined trial error. According to the change of the environment the order must be adjusted as well as the coefficients of the filter.

3 Error Back-Propagation Algorithm

The error back-propagation (BP) algorithm has been well-known since it was proposed by Rumerhart et al. [6] in 1985. The self-tuning PID being described in detail later is based on the derivation of this algorithm. First,

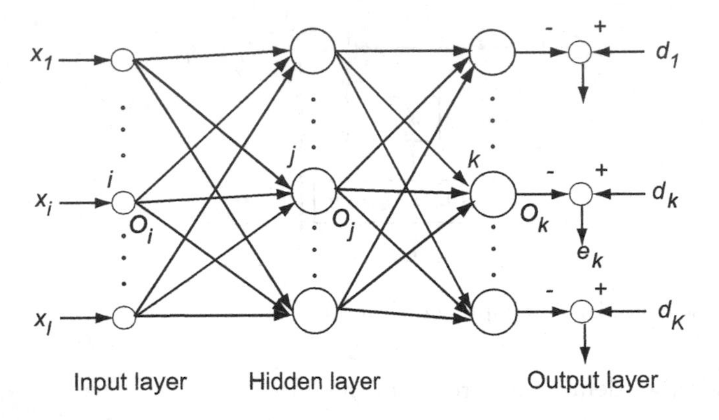

Fig. 3 Structure of a layered neural network.

we will explain the derivation of the BP algorithm in compact way. The form of a neural network described by Fig. reflayerednn is called a layered neural network since they have more than three layers which are called input layer, hidden layer, and output layer. Outputs of neurons in the input layer are the input data which should be processed. We assume that numbers of neurons in the input, hidden, and output layers are I, J, and K, respectively.

In Fig. 3, large circles denote neurons and each neuron, for example, neuron j can be described by the following nonlinear input-output relation:

$$O_j = f(\text{net}_j), \qquad \text{net}_j = \sum_{i=1}^{I} w_{ji} O_i - \theta_j, \qquad f(x) = \frac{1}{1 + \exp(-x)} = sigmoid(x). \tag{7}$$

where O_j denotes the output of neuron j, w_{ji} denotes the connection weight from a neuron i to a neuron j, θ_j is a threshold value of neuron j.

Note that the output of a neuron is limited within 0 to 1 since $f(x) \in [0,1]$. If we assume that $O_0 = -1$ and $w_{j0} = \theta_j$, then we can rewrite net_j as follows:

$$O_j = f(\text{net}_j), \qquad \text{net}_j = \sum_{i=0}^{I} w_{ji} O_i, \qquad f(x) = \frac{1}{1 + \exp(-x)}. \tag{8}$$

Then we will compare the output $\{O_k\}$ with the desired value $\{d_k\}$ for each $k, k = 1, 2, \ldots, K$ and if there are large discrepancies, we will correct the weighting functions, $w_{ji} and w_{kj}$ such that the following error function E will be decreased.

$$E = \frac{1}{2} \sum_{k=1}^{K} e_k^2, \qquad e_k = d_k - O_k. \tag{9}$$

Using the gradient search, the minimizing cost of E is given by the following relation(the error back-propagation algorithm):

$$\delta_k = e_k O_k (1 - O_k). \tag{10}$$

$$\delta_j = \sum_{k=1}^{K} \delta_k w_{kj} O_j (1 - O_j) \quad k = 1, 2, \ldots, K, \quad j = 0, 1, \ldots, N \tag{11}$$

$$\Delta w_{kj}(\text{new}) = \eta \delta_k O_j + \alpha \Delta w_{kj}(\text{old}) \ , \ j = 0, 1, \ldots, N, \ k = 1, 2, \ldots, K \tag{12}$$

$$\Delta w_{ji}(\text{new}) = \eta \delta_j O_i + \alpha \Delta w_{ji}(\text{old}) \ , \ i = 0, 1, \ldots, M, \ j = 1, 2, \ldots, N \tag{13}$$

where the first term and second term of (12) and (13) are called the learning term and the momentum terms, respectively and η and α are called learning rate and momentum rate, respectively.

4 Acoustic Data from a Bank Machine

We will consider classification of new and fatigued bills based on the acoustic data which have been measured near the bank machine as shown in Fig. 4. In this case, there are various kinds of noises in the data since the bank note machine includes many gears and transmission connecters derived by a motor as well as background noise [5] as shown in Fig. 5.

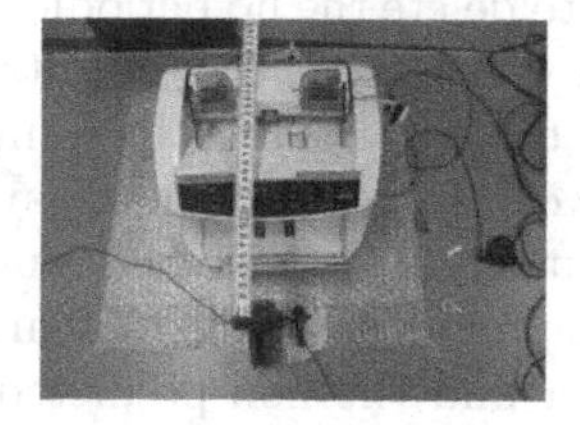

Fig. 4 Bank machine used here.

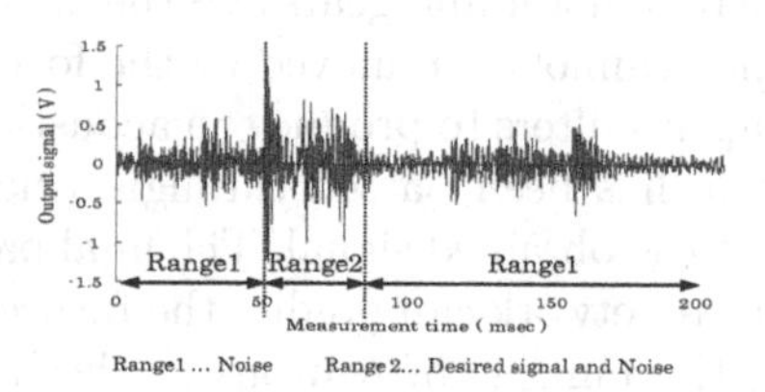

Fig. 5 Acoustic data from a bank machine.

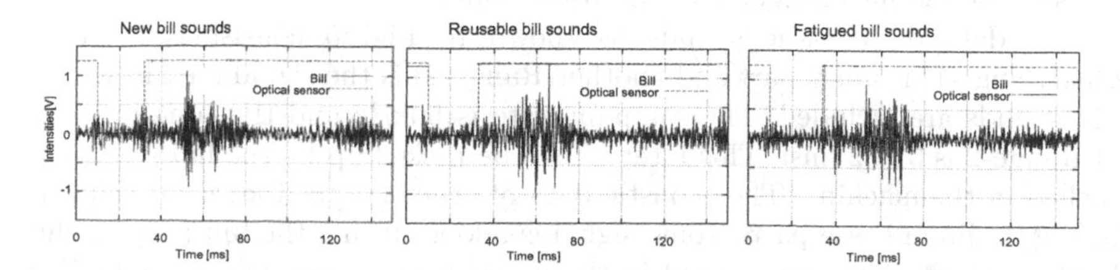

Fig. 6 Various acoustic data from a bank machine.

5 Noise Reduction Method

In order to classify the bill into new and fatigued ones based on the acoustic data, we must delete those noises. We propose a sub-band adaptive digital

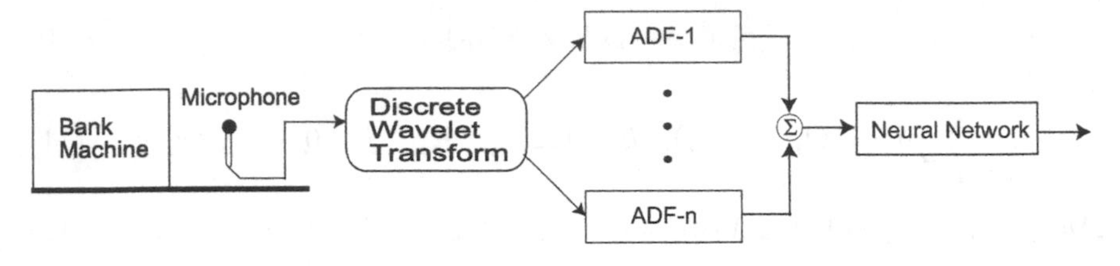

Fig. 7 Noise reduction method.

filter to reduce those noises. The proposed algorithm is shown in Fig. 7 where two microphones are located front and back places of the bank machine, ADF is adaptive digital filter, and NN is a neural network.

From the bank machine the acoustic sound such as Fig. 5 has been measured from each microphone. In order to construct the adaptive digital filter, we need many trials to determine the order of the filter since the acoustic data has broad-band frequencies and it is difficult to tune the parameters of only one adaptive digital filter. Thus, we propose to divide the measurement signals into some frequency bands by using discrete wavelet transform where Daubechies basis functions were adopted. After that for each spectral band we construct two types adaptive digital filters. One is prediction type and another is noise cancelling type. Former is used to reduce periodic-like noises due to a motor and gears and the latter is used to delete the no periodic noises which cannot be removed by the former. Then we compose the outputs from adaptive filters to produce an acoustic sound of the bill. Finally, the composed signal has been passed through a neural network to remove the noisy part from the obtained signal. The final part is to extract pure noise part since the neural network can predict the future values as precise as possible by increasing the number of neurons in the hidden layer and the non-predicted part could be regarded as noise. The bank machine(GFR-X120) made in Japan in Fig. 4 is able to classify the bills into several kinds of banknotes with thirty pieces per second based on the neuro-pattern algorithm although it could not classify the bill into fatigue one and usable one.

The data of acoustic sounds as shown in Fig. 5 consist two parts. One(Range-1) is noisy part and another(Range-2) is the signal part in which bill sounds are included. The sampling rate is 50kHz and the time interval of Range-2 is 51-82 [ms]. This interval can be picked up by the hardware attached in the machine. The acoustic data of other fatigued levels are shown in Fig. 6. From these parts, some signal is added during the bill passing the bank machine. The noise added in the measurement data might be similar in Range-1 although they are not the same because the bank machine will produce a little bit different sounds by the rotation change under loads. As we could expect that the signals in Range-1 and Range-2 are correlated each other, we adopted the noise cancelling technique in this paper. But measurement data include various noises due to the motor and gears, which could be assumed to be periodic and the adaptive digital filter could reduce such components very easy. From these facts, we use the following steps to reduce the noises in the measurement data of bill sounds:

Step 1. Decompose the measurement data into low frequency part and high frequency part by using discrete wavelet transform.

Step 2. Construct two-type adaptive digital filters: first, the prediction type and second, noise castellation type for each frequency band.

Step 3. Compose the filter output from each frequency band.

Step 4. Construct the nonlinear predictor using a neural network to reduce nonlinear noise component.

6 Experimental Results

In order to classify the bill into new and fatigued ones based on the acoustic data, we must delete those noises. We propose a sub-band adative digital filter to reduce those noises. The parameters used for the simulation is given in Table 1.

Comparing spectrum of measurement data between Range-1 and Range-2 shown in Fig. 5, we can see that both ranges have similar spectrum, which means that the separation between noise and bill sounds is a difficult work. Results by the method by Kang et al.([2]) and the proposed method are shown in Fig. 8 and Fig. 9, respectively. The conventional method by Kang

Table 1 Various parameters.

Specification	Values
learning times of ADF [s]	4,700
order of ADF	15
order of input delay for NN	1
learning time of NN [s]	100,000
No. of neurons(input layer)	12
No. of neurons(hidden layer)	6
No. of neurons(output layer)	1
learning coefficient of NN	0.002
momentum coefficient	0.01

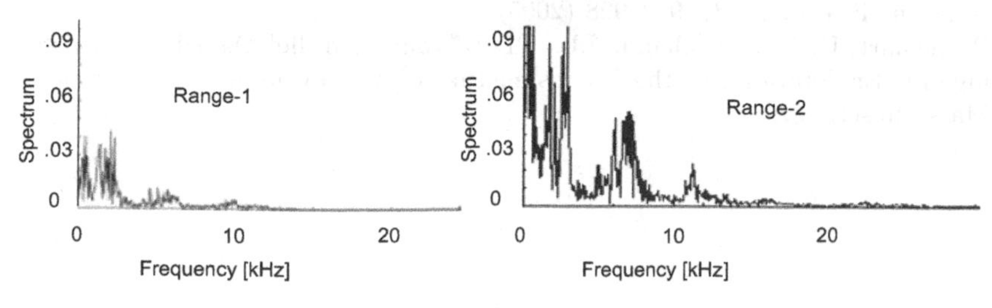

Fig. 8 Range spectrum.

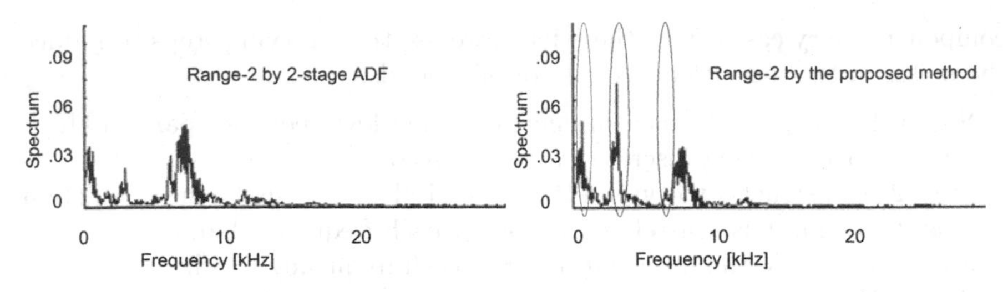

Fig. 9 Comparison of spectrum.

et al.([2]) is based on two-stage adaptive digital filters. This is effective when a simple noise is added to the signal. But for the present situation with many noise factors is not so good compared with the proposed method.

7 Conclusions

We have considered the noise reduction method from very noisy acoustic data obtained from a bank machine based on the proposed method. The simulation results show the effectiveness to reduce the noise from the noisy measurement data. From now on, we consider how much improvement of the classification rate of the new and fatigued bills could be achieved.

References

1. Teranishi, M., Omatu, S., Kosaka, T.: Continuous Fatigue Level Estimation for the Classification of Fatigued Bills Based on an Acoustic Signal Feature by a Supervised SOM. Artificial Life and Robotics 13(2), 547–550 (2009)
2. Kang, D., Omatu, S., Yoshioka, M.: New and Used Bills Classification Using Neural Networks. IEICE Trans. on Fundamentals of Electronics, Communications, and Computer Sciences E82A(8), 1511–1516 (1999)
3. Widrow, B., Stearns, S.D.: Adaptive Signal Processing. Prentice-Hall, New Jersey (1985)
4. Mallat, S.: Wavelet Tour of Signal Processing. Acaemic Press, New York (1998)
5. Wang, B., Omatu, S., Abe, T.: Identification of the Defective Transmission Devices Using the Wavelet Transfom. IEEE Trans. on Pattern Analysis and Machine Intelligence 27(3), 919–928 (2005)
6. Rumelhart, D.E., McClelland, J.L.: PDP Group: Parallel Distributed Processing. In: Explorations in the Microsteucture of Cognition, vol. 1. MIT Press, Massachusetts (1987)

Otoliths Identifiers Using Image Contours EFD

R. Reig-Bolaño, Pere Marti-Puig, S. Rodriguez, J. Bajo,
V. Parisi-Baradad, and A. Lombarte

Abstract. In this paper we analyze the characteristics of an experimental otolith identification system based on image contours described with Elliptical Fourier Descriptors (EFD). Otoliths are found in the inner ear of fishes. They are formed by calcium carbonate crystals and organic materials of proteic origin. Fish otolith shape analysis can be used for sex, age, population and species identification studies, and can provide necessary and relevant information for ecological studies. The system we propose has been tested for the identification of three different species, *Engraulis encrasicholus*, *Pomadasys incisus* belonging to the different families (Engroulidae and Haemolidae), and two populations of the species *Merluccius merluccius* (from CAT and GAL) from the family Merlucciidae. The identification of species from different families could be carried out quite easily with some simple class identifiers -i.e based on Support Vector Machine (SVM) with linear Kernel-; however, to identify these two populations that are characterized by a high similarity in their global form; a more accurate, and detailed shape representation of the otoliths are required, and at the same time the Otolith identifiers have to deal with a bigger number of descriptors. That is the principal reason that makes a challenging task both the design and the training of an otolith identification system, with a good performance on both cases.

Keywords: Otoliths Analysis, Elliptic Fourier Descriptors (EFD), Shape Analysis, Image Processing, Pattern Recognition, Pattern Classification.

R. Reig-Bolaño and Pere Marti-Puig
Department of Digital Information and Technologies, University of Vic (UVIC)
C/ de la Laura, 13, E-08500, Vic, Catalonia, Spain
e-mail: ramon.reig@uvic.cat, pere.marti@uvic.cat

S. Rodriguez and J. Bajo
BISITE Group (USAL), Pza. de la Merced s/n, 370008 Salamanca, Spain

V. Parisi-Baradad
Department of Electronic Engineering , Politechnical Univerity of Catalonia (UPC)

A. Lombarte
ICM-CSIC, Passeig Marítim 37-49, 08003, Catalonia, Spain

A.P. de Leon F. de Carvalho et al. (Eds.): Distrib. Computing & Artif. Intell., AISC 79, pp. 9–16.
springerlink.com

1 Introduction

Otoliths are found in the inner ear of fishes. They are formed by calcium carbonate crystals and organic materials of proteic origin. The shapes and proportional sizes of the otoliths vary with fish species. In general, fish from highly structured habitats such as reefs or rocky bottoms will have larger otoliths than fish that spend most of their time swimming at high speed in straight lines in the open ocean. Flying fish have unusually large otoliths, possibly due to their need for balance when launching themselves out of the water to "fly" in the air. Often, the fish species can be identified from distinct morphological characteristics of an isolated otolith. It is also widely accepted that fish otolith shape analysis can be used for sex, age, population and species identification studies, and can provide necessary and relevant information for ecological studies. Successful stock discriminations using otolith shape analyses have been reported by Casselman et al. (1981), Bird et al. (1986), Campana & Casselman (1993). Moreover the variability in the left sagitta otolith shape has been related to genetic, ontogenetic and environmental factors (Lombarte et al. 2003). Several methods are used to describe and compare form in morphological studies, such as ratios of linear dimensions, Euclidean distance matrix analysis, eigenshape analysis, and several variations of Fourier analysis (Chen et al. 2000; Iwata 2002). The traditional approach to contour feature extraction is based on expanding the contour into a two-dimensional series by means of elliptic Fourier analysis (Kuhl and Giardina 1982). The data are reduced by selecting only a set of coefficients for the expansion. Appropriate feature extraction and compaction is essential for obtaining good results in automatic classification systems. One of the major problems when complex data sets are classified is the number of variables involved. In pattern recognition and image processing, feature extraction is a form of dimensionality reduction. Although Elliptic Fourier Descriptors (EFD) can represent any outline when a large number of coefficients are used, in practical applications the number of harmonics is limited to a certain value, i.e 40 coefficients (Parisi et al. 2005; Tracey et al. 2006); therefore, EFDs only represent an approximation to the original contour. This problem comes to light when there is a close similarity between different species or populations from the same species. In these cases the design and the training of thc identifier is a challenging task. In this paper we will show some results on both cases.

2 Materials and Methods

The test material was taken from the AFORO database (http://aforo.cmima.csic.es), a web based environment for shape analysis of fish otoliths (Lombarte et al. 2006). The database is regularly updated and at present (05/20/2010) it contains a total of 2874 high resolution images corresponding to 841 species and 168 families from the Western Mediterranean Sea, Weddell Sea and Antartic Peninsula, Southwestern Atlantic (Uruguay and Argentina, Northwestern Atlantic and Gulf of Mexico), Northeastern Atlantic (From Senegal to North Sea), Southeastern Atlantic (Namibia), Indic (Tanzania) and Pacific (New

Caledonia, Alaska, Canada, Perú, and Chile). The sagitta is the otolith with the largest morphologic variability and therefore is the most studied. This database provides an open online catalogue of otolith images, and is associated with a shape analysis module that uses pattern recognition techniques and applies Fourier transform (FT), curvature scale space (CSS) and wavelet transform (WT). The site also implements a search in an identification engine, using query images of otoliths. The method presented in this paper has been tested with otoliths from three different species, *Engraulis encrasicholus*, *Pomadasys incisus* belonging to the different families (Engroulidae and Haemolidae), and two populations of the species *Merluccius merluccius* (from CAT and GAL) from the family Merlucciidae. These groups are characterized by having high similarity between species; therefore, detailed shape analyses of their otoliths can help to identify and discriminate morphologically close species. But, at the same time, this makes the identifier more difficult to design and train.

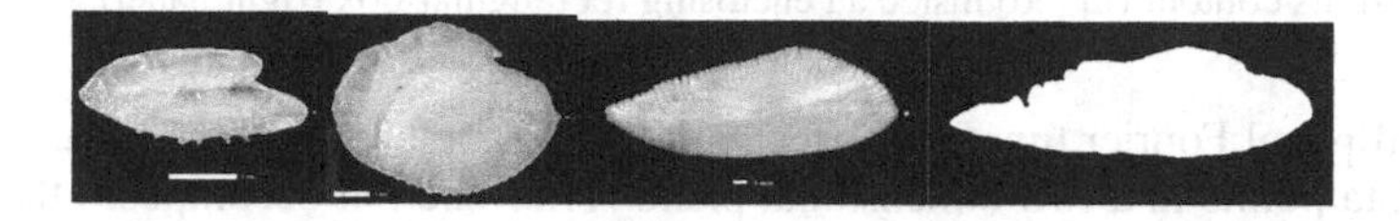

Fig. 1 (from left to right) Left sagitta Otoliths from *Engraulis encrasicholus, Pomadasys incisus, Merluccius merluccius* CAT (Catalonia) and *Merluccius merluccius* CAN (Canada)

The identifiers are based on a simple Support Vector Machine (SVM) with a linear kernel, like the one described in the tutorial of Burges (1998), trained for each one of the classes we want to classify, the output of the system will be one of the following classes: ENG *–Engraulis-*, POM *–Pomadasys-*, CAT *–Merluccius* CAT-, GAL *–Merluccius* GAL-, NOT –not determinate-. The output will be decided from the binary outputs of the SVMs. The descriptors used in the identifiers will be obtained from the EFD of the image contour under test.

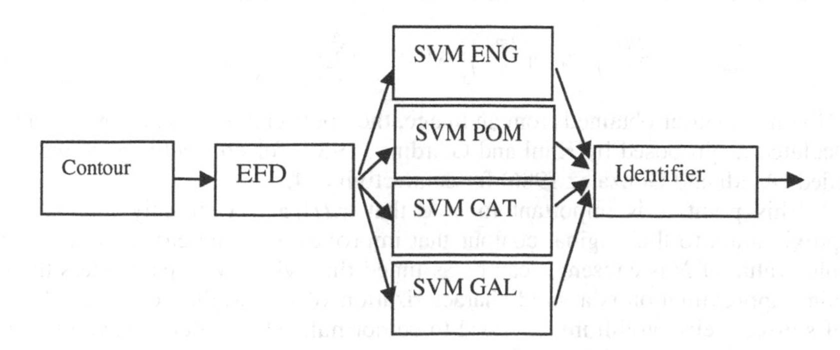

Fig. 2 Schema for the identification system studied in this paper. In some cases de EFD analysis would be independent for each SVM.

2.1 Contours Extraction and Elliptic Fourier Descriptors (EFD)

The first step of the system following the schema of Fig.2 is the contour extraction from the otolith images. In AFORO database all the images are highly contrasted respect to the background, so any contour extractor works properly; in our case we have used a morphological contour extractor (Serra 1982); it is applied on a segmented binary image like the last one that is represented in Fig.1, and the result is shown in Fig 3.

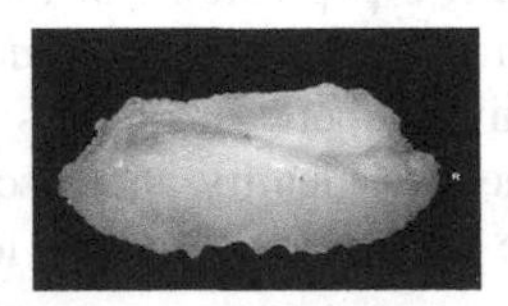

Fig. 3 Otolith contour extraction from a normalized image. Normalized otolith image (left panel) with its contour (in red) inside an enclosing rectangular box (right panel).

An elliptical Fourier function perfectly describes a closed curve with an ordered set of data points in a two-dimensional plane. The function decomposes the curve orthogonally into a sum of harmonically related ellipses (Kuhl and Giardina 1982). These ellipses can be combined to reconstruct an arbitrary approximation to the closed curve.

A closed contour with period T is defined by the evolution of its coordinates $x(t)$ and $y(t)$ along the variation t within the length of this contour, and can be expanded as follows:

$$x(t) = a_0 + \sum_{n=1}^{\infty}\left[a_n \cos\left(\frac{2\pi nt}{T}\right) + b_n \sin\left(\frac{2\pi nt}{T}\right)\right] \quad y(t) = c_0 + \sum_{n=1}^{\infty}\left[c_n \cos\left(\frac{2\pi nt}{T}\right) + d_n \sin\left(\frac{2\pi nt}{T}\right)\right] \tag{1}$$

The coefficients a_n, b_n, c_n and d_n are an alternative way of perfectly describing the contour. The signal is characterized and the information reduced by selecting a reduced set of 4N coefficients, which leads to the expression:

$$x_N(t) = a_0 + \sum_{n=1}^{N}\left[a_n \cos\left(\frac{2\pi nt}{T}\right) + b_n \sin\left(\frac{2\pi nt}{T}\right)\right] \quad y_N(t) = c_0 + \sum_{n=1}^{N}\left[c_n \cos\left(\frac{2\pi nt}{T}\right) + d_n \sin\left(\frac{2\pi nt}{T}\right)\right] \tag{2}$$

Given a contour obtained from an image, the coefficients a_n, b_n, c_n and d_n can be calculated as proposed by Kuhl and Giardina (1982) for connectivity -8 and simplified (Abidi and Gonzalez 1986) for connectivity -4.

At this point it is important to note that $x_N(t)$ and $y_N(t)$ only represent an approximation to the original contour that improves as N increases. When a particular value of N is chosen, it can be assumed that with $4N+2$ parameters the resulting approximation is a valid characterization of the original contour. To use EFDs effectively, otolith images need to be normalized in order to have the same image acquisition conditions. The otolith contour is usually normalized for size and orientation: the coefficients a_0 and c_0 are taken as 0 (this locates the centre of mass at the origin), the rest of the coefficients are normalized so that a_1=1, and the main radial is normalized to 1.

To complete the EFD computation, we obtain the Fourier series expansion of one period T' (of K points) of the discrete periodical functions *y(k)* and *x(k)*. Taking into account that the range of the digital frequencies (ω_n) that can be represented in the discrete domain is limited to the set of harmonics given by the expression (Proakis 2007), then the discrete Fourier series expansion of *y(k)* takes the form:

$$\omega_n = \frac{2\pi}{T'} n = \frac{2\pi}{K} n \quad n = 0,..,K-1$$

$$y(k) = c_0 + \sum_{n=1}^{K}\left[c_n \cos\left(\frac{2\pi}{K}nk\right) + d_n \sin\left(\frac{2\pi}{K}nk\right)\right] \quad \begin{matrix} k = 0,..,K-1 \\ n = 0,..,K-1 \end{matrix} \tag{3}$$

The y_N approximation is carried out by selecting a reduced set of *2N* coefficients, with *N<K*.:

$$y_N(k) = c_0 + \sum_{n=1}^{N}\left[c_n \cos\left(\frac{2\pi}{K}nk\right) + d_n \sin\left(\frac{2\pi}{K}nk\right)\right] \quad \begin{matrix} k = 0,..,K-1 \\ n = 1,..,N \end{matrix} \tag{4}$$

The coefficients of the one-dimensional Fourier series can be obtained very quickly with the FFT (Fast Fourier Algorithm). The complex values provided by the FFT are a compact representation of the two real coefficients given in the Fourier series expansion (Cooley and Tukey 1965). And the same could be done for the *x(k)* function.

2.2 Support Vector Machine (SVM) and Identifier

The identifiers are based on a set of SVM with linear kernel, like the one described in the tutorial of Burges (1998). During the training phase, the linear Kernel (eq. 8) measures the Euclidean distance between the annotated descriptors S_I and S_J , and finds the optimal α_l set values to separate one class from the others, which corresponds to the selection of an hyperplane in the L-dimensional space of the descriptors, to separate one class from the others.

$$K(S_I, S_J) = \sum_{l \epsilon L} \alpha_l d_l(S_I, S_J) \tag{5}$$

The design point from our approach was the selection of the descriptors used in the training and subsequently into the test phase.

Once we have trained each one of the SVM corresponding to the classes, during the test phase the Identifier is fed with those results, and the output of the system should be one of the following classes: ENG *-Engraulis-*, POM *-Pomadasys-*, CAT *-Merluccius* CAT-, GAL *-Merluccius* GAL-, NOT -not determinate-. In the case that more than one is activated, the output would be NOT, except for the CAT and GAL case, as they correspond to the same species.

3 Implementation and Results

This experimental Identification system has been implemented with Octave 3.2 software under Linux 10.04 on a Core 2 Duo T7100 / 1.8 GHz Laptop with 2 GB

RAM. We used a DATASET extracted from AFORO database: 10 specimens of *Engraulis encrasicholus*, from the family Engroulidae, 14 specimens of *Pomadasys incisus* belonging to the family Haemolidae, and 44 specimens from different populations of *Merluccius merluccius* from the family Merlucciidae. The database is available from http://aforo.cmima.csic.es for research purposes. For the training step corresponding to the SVM of ENG, POM, CAT, GAL we divide and annotate the population with 10 specimens of each group. During the training we select randomly 50% of this population, with a minimum of 20% from the corresponding class. The test phase was done using the complete population. In all the cases we use a retraining phase with part of the training individuals.

The main parameters analyzed for the design of the system are two: the number of points of the contours (K), it is used to normalize the contours of different specimens. The value K could go from 2 to 1024, with K very low we only represent values related to basic diameters of the contour. But with K getting high, the number of parameters increases. Another value is L the total number of dimensions on the classification space, it is related with the order of the EFD (N), with the equation $L=4\cdot(2\cdot N)-2$, moreover it is necessary that $N \leq K/2$. With L low, the contour is represented with basic ellipsoids, not having into account the detail of the contours, but at the same time, the classifiers have a low dimensional space (easier to train and generalize). When L goes higher, the contour is represented with more details, however, the classifier has higher dimension and becomes instable and sometimes could not be generalized.

The plotted results are for medium values of K=64, and N=32, which leads to a decision space of L=254 parameters. To represent the results we plot the true

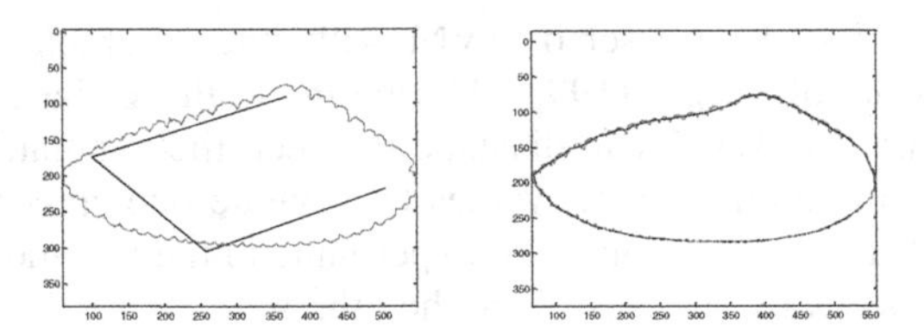

Fig. 4 Results for CAT contour (in red), respect to the undersampled contour (in black) with K=4 and N=2 in the left panel, and the same with K=64 and N=32 in the right one.

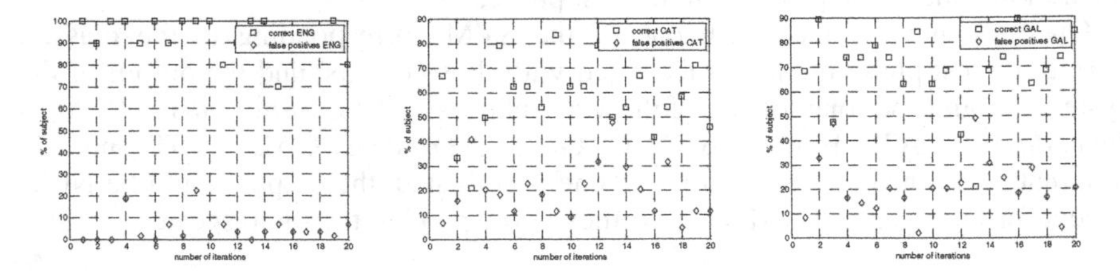

Fig. 5 Results for ENG, CAT, GAL identification using K=64 and N=32 (L=254), the mean value of the correct identification is 93% , 57% and 68%; The % of false positives is 4.6 %, 20% and 21%.

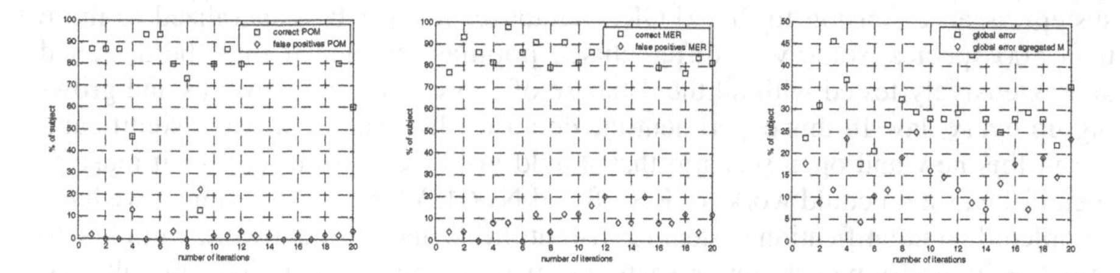

Fig. 6 From lest to right results for POM and MER identification, using K=64 and N=32 (L=254), the mean value of the correct identification is 79% and 86% and the false positives is 3% and 9%. Results for Global error identification (the right panel) we compare the error without aggregating CAT and GAL, and with their aggregation as MER. The mean error is 29% in one case and 14% in the other

classifications results (in %), represented with black squares; and in the same plot we draw the value of false postitives for each one of the classes (in %), represnted by blue romboids.

The results for CAT and GAL have a great number of false positives, as there is a great similarity between the two populations of the same specie. If we aggregate the results as the specie *Merluccius merluccius* (MER), the results are similar to the other species, as we can see in Fig 6.

4 Discussion

For the same K=64, we have compared the results for N=8, 16 and are quite similar, although the CAN, GAL errors become greater, for N=4 (L=30) the global error reaches 51% without aggregation and 29% with aggregation. Meaning that the class classification could be acceptable with a low number of descriptors, but the population cannot. There is a somehow surprising result if we maintain N=4 (L=30) but K=16, then the results are better, global error 30% and 17%, and the partial result similar to those of the section 4, and the same happens with K=8. These could be seen as an effect of having more frequencies information in less coefficients as the number of points in the contour decreases; however, the main difference with initial values K=64 is that not all the trains are valid, and the standard deviation of the measures are greater in all the graphics. With K=4 and N=2 (L=14), the global errors are 36% and 16% respectively but the identification errors are greater in more training essays. With K=2 and N=1 we have a lot of errors. With K=128 or greater, the system becomes more difficult to train and the system gets worse results. Another interesting point is the results for the classification of populations, but they have to be more accurately studied, in deeper detail with a wider set of otoliths, and selecting by hand the best classifiers of the training set.

5 Conclusion

The proposed method for otoliths identification could be a good method analyzing these preliminary results. But to be used as a classification method in an automatic system as an extension to the AFORO database, it must be generalized to more than 800 species, some with differentiated populations, and it must be designed and extensively tested with all the database of more than 2800 otoliths and growing up every day. In this experiment the detail of the contour was not significant, but in this new and open scenario that should not be always true, then it must be seen if the system could work for K>128 and N>64. With this experiment we have completed an identification system for fish otoliths, and our next work would be to improve its performance, tuning each one of the elements and analyzing alternative solutions for all of them.

Acknowledgements. This work has been supported by *'Ajuts per a estades de recerca fora de Catalunya 2009-10'* and the Digital Technologies Research Group from the University of Vic: cost center R008-R0904.

References

Abidi, M.A., Gonzalez, R.C.: Shape Decomposition Using Elliptic Fourier Descriptors. In: Proc. 18th IEEE Southeast Sympo. Sys. Theory, Knoxvlle, TN USA, pp. 53–61 (1986)
Bird, J.L., Eppler, D.T., Checkley, D.M.: Comparisons of herring otoliths using Fourier-series shape-analysis. Can. J. Fish. Aquat. Sci. 43(6), 1228–1234 (1986)
Burges, C.: A Tutorial on Support Vector Machines for Pattern Recognition. Data Mining and Knowledge Discovery 2, 121–167 (1998)
Campana, S.E., Casselman, J.M.: Stock discrimination using otolith shape analysis. Can. J. Fish. Aquat. Sci. 50, 1062–1083 (1993)
Casselman, J.M., Collins, J.J., Crossman, E.J., Ihssen, P.E., Spangler, G.R.: Lake whitefish (Coregonus clupeaformis) stocks of the Ontario waters of Lake Huron. Can. J. Fish. Aquat. Sci. 38, 1772–1789 (1981)
Chen, S.Y.Y., Lestrel, P.E., Kerr, W.J.S., McColl, J.H.: Describing shape changes in the human mandible using elliptical Fourier functions. Eur. J. Orthodont. 22, 205–216 (2000)
Cooley, J.W., Tukey, J.W.: An Algorithm for the Machine Calculation of Complex Fourier Series. Math. Computat. 19, 297–301 (1965)
Iwata, H., Ukai, Y.: SHAPE: a computer program package for quantitative evaluation of biological shapes based on elliptic Fourier descriptors. J. Hered. 93, 384–385 (2002)
Kuhl, F.P., Giardina, C.R.: Elliptic Fourier features of a closed contour. Comput. Graph. Image Process. 18, 236–258 (1982)
Lombarte, A., Torres, G.J., Morales-Nin, B.: Specific Merluccius otolith growth patterns related to phylogenetics and environmental. J. Mar. Biol. Ass. U.K. 83, 277–281 (2003)
Lombarte, A., Chic, O., Parisi-Baradad, V., Olivella, R., Piera, J., García-Ladona, E.: A web-based environment for shape analysis of fish otoliths. The AFORO database. SCI. MAR. 70(1), 147–152 (2006)
Parisi-Baradad, V., Lombarte, A., García-Ladona, E., Cabestany, J., Piera, J., Chic, Ò.: Otolith shape contour analysis using affine transformation invariant wavelet transforms and curvature scale space representation. Mar. Freshw. Res. 56, 795–804 (2005)
Proakis, J.G., Manolakis, D.G.: Digital Signal Processing, 4th edn. Pearsons Education (2007)
Rosenfeld, A., Pfaltz, J.L.: Distance functions on digital pictures. Pattern Recognition 1(1), 33–61 (1968)
Serra, J.: Image Analysis and Mathematical Morphology. Academic Press, New York (1982)
Tracey, S.R., Lyle, J.M., Duhamelb, G.: Application of elliptical Fourier analysis of otolith form as a tool for stock identification. Fisheries Research 77, 138–147 (2006)

Semantic Based Web Mining for Recommender Systems

María N. Moreno García, Joel Pinho Lucas, Vivian F. López Batista, and María Dolores Muñoz Vicente

Abstract. Availability of efficient mechanisms for selective and personalized recovery of information is nowadays one of the main demands of Web users. In the last years some systems endowed with intelligent mechanisms for making personalized recommendations have been developed. However, these recommender systems present some important drawbacks that prevent from satisfying entirely their users. In this work, a methodology that combines an association rule mining method with the definition of a domain-specific ontology is proposed in order to overcome these problems in the context of a movies' recommender system.

Keywords: Semantic Web Mining, Recommender Systems, Associative Classification.

1 Introduction

Endowing Web systems with mechanisms for selective recovery of the information is nowadays a highly demanded requirement. Many Web applications, especially e-commerce systems, already have procedures for personalized recommendation that allow users to find products or services they are interested in. However, in spite of the advances achieved in this field, the recommendations provided by this type of systems have some important drawbacks, such as low reliability and high response time. Therefore, it is necessary to research in new recommender methods that join precision and performance as well as solving other usual problems in this kind of systems. Data mining is one of the techniques providing better results in these two important aspects but it cannot deal whit other drawbacks such as *first-rater*, *gray-sheep* problem, etc. The process of applying data mining techniques on web data is known as web mining. Patterns extracted from web data can be applied to web personalization applications.

María N. Moreno García, Joel Pinho Lucas,
Vivian F. López Batista, and María Dolores Muñoz Vicente
Dept. of Computing and Automatic, University of Salamanca, Plaza de los Caídos s/n,
37008 Salamanca

A.P. de Leon F. de Carvalho et al. (Eds.): Distrib. Computing & Artif. Intell., AISC 79, pp. 17–25.
springerlink.com

In the last years new strategies using ontology are being tested. The idea is to enrich the data to be mined with semantic annotations in order to produce more interpretable results and to obtain patterns at different levels of abstraction that allow to solve the problems mentioned before. In this paper, a semantic based web mining technique is proposed. It is a hybrid methodology for personalized recommendation applicable to e-commerce systems. Recommendations are obtained by means of applying an associative classification data mining algorithm to data annotated with semantic metadata according to a domain-specific ontology.

The rest of the paper is organized as follows: Sections 2, 3 and 4 include a brief description of the state of the art of recommender systems, associative classification and semantic web mining respectively. In section 5 the proposed methodology and its application in a specific recommender system is presented. Finally, the conclusions are given in section 6.

2 Recommender Systems

There are a great variety of procedures used for making recommendation in the e-commerce environment. They can be classified into two main categories (Lee et al., 2001): *collaborative filtering* and a *content-based approach*. The first class of techniques was based initially on nearest neighbor algorithms. These algorithms predict product preferences for a user based on the opinions of other users. The opinions can be obtained explicitly from the users as a rating score or by using some implicit measures from purchase records as timing logs (Sarwar et al., 2001). In the content based approach text documents are recommended by comparing between their contents and user profiles (Lee et al., 2001). The main shortcoming of this approach in the e-commerce application domain is the lack of mechanisms to manage web objects such as motion pictures, images, music, etc. Besides, it is very difficult to handle the big number of attributes obtained from the product contents. Currently there are two approaches for collaborative filtering, *memory-based* (*user-based*) and *model-based* (*item-based*) algorithms. Memory-based algorithms, also known as nearest-neighbor methods, were the earliest used (Resnick et al., 1994). They treat all user items by means of statistical techniques in order to find users with similar preferences (neighbors). The prediction of preferences (recommendation) for the active user is based on the neighborhood features. A weighted average of the product ratings of the nearest neighbors is taken for this purpose. The advantage of these algorithms is the quick incorporation of the most recent information, but they have the inconvenience that the search for neighbors in large databases is slow (Schafer et al., 2001). Model-based collaborative filtering algorithms use data mining techniques in order to develop a model of user ratings, which is used to predict user preferences.

Collaborative filtering, specially the memory-based approach, has some limitations in the e-commerce environment. Rating schemes can only be applied to homogeneous domain information. Besides, sparsity and scalability are serious weaknesses which would lead to poor recommendations (Cho et al., 2002). Sparsity is due to the number of ratings needed for prediction is greater than the number of the ratings obtained because usually collaborative filtering requires user explicit expression of personal preferences for products. The second limitation is related to performance problems in the search for neighbors in memory-based algorithms. These problems are caused by the necessity of processing large amount of information. The computer time grows linearly with both the number of customers and the number of products in the site. The lesser time required for making recommendations is an important advantage of model-based methods. This is due to the fact that the model is built off-line before the active user goes into the system, but it is applied on-line to recommend products to the active user. Therefore, time spent in building the model has no effects in the user response time since little process is required when recommendations are requested by the users, contrary to the memory based methods that compute correlation coefficients when user is on-line. Nevertheless, model based methods present the drawback that recent information is not added immediately to the model but a new induction is needed in order to update the model.

3 Associative Classification

Associative classification methods obtain a model of association rules that is used for classification. These rules are restricted to those containing only the class attribute in the consequent part. This subset of rules is named class association rules (CARs) (Liu et al., 1998).

Since Agrawal and col. introduced the concept of association between items (Agrawal et al., 1993a)(Agrawal et a l., 1993b) and proposed the Apriori algorithm (Agrawal and Srikant, 1994), association rules have been the focus of intensive research. Most of the efforts have been oriented to simplify the rule set and improve the algorithm performance. The number of papers in the literature focused in the use of the rules in classification problems is lesser. A proposal of this category is the CBA (Classification Based on Association) algorithm (Liu et al., 1998) that consists of two parts, a rule generator based on Apriori for finding association rules and a classifier builder based on the discovered rules. CMAR (Classification Based on Multiple Class-Association Rules) (Li et al., 2001) is another two-step method, however CMAR uses a variant of FP-growth instead Apriori. Another group of methods, named integrated methods, build the classifier in a single step. CPAR (Classification Based on Predictive Association Rules) (Yin and Han, 2003) is the most representative algorithm in this category.

4 Semantic Web Mining

Semantic Web Mining is a new research field where converge the Semantic Web and the Web mining, which are two areas that are evolving very rapidly. Both are contributing to the latest challenges of the WWW in a complementary way. The lack of structure of most of the data in the Web can be only understood by humans; however, the huge amount of information can only be processed efficiently by machines. The semantic Web addresses the first problem by enriching the Web with machine-understandable information, while web mining addresses de second one by automatically extracting the useful knowledge hidden in web data. Semantic web mining aims at improving the results of web mining by exploiting the semantic structures in the web as well as building the semantic web making use of web mining techniques (Stumme et al., 2006).

We will focus in the first case since it is the target of the research presented in this paper. In this area, ontology is used to describe explicitly the conceptualization behind the knowledge represented in the data to be mined. An ontology is "an explicit formalization of a shared understanding of a conceptualization". Most of them include a set of concepts hierarchically organized and additional relations between them. This formalization allows integrating heterogeneous information as a preprocessing step of web mining tasks (Liao et al., 2009).

Taxonomic abstraction provided by an ontology is often essential for obtaining meaningful results. In addition, taxonomies allow inducing patterns at more abstract level, that is, regularities can be found between categories of products instead of between specific products. These patterns can be used in recommender systems for recommending new products that still have not been rated by the users (Huang and Bian, 2009). This is a major problem since new products introduced in the catalog cannot be recommended if classical collaborative filtering algorithms are used because the induced models do not include these products. However, when taxonomies are used, new products can be classified into one or more categories and recommendations can be done from models enclosing more general patterns, which relate user profiles with categories of products.

On the other hand, applications of traditional data mining methods can be extended and tailored to web systems in order to obtain similar benefits such as those provided by using association rules and clustering techniques in market basket analysis and in cross selling promotions. In (Mossavi et al., 2009) concepts of classical market analysis are extended by means of ontological techniques in order to enhance electronic market. Substitution is one of these concepts which represent the similarity degree between related products (Resnik, 1999). Usually, techniques for computing the similarity degree are based on finding product properties (Resnik, 1999), (kanappe, 2005), (Ganjifar et al., 2006). Complement is another useful concept , mainly used for designing marketing strategies, since it provide them and additional value. In some circumstances recommendation of similar products do not give the desired results, but recommendation of complementary products does. Mossavi et al. (Mossavi et al., 2009) apply an ontological technique to determine complement products and the OWL language to model types of products. In this direction, Liao et al. (2009) have updated the marketing concept of branding, which is traditionally used in the business field for making a brand by means of differentiating the brand products from those of the competitors (Baker, 1996). A brand is "a combination of features (what the product is), customer benefits (what needs and wants the product meets) and values (what the customer associates with the product)". Liao et al (2009) analyze a specific market segment by means of brand spectrums depicted from data relating to consumer purchase behaviors and beverage products. They develop a set of ontologies for describing the integrated consumer behavior and a set of databases related to these ontologies. In further steps two data mining techniques are applied. First, a clustering algorithm is used for segmenting the data according to customer information, lifestyles, and purchase behavior. Then, the relationship among the clusters is analyzed by means of the Apriori algorithm for association rule induction.

5 Recommendation Methodology

The procedure proposed in this work aims to overcome some frequent drawbacks of recommender systems:

- *Sparsity*: Caused by the fact that the number of ratings needed for prediction is greater than the number of the ratings obtained from users.
- *Scalability*: Performance problems presented mainly in memory-based methods where the computation time grows linearly with both the number of customers and the number of products in the site.
- *First-rater problem*: Takes place when new products are introduced. These products, never have been rated, therefore they cannot be recommended.
- *New users*: They cannot receive recommendations since they have no evaluations about products.

In order to overcome these problems we present a methodology that combines an associative classification data mining method with the definition of a domain-specific ontology.

The study was carried out with data from MovieLens, a well known recommender system used in may research works. The database contains user demographic information and user rating about movies, collected through the MovieLens Web site (movielens.umn.edu) during a seven-month period. All movies belong to 18 different genres. User ratings were recorded on a numeric five point scale. Users who had less than 20 ratings or did not have complete demographic information were removed from the data set. It consists of 100,000 ratings (1-5) from 943 users on 1682 movies. Since the rating attribute is used to decide if a movie is going to be recommended to a user, we changed such attribute in order to have only two values: "Not recommended" (score 1 or 2) and "Recommended" (score 3, 4 or 5). This new attribute, *rating_bin*, will be the label attribute to be predicted. In this way, the classification is simplified and no further transformation is needed for making the recommendations to the user. Figure 1 illustrates this transformation by showing the data distribution into the initial five classes and the distribution into the final two ones. The *age* attribute was discretized into five age ranges. The user's *occupation* attribute is a nominal variable with 21 distinct values. The MovieLens file originally contained 19 binary attributes related to movie genres. an instance with value 1 expressed that the movie belongs to a specific gender and 0 otherwise. The consistency of the association model would be compromised if 19, among the 23 attributes available, were binaries. Thus, the 19 binary attributes were reduced to just one attribute representing the movie genre's name. After data pre-processing and transformation, 14587 records were remained in the input file for the algorithms used in this study.

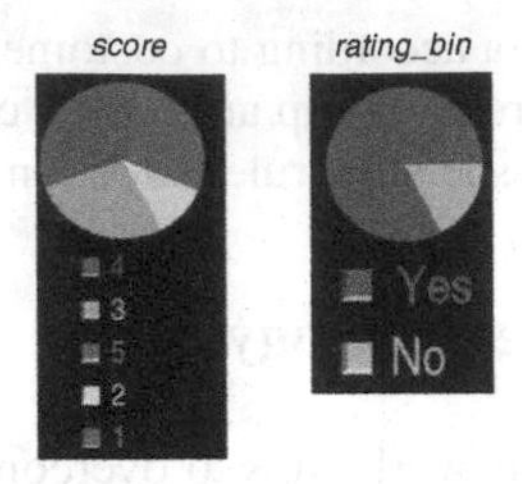

Fig. 1 "Rating" attribute transformation

The next step was the development of an ontology for the available data. An ontology consist of abstract concepts (classes) and relationships defined to be shared and reusable in a particular application domain. Web data can be considered the instances of ontology entities when they are classified according to a specific ontology. In this way, web mining techniques can be applied to these instances giving more meaningful knowledge to the user (Lim and Sun, 1996).

In this work, ontology is used to improve recommender systems and overcome their main drawbacks, previously commented. Data from MovieLens have been selected and annotated with semantic metadata according to a domain-specific ontology. We have adapted a public ontology about movies and cinema developed by Monika Solanki of De Montfort University. The ontology is part of the DAML repository (http://www.daml.org/ontologies), which aims at developing a language and tools to facilitate the concept of the Semantic Web. It defines classes and properties for movies. The classes are: Cinema, Genre, Movie and shows. The properties are: actor, actress, address, cinemaname, director, duration, email, has_genre, has_shows, movieName, musicdirector, producer, screenplay, showing_at, synopsis, telephonenumber. Since this information is not available in the MovieLens database, the ontology has been simplified.

On the other hand, the database contains demographic and rating information from users, which is used for making the recommendations. Therefore, we have organized this data according a different ontology related to user characteristics, such as gender, age and occupation.

Apart from the previous considerations, the proposed ontology for our application domain must take into account the available data: User (*id_user*, *gender*, *age*, *occupation*, *zip*), movie (*id_movie*, *title*, *genre*) and rating (*ide_user*, *id_movie*, *score*, *rating_bin*). The definition of the proposed ontology is showed in figure 2. The database used for applying the web mining methods and predict user preferences was designed following the structure given by the defined ontology.

Classes = {Movie, User, Rating, Genre}
Relationships = {evaluates (User, Movie), scores (Rating, evaluates (User, Movie)), belongs (Movie, Genre}
Properties = {User (gender, age, occupation, zip), Movie (title), Rating (score, rating_bin), Genre (name_genre)}

Fig. 2 Ontology definition

We propose a recommendation procedure to apply to data annotated with semantic information given by the defined ontology. The target of the methodology is to predict in an efficient way the user preferences in order to recommend him products he is interested in. To overcome scalability problems a model based approach was applied. The predictive models are induced off-line (before the entry of the users in the system) but are used on-line (when active user is connected to the system). Therefore, the time spent in building the models does not influence the user response time. This is the main advantage of this approach that avoids problems associated with traditional memory-based techniques.

Predictive models were induced by means of a data mining algorithm, specifically an associative classification method. Traditionally, association analysis is considered an unsupervised technique, so it has been applied in knowledge discovery tasks. Recent studies have shown that knowledge discovery algorithms, such as association rule mining, can be successfully used for prediction in classification problems. In this case, the induced association rules contain the class as the consequent part. The choice of an associative classification algorithm is due to the better behavior of these methods in sparse data contexts (Moreno et al., 2009). Consequently more reliable recommendation can be obtained with a lesser number of ratings. This is the way of dealing with the sparsity problem.

The class association rule algorithm is applied in two different abstraction levels. Semantic annotations provided to the data following the defined ontology allow inducing patterns at a more abstract level. Therefore, regularities between types of movies and user profiles can be obtained instead of between particular users or specific movies. These patterns can be used for recommending new products that still have not been rated by the users. In a similar way, new users can receive recommendations according his profile, which is defined from properties and relations given by the ontology. Figure 3 shows some of the rules obtained with the predictive Apriory algorithm at this level of abstraction. We can observe the high precision of the rules (label *acc* in the right side) in spite of the sparsity of

```
Best rules found:

 1. age=[50 - 55] occupation=executive genre=drama 45 ==> rating=yes 45    acc:(0.99347)
 2. age=[35 - 44] occupation=programmer genre=drama 81 ==> rating=yes 80    acc:(0.99272)
 3. age=[25 - 34] occupation=salesman genre=drama 39 ==> rating=yes 39    acc:(0.9927)
 4. gender=F occupation=salesman genre=drama 34 ==> rating=yes 34    acc:(0.99164)
 5. age=[35 - 44] occupation=lawyer genre=drama 33 ==> rating=yes 33    acc:(0.99135)
 6. age=[35 - 44] occupation=student genre=drama 27 ==> rating=yes 27    acc:(0.98878)
 7. age=[50 - 55] gender=F occupation=educator 24 ==> rating=yes 24    acc:(0.98665)
 8. age=[18 - 24] occupation=none genre=comedy 23 ==> rating=yes 23    acc:(0.98575)
 9. age=[35 - 44] gender=F occupation=educator genre=drama 94 ==> rating=yes 92    acc:(0.98548)
10. age=[35 - 44] gender=M occupation=librarian genre=drama 51 ==> rating=yes 50    acc:(0.98383)
11. occupation=writer genre=documentar 19 ==> rating=yes 19    acc:(0.9808)
12. age=[25 - 34] occupation=student genre=sci_fi 19 ==> rating=yes 19    acc:(0.9808)
13. age=[35 - 44] occupation=artist genre=comedy 19 ==> rating=yes 19    acc:(0.9808)
14. age=[50 - 55] genre=sci_fi 18 ==> rating=yes 18    acc:(0.97912)
15. age=[35 - 44] gender=F genre=musical 18 ==> rating=yes 18    acc:(0.97912)
```

Fig. 3 A portion of the predictive rule model

the data. In the lowest abstraction level the rules induced by the associative classification method relate particular users and their characteristics with specific movies. This model is used for users who have enough numbers of evaluated movies and for rated movies.

6 Conclusions

In this work a methodology for recommender systems is proposed. The aim is to overcome some usual drawbacks such as sparsity, scalability, first rater problem and new users. The methodology consists on combining a web mining method and the definition of a specific ontology for the application domain to be studied.

An associative classification algorithm is used to generate de predictive models used for making recommendations. These models can be generated in two abstraction levels. The lowest level relates users, movies and ratings for making the recommendations. Web data are annotated with semantic information according to the defined ontology. This allows generating patterns at high level abstraction by means of the associative classification algorithm. In this level, the rules relate types of movies and user profiles instead of specific movies or particular users. This model is used for recommender not rated movies or for making recommendation to new users.

References

Agrawal, R., Srikant, R.: Fast algorithms for mining association rules in large data-bases. In: Proc. of 20th Int. Conference on Very Large Databases, Santiago de Chile, pp. 487–489 (1994)

Agrawal, R., Imielinski, T., Swami, A.: Database mining. A performance perspective. IEEE Trans. Knowledge and Data Engineering 5(6), 914–925 (1993a)

Agrawal, R., Imielinski, T., Swami, A.: Mining associations between sets of items in large databases. In: Proc. of ACM SIGMOD Int. Conference on Management of Data, Washinton, D.C., pp. 207–216 (1993b)

Baker, M.: Marketing-An introductory text. Macmillan Press, London (1996)

Cho, H.C., Kim, J.K., Kim, S.H.: A Personalized Recommender System Based on Web Usage Mining and Decision Tree Induction. Expert Systems with Applications 23, 329–342 (2002)

Ganjifar, Y., Abolhasani, H., et al.: A similarity measure for OWL-S annotated web services, Web Intelligence, pp. 621–624 (2006)

Huang, Y., Bian, L.: A Bayesian network and analytic hierarchy process based personalized recommendations for tourist attraction over the Internet. Expert Systems with Applications 36, 933–943 (2009)

Kanappe, R.: Measures of semantic similarity and relatedness for use in ontology-based information retrieval. Thesis of Doctor. Denmark, Roskilde University (2005)

Lee, C.H., Kim, Y.H., Rhee, P.K.: Web Personalization Expert with Combining collaborative Filtering and association Rule Mining Technique. Expert Systems with Applications 21, 131–137 (2001)

Li, W., Han, J., Pei, J.: CMAR. Accurate and efficient classification based on multiple class-association rules. In: Proc. of the IEEE International Conference on Data Mining (ICDM 2001), California, pp. 369–376 (2001)

Liao, S., Ho, H., Yang, F.: Ontology-based data mining approach implemented on exploring product and brand spectrum. Expert Systems with Applications 36, 11730–11744 (2006)

Lim, E., Sun: A.Web Mining – the Ontology Approach. In: Proc. of International Advanced Digital Library Conference, IADLC 2005 (2005)

Liu, B., Hsu, W., Ma, Y.: Integration classification and association rule mining. In: Proc. of 4th Int. Conference on Knowledge Discovery and Data Mining, pp. 80–86 (1998)

Moosavi, S., Nematbakhsh, M., And Farsani, H.K.: A semantic complement to enhance electronic market. Expert Systems with Applications 36, 5768–5774 (2009)

Moreno, M.N., Pinho, J., Segrera, S., y López, V.: Mining Quantitative Class-association Rules for Software Size Estimation. In: Proc. of IEEE 24th International Symposium on Com-puter and Information Sciences (ISCIS 2009), pp. 199–204. IEEE, Northern Cyprus (2009)

Resnick, P., Iacovou, N., Suchack, M., Bergstrom, P., Riedl, J.: Grouplens: An open architecture for collaborative filtering of netnews. In: Proc. Of ACM CSW 1994 Conference on Computer. Supported Cooperative Work, pp. 175–186 (1994)

Resnik, P.: Semantic similarity in a taxonomy: An information based measure and its application to problems of ambiguity in natural language. Journal of Artificial Intelligence, 94–130 (1999)

Sarwar, B., Karypis, G., Konstan, J., Riedl, J.: Item-based Collaborative Filtering Recommendation Algorithm. In: Proceedings of the tenth International World Wide Web Conference, pp. 285–295 (2001)

Schafer, J.B., Konstant, J.A., Riedl, J.: E-Commerce Recommendation Applications. Data Mining and Knowledge Discovery 5, 115–153 (2001)

Stumme, G., Hotho, A., Berendt, B.: Semantic Web Mining. State of the art and future direction. Journal of Web Semantics 4, 124–143 (2006)

Yin, X., Han, J.: CPAR. Classification based on predictive association rules. In: SIAM International Conference on Data Mining (SDM 2003), pp. 331–335 (2003)

Classification of Fatigue Bills Based on K-Means by Using Creases Feature

Dongshik Kang, Satoshi Miyaguni, Hayao Miyagi, Ikugo Mitsui, Kenji Ozawa, Masanobu Fujita, and Nobuo Shoji

Abstract. The bills in circulation generate a large amount of fatigue bills every year, causing various types of problems, such as the paper jam in automatic tellers due to overwork and exhaustion. A highly advanced bill classification technique, which distinguishes whether a bill is a reusable bill specifying the level of fatigue, is greatly required in order to comb out these problematic bills. Therefore, a purpose of this paper is to suggest a classification method of fatigue bills based on K-means with bill image data. The effectiveness of this approach is verified by the bill discriminant experimentation.

1 Introduction

In recent days, we often encounter several bills that, in process of circulation, have been discolored or worn out. These bills can be regarded as invalid, and some audit machines and classifying devices have been put in order to eliminate them.

One of those invalid bills are ones that are low in intensity from the circulation (hereinafter called 'fatigue bills'), and be distinguished from valid ones (hereinafter called 'new bills'). The purpose of this distinction is the prevention of paper jam in automatic cash machines such as ATM, CD(Cash Dispenser) and vending machine. The methods high-quality of distinction are demanded.

Nowadays the study of sound signal distinction, which distinguishes those fatigue bills and new bills from their sounds, is reported. It deals with the sounds which the bills emit, corrects the sounds into data, and analyzes them with FLVQ and ICA. This sound-signal distinction is one of the most well-known methods of bill distinction, and has reported high performance among them[1-6].

However, this method is unavailable with noises around; this method demands precise information of sounds, and those noises can be a hindrance when correcting the sounds[7].

Dongshik Kang, Satoshi Miyaguni, and Hayao Miyagi
University of the Ryukyus, Okinawa, Japan

Ikugo Mitsui, Kenji Ozawa, Masanobu Fujita, and Nobuo Shoji
Japan Cash Machine Co., Ltd., Osaka, Japan

A.P. de Leon F. de Carvalho et al. (Eds.): Distrib. Computing & Artif. Intell., AISC 79, pp. 27–33.
springerlink.com

Then, in this study we focus in image datum of paper money from CIS scanners[9,10]. The image compensation is carried out in order to extract a wrinkle of the paper money image and features by the discoloration. Furthermore, the bill of two categories was prepared, and the discrimination experiment was performed using the K-means method.

2 Basic Concept

This chapter briefly explains three concepts, gamma compensation, color histogram and K-means that are introduced in this study.

2.1 Length

The procedure of gamma compensation on image processing is how the brightness changes from its highest to lowest. As it were, it changes the additive color. When turning it bright γ▯ ▯ , and when turning it dark γ▯ ▯ . The color information ranges from 0 to 255, so the equation (2.1) is used.

$$OutPixel = 255 * (InPixel / 255)^{1/\gamma} \tag{2.1}$$

2.2 K-Means Clustering

In statistics and machine learning, K-means (MacQueen, 1967) is one of the simplest unsupervised learning algorithms that solves the well known clustering problem[8]. The procedure follows a simple and easy way to classify a given data set through a certain number of clusters (assume k clusters) and fixed a priori. The main idea is to define k centroids, one for each cluster. These centroids shoud be placed in a cunning way because different locations cause different results. So, the better choice is to place them as far as possible away from each other. The next step is to take each point belonging to a given data set and associate it to the nearest centroid. When no point is pending, the first step is completed and an early groupage is done. At this point we need to recalculate k's new centroids as barycenters of the clusters resulting from the previous step. After we have these k new centroids, a new binding has to be done between the same data set points and the nearest new centroid. Now a loop has been generated. As a result of this loop we may notice that the k centroids change their location step by step until no more changes are done. In other words, centroids do not move any more.

Finally, this algorithm aims at minimizing an objective function, in this case a squared error function. The objective function

$$V = \sum_{j=1}^{k}\sum_{i=1}^{n} \| x_i^{(j)} - c_j \|^2 ,$$

where $\| x_i^{(j)} - c_j \|^2$ is a chosen distance between a data point $x_i^{(j)}$ and c_j the cluster centre, is an indicator of the distance of the n data points from their respective cluster centres.

K-means is a simple algorithm that has been adapted to many problem domains. As we are going to see, it is a good candidate for extension to work with fuzzy feature vectors.

3 Feature Extraction

In this study, the histogram's change due to discoloration and creases is used as the feature. For extraction of the feature, gamma compensation is performed to the image of the fatigued bills for gamma value 0.2, and the prepared image shows remarkable creases. For the new bills, the compensation of the same condition is done. In this study, the setting of gamma values for the image compensation was repeated by changing gamma value, and the comparison of the images decide the gamma value. This prepared image is changed into gray scale, and is then generated into a color histogram. Figure 1 is the fatigued bills after gamma compensation.

In addition, as feature extraction in the full color image and compression from a practical application standpoint, the color histogram is divided into 8 section and the color decreases in eight levels from 256. The provided feature is used in the classification experiment as the bill feature.

Figure 2 (a), (b) show the RGB histogram of the new bill and fatigue bill respectively, and the segmented region is a range delimited in the line. In this study, the range is divided into eight. In addition, the values were integrated within the range of division, and were used for the verification experiment as an RGB feature.

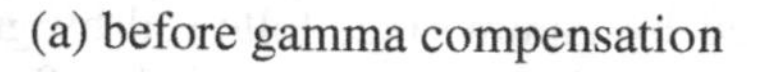

(a) before gamma compensation (b) after gamma compensation

Fig. 1 The fatigued bill

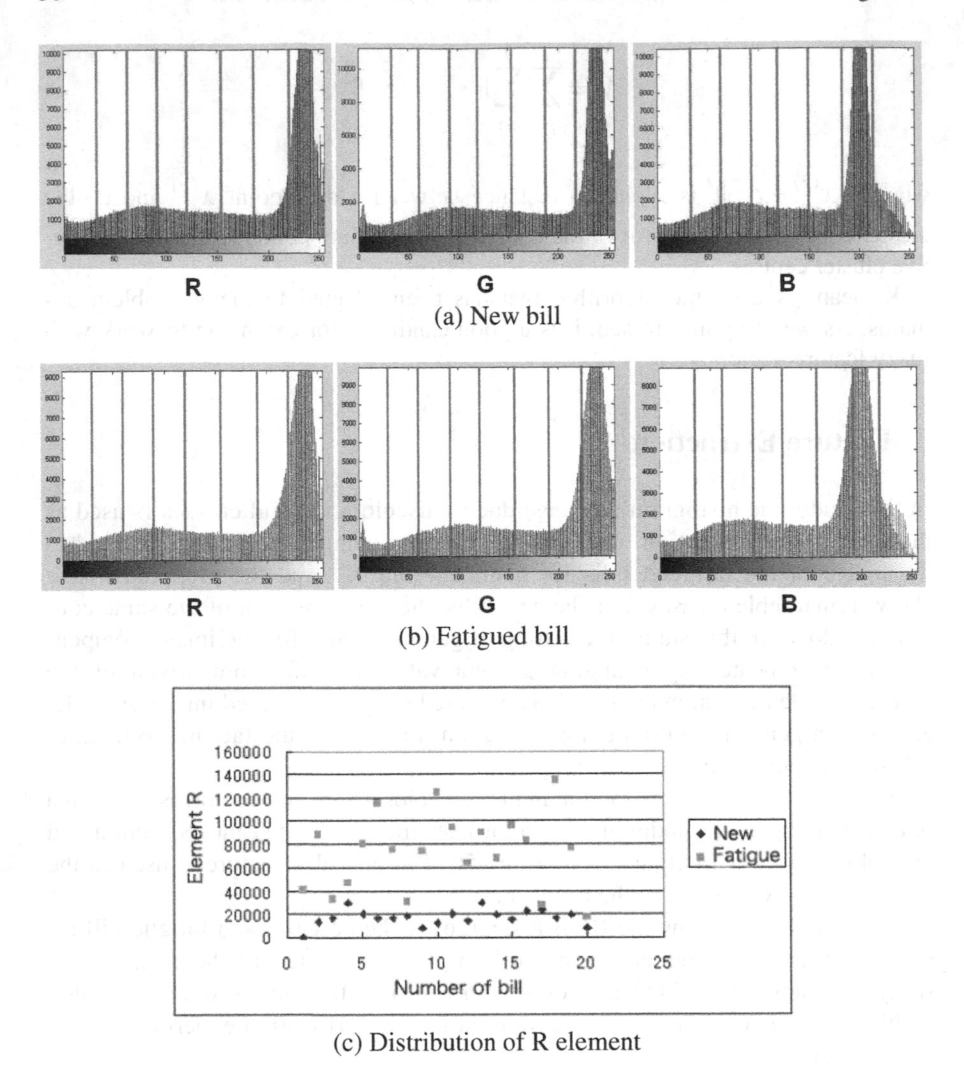

(a) New bill

(b) Fatigued bill

(c) Distribution of R element

Fig. 2 Example of RGB feature

4 Distinction Experiment

This experiment was performed using American 1 doller bills where 120 bills were used in total. The size of the image datum of paper money was 520*1230, and the measurement was carried out by the CIS scanner.

In the first step, the bills are compensated by gamma 0.2, followed by gray scaling and color histogram conversion. The color histogram is divided to 8 and the color decreases in eight levels starting around 256. And the RGB feature is input to K-means method.

In the next step, 3 new bills and 3 fatigue bills, respectively, have random supervised data selected as a distinguish method, and is clustered by the K-means method. A center point of two classes is obtained by the K-means.

Finally, the test data is inputed, and the distance from the center of each cluster is measured, and it is determined whether it is the new bill or the fatigue bill.

The comparison experiment using the image (original image) before the compensation and the image after the compensation was carried out in order to verify the effectiveness of the discriminant technique of this study. The distribution of the amount of feature before gamma compensation is corrected to Figure 4 (a) as shown, and the distribution of the amount of characteristic after gamma compensation is corrected to Figure 4 (b) as shown. Here, a black point and a white point under distribution show new bills and fatigue bills, respectively.
Figure 4 shows that distance among the color histogram of the new bills is narrow, and, compared with the bills before gamma compensation, the bills after gamma compensation had better accuracy. This experiment showed that the gamma compensation was effective.

In this experiment, both bills before compensation and bills after compensation were able to perform high accurate classification altogether. Especially, in the fatigued bills, a discrimination accuracy over 90% was obtained.

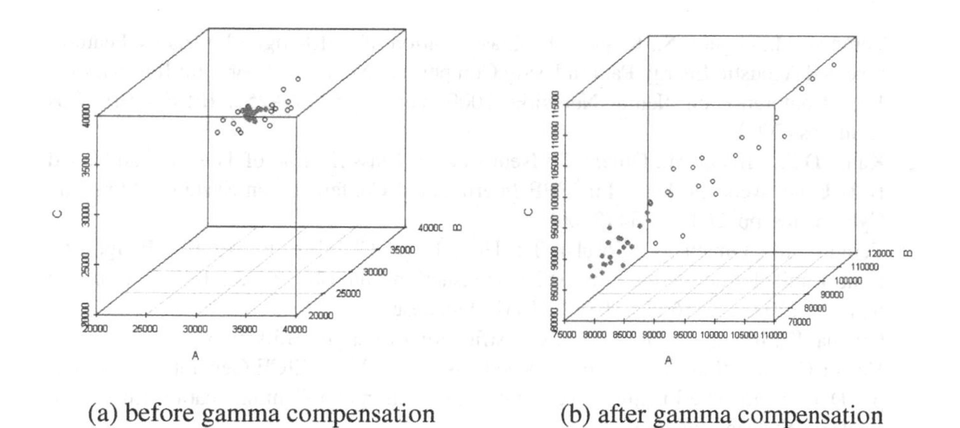

(a) before gamma compensation (b) after gamma compensation

Fig. 3 Result of K-means method

Table 1 Classification Result

Bill	Result
New Bill	50/50 (100%)
Fatigue Bill	187/200 (94%)

5 Conclusion

In the present study, we paid attention to the change in the histogram by the wrinkle and discoloration in the image data. The distinction between fatigued bills and new bills were done by observation the change of histograms. The experiments from various points of view were done; using horizontal parts considering where the actual automatic cash machines sensor were on one hand, and using vertical parts considering the stronger influence of creases on the other. Additionally, the distinction experiment using the Euclid distance method of color histogram was performed.

As a result, it is thought that the effectiveness of the fatigue bill identification by this proposal technique was able to be shown through that uses the change in the histogram by the wrinkle and discoloration in the image data.

For the future, the new feature and identification technique that aim at the improvement of the identification rate are will be examined.

Acknowledgments. This work has been supported by the Japan Cash Machine Co., Ltd. in 2-3-15 Nishiwaki, Hirano-ku, Osaka 547-0035 Japan.

References

1. Teranisi, M., Omatu, S., Kosaka, T.: Classification of Bill Fatigue Levels by Feature-Selected Acoustic Energy Pattern Using Competitive Neural Network. In: International Joint Conference on Neural Networks 2000, vol. 6, pp. 249–252. IEEE Press, Los Alamitos (2000)
2. Kang, D.S., Miyagi, H., Omatu, S.: Neuro-Fuzzy Classification of The New and Used Bills Using Acoustic Data. In: IEEE International Conference on Systems, Man, and Cybernetics, pp. 2649–2654 (2000)
3. Teranisi, M., Omatu, S., Kosaka, T.: Three-Level Classification of Bill Fatigue by Band-Limited Energy Patterns. IEEJ Transactions on Electronics, Information and Systems 120-C, 11, 1602–1608 (2000) (in Japanese)
4. Oyama, K., Kang, D., Miyagi, H.: Classification of Fatigue Bills by a Fuzzy Learning Vector Quantization Method. In: Proceedings of the 2005 IEICE General Conference, pp. D-12–D-45. The Institute Electronics, Information and Communication Engineers, Osaka (2005) (in Japanese)
5. Uehara, R., Kang, D., Miyagi, H.: Classification of Fatigue Bill with Independent Component Analysis. Technical Report of IEICE, PRMU, 107(384), 71–75 (2007) (in Japanese)
6. Ishigaki, T., Higuchi, T.: Detection of Worn-out Banknote by Using Acoustic Signals ~Time-varying Spectrum Classification by Divergence based Kernel Machines~. Journal of SICE 44(5), 444–449 (2008) (in Japanese)
7. Motooki, T., Omatu, S., Yoshioka, M., Teranishi, M.: Noise Reduction of Acoustic Data of Bill under Noisy Environment Using Adaptive Digital Filter and Neural Network. IEEJ Transactions on Electronics, Information and Systems 129-C, 9, 1724–1729 (2009) (in Japanese)
8. Yamamoto, K., Murakami, S.: A Study on Image Segmentation by K-Means Algorithm. Technical Report of IEICE, PRMU, Vol. 103(514), 83–88 (2003) (in Japanese)

9. Miyaguni, S., Kang, D.: Classification of Fatigue Bills Using by feature quantity of Wrinkles. Record of 2007 Joint Conference of Electrical and Electronics Engineers, 234–234 (2007) (in Japanese)
10. Miyaguni, S., Kang, D., Miyagi, H.: Classification of Fatigue Bills Using by Creases Feature. In: Proc. of the 8th International Conference on Applications and Principles of Information Science (APIS 2009), pp. 263–266 (2009)

A Classification Method of Inquiry e-Mails for Describing FAQ with Self-configured Class Dictionary

Koichi Iwai, Kaoru Iida, Masanori Akiyoshi, and Norihisa Komoda

Abstract. Recently the number of interactions between a company and its customers has been increased and it has taken a lot of time and cost of help desk operators. Companies construct FAQ pages in their web site and try to provide better services for their customer, however it takes surplus costs to analyze stored inquiries and extract frequent questions and answers. In this paper the authors propose a classification method of inquiry e-mails for describing FAQ (Frequently Asked Questions). In this method, a dictionary used for classification of inquiries is generated and updated automatically by statistical information of characteristic words in clusters, and inquiries are classified correctly to a proper cluster. This method achieved 70 percent precision of inquiry classification in an experiment with practical data stored in the registration management system for a sports association.

Keywords: Natural language processing, FAQ, Help desk, Clustering.

1 Introduction

Recently company services have become more diversified, and inquiries from the customers have also become more complicated. Companies provide a help desk

Koichi Iwai
Osaka University, Yamadaoka 2-1 Suita Osaka
e-mail: iwai.koichi@ist.osaka-u.ac.jp

Kaoru Iida
Osaka University, Yamadaoka 2-1 Suita Osaka
e-mail: iida.kaoru@ist.osaka-u.ac.jp

Masanori Akiyoshi
Osaka University, Yamadaoka 2-1 Suita Osaka
e-mail: akiyoshi@ist.osaka-u.ac.jp

Norihisa Komoda
Osaka University, Yamadaoka 2-1 Suita Osaka
e-mail: komoda@ist.osaka-u.ac.jp

A.P. de Leon F. de Carvalho et al. (Eds.): Distrib. Computing & Artif. Intell., AISC 79, pp. 35–43.
springerlink.com

as a contact point to their customers to deal with inquiries and claims, cultivate new demands and develop new services. In order to make the task of help desk effective, useful methods have been proposed[1, 2, 3, 4, 5]. Company also uploads FAQ(Frequently Asked Questions) and cuts the cost to deal with such inquiries. The researches related to FAQ have been also developed recently[6, 7, 8, 9, 10].

However, it is hard to add a new question and answer set into an existing FAQ because it takes a lot of costs to investigate a huge number of inquiries and replies correctly. In this paper the authors propose a method to display candidate question and answer sets to help desk operators. The candidate Q&As are extracted as a cluster of stored inquiries and replies similar to each other. The difficulties are what kind of clusters are proper for help desk operators as candidate Q&As and how to create such clusters from stored inquiries and replies. Some conventional clustering methods are not able to extract proper candidate Q&As because inquiries from customers include various expressions and lengths of sentences. To solve this problem a clustering method is proposed that makes and extends clusters in several stages. In the first stage, small clusters are constructed under strict condition and dictionary is generated to make clusters from the rest of inquiries and replies. In the second stage, the clusters constructed in the first stage are expanded under less strict condition. In the third stage, the clusters are refined as a candidate Q&A.

In section 1, the outline of this paper is described. In section 2, a conventional method of clustering is analyzed and the problems of that are made clear. In section 3, the authors propose a clustering method and describe the details. In section 4, an experiment to improve that the proposed method is efficient is described and the result of the experiment is analyzed. In section 5, the authors summarize this proposal.

2 Document Clustering for Making Candidate Q&A

2.1 Related Work

Document clustering[11] is a method making groups where documents are similar to each other. It has been used in a lot of different ways, such as improvement of effectiveness on information retrieval and visualization of search result. In this research document clustering is supposed to be effective because the purpose of the research is extracting frequently asked questions and answers of those from stored inquiries and replies.

There are some steps below generally in document clustering. First step is dividing document into words by morphological analysis. Japanese sentences have to be separated into words by morphological analysis, which is a method dividing natural language into minimum unit by predicting word class of each word. Second step is generating word vector from document. Word vector consists of weighted words and reflects feature of document. The weight of words is generally calculated by word frequency and *tf-idf(term frequency - inverse document frequency)* method. Third step is definition of similarity between documents. Cosine similarity is generally used, which means cosine value of word vectors. Fourth step is making clusters. Clusters are made based on the similarity index defined the step above.

2.2 Conventional Clustering Method and the Problem

In this section the problems are described when the conventional clustering method is applied to making candidate Q&A for generating new FAQ. First, one of challenging things is how to select words used in word vector. Documents used in this research are sentences a lot of people write under no rules and a lot of words are spelled in several different ways. Grammatical error and redundant expressions are included in a lot of sentences. Therefore it is important to select words used as a dimension of word vector. Second, it is also important to decide which words should be weighted and how they are weighted. *tf-idf* method, which is generally used in a lot of cases, is a statistics index and useful when there are a lot of document data and documents include a lot of words. The documents used in this research, however, have various patterns in the length of sentence and some inquiries can include a few words. Therefore it is difficult to make word vector that reflects the feature of document.

Finally, clustering result of conventional method depends on the threshold value and it is quite difficult to decide proper threshold value. When it is high, a lot of small clusters are made because documents do not tend to be combined. On the other hand, big clusters can be made when the threshold is low, however the preciseness of clusters may become low. In this research big and precise clusters are needed because the clusters show the frequency of questions and operators extract a frequent question from each cluster. The proposed method have to achieve such a trade-off criteria in constructing clusters.

3 Proposed Clustering Method with Self-configured Class Dictionary

3.1 Approach

The purpose of our research is clustering of documents that are a pair of inquiry and reply, which is called "thread", and making groups of frequent question and answer as a new Q&A in FAQ. The features of clusters that have to be made in the research are two points, cluster size is large and preciseness of cluster is high.

Large clusters, which mean there are great deal of inquiries similar to each other, should be generated because the size of clusters means the frequency of inquiries. It is needed to make the threshold value low to make large clusters. But in this case the clusters have to be precise. It means that threads in a cluster have to be similar to each other because operators make a new Q&A from a cluster. To achieve both conditions, the process of making clusters is divided into three parts, which are making core clusters, expanding clusters and sophistication of clusters, and the threshold values are arranged properly in each step.

3.2 Outline of the Clustering

Our system outputs candidate Q&A clusters via mainly three steps shown in Fig. 1.

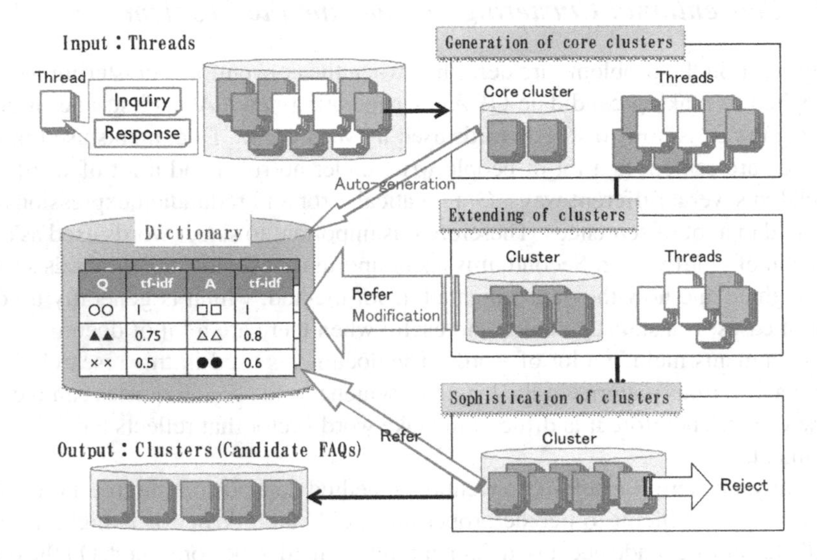

Fig. 1 An overview of Generation on Candidate FAQ Clusters

The first step is to construct core clusters that are small clusters but include quite similar threads. And then the dictionary is constructed based on the core clusters. The dictionary have weighted values of words for each cluster.

The second step is expansion of clusters. The threads that are not clustered in the first step is added to core clusters and the clusters are expanded by using the dictionary constructed in the first step.

Final step is sophistication of clusters. In this step all the threads are checked whether the cluster that they are included in is really proper. If threads are contained in a wrong cluster, they are removed from the cluster. Each step is described in the later sections.

3.3 Construction of Core Clusters

Core clusters are needed to be constructed precisely because they should involve statistical information for making the dictionary. Therefore core clusters have to be constructed with strictly similar threads to each other. The similarity index is calculated from the sum of the similarity between inquiries of threads and the similarity between replies of threads. The similarity index is defined as shown in the below and α is a constant value from 0 to 1.

$$Sim(Doc_i, Doc_j) = (1-\alpha) \times cosSimQ + \alpha \times cosSimA \tag{1}$$

$$cosSimQ = \frac{\mathbf{Q_i} \cdot \mathbf{Q_j}}{|\mathbf{Q_i}| \times |\mathbf{Q_j}|}, \quad cosSimA = \frac{\mathbf{A_i} \cdot \mathbf{A_j}}{|\mathbf{A_i}| \times |\mathbf{A_j}|}$$

$\alpha : Constant$

$\mathbf{Q_{i,j}} : Word\ vector\ of\ inquiry$

$\mathbf{A_{i,j}} : Word\ vector\ of\ reply$

Self-configured Class Dictionary

Self-configured class dictionary is constructed automatically from core clusters and provides the feature of each cluster. The feature of clusters is as a word vector. The each element of vector means the feature value of the element word. In this research, *tf-idf(term frequency inverse document frequency)* is used as the feature value. It is one of the most popular index to calculate the importance of words in a thread. Words having a high score on *tf-idf* rule means important one in a cluster because they are frequently appearing words in a cluster and the number of clusters including them is small.

As the clusters are expanded, the dictionary is modified. As the number of threads in a cluster increases, the statistics of words in the cluster changes. At the same time the value of *tf-idf* changes.

3.4 Expansion of Clusters

In this step core clusters generated in the first step are expanded. There are two types of cluster expansion with the dictionary, adding a thread to a cluster and combining two clusters.

Adding a thread to a cluster is a process of making a thread such as inquiries and responses included in a cluster by referring to the dictionary. Combining two clusters is also needed to make large clusters. Similar inquiries and replies have to be merged to display proper candidates. Expanded clusters by the processes above modify the dictionary. The system keeps expanding clusters and the clusters that are not able to be expanded are regarded as a candidate FAQ.

Adding a thread to a cluster

In order to decide the cluster that a thread have to be included in, feature words in clusters and the threads are compared. All the words in clusters are scored in the *tf-idf* manner and a feature vector is created in each cluster. The feature vector is a weighted word vector that has the *tf-idf* scores in the dictionary as the weight of words. At the same time the feature of thread is described in a word vector. The element value is defined as the frequency of each word occurrence. The similarity between a cluster and a thread is decided by the cosine value between the feature vector of the cluster and the word vector of the thread. And the most similar cluster is decided by comparing all the similarities between clusters and a thread. If the highest similarity is over the threshold value decided in advance, the thread is included into the cluster the most similar to it. The expression are described as follows.

$$Sim(Cluster_m, Doc_j) = (1-\alpha) \times cosSimQ + \alpha \times cosSimA \tag{2}$$

$$cosSimQ = \frac{\sum_{i=1}^{n} cosSimQ_i}{n}, \quad cosSimA = \frac{\sum_{i=1}^{n} cosSimA_i}{n}$$

$$cosSimQ_i = \frac{\mathbf{tfidf_{Q_m}(Q_i) \cdot Q_j}}{|\mathbf{tfidf_{Q_m}(Q_i)}| \times |\mathbf{tfidf_{Q_m}(Q_j)}|}$$

$$cosSimA_i = \frac{\mathbf{tfidf_{A_m}(A_i) \cdot A_j}}{|\mathbf{tfidf_{A_m}(A_i)}| \times |\mathbf{tfidf_{A_m}(A_j)}|}$$

$\alpha : Constant$

$n : The\ number\ of\ threads\ in\ cluster_m$

$\mathbf{Q} : Word\ vector\ of\ inquiry$

$\mathbf{A} : Word\ vector\ of\ reply$

$\mathbf{tfidf}() : Feature\ vector\ of\ cluster_m$

Combining two clusters

The threshold value in the step of making core clusters is a high value. Therefore a lot of small clusters can be made. Our purpose of making clusters is displaying candidate Q&A to operators, so that similar clusters have to be combined and large clusters are needed to be constructed. The similarity between clusters is defined as the cosine similarity between feature vectors of the clusters.

The number of feature vector elements used in this process is k. This is why that typical words in a cluster discriminating the cluster tend to have high score of *tf-idf*, therefore so many words are not needed to be taken in consideration for deciding similarity. It means that upper some words just individualize the cluster they are included in. The general expression of the similarity between clusters is shown as follows.

$$\begin{aligned} &Sim(Cluster_m, Cluster_n) \\ &= \mathbf{tfidf(k)_m \cdot tfidf(k)_n} \\ &= \frac{\mathbf{tfidf(k)_{Q_m} \cdot tfidf(k)_{Q_n} + tfidf(k)_{A_m} \cdot tfidf(k)_{A_n}}}{2} \end{aligned} \tag{3}$$

$\mathbf{tfidf(k)} : Feature\ vector$

3.5 Sophistication of Clusters

The end of processes in the proposed method is sophistication of clusters. Sophistication of clusters means that threads meaning another content in a cluster are removed. This process is needed because the pre-process may add threads to a wrong cluster. It occured because the threshold value is low for expanding clusters. In this process such threads in a wrong cluster are removed.

Whether threads in a cluster should be removed is decided by the value of characterized word including index($C(i)$). The index shows how much the threads in a cluster have the words characterizing the cluster. The expression is as follows.

$$C(i) = \sum_{w_i \in Dict(Cluster_m)} f(w_i) \tag{4}$$

$$f(w_i) = \frac{Word(w_i)(tf\text{-}idf)}{Number\ of\ words}$$

$$Dict(Cluster_m) : Words\ of\ cluster_m\ in\ the\ dictionary$$

If the index of threads is lower than the threshold value, the threads are removed from the cluster.

4 Evaluated Experiment

In the experiment threads of inquiries and replies about a sport membership administration Web site were used. The number of threads that are a pair of an inquiry and a reply is 1350. The number of clusters made by hand is 327 and the clusters having over 50 threads are defined as a candidate FAQ.

Evaluation criterions are size of cluster and precision of clustering. Size of a cluster is defined as the number of threads in the cluster. Candidate FAQ have to be a big cluster to help picking up frequent inquiries with operators. And precision of clustering is the rate of threads classified correctly in a cluster. This value was evaluated the way that threads in a cluster on the proposed method and those in a correct cluster made by hand are compared. These items were compared with the two of conventional methods; simply use of Jaccard coefficient and Cosine similarity.

4.1 Result of Experiment

The result of the experiment is shown in Fig. 2. As for size of cluster, the proposed method could generate about over three times bigger clusters than the conventional methods in the results of cluster no.1 and 2. In the conventional methods, the size of cluster no.1 is as same as that of cluster no.2. It means that operators can not judge which cluster should be more proper candidate FAQ by the conventional methods. On the other hand, the proposed method makes operators choice clusters that should

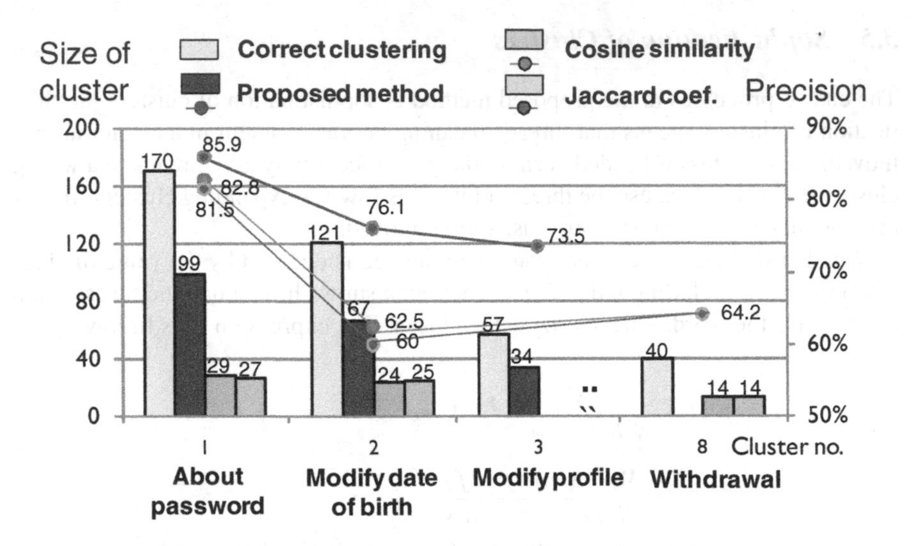

Fig. 2 Size of clusters and precision

be become a FAQ at a glance. In addition, the proposed method could make a cluster about cluster no.3 even though the conventional methods could not generate that.

As for precision of clustering, the proposed method could work more than the conventional methods. The precision in the proposed method keep over 70 percent.

5 Conclusion

The authors propose a clustering method for displaying candidate FAQs to operators at help desk. Threads consist of an inquiry from a customer and response of an operator and they are clustered correctly in the proposed method. The process of clustering is divided into three steps. The first step is making core clusters in a strict condition and documents in the clusters are related to each other. The second step is expansion of the clusters. The third step is sophistication of clusters and some threads are removed from their cluster, and the clusters become more precisely. An experimental result shows the effectiveness of the proposed method. The size of clusters in the proposed method was three times as big as those of conventional methods. And the precise of the clusters keep 70 percent in the proposed method.

References

1. Morimoto, Y., Mase, H., Hirai, C., Kinugawa, K.: Operator Support for Help Desk System (Information Retrieval). Transactions of Information Processing Society of Japan 44(7), 1731–1739 (2003)

2. Enoki, H., Tatsumi, S.: Workflow Management with a Multi-agent System and an Application to a Helpdesk System(Special Issue on Next Generation Mobile Communication Networks and their Applications). Transactions of Information Processing Society of Japan 43(12), 4023–4033 (2002)
3. Nasukawa, T.: Text Mining Application for Call Centers(Special Issue: Text Mining). Journal of Japanese Society for Artificial Intelligence 16(2), 219–225 (2001)
4. Takayama, Y., Aikawa, T., Suzuki, K.: Problem Solving Function in Helpdesk Support Systems. In: Proceedings of the IEICE General Conference, Information system (1), p. 61 (2000)
5. Shimazu, H., Ito, M.: Trends, Technologies and Strategies for Customer Support. Journal of Information Processing Society of Japan 39(9), 912–917 (1998)
6. Sneiders, E.: Automated FAQ Answering with Question-specific Knowledge Representation for Web Self-service. In: 2nd Conference on Human System Interactions (HSI 2009), pp. 298–305 (2009)
7. Sheng-Yuan, Y.: Developing an Ontological FAQ System with FAQ Processing and Ranking Techniques for Ubiquitous Services. In: First IEEE International Conference on Ubi-Media Computing, pp. 541–546 (2008)
8. Chih-Hao, H., Song, G., Rung-Chin, C., Shou-Kuo, D.: Using Domain Ontology to Implement a Frequently Asked Questions System. In: World Congress on Computer Science and Information Engineering, vol. 4, pp. 714–718 (2009)
9. Harksoo, K., Jungyun, S.: Cluster-Based FAQ Retrieval Using Latent Term Weights. IEEE Intelligent Systems 23(2), 58–65 (2008)
10. Hammond, K., Burke, R., Martin, C., Lytinen, S.: FAQ finder: a case-based approach to knowledge navigation. In: 11th Conference on Artificial Intelligence for Applications, pp. 80–86 (1995)
11. Li, Y., Chung, S.M., Holt, J.D.: Text document clustering based on frequent word meaning sequences. Data & Knowledge Engineering 64(1), 381–404 (2008)

A Recommendation System for the Semantic Web

Victor Codina and Luigi Ceccaroni

Abstract. Recommendation systems can take advantage of semantic reasoning-capabilities to overcome common limitations of current systems and improve the recommendations' quality. In this paper, we present a personalized-recommendation system, a system that makes use of representations of items and user-profiles based on ontologies in order to provide semantic applications with personalized services. The recommender uses domain ontologies to enhance the personalization: on the one hand, user's interests are modeled in a more effective and accurate way by applying a domain-based inference method; on the other hand, the matching algorithm used by our content-based filtering approach, which provides a measure of the affinity between an item and a user, is enhanced by applying a semantic similarity method. The experimental evaluation on the Netflix movie-dataset demonstrates that the additional knowledge obtained by the semantics-based methods of the recommender contributes to the improvement of recommendation's quality in terms of accuracy.

Keywords: Recommendation systems, Semantic Web, Ontology-based representation, Semantic reasoning, Content-Based filtering, Services Orientation.

1 Introduction

Most common limitations of current recommendation systems are: *cold-start*, *sparsity*, *overspecialization* and *domain-dependency* [4]. Although some particular combination of recommendation techniques can improve the recommendation's quality in some domains, there is not a general solution to overcome these limitations. The use of *semantics* to formally represent data [1] can provide several advantages in the context of personalized recommendation systems, such as the dynamic contextualization of user's interests in specific domains and the guarantee of interoperability of system resources. We think that the next generation of

Victor Codina and Luigi Ceccaroni
Departament de Llenguatges i Sistemes Informàtics (LSI),
Universitat Politècnica de Catalunya (UPC),
Campus Nord, Edif. Omega, C. Jordi Girona, 1-3, 08034 Barcelona, Spain
e-mail: {vcodina,luigi}@lsi.upc.edu

A.P. de Leon F. de Carvalho et al. (Eds.): Distrib. Computing & Artif. Intell., AISC 79, pp. 45–52.
springerlink.com

recommenders should focus on how their personalization processes can take advantage of semantics as well as social data to improve their recommendations. In this paper, we describe how the accuracy of recommendation systems is higher when semantically-enhanced methods are applied.

The structure of the paper is as follows: in section 2 we present the state of the art of recommendation systems and semantic recommenders; in section 3 we describe a new domain-independent recommendation system; and in section 4 we presentan experimental evaluation of the recommender.

2 Related Work

Different recommendation approaches have been developed using a variety of methods. A detailed review of the traditional approaches based on user and item information, and also a description of the current trend in systems that try to incorporate contextual information to the recommendation processis presented in section2.3 of Codina [4]. *Semantic* recommendation systems are characterized by the incorporation of semantic knowledge in their processes in order to improve recommendation's quality.

Most of themaim to improvethe user-profile representation(*user modeling-*stage), employing a concept-based approach and using standard vocabularies and ontology languages like OWL. Two different methods can be distinguished:

- Approaches employing *spreading activation* to maintain user interests and treating the user-profile as a semantic network. The interest scores of a set of concepts are propagated to other related concepts based on pre-computed weights of concepts relations. A news recommender system [3] and a search recommender [8] employ this method.
- Approaches that apply *domain-based inferences*, which consist of making inferences about user's interests based on the hierarchical structure defined by the ontology. The most commonly used is the *upward-propagation*, whose main idea is to assume that the user is interested in a general concept if he is interested in a given percentage of its direct sub-concepts. This kind of mechanisms allows inferring new knowledge about the long-term user's interests and therefore modeling richer user-profiles. *Quickstep* [7], a scientific-paper recommender, and *Travel Support System* [6], a tourism-domain recommender, employ an *upward-propagation* method to complete the user profile.

Other recommenders focus onexploiting semantics to improve the*content adaptation* stage. Most of them make use of *semantic similarity* methods to enhance the performance of a *content-based*approach (CB), although there are also some recommenders using semantics to enhance the user-profile matching ofa *collaborative filtering*approach. The only recommender that makes use of semantic reasoning methods in both stages of the personalization process is *AVATAR*[2], a TV recommender that employs *upward-propagation*and *semantic similarity* methods.

3 A Semantic Recommendation System

In this section we present the main components and characteristics of the semantic recommendation system we developed, which makes use of semantics-based methods to enhanceboth stages of the personalization process.

3.1 *Architectural Design*

In order to develop a domain-independent recommender, it is necessary to decouple the recommendation engine from the application domains. For this reason, we designed the system as a service provider following the well-known *service oriented architecture* (SOA) paradigm.In**Fig. 1**, the abstract architectural design is represented. Using this decoupled design, each Web-application or domain has to expose a list of items to be used in the personalization process; items has to be semantically annotated using the hierarchically structured concepts of the domain ontology, which is shared with the recommender. Thus, the recommendation engine can work as a personalization service, providing methods to generate personalized recommendations as well as to collect user feedback while users interact with Web-applications. In order to facilitate the reuse of user profiles as well as the authentication process we employ the widely used FOAF vocabulary as the basis of our ontologically extended user profiles, which is compatible with the *OpenID* authentication [http://openid.net/].

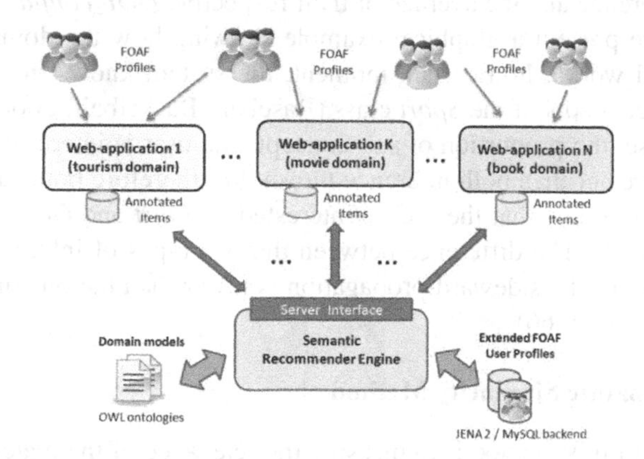

Fig. 1 General architecture design

3.2 *Semantic Reasoning Method*

Our semantic recommender employs the typical weighted overlay approach, used in ontological user profiles to model user's interests, that consists of mapping collected feedback about semantically annotated items to the corresponding concepts of the domain; the association is done with a weight, which indicates the degree of interest (*DOI_weight*) of the user. In combination with the weight value, we use a measure of how trustworthy is the interest prediction of the particular concept (*DOI_confidence*) to reduce/increase its influence during the recommendation. The recommender takes advantage of this ontological representation in the two stages of the personalization process:

- The user-profile learning algorithm, responsible for expanding and maintaining up-to-date the long-term user's interests, employs a *domain-based inference* method in combination with other relevance feedback methods to populate more quickly the user profile and therefore reduce the typical cold-start problem.
- The filtering algorithm, which follows a CB approach, makes use of a *semantic similarity* method based on the hierarchical structure of the ontology to refine the item-user matching score calculation.

3.2.1 The Domain-Based Inference Method

The domain-based inference method we used is an adaptation of the approach presented in [5] and consists of inferring thedegree of interestfor a concept using subclass or sibling relations (upward or sideward propagation) when the user is also interested in a minimum percentage (the inference threshold) of direct subconcepts or sibling concepts. The predicted weight is calculated as the *DOI_weight* average of the sub-concepts or sibling concepts the user is interested in, and the confidence value is based on the percentage of sub-concepts or siblings used in the inference and the average of their respective *DOI_confidence* values.

In Fig. 2, we present a graphical example showing how the domain-based inference method works. In a certain moment, the system knows the user is interested in 4 sub-concepts of the *Sport* class (Baseball, Basketball, Football and Tennis). In this case, the proportion of sub-concepts the user is interested in (4 out of 5, i.e., 0.8) is greater than both inference thresholds, therefore both can be applied. Thus, the system infers that the user is interested in *Sport* and *Golf* with the same *DOI_weight* (0.62). The difference between the two types of inference is that the *DOI_confidence* of the sideward-propagation is lower than the one of the upward-propagation (0.5 vs. 0.66).

3.2.2 The Semantic Similarity Method

The basic idea of this method is to measure the relevance of the matching between a particular concept the user is interested in and a concept describingthe item. (In Fig. 3, two examples are shown, in which the user's interest is the parent of the item concept.) We can distinguish two types of matching:

- The item concept is one of the user's interests, so the matching is perfect and the similarity is maximum (1).
- An ancestor of the item concept (e.g., the direct parent) is one of the user's interests. In this case the similarity is calculated using the following recursive function whose result is always a real number(lower than 1).

1. $SIM_n = SIM_{n-1} - K * SIM_{n-1} * n$ (partial match, n>0)
2. $SIM_0 = 1$ (perfect match, n=0)

Where:

- ***n*** is the distance between the item concept and the user's interest (e.g., when it is the direct parent, ***n*** = 1);
- ***K*** is the factor that marks the rate at which the similarity decreases (the higher ***n***, the higher the decrement). This factor is calculated taking into account the depth of the item concept in the hierarchy and is based on the assumption that semantic differences among upper-level concepts are bigger than those among lower-level concepts.

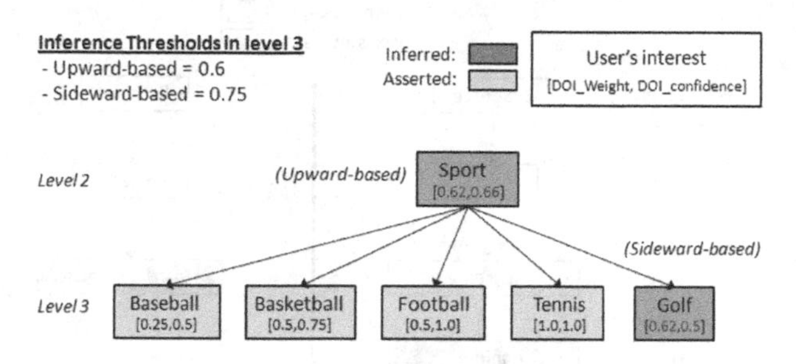

Fig. 2 An example of how new interests are inferred

4 Experimental Evaluation

In this section the undertaken experimental evaluation of the recommender is presented.

The main goal of the experiments is to demonstrate how the recommendation's quality of a CB approach is improved when semantically-enhanced algorithms are employed. We employ the well-known Netflix-prize movie dataset in order to evaluate the recommendation's quality of the recommender in terms of accuracy of rating predictions. TheNetflix dataset consists of 480,000 users, 17,700 movies and a total of 100,480,507 user's ratings ranging between 1 and 5. We employ the same predictive-based metric used in the contest, the *root mean square error* (RMSE).

4.1 Experimental Setup

To evaluate how the semantically-enhanced algorithms contribute to improve the recommendation's quality in terms of accuracy, we compare the prediction results obtained executing the recommender in three different configurations:

- *CB*. It represents the traditional CB approach; therefore the methods that take advantage of the ontology representation are disabled.In this case, the item-user matching only takes into account the concepts that perfectly match.

- *Sem-CB*. It employs the semantics-based methods presented in section 3 using, as domain ontology and movie indexation, the same taxonomy of three levels of depth used by Netflix and publicly available[http://www.netflix.com/AllGenresList].
- *Sem-CB+*. It employs the semantic-based methods using, as domain ontology, an adaptation of the Netflix taxonomy, with a concepts hierarchy of four levels of depth (see Fig. 4). We also changed the indexation for concepts referring to two or more other concepts (i.e., we indexed movies related toNetflix's concept "*Family Dramas*"separately under"*Family*" and "*Drama*") in order to reduce the ontology size.

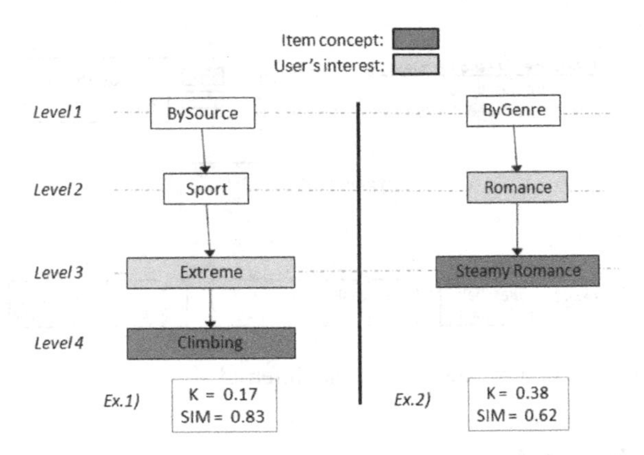

Fig. 3 How the similarity method works

4.2 Results

The errorof the predictions generatedby the system (see Table 1) demonstratesthat, when semantics is used, the recommendation's accuracy improves with respect to the *CB* configuration. The accuracy of *Sem-CB+* is not better than *Sem-CB* when the parameters of the algorithms are properly adjusted (see Ex. 3 in Table 2). We compare both configurations using the same inference thresholds and the value of the K factor which provides the best accuracy in each case. In the case of *Sem-CB*: K=0.12 when the concept level is 3; K=0.31 when the level is 2. In the case of *Sem-CB+*: K=0.30 when the level is 4; K=0.40 when the level is 3; and K=0.50 when the level is 2. It can be observed that the improvement of accuracy is strongly related with the upward-inference threshold (the higher the number of upward-propagations, the better the results). For example, for Sem-CB+: 1.0443 – 1.0425 – 1.0397.

For comparison, a trivial algorithm that predicts for each movie in the quiz set its average grade from the training data produces an RMSE of 1.0540. Netflix's *Cinematch*algorithm uses "straightforward statistical linear models with a lot of data conditioning". Using only the training data, Cinematch scores an RMSE of 0.9514 on the quiz data, roughly a 10% improvement over the trivial algorithm.

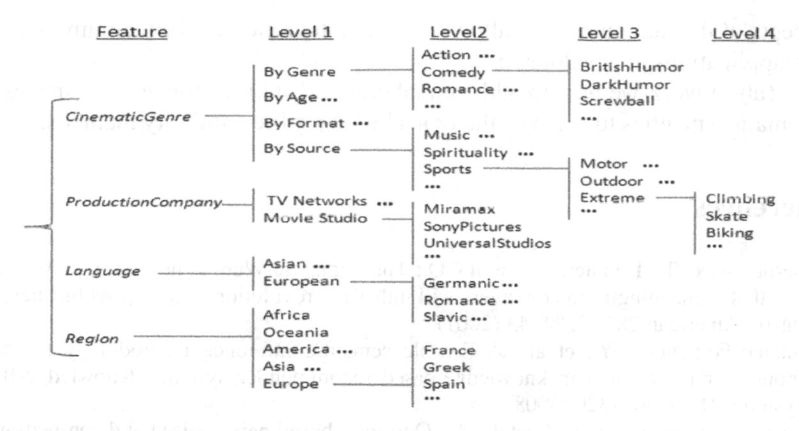

Fig. 4 Partial representation of the adapted movie taxonomy

Table 1 Global prediction-error (RMSE) results

Configuration	RMSE
CB	1.0603
Sem-CB	1.0391
Sem-CB+	1.0397

Table 2 Comparison of semantic-based configurations

Execution (*Upward – Sideward*) thresholds		Avg. Upward propagations	Avg.Sideward propagations	RMSE
Ex. 1 (0.60-0.75)	*Sem-CB*	4.32	2.87	1.0482
	Sem-CB+	6.01	3.83	1.0443
Ex.2 (0.40-0.75)	*Sem-CB*	8.89	3.85	1.0440
	Sem-CB+	9.99	3.89	1.0425
Ex.3 (0.20-0.85)	*Sem-CB*	13.84	2.88	**1.0391**
	Sem-CB+	17.73	3.30	**1.0397**

5 Conclusions and Future Work

This paper describes howthe accuracy of recommendation systemsis higher when semantically-enhanced methods are applied. In our approach, we make use of semantics by applying two different methods. A domain-based method makes inferences about user's interests and a taxonomy-based similarity method is used to refine the item-user matching algorithm, improving overall results.

The recommender proposed is domain-independent, is implemented as a Web service, and uses both explicit and implicit feedback-collection methods to obtain information on user's interests. The use of a FOAF-based user-model linked with concepts of domain ontologies allows an easy integration of the recommender into Web-applications in any domain.

As future work we plan to add a collaborative-filtering strategy that makes use of domain semantics to enhance the typical user-profile similarity methods.

References

1. Berners-Lee, T., Hendler, J., Lassila, O.: The Semantic Web. A new form of Web content that is meaningful to computers will unleash a revolution of new possibili-ties. Scientific American 284(5), 34–43 (2001)
2. Blanco-Fernández, Y., et al.: A flexible semantic inference methodology to reason about user preferences in knowledge-based recommender systems. Knowledge-Based Systems 21(4), 305–320 (2008)
3. Cantador, I., Bellogín, A., Castells, P.: Ontology-based personalised and con-text-aware recommendations of news items. In: Proc. of IEEE/WIC/ACM International Conference on Web Intelligence and Intelligent Agent Technology, pp. 562–565 (2008)
4. Codina, V.: Design, development and deployment of an intelligent, personalized recommendation system. Master Thesis. Departament de Llenguatges i Sistemes Informàtics, Universitat Politècnica de Catalunya. 101 pp (2009)
5. Fink, J., Kobsa, A.: User Modeling for Personalized City Tours. Artificial In-telligence Review 18(1), 33–74 (2002)
6. Gawinecki, M., Vetulani, Z., Gordon, M., Paprzycki, M.: Representing users in a travel support system. In: Proceedings - 5th International Conference on Intelligent Systems Design and Applications 2005, ISDA 2005, art. no. 1578817, pp. 393–398 (2005)
7. Middleton, S.E., De Roure, D.C., Shadbolt, N.R.: Capturing Knowledge of User Preferences: ontologies on recommender systems. In: Proceedings of the First International Conference on Knowledge Capture (K-CAP 2001), Victoria, B.C. Canada (October 2001)
8. Sieg, A., Mobasher, B., Burke, R.: Ontological user profiles for personalized Web search. In: AAAI Workshop - Technical Report WS-07-08, pp. 84–91 (2007)

Natural Scene Segmentation Method through Hierarchical Nature Categorization

F.J. Díaz-Pernas, M. Antón-Rodríguez, J.F. Díez-Higuera, M. Martínez-Zarzuela, D. González-Ortega, D. Boto-Giralda, and I. de la Torre-Díez

Abstract. In this paper we present a hierarchical learning method to segment natural colour images combining the perceptual information of three natures: colour, texture, and homogeneity. Human knowledge is incorporated to a hierarchical categorisation process, where each nature features are independently categorised. Final segmentation is achieved through a refinement process using the categorisation information from each segment. Experiments are performed using the Berkeley Segmentation Dataset achieving good results even when comparing them to other significant methods.

Keywords: Natural colour image segmentation; supervised categorisation; hierarchical neural network; Adaptive Resonance Theory, pattern refinement; Berkeley segmentation dataset.

1 Introduction

This paper considers the problem of segmentation of natural scenes defined by multi-nature features based on colour, texture and homogeneity. Great advances have been performed in colour image segmentation [5]. Natural scenes are formed of perceptual significance segment such as "sky", "water", "mountain", etc. Fig. 3-left shows manually segmented images of Berkeley Segmentation DataSet (BSDS) [13]. Segmentation of natural scenes is particularly difficult since the segment low-level features are not well defined and are not uniform due to the effects of lighting, perspective, scale, changes, etc. [4]. The use of human segmented images to add human knowledge to the segmentation of natural images has been widely employed. Pyun et al. [12] used the manually segmented images in the supervision process for aerial images. Martin et al. [9] proposed an interesting

F.J. Díaz-Pernas, M. Antón-Rodríguez, J.F. Díez-Higuera, M. Martínez-Zarzuela, D. González-Ortega, D. Boto-Giralda, and I. de la Torre-Díez
Department of Signal Theory, Communications and Telematics Engineering
Telecommunications Engineering School. University of Valladolid. Valladolid. Spain
e-mail: {pacper,mirant,josdie,marmar,davgon,danbot}@tel.uva.es, isator@tel.uva.es

A.P. de Leon F. de Carvalho et al. (Eds.): Distrib. Computing & Artif. Intell., AISC 79, pp. 53–60.
springerlink.com

colour-texture gradient. They trained a classifier using human labelled images as ground truth. Hanbury and Marcotegui [8] proposed a method based on the distance function and also used the human segmented images as ground truth for boundaries learning. Grossberg and Williamson suggest that categorization processes take part in the human segmentation system [7]. In this paper we propose a hierarchy of categorisation processes to segment natural scenes through a supervised learning according to the knowledge of human perception. Proposed hierarchy of neural networks is based on the supervised and unsupervised Fuzzy models of the Adaptive Resonance Theory [3,2], which shows the high potentiality of this theory to develop complex recognition architectures. Another example of this feature is shown in the recently released work of Grossberg and Huang [6], who propose a neural system for natural scene classification, ARTSCENE.

The segmentation method presented is based on hierarchical categorization with pattern refinement feedback and variable similarity measure to achieve a better clustering. In a previous work [1,10], we proposed a neural architecture for texture recognition, which proved satisfactory processing images from the VisTex texture dataset [14]. Based on that experience and using part of the development performed, we now advance a hierarchical extension to segment natural scenes.

2 Proposed Approach Structure

The structure of the proposed approach is shown in Fig. 1. The architecture have four main stages: colour and texture feature extraction, nature categorisation stage, global pattern categorization and region merging stage.

2.1 Colour and Texture Feature Extraction

The proposed approach starts from three feature patterns of different natures: texture (tex), low-level homogeneity (llh), and opponent colour (col). These features are extracted using a multi-scale neural architecture with opponent colour codification, perceptual contour extraction, and diffusion processes. This neural architecture has been previously proposed within a coloured and textured image classification system, achieving very good results [1,10]. Fig. 2 displays a processing example of this architecture, input to the categorisation processes. The texture pattern is expressed following equation (1).

$$\mathbf{T}_{ij} = \{t_i\} = \left\{(\mathrm{Re},\mathrm{Im})_0^1, (\mathrm{Re},\mathrm{Im})_0^2, (\mathrm{Re},\mathrm{Im})_0^3, \ldots, (\mathrm{Re},\mathrm{Im})_{K-1}^1, (\mathrm{Re},\mathrm{Im})_{K-1}^2, (\mathrm{Re},\mathrm{Im})_{K-1}^3\right\} \quad (1)$$

where $(\mathrm{Re},\mathrm{Im})_k^s$ is the pair corresponding to the real and imaginary parts of the Gabor filtering for the scale s $\{s=1,2,3\}$ and the orientation k $\{k=0,\ldots K-1\}$ referring to K=6 orientations $\theta=\{0°,30°,60°,90°,120°,150°\}$.

The homogeneity pattern has two components $\{h_i, i=1,2\}$ corresponding to the diffusion signals of the two opponent channels (see Fig. 2, last row). Finally, the 6-dimensional pattern includes the signals from the two colour opponent channels (red-green and blue-yellow) for the three scales $\{c_i, i=1,\ldots,6\}$ (see Fig. 2, third row).

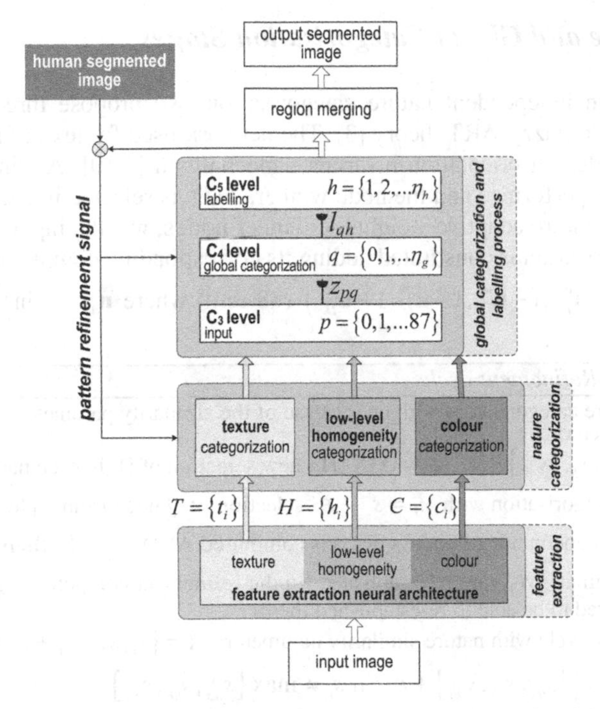

Fig. 1 Proposed segmentation method structure.

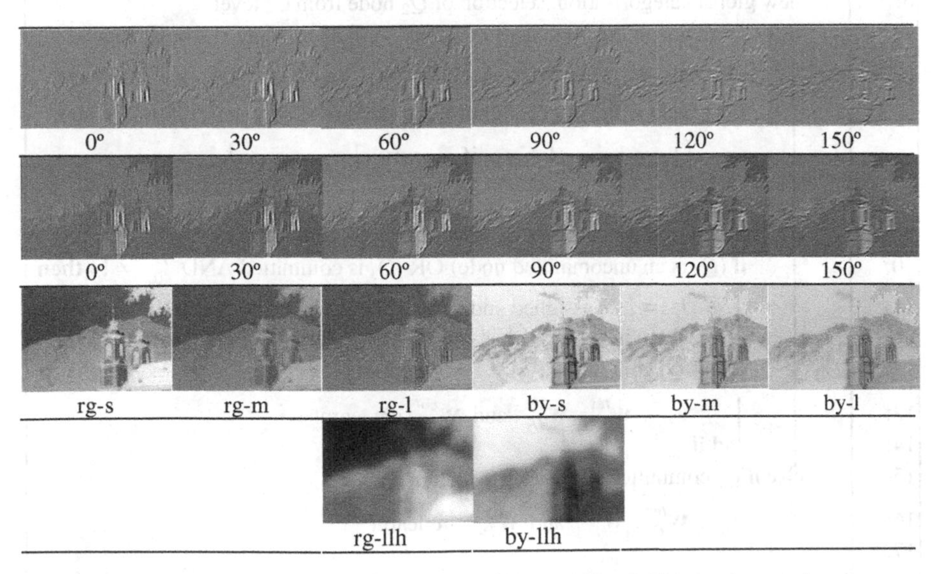

Fig. 2 Colour and texture feature images of an image of Fig. 3. First and second rows: texture components, imaginary and real components of the Gabor filtering for the small scale and the 6 orientations used; Third row: colour opponent components of the red-green and blue-yellow channels for the three scales (s,m,l). Last row: low-level homogeneity components, channels red-green and blue-yellow.

2.2 Nature and Global Categorisation Stages

To achieve an independent nature categorisation, we propose three neural networks based on Fuzzy-ART theory [3]. The network used for texture nature classification includes an orientation-invariance mechanism [1,10]. A winner takes all competition is performed and the node winner, *d=D*, is selected in each network.

Using the nature adaptive weights of winner nodes, $\mathbf{w}_D$, the input pattern to the global categorisation is constituted adding its corresponding complementary code: $\mathbf{P}_{ij} = \left(\mathbf{u}_D^{tex}, \mathbf{u}_D^{llh}, \mathbf{u}_D^{col}, 1-\mathbf{u}_D^{tex}, 1-\mathbf{u}_D^{llh}, 1-\mathbf{u}_D^{col}\right)$ (88-dim) where $\mathbf{u}_D = \min\left(\mathbf{w}_D^P, 1-\mathbf{w}_D^C\right)$

Algorithm 1: Refinement cycle

1: new nature categorisation with an increase of the similarity parameter: nature similarity parameters $\lambda = \{s_{tex}, s_{llh}, s_{col}\} + \varepsilon$, $\varepsilon = 0.0001$, new selection of D_2 in each nature.
2: global categorisation with $\lambda_l = s_g + \varepsilon$, selection of Q node from C_4 level
3: **if** (Q is an uncommitted node) OR (Q is committed AND $l_{Qh} \neq 1$) **then**
4: originals $\mathbf{w}_D$ are re-established (in the refinement compute, originals $\mathbf{w}_D$ are stored to be able to re-established them)
5: new cycle with nature similarity parameters $\lambda = \{s_1, s_2, s_3\} + \varepsilon$ where

$$s_i = \begin{cases} \{s_{tex}, s_{llh}, s_{col}\} + \varepsilon & \text{if } s_i \neq \max\{s_{tex}, s_{llh}, s_{col}\} \\ 0 & \text{if } s_i = \max\{s_{tex}, s_{llh}, s_{col}\} \end{cases}$$

6: new global categorisation, selection of Q_2 node from C_4 level
7: **if** (Q_2 is an uncommitted node) OR (Q_2 is committed AND $l_{Q_2l} \neq 1$) **then**
8: originals $\mathbf{w}_D$ are re-established
new cycle with nature similarity parameters $\lambda = \{s_1, s_2, s_3\} + \varepsilon$ where

$$s_i = \begin{cases} \{s_{tex}, s_{llh}, s_{col}\} + \varepsilon & \text{if } s_i = \min\{s_{tex}, s_{llh}, s_{col}\} \\ 0 & \text{if } s_i \neq \min\{s_{tex}, s_{llh}, s_{col}\} \end{cases}$$

9: new global categorisation, selection of Q_3 node from C_4 level
10: **if** (Q_3 is an uncommitted node) OR (Q_3 is committed AND $l_{Q_3l} \neq 1$) **then**
11: $l_{Ql} = 1$ is assigned and $\mathbf{z}_Q$ is learnt (Q is the node selected in step 2) There is not any other favourable situation in the relaxation cycle.
12: **else if** Q_3 committed AND $l_{Q_3l} = 1$ **then**
13: $\mathbf{z}_{Q_3}$, $\mathbf{w}_{D_2}^{tex}$, $\mathbf{w}_{D_2}^{llh}$, and $\mathbf{w}_{D_2}^{col}$ are learnt
14: **end if**
15: **else if** Q_2 committed AND $l_{Q_2h} = 1$ **then**
16: $\mathbf{z}_{Q_2}$, $\mathbf{w}_{D_2}^{tex}$, $\mathbf{w}_{D_2}^{llh}$, and $\mathbf{w}_{D_2}^{col}$ are learnt
17: **end if**
18: **else if** Q committed AND $l_{Qh} = 1$ **then**
19: $\mathbf{z}_Q$, $\mathbf{w}_{D_2}^{tex}$, $\mathbf{w}_{D_2}^{llh}$, and $\mathbf{w}_{D_2}^{col}$ are learnt
20: **end if**

The global pattern categorisation accomplishes the supervised labelling of the patterns. Hence, a supervised network comprised of three neural levels is proposed (see Fig. 1): input (C_3), global categorisation (C_4), and labelling levels (C_5).

The winner node activation, the process selection, and the similarity criterion procedure are similar to the nature categorisation network model [1,10]. In this work, we propose an adaptive labelling weights, l_{qh}, to link the winner node C_4, Q, to the C_5 node h, with h $\{h=0,1,\ldots,\eta_h\}$ the human supervised label, activating the associated adaptive weight, i.e. $l_{Qh}=1$.

3 Segmentation Approach with Pattern Refinement Cycle

The proposed segmentation approach involves learning all the variability of the features included in segments with perceptual significance. The segmentation method is supervised by the labels assigned to the human segmented images of the BSDS [13]. Our proposal has two operating modes: training and testing.

In the training mode, 2000 random samples of each image are chosen. These images are processed by the low-level feature extraction neural architecture, generating the input pattern to each nature from the samples chosen. Next, a global categorization is performed. If the label associated to the global selected node and the human supervision label are equal, the patterns (nature and global networks) are learnt. If there is a mismatch, the similarities computed will not be enough, so a new search with more strict similarity criteria will be needed. The cycle refinement included in this approach is established (see algorithm 1). In order to find the best categorisation, we place the more strict feasible criteria. If the linked label is equal to the supervision label, a favourable situation is achieved, so the patterns are learnt. If there is another mismatch, the chosen criterion is relaxed.

In the test mode, the label linked to the global selected node will represent the learnt label for the input pattern of each point in the image. These categorisation labels shape the hierarchical categorization segmented image, $\mathbf{S}_{ij}$.

3.1 Region Merging Stage

Since the categorisation performed is positional, small segments appear in the output image, which can be eliminated by a region merging process. The input is the categorisation label image, $\mathbf{S}_{ij}$, and the adaptive weights, $\mathbf{z}_{pq}$. The merging process is performed in two sequential stages: eliminating the small segments (size<50) and merging those with very high similarity measure (>95%). The process starts with the highest similarity union and it continues following a descending order of similarities, while the segments merging can be possible. The similarity measure of segments R_1 and R_2 with weights $\mathbf{z}_{p1}$ and $\mathbf{z}_{p2}$ $\{p=0,\ldots,87\}$ and updated weights

$$s(R_1,R_2)=\frac{\frac{1}{3}\left(\sum_{tex}\min(\mathbf{z}_{p1},\mathbf{z}_{p2})+\sum_{llh}\min(\mathbf{z}_{p1},\mathbf{z}_{p2})+\sum_{col}\min(\mathbf{z}_{p1},\mathbf{z}_{p2})\right)}{\min\left(\sum_{p}\mathbf{z}_{p1},\sum_{p}\mathbf{z}_{p2}\right)} \text{ and } \mathbf{z}_{p1\cup 2}=1/2\left(\mathbf{z}_{p1}+\mathbf{z}_{p2}\right).$$

PRI: 0.9128, VoI: 0.897,
GCE: 0.128, BDE: 3.3851,
F-value: 0.6765
PRI: 0.9368, VoI: 0.8499,
GCE: 0.1101, BDE: 4.6927,
F-value: 0.6762
PRI: 0.8448, VoI: 1.9143,
GCE: 0.1997, BDE: 4.0482,
F-value: 0.7002
PRI: 0.9292, VoI: 1.5463,
GCE: 0.1934, BDE: 3.5761,
F-value: 0.709
PRI: 0.9331, VoI: 1.3085,
GCE: 0.1719, BDE: 5.892,
F-value: 0.6836
PRI: 0.9072, VoI: 0.5224,
GCE: 0.085, BDE: 3.9874,
F-value: 0.6363

Fig. 3 Examples of the segmentation achieved with the proposed approach (right) and a comparison to the human segmented image (left). The average values of PRI, VoI, GCE and BDE of the entire Berkeley dataset (human segmented images) are 0.87, 1.1, 0.08, and 4.99 respectively [15]; and the F-measure average value is 0.79 [8]

The approach output is the segmented image in the iteration when there is not any possibility to join any segments, that is, when the merging process has ended.

4 Experiments

We compare our method against two significant algorithms, [15] and [8], that have quantitatively measured their results. The comparison to [15] is performed according to the four quantitative measures they used: the Probabilistic Rand Index (PRI), the Variation of Information (VoI), the Global Consistency Error (GCE), and the Boundary Displacement Error (BDE). The comparison to [8] is accomplished with the GCE parameter and with the boundary-based error, F-measure, which are the measures included in the paper. The measures computation has been made using the Matlab code supplied in their web pages [13,11].

We have used 114 images from the BSDS [13]. Their human segmented images were taken as the ground truth to accomplish the learning process, using 2000 random points from each image. In Table 1, we can see the achieved average of the five quantitative measures. In Fig. 3 we can visually compare some of our segmentation results with the corresponding human segmented images. The mean time per image has been 248 seconds. With the aim of accelerating the procedure, we work on parallel processing through GPUs (Graphic Processing Units) [10].

We compare our approach with two methods based on a supervised learning [8] and on a clustering process of Gaussian mixture models [15]. Hanbury and Marcotegui [8] proposed two segmentation methods with colour-texture gradient based on two hierarchical approaches: watershed using volume extinction values and the waterfall algorithm, respectively. They used the colour-texture gradient proposed by Martin et al. [9] which uses boundaries learning. In Table 1, we present their results, taken from their papers, along with our proposal measures.

Observing these data we can highlight that our approach achieves better global results. It accomplishes the higher value regarding the numerical boundary measure, F-value, 0.56 versus 0.55 and 0.44. It is a wide difference concerning the Waterfall algorithm, which achieves a slightly better GCE value. Hence, the Waterfall method worse determines the segment boundaries but generates a slightly more coherent segmentation regarding to the human one.

Yang et al. [15] modelled the texture using a mixture of Gaussians distributions and performed a clustering allowing the component degeneration. They used the colour metric L*a*b*. The features were generated through Gaussian convolutions of size 7x7, to be projected over the eight-dimensional space by the PCA method. Comparing both methods results, (see Table 1) we can observe that we achieve better results in three of the four measures (PRI, VoI, and BDE) measures. PRI and VoI are measures more correlated with the human segmentation in terms of visual perception [15]. GCE and BDE measures penalise the under-segmentations more heavily than the over-segmentations; in fact, the GCE value does not penalise the over-segmentations. Hence, our method perceptually achieves better results, tending towards the under-segmentation.

Table 1 Comparative results of the methods proposed in [8], [15] and our approach. Results correspond to the average value of all the segmentations performed.

Method	**PRI** *better when higher (0.87)*	**VoI** *better when lesser (1.1)*	**GCE** *better when lesser (0.08)*	**BDE** *better when lesser (4.99)*	**F-value** *better when higher (0.79)*
Waterfall level 2 [8]	-	-	0.19	-	0.44
Watershed using volume extinction values [8]	-	-	0.22	-	0.55
Yang et al. [15]	0.7627	2.036	0.1767	9.4211	-
Method proposed	0.8126	1.8434	0.2034	9.1339	0.5646

5 Conclusions

A method for colour and texture segmentation of natural images is proposed in this work. This approach incorporates the knowledge of human segmentation of natural scenes in hierarchical learning processes. Our method supervisedly learns the texture, colour, and homogeneity features of the human-generated segments. This knowledge determines the final segmentation through pattern refinement cycles. Based on the experience and using part of the development performed in a previous work [1,10], we have advanced a hierarchical extension to segment natural scenes. In this work, the feature extraction architecture is taken from that previous work. This reutilisation proves the architecture to be independent from the processing objective, similarly as it occurs in the human visual system.

Two comparisons with other significant methods for segmenting natural scenes have been included; in a global evaluation, our approach achieves better results. It shows more perceptual features and a higher tendency to the under-segmentation.

References

1. Antón-Rodríguez, M., Díaz-Pernas, F.J., Díez-Higuera, J.F., Martínez-Zarzuela, M., González-Ortega, D., Boto-Giralda, D.: Recognition of coloured and textured images through a multi-scale neural architecture with orientational filtering and chromatic diffusion. Neurocomputing 72(16-18), 3713–3725 (2009)
2. Carpenter, G.A.: Default ARTMAP. In: Proceedings of the International Joint Conference on Neural Networks (IJCNN 2003), pp. 1396–1401 (2003)
3. Carpenter, G.A., Grossberg, S., Rosen, D.B.: Fuzzy ART: Fast stable learning and categorization of analog patterns by an adaptive resonance system. Neural Networks 4(6), 759–771 (1991)
4. Chen, J., Pappas, T.N., Mojsilovic, A., Rogowitz, B.E.: Adaptive perceptual color-texture image segmentation. IEEE Trans. on Image Processing 14(10), 1524–1536 (2005)
5. Cheng, H.D., Jiang, X.H., Sun, Y., Wang, J.: Color image segmentation: advances and prospects. Pattern Recognition 34(12), 2259–2281 (2001)
6. Grossberg, S., Huang, T.: ARTSCENE: A neural system for natural scene classification. Journal of Vision 9(4), 1–19 (2009)
7. Grossberg, S., Williamson, J.R.: A self-organizing neural system for learning to recognize textured scenes. Vision Research 39, 1385–1406 (1999)
8. Hanbury, A., Marcotegui, B.: Morphological segmentation on learned boundaries. Image and Vision Computing 27(4), 480–488 (2009)
9. Martin, D.R., Fowlkes, C., Malik, J.: Learning to Detect Natural Image Boundaries Using Local Brightness, Color, and Texture Cues. IEEE Transaction on Pattern Analysis and Machine Intelligence 26(5), 530–549 (2004)
10. Martínez-Zarzuela, M., Díaz-Pernas, F.J., Antón-Rodríguez, M., Díez-Higuera, J.F., González-Ortega, D., Boto-Giralda, D., López-González, F., De la Torre-Díez, I.: Multi-scale Neural Texture Classification using the GPU as a Stream Processing Engine. Machine Vision and Applications (2010), doi:10.1007/s00138-010-0254-3
11. Yang, et al.: MATLAB Toolboxes and Segmentation Results from the BSDS (2008), `http://www.eecs.berkeley.edu/~yang/software/lossy_segmentation/` (last visited: December 2009)
12. Pyun, K., Lim, J., Won, C.S., Gray, R.M.: Image segmentation using hidden Markov Gauss mixture models. IEEE Trans. on Image Processing 16(7), 1902–1911 (2007)
13. The Berkeley Segmentation Dataset and Benchmark, `http://www.eecs.berkeley.edu/Research/Projects/CS/vision/grouping/segbench/` (last visited December 2009)
14. VisTex: Vision texture database, Massachusetts Institute of Technology, `http://vismod.media.mit.edu/vismod/imagery/VisionTexture/vistex.html` (last visited: December 2009)
15. Yang, A.Y., Wright, J., Ma, Y., Sastry, S.S.: Unsupervised segmentation of natural images via lossy data compression. Computer Vision and Image Understanding 110(2), 212–225 (2008)

First Steps towards Implicit Feedback for Recommender Systems in Electronic Books

Edward R. Núñez V., Oscar Sanjuán Martínez, Juan Manuel Cueva Lovelle, and Begoña Cristina Pelayo García-Bustelo

Abstract. Currently, a variety of eBooks with some intelligent capabilities to store and read digital books have been developed. With the use of these devices is easier to interact with the content available on the Web. But in some way access to such content is limited due to data overload problems. Trying to resolve this problem have been developed some techniques for information retrieval, among which are the recommender systems. These systems attempt to measure the taste and interest of users for some content and provide information relating to your profile. Through the feedback process attempts to collect the information that a recommendation system needs to work; but often this process requires the direct intervention of users, so that sometimes it is tedious and uncomfortable for users. For what we believe is necessary for a recommender system should be able to capture and measure implicitly the interaction parameters of a user with content in an eBook. Considering this need, we present a series of parameters that can be measured implicitly and how they will measured in the feedback process so that a recommender system to be reliable in electronic books.

Keywords: Electronic books, recommender system, feedback, implicit feedback.

1 Introduction

Due to the large amount of information found on the Internet, it is sometimes difficult for users to find content that they really need in a quick and easy way, so the user tends to seek guidance from others who may recommend some content that meets their needs or select those objects that come closest to what they want.[1]

Through the use of recommender system as technical of information retrieval attempts to solve the problem of data overload; and that through these you may

Edward R. Núñez V., Oscar Sanjuán Martínez, Juan Manuel Cueva Lovelle, and Begoña Cristina Pelayo García-Bustelo
University of Oviedo, Department of Computer Science, Sciences Building, C/ Calvo Sotelo s/n 33007, Oviedo, Asturias, Spain
e-mail: nuñezedward@uniovi.es, osanjuan@uniovi.es, cueva@uniovi.es, crispelayo@uniovi.es

A.P. de Leon F. de Carvalho et al. (Eds.): Distrib. Computing & Artif. Intell., AISC 79, pp. 61–64.
springerlink.com

filter the information available on the web and find information more interesting and more valuable to users. [2, 3, 4]

To make an effective recommender system we believe that is necessary to make a feedback process that does not require direct user intervention, but is able to capture as much information as possible related to the user profile implicitly.

This article describes the different implicit interaction parameters than can be collected and measured during the interaction of a user with an Electronic book and how we collect and measured these values. This process could determine whether implicit feedback gives good results for a recommender system in electronic books.

2 Problems

To capture and measure efficiently the interaction parameters of a user with an electronic book and implement a recommender system suitable for these types of devices, we must take into consideration a number of problems. In general we can say that there are three major problems associated with this subject:

- Information Overload.
- Feedback problems.
- Limited computing capability in the electronic books.

3 State of Art of the Recommender System

A recommender system is "*A system that has as its main task, choosing certain objects that meet the requirements of users, where each of these objects are stored in a computer system and characterized by a set of attributes.*" [5]

These consist of a series of mechanisms and techniques applied to the retrieval of information to try to resolve the problem of data overload on the Internet. These help users to choose the objects that can be useful and interesting, these objects can be any type, such as books, movies, songs, websites, blogs. [6]

Through the feedback process recommender systems collect user information and stored in your profile in order later to reflect your tastes and make recommendations. The feedback process can be two types: [7, 8, 9]

- Explicit feedback. Through a survey process, the user evaluates the system by assigning a score to an individual object or a set of objects.
- Implicit feedback. This process is that the objects are evaluated in ways transparent to the user. Namely, that the evaluation is done without the user being aware through the actions user performs while interacting with the system.

4 Case Study

The fundamental basis of recommender system is the ability of the system to collect the data necessary to meet efficiently the feedback process. We believe that with capture implicit parameters, we can measured the user interaction with electronic books.

To achieve an approach to the solution of the explicit feedback, we are developing an application that contains a series of content that users will be evaluate explicitly; and secondly, transparently to users we will record the user behavior during the whole period of interaction with the application, to capture the implicit parameters. In this process we measure series implicit parameters, which later will be compared with explicit user ratings. Through this process we can determine whether the feedback implicit in a recommendation system could discover the tastes and needs of users in electronic books.

This study will be done in a web application for electronic books, which involved 100 users with different levels of knowledge, different ages, without prior knowledge of the contents and chosen at random. We will make this study in the electronic book that is being developed in the project [einkPlusPlus][1] in which this research is based.

The evaluation mechanism will be stored in a repository the data retrieved from every action taken by the user with the ebook. These data will be sent to a server for the comparison and analysis. Finally the results will be integrated into the system recommender who later sent the recommendations to the electronic book.

4.1 Implicit Parameters to Measure

As Nielsen [10] presents a series of indicators to measure Web usability, we chose a set of parameters that we believe could help discover the interest and tastes of users implicitly.

The different parameters that we measure in this process are described below:

- Session duration / content size. With the evaluation of this parameter indicates the user's connection time, allowing the user to know how much it took to evaluate and interact with content accessed by the user.
- Number of Clicks. This will determine how many click the user needed to evaluate the content.
- Reading time. With this parameter will determine how long a user takes a reading or viewing content. this parameter is important because it could determine the user's interest based on the average time reading or viewing of contents
- Complete reading time: This parameter determines the time that the user needs to view a specific section or category.
- Number of accesses to a specific section or category: this parameter will determine how many times the user accessed a specific section.
- Number of iterations: This parameter determines the number of times a user read or viewed content. It may determine that a larger number of repetitions, more interested by the content.
- Number of comments: With this parameter we could determine the general interest by a specific content, according to the amount of comments that have content.

---

[1] www.einkplusplus.com

- Number of recommendation. With this parameter we could determine the general interest in specific content, according to the number of people who have been recommended content.

5 Conclusion and Future Work

With the development of this application will get the values of the implicit parameters, and through comparative analysis and the search for correlations determine the value of these implicit parameters to capture user interest and to make good recommendations in electronic books.

We believe that the use of these parameters could build recommender systems more effective and does not need the explicit capture parameters for the feedback.

In future work we will focus on comparative analysis of the data collected and then according to the results obtained in this study.

References

1. Sanjuan, O.: Using Recommendation System for E-learning Environments at degree level. International Journal of Artificial Intelligence and Interactive Multimedia 1(2) (2009) ISSN - 1989-1660
2. Taghipour, N., Kardan, A.: A Hybrid Web Recommender System Based on Q-Learning. In: SAC 2008: Proceedings of the 2008 ACM symposium on Applied computing, Fortaleza, Ceara, Brazil, pp. 1164–1168 (2008) ISBN:978-1-59593-753-7
3. O'Donovan, J., Smyth, B.: Trust in recommender systems. In: Proceedings of the 10th international conference on Intelligent user interfaces, pp. 167–174. ACM, San Diego (2005) ISBN:1-58113-894-6
4. Noor, S., Martinez, K.: Using Social Data as Context for Making Recommendations: An Ontology based Approach. In: ACM International Conference Proceeding Series, Proceedings of the 1st Workshop on Context, Information and Ontologies, Heraklion, Greece. ACM, New York (2005) ISBN:978-1-60558-528-4
5. Wang, P.: Why recommendation is special? In: Papers from the 1998 Workshop on Recommender Systems, part of the 15th National Conference on Artificial Intelligence (AAAI 1998), Madison, Wisconsin, EUA, pp. 111–113 (1998)
6. Gonzalez, R., Sanjuan, O., Cueva, J.M.: Recommendation System based on user interaction data applied to intelligent electronic books (2010)
7. Resnick, P., Varian, H.R.: Recommnder Systems. Communications of the ACM, 56–58 (1997)
8. Adonomavicius, G., Sen, S., Tuzhilin: A. Incorporating Contextual Information in Recommender Systems Using a Multidimensional Approach. ACM Transactions on Information Systems (TOIS) 23(1), 103–145 (2005)
9. Ziegler, C.-N., et al.: Improving Recommendation Lists Through Topic Diversification. In: International World Wide Web Conference. Proceedings of the 14th international conference on World Wide Web, Chiba, Japan, pp. 22–32. ACM, New York (2005) ISBN:1-59593-046-9
10. Nielsen, J.: Usability Metrics: Tracking Interface Improvements. IEEE Software 13(6), 12–13 (1996)

A Trust-Based Provider Recommender for Mobile Devices in Semantic Environments*

Helena Cebrián, Álvaro Higes, Ramón Hermoso, and Holger Billhardt

Abstract. Semantic web services have been studied during last years as an extension of Service-Oriented Computing on the Web 2.0. A lot of effort has been made to address some problems such as service discovery, matchmaking or service composition. Nevertheless, there is not much work in the literature about how to integrate trust into the process of selecting service providers. In this paper, an abstract architecture with a trust-based recommender module is presented, and a case study is put forward explaining how to apply the architecture to implement an agent in a mobile device.

Keywords: Semantic web services, trust models, mobile devices.

1 Introduction

Web Services (WS) is a technology based on the concept of Service-Oriented Computing (SOC) with a very promising horizon [9]. WSs provide the foundations for the development of business processes distributed on the Web, and available via standard interfaces and protocols.

The recently more and more development of Semantic Web (SW) [6] allows providing web services with semantic information that facilitates the automation of certain tasks, such as *discovery*, *composition* or *invocation*. Semantic Web Services (SWS) apply knowledge from the SW to web services by using semantic descriptions using languages such as OWL-S [3].

Helena Cebrián, Álvaro Higes, Ramón Hermoso, and Holger Billhardt
Centre for Intelligent Information Technologies (CETINIA)
University Rey Juan Carlos, Madrid (Spain)
e-mail: {helena.cebrian,alvaro.martinez,ramon.hermoso}@urjc.es,
holger.billhardt@urjc.es

* The present work has been partially funded by the Spanish Ministry of Education and Science under project OVAMAH-TIN2009-13839-C03-02 and "Agreement Technologies" (CONSOLIDER CSD2007-0022, INGENIO 2010) and by CDTI under project "miO!" (CENIT-2008 1019).

A.P. de Leon F. de Carvalho et al. (Eds.): Distrib. Computing & Artif. Intell., AISC 79, pp. 65–72.
springerlink.com

In SWS community [7][8] there are four axis of researching to deal with: i) service *discovery*; ii) service *matchmaking*; iii) service *composition*; and iv) service *provider selection*. The first three aspects have been deeply studied [8][4][1]. Nevertheless service provider selection is often situated as task of the matchmaking process. We claim in this paper that provider selection may be added as an extra-step when an entity is searching a particular service.

In this paper we present an abstract agent's architecture that incorporates a semantic web services recommender module that encapsulates discovery and matchmaking processes. Furthermore, we endow it with a trust model that allows tackle the provider selection problem.

Ubiquitous computing has become extremely popular over the last years. Mobile devices, such as mobile phones are tools with a huge potential from a computing point of view. Such is the case of semantic web services, since allow users to use their functionality in their everyday lives. The second part of the paper attends to show the implementation of the abstract architecture that has been developed for its use in mobile phones. A case study will be presented to illustrate the details.

The paper is organised as follows: Section 2 describes in detail the abstract architecture for the recommender; Section 3 presents a case-study that shows how the recommender has been implemented, using a specific scenario for that. Finally, Section 4 sums up our approach and outlines some future work.

2 An Agent Abstract Architecture for Semantic Service Selection

As aforementioned, we intend to build an agent architecture that is endowed with a recommender module for searching service providers and that encapsulates discovery and matchmaking processes. Furthermore, we endow it with a trust model that allows tackle the provider selection problem. In this section we present the architecture depicted in Figure 1. Although architectures are usually detailed in a formal way (for instance, following Kruchten's approach [5]), due to lack of space we put forward the present architecture in a more colloquial manner.

From this schema can be observed that there exist two different parts in the agent's architecture. On the one hand the recommender module encapsulates the automatic process of recommending, from an agent's goal, a set of potential service providers able to make the agent be satisfied regarding the goal she wanted to achieve, and ordered by trust evaluations. On the other hand, also inside the agent's architecture, we find some modules that correspond to intermediate actions that the individual must implement in order to interact with the recommender. Such modules are *template selection*, *instance selection* and *Rating*. Finally, for the sake of simplicity, we include the service platform in the schema, that interacts with the recommender module in the way we will put forward in next paragraphs.

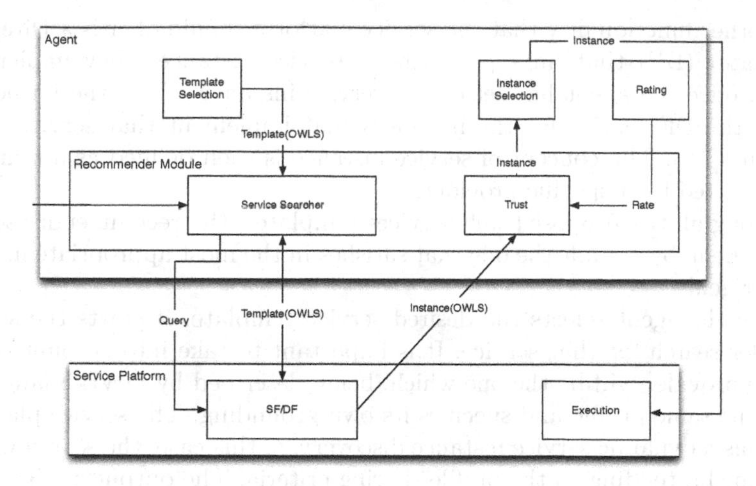

Fig. 1 Recommender agent architecture

Goals are usually generated by the agent to cope with the satisfaction of her own preferences. In a service-oriented environment, goals entail the search and the immediately after execution of services that achieve the aforesaid goals. The architecture approach we present intends to show how to encapsulate this process in the agent's core. Notice we only deal with the process of, once the agent knows which is the goal to achieve first, searching and selecting a service and its corresponding execution, but not with the previous reasoning through which the agent manages goals, preferences, desires, etc. For that reason this approach should be taken into account as a complementary abstraction that may be included in a more complex architecture, such as, for instance, a BDI architecture.

The *service search* module has as input a goal or a list of goals[1]. It will communicate with the Service Platform, more precisely with a Service Facilitator (SF) that maintains a registry of which services are offered in the environment. This entity should encapsulate the service discovery and the matchmaking process as well. As an abstract architecture, each of this entities may be instantiated in several ways, implementing different techniques for discovery and matchmaking. It is not in the scope of this paper to go into it in depth. Thus, goals could be specified using a description language such as OWL, that could also be understandable for the SF.

The proposed architecture makes the distinction between service template and service instance. Therefore, from the process of service searching the agent obtains a list of service templates that semantically match in some degree to the agent's original goal (*matchmaking*). Using this list, the agent – according the implementation in *template selection* module – selects the most suitable service template for her interests.

[1] As we show in Section 3.

Another functionality that the service platform should offer is a Directory Facilitator (DF) that links providers to service instances they implement. Hence, once the agent has selected a service instance s_i^{temp}, the former requests the DF service for the instances that implement that service (from now on s_i^{inst}). The concept of service instance is then defined as a template implemented by a specific provider.

Among all the proposed web services templates, the recommender selects one of them, by default the one that satisfies in the most appropriate manner the her goal.

Once the agent selects the desired service template, it starts the service provider search for that service. It is important to take into account that a service provider will be the one which, being described by OWL-S language, share the same profile and specifies its own grounding. The service platform performs a dynamic service instance discovery, in this case, the search will be performed attending to the profile sharing criteria. The output at this point consist on a not ordered list of service providers.

In order to improve their performance, clients in Service-Oriented Computing (SOC) commonly deal with the problem of deciding appropriate service providers for their requests according to their own interests, probably modelled with beliefs or goals. Trust models, most of them based on reputation mechanisms, have been proved to be worth techniques to estimate expectations on other individuals in order to better select interaction partners. We adhere to the work of Billhard et al. [2] in which authors claim that trust and reputation mechanisms should be added as an additional step after traditional phases in service environments. This is described in Figure 2. Following the schema, trust is presented to fill the gap between service matchmaking and eventually provider selection, since amongst the set of individuals that can perform a service, there could exist trust relationships with the client that may allow the latter to better select a provider. Here trust do not substitute current techniques of service discovery or matchmaking processes, but are complementary. Advantages of this model are defined in [2]. At this point, it is important to note that the proposed architecture is designed to accept any other trust model.

In the architecture, the trust module receives a list of service instances of the service template selected by the agent to fulfil the goal. This module should implement a trust model that has to be able to evaluate trust on different potential providers. This model could base its assessments on either local own information derived from past interactions with different providers or information gathered from third parties.

Once the list of service providers that satisfies the initial goal are discovered, the recommender module (and more precisely the trust module) is in charge of evaluating trust on different service instances (and their service providers), as well as building a ranking with that information, so allowing the agent to have more information to make a decision. This recommendation is done conforming to the [2] model. This model bases on the concept

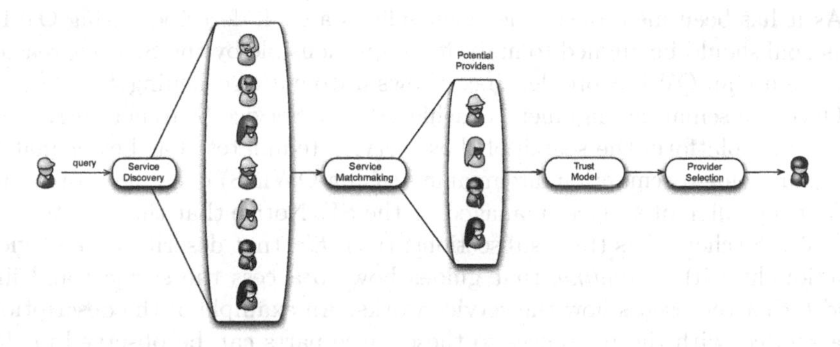

Fig. 2 Trust integration in a service-oriented domain

of similarity between situation that the agent had in the past and situations that she is involved in in the present. A situation in [2] is defined as a tuple $\langle s_i, p_j, t \rangle$ where s_i^{inst} is the service instance that the agent used, and p_j is the provider that performed s_i^{inst}. Similarity between service instances is measured with a closeness function based on the distance of concepts in a service taxonomy, previously existing[2].

The trust module returns an ordered list – a ranking – of the evaluation of different service instances. Then, the agent selects the desired instance – implementing her *instance selection* module – and invokes its execution on the service platform. This execution is made attending to the established specification in the OWL-S grounding.

After the execution, the agent rates the service execution, which is used as feedback for the recommendation module. This rating is aggregated with her past interactions with the aim of presenting more suitable recommendations in the future for the agent's profile.

3 Case Study

So far, we have put forward an architecture in a theoretical level. In this section we attempt to present a real scenario where the architecture has been instantiated on a mobile device – a mobile phone – running over the Android platform [3]. In the current case-study Paula is visiting Córdoba and will use her mobile phone to obtain recommendations about what to do in the city. An agent implementing the architecture presented in Sectionsec:recommender will be available in the mobile phone. The scene comes into being when Paula arrives at the town and decides to start visiting. Thus, the agent in the phone receives (via Paula) the following goal: **Visit the town of Córdoba**.

[2] Due to lack of space a wider explanation about how to calculate similarities may be looked up on [2].

[3] http://www.android.com

As it has been mentioned, the agent selects a goal, described using OWL. This goal should be aligned to make it comprehensible by the SF/DF, reason why we use an OWL-S profile, since allows automatic reasoning [4].

Once the semantic alignment is achieved, the Service Searcher entrust to the service platform the search of those service templates that better match the goal, doing a semantic matchmaking (using OWL-S) over the postcondition of the different services managed by the SF. Notice that the structure of an OWL-S scheme has three subsections: i) *profile*, that describes the service functionality; ii) *grounding*, that guides how to access the service; and iii) *model*, that represents how the service works. An example of the description of a service, with the references to these three parts can be observed in the following representation in OWL-S:

```
<service:Service rdf:ID="Tourist Guide">
    <!-- Reference to the Tourist Guide Profile -->
    <service:presents rdf:resource="profile";
                        Profile Tourist Guide">
    <!-- Reference to the Tourist Guide Process Model -->
    <service:describedBy rdf:resource="process;
    Tourist Guide ProcessModel"/>
    <!-- Reference to the Tourist Guide Grounding -->
    <service:supports rdf:resource="grounding;
    Grounding Tourist Guide"/>
</service:Service>}
```

In our scene, the recommender module receives four possible service descriptions (OWL-S profiles) with a high enough matching degree that was previously specified by the user. With the inclusion of an affinity threshold, we guarantee that all discovered services satisfy at least in a minimum degree the user expectations. This unordered service templates list is received by the Service Searcher which translates the semantic language to a friendly format for Paula. The received list contains the following service templates: [*Cordoba Street Searcher* that offers a web service to access to street directory; *Tourist Guide of Córdoba* that offers tourist information of Córdoba; *Gastronomic Route* that offers A route offering places to enjoy wines and typical food; *Cordoba on wheels* that offers an on-line booking service for tourist vehicles].

Paula chooses the one that better fits her insterests. In this case her choice is **Córdoba on wheels**. Once the service has been selected, the Service Searcher communicates with the DF for discovering all available service providers for that template. Thanks to the OWL-S description format, this search just consists on finding all those providers that have been instantiated in the consulted service grounding. The DF then returns an unsorted list with all the available service instances (and its providers) for that service. This list will be sorted by the recommender's trust module straight afterwards. The received list (only service provider are shown) together with the

[4] Due to lack of space semantic alignment and matchmaking processes are not detailed.

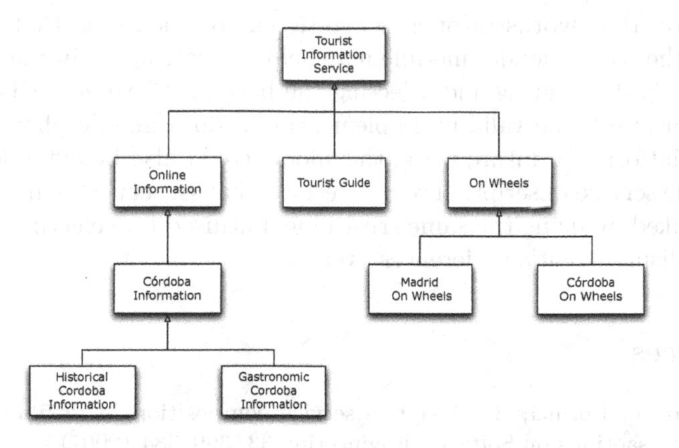

Fig. 3 Service Taxonomy

profile description, is as follows: i) Córdoba's Tourist Board (from now on CTB). One of the available services of this provider is to manage a tourist train service; ii) Córdoba Tours. Offers group excursions by bicycle over surroundings; iii) Visio Bus. Offers a city tour by bus. Paula has never used the *Córdoba On Wheels* service, that is the reason why the trust module of the recommender has no past information in its internal structures. To solve this situation, the trust module uses trust model of Billhardt et al [2]. Thus, it estimates a value for each received tuple ($(\langle S, P \rangle)$), based on the services offered by the same provider. More precisely, Paula has already had previous interaction with the providers: CTB and Visio Bus. However, she has no previous experience with Córdoba Tours provider. The similarity between the actual service, *Córdoba On Wheels*, and the past services offered by the CTB and Visio Bus, Cordoba Online Information and Vision Bus Madrid respectively, is calculated by the SF/DF according to the internal service taxonomy that it keeps (Figure 3). Part of this taxonomy, is represented in Figure 3. As it is suggested in the [2], the similarity between two services would be calculated as a closeness function between the past known services, and the current one, so estimating how good it will be in future interactions.

Then, an ordered ranking is returned:[1. Visio Bus; 2. CTB; 3. Córdoba Tours]. Thus, after applying trust model in [2] Paula will finally choose Visio Bus service. After the service execution, the agent requests her to rate de service. This information will be used as feedback for the trust module. Finally, Paula rates the service in the way the system proposes (typically by giving a feedback value in a pre-defined range).

4 Conclusions

In this paper we have presented an abstract agent architecture with a trust-based recommender module. The underlying idea of the proposed architecture, that works in open semantic environments, is that according to a goal, the recommender module may help the user agent in the process of automatically discovering and selecting the best $\langle S, P \rangle$ tuple for her goal. It has been proved to be valid by implementing it on a mobile phone with the Android platform. In future work, this model could also be extended to personalize the service description search result. That is, service templates could also be ranked by using the same trust model utilised for selecting providers, so personalising agent's preferences over time.

References

1. Ardagna, D., Pernici, B.: Adaptive service composition in flexible processes. IEEE Transactions on Software Engineering 33, 369–384 (2007)
2. Billhardt, H., Hermoso, R., Ossowski, S., Centeno, R.: Trust-based service provider selection in open environments. In: SAC 2007: Proceedings of the 2007 ACM symposium on Applied computing, pp. 1375–1380. ACM, New York (2007)
3. Burstein, M., Hobbs, J., Lassila, O., Mcdermott, D., Mcilraith, S., Narayanan, S., Paolucci, M., Parsia, B., Payne, T., Sirin, E., Srinivasan, N., Sycara, K.: Owl-s: Semantic markup for web services. W3C Member Submission (November 2004)
4. Czerwinski, S.E., Zhao, B.Y., Hodes, T.D., Joseph, A.D., Katz, R.H.: An architecture for a secure service discovery service. In: MobiCom 1999: Proceedings of the 5th annual ACM/IEEE international conference on Mobile computing and networking, pp. 24–35. ACM, New York (1999)
5. Kruchten, P.: The 4+1 view model of architecture. IEEE Software 12(6), 42–50 (1995)
6. Lee, B.T., Hendler, J., Lassila, O.: The semantic web. Scientific American (May 2001)
7. McIlraith, S.A., Martin, D.L.: Bringing semantics to web services. IEEE Intelligent Systems 18(1), 90–93 (2003)
8. Paolucci, M., Kawamura, T., Payne, T., Sycara, K.: Semantic matching of web services capabilities. pp. 333–347 (2002)
9. Weerawarana, S., Curbera, F., Leymann, F., Storey, T., Ferguson, D.F.: Web Services Platform Architecture: SOAP, WSDL, WS-Policy, WS-Addressing, WS-BPEL, WS-Reliable Messaging, and More. Prentice Hall PTR, Englewood Cliffs (2005)

Quality of Service and Quality of Control Based Protocol to Distribute Agents

Jose-Luis Poza-Luján, Juan-Luis Posadas-Yagüe, and José-Enrique Simó-Ten

Abstract. This paper describes an agent's movement protocol. Additionally, a distributed architecture to implement such protocol is presented. The architecture allows the agents to move in accordance with their requirements. The protocol is based on division and fusion of the agents in their basic components called Logical Sensors. The movement of the agents is based on the quality of services (QoS) and quality of control (QoC) parameters that the system can provides. The protocol is used to know the impact that the movement of the agents may have on the system and obtain the equilibrium points where the impact is minimal.

1 Introduction

Distributed control architectures based on multi-agent systems require a middleware to hide the complexity of the communications to the agents. The middleware can provide the data about QoS to the system, and system can provide this information to the agents. On the other hand, when an agent needs to move between the control nodes, the arrival of the agent have an impact on the global performance of the control node. To provide the support to agents an architecture, called Frame-Sensor-Adapter with Control support (FSACtrl), has been developed. This architecture is based on the model FSA, widely tested on mobile robot, and home automation systems [1]. The middleware of the architecture is based on the Data Distribution Service (DDS) model proposed by The Object Management Group (OMG) [2].

When an agent arrives to a node, competes with the other agents for the resources. The effect of the new agent on the node performance is determined by the Liebig's law of the minimum [3]: similar species competing for the same resources minimizing the availability of the global resources. To minimize the impact of the agent movement the architecture can move and evaluate only fragments of an agent. How the architecture can provide the necessary information to know the impact of the agent movement is the aim of this article.

The article has been organized as follows: Next section introduces the theoretical Concepts that are the base of the architecture on which the protocol shown is

Jose-Luis Poza-Luján, Juan-Luis Posadas-Yagüe, and José-Enrique Simó-Ten
University Institute of Control Systems and Industrial Computing (ai2).
Universidad Politécnica de Valencia. Camino de vera, s/n. 46022 Valencia (Spain)
e-mail: {jopolu,jposadas,jsimo}@ai2.upv.es

A.P. de Leon F. de Carvalho et al. (Eds.): Distrib. Computing & Artif. Intell., AISC 79, pp. 73–80.
springerlink.com

developed. The third section describes the components of the architecture and the quality-based supply and demand cycle. The fourth section shows the phases of the protocol. Finally, presents concluding remarks.

2 Middleware, Quality of Service and Quality of Control

There are different paradigms of communication with support to quality of service, among them publish-subscribe model is one of the most suitable [4]. DDS provides a platform independent model that is aimed to real-time distributed systems. DDS is based on publish-subscribe communications paradigm. Publish-subscribe components connect information producers (publishers) and consumers (subscribers) and isolate the publishers and the subscribers in time, space and message flow [5].

When a producer (component, agent or application) wants to publish some information, should write it in a Topic by means of a component called Data Writer which is managed by another component called Publisher. Both components, Data Writer and Publisher, are included in another component called Domain Participant. On the other hand, a Topic cans delivery messages to both components: Data Readers and Listeners by means of a Subscriber. Data Reader provides the messages when the application requires. Instead of a Listened sends the messages without waiting for the application.

Quality of Service is defined as the collective effect of service performance, which determines the degree of satisfaction of a user of the service [6]. FIPA

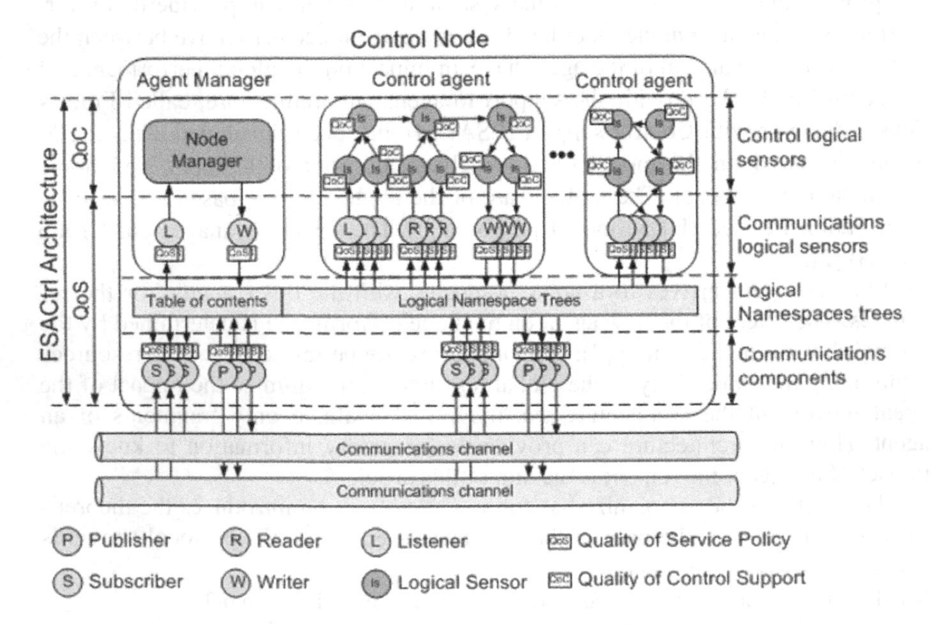

Fig. 1 Components of FSACtrl architecture

defines a set of QoS parameters mainly based on temporal issues of the messages [7]. DDS specification proposes 22 different QoS policies that cover all aspects of communications management: message temporal aspects, data flow and metadata. For example, by means the "Deadline" policy, that determines the maximum time for the message arrival, and the "Time-Based-Filter" policy, that determines the minimum time between two messages, a component can establish a temporal window to receive messages from other components.

In the same way that to evaluate the efficiency of communications uses QoS parameters; control must provide the corresponding parameters, known as quality control parameters. It's considered a perfect control when the signal is sent to the actuator causes the signal measured by the sensor is identical to a reference signal, therefore there is no error between the measured signal and reference signal. Control error is used to modify the signal sent to actuator. The most commonly used QoC parameters are the value of the Integral Absolute value of Error (IAE) and the Integral of the Time and the Absolute value of Error (ITAE). Both parameters allow the system to know how evolve the error, and predict the new action. [8].

3 Distributed Architecture

3.1 Componentes

Figure 1 shows the main components of the architecture FSACtrl [9]. Each control node has a manager agent, the necessary control agents, the communications components to provide support at control agents, and a set of topics to connect the control agents with the communications components, Topics are organized in an ontology called Logical Namespace Tree.

Each control node contains a special ontology called Table of Contents (TOC). TOC describes the control node to the other control nodes of the system. The communications components are the components proposed by the DDS standard. Publishers and Subscribers are the control node communications components and they are accessible to all control agents, whereas Data Writers, Data Readers and Listeners are the control agent communications components and they are exclusive to each control agent. Control algorithms are implemented by the components called Control Logical Sensors that provides the QoC parameters.

Manager agent provides support to all functions of the MAS: processes the request of the control agents, both from within and outside the control node, manage the ontology to connect successfully communications components and control components and mediates in the negotiation based on the QoS and QoC parameters. System manager and LNT are similar to other MAS system as JADE [10].

3.2 QoS and QoC Based Supply and Demand Cycle

Figure 2, shows the location of the parameters in the control loop based on a client-server model. Based on the QoS cycle [11], the Quality-based Supply and

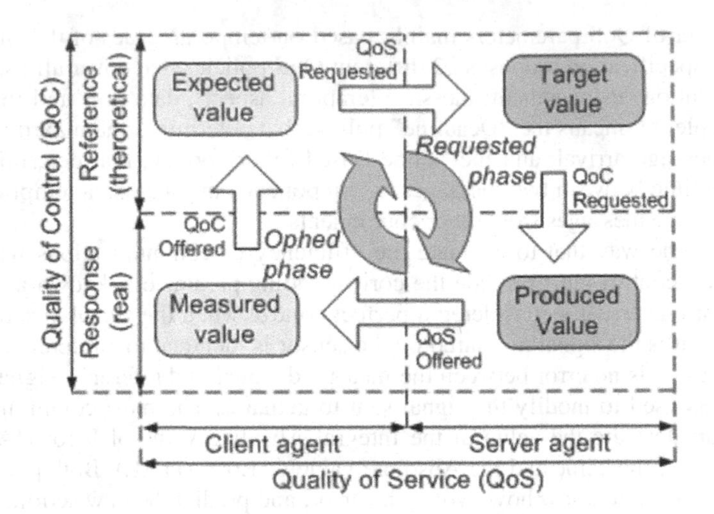

Fig. 2 Both qualities, quality of service and quality of control determine the quality-based supply and demand cycle

Demand Cycle adds the QoC. The QoS parameters are associated with the client-server communication and the QoC parameters characterize the relationship between the theoretical expectations and the real results of the control actions, both the client and the server side.

DDS offers a protocol for negotiating the QoS parameters by means the QoS policies. However, when the QoC is included in the negotiation process, it is necessary that the control agent can provide the appropriate protocol and parameters to measure the optimization level.

As seen in figure 2, the QoS and QoC have a fist service request phase to communicate the value needed of the parameters. The second phase, communicates the values offered. Consequently the cycle consists in two phases: the QoS and QoC request, and the QoS and QoC offer. This cycle is known as “quality-based Supply and Demand Cycle”.

4 Moving or Cloning Agents in the Distributed System

4.1 Search Destiny Control Node Phase

FSACtrl architecture allows to the agents to perform the same actions than the rest of the MAS [12]. Move or clone an agent involves a set of operations like insert or remove agents. In FSACtrl, insert an agent in a node is done inserting the agent logical sensors. The QoS and the QoC of an agent is easy to calculate because it depends from its internal components. However, when an agent needs change the control node, it’s difficult to know the effect of the change in the agent’s behaviour.

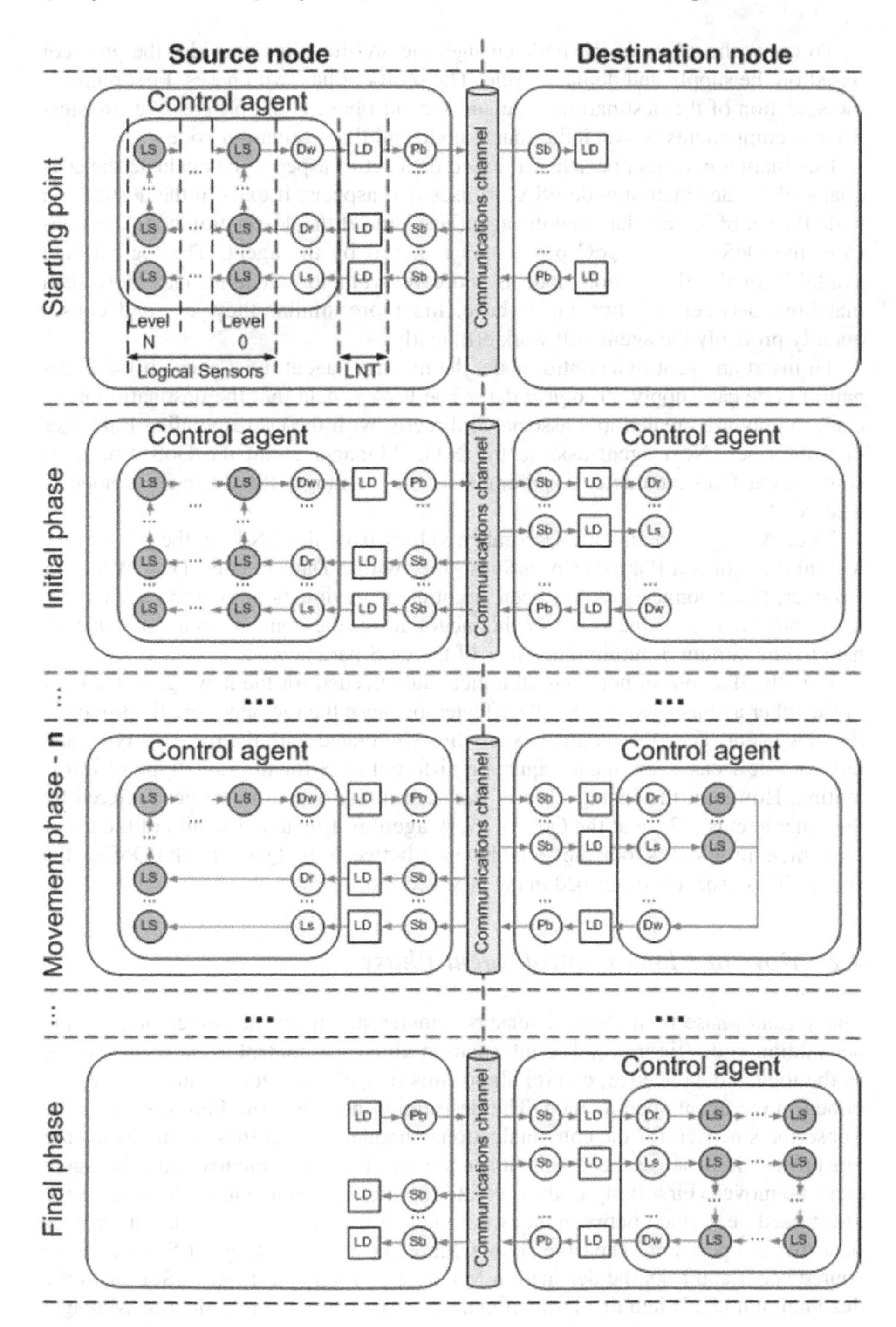

Fig. 3 All phases of the process of an agent movement

To know the effect of the node change, the architectures provides the protocol based on the supply and demand cycle. The protocol has two phases. First phase is the selection of the destination node and second phase is the progressive transference of components between the source node and the destination node.

Destination node can be selected based on a set of aspects. To evaluate the adequacy of the destination node FSACtrl uses two aspects: if exits in the destination node the set of logical data that the agent need and if the destination node can conform the QoS and the QoC parameters required by the agent. The logical data available in the destination node is more relevant to select it. The more data matching between the two nodes have, the more similar they are, and consequently probably the agent will work efficiently.

To insert an agent in a control node, the manager agent must check if the destination node can supply all logical data. The logical data that the destination node can't supply are created and associated directly with the corresponding Publisher or Subscriber. Next agent asks to the Node Manager about the QoS ranges. If some agent QoS parameter requirement is out of range, the destination node is discarded.

Node Manager obtains the QoS range values from the LNT. In the LNT every logical data joins a Publisher or a Subscriber with a Data Reader, Data Writer or Listener, these connections has been negotiated previously among the agents and the control node. As the result of this negotiation, the control node can determinate the maximum or minimum values of the QoS parameters.

Usually, if a control node has all logical data needed for the new agent, the QoS of the other agents was only hardly affected because the messages are the same for the new agent. The QoS parameters are directly related with the message type, and only in a few cases, an agent requires a different QoS for the same type of information. However, the Liebig's law of the minimum affects the services offered by the other agents [13], and the QoC of every agent is affected. The aim of the agent movement protocol is to achieve the balance between the QoC and the QoS of the agents. This aspect is described in the next section.

4.2 Move or Clone Control Agent Phase

The second phase is the logical sensors transference from the source node to the destination node (figure 3). The information about the control agent composition, as the logical data needed, control algorithms or QoS and QoC parameters is contained in the agent specification. The destination node has the Publishers and the Subscribers needed for the communications channel used, if these components are not present is because the destination node can't use this channel and the agent can't be moved. First thing to do is create the destination agent, in the case of the agent need be divided between the two nodes, or use an existing agent, in the case of a merge operation. Then the source Manager selects a Logical Sensor in the source agent and asks the destination Manager to insert the Logical Sensor in the destination node. When the destination agent is inserted, both destination Manager and source Manager creates the same Logical Data to communicate with the Logical Sensors of the source agent, finally launch the new Logical Sensor and stops the old Logical Sensor. The key of the protocol is the running of the Quality-based supply and demand cycle to every Logical Sensor movement phase. Once a cycle has finished, Manager knows the new QoS and QoC values of all the agents in the destination node, and can decide to move other Logical Sensor or undo the last action.

The Logical Sensors that has not been translated send (or receive) the messages to (or from) the moved agents through the LNT. The Manager Agent creates the Data Writer, Data Reader or Listener and the Logical Data needed to supply the old direct connection between the Logical Sensors [14]. Logical Sensors are being moved one to one between the source and the destination node. For every Logical Sensor inserted the QoS and the QoC of the destination node changes, generally global quality decreases. The evolution of the QoS and QoC parameters allows node and agents to regulate the balance between the QoS and the QoC. The agent moved, can decide stop the process at the moment when the QoS and QoC parameters are optimal. The destination node can stop the process if the last Logical Sensor exceeds the QoS and QoC limits determined by the other agents.

5 Conclusions

This article has presented a protocol to move agents between nodes along a distributed system based on an architecture called FSACtrl. The main contribuition is the joint use of QoS and QoC parameters to determine, gradually, the impact that an agent has in a control node during the process of movement.

The protocol is used to balance the agent components, usually different control algorithms, between the control nodes. From the location of the agents in the control node, it is possible to determine the optimal composition of the agents to a specific environment.

Currently the formulas to measure the impact in the node and in the agent, in terms of QoS and QoC are being developed. One of the aims of the formulas is provide global the Quality of Node (QoN) and the Quality of Agent (QoA) can be determined by means the combination of the QoS and QoC parameters. With the QoA, the continuous process of divisions and merges of Logical Sensors can stop in the instant that QoS and QoC of all agents are optimal.

Acknowledgments. The architecture described in this article is a part of the coordinated project SIDIRELI: Distributed Systems with Limited Resources. Control Kernel and Coordination. Education and Science Department, Spanish Government. CICYT: MICINN: DPI2008-06737-C02-01/02.

References

1. Posadas, J.L., Poza, J.L., Simó, J.E., Benet, G., Blanes, F.: Agent Based Distributed Architecture for Mobile Robot Control. In: Engineering Applications of Artificial Intelligence, vol. 21(6), pp. 805–823. Pergamon Press Ltd., Oxford (2008)
2. Object Management Group (OMG): Data Distribution Service for Real-Time Systems, v1.1. Document formal / 2005-12-04 (2005)

3. Odum, E.P.: Fundamentals of Ecology, 3rd edn. W.B. Saunders Company, Philadelphia (1971)
4. Aurrecoechea, C., Campbell, A.T., Hauw, L.: A Survey of QoS Architectures. ACM/Springer Verlag Multimedia Systems Journal, Special Issue on QoS Architecture 6(3), 138–151 (1998)
5. Pardo-Castellote, G.O.: Data-Distribution Service: architectural overview. In: Proceedings of 23rd International Conference on Distributed Computing Systems Workshops, vol. 19-22, pp. 200–206 (2003)
6. International Telecommunication Union (ITU). Terms and Definitions Related to Quality of Service and Network Performance Including Dependability. ITU-T Recommendation E.800 (0894) (1994)
7. Foundation for Intelligent Physical Agents. FIPA Quality of Service Ontology Specification, Experimental Doc: XC00094 (2002)
8. Dorf, R.C., Bishop, R.H.: Modern Control Systems, 11th edn. Prentice Hall, Englewood Cliffs (2008)
9. Poza, J.L., Posadas, J.L., Simó, J.E.: Middleware with QoS Support to Control Intelligent Systems. In: 2nd International Conference on Advanced Engineering Computing and Applications in Sciences, ADVCOMP, pp. 211–216 (2008)
10. Bellifemine, F., Poggi, A., Rimassa, G.: Jade: A FIPA-compliant agent framework. In: Proceedings of PAAM 1999, pp. 97–108 (1999)
11. Poza, J.L., Posadas, J.L., Simó, J.E.: From the Queue to the Quality of Service Policy: A Middleware Implementation. In: Omatu, S., Rocha, M.P., Bravo, J., Fernández, F., Corchado, E., Bustillo, A., Corchado, J.M. (eds.) IWANN 2009. Part II. LNCS, vol. 5518, pp. 432–437. Springer, Heidelberg (2009)
12. Foundation for Intelligent Physical Agents. FIPA Agent Management Specification, Doc: FIPA00023 (2000)
13. Jeong, B., Cho, H., Kulvatunyou, B., Jones, A.: A Multi-Criteria Web Services Composition Problem. In: Proceedings of the IEEE International Conference on Information Reuse and Integration, 2007 (IRI 2007), pp. 379–384. IEEE, Los Alamitos (2007)
14. Poza, J.L., Posadas, J.L., Simó, J.E., Benet, G.: Distributed Agent Specification for an Intelligent control Architecture. In: 6th International Workshop on Practical Applications of Agents and Multiagent Systems. IWPAAMS (2007) ISBN 978-84-611-8858-1

Smart Communication to Improve the Quality of Health Service Delivery

Rosa Cano, Karla Bonales, Luis Vázquez, and Sandra Altamirano

Abstract. Nowadays, the society demands more and better products and services, and as a result organizations need to count on business strategies in order to provide a way for their employees communicate and fulfil their clients and work mates' expectations more efficiently. This article presents the Hippocrates of Cos Multi-Agent System 1.0 (HdeC-MAS 1.0) a multi-agent system (MAS) based on virtual organizations. The system can access information and services ubiquitously from either land or mobile devices connected to wired or wireless networks. HdeC-MAS 1.0 supports the communication process for the healthcare services of a hospital. This paper presents the results obtained from the implementation of the system and demonstrates the advantages produced by the use of this new technology.

Keywords: agent organizations, open MAS, medical informatics, HCE, HL7, DICOM, ISO: 9000 and 15489 and organizational communication.

1 Introduction

The internal communication process in an organization is a key factor for its members being able to interact, because by means of such interactions it is possible to achieve the organization's strategy goals. The vast amount of information that is generated in an organization, product or its domestic efforts of the

Rosa Cano
Departamento de Sistemas y Computación, Instituto Tecnológico de Colima
Av. Tecnológico, 1, 28976, Villa de Álvarez, Colima, México
e-mail: rdegca@gmail.com

Karla Bonales
Dirección de I+ D+I, ABC Technology
Nicolás Bravo, 367, 28000, Colonia Centro, Colima, Colima, México
e-mail: kbobales@gmail.com

Luis Vázquez
Dirección de Innovación, Gobierno del Estado de Colima
Av. Ejército Mexicano y 3er Anillo Periférico, 28010, El Diezmo, Colima, México
e-mail: vazquezornelas@gmail.com

Sandra Altamirano
Colegio de Estudios Científicos y Tecnológicos del Estado de Jalisco, Plantel 19
Camino viejo a Santa Lucia, 1, 45220, Nextipac, Zapopan, Jalisco, México
e-mail: saltamirano@gmail.com

A.P. de Leon F. de Carvalho et al. (Eds.): Distrib. Computing & Artif. Intell., AISC 79, pp. 81–88.
springerlink.com

interaction it has with other outside entities, usually is generated and stored on paper, the main problems associated with this are among others: unknown the amount of information that is generated and the lack of standards for: development, classification, management and retrieval of documents. Through computer systems it is possible to resolve these problems, which allows formalize and systematize processes associated with the generation, use, storage and retrieval of documents.

Communication is one of the fundamental processes in which individuals engage throughout their lifetime. It is the tool that society uses to advance towards the compliance of the demands and needs of its members. As with specific societies, business organizations require the tools, systems and models that allow them to adapt to the new knowledge-based society on a daily basis. They need to adapt their strategies in order to become the type of business organization that learns and focuses on teamwork, the optimal use of human resources, flexibility, and the involvement of professionals in an innovative corporate culture that is fully integrated within a society. Such organizations emphasize just-in-time objectives, quality, efficiency and the continual improvement of processes [1].

Among the most prominent MAS objectives is the ability to build systems capable of autonomous and flexible decision-making, and to cooperate with other systems within a "society" [2] that meets the following requirements: distribution, continual evolution, flexibility to allow members (agents) to enter and exit, the appropriate management of an organizational structural, and the ability to execute agents in both multi devices and other devices with limited resources. It is possible to meet these requirements with the use of an open system and virtual organization paradigm [3].

Open MAS [4] can be thought of as distributed systems in which the agent population interacts and exhibits various behaviors. Unlike distributed systems, the cooperation between agents is not determined during the design process, rather it emerges upon execution.

The objective of this paper is to present the HdeC-MAS 1.0 system, an open MAS that can manage the communication process within healthcare environments. The system is based on the PAINALLI architecture, which is used for developing open MAS for virtual organizations. This makes it easier to develop open MAS systems that can optimize the communication process in business organizations by using an agent platform, which can be accessed via fixed and mobile devices such as PDAs (Personal Digital Assistant), smart phones and personal computers connected to wired and wireless networks.

The majority of the methodologies used in the development of MAS are an extension of existing methodologies from other fields [5]. Some are based on specific agents and architectures [6], while others are geared towards organizations, Agent-Group-Role [7], Tropos [8], MOSEInst [9], OMNI [10], and E-institutions [11]. There does not currently exist a methodology for developing MAS that is based on organizational structures and that would allow direct communication between the various communicative aspects within an organization. Nor is there a methodology that allows a complete specification of the system´s social structure. PAINALLI is based on THOMAS (MeTHods, Techniques and Tools for Open

Multi-Agent Systems) [12] [13], a system that strengthens the definition for the organizational model and the specific methods for the development of open MAS focused on the concept of organization.

Poor internal communication is one of the weaknesses that most affects the development of an organization. For this reason, it is necessary to create a new business model in which communication assumes a critical role [14] [15].

The following section describes the components of the architecture. Section 3 presents the HdeC-MAS 1.0 system, and section 4 offers the results and conclusions.

2 PAINALLI Architecture

PAINALLI is an architecture that is used for developing open MAS based on virtual organizations. It focuses specifically on the communication process in business organizations. The communication process in PAINALLI is based on the use of templates and metadata that are associated with attached files. The input methods are based on international standards: ISO 15489 e ISO 9000:2000.

The goals for the architecture were rooted in the exchange of messages and attached files as a support mechanism for an organization´s communication process. The primary objectives are to manage personal and business related messages, documents, user accounts, user groups, and agendas. All of the information is stored and transferred with encryption algorithms.

Users are able to manage their own messages, which are created from templates: scheduling, information, question and request. The sender is notified once the message has been read. A conversation is a set of messages with a common theme, which is "owned" by the original sender. Only the original sender can eliminate messages and conversations.

The architecture administrator is responsible for managing the templates and the input methods, both of which can be tailored to the specific needs of the organization. The administrator manages the users, who can be grouped according to the organizational structure.

PAINALLI is composed of: applications and services, communication protocol, and an agent platform.

➢ **Applications and services**

The combined set of applications and services proposed in PAINALLI constitute the foundation that support the functionalities given to the users and developers who can now benefit from web services such as: web interoperability, offering services through an open exchange of applications and data, offering remote procedure services, and requesting procedure services through the web.

The programs required for accessing the system functionalities are the applications. They must be dynamic and able to adapt to their context, and capable of reacting differently to particular situations and the type of service requested. The applications make it possible for the services to be executed locally or remotely, through computers or mobile devices. The low processing capability of the applications is irrelevant since computing tasks are carried out by the agents and services located in the devices specifically intended for these activities.

Services constitute a set of activities that are meant to respond to the needs of the requester. They are programs that provide methods for accessing databases, manage connections, analyze data, obtain information from external devices, publish information, or even use other services to carry out a particular task.

PAINALLI uses service directories that can be managed dynamically according to the needs of the developers and the demands of the users. Both services and applications in PAINALLI are reusable and function independently from the system to which they are offering their functionalities and from the programming language used by the platform agents [16].

➢ **Communication protocol**

Messages are transmitted in PAINALLI through a communication protocol that allows applications and services to communicate directly with the agent platform. The SOAP (Service Oriented Architecture Protocol) [17] specification serves as a reference for establishing the communication protocol in PAINALLI and allows the programming language to function independently. Agents use the ACL (Agent Communication Language) specification in FIPA [18] to communicate. ACL messages are objects and require protocols to allow their transport, which is provided by RMI (Remote Method Invocation). This specification is very useful in the event that the applications are executed from mobile devices, which have limited processing capabilities.

➢ **Agent platform**

The set of roles that comprise the agent platform in PAINALLI can control and manage all of the architecture´s functionalities: applications, services, communication, output, reasoning capability and decision-making. Any modifications on agent behavior are carried out according to user preferences, knowledge obtained from previous interactions, and the available options for responding to a particular situation.

In PAINALLI, the agents take on the following roles, which will subsequently help the external agents in the case study: IA: Interface agent, ACA: application communication agent, CSA: communication services agent, DAA: document administrator agent, MAA: message administrator agent, DiA: service directory manager agent, SeA; security manager agent, SuA: supervisor agent, KEA: knowledge extraction agent, MA: manager agent

With both PAINALLI and web services, security has to control the users and the access given to each of the operations exposed through WSDL (Web Services Description Language) to all other users. It is also necessary for security to provide a set of bookstores that make it easy to work with security APIs for both the client and the server. When using SOAP protocol, one way of dealing with security is through WS-Security (Web Service Security), which allows the exchange of security Tokens between client and server (End-to-End security). This exchange of Tokens, as well as the mechanisms associated with encryption and signature makes it possible to ensure the authentification, integrity and confidentiality of the operations that are carried out.

3 HdeC-MAS 1.0 System

The HdeC-MAS 1.0 system supports the administrative and medical communication process in a hospital. A hospital is an organization in which a great deal of information is generated and used, and where the key players in the process can be grouped into three categories: sources, recipients and intermediaries.

The information system in a healthcare environment, particularly that of a hospital, must be able to adopt the messages, formats, code and structure of medical histories in order to allow interoperability within the system.

The use of standards permits an increase in security, lowers costs and encourages market development. The implementation of standards and norms for all users, manufacturers and service providers, promotes the creation of more economical and stable solutions.

A document is one type of standard that is established by consensus, approved by a recognized entity, provides rules, guides or characteristics needed to carry out activities [19]. The use of standards makes it possible to achieve interoperability within the system and its components.

Interoperability is the ability of two or more systems or components to exchange information and use the information that has been exchanged. Syntactic interoperability (operative or functional) is the structure of communication, the equivalent of spelling and grammar rules. H7 (Health Level Seven) is one type of method that can be used to exchange messages or data within a health environment. Semantic interoperability contains the meaning of the communication, the equivalent of a dictionary or thesaurus. The CDA (Clinical Document Architecture) within HL7 is a structural and semantic. It is based on XML language, and SNOMED (Systematized Nomenclature of Medicine) terminologies and LOINC (Logical Observation Identifiers Names and Codes), both of which are examples of semantic standards.

For messaging and data exchange, HdeC-MAS 1.0 uses one of the HL7 standards, while for image exchange it uses DICOM. In order to best integrate the data and the HdeCE structure, the CDA standard proposed by HL7 is followed.

The following section presents the various roles that HdeC-MAS 1.0 needs to execute all the actions described in PAINALLI, in addition to the communication process for a hospital.

3.1 Agent Platform

After analyzing the communication process, the information, and user requests made by hospital personnel, it is possible to see how these elements can form an adaptive virtual organization whose administration depends on the variability of its products, users, etc. [20]. In order to completely satisfy the demands of the users in the communication process of a hospital, as well as those of the agent roles presented in PAINALLI, new roles can be added using the system´s open architecture. The new roles allow the system to conform to HCE by managing patient data, appointments, lab analyses, clinical tests, diet and patient care.

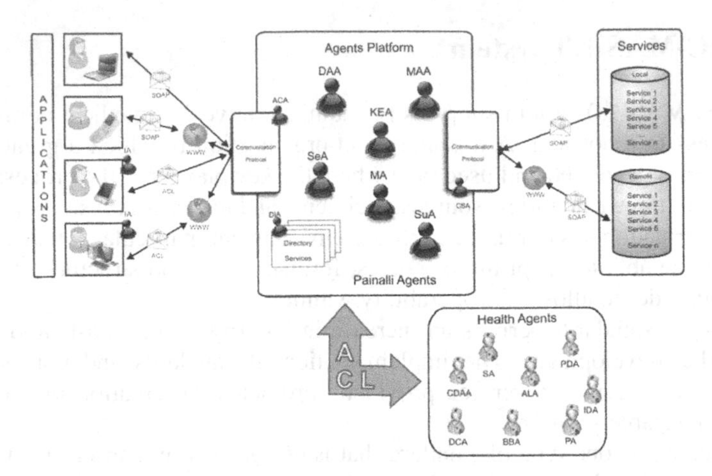

Fig. 1 Scenario of the communication process in a hospital

Figure 1 shows the PAINALLI architecture and the health agents that together manage and formalize the communication process in a hospital.

Health agents take on the following roles: PDA (Patient Data Agent): manages the administrative and personal data of the patients; SA (Scheduling Agent): manages appointments; ALA (Analysis and lab tests agent): handles order forms for analyses and clinical studies; IDA: (Image diagnostic agent): keeps track of messages for managing hemoderivative transfusion devices; PA (Pharmaceutical agent): controls the templates for medical prescriptions; CDAA (Clinical document architecture agent): assumes the same responsibility as the AG in PAINALLI, but with the roles corresponding to health agents.

4 Results and Conclusions

HdeC-MAS was installed in a hospital in Salamanca, Spain. Several interviews were carried out in early 2008. During the same period a number of hospital activities were timed. The goal of both activities was to have reference data to compare against the results obtained after installing the system. Because the purpose of this paper is not to provide a full description of the various quality indicators, only a few descriptions will be identified. The results presented in this section are divided into 3 indicator groups. For each group there will be 2 variables described in detail, which makes it possible to gain an insight on: doctors, nurses, managers, patients and caregivers.

➢ **Assistance indicators**

This information allows us to understand how much time doctors and nurses invest in writing reports and connected to the system.

Medical histories are the most important documents in a health system since they contain all of the information related to each of the patients. The indicator shows a 20% decrease in preparing these documents, which is due to the fact that

HdeC-MAS can "write" the information directly in an electronic document. The indicator that most significantly demonstrates the effectiveness of using HdeC-MAS 1.0 is the interconnectivity that doctors can have with other doctors or clinical services in the hospital. What previously took on average 5 days to complete can now be done immediately.

➢ **Management indicators**

Confirming patient eligibility is an important step towards ensuring that services are rendered as quickly as possible, and that whoever receives treatment is eligible. A decrease of 32.44% in the waiting period involved in confirming eligibility makes it easier to grant services and more importantly ensures that 100% of the patients receiving services are in fact eligible. The indicator for managing the costs involved in image diagnostics is reduced significantly (90%) because with HdeC-MAS 1.0 it is no longer necessary to print images.

➢ **Patient and caregiver satisfaction indicators**

Some of the indicators that demonstrate the level of satisfaction are: (i) the waiting time between appointments, which fell by almost 28% making patients feel that they are being better attended; (ii) the number of complaints, which fell from 84 to 35, resulting in 49 fewer complaints. To support this indicator, we analyzed the complaints. Of the 35 that were brought forth, 20 were related with variables from HdeC-MAS and 15 were related to aspects, such as administration and billing that will be addressed in version 2.0.

The indicators reflect a significant decrease in time, which can now be spent on activities related to the professional and personal development of the employees.

One of the advantages of PAINALLI is the flexibility for adapting to different systems. HdeC-MAS 1.0 takes the agent platform architecture and adds new agents to solve the problem of communications in public or private healthcare environments.

References

[1] Garcia Echevarría, S., Val Núñez, M.T.: New perspective in business organizations: lean Management. IDOE, Alcalá de Henares (1995)

[2] Corchado, J.M., Bajo, J., Abraham, A.: GERAmI: Improving the delivery of health care. IEEE Intelligent Systems, Special Issue on Ambient Intelligence 3(2), 19–25 (2008)

[3] Corchado, J.M., Tapia, D.I., Bajo, J.: A Multi-Agent Architecture for Distributed Services and Applications. Computational Intelligence (2009) (in press)

[4] Jennings, N.R., Sycara, K., Woldridge, M.: A roadmap of agent research and development. Autonomous agents and multiagente systems (1998)

[5] Iglesias, C.A., Garijo, M., González, J.C.: A survey of agent-oriented methodologies. In: Rao, A.S., Singh, M.P., Müller, J.P. (eds.) ATAL 1998. LNCS (LNAI), vol. 1555, pp. 317–330. Springer, Heidelberg (1999)

[6] Bussmann, S.: Agent-oriented programming of manufacturing control tasks. In: Proc. 3rd Int. Conference on Multi-Agent Systems, ICMAS 1998 (1998)

[7] Ferber, J., Gutkenecht, O., Michel, F.: From agents to organizations: an organizational view of multi-agent systems. In: Proc. AAMAS 2003 – Agent-Oriented Software Engineering Workshop, AOSE (2003)

[8] Castro, J., Kolp, M., Mylopoulos, J.: A requirements-driven software development methodology. In: Conference on Advanced Information Systems Engineering (2001)
[9] Gateau, B., Boissier, O., Khadraoui, D., Dubois, E.: MOISE-Inst: An Organizational model for specifying rights and duties of autonomous agents. In: der Torre, L.V., y Boella, G. (eds.) 1st International Workshop on Coordination and Organization (2005)
[10] Vazquez-Salceda, J., Dignum, V., Dignum, F.: Organizing Multiagent systems. Technical Report UU-CS-2004-015, Institute of Information and Computing Sciences. Utrecht University (2004)
[11] Esteva, M., Rodriguez, J., Sierra, C., García, P., Arcos, J.: On the formal Specification of Electronic Institutions. Agent Mediated Electronic Commerce (2001)
[12] Argente, E., Botti, V., Carrascosa, C., Giret, A., Julian, V., Rebollo, M.: An Abstract Architecture for Virtual Organizations: The THOMAS project DSIC-II/13/08 (2008)
[13] Carrascosa, C., Giret, A., Julian, V., Rebollo, M., Argente, E., Botti, V.: Service Oriented MAS: An open architecture. Actas del AAMAS (2009) (in press)
[14] Daft, R.: Teoría y diseño organizacional, 6th edn. International Thomson, México (1998)
[15] Fox, M.: An organizational view of distributed systems. IEEE Trans. on System, Man and Cybernetics (1981)
[16] Cano, R., Corchado, J.M.: Total Versatility: Intelligent Agents in Organizations. In: Antunes, L., Moniz, L., Pavón, J. (eds.) IBERAGENTS 2008 7th Ibero-American Workshop in Multi-Agent Systemas, Lisbon, Portugal (2008)
[17] Cerami, E.: Web Services Essentials Distributed Applications with XML-RPC, SOAP, UDDI & WSDL, 1st edn. O'Reilly & Associates, Inc., Sebastopol (2002)
[18] FIPA. Foundation for Intelligent Physical Agents (2005), http://www.fipa.org (retrieved 15-02-2008)
[19] van Bemmel, J.H., Musen, M.A.: Handbook Of medical Informatics. Springer, Heidelberg (1997)
[20] Camacho, D., Borrajo, D., Molina, J.M.: Intelligence Travel Planning: a Multiagente planning system to solve web problems in the e-tourism domain. International Journal on Autonomous Agents and Multiagent systems (2001)

APoDDS: A DDS-Based Approach to Promote Multi-Agent Systems in Distributed Environments

Raul Castro-Fernandez and Javier Carbo

Abstract. Multi-agent systems (MAS) paradigm emerged as an innovative technology that seemed to be applicable to a large number of distributed problems. However, during these years, ubiquitous computing and ambient intelligence among other distributed paradigms have proposed problems that are currently coped with other technologies. MAS have remained in research environments without establishing themselves in the distributed computing field, despite the benefits it could provide to it. In this paper, the key factors that have produced this situation are pointed, and solutions in order to fix it are proposed. *APoDDS* is a platform which collects all these solutions and merge them. Finally, a comparison between the new approach and the well-known agent platform *JADE* is made in order to evaluate the proposal.

Keywords: Multi-Agent systems, Distributed Computing, Communication Middleware.

1 Introduction

During the last two decades, several distributed computing models have emerged. Ubiquitous and pervasive computing, context-aware computing, ambient intelligence and cyber-physical systems are some of these models. The trend is to get more and more distributed systems around us. Avionics, automotive systems, health-care, monitoring, control systems and entertainment are just a few examples of current and future distributed applications.

In the 90's and because of their properties, Multi-Agent paradigm seemed to fit perfectly with all these distributed environments. MAS are composed by agents, which are defined [9] as independent computational entities allocated in any environment, which are able to act autonomously within that scenario in order to reach its objectives. They have mainly the following two properties [8]. Firstly, they are

Raul Castro-Fernandez and Javier Carbo
Computer Science department, Universidad Carlos III de Madrid
e-mail: rcfernan@pa.uc3m.es, jcarbo@inf.uc3m.es

A.P. de Leon F. de Carvalho et al. (Eds.): Distrib. Computing & Artif. Intell., AISC 79, pp. 89–96.
springerlink.com

autonomous. They can decide the next action to do to accomplish its goal. Secondly, the communication capability. They must be able to communicate with other agents or entities for solving the problem. In distributed computing models both autonomy of the elements which form the distributed environment and communication are key features. The relation is direct, MAS seems to fit in distributed applications, and this symbiosis is an advantage since it allows developers to think clearly about the final goal and use the concept of agent in order to accomplish it.

However, despite the many benefits that MAS can give to distributed computing models, they remain constrained to research environments. Currently, distributed problems are solved by means of other technologies, like Service-Oriented Architectures. MAS can be deployed alongside other technologies in order to improve the solution to the distributed environment, bringing several advantages to distributed environments. Therefore, in this paper, the main factors that have disabled MAS from being used in distributed environments are pointed. Then, solutions in order to tackle these problems are proposed. *APoDDS* takes all these solutions and establishes them as design requirements, then a new platform is implemented and it is compared to the well-known agent platform *JADE* [1]. The remainder of this paper is organized as follows. The main problems of MAS are collected through studying the state of the art in Sect. 2. In Section 3, the design of *APoDDS* is described, and its advantages are shown. A comparison between *APoDDS* and *JADE* is done in Sect. 4. Finally, conclusions and future work are presented in Sect. 5.

2 State of the Art Analysis

This section is organized as follows. Firstly, the state of the art regarding Multi-Agent systems applied to distributed environments is investigated, and the problems that have the different approaches are shown. Then, these problems are defined and specific solutions are proposed.

2.1 Problems of Multi-Agent Platforms

In order to make an accurate investigation, the state of the art is focused from two different perspectives.

Versatility and scalability. There are a large number of different distributed applications, with their own requirements and needs. Some properties are common, like autonomy and communication, but other ones are domain-specific. Hence, it is necessary for the technology that is going to tackle the problem to be versatile enough. Multi-Agent paradigm is very versatile, but this is not the general case of current agent platforms which implement the paradigm. The FIPA standard [10] was designed in order to homogenize Multi-Agent systems. It offers a defined architecture and several facilities. While this can suppose various advantages, a FIPA-compliant Multi-Agent platform is constrained to this standard. In FIPA architecture, it is defined an entity called *Directory Facilitator*, which is a directory agent in charge of managing the information and functionalities of other agents in the platform. This can be useful when the domain is small, but when the system grows, a centralized entity affects severely the system scalability. Moreover, having a defined architecture leads to a poor versatility, reducing the applicability of MAS. Some platforms which are not FIPA-compliant appeared, like *Cougaar* [3] and *Tryllian* [11]. The problem of *Tryllian* is the platform performance due to the communication model that it implements, which leads to scalability problems, like it can be seen in [2]. On the other hand, while Cougaar does not follow the FIPA standard, it has its own architecture that leads to the same problem.

Heterogeneity of devices, operating systems, and programming languages is another problem in distributed environments. Most of the agent platforms do not deal with this problem, while current middleware deal and solve it. An agent platform [6] has been developed in order to deal with this problem, by using the middleware *ZeroC ICE* [12], but this platform is FIPA-compliant.

Agent platforms are often very complex, leading to a large overhead in final applications. When tackling with distributed systems, it is desirable to have a lightweight platform, which can be adjusted to the concrete application. Moreover, the less complex, the easier to use it. Cougaar is a lightweight platform, although it presents the aforementioned problems. *JADE* is very well designed, allowing the developer to adjust it to the specific application. But it is FIPA-compliant offering a constrained architecture.

Communication model. Due to the distributed nature of MAS, the earliest platforms integrated a communication middleware which offered high-level primitives to the agent developers. Several communication middleware have been used like *CORBA*, *Java RMI*, and *JMS* more recently. The communication technology used in the first days of agent platforms is often the same as nowadays. One of the main properties of the agents is communication. Therefore, the communication middleware must be a key technology when developing Multi-Agent platforms. Scalability, versatility, lightness and autonomy are features that depend directly on the communication middleware. *JADE* and *Cougaar* have paid attention to this design decision, although the communication middleware that they use is not maybe the best one. Other platforms are built from scratch, leading to robustness, complex, and versatility problems.

2.2 State of the Art Conclusions and Design Requirements

This subsection proposes the specific solutions for each of the problems pointed above. These solutions are also the requirements of *APoDDS*.

Firstly, problems emerging from adjusting the platforms to a specific architecture have been shown. *APoDDS* must not have a predefined architecture in order to be adjustable to the largest range of distributed applications possible.

Moreover, due to the increasing heterogeneity that distributed systems present nowadays, the platform must provide means for making that heterogeneity transparent. In this way, it must be allowed the agents to execute independently of hardware architecture or programming language.

Sometimes, platforms are very complex, leading to large overheads and to complex usage. *APoDDS* must be a light platform. In this way, it must be easy to port to other programming language and it must be easy to use. A developer must be aware of agent theory but not about specific platform internals, because distributed applications are already complex enough.

Communications models used by current platforms are usually language dependent, like in the case of *RMI* or *JMS*, or they are very complex, like *CORBA*. In this way, it is necessary to draw upon an independent-platform communication middleware. The technology must fit with the Multi-Agent paradigm. This is the reason why middleware choice is perhaps the most important decision that should be done. There is a current trend in confer more importance to the communication middleware explained in [7] and [5], where the relationship between the agent environment and the middleware is described. Furthermore, in these recent works is established, as future research, the vertical integration between agent specific functions and communication middleware. *APoDDS* ought to be built upon these guidelines. It must integrate vertically the agent functions with the communication middleware. It has to facilitate the autonomous property of agents, and it should provide scalability by avoiding centralized entities.

3 APoDDS Design Principles

Once the main problems that keep current agent platforms have been pointed, and the proposed solutions have been explained, in this section it is shown the design of *APoDDS*. After exploring the communication models, the communication middleware chosen has been the *OMG* standard *Data Distribution Service DDS* [13]. This middleware implements a publish-subscribe communication model. Despite its relative youth, it has been deployed on critical systems like military and avionics environments. Thus, it is a robust technology that presents several features suitable for *APoDDS*. It is presented the design decisions regarding the development of *APoDDS* below:

- It has been stated in [7] the importance of providing a vertical integration between the agent environment and the communication middleware. *APoDDS* aims to provide this integration by offering a high level API at the top, which is directly related to the primitives of the communication middleware *DDS*.
- The publish-subscribe model allows an agent to publish some kind of information regardless the localization of other agents, because the *DDS* service is the entity in charge of delivering the information to the correct agents, see Fig. 1. In this way, an agent has just to know what information has to be sent, not to what agent. This feature helps to encourage the agents autonomy even in the communication level, which is a key factor in order to grant scalability.

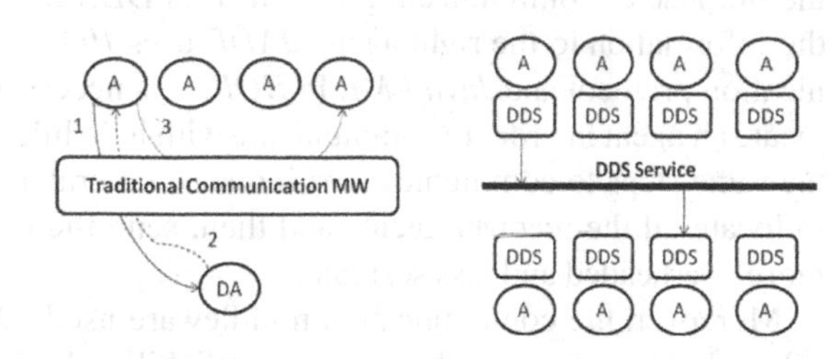

Fig. 1 On the left, the agent needs to locate the destiny agent and then it can send the data. On the right the agent just need to send the data, and the DDS entity is in charge of delivering the information to the correct agent.

- *DDS* is platform and language independent, and it can be installed both on small embedded devices and on large computers. It supports the main programming languages, so it covers the heterogeneity problems that actual distributed systems have.
- *DDS* is a lightweight middleware. Furthermore, it has been tested on critical environments, so it is highly optimized. This leads to a very good performance. The integration between the agent library and *DDS* ensures a lightweight and optimized platform. The light agent library, merged with *DDS*, makes an easy to use platform. Moreover, the overhead introduced by the agent library is small, and its underlying communication middleware helps to palliate it.

In this way, *APoDDS* deals with the problems that current platforms present. Both the communication middleware and the library implementation play a main role in this achievement. The publish-subscribe model allows a decoupled application development, and the agent library provides the developer with a task-oriented programming style. However, the developer is not drafted to use this style, but anyone that could be necessary. In this way, *APoDDS* presents a versatile enough architecture, which can cope with many different distributed applications. Moreover, the vertical integration between middleware and agent environment provides autonomy and scalability to the application. The learning curve is almost immediate because of the facilities provided by *APoDDS*, so the developers can spend their time thinking about the application problems, and not trying to make the platform works.

4 Evaluation

Two different tests have been carried out in order to evaluate *APoDDS*. A qualitative test in which advantages of *APoDDS* are underlined, and a quantitative test, where specific results are shown. In the quantitative test, some experiments have been done to compare *APoDDS* with *JADE*. *JADE* is nowadays the most well-known agent platform, and it is proved to be the current most scalable platform [4].

Qualitative evaluation. There are some remarkable points present in *APoDDS*. These are remarkable because these features are crucial in order to make *APoDDS* to accomplish the design requirements.

On one hand publish-subscribe communication model provides real autonomy to agents. An agent does not need to be aware of the localization of another agent for the purpose of communicating with it. It is DDS the entity in charge of delivering the information to the right agent. *JADE* uses *IIOP*, which is the *CORBA* communication protocol and *Java RMI*. In *IIOP*, it is necessary to know previously how to locate an agent in order to communicate with it. While in *APoDDS* it requires exactly the same steps to communicate with one or several agents, in *JADE*, it is necessary to locate all the receiver agents and then, send them the data, which makes *JADE* more overheaded and less scalable.

Moreover, the communication middleware used, *DDS*, comes from critical distributed environments, which ensures reliability, high performance and robustness. Furthermore, *DDS* includes several Quality of Service (*QoS*) parameters, which confers it with several non-functional facilities regarding time constraints, reliability, fault-tolerance mechanisms, etc.

Quantitative evaluation. For the quantitative test, two different experiments have been performed. The first one consists on measuring the Round Trip Time (RTT), which aims to show how beneficial the vertical integration is. *DDS* performance is also shown with the first experiment. The second experiment aims to show how scalable is *APoDDS* versus *JADE*. Both tests have been done in a *Intel Core 2 Duo@3Ghz*, with *3Gb RAM@800Mhz*, using *Ubuntu 8,04 LTS* as operating system and the Java programming language with *JVM 1.6*.

RTT Experiment. There exist two agents. Agent *A* sends some data to agent *B*. When agent *B* receives data from agent *A*, it forwards this data to agent *A* again. Agent *A* measures time when it sends the data, and when it receives it back. In *JADE*, it is necessary firstly to look for agent *B* (by means of the directory agent), and then send it the data. In *APoDDS*, it is just necessary to send data regardless the actual localization of agent *B*. Results are very favorable to *APoDDS* which performs about 20 times better regarding time than *JADE*. In Fig. 2 it is shown twenty different measurements and finally, the mean value for *JADE* and *APoDDS*.

The large performance difference is mainly due to two reasons. Firstly, *DDS* has been proven to be more optimized than *IIOP* and *RMI*. And secondly, *DDS* is totally decoupled, and it is not necessary to look for the receiver, like in the *JADE's* case. Therefore, while in *JADE* four messages are necessary to calculate the RTT, in *APoDDS* are only necessary two. The exchange of data is made in shared memory. However, if the experiments would had been done in a distributed environment, the latency that the network would introduce in the communication would affect *JADE* more than *APoDDS*, because of the number of messages exchanged.

	Jade(ns)	ApoDDS(ns)		Jade(ns)	ApoDDS(ns)		Jade(ns)	ApoDDS(ns)		Jade(ns)	ApoDDS(ns)
Exp. 1	2909229	187199	Exp. 6	4844281	189761	Exp. 11	6720202	192464	Exp. 16	3183317	183488
Exp. 2	2886722	188654	Exp. 7	4287204	190226	Exp. 12	3261990	190562	Exp. 17	8473780	192716
Exp. 3	3453842	188389	Exp. 8	2614491	193775	Exp. 13	3584311	184217	Exp. 18	2272435	182861
Exp. 4	3417996	187460	Exp. 9	7993185	190856	Exp. 14	6793363	189773	Exp. 19	4182101	189767
Exp. 5	2980128	190991	Exp. 10	2662484	186560	Exp. 15	4071759	189673	Exp. 20	2224184	185196

	Jade(ns)	ApoDDS(ns)
MEAN	4140850,2	188729,4

Fig. 2 RTT test. 20 experiments which show RTT's under *JADE* and under *APoDDS*. Finally the mean RTT for both *JADE* and *APoDDS*

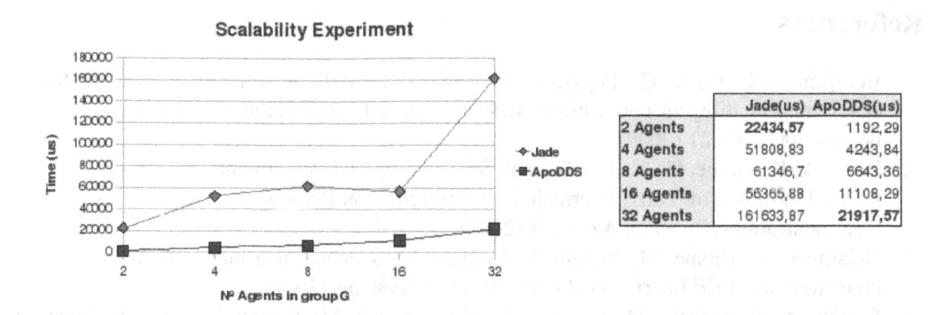

	Jade(us)	ApoDDS(us)
2 Agents	22434,57	1192,29
4 Agents	51808,83	4243,84
8 Agents	61346,7	6643,36
16 Agents	56365,88	11108,29
32 Agents	161633,87	21917,57

Fig. 3 Scalability Test. On the left picture, the graph represents times from *JADE* and *APoDDS* while increasing agents in group G. On the right one the numerical values presented in the graph

Scalability experiment. In this experiment, shown in Fig. 3, an agent *A* sends data to a variable group of agents *G*. Each agent in *G* must respond to agent *A*. Agent *A* measures the time between it sends the data and it receives the answers from all agents in *G*. *APoDDS* obtains better times than *JADE*, which is easy to understand because of the previous experiment. The relevant fact in this experiment is the growth of the times, which represents the scalability of both platforms. While in *APoDDS* times grow linearly, in *JADE*, once the number of agents in *G* is greater than 16, times grow above linear behavior. It is worth realising that in a distributed environment it is expected that *JADE* would be more affected by the network latency because of the centralized entity (the *Directory Facilitator*), while in the publish-subscribe model this would not suppose a problem.

5 Conclusions and Future Work

In this paper it has been presented the main problems that have disabled MAS to coexist with other technologies in distributed environments. MAS fit well with distributed paradigms, and this is what has motivated this research work. Specific solutions have been proposed for the aforementioned problems, and a new platform called *APoDDS* has been implemented for the purpose of merging all the solutions. A comparison has been made between *APoDDS* and *JADE* in order to validate the new approach, and the results are promising. It is worth noting, however, that *APoDDS* is an agent platform focused on applying MAS to distributed environments.

As future work, it is necessary to make more exhaustive experiments, and do them in distributed environments. Moreover, it is necessary to include *DDS QoS* support in *APoDDS* in order to provide the developers with new facilities. It is also necessary to improve the modular architecture of *APoDDS* in order to allow users to include their own functionality modules, like ontologies support or reason engines. Finally, taking advantage of real-time features of *DDS* could be a very desirable feature in modern and future distributed systems.

References

1. Bellifemine, F., Caire, G., Poggi, A., Rimassa, G.: JADE: A software framework for developing multi-agent applications. Lessons learned Information and Software Technology 50(1), 10–21 (2008)
2. Burbeck, K., Garpe, D., Nadjm-Tehrani, S.: Scale-up and Performance studies of Three Agent Platforms. In: IEEE International Conference on Performance, Computing, and Communications, Phoenix, AZ, USA (2004)
3. Helsinger, A., Thome, M., Wright, T.: Cougaar: A scalable, distributed multi-agent architecture. In: IEEE International Conference on Systems (2004)
4. Chmiel, K., Gawinecki, M., Kaczmarek, P., Szymczak, M., Paprzyckib, M.: Efficiency of JADE agent platform. Scientific Programming. IOS Press, Amsterdam (2005)
5. Platon, E., Mamei, M., Sabouret, N., Honiden, S., Van Dyke Parunak, H.: Mechanisms for Environments in Multi-Agent Systems Survey and Opportunities. Kluwer Academic Publishers, Norwell (2006)
6. Vallejo, D., Albusac, J., Mateos, J.A., Glez-Morcillo, C., Jimenez, L.: A modern approach to multiagent development. The Journal of Systems and Software (2009)
7. Weyns, D., Helleboogh, A., Holvoet, T., Schumacher, M.: The agent environment in multi-agent systems: A middleware perspective. Multiagent and Grid Systems (2009)
8. Wooldridge, M.: An introduction to MultiAgent Systems. Wiley, Chichester (2009)
9. Wooldridge, M., Jennings, N.R.: Intelligent Agents: Theory and practice. The knowledge engineering review 10(2), 110–152 (1995)
10. The foundation of intelligent agents, http://www.fipa.org/
11. Tryllian Agent Development Kit, http://www.tryllian.org/
12. ZeroC ICE, http://www.zeroc.com/
13. Data Distribution Service for Real-Time Systems, http://www.omg.org/technology/documents/formal/data_distribution.htm

Architecture for Multiagent-Based Control Systems

D. Oviedo, M.C. Romero-Ternero, M.D. Hernández, A. Carrasco,
F. Sivianes, and J.I. Escudero

Abstract. This paper presents a multiagent architecture that covers the new requirements for the new control systems such as the distribution and decentralisation of system elements, the definition of communications between these elements, the fast adaptation in the control and organizational changes. The agents in this architecture can cooperate and coordinate to achieve a global goal, encapsulate the hardware interfaces and make the control system easily adapt to different requirements through configuration. Finally, the proposed architecture is applied to a control system of a solar power plant, obtaining a preliminary system that achieve the goals of simplicity, scalability, flexibility and optimization of communications system.

1 Introduction

The control systems have evolved to a complex system that employs more and more equipment and sensors. The hardware diversity of the sensors and actuators greatly affects the portability of the control system and the complexity may differ from each other. A goal of control systems is to enable the integration of different types of devices in a scalable and flexible system. The problem is how communications among the different parts of the system is organised and optimized [1]. To resolve problems in the control system it is necessary have knowledge and experience working in the field. In this paper, the solutions to control problems will be the responsibility of different agents in the system. The agents should handle the problem domain knowledge and be able to communicate this knowledge in order to provide efficient solutions and recommendations. An architecture is proposed that presents an organizational scheme of agents and a global communication protocol. The architecture was constructed to support a wide range of control systems; however, this paper focuses on its application in the control of a solar power plant.

D. Oviedo, M.C. Romero-Ternero, M.D. Hernández, A. Carrasco,
F. Sivianes, and J.I. Escudero
Departamento de Tecnología Electrónica, Universidad de Sevilla, Spain
e-mail: oviedo@dte.us.es, mcromerot@us.es, marilohv@dte.us.es,
acarrasco@us.es, sivianes@dte.us.es, ignacio@us.es

A.P. de Leon F. de Carvalho et al. (Eds.): Distrib. Computing & Artif. Intell., AISC 79, pp. 97–104.
springerlink.com

It has a distributed input-output agent architecture to accomodate the changing requirements. This distributed intelligent agent architecture provides flexible and scalable ways to integrate the different sensors and actuators. The design goal is to create a system architecture that is general enough to support many different kinds of sensors and actuators, while being distributed and scalable.

Finally, to illustrate the application of the architecture, an example system is built (a solar power plant control system). This preliminary implementation of system gives an idea of the spread of knowledge flows in multiagent-based control systems. The spread of this knowledge in the system will not be free since it is determined by the architecture of the system.

2 Problem Domain

The problem discussed here belongs to the field of distributed control systems. An introduction to the subject and issues can be found at [3], [8]. Control systems based on the theory of multiagents, and restricted to specific domains have been developed. For instance, one of the early works by Junpu et al. [7] discusses the feasibility of agent-based distributed hierarchical intelligent control. Another study [5] examines the modeling of multiagent control for energy infrastructures that presents an agent-based control system for distributed energy resources in low voltage power grids.

Other important aspect that is discussed here concerns the spread of knowledge in control systems based on multiagents. In [4] the study of knowledge transfer and action restriction among agents in multiagent systems founded on the definition of patterns of dialogues between groups of agents, are expressed as protocols. This paper is focused in the organization of the elements of the system and the flow of information within the system, with the aim of creating a simple and optimized model.

Finally, the aim is to build a remote and automatic control system for a solar power plant (CARISMA Project) through the use of multi-agent technology, where a set of performance areas controlled by a variable number of agents is defined in the system. These areas will cover some agents assigned to allow coordination among areas, with the purpose of making joint control actions on different areas. This system should be simple and streamline communications, as well as being flexible and robust.

3 Architecture Proposal

It is proposed an architecture divided into three layers looking to optimize the number of messages sent through the network to communicate knowledge and to establish a structure that allows some agents to perform certain recommendations or make decisions based on knowledge dispersed over different parts of the network. Figure 1 shows a diagram of the architecture.

As shown in figure 1, the first layer, called the *input layer* comprises all agents that allow for the entry of new knowledge in the system introduced by a user or generate changes in the knowledge base of the system from information sent by lower layer agents. The number of agents that may belong to this layer is not limited, but the simplest is formed by a single agent (decisions and recommendations centralized generator). The agents of this layer only communicate with the agents of intermediate layers, but allow direct communication with final layer agents when the response speed requirements are high (ocasionally).

The *middle layer* comprises all agents that have the ability to generate decisions or recommendations based on knowledge acquired from different points or areas of the multi-agent system and coordinate the agents belonging to lower layers. This layer can be subdivided into many middle layers as desired, allowing for scale and creating a hierarchy multiagent system for decision or recommendation generation. The number of agents in this layer is not limited. It is usually a number greater than one and less than the number of agents of the bottom layer that controls. In the communication level, the agents of this layer are able to coordinate (horizontal comunication) and disseminate knowledge in the layers in both directions (upper-layer and lower-layer).

The *final layer* comprises all terminal agents that have the ability to obtain environmental data or act on it. The number of agents in this layer is not limited, it is usually a number greater than the nummbers in the rest of layers. These agents can be organized into zones or regions of coverage controlled by a set of agents from the upper layer. The agents of this layer only communicate with the agents of the intermediate layers, but allow direct communication with agents from the input layer when the response time requirements are high (ocasionally).

3.1 Architecture Overview

A general system architecture is proposed that is composed of four types of agents: Teleoperator Agent (TA), Coordinator Agent (CA), Operator Agent (OA) and Device Agent (DA). This architecture is the basis of the CARISMA project. The number of coordinator, operator and device agents is free (specified in the configuration of the system), while there is only one teleoperator agent. The network can expand or shrink according to the number of the solar power plants and the complexity of the control system. The architecture of individual agents is based on the paradigm

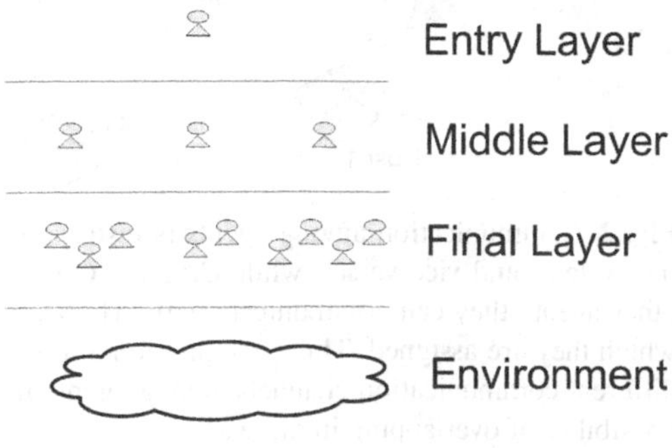

Fig. 1 Our model proposes the organization of a multiagent system in at least three layers. This is in the capabilities knowledge management and communication with agents (the fact that an agent belongs to one layer or another will depend primarily on the behaviors that it implements).

Belief-Desire-Intention [2]. An example of the network topology including the three zones is shown in figure 2.

The Teleoperator Agent is the entry point into the system, providing a user interface that allows configuration, deployment and knowledge input to the platform. The main responsabilities of the TA are to:

- Define the hosts that make up the platform, and define the CAs and OAs and also define the areas
- Register agents in the corresponding hosts and areas
- Generate the specific DAs for the sensors and actuators to be controlled and their asociation with the corresponding OA
- Provide user GUIs and interfaces to manage the configuration
- See agent status and view data acquired by sensors
- Introduce new knowledge (insert, modify or delete rules, concepts, predicates and actions)

The Coordinator Agents goals are to coordinate global solutions for a state of failure or alarm detected from multiple points in different areas. The main responsibilities of the CAs are to:

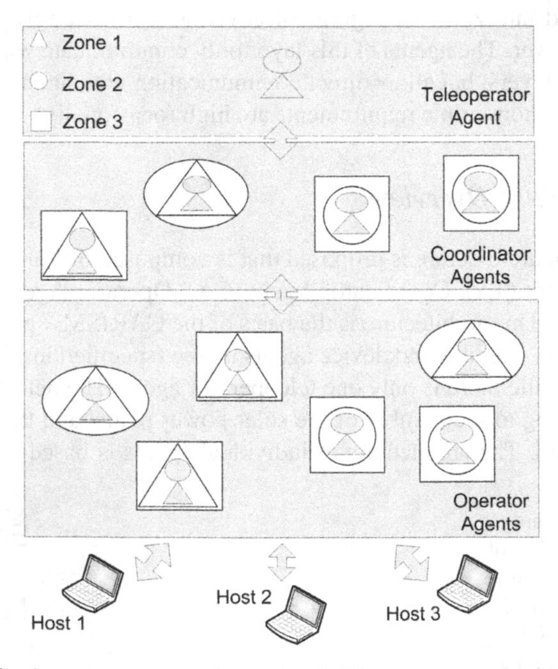

Fig. 2 Communication among agents is restricted: TA can communicate with any agent of the system (and vice versa), while CAs and OAs will have specific information about what other agents they can communicate with. The DAs may only communicate with the OA to which they are assigned. This configuration allows us to define flexible areas, by supporting different communication channels among agents living in the system, which can lead to the possibility of overlapping in these areas.

- Control the OAs in their area
- Manage the different rules of global action
- Coordinate with the TA to give advice or solutions in a global alarm
- Serve as an intermediate point of communication of new knowledge among the TA and OAs

The Operator Agents are responsible for controlling the various DAs, and if they detect failures or local alarms in accordance with information received from the DAs, they communicate them to the rest of the system.An example of communication among OAs and DAs is shown in figure 3. The main responsibilities of the OAs are to:

- Obtain data and comunicate operation commands to DAs that have allocated
- Detect local fails or alarms
- Perform direct actions to address failures in case has the necessary knowledge
- Communicate recommendations to other agents

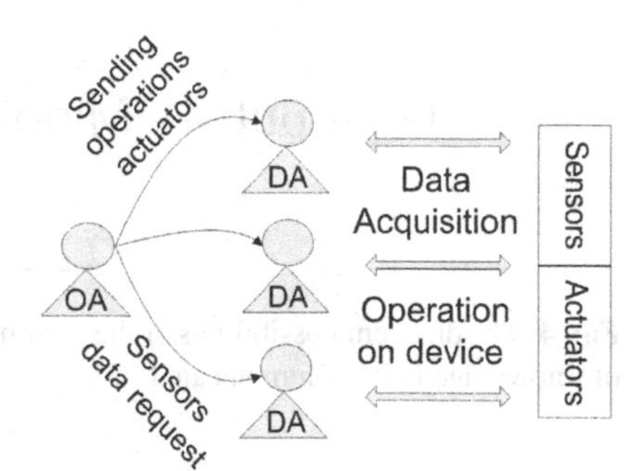

Fig. 3 The Dispositive Agents are hybrid agents (Cognitive and Reactive), that have the ability to alert the OA or act directly in case of changes in the state of a device. Each DA will have a concrete implementation intended to obtain data from a particular sensor or perform actions on a given actuator.

In terms of hardware, there are no restrictions on the number or type agents that can reside in a device or style of devices that compose the system, but these have to be capable of computation. Generally, the agent node device consists of an embedded system with support for various transmission technologies (RF, Ethernet, Bluetooth, ...). One agent node is attached various sensors and actuator devices. The sensors include thermal sensors, humidity sensors, CO2 sensors, sensors for signal from solar plant appliances and various intelligent meters such as solar irradiation and video control [6]. The attached actuators include various valves, motors, and switchers for heating system, ventilation, humidity control, screen control, etc.

3.2 *Knowledge Spreading in the Agents Network*

Another important aspect to consider in a multiagent system is the spread of knowledge (represented by ontologies, [10]) in the system. As shown in Figure 4 the possibilities that can be implemented using expert systems [11] are:

(A) Communication of new knowledge by a human user, or propagation of an action in the system from a global knowledge of system.

(B) Communication of local knowledge: In this case, knowledge spreads from the final agents to agents of the *middle layer* or *input layer*. In the latter case, the spread is usually done through the middle layer, and it can be performed directly among agents from the final layer and the input layer (dotted lines) if response time requirements are high.

(C) Communication of knowledge to the input layer or final layer by an agent of the middle layer, from a partial knowledge of the system.

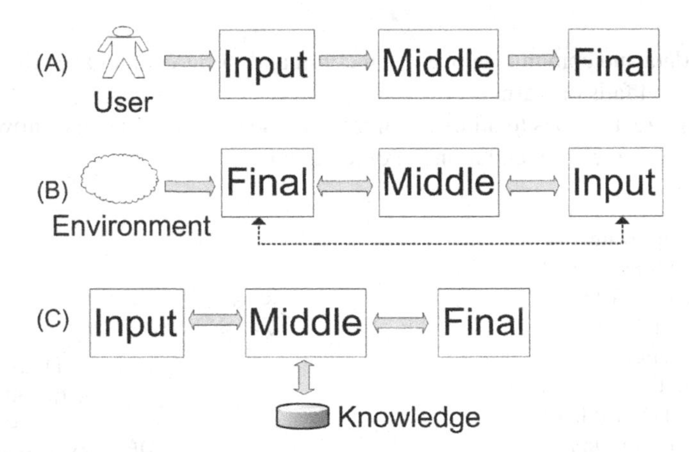

Fig. 4 The different possibilities of the communication flow modeled, according to the source of knowledge to be communicated

4 Implementation

When applying the architecture proposed in section 3, an automatic control and decision support system which is very simple, reliable and scalable can be implemented. For instance, figure 5 shows an example of spreading knowledge in CARISMA, for the B case seen in figure 4 (generation of knowledge in the final layer).

Note that information travels in phases throughout the system and if the necessary knowledge is not available, then a human is requested. This allows for the optimization of the information or knowledge exchanged in the system, and easily increases the knowledge base of system, with scalability.

4.1 First Results

From exposed architecture, a preliminary implementation of the system CARISMA has been carried out. This implementation was performed using the JADE development platform [9] and using hardware devices such as embedded systems and

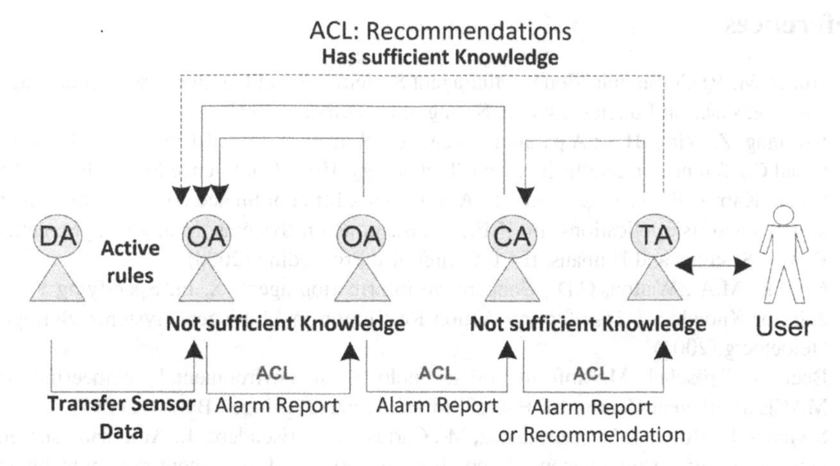

Fig. 5 OAs and DA are in the final layer of the system. CA represents the middle layer and TA of the input layer.

PCs. The base case system developed is composed of 19 agents (among AOs, ACs, ADs and AT) distributed in four areas. The first significant results obtained were: (A) Simple system design, (B) Short implementation time (less than two months), (C) Main features of the control system were covered, based only on the premises exposed architecture, (D) The initial knowledge base of multi-agent system was extended in a quick and easy way by a user without programming knowledge or agents.

5 Conclusions

Traditionally, control systems have had difficulties as far as their design because they must meet high requirements. Using our model of architecture in the design of a multiagent-based control system, we can obtain many benefits, such as simplicity. Based on the organization of agents in layers exposed, we can easily program control systems. Other problems in control systems are scalability and flexibility: this architecture does not limit the number and types of agents and inclusion / exclusion of layers is possible. It also allows for dynamic configurability, so we can dynamically change parts of the system since agents can move from one layer to another, or new types of agents can be added. It is even possible to dinamically add new layers with certain restrictions, only by changing the behavior of agents.

Acknowledgements. The work described in this paper has been funded by the Consejería de Innovación, Ciencia y Empresas (Junta de Andalucía) with reference number P08-TIC-03862 (CARISMA Project).

References

1. Huget, M.-P.: Communication in Multiagent Systems: Agent Communication Languages and Conversation Policies, 1st edn. Springer, Heidelberg (2003)
2. Huiliang, Z., Ying, H.S.: A parallel bdi agent architecture. In: IEEE/WIC/ACM International Conference on Intelligent Agent Technology. IEEE Conference Proceeding (2005)
3. Li, H., Karray, F., Basir, O., Song, I.: A framework for coordinated control of multiagent systems and its applications. In: IEEE Transactions on Systems, Man and Cybernetics, Part A: Systems and Humans. IEEE Conference Proceeding (2008)
4. Grando, M.A., Walton, C.D.: Cooperative information agents x. In: Specifying Protocols for Knowledge Transfer and Action Restriction in Multiagent Systems. Springer, Heidelberg (2006)
5. Beer, S., Tröschel, M.: Information technologies in environmental engineering. In: MACE Multiagent Control for Energy Infrastructures. Springer, Berlin (2009)
6. Sivianes, F., Romero, M., Hernandez, M., Carrasco, A., Escudero, J.: Automatic surveillance in power system telecontrol applying embedded and multiagent system technologies. In: ISIE 2008. IEEE International Symposium on Industrial Electronics. IEEE Conference Proceeding (2008)
7. Junpu, W., Chen Hao, X.Y.L.S.: An architecture of agent-based intelligent control systems. In: Proceedings of the 3rd World Congress on Intelligent Control and Automation. IEEE Conference Proceeding (2000)
8. Yang, J., Yang, L., Zhao, T., Jia, Z.: Automatic control system of water conservancy project model based on multi agent. In: WKDD 2009. Second InternationalWorkshop on Knowledge Discovery and Data Mining. IEEE Conference Proceeding (2009)
9. Bellifemine, F., Caire, G., Poggi, A., Rimassa, G.: JADE - A white paper. In: EXP in search of innovation - Special Issue on JADE. TILAB Journal (2003)
10. Colomb, R.M.: Ontology and the Semantic Web, 1st edn. IOS Press, Amsterdam (2007)
11. Dongliang, L., Kanyu, Z., Xiaojing, L.: ECA Rule-based IO Agent Framework for Greenhouse Control System. In: ISCID 2008. International Symposium on Computational Intelligence and Design. IEEE Conference Proceeding (2008)

SithGent, an Ill Agent

L. Otero Cerdeira, F.J. Rodríguez Martínez, and T. Valencia Requejo

Abstract. In this paper we introduce our future lines of investigation on the field of Intelligent Agents, focusing on analysing why agents which were well defined according to the guidelines of Agent Oriented Software Engineering, do not show the expected behaviour when they are introduced on a real environment, causing the system to collapse. When addressing to agents malfunctions we will use a metaphor in which agents are considered as patients waiting to be diagnosed. Problems agents show would be equivalent to diseases patients suffer from, and the diagnostic process in agents will also be compared to the procedure followed by physicians when they need to heal patients. Therefore we introduce the concepts of SithGent and SithGent Diagnose, to refer to, sick agents and its diagnostic process.

1 Introduction

Russell and Norvig [1] define an agent as anything that can be viewed as perceiving its environment through sensors and acting upon that environment through effectors. Wooldridge and Jennings [2] present an agent as a hardware or (more usually) software-based computer system that enjoys the properties of autonomy, social ability, reactivity and pro-activeness. Other authors offer variations on this theme, but there is the consensus that autonomy, the ability to act without the intervention of humans or other systems, is a key feature of an agent. [3]

So far, we have seen that agents behave and interact with their environment, as humans could do, the next step would be identifying the main characteristics an intelligent agent should accomplish.

1.1 Properties of Intelligent Agents

In the IA context, Wooldridge and Jennings [2], in their definition of agents, mention following properties:

Lorena Otero Cerdeira, Francisco Javier Rodríguez Martínez, and Tito Valencia
Universidad de Vigo, España
e-mail: locerdeira@uvigo.es, franjrm@uvigo.es, tvalencia@uvigo.es

A.P. de Leon F. de Carvalho et al. (Eds.): Distrib. Computing & Artif. Intell., AISC 79, pp. 105–111.
springerlink.com

- **Autonomy:** Agents operate without the direct intervention of humans or others, and have some kind of control over their actions and internal state. When an agent lacks of autonomy we could reach the case when its acts are leaded by another entity, and so, some objectives may never be accomplished.
- **Social ability:** Agents interact with other agents (and possibly humans) via some kind of agent-communication language. If this social ability is reduced, the agent will not be able to interact with the other entities and then it will isolate itself, this will have severe consequences to the whole environment since agents need to cooperate in order to reach the design objectives.
- **Reactivity:** Agents perceive their environment, and respond in a timely fashion to changes that occur in it. When an agent has its sensors spoilt, they could offer mistaken information, which will cause the agent to choose a wrong sequence of actions, the system will end up being unpredictable.
- **Pro-activeness:** Agents do not simply act in response to their environment. They are able to exhibit goal-directed behaviour by taking the initiative. If an agent is not pro-active, the intelligent behaviour could never be emulated, and the agent will turn into a vending machine.

Other properties [2] mentioned in the context of intelligent agents include:

- **Mobility:** Ability of an agent to move around an electronic network. When an agent has mobility issues, it may not be able to accomplish its design objectives, since it will be stuck without being able to reach the places where it is needed.
- **Veracity:** Assumption that an agent will not knowingly communicate false information. If an agent lies, all the actions triggered by this wrong piece of information, may cause the system to reach an undesirable state, since others agents actions, may take this information as truthful.
- **Benevolence:** Assumption that agents do not have conflicting goals, and that every agent will therefore always try to do what is asked of it. If an agent has colliding objectives, it will get stuck in a liveblock, it will never choose an action to execute, since it will not be able to decide which goal should be persecuted first.
- **Rationality:** Assumption that an agent will act in order to achieve its goals, and will not act in such a way as to prevent it goals being achieved at least insofar

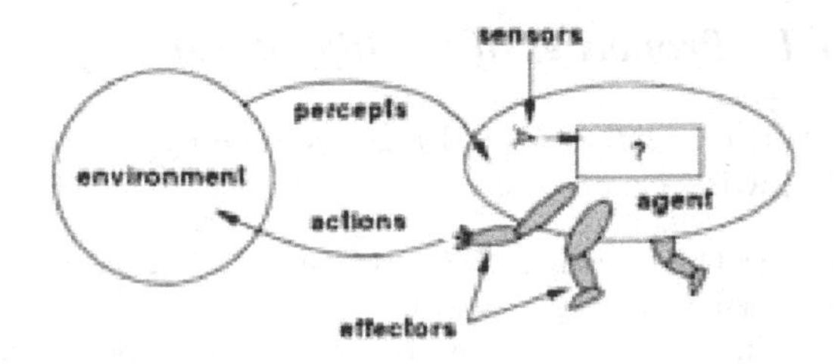

Fig. 1 *Agent interacting with its environment*[1]. It shows a high-level view of an agent within its environment. An agent receives an input from its environment through its sensors and reacts to the input doing something (from a list of possible actions) using its effectors.

as its beliefs permit. If an agent prevents itself from reaching its goals, the whole system will be compromised.

- **Learning/Adaptation [3]:** Agents improve performance over time. If an agent does not learn, it may execute actions that have already been proved to be misleading.

We have already introduced when describing these properties what would happen when an agent does not satisfy them. Diseases we plan on describing will have its basis on the non-accomplishment of these features.

These defining features may be seen as organs in human beings. A human disease strokes an organ, an agent disease strokes a key feature.

By analysing the main features an agent should have we can conclude that agents are social beings, which, far from acting alone, are usually integrated on a multi-agent system and in an environment.

1.2 Environments

The critical decision [3] an agent faces is determining which action to perform to best satisfy its design objectives. Russell and Norvig [1] propose different properties that can affect the complexity of the agents decision-making process to classify agent environments:

- **Accessible vs. inaccessible:** An accessible environment is one in which the agent can obtain complete, timely and accurate information about the state of the environment. The more accessible an environment, the less complicated it is to build agents to operate within it. Most moderately complex environments are inaccessible.
- **Deterministic vs. non-deterministic:** Most reasonably, complex systems are non-deterministic the state that will result from an action is not guaranteed even when the system is in a similar state before the action is applied. This uncertainty presents a greater challenge to the agent designer.
- **Episodic vs. non-episodic:** In an episodic environment, the actions of an agent depend on a number of discrete episodes with no link between the performance of the agent in different scenarios. This environment is simpler to design since there is no need to reason about interactions between this and future episodes; only the current environment needs to be considered.
- **Static vs. dynamic:** Static environments remain unchanged except for the results produced by the actions of the agent. A dynamic environment has other processes operating on it thereby changing the environment outside the control of the agent. A dynamic environment obviously requires a more complex agent design.
- **Discrete vs. continuous:** If there are a fixed and finite number of actions and percepts, then the environment is discrete. A chess game is a discrete environment while driving a taxi is an example of a continuous one.

To define the environments where agents interact shows up the need of a new approach, which would be Agent Oriented Software Engineering.

1.3 Agent Oriented Software Engineering

With intelligent agents, we have a new main character in the list of software design concepts; and with it, a new paradigm: *Agent Oriented Paradigm*. Agents have little differences with traditional objects that we are used to model. In general, objects are considered as "something" with attributes that only take values and from that a life cycle, well defined and simply in most cases, is developed. In the other hand, agents are intangible abstractions whose most important elements are the desired goals and its adaptability to achieve those goals. To face this new concept, classical methodologies aren't enough and that's the reason of the birth of new generation of software design development: AOSE (Agent Oriented Software Engineering). There are many alternatives to solve the agent oriented development problems in design terms. Some of them try to convert classical methodologies to a new paradigm by extending some concepts and elements from models to satisfy new needs. One sample is INGENIAS that uses AUML, an extension of classic UML, to agent modelling. Perhaps experts talk about big differences between agents and objects, so the best way would be to try a new approach from zero, starting from the intelligent agent concept and its life cycle without emulating an object. Sample of this is Tropos that uses an intuitive language to modelling systems and organizations.

Even when agents and their environments are designed following a methodology or a specific language for agent modelling, it does not necessary implies that the features presented above are achieved, at that point problems and system failures show up.

2 Possible Problems

Agent-based solutions are increasing their use in complex and critical areas, so, there is a huge need to analyze, comprehensively the behaviour of such systems, in order to avoid system malfunctioning. Hence, it is not surprising to find that formal verification techniques adapted to agent-based systems are an area that is now attracting a huge amount of attention.

While program verication is well advanced, verication of agent-oriented programs sets up new challenges that have not yet been adequately addressed.

In agent verification, we have to test not only what the agent does, but why it chose an specific course of action, what the agent believed that made it choose to act that way, and what its intentions were by doing so. [4]

If we want to be able to ensure that a system behaves exactly as it is supposed to do we need to compare its actions to its theoretical behaviour described in system's specification. In order to achieve this we need system's specification to be formal.

2.1 SithGent and SithGent Diagnose

At the beginning of this paper, we have set that an agent behaves almost like a human being, they perceive the environment through sensors and act upon that environment through effectors, they communicate with others, and have goals to achieve. Since we are able to stablish this behavioural resemblance between human beings and agents, we have tried to spread out this likeness to other areas, so, as long as human beings get sick, why wouldn't agents? And further, why wouldn't human diseases also affect intelligent agents? Always bridging the gap between humans and agents. From now on we will name an ill intelligent agent as *SithGent*. We have noticed by analysing the diseases that can affect a human being, that most of them can also affect an intelligent agent, at least, there is a clear parallelism between human diseases and agent diseases.As an example, we can think of a person that due to a neuronal disease or an accident, needs help to carry out his regular tasks. This person would be functionally dependent. When an agent needs help to carry out tasks that it should be able to accomplish acting on its own, it would also be funcionally dependent, just like the human We have also found that illnesses agents can suffer from, as happens with human beings, could be grouped under several categories like, miscommunication, environment maladjustment, or poorly functional behaviour, among others. In SithGent Diagnose we face this kind of situations, where a non functional agent obstructs the global behaviour of the multi-agent system, and we will try to supply with two types of solutions to this problem. First solution, from now on, palliative treatment, focuses on providing methods and mechanisms to the platform, that will aid SithGent's identification and SithGent's isolation, in order to prevent the system from collapsing. This treatment sets its basis on symptom detection. The other solution, from now on, chirurgical treatment, focuses on providing tips to the programmers and designers of the system, so they can fix the problems caused by the SithGent.

In conclusion, one of ours lines of investigation will be addressed to symptom indentification of human diseases and how these could be linked to symptoms in agent's ailments. Besides, once a non-desired situation is identified, we will explore how to provide means to solve it.

2.2 *Agent Oriented Design vs. SithGent Diagnose*

As the concept of Agent Oriented Software Engineering gains wider interest, it has become clear that, situations where all this software engineering can not prevent the agents from showing unsuccessful or unpredictable behaviours, also arise. Our work focus on examining repeated failures multiagent systems show and then provide two different approaches to solve the situation, in order to transform the problem into a more desirable situation. In one side, our work will provide tips so an agent, that is not acting as supposed, could be easily detected and in the other side, how this malfunctioning agent can be repaired. Even though SithGent Diagnose was developed from the notion of Agent Oriented Design, there are significant differences between the two concepts. SithGent Diagnose starts from a multiagent system that is not working as it should in the "real-world", this is quite different from a Design problem, which starts from a system's specification and works towards an abstract solution.

Table 1 Comparison between Agent Oriented Design and SithGent Diagnose.

	A.O Design	SithGent Diagnose
Focuses on	Designing Agents	Agent Malfunction
Starts From	Well-defined Specifications	Real-World Problem
Solution proposes	Design Patterns	Tips to identify the problem. Ways to approach.

"Focuses On": Agent Oriented Design authors study successful solutions to articulate a repeatable path from the problem to the solution. In SithGent Diagnose we study failed solutions, working to determine common mistakes made and ways to solve them. *"Starts From"*: Also, users of Agent Oriented Design typically begin with a well-defined problem and in finding the relating pattern, identify the best path to take to the solution. We, in SithGent Diagnose begin with a situation where an agent is not acting as supposed, it means, a failed solution. In a Multi-agent System identifying which one of the agents in the system is a SithGent, how it is affecting the entire system and which illness it suffers from, can be very challenging, because agent's illness are rarely isolated. Agents with similar symptoms suffer from different diseases or even more, the same disease can cause different symptoms. *"Solution proposes"*: In Agent Oriented Design, there are structures and patterns that propose a design to solve the problem. These structures have been reached due to mapping of typical problems, within their proved solutions. In SithGent Diagnose we provide tips to identify when a problem is going on with an agent on a multi-agent system, and possible ways to face the problem, without providing unique solutions or designs.

3 Conclusions and Future Work

Agent-based systems technology is a vibrant and rapidly expanding field of academic research and business world applications. Agent-based solutions are increasing their use in complex and critical areas, so, there is a huge need to analyze, comprehensively the behaviour of such systems, in order to avoid system malfunctioning. For that, we project our future work, as follows, first of all would be formalize the concepts, SithGent and SithGent Diagnose, introduced in this paper. Once the specification is complete, we plan on analysing different cases from real world, where a problem has been detected, in order to distinguish the symptoms that are causing the system failure, and how these symptoms can be also easily identified on a human being. By identifying the disease on humans and its treatment we plan on providing treatment for the agents disease. This cure will be composed of two kinds of treatments, one addressed to identifying and preventing the disease and the other addressed to healing the agent.

Acknowledgements. This work has been supported by the project *XUGA 08SIN009305PR* supported by Xunta de Galicia.

References

1. Russell, S., Norvig, P.: Articial Intelligence: A Modern Approach. Prentice-Hall, Inc., Englewood Cliffs (1995)
2. Wooldridge, M., Jennings, N.R.: Intelligent Agents: Theory and Practice. The Knowledge Engineering Review 10(2), 115–152 (1995)
3. Rudowsky, I.: Intelligent Agents. In: Proceedings of the Americas Conference on Information Systems, New York (August 2004)
4. Dennis, L.A., Farwer, B., Bordini, R.H., Fisher, M.: A Flexible Framework for Verifying Agent Programs (Short Paper). In: Padgham, Parkes, Mller, Par-sons (eds.) Proc. of 7th Int. Conf. on Autono-mous Agents and Multiagent Systems (AAMAS 2008), Estoril, Portugal, May 12-16 (2008)
5. Wooldridge, M.: An Introduction to MultiAgent Systems. Wiley, England (2002)
6. Franklin, S.P., Graesser, A.G.: Is It an Agent, or Just a Program? A Taxonomy for Autonomous Agents. In: Muller, J.P., Wooldridge, M.J., Jennings, N.R. (eds.) Intelligent Agents III. Springer, Heidelberg (1996)
7. Sycara, K.P.: Multiagent Systems, pp. 79–92, American Association for Artificial Intelligence (Summer, 1998)
8. Bradshaw, J.: Software Agents, ch. 1. AAAI Press, Menlo Park (1997)
9. Brenner, W., Wittig, H., Zarnekow, R.: Intelligent Software Agents: Foundations and Applications. Springer-Verlag New York, Inc., Secaucus (1998)
10. Murch, R., Johnson, T.: Intelligent software agents. Prentice Hall PTR, Upper Saddle River (1998)
11. Henderson-Sellers, B., Giorgini, P.: Agent-oriented Methodologies. Idea Group Inc., USA (2005)
12. Jennings, N.R., Wooldridge, M.: Agent technology: foundations, applications, and markets. Springer-Verlag New York, Inc., Secaucus (1998)
13. Alonso, E.: AI and agents: state of the art. AI Magazine 23(3), 25–29 (2002)

Trends on the Development of Adaptive Virtual Organizations

Sara Rodríguez, Vicente Julián, Angel L. Sánchez, Carlos Carrascosa,
Vivian F. López, Juan M. Corchado, and Emilio Corchado

Abstract. Nowadays there is a clear trend towards using methods and tools that can develop multi-agent systems (MAS) capable of performing dynamic self-organization when they detect changes in the environment. Moreover, the ideas that model the interactions of a multi-agent system cannot be related only to the agents and their communication skills in gaining strength, but it is necessary to use the concepts of organizational engineering. This paper presents a comprehensive list of development issues for primary adaptive virtual organizations (AVO). These issues will be a starting point for defining a complete list of AVO development requirements. From these requirements it could be possible to define an abstract architecture specifically addressed to the design of open multi-agent systems and virtual organizations.

Keywords: Multi-agent systems, Virtual Organizations, Open multi-agent systems, Adaptive Systems.

1 Introduction

Nowadays, one of the goals of multi-agent systems (MAS) is to construct systems capable of autonomous and flexible decision-making, and of cooperating with other systems within a "society". This "society" should take certain characteristics into account, such as distribution, constant evolution, or a flexibility that allows its members (agents) to enter and exit at will, a correct organizational structure, and the ability for the system to be executed on different types of devices. Each of these characteristics can be achieved through the virtual organization paradigm. This paradigm was conceived as a solution to the management, coordination and control of the agents' behavior.

Vicente Julián and Carlos Carrascosa
Dept. Sistemas Informáticos y Computación, Universidad Politécnica de Valencia
Valencia, Spain
e-mail: {vinglada,carrasco}@dsic.upv.es

Sara Rodríguez, Angel L. Sánchez, Vivian F. López, Juan M. Corchado,
and Emilio Corchado
Departamento Informática y Automática, Universidad de Salamanca, Salamanca, Spain
e-mail: {srg,als,vivian,corchado,escorchado}@usal.es

A.P. de Leon F. de Carvalho et al. (Eds.): Distrib. Computing & Artif. Intell., AISC 79, pp. 113–121.
springerlink.com

Virtual Organizations (VO)[10] are a means of understanding system models from a sociological perspective. From a business perspective, a virtual organization model is based on the principles of cooperation among businesses within a shared network. It takes advantage of the distinguishing elements that provide flexibility and a quick response capability and these form a strategy aimed at customer satisfaction. Virtual Organizations have been usefully employed as a paradigm for developing agent systems [2][10] One of the advantages of organizational development is that systems are modeled with a high level of abstraction, so the conceptual gap between real world and models is reduced. Also this kind of system offers facilities to implement open systems and heterogeneous member participation [20]. Organizations should not only be able to describe structural composition (i.e. functions, agent groups, interaction and relationship patterns between roles) and functional behavior (i.e. agent tasks, plans or services), but should also be able to describe behavioral norms of agents, the entry and exit of dynamic components, and formation, which is also dynamic, for agent groups.

Given the characteristics of these open environments, particularly their dynamism, it is essential to find a new approach to support the evolution of these systems and to facilitate their growth and update at execution time.

In general, it is necessary to define the standards and platforms required for the interoperability and adaptability of the agents that meet these requirements. This article attempts to identify and analyze the research topics that touch on the development of adaptive virtual organizations (AVO). Additionally, it will present a high level abstract architecture applied with the specific intent of addressing the design of open multi-agent systems and virtual organizations.

The article is structured as follows: Section 2 describes the panorama of Virtual Organizations and presents the challenges for adaptive systems; Sections 3 introduces an abstract architecture specifically adapted to the design of open multi-agent systems and virtual organizations (OVAMAH); finally some conclusions are given in Section 4.

2 Background of Virtual Organizations

MAS agents based on organizational concepts coordinate and exchange services and information; they are capable of negotiating and coming to an agreement; and they can carry out other more complex social actions. At present, research focused on the design of MAS from an organizational perspective seems to be gaining the most ground. The emergent thought is that modeling interactions in a MAS cannot be related exclusively to the actual agent and its communication capabilities; instead, organizational engineering is also necessary. The concepts of rules [1], norms and institutions [9] and social structures [22] are rooted in the idea of needing a higher level of abstraction, independent from the agent that explicitly defines the organizations in which the agents reside.

The purpose of this section is to make an overall research of the needs that arise when developing OV, relying more on concepts than in a specific terminology. There are specific organizational concepts that must also be considered in the OV development process:

Social Entity: The organisations are formed by members or social entities that can in turn be composed of a specific number of members or agents. These entities, according to [23]: (i) have some responsibilities, i.e., a set of sub-tasks to carry out, included within the objectives of the organization; (ii) have and consume resources. The members have certain resources with which they can perform their tasks. The resources required by a member will depend on the role that they are playing in the organization at that moment; (iii) they are structured according to fixed patterns of communication; (iv) they attempt to achieve the global objectives of the organization and (v) are regulated by rules and restrictions.

Structure: can be defined as the distribution, order and interaction of the different parts which compose the organization. In that structure, the agents will be arranged and communicated with depending on the topology that defines the structure. Different types of organizational topologies exist, including (hierarchies, coalitions, congregation groups, federations, matrix organisations) [25].

Functionality: is determined by the mission, i.e., by the global objectives that describe the purpose of its existence. The mission defines the strategy, the functional requirements (what the organisation does) and interaction (how it does it).

Normative: represents the set of norms that control the organization. Norms facilitate the mechanisms to drive the behavior of agents, especially in those cases when their behavior affects other agents [19]. A wide variety of norms can be found[19] (i) deontic (Obligations, Permissions and Prohibitions); (ii) legislatives, for creating, modifying or revoking norms; (iii) reinforcement, for controlling and penalizing; (iv) rewards.

Environment: defines what exists around the system: resources, applications, objects, assumptions, restrictions, stakeholders (suppliers, clients, beneficiaries). Defining the environment; the connection of roles with respect to the environmental elements is established. Access mode (reading, interaction, information extraction), access permission, etc.

Dynamic: The Organizational Dynamic is related to the input/output of agents, with the adoption of roles by part of the same, the creation of groups and with the control of behaviour. In the definition of the dynamic of an organization we must specify: (i) with respect to the entry of agents: when agents are permitted to enter the organization; what their position in the organization will be; the process of expulsion of agents with anomalous behaviours; (ii) with respect to the adoption of roles: how the agents will adopt a specific role; the association of agents with one or more roles; (iii) with respect to the dynamic creation of groups: the definition of federations, coalitions, congregations, etc and (iv) ultimately, with respect to the control of behaviour: how to control the compliance of agents' behaviour with the rules of society.

Adaptation. For a society, the idea of adaptation is its capacity to get involved with the environment and to be part of this simbiosis which allows both to work with the other. And it cannot only be considered as a capacity, but also a necessity that it gets involved with the environment to maximize the learning needs of each individual and ensure that the system succeeds in acquiring significant knowledge.

Social learning: Social learning is perceived as a process in which, thanks to an environment or element in common, different agents can interact and evaluate their experiences and information. An agent that learns from another agent is saving a lot of effort as the information that it has gained has been obtained through the effort of others.[6]. The two most relevant types of learning in society are social facilitation [21] and imitative learning [6]. Learning is intrinsically connected to adaptation. Both social facilitation and imitation are techniques that allow us to learn from new situations in order to be able to act in a certain way in the future. In short, it can be considered to be a process of adaptation to new situations. As can be seen in the following subsection, different approaches exist for adaptation in society and amongst them, imitation is one of the most used. [6].

2.1 Adaptive Issues

Agents in a VO can jointly coordinate their knowledge, goals, abilities, tasks and plans to achieve a global objective. To perform tasks, an agent may need system resources, in which case it would have to coordinate with other system agents that need the same resource. Coordination allows the agents to consider all of the tasks and regulate them so as to avoid executing undesirable actions such as: (i) the agents not generating or communicating sub-solutions that would lead to the solution of a problem; (ii) the agents generating and communicating redundant solutions; (iii) an inappropriate distribution of workload amongst agents.

This coordination is related to the action planning required for solving tasks: (i) high level understanding and ability to predict the behavior of other system agents; (ii) exchange intermediate results that progress towards the solution of a global task; (iii) avoid redundant actions, if they are undesirable.

Agent organizations depend on the type of coordination and communication among agents, as well as the type of agents that the group is comprised of. There are several different organizational approaches and platforms: JADE[1] , MOISE[2], OperA[7], RETSINA[12], Jack [15], EIDE [9], RICA-J [28], S-Moise+ [16], Jack Teams [0], SIMBA [4] y SPADE [8]. However, while these studies provide mechanisms for creating coordination among participants, there is much less work focused on adapting organizational structures at execution time, or on norms defined at design time. For example [16] proposes a model for controlling adaption by creating new norms. [11] propose a distributed model for reorganizing their architecture. [1] requires agents to follow a protocol to adapt the norms. Each of these studies focuses on the structure and/or norms based on adapting the coordination among participants. There are other approaches, such as those based on social norms (on the role of a social group and the effect on the appearance and support of social norms) [14][18].

Another possibility is the development of a MAS that focuses on the concept of organization/institution. Human interactions are regulated by institutions that represent the rules of the game in a society, and define what the individuals can and cannot do and under what conditions. This perspective defends the adoption of a mimetic strategy whose objective is to tackle the complexity of the development of open MAS from an organizational concept. One electronic institution [9] should

---

[1] http://jade.tilab.com/

be considered as a social middleware between the external participating agents and the selected communication layer r responsible for accepting or rejecting agent actions. There are also different studies that focus this theme on the definition of "Automatic Electronic Institutions", an extension of electronic institutions with self-adaptive capabilities[3]. The primary difference with the other proposals is that the adaptation is carried out by the institution instead of by the agents.

Other points of vicw [29][17] focus on mechanisms for regulating behavior. Here the key factor is "imitation", which refers to the social conventions that occur through propagation and contagion [6].

Lastly, there are approaches in which the agents use the information obtained du ring their interactions [27]. In these studies, social norms (as special types of social conventions) are used by the agents to self-regulate their behavior according to s social information: prior history, reputation, etc. The research focuses on social group mechanisms based on the social information gathered during interactions [30]. They study the effects of information transmission algorithms in recognizing and forming social groups.

None of these approaches is capable of coordinating tasks for member agents of the organization to solve a common problem, nor do they consider that task planning should adapt to changes in the environment. The architecture shown in this study is OVAMAH (Multiagent-based Adaptive Virtual Organizations). OVAMAH is based on THOMAS (MeTHods, Techniques and Tools for Open Multi- Agent Systems) [5][13], which focus on defining structure and norms. The adaptation feature used in these models is based on coordination among organization participants. The following section presents OVAMAH, the evolution of THOMAS and shows where the necessary modifications have been made so that the system can be used as a model for adaptive virtual organizations.

3 OVAMAH Architecture

The most well-known agent platforms (like Jade) offer basic functionalities to agents, such as AMS (Agent Management System) and DF (Directory. Facilitator) services; but designers must implement nearly all organizational features by themselves, like communication constraints imposed by the organization topology. In order to model open and adaptive virtual organizations, it becomes necessary to have an infrastructure than can use agent technology in the development process and apply decomposition, abstraction and organization techniques, while keeping in mind all of the requirements cited in the previous section. OVAMAH is the name given to an abstract architecture for large-scale, open multi-agent systems. It is based on a services oriented approach and primarily focuses on the design of virtual organizations. The architecture is essentially formed by a set of services that are modularly structured. It uses the FIPA architecture, expanding its capabilities with respect to the design of the organization, while also expanding the services capacity. The architecture has a module with the sole objective of managing organizations that have been introduced into the architecture, and incorporates a new definition of the FIPA Directory Facilitator that is capable of handling services in a much more elaborate way, following service-oriented architecture directives.

OVAMAH consists of three principle components: Service Facilitator (SF), Organization Manager Service (OMS) and Platform Kernel (PK). The advantages of using these components for an adaptive and decentralized open system are:

- Allowing dinamicity: new organizations can be created at execution time, allowing the development of MAS which emerge or change dynamically. Moreover, the organizations may also be destroyed when its purpose is reached.
- Improving the way emergent behaviors such as composed services may arise: new and relevant, complex services can be composed at runtime, composing the new registered services with the existing ones.
- Improving the localization techniques and composition of services: entities may publish the services they demand (not only ones they offer) so that, due to the dynamic of an open system, when an entity arrives at the system and discovers that it is able to provide this demanded service, it registers as a provider.
- Allowing to express a normative control: the OMS is in charge of controlling the role enactment process. It also stores all norms defined in the system and provides some services for adding or deleting norms.

The SF primarily provides a place where autonomous entities can register service descriptions as directory entries. The *OMS* component is primarily responsible for specifying and administrating its structural components (role, units and norms) and its execution components (participating agents and the roles they play, units that are active at each moment). In order to manage these components, OMS handles the following lists: *UnitList*: maintains the relationship between existing units and the immediately superior units (SuperUnit), objectives and types; *RoleList*: maintains the relationships between existing roles in each unit, which roles the unit inherits and what their attributes are (accessibility, position); *NormList*: maintains the relationship between system rules; *EntityPlayList*: maintains the relationship between the units that register each agent as a member, as well as the role that they play in the unit. Each virtual unit in OVAMAH is defined to represent the "world" for the system in which the agents participate by default. Additionally, the roles are defined in each unit. The roles represent the functionality that is necessary for obtaining the objective of each unit. The *PK* component directs the basic services on a multi-agent platform and incorporates mechanisms for transporting messages that facilitate interaction among entities.

The complexity of open-systems is very high and current technology to cover all the described functionalities is lacking. There are some new requirements that still need to be solved. These requirements are imposed mainly by: (i) computation as an inherently social activity; (ii) emergent software model as a service; (iii) a non-monolithic application; (iv) computational components that form virtual organizations, with an autonomous and coordinated behavior; (v) distributed execution environments; (vi) multi-device execution platforms with limited resources and (vii) security and privacy policies for information processing. In order to satisfy all of those requirements, the architecture must provide interaction features between independent (and usually intelligent) entities, that can adapt, coordinate and organize themselves [24]. From a global perspective, the architecture offers total integration, enabling agents to transparently offer and request services from other agents or entities and at the same time allowing external entities to interact with agents in thc architecture by using the services provided. Reorganization and adaptation features in the agent´s behavior are necessary for this platform, for which we have proposed a social planning model [25][26]. This social planning model offers the possibility of deliberative and social behavior. It is worth mentioning that this is a unique model that incorporates its own reorganization and social adaptation mechanism. The architecture facilitates the development of MAS in an organizational paradigm and the social model adds reorganization and adaptation functions.

4 Conclusions

In this study, a list of basic concepts and questions is presented for the development of adaptive virtual organizations. These questions are defined from a detailed study of state of the art development methods, works and tools associated with virtual organizations and their adaptation.

It is well known that VO are specially suited for open and large systems in which organizational structures are required in order to manage system complexity and its environmental adaptivity. These facts make it compulsory to use principles, methods and techniques in the entire development cycle of this kind of system. Many research efforts have been developed in this field. This paper has shown that many of the fundamental AVO issues are not completely covered by these works. The essential development issues presented in this study are a starting point for developing a complete requirement list for AVO development. The presented architecture, currently in development, takes into account the viewed organizational concepts as well as the specific requirements of an adaptive organisation. The final goal of our research is the definition of a general and complete architecture for AVO.

Acknowledgments. This work has been partially supported by the MICINN project TIN2009-13839.

References

[1] Agent Oriented Software, Ltd., JACK Intelligent Agents-Agent Practicals, 4 (2004)
[2] Artikis, A., Kaponis, D., Pitt, J.: Multi-Agent Systems: Semantics and Dynamics of Organisational Models. In: Dynamic Specications of Norm-Governed Systems, IGI Globa (2009)
[3] Boissier, O., Gateau, B.: Normative multi-agent organizations: Modeling, support and control. In: Normative Multiagent Systems (2007)

[4] Bou, E., López-Sánchez, M., Rodríguez-Aguilar, J.A., Sichman, J.S.: Adapting Autonomic Electronic Institutions in Heterogeneous Agent Societies. In: Vouros, G., Artikis, A., Stathis, K., Pitt, J. (eds.) OAMAS 2008. LNCS (LNAI), vol. 5368, pp. 18–35. Springer, Heidelberg (2009)

[5] Carrascosa, C., Rebollo, M., Soler, J., Julian, V., Botti, V.: SIMBA Architecture for Social Real-Time Domains EUMAS. In: 1st E. Workshop Multi-Agent Systems (2003)

[6] Carrascosa, C., Giret, A., Julian, V., Rebollo, M., Argente, E., Botti, V.: Service Oriented MAS: An open architecture (Short Paper). In: Decker, Sichman, Sierra, Caste franchise (eds.) Proc. of 8th Int. Conf. on Autonomous Agents and Multiagent Systems (AAMAS 2009), Budapest, Hungary, pp. 1291–1292 (2009)

[7] Conte, R., Policy, M.: Intelligent social learning. Artificial Society and Social Simulation 4(1), 1–23 (2001)

[8] Dignum: A model for organizational interaction: based on agents, founded in logic, PhD. Thesis (2004)

[9] Escrivá, J., Palanca, G., Aranda, A., Fornes, G., et al.: A Jabberbased multiagent system platform. In: Proc. of AAMAS 2006, pp. 1282–1284 (2006)

[10] Esteva, M., Rodríguez, J., Sierra, C., Garcia, P., Arcos, J.: On the formal specific ations of electronic institutions. In: Dignum, F., Sierra, C. (eds.) Agent-mediated Electronic commerce (European AgentLink Perspective). LNCS (LNAI), pp. 126–147. Springer, Heidelberg (2001)

[11] Ferber, J., Gutknecht, O., Michel, F.: From Agents to Organizations: an Organizational View of Multi-Agent Systems. In: Giorgini, P., Müller, J.P., Odell, J.J. (eds.) AOSE 2003. LNCS, vol. 2935, pp. 214–230. Springer, Heidelberg (2004)

[12] Gasser, L., Ishida, T.: A dynamic organizational architecture for adaptive problem solving. In: Proc. of AAAI 1991, pp. 185–190 (1991)

[13] Giampapa, J.A., Sycara, K. (2002) Team-Oriented Agent Coordination in the RETSINA Multi-Agent System. Tech. Report CMU-RI-TR-02-34, Robotics Institute, Carnegie Mellon University, December 2002. Presented at AAMAS (2002)

[14] Giret, A., Julian, V., Rebollo, M., Argente, E., Carrascosa, C., Botti, V.: An Open Architecture for Service-Oriented Virtual Organizations. In: Seventh international Workshop on Programming Multi-Agent Systems. PROMAS 2009, pp. 23–33 (2009)

[15] Hales, D.: Group Reputation Supports Beneficent Norms. The Journal of Artificial Societies and Social Simulation (JASSS) 5(4) (2002)

[16] Howden, N., et al.: JACK intelligent agents-summary of an agent infrastructure. In: Proceedings of IEEE international conference on autonomous agents, Montreal (2001)

[17] Hubner, J., Sichman, J., Boissier, O.: S-Moise+:A middleware for developing organized multi-agent systems. In: Boissier, O., Padget, J., Dignum, V., Lindemann, G., Matson, E., Ossowski, S., Sichman, J.S., Vázquez-Salceda, J. (eds.) ANIREM 2005 and OOOP 2005. LNCS (LNAI), vol. 3913, pp. 64–78. Springer, Heidelberg (2006)

[18] Kittock, J.E.: Emergent conventions and the structure of multi-agent systems. In: Lectures in Complex systems: Complex systems summer school, Santa Fe Institute Studies in the Sciences of Complexity Lecture, vol. VI, pp. 507–521. Addison-Wesley, Reading (1993)

[19] Lakkaraju, K., Gasser, L.: Norm Emergence in Complex Ambiguous Situations. In: Proceedings of the AAAI Workshop on Coordination, Organizations, Institutions and Norms AAAI, Chicago (2008)

[20] Lopez, F., Luck, M., d'Inverno, M.: A normative framework for agent-based systems. Computational and Mathematical Organization Theory 12, 227–250 (2006)

[21] Mao, X., Yu, E.: Organizational and social concepts in agent oriented software engineering. In: Odell, J.J., Giorgini, P., Müller, J.P. (eds.) AOSE 2004. LNCS (LNAI), vol. 3382, pp. 184–202. Springer, Heidelberg (2005)

[22] Mataric, M.J.: Learning Social Behavior. Robotics and Autonomous Systems 20, 191–204 (1997)
[23] Van Dyke Parunak, H., Odell, J.J.: Representing Social Structures in UML. In: Wooldridge, M.J., Weiss, G., Ciancarini, P. (eds.) AOSE 2001. LNCS, vol. 2222, p. 1. Springer, Heidelberg (2002)
[24] Pattison, H.E., Corkill, D., Lesser, V.R.: Instantiating Descriptions of Organizational Structures. In: Distributed Artificial Intelligence, pp. 59–96. Pitman Publishers (1987)
[25] Rodríguez, S., Pérez-Lancho, B., De Paz, J.F., Bajo, J., Corchado, J.M.: Ovamah: Multiagent-based Adaptive Virtual Organizations 12th International Conference on Information Fusion, Seattle, Washington, USA (2009)
[26] Rodríguez, S.: Modelo Adaptativo para Organizaciones Virtuales de Agentes, PhD on Computers and Automation, Universidad de Salamanca (2010)
[27] Rodríguez, S., Pérez-Lancho, B., Bajo, J., Zato, C., Corchado, J.M.: Self-adaptive Coordination for Organizations of Agents in Information Fusion Environments. In: SS Information Fusion: Frameworks and Architectures HAIS 2010 (2010)
[28] Sen, S., et al.: Emergence of Norms through Social Learning. IJCAI, 1507–1512 (2007)
[29] Serrano, J.M., Ossowski, S.: RICA-J -A Dialogue-Driven Software Framework for the Implementation of Multiagent Systems. In: JISBD Taller en Desarrollo de Sistemas Multiagente (DESMA 2004), Málaga, pp. 48–61 (2004)
[30] Shoham, Y., Ennenholtz, M.: On the emergence of social conventions: Mode ling, analysis, and simulations. Artificial Intelligence 94(1-2), 139–166 (1997)
[31] Villatoro, D., Sabater-Mir, J.: Mechanisms for Social Norms Support in Virtual Societies. In: Proceedings of the Fifth Conference of the European Social Simulation Association, ESSA 2008 (2008)
[32] Zambonelli, F.: Abstractions and Infrastructures for the Design and Development of Mobile Agent Organizations. In: Wooldridge, M.J., Weiss, G., Ciancarini, P. (eds.) AOSE 2001. LNCS, vol. 2222, pp. 245–262. Springer, Heidelberg (2002)

Using Case-Based Reasoning to Support Alternative Dispute Resolution

Davide Carneiro, Paulo Novais, Francisco Andrade, John Zeleznikow, and José Neves

Abstract. Recent trends in communication technologies led to a shift in the already traditional Alternative Dispute Resolution paradigm, giving birth to the Online Dispute Resolution one. In this new paradigm, technologies are used as a way to deliver better, faster and cheaper alternatives to litigation in court. However, the role of technology can be further enhanced with the integration of Artificial Intelligence techniques. In this paper we present UMCourt, a tool that merges concepts from the fields of Law and Artificial Intelligence. The system keeps the parties informed about the possible consequences of their litigation if their problems are to be settled in court. Moreover, it makes use of a Case-based Reasoning algorithm that searches for solutions for the litigation considering past known similar cases, as a way to enhance the negotiation process. When parties have access to all this information and are aware of the consequences of their choices, they can take better decisions that encompass all the important aspects of a litigation process.

Keywords: Online Dispute Resolution, Case-based Reasoning, Multi-agent System.

1 Introduction

The shift of already traditional Alternative Dispute Resolution (ADR) methods from a physical to virtual place [1] led to a new paradigm called Online Dispute Resolution (ODR). Using this new technology-based approach, disputant parties have an easier, simpler and faster course than litigation in court, saving both temporal and monetary costs [2]. In that sense, several methods of ADR and ODR may be considered, from negotiation and mediation to modified arbitration or modified jury proceedings [3].

Davide Carneiro, Paulo Novais, Francisco Andrade, and José Neves
University of Minho, Braga, Portugal
e-mail: dcarneiro@di.uminho.pt, pjon@di.uminho.pt,
fandrade@direito.uminho.pt, jmn@di.uminho.pt

John Zeleznikow
Victoria University, Melbourne, Australia
e-mail: John.Zeleznikow@vu.edu.au

A.P. de Leon F. de Carvalho et al. (Eds.): Distrib. Computing & Artif. Intell., AISC 79, pp. 123–130.
springerlink.com

The process of developing ODR systems frequently consists on the development of tools that provide legal advice to the disputing parties. Here, it must be considered the role of the BATNA or Best Alternative to a Negotiated Agreement [4]. In fact, when parties enter into a negotiation process, they expect to achieve better results than it would otherwise occur. It is of utter importance that, during this negotiation process, the parties are aware of the possible results if the negotiation is unsuccessful. In fact, failing to do so may drive the parties into accepting an agreement that they would better of rejecting or rejecting one that they would better of enter into. Likewise, the WATNA, or the Worst Alternative to a Negotiated Agreement is equally important. Looking at these two elements, parties can definitively improve their negotiation by looking at the whole picture. ODR platforms that embody such concepts as BATNAs and WATNAs can help parties take better decisions [5]. The so-called second generation ODR platforms [6] have a considerable degree of autonomy and can be used for idea generation, planning, strategy definition and decision making processes. The development of such platforms, in which an ODR system might act as an autonomous agent [6] is an appealing way for solving disputes. Moreover, the architecture of such systems needs to be expansible, modular and compatible. Thus using Case-based Reasoning (CBR) [7], Multi-agent Systems and Rule-based Systems is appropriate.

In this paper we present a hybrid system that merges the versatility of an agent-based architecture with the completeness of CBR and the simplicity and efficiency of rules. This allows the system to look at past cases, select the most similar ones and adapt the solutions to the current problem. Furthermore, rules are used to define the simpler tasks and secure the whole process. This work is being applied in UMCourt, an ODR platform in the context of the Portuguese Labour Law. This platform is part of the TIARAC project - Telematics and Artificial Intelligence in Alternative Conflict Resolution, a project funded by FCT – the Portuguese Science and Technology Foundation.

2 Related Work

Given the fact that legal practice is largely based on the concept of precedence and the notion of case, it is important to investigate CBR [18]. The many successful cases of application of CBR techniques to the legal domain attest the ability of this technique to deal with this knowledge-based domain and support our efforts to enhance the dispute resolution process. In MEDIATOR [8], CBR is used to look at past cases in order to find solutions for problems in the context of international disputes. JUDGE [9] in the other hand, focuses on criminal sentencing. We can also mention HYPO [10] that addresses patent law, and CABARET [11], the result of improving HYPO with Rule-based reasoning. BEST [16] is a project that specifically looks at the use of the BATNA in a semantic web context and INSPIRE [17] focuses on the study of negotiation processes. Whilst BEST and INSPIRE support ADR, most decision support tools have one thing in common: they attempt to help lawyers to win cases in a trial. This means that parties will engage in potentially time-consuming and expensive processes, in which both parties will lose something (e.g. time, privacy, reputation, money) even if one of them eventually wins the litigation.

In this paper we propose an alternative approach to handle this problem, based on two key ideas. On the one hand, we look at past cases and suggest a possible outcome to the parties. In order to propose this outcome, we use the CBR algorithm to determine the MLATNA - Most Likely Outcome for a Negotiated Agreement [19]. This concept denotes the most likely outcome scenario if the negotiation process fails. If the parties do not agree on the suggestion, they can start gradually working out a more satisfactory one, using as a starting point the MLATNA. On the other hand, we warn the parties about the possible and potential consequences of solving the dispute in court. In this sense, parties are able to take their decisions while encompassing the whole picture. With this approach we plan to diminish the number of cases that have to be solved through litigation.

3 System Architecture

As being said in the introductory section, the architecture is based on the multi-agent paradigm [12]. Specifically, we are using Jade, in compliance with FIPA specifications. This allows us to develop the application layer in a highly modular fashion which makes it possible to build an architecture that is highly expansible and extensible. The development of the architecture is depicted in [13] and is out of the scope of this paper. However, we will briefly describe the agents that make up the architecture as well as their roles.

Table 1 The Main Agents that implement the CBR process.

Agent Name	Role
Coordinator	Receives task requests from other agents and takes the necessary steps in order to execute them. This agent maintains a list of active tasks and has access to a list of automata that define the next action for each task.
Retriever	Retrieves the more similar cases. It has the autonomy to change, in real-time, search settings, similarity parameters and retrieve algorithms.
Reuse	Performs the necessary actions to adapt a known case to a new context, so that it can be used.
Reviser	Looks at a group of cases in order to select an outcome/solution. Proposes the outcome to the coordinator as well as the corresponding justification, waits for the outcome and compiles a list of possible reasons for the failure, if necessary.
Learning	Has the autonomy to make changes to the knowledge base and to the rules that reflect the result of the actions of the system.

The agents are organized in two groups. The *Main Agents* group is populated by agents that have a major and autonomous role in the CBR process. These are detailed in Table 1. In Table 2 the agents of the *Secondary Agents* group are listed which have no autonomy, having as its foremost objective to support the actions of the main agents. This departure between main and secondary agents has been

Table 2 The Secondary Agents that support the actions of the Main ones.

Name	Role
FSA	Contains a list of Jade FSM behaviours that describe the guidelines or steps necessary for an agent to implement specific actions.
Selector	Multiple instances of this agent exist that implement different pre-selected algorithms (e.g. Template Retrieval, Clustering).
Similarity	This agent is able to compute the values of similarity between two cases, according to the desired rules.
Settings	Defines several search and similarity settings according to which retrieve parameters can be changed.
Database	Implements an application layer that surrounds the database of cases, that caters for all the actions to be applied to the cases stored.
Rules	Embodies rules of type if condition then action that provide the basic reactive actions for guiding some of the remaining agents.
ATNA	Computes the BATNA and WATNA in a given context using a set of logical rules defined after the Portuguese labour law.
Loader	Loads the information of cases from XML files and provides it as a Java object maintaining and updating loaded cases.
Indexer	Indexes each new case in the database according to the rules defined.
Parser	Checks the validity and parses XML files according to the defined schemas.
Process Validity	Verifies the validity of a case in terms of the dates and the corresponding statutory periods.
Roles	Contains information about the roles of registered external agents. This is used to decide which actions each external agent can perform.

performed in order to simplify the main agents but also increase code (thus functionalities) reuse. The services that these agents provide can be individually used by external agents or can be used in specific sequences in order to implement more complex tasks. Thus, this architecture can extend external systems and it can be extended either by making use of external services or by means of new agents.

4 The CBR Process

The basic unit of information in the CBR paradigm is the case. It represents a past experience that took place under a context that is also considered in the case. It is therefore a contextualized piece of knowledge. This allows estimating the outcome of an experience that we are now living by looking at a past similar one and its respective outcome. Cases in UMCourt contain the description of the problem, the solution adopted and the verified outcome [20]. Part of this information is indexed in the database, whereas the remaining is stored in XML files. However, purely storing information is not enough. In this section we define the processes that acquire and use this information, defined after the work of [7, 14] (Figure 1).

Retrieve. The retrieve process is used to select a group of the most similar cases that are of relevance for solving a given problem by means of a similarity measure. Unlike database searches, retrieval of cases from the case-base must be equipped with heuristics that perform partial matches, since in general there is no existing case that exactly matches the new case [15]. We combine a template algorithm (similar to SQL queries) with a nearest neighbour one, resulting in a hybrid approach. Thc key idea is for the template retrieval algorithm to narrow the search space so that the nearest neighbour algorithm performs quicker.Then, the nearest neighbour algorithm is applied only to this set of cases instead of applying it to all the cases in the case memory, thus increasing efficiency.

$$\frac{\sum_{i=1}^{n} W_i * fsim_i(Arg_i^N, Arg_i^R)}{\sum_{i=1}^{n} W_i} \tag{1}$$

Equation 1 shows the closest neighbour algorithm. In this equation, n represents the number of elements being considered; W_i denotes the weight of element i with respect to the overall similarity; $Fsim$ speaks for the similarity function for element i; Arg refers to the arguments for the similarity function representing the values of the element i for the new case and the retrieved case, respectively N and R. It is also important to detail the information of the case that is considered to be relevant for the computation of the similarity, i.e., the components. According to the scope of our application, we consider three types of information: (1) the objectives stated by each party at the beginning of the dispute; (2) the norms addressed by parties and witnesses; (3) the date of the dispute. The norms addressed and the objectives are lists of elements, thus the similarity function consists in comparing two lists (equation 2). Concerning the date, the similarity function verifies if the two dates are within a given time frame.

$$fsim_{list} = \frac{|L_N \cap L_R|}{n}, n = \begin{cases} |L_N|, & |L_N| \geq |L_R| \\ |L_R|, & |L_N| < |L_R| \end{cases} \tag{2}$$

Furthermore, the agent is able to dynamically change the metrics of the search and similarity algorithms. Specifically, the agent is able to choose what components to use. Thus, the agent can decide to compute the similarity of the norms addressed concerning only the article of each law (thus retrieving more laws but with less expected similarity). If the agent retrieves a significant amount of cases, it can be more precise and select the cases concerning also the number and even the item of each norm addressed, retrieving less cases, but with a higher similarity.

Reuse. Having a list of cases with associated values of similarity, the system can present the users with the solutions that are most likely to occur, among other useful operations. This phase consists of adapting the solutions of the selected cases to match the context of the new case. Solutions are lists of steps that the parties take in order to achieve the outcome (typically trade-offs). This information is structured in a way that makes it possible to adapt it to other cases. The information considered is organized as a list of actions, each action containing a unique

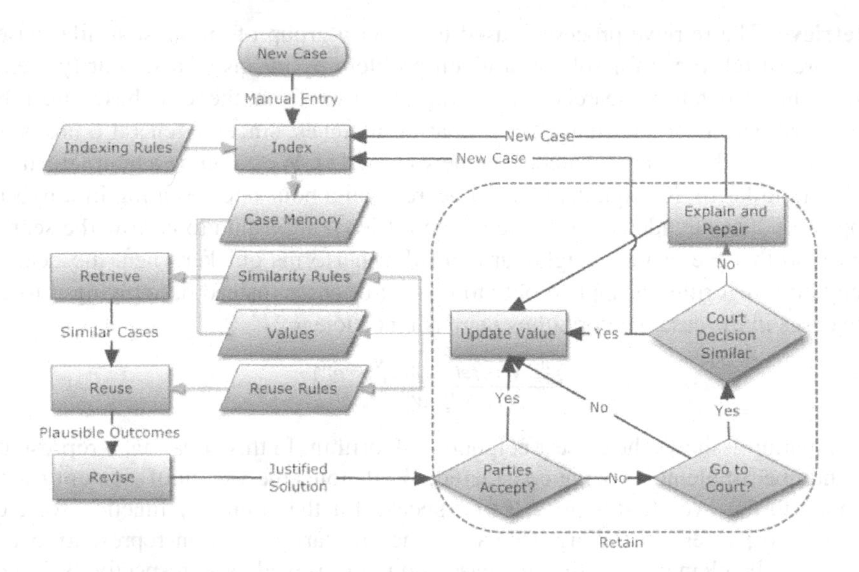

Fig. 1 A Flowchart depicting the major steps in the CBR process. Gray arrows represent access to information structures while black arrows represent the flow of control. Rectangles represent processes, parallelograms represent information structures and rhombuses represent decisions.

identifier and a structured textual description. These actions include demanding or offering an item, abdicating a given right in return for another item, among others. The items may include components of the indemnity, rights or sums of money. Thus, in this phase, the solution of the retrieved cases is adapted by changing the necessary fields (e.g. names, dates, locations). The resulting solution can then be presented to the users in the form of plain text, in their natural language.

Revise. In the revise phase, the parties become aware of the outcome of the most similar case as a first solution for the dispute, which is the MLATNA. The parties can afterwards analyze the proposal, decide what each one should give and take and whether to accept each other decision or not. If the parties do not agree on the suggestion, they can search for another solution starting from the suggested one, by adding or removing actions to the original list. Furthermore, parties can state how much they value each item in dispute and let the system make suggestions that are aimed at the maximization of the mutual gain. At the same time, parties can analyze their BATNA and WATNA and know what they are risking if they decide not to accept an agreement. Moreover, the parties can see other similar cases as well as their respective solutions, which can be used to assist in the establishment of their own outcome. Each of these selected cases that the parties have access to are accompanied by a justification, stating why they are considered similar. This will help the parties determine how much attention they should devote to each case, according to their personal expectations.

Retain. The last phase is the one in which the system embodies the changes that occurred during the whole CBR process. These changes include but are not limited to: changes to the value of cases, new cases learned, changes to thresholds or changes to rules. It is in this phase that the system actually adapts to new situations and learns new experiences, thus enriching its ability to deal with future problems. As an example, if the parties do not accept a proposed solution, the property of the corresponding case that denotes the acceptance rate will be decreased and vice-versa. Furthermore, if the parties decide to go into court, the system will also learn the new case that will be decided by a judge. Moreover, the system records the search and similarity settings for each type of case, and the ones that are more successful will be preferred in future iterations.

5 Results and Conclusions

In this paper we have focused in the description of the work done on the labour law domain. Nevertheless, the architecture is highly modular and can be applied in other domains, namely by changing secondary agents or making use of external services. In that sense, we are applying this architecture in two other prototypes in the fields of consumer law and property division. The common innovation that stems from these three prototypes is the use of three key concepts, namely BATNA, WATNA and MLATNA, as the way to define the negotiation process. Indeed, BATNA and WATNA define the Zone of Potential Agreement, i.e., the boundaries of the solution zone. Simultaneously, MLATNA constitutes the starting point for the negotiation process, representing the most likely outcome. Starting from this point, parties have a better chance of reaching a satisfactory outcome. We do not aim to propose solutions to solve a dispute in court. We rather aim at preventing the parties from going into litigation, avoiding unnecessary costs. Knowledge-based domains such as the legal one are usually complex to model. Nevertheless, the role of the actors can be significantly improved with the use of Intelligent Systems based techniques. In UMCourt, parties can intuitively look at past cases and their solutions in an attempt to find a mutually satisfactory solution. At the same time, they can be aware of what may happen if they decide not to negotiate and go into a court. Moreover, the system is not static. It is dynamic as it evolves with the results of its application to particular disputes, in an attempt to adapt to the desires of the parties and to the eventual legal changes.

Acknowledgments. The work described in this paper is included in TIARAC - Telematics and Artificial Intelligence in Alternative Conflict Resolution Project (PTDC/JUR/71354/2006), which is a research project supported by FCT (Science & Technology Foundation), Portugal.

References

1. Bellucci, E., Lodder, A., Zeleznikow, J.: Integrating artificial intelligence, argumentation and game theory to develop an online dispute resolution environment. In: 16th IEEE International Conference on Tools with AI, pp. 749–754 (2004)

2. Klaming, L., Van Veenen, J., Leenes, R.: I want the opposite of what you want. In: Expanding the horizons of ODR. In: Proceedings of the 5th International Workshop on Online Dispute Resolution (ODR Workshop 2008), Firenze, Italy, pp. 84–94 (2008)
3. Goodman, J.W.: The pros and cons of online dispute resolution: an assessment of cyber-mediation websites. In: Duke Law and Technology Review (2003)
4. Notini, J.: Effective Alternatives Analysi. In: Mediation: BATNA/WATNA Analysis Demystified (2005), http://www.mediate.com/articles/notini1.cfm
5. De Vries, B.R., Leenes, R., Zeleznikow, J.: Fundamentals of providing negotia-tion support online: the need for developing BATNAs. In: Proceedings of the Second International ODR Workshop, pp. 59–67. Wolf Legal Publishers, Tilburg (2005)
6. Peruginelli, G., Chiti, G.: Artificial Intelligence in alternative dispute resolution. In: Proceedings of the Workshop on the law of electronic agents – LEA (2002)
7. Aamodt, A., Plaza, E.: Case-based reasoning: foundational issues, methodological variations, and system approaches. AI Communications 7(1), 39–59 (1994)
8. Simpson, R.L.: A Computer Model of Case-Based Reasoning in Problem Solving: An Investigation in the Domain of Dispute Mediation. Technical Report GIT-ICS-85/18, Georgia Institute of Technology, School of Information and CS, Atlanta, US (1985)
9. Bain, W.M.: Case-Based Reasoning: A Computer Model of Subjective Assessment. Ph.D. Thesis, Yale University, Yale, CT, US (1986)
10. Ashley, K.D.: Arguing by Analogy in Law: A Case-Based Model. In: Analogical Reasoning: Perspectives of AI, Cognitive Science, and Philosophy, D. Reidel (1988)
11. Rissland, E.L., Skala, D.B.: Combining case-based and rule-based reasoning: A heuristic approach. In: Eleventh International Joint Conference on Artificial Intelligence, IJCAI 1989, Detroit, Michigan, pp. 524–530 (1989)
12. Wooldrige, M.: An Introduction to Multiagent Systems. John Wiley & Sons, Chichester (2002)
13. Andrade, F., Novais, P., Carneiro, D., Zeleznikow, J., Neves, J.: Using BATNAs and WATNAs in Online Dispute Resolution. In: JURISIN 2009 - Third International Workshop on Juris-informatics, Tokyo, Japan, pp. 15–26 (2009)
14. Riesbeck, C., Bain, W.: A Methodology for Implementing Case-Based Reasoning Systems, Lockheed (1987)
15. Watson, I., Marir, F.: Case-based reasoning: A Review. Knowledge Engineering Review 9, 327–354 (1994)
16. Wildeboer, G., Klein, M., Uijttenbroek, E.: Explaining the Relevance of Court Decisions to Laymen. In: Lodder, A.R., Mommers, L. (eds.) Proceedings of JURIX 2007, pp. 129–138. IOS Press, Amsterdam (2007)
17. Kersten, G.E., Noronha, S.J.: Negotiation via the World Wide Web: A Cross-cultural Study of Decision Making. Group Decision and Negotiation 8, 251–279 (1999)
18. Ashley, K.D.: Case-based reasoning and its implications for legal expert systems. Artificial Intelligence and Law 1, 113–208 (1992)
19. Guasco, M.P., Robinson, P.R.: Principles of negotiation. Entrepreneur Press (2007)
20. Carneiro, D., Novais, P., Andrade, F., Balke, T., Neves, J.: TIARAC Technical Report, http://tiaracserver.di.uminho.pt/tiarac/

Statistical Machine Translation Using the Self-Organizing Map

V.F. López, J.M. Corchado, J.F. De Paz, S. Rodríguez, and J. Bajo

Abstract. The paper describes a contextual environment using the Self-Organizing Map, which can model a semantic agent (SOMAgent) that learns the correct meaning of a word used in context in order to deal with specific phenomena such as ambiguity, and to generate more precise alignments that can improve the first choice of the Statistical Machine Translation system giving linguistic knowledge.

1 Introduction

For more than half a century, various aspects of translation have been studied and considered in order to develop Machine Translation (MT). However, it is well-known that MT is a very difficult task. The more general the domain or complex the style of the text, the more difficult it is to achieve a high quality translation. Today there is a wave of optimism that is spreading throughout the MT research community, one that has been caused by the revival of statistical approaches to MT. Very specifically, we refer to the birth of Statistical Machine Translation (SMT). In contrast to previous approaches based on linguistic knowledge representation, SMT is based on large amounts of human-translated example sentences (parallel corpora) from which it is possible to estimate a set of statistical models describing the translation process [9].

The incorporation of syntactic information in SMT is a current research topic. It is based on both syntax and on hierarchy of phrases. To this end, in [9, 19] there appears the need to introduce alternative techniques to include information on morphology derivation and verb group information into word alignment algorithms.

In this paper, we study improvements in translation quality that can be achieved by using the open-source Syntax Augmented Machine Translation (SAMT). By preprocessing with a multi-agent system, we experimented with different degrees of linguistic analysis from the lexical level to a syntactic or semantic level in order to generate a more precise alignment. We developed a contextual environment using

V.F. López, J.M. Corchado, J.F. De Paz, S. Rodríguez, and J. Bajo
Dept. Informática y Automática. University of Salamanca,
Plaza de la Merced S/N, 37008. Salamanca
e-mail: vivian@usal.es

A.P. de Leon F. de Carvalho et al. (Eds.): Distrib. Computing & Artif. Intell., AISC 79, pp. 131–138.
springerlink.com

the Self-Organizing Map where we model a semantic agent (SOMAgent) that learns the correct meaning of a word used in a particular context in order to deal with specific phenomena such as ambiguity and to generate more precise alignments that can improve the first choice of the SMT system.

The Machine Translation and the Statistical Approach are further described in Section 2. The SAMT is presented in Section 3. The Word Alignment with SOMAgent and our system is further described in Section 4 and the conclusions are briefly outlined in Section 5.

2 Machine Translation and the Statistical Approach

SMT as a research area started in the late 1980s with the Candide Project at IBM, which included the classic IBM word-based model. Their estimation of a parallel corpus can be found in [1]. When IBM researchers presented the statistical approach to MT, the interest among both natural language and speech processing research communities increased. The IBM model included the possibility of working towards a level of phrases. The evolution from word-based models to phrase-based models is described in [10] and Moses MT (*http://www.statmt.org/moses/*). Marcu [14] introduced a joint-probability model for phrase translation. As a result, most competitive SMT systems, such as the CMU, IBM, ISI, and Google systems, to name just a few, use phrase translation. Phrase-based systems came out ahead of the participation list at a recent international MT competition (DARPA TIDES Machine Translation Evaluation 2003-2006 on Chinese-English and Arabic-English). They also appear the SMT model based on tuple N-grams [15], or Ngram-based SMT. This approach is an evolution of a previous Finite-State Transducer implementation of X-grams [2], which adapted speech recognition tools for speech-oriented MT. The result is a competitive SMT model whose basic unit is the tuple, composed by one or more words of the language source and for one or more words of the target language.

In the last year, many efforts have been devoted to building syntax-based models that use either real syntax trees generated by syntactic parsers, or tree transfer methods motivated by syntactic reordering patterns. This statistical approach had considerable success. Several other strategies have been followed, including systems based on syntax [16], and those based on the hierarchy of phrases [5].

3 Syntax Augmented Machine Translation

Defined in [22] as a specific parameterization of the probabilistic synchronous context-free grammar (PSCFG) approach to MT. It takes advantage of nonterminal symbols, as in monolingual parsing, to generalize beyond purely lexical translation. [6] extends SAMT to include nonterminal symbols from target language phrase structure parse trees. Each target sentence in the training corpus is parsed with a stochastic parser [4] to produce constituent labels for target spans. PSCFG are defined by a source vocabulary T_s, a target vocabulary T_t, and a shared non-terminal set N, and induce rules of the form

$$X = < \gamma, \alpha, \iota, \psi > \tag{1}$$

Where $X \in N$ is a nonterminal (initial rule), $\gamma \in (NUT_s)^*$ is a sequence of nonterminals and source terminals, $\alpha \in (NUT_t)^*$ is a sequence of nonterminals and target terminals, ι is a one to one mapping from nonterminal tokens in γ to nonterminal tokens in α, and ψ is a non negative weight assigned to the rule.

PSCFG models define weighted transduction rules that are automatically learned from parallel training data. As in monolingual parsing, such rules make use of non-terminal categories to generalize beyond the lexical level. These rules seem considerably more complex than weighted word-to-word rules [1], or phrase-to-phrase rules [10] but can be viewed as natural extensions to these well established approaches. In [6] it is pointed out a procedure to learn PSCFG rules from word-aligned parallel corpora, using the phrase-pairs as a lexical basis for the grammar.

The translation quality is represented by a set of the functions for every rule, that are trained via Minimum Error Rate (MER) [17] to maximize translation quality according to a user specified automatic translation metric, like BLUE Papineni et al. [18] or NIST [8]. The weights of the functions are computed on the basis of the maximization of the BLUE measure.

4 Word Alignment

In this application it is intended to demonstrate that Kohonen Maps [12][11] can be applied to introducing linguistic information, other than the lexical units, to the process of building word and phrase alignments. We consider that linguistic information may be helpful to built better translation models. The alignment model as part of a whole translation scheme can also be defined as an independent Natural Language Processing task. In fact, most of current new generation translation models treat word alignment as an independent result from the translation model. In [19] the task of automatic word alignment focuses on detecting, given a parallel corpus, which tokens or sets of tokens from each language are connected together in a given translation context, revealing thus the relationship between these bilingual units.

4.1 The Word Alignment with SOMAgent

Our approach exploits the possibility of working with alignments at different levels of granularity, from the lexical to the semantic level, as suggests [19]. Therefore, assuming we are able to extract a set of tuples from a given parallel text, we can use a multi-agent system (SOMAgent) [13] to estimate the bilingual model and, to perform a corpus preprocessing, for SMT in a prototype of an Automatic German-Spanish Translator.

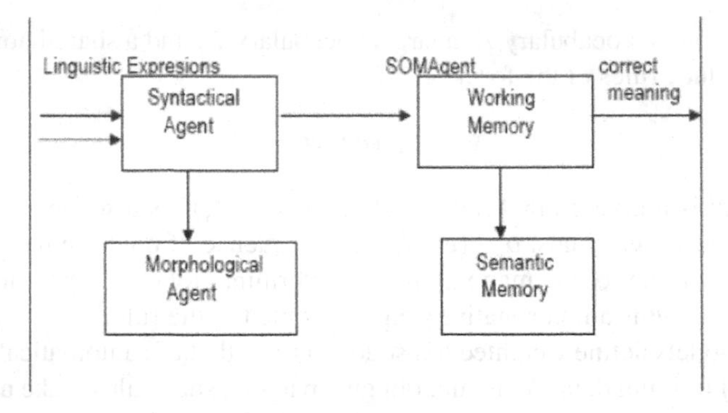

Fig. 1 Architecture of multi-agent system

The overall architecture of multi-agent system is presented in Figure 1. The SOMAgent receives perceptual inputs: linguistic expressions. There are potential actions: the agent can disambiguate an expression. The perceptions words are primarily stored in the working memory. The semantic memory associates contextual information and gives the correct meaning. Communication between the agents is motivated by the exchange of information related to linguistic expressions: morphological, syntactical and semantic information about the lexical items that are necessary for the resolution of specific tasks.

In order to implement this model, grammar knowledge comprises the initial tree models, which represent the structure of German sentences and the lexicalization dictionary forming the Syntactical Agent knowledge. This agent can be seen as a subsociety [20], formed by agents handling simpler task or information associated with the features (e.g. complements) used in the parsing. This subsociety can be dynamically organized according to the problem it is expected to solve: to assist in a best alignment. The Syntactical agent [13] divides the sentence into subject, verb, object and enrich tokens with features further than lexical such as part-of-speech (PoS),lemma,and chunk IOB label.In cases where syntactic-semantic analysis of the society of agents is insufficient to resolve a lexical ambiguity so that it should be solved by context reference. The network is trained using the SOMAgent with a large set of sentences that reflects every type of context in the corpus. These sentences, following the steps of the general algorithm [13], form a file of input data vectors for doing the training, creating the semantic memory (a trained network) with the semantic classes specified.

To study semantic relationships in their pure form, it is recognized that semantic value should not be inferred from any semantic pattern used for the encoding of individual words but only from the context where each word appears. In the self-organizing process, the inputs consist of sequences of three words selected from certain patterns of contexts. Such class patterns are defined off-line. With sentence patterns generated based on this contexts, sentences are created covering every possible context combination, for example: *Peter spielt Fußball (Peter plays football), Peter spielt Karten (Peter plays cards) or Peter spielt Schach (Peter plays chess).*

Table 1 A case of word alignment possibilities on a top of lexical units (a) and linguistic data (b)

Peter spielt Fu*β*ball, Peter spielt Gitarre, and Peter liebt romantische Spiel
↙ ↙ ↙ ↙ ↙ ↙ ↙ ↙ ↙ ↙ ↙
Peter juega futball, Peter toca guitarra, and Peter loves romantic play

The training phase consists of the sequential presentation of semantically correct sentences until the network converges. After training, the network becomes topologically ordered, and it can be verified what units of the map are active for each input vector and are then labeled, with the principal semantic classes, taking the best answer for conducting via automatic model clustering to reduce the ambiguity. For those cases, the SOMAgent is called to collaborate in solving the ambiguity, the agent takes as its input the results of the previous agents: the semantic agent searches for meanings associated with each word, forming key sentences with the combination of words in German which could not disambiguate, these feed the network input, which should be able to classify it within the active classes, taking the best answer as the correct meaning and the best alignment.

For example, suppose the case illustrated in table 1, for the sentence *Peter spielt Fuβball* we take, the German verb *spielen* (to play) that has two meanings represented by different Spanish verbs: either *tocar*, which appears in the context of playing musical instruments Klavier, Gitarre, Flöte, or *jugar* which appears in the context of games, Fu*β*ball, Karten or Schach. In addition, the lexical item *spielen* is seen acting as a verb and as a noun. Considering these two words, with the same lexical realization, as a single token adds noise to the word alignment process. The Syntactical Agent represent this information, by syntactic label (by means of linguistic data views), as *spielen VBZ* and *spielen NNS* would allow us to distinguish between the two cases. For those cases where the Semantic Agent collaborates in solving the semantic ambiguity, the agent takes as its input the results of the previous agents, searches for meanings associated with each word, forming key sentences with the combination of words in German which could not disambiguate, these feed the network input, which should be able to classify it within the active semantic classes, taking the best answer as the correct meaning. For example for the sentence *Peter spielt Fuβball*, the network find the true meaning of German verb *spielt*, alignment this entry inside the active classes, in this case, the class 2 [13](to play) whose meaning is *jugar* in Spanish.

4.2 Experimental Work

We present the experimental results for Germanto to Spanish translation task, based on a set of sentences of the full DWDS corpus (*http://utils.mucattu.com/*) of the domain of news. The results were obtained using only the first 40K lines of the corpus. The statistical data set of the corpus can be seen in the table 2.

Table 2 Training set

	Spanish	German
Sentences	40 K	40 K
Words	1,31	1,47
Length average	18,10	31,11
Vocabulary	41,12	21,10

For phrase extraction we have used MOSES MT. The number of phrases of the style Moses extracted with the system based on phrases was 4,8M. The first preliminary step requires the preprocessing of the parallel data using SOMAgent, so that it is sentence aligned and tokenised. It has as aim to deal with specific phenomena such as ambiguity and to generate more precise alignments. The output of the tokenised is formed from words that are meaningful within a particular context (or domain). For dimensionality reduction it excludes words which are meaningless because they are independent of the domain and they belong to categories such as articles, prepositions, conjuntions and pronouns. This allows the network to be trained with a smaller range of errors. The training data were provided for the sentence aligned (one sentence per line), in two files, one for the German sentences, one for the Spanish sentences. A phrase-based translation models was built of the output of the multi-agent systems to extract the purely lexical phrases, which later were used to create the grammar of the SAMT. Then, running the script that forms part of the Moses MT System grow-diag-final aligned as well as was computed the word-to-word lexical relative frequencies[6] were created. To continue with the experiments we follow the directive, available on-line in open-source SAMT system,($http://www.cs.cmu.edu//zollmann/samt$) that consists of three parts:

1. Extraction of statistical translation rules from a training corpus: to extract purely lexical phrases by SOMAgent, which later were used to creat the grammar of the SAMT.
2. Cocke-Kasami-Younger (CKY+) [3] style chart-parser employing the statistical translation rules to translate test sentences.
3. A MER optimization and scoring tool (integrated into the chart parser) to tune the parameters of the underlying log-linear model on a held-out development corpus.

The target set of the training corpus was processing by the Penn Treebank parser of Charniak [4]. The size of the vocabulary of Penn Treebak is 61 elements.

We train the language model by using the beam-search decoder engine MER, in order to fit the weights of the characteristic functions and to generate the translations N-best and 1-best [21]. In the optimization process, the iterations number is limited to 10 and the 1000-best list was extracting. We used the measure BLUE like criterion of optimizations for maximize translation quality. Finally we did other sets of experiments with a phrase-based translation models using the same sentences but without preprocessing. The results for the system SAMT appear in the table 3.

Table 3 Evaluation of the translation for German to Spanish using SAMT

	BLUE
SAMT	42,20
SAMT-SOMAgent	63,11

5 Conclusions

The diagram described in the paper was created by using a MAS to apply a corpus preprocessing, which enabled the used of an open source SAMT.We applied the SomAgent to estimate the bilingual model. We experimented with different degree of linguistic analysis, from the lexical level to syntactic or semantic level, in order to generate a more precise alignment. Our work confirms the feasibility of the SOMAgent to automatically determinate the correct meaning of a word used in context and to collaborate in the use a word alignment to learn a phrase translation table. This approach confirms the idea that the linguistic information may be helpful, specially when the target language has a rich morphology (e.g. Spanish). Nevertheless this model offers a methodology that also illustrates the formation of a terminological mapping between two languages through an emergent conceptual space, and that can improve the first choice of the translator.

We have obtained interesting comparative results with regard to the measures BLUE: the SAMT system with SOMAgent overcomes his rival in 20 percent.

Acknowledgements. This work has been partially supported by the MICINN project TIN 2009-13839-C03-03.

References

1. Brown, P.F., Della Pietra, V.J., Della Pietra, S.A., Mercer, R.: The mathematics of statistical machine translation: parameter estimation. Comput. Linguist. 19(2), 263–311 (1993)
2. Casacuberta, F., Vidal, E., Vilar, J.M.: Architectures for speech-to-speech translation using finite-state models. In: Proceedings of the Workshop on Speech-to-Speech Translation: Algorithms and Systems, pp. 39–44 (2002)
3. Chappelier, C., Rajman, M.: A generalized CYK algorithm for parsing stochastic CFG. In: First Workshop on Tabulation in Parsing and Deduction (TAPD 1998), Paris, pp. 133–137 (1998)
4. Charniak, E.: A maximum entropyinspired parser. In: Proceedings of NAACL 2000, pp. 132–139 (2000)
5. Charniak, J.: Learning non-isomorphic tree mappings for machine translation. In: Proceedings of ACL 2003, (Compain Volume) pp. 205–208 (2003)
6. Chiang, D.: A hierarchical phrasebased model for statistical machine translation. In: Proceedings of ACL 2005, pp. 263–270 (2005)
7. Chiang, D.: Hierarchical phrase based translation. Computational Linguistics (2007)
8. Doddington, G.: Automatic evaluation of machine translation quality using n-gram cooccurrence statistics. In: Proceedings ARPA Workshop on Human Language Technology (2002)

9. Honkela, T.: Philosophical Aspects of Neural, Probabilistic and Fuzzy Modeling of Language Use and Translation. In: International Joint Conference on Neural Networks, IJCNN 2007, pp. 2881–2886 (2007)
10. Koehn, P., Och, F.J., Marcu, D.: Statistical phrase-based translation. In: NAACL 2003: Proceedings of the 2003 Conference of the North American Chapter of the Association for Computational Linguistics on Human Language Technology, pp. 48–54. Association for Computational Linguistics, Morristown (2003)
11. Kohonen, T.: Self-organized Formation of Topologically Correct Feature Maps. In: Neurocomputing, pp. 511–522. The MIT Press, Cambridge (1990)
12. Kohonen, T.: Self-organized Maps. Proceedings of the IEEE 78(9), 1464–1480 (1990)
13. López, V., Alonso, L., Moreno, M.: A SOMAgent for Identification of Semantic Classes and Word Disambiguation. In: 7th International Conference on Practical Applications of Agents and Multi-Agent Systems (PAAMS 2009). Advances in Intelligent and Soft Computing, vol. 55, pp. 207–215 (2009) ISBN: 978-3-642-00486-5
14. Marcu, D., Wong, W.: A Phrase-Based, Joint Probability Model for Statistical Machine Translation. In: Proceedings of the Conference on Empirical Methods in Natural Language Processing (EMNLP), Philadelphia, pp. 133–139 (2002)
15. Mariño, J.B., Banchs, R.E., Crego, J.M., Gispert, A., de Lambert, F.P., Costa-jussá, M.R.: N-gram based machine translation. Computational Linguistics 32(4), 527–549 (2006)
16. Melamed, I.D.: Statistical machine translation by parsing. In: Proceedings of ACL 2004, pp. 111–114 (2004)
17. Och, F., Ney, H.: A systematic comparison of various statistical alignment models. Computational Linguistics 29(1), 19–52 (2003)
18. Papineni, K., Roukos, S., Ward, T., Zhu, W.-J.: BLUE: a method for automatic evaluation of machine translation. In: Proceedings of the Annual Meeting of the Association for Compuational Linguistics, ACL (2002)
19. Picó, D.: Combining Statistical and Finite-State Methods for Machine Translation. Thesis for the degree of doctor.Universitat Politécnica de Valéncia. Departament de Sistemes Informátics I Computació. Spain (2005)
20. Strube, V.L., Carneiro, P.R., Filho, I.: Distributing linguistic knowledge in a multiagent natural language processing system: re-modelling the dictionary. Procesamiento del lenguaje natura 23, 104–109 (1998)
21. Venugopal, A., Zollmann, A., y Vogel, S.: An Efficient Two-Pass Approach to Synchronous-CFG Driven Statistical MT. In: Proceedings of HLT/NAACL 2007, pp. 500–507 (2007)
22. Zollmann, A., Venugopal, A.: Syntax augmented machine translation via chart parsing. In: Proceedings of NAACL 2006 (2006)
23. Yamada, K., Knight, K.: A decoder for syntax-based statistical MT.In: Annual Meeting of the ACL. Proceedings of the 40th Annual Meeting on Association for Computational (2001)

Virtual Organisations Dissolution

Nicolás Hormazábal and Josep Lluís de la Rosa

Abstract. Virtual organisations are created to satisfy requests for complex services, after the creation phase they operate usually until they fulfil their objectives and dissolve the organisation freeing its members from their resource commitment towards the organisation; this is a common virtual organisation life-cycle. In some environments, the services requests may vary over time, having high numbers of requests at some periods requiring more organisations to cover them, resulting on high number of virtual organisations formation processes. But besides the fulfilment, other dissolution causes can be considered. In this paper we present other causes that should be considered, and explain how they can affect on the overall performance regarding the formation costs and services requests assignment. In addition, we present a virtual organisation test platform (VOCODIT, Virtual Organisation and COalition DIssolution Test platform) for evaluate this approach.

1 Introduction

In multi-agent systems, the agents, being autonomous, usually pursue individual goals, but in some cases, these goals can be achieved with better performance or higher benefits inside a cooperative environment with other agents, where the resulting organisation can even offer new services through complementary abilities combination. Virtual organisations (VOs or organisations for short from now on) are temporary organisations of different partners that come together in response to or in anticipation of new requests of complex services that require combined resources, strengths and skills that no partner can fulfil by itself [1].

VOs' life-cycle is often described as a three-phase life-cycle, which are *formation*, *operation* and *dissolution*. Usually, the dissolution should happen after the VO

Nicolás Hormazábal
Agents Research Lab, Universitat de Girona, Av. Lluis Santaló S/N,
Campus Montilivi, Edifici PIV 17071, Girona Spain
e-mail: nicolash@eia.udg.edu

Josep Lluís de la Rosa
Agents Research Lab, Universitat de Girona
e-mail: peplluis@eia.udg.edu

A.P. de Leon F. de Carvalho et al. (Eds.): Distrib. Computing & Artif. Intell., AISC 79, pp. 139–146.
springerlink.com

has fulfilled its objectives [2], but there are more causes to be considered for the VO and other coalitions dissolution [3]. How the dissolution is managed could affect on the overall environment where the VOs operate. VOs have costs related to each of their life-cycle's phases; during the formation phase, the task allocation, the agent finding process for the VO and other activities lead to different types of costs (time, computer cycles, money), a good management on when the dissolution should happen could save formation-related costs by reusing existing VO for new services requests.

The next section will overview the dissolution causes. Then a platform oriented to test and evaluate how the VO dissolution can affect on the other phases of their life-cycle and the overall performance is presented, at the following section an experiment using this platform regarding one dissolution cause is presented. Then conclusions and future work are described.

2 Dissolution Causes

Besides the fulfillment, there are other dissolution causes that should be considered. In [4], 7 different dissolution causes are listed, classified in the causes that once identified dissolve the organisation, and the causes that once identified allows the organisation to continue operating waiting for the VO members' confirmation for the dissolution, these classifications are named as *sufficient* and *necessary* causes, as the last ones are necessary for the dissolution but not sufficient.

VO operate in environments where services requests and available services vary dynamically over time [5], in where during the VO formation services providers are assigned depending on the emerging requirements. Sometimes, during its operation VOs may be under-performing and should be modified and in some cases after their goals fulfillment, similar services requests could be satisfied by the already created VO by reassigning its members, these reconfiguration actions are part of an additional phase during the operation (evolution) in the VO life-cycle [6]. This is why some causes shouldn't necessarily dissolve the VO but allow them to reconfigure themselves.

As hinted before, when a VO has fulfilled its objectives, it doesn't have to dissolve (as it is not a sufficient cause). In case the VO continues existing, it can satisfy new services requests in a dynamic environment if the new services requests match the services provided by the VO. The main and most obvious drawback of this approach is that if a VO continues indefinitely, it looses its dynamic property, and the services providers (the VO members) remain attached to the VO making more difficult to create new ones when needed.

Sufficient causes, once identified, are sufficient for the automatic dissolution of the VO. Necessary causes are necessary, but not sufficient. For being sufficient, the the partners' agreement to dissolve the VO is needed. In other words, the partners have to take actions to prevent the dissolution. The dissolution causes are described in [4].

2.1 Virtual Organisation and Coalition Dissolution Test Platform

In order to evaluate how the different dissolution causes can affect on the VOs and their overall performance, the Virtual Organisation and Coalition Dissolution Test Platform (VOCODIT) is currently in development. The main goal is to provide an environment where different services requests (containing a different services combination) emerge and autonomous agents that offer services create coalitions for attending them. As it is a platform for evaluate the dissolution, the formation process is simplified and any type of negotiation is avoided.

VOCODIT is being developed using the Repast Simphony agent-based modeling toolkit (available at http://repast.sourceforge.net/) that includes built-in results logging and graphing tools as well as automated connections to a variety of optional external tools. VOCODIT is written in Java.

The performance criteria used will be based on cost, quality and time for each attended service, which usually are the parameters used for operational efficiency and as services attributes in VOs scenarios [7] and [8].

2.2 Services Requests

In the environment, a limited number of different services can be requested and performed. Each service have an average cost for each work time unit (e.g. minutes, seconds, etc), so:

$$S = \langle SID, C \rangle \tag{1}$$

Where

- SID is the unique identifier for the service.
- C is the average cost per work unit to provide the service.

Periodically services requests are created, containing a random number of different services whose quantity is between 2 and the total different services that exist. Each service request is made up by a combination of different services, which each one of them require a specific amount of work units for this service request.

$$SR = \{\langle S_1, W_1 \rangle, \langle S_2, W_2 \rangle, ..., \langle S_n, W_n \rangle\} \tag{2}$$

Where

- S_i is a service.
- W_i is the required amount of work units for the service S_i.
- n is a number between 2 and the maximum number of services that exist in the environment.

At each timestep, a random number of services requests are created, with an average of T, which varies depending on the experiment.

2.3 Agents

The agents in the platform can provide only one service, so for attending the services requests, it is needed more than one agent. Each one of them have a fixed amount of workload available for each timestep, and they provide the service with a given quality and price. This way, there are agents that provide services better than others in terms of quality and cost. The cost they provide a service is the average cost of the service plus a random value from a normal distribution:

$$PC_{a,s} = C_s + N \tag{3}$$

Where

- $PC_{a,s}$ is the associated cost for an agent a when providing the service s.
- C_s is the average cost per work unit to provide the service, taken from equation 1.
- N is a random number taken from a normal distribution with average 0, and variance 0,5.

This way, each agent A is defined by:

$$A = \langle AID, W, S, PC, Q \rangle \tag{4}$$

Where:

- AID is the unique identifier of the agent.
- W is the total workload by timestep that the agent can perform.
- S is the service that the agent can attend.
- CP is the agent's associated cost when providing the service S, (first calculated in equation 3).

2.4 Virtual Organisations

As each agent can only provide one unique service, in order to attend the services requests, they have to create organisations. The formation phase is simplified, selecting random agents for the service request.

Once an organisation is created, the system assigns a cost based on the number of agents that join the organisation.

$$CF = \overline{A} * I \tag{5}$$

Where:

- CF is the organisation's formation cost.
- $\overline{A}$ is the total number of agents in the organisation.
- I is the individual cost for adding an agent to the organisation.

In future VOCODIT releases, the formation phase will use information taken from dissolution reports as explained in [4], so it will use a trust measure depending on the performance (quality, price and time) on past VOs.

During the formation process, a contract detailing the commitment for each agent towards the organisation is created. The contract structure is inspired by the contract's cooperation effort outlined in [9]. Each organisation has a cooperation effort detailing the commitment for each one of its members.

$$CoopEff = \{\langle A_i, WL_i \rangle\} \tag{6}$$

$$WL_i = \langle MinQt, MaxQt, Freq, CP \rangle \tag{7}$$

Where:

- $CoopEff$ represents de cooperation effort inside the organisation for each agent A_i.
- WL_i is the workload that the agent A_i is willing to perform.
- $MinQt$ and $MaxQt$ is minimum and maximum workload commited, the agent will perform between these values.
- $Freq$ is the frequency that the agent will perform, specified in timesteps. So each timestep, the agent will perform at most $MaxQt/Freq$.
- CP is the cost associated to the workload (from equation 4). So each timestep, the agent will spend at most $(MaxQt/Freq) * CP$.

At each timestep, the agents can perform a maximum of W units of workload (tagen from equationeq:agents), so when requested for joining a new organisation, the agent first checks its availability to join or not.

In case that a VO can be reused, the system first checks if there are organisations available to attend the service request instead of creating a new one.

When an organisation is being reused, its cooperation effort is changed accordingly to the new assigned service request. For considering that an organisation can attend a new service request, first the platform checks if the services that the organisation can attend matches the services that are part of the service request, then the system checks the organisation members workload availability.

A service request can only be attended by one organisation. If an organisation is dissolved, the unfulfilled services requests can then be attended by another organisation.

2.5 Dissolution

VOCODIT is still in development, but currently supports two dissolution causes: *Fulfillment* and *Inactivity*. After the dissolution, the organisation stores a report, specifying the dissolution cause, the organisation's members, the creation and dissolution time (which timestep), the total costs, the assigned services requests and the average quality each service has been attended. Future development expects to add additional data in the dissolution report.

- Fulfillment: Each timestep, each organisation checks the status of the services requests it has assigned. If all the services contained in the services request are fulfilled (this is, that all the required workload has been done), then the *fulfillment* dissolution cause is identified.
- Inactivity: VOCODIT monitors the activity in the environment. How many services requests are created, when are created, how many of these are without an organisation, how many organisations, etc. Also the organisations report to the monitor when one of their members has performed a workload amount, indicating that the organisation is active and providing services. The monitor tracks when was the last time each organisation has reported activity.

There are different normative layers related to the organisations activities, at the higher level there are the norms that govern all organisations in the environment, and the norms that are specific for each organisation, here represented (in a simplified version) by the cooperation effort from equation 7. The higher level norms, named institutional norms inspired from [10] as they are common for the whole electronic institution the organisations operate in, may contain some parameters that among other goals, give support to the dissolution [4]. One of these parameters is the time of inactivity for the dissolution; once the time of inactivity has reached this value, the inactivity dissolution cause is identified.

3 Inactivity and Fulfillment, Experiments and Results

In order to evaluate how the inactivity dissolution cause affects in the overall performance, three experiments have been performed using the VOCODIT platform. The results presented here are preliminary and it is expected to run further experimentation regarding inactivity and other dissolution causes in the future. The three experiments have in common:

- 10 different services can be part of a service request.
- 100 agents available.
- 100 units of workload available for each agent for timestep.
- the agents agree to perform a minimum of 50 workload units per timestep, and a 100 as maximum.
- 1 average service request per timestep.
- a random value between 200 and 800 of workload required for the services in the services requests (so an agent could fulfill a service in between 2 and 8 timesteps if it assigns all its resources to the organisation).

The experiments differ on when the VOs dissolve:

1. Fulfillment dissolution cause: If the organisation has fulfilled its objectives, it dissolves.
2. Inactivity and fulfillment dissolution cause: If the organisation has fulfilled its objectives, and it has been inactive for a specific period of time (20 timesteps, time for an average of 20 new services requests), it dissolves.

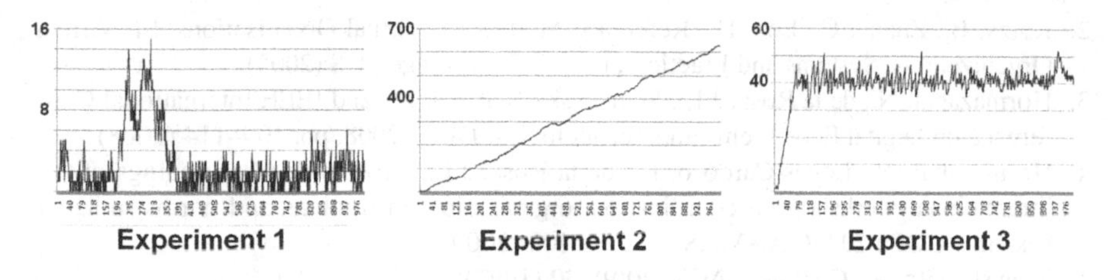

Fig. 1 The total number of Services Requests without a VO per timestep

3. Inactivity (dynamic) and fulfillment dissolution cause: Like the experiment above, but dynamically varying the required time of inactivity for keeping the unattended services requests below a maximum value.

For evaluate the performance, the number of services requests without organisation (or pending service requests to be fulfilled) will be measured, which is the total number of services requests that are not being attended by any organisation. The results can be seen in fig 1.

Experiment 2 shows that not dissolving the VOs could lead to an significant increase of pending services requests as the number of agents available to create new organisations for the emerging service requests weren't enough. Experiment 3 shows that it is possible to control the number of pending requests while old VOs are reused instead of creating new ones when possible.

4 Conclusions and Future Work

After defining the dissolution causes, current work is focused on the VOCODIT[1] development, a simple platform for evaluating how the dissolution can affect virtual organisations and for future dissolution-related experiments. Some preliminary results regarding the inactivity dissolution cause have been presented.

The dissolution can affect all the other phases in the virtual organisation lifecycle, as an early dissolution could prevent to reuse a virtual organisation for future requests, or an inefficient organisation should be dissolved even if it hasn't fulfilled its objectives in order to free its resources for new better organisations. Future work will focus on improve the VOCODIT platform, adding more features to the organisations and the agents, such as the ability to dynamically reconfigure the organisations allowing to add or remove agents, a voting system for the dissolution, trust and reputation based on dissolution reports and the other dissolution causes support. It is expected then, to run new experiments while the platform continues to improve.

References

1. Dignum, V., Dignum, F.: PROVE 2002: Proceedings of the IFIP TC5/WG5.5 Third Working Conference on Infrastructures for Virtual Enterprises, pp. 363–370. Kluwer, B.V., Deventer (2002)

[1] More info at http://arlab.udg.edu/wiki/index.php/VOCODIT

2. Katzy, B., Zhang, C., Loh, H.: Reference Models for Virtual Organisations. In: Virtual Organizations, Systems and Practices, pp. 45–58. Springer, US (2005)
3. Hormazabal, N., de la Rosa, J.L., Stanoevska-Slabeva, K.: 2nd IEEE International Conference on Digital Ecosystems and Technologies, DEST 2008, pp. 109–114 (2008)
4. Hormazabal, N., Lopes-Cardoso, H., de la Rosa, J.L., Oliveira, E.: Proceedings of the international workshop on coordination, organization, institutions and norms in agent systems (COIN 2009@AAMAS), pp. 93–108 (2009)
5. Mowshowitz, A.: Commun. ACM 40(9), 30 (1997), urlhttp://doi.acm.org/10.1145/260750.260759
6. Camarinha-Matos, L.: Proceedings of IEEE Conference on ETFA 2003 Emerging Technologies and Factory Automation, vol. 2, pp. 405–414 (2003), doi:10.1109/ETFA.2003.1248728
7. Katzy, B.: Hawaii International Conference on System Sciences, vol. 4, p. 0142 (1998), http://doi.ieeecomputersociety.org/10.1109/HICSS.1998.655269
8. Norman, T.J., Preece, A., Chalmers, S., Jennings, N.R., Luck, M., Dang, V.D., Nguyen, T.D., Deora, V., Shao, J., Gray, W.A., Fiddian, N.J.: Knowledge-Based Systems. In: AI 2003, The Twenty-third SGAI International Conference on Innovative Techniques and Applications of Artificial Intelligence, vol. 17(2-4), p. 103 (2003), http://www.sciencedirect.com/science/article/B6V0P-4C4VYG9-1/2/dedd161f4efe5c718a20b4d21e235328, doi:10.1016/j.knosys.2004.03.005
9. Lopes-Cardoso, H., Oliveira, E.: Virtual Enterprise Normative Framework Within Electronic Institutions. In: Gleizes, M.-P., Omicini, A., Zambonelli, F. (eds.) ESAW 2004. LNCS (LNAI), vol. 3451, pp. 14–32. Springer, Heidelberg (2005)
10. Lopes-Cardoso, H., Oliveira, E.: Artif. Intell. Law, vol. 16(1), p. 107 (2008), http://dx.doi.org/10.1007/s10506-007-9044-2

Cloud Computing in Bioinformatics

Javier Bajo, Carolina Zato, Fernando de la Prieta, Ana de Luis, and Dante Tapia

Abstract. Cloud Computing presents a new approach to allow the development of dynamic, distributed and highly scalable software. For this purpose, Cloud Computing offers services, software and computing infrastructure independently through the network. To achieve a system that supports these characteristics, Service-Oriented Architectures (SOA) and agent frameworks exist which provide tools for developing distributed and multi-agent systems that can be used for the establishment of Cloud Computing environments. This paper presents a CISM@ (Cloud computing Integrated into Service-oriented Multi-Agent) architecture set on top of the platforms and frameworks by adding new layers for integrating a SOA and Cloud Computing approach and facilitating the distribution and management of functionalities. CISM@ has been applied to the real case study consisting of the analysis of microarray data and has allowed the efficient management of the allocation of resources to the different system agents.

Keywords: Cloud Computing, SOA, Bioinformatics, Microarray, Multi-Agent Architecture.

1 Introduction

One of the recent developments in terms of Web Architectures referred to is denoted by the term Cloud Computing [17]. This new architectural concept offers different advantages with respect to preceding architectures as it has the capacity to offer the same level of traditional services, from complete software packages to infrastructural hardware. They are dynamic, distributed and scalable systems that provide different services on demand. These types of software usually require the creation of increasingly complex and flexible applications, so there is a trend toward reusing resources and sharing compatible platforms or architectures. In some cases, applications require similar functionalities already implemented into other systems, which are not always compatible. Microarray Data Analysis [0][1] consists of the expression study of different levels of expression in different genes. For this, statistical techniques which are widely used in various fields are carried out so that the functionality is largely reusable.

Javier Bajo, Carolina Zato, Fernando de la Prieta, Ana de Luis, and Dante Tapia
Departamento de Informática y Automática, Universidad de Salamanca,
Plaza de la Merced, s/n, 37008, Salamanca, Spain
e-mail: {jbajope,carol_zato,fer,adeluis,dantetapia}@usal.es

A.P. de Leon F. de Carvalho et al. (Eds.): Distrib. Computing & Artif. Intell., AISC 79, pp. 147–155.
springerlink.com

It is necessary to develop innovative solutions that integrate different approaches in order to create flexible and adaptable systems, especially for achieving higher levels of reutilization of developed algorithms with independence from the architecture used. It is therefore necessary to develop new functional architectures capable of providing adaptable and compatible frameworks and allowing access to services and applications regardless of time and location restrictions. There are Service-Oriented Architectures (SOA) and agent frameworks [3] [4] [5], which provide tools for developing distributed systems and multi-agent systems [6] [7] [8] that can be used for the establishment of cloud computing environments. However, these tools do not solve the development requirements of these systems by themselves.

The main purpose of this research is to design CISM@ (Cloud Computing Integrated on Service-oriented Multi-Agent) architecture with several features capable of being executed in dynamic and distributed environments to provide interoperability in a standard framework. These features can be implemented in devices with limited storage and processing capabilities. CISM@ architecture is set on top of the platforms and frameworks by adding new layers for integrating a SOA and Cloud Computing approach and facilitating the distribution and management of functionalities. A distributed agent-based architecture provides more flexible ways to move functions to where actions are needed. Additionally, the programming effort is reduced because it is only necessary to specify global objectives so that agents cooperate in solving problems and reaching specific goals, thus giving the systems the ability to generate knowledge and experience.

CISM@ has been applied in the analysis of microarray data. CISM@ integrates intelligent agents with a service-oriented philosophy that allows analysis of microarray data through the integration of different services in CISM@ and for the expression analysis of different genetic characteristics. An expression analysis consists of three stages: normalization and filtering; clustering and classification; and extraction of knowledge. The context that provides a Cloud Computing-based architecture is ideal for the treatment and analysis of bioinformatics data. It allows the exchange of great quantities of data, covers the required computational needs at the execution time of different algorithms on the aforementioned data and provides an adequate software environment for the display and study of the results obtained.

In the next section, the problem of microarray analysis is briefly explained. The specific characteristics and the agent-based architecture will be described in section 3. Finally, section 4 will present the results and the conclusions obtained.

2 Microarray Analysis

The use of microarrays, and more specifically expression arrays, enables the analysis of different sequences of oligonucleotides [0][1][0]. Microarrays have become an essential tool in genomic research, making it possible to investigate global gene expression in all aspects of human disease. [8]. Microarray technology is based on a database of gene fragments called ESTs (Expressed Sequence Tags), which are used to measure target abundance using the scanned fluorescence intensities from tagged molecules hybridized to ESTs [10]. Specifically, the HG U133 plus 2.0 [9] are chips used for expression analysis. These chips analyze the expression level of over 47,000 transcripts and variants, including 38,500 well-characterized human genes. It is comprised of more than 54,000 probe sets and 1,300,000 distinct oligonucleotide features. The HG U133 plus 2.0 provides multiple, independent measurements for each transcript.

Simply put a microarray is an array of probes that contains genetic material with a predetermined sequence. These sequences are hybridized with the genetic material of patients, thus allowing the detection of genetic mutations through the analysis of the presence or absence of certain sequences of genetic material. The analysis of expression arrays is called expression analysis. An expression analysis basically consists of three stages: normalization and filtering; clustering and classification; and extraction of knowledge. These stages are carried out from the luminescence values found in the probes.

3 CISM@ Architecture

CISM@ is a novel architecture which integrates a Cloud Computing approach with SOA and intelligent agents for building a system that needs be dynamic, flexible, robust, adaptable to changes in context, scalable and easy to use and maintain. The architecture proposes a new and easier method to develop distributed intelligent systems, where cloud services can communicate in a distributed way with intelligent agents, even from mobile devices, independent of time and location restrictions. The functionalities of the systems are not integrated into the structure of the agents; they are modeled as distributed services and applications that are invoked by the agents acting as controllers and coordinators. Another important functionality is that, thanks to the agents' capabilities, the systems developed can make use of reasoning mechanisms to handle cloud services according to context characteristics, which can change dynamically over time.

CISM@ is based on agents because of their characteristics (autonomy, reasoning, reactivity, pro-activity, mobility and organization), which allow them to cover several needs for highly dynamic environments, especially ubiquitous communication and computing and adaptable interfaces. CISM@ combines a cloud computing approach built on top of Web Services and intelligent agents to obtain an innovative architecture, facilitating high levels of human-system-environment interaction. It also provides an advanced flexibility and customization to easily add, modify or remove services on demand. The main goal in CISM@ is not only to distribute services, but also to promote a new way of developing highly dynamic systems focusing on ubiquity and simplicity. It provides the systems with a higher ability to recover from errors and a better flexibility to change their behavior at execution time.

CISM@ is set on top of existing agent frameworks by adding new layers to integrate a cloud computing approach and facilitate the provision and management of services at two different levels, Software as a Service (Seas) and Platform as a Service (Peas) [12]. Therefore, the CISM@ framework has been modeled following the Cloud Computing model based on SOA, but has added the applications block which represents the interaction with users. These blocks provide all the functionalities of the architecture. CISM@ adds new features to common agent frameworks, such as OAA, RETSINA and JADE and improves the services provided by these previous architectures. These aforementioned architectures have limited communication abilities.

As can be seen in Figure 1, CISM@ defines four basic blocks:

1. **PaaS (Platform as a Service).** This involves all the custom applications that can be used to take advantage of the system functionalities. Applications are dynamic and adaptable to context, reacting differently according to the particular situation. They can be executed locally or remotely, even on mobile devices with limited processing capabilities, because computing tasks are largely delegated to the agents and services.
2. **Agent Platform.** This is the core of CISM@. The set of agents contains agents predefined by the CISM@ architecture and virtual organisation for massive data analysis. The virtual organisation of the agents is established in function of the case studies, so that for the case of microarray data analysis an organisation that simulates the behaviour of laboratory personnel is generated.
3. **SaaS (Software as a Service).** These represent the activities that the architecture offers. Services are designed to be invoked locally or remotely and they can be organized as local services, Web Services, Cloud services, or even as individual stand-alone services. Services can make use of other services to provide the functionalities that users require. CISM@ has a flexible and scalable directory of services, so they can be invoked, modified, added, or eliminated dynamically and on demand. As well as the Agent Platform it includes the services of the specific case study.
4. **Communication Protocol.** This allows applications and services to communicate directly with the agent platform. The protocol is completely open and independent of any programming language, facilitating ubiquitous communication capabilities. This protocol is based on SOAP specification to capture all messages between the platform and the services and applications [13]. All external communications follow the same protocol, while the communication amongst agents in the platform follows the FIPA Agent Communication Language (ACL) specification.

One of the advantages of CISMA@ is that the users can access the system through distributed applications, which run on different types of devices and interfaces. The agents in the platform handle all requests and responses. The agents analyze all requests and invoke the specified services either locally or remotely. Services process the requests and execute the specified tasks. Then, the services send back a response with the result of the specific task.

The Web Services Architecture model uses an external directory, known as UDDI (Universal Description, Discovery and Integration), to list all available services. Each service must send a WSDL (Web Services Description Language) file to the UDDI to be added to the directory. Applications consult the UDDI to find a specific service. These services are grouped in accordance with their functionality, to facilitate their selection. However, CISM@ does not include a service discovery

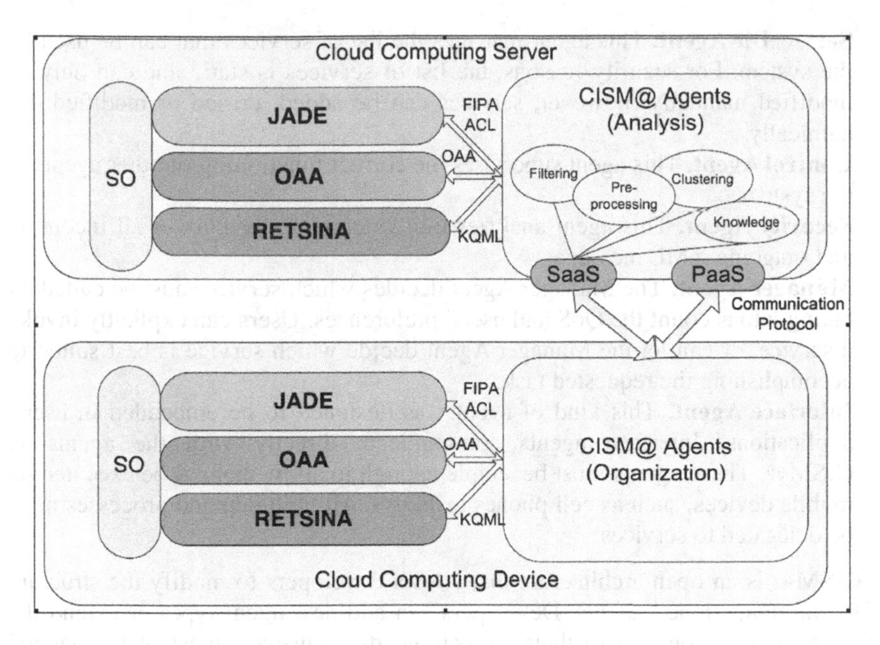

Fig. 1 Cism@ Architecture

mechanism, so applications must use only the services listed in the platform. In addition, all communication is handled by the platform, so there is no way for direct interaction between applications and services. Moreover, the platform makes use of deliberative agents to select the optimal option to perform a task, so users do not need to find and specify the service to be invoked by the application. These features have been introduced in CISM@ to create a secure communication between applications and services.

CISM@ is a modular multi-agent architecture, where services and applications are managed and controlled by deliberative BDI agents. There are different kinds of agents in the architecture, each one with specific roles, capabilities and characteristics. This fact facilitates the flexibility of the architecture when incorporating new agents. However, there are pre-defined agents, which provide the basic functionalities of the architecture.

The pre-defined CISM@ agents are described next:

1. **PaaS Agent.** This agent is responsible for all communications between applications and the agent platform. It manages incoming requests from the applications to be processed by services. It also manages responses from services (via the platform) to applications.
2. **SaaS Agent.** This agent is responsible for all communications between services and the agent platform. The functionalities are similar to PaaS Agent but the other way around. All messages are sent to the Security Agent for their structure and syntax to be analyzed. This agent also periodically checks the status of all services to know if they are idle, busy, or crashed. The analysis of messages is carried out through the use of previously implemented services.

3. **ServiceDir Agent.** This agent manages the list of services that can be used by the system. For security reasons, the list of services is static and can only be modified manually; however, services can be added, erased or modified dynamically.
4. **Control Agent.** This agent supervises the correct functioning of other agents in the system.
5. **Security Agent.** This agent analyzes the structure and syntax of all incoming and outgoing XML messages.
6. **Manager Agent.** The Manager Agent decides which service must be called by taking into account the QoS and users' preferences. Users can explicitly invoke a service, or can let the Manager Agent decide which service is best suited to accomplishing the requested task.
7. **Interface Agent.** This kind of agent was designed to be embedded in users' applications. Interface agents communicate directly with the agents in CISM@. These agents must be simple enough to allow them to be executed on mobile devices, such as cell phones or PDAs. All high demand processes must be delegated to services.

CISM@ is an open architecture that allows developers to modify the structure of the aforementioned agents. Developers can add new agent types or extend the existing ones to conform to their projects needs. However, most of the agents' functionalities should be modeled as services, releasing them from tasks that could be performed by services.

4 CISM@ Architecture in Expression Analysis

CISM@ Architecture has been adapted to the analysis of microarray expression, since it has been necessary to include agents that simulate the behaviour of a laboratory and the necessary services in the Agent Platform in order to carry out analysis.

As well as the predefined agents, the Agent Platform includes agents that simulate the roles associated with the case study. Figure 1 shows two types of agent layers:

- **Organization.** The organization agents run on the user devices or on servers. The agents installed on the user devices create a bridge between the devices and the system agents which perform data analysis. The agents installed on servers will be responsible for conducting the analysis of information following the CBP-BDI [16] reasoning model. The agents from the organizational layer should be initially configured for the different types of analysis that will be performed. Because these analyses vary according to the available information and the search results, it is imperative to establish a previous workflow configuration at the analysis layer.
- **Analysis.** The agents in the analysis layer are responsible for selecting the configuration and the flow of services best suited to the problem that needs to be solved. They communicate with Web services to generate results. The agents of this layer follow the CBP-BDI [16] reasoning model. The workflow

and configuration of the services to be used are selected with a Bayesian network and graphs, using information that corresponds to previously executed plans. The agents at this layer are highly adaptable to the case study to which they are applied. Specifically, the microarray case study includes the required agents to carry out expression analysis.

On the other hand, the services necessary to carry out expression analysis must be implemented within SaaS (Software as a Service) [0][1]. These services are those used by agents from the analysis layer to carry out data analysis.

- **Pre-processing Service.** This service implements the RMA (Robust Multi-array Average) algorithm which is frequently used for pre-processing Affymetrix microarray data.
- **Filtering Service.** The filtering service eliminates variables that do not allow classification of patients by reducing the dimensionality of the data. Three services are used for filtering: Variability, Uniform Distribution and Correlations.
- **Clustering Service.** It addresses both clustering and association of a new individual to the most appropriate group. The services included in this layer are: the ESOINN neural network. Additional services for clustering in this layer are the Partition around medoids (PAM) and demdrograms.
- **Knowledge Extraction.** The knowledge extraction technique applied has been the CART (Classification and Regression Tree) [13] algorithm and C 4.5 [14].

As shown in Figure 1, the agents from the different layers interact to generate the plan for the final analysis of data. The different system agents are distributed according to the layers and the connections that each type of agent can make with the other types of system agents and services. For example, in order to carry out its task, the Diagnosis agent at the organizational layer uses a specific sequence to select agents from the analysis layer. In turn, the analysis layer agents select the services that are necessary to carry out the data study: the filtering agent at the analysis layer selects, from the services and workflow available, those that are most suitable for the data.

The agents at the organizational layer are CBP-BDI agents with the ability to generate plans automatically based on previously existing plans in the system. Each of the CBP-BDI agents handles its own case memory in which it stores past experiences related to the specific tasks assigned to the agent. As a result, each CBP-BDI agent manages its own case memory, which is updated each time a global plan is carried out.

5 Conclusions

CISM@ facilitates the development of dynamic and intelligent multi-agent systems. Its model is based on a Cloud Computing approach where functionalities are implemented using Web Services. The architecture proposes an alternative where agents act as controllers and coordinators. CISM@ takes advantage of the agents' characteristics to provide a robust, flexible, modular and adaptable solution that can cover most requirements of a wide diversity of distributed systems. All functionalities, including those of the agents, are modelled as distributed services and applications allowing the decoupling of functionality of agents, which contributes better system integrity with regard to failure of the agents. The decoupling of functionality also gives the system greater reutilization and adaptation to new information processing.

One of the objectives of the research activity was testing the application of Cloud Computing and Cloud services to systems and platforms oriented to the analysis of large volumes of information. The architecture has enabled the quick and efficient integration of a case study and made the inclusion of new case studies possible with a simple rearrangement of the Agent Platform, based on the needs of the problem and the definition of new services where necessary.

As a conclusion we can say that although CISM@ is still under development, preliminary results demonstrate that it is adequate for building complex systems and taking advantage of composite services. However, services can be any functionality (mechanisms, algorithms, routines, etc.) designed and deployed by developers. CISM@ has laid the groundwork to boost and optimize the development of future projects and systems that combine the flexibility of a Cloud Computing approach with the intelligence provided by agents. CISM@ makes it easier for developers to integrate independent services and applications because they are not restricted to programming languages supported by the agent frameworks used (e.g. JADE, OAA, RETSINA). The distributed approach of CISM@ optimizes usability and performance because it can obtain lighter agents by modelling the systems' functionalities as independent services and applications outside of the agents' structure, thus these may be used in other developments.

Acknowledgments. This research has been partially supported by the project PET2008_0036.

References

[1] Lina, K.S., Chien, C.F.: Cluster analysis of genome-wide expression data for feature extraction. Expert Systems with Applications 36(2-2), 3327–3335 (2009)

[2] Stadlera, Z.K., Come, S.E.: Review of gene-expression profiling and its clinical use in breast cancer. Critical Reviews in Oncology/Hematology 69(1), 1–11 (2009)

[3] Maamar, Z., Kouadri, S., Yahyaoui, H.: Toward an Agent-Based and Context-Oriented Approach for Web Services Composition. IEEE Transactions on Knowledge and Data Engineering 17(5), 686–697 (2005)

[4] Buhler, P., Vidal, J.M.: Integrating Agent Services into BPEL4WS Defined Workflows. In: Proceedings of the 4th International Workshop on Web-Oriented Software Technologies, pp. 244–251 (2004)

[5] Fuentes-Fernández, R., García-Magariño, I., Gómez-Sanz, J.J., Pavón, J.: Integration of Web Services in an Agent-Oriented Methodology. International Transactions on Systems Science and Applications 3, 145–161 (2007)

[6] Martin, D.L., Chever, A.J., Moran, D.B.: The Open Agent Architecture: A framework for Building Distributed Software Systems. Applied Artificial Intelligence 13, 91–128 (1999)

[7] Sycara, K., Paolucci, M., Van Velsen, M., Giampapa, J.: The RETSINA MAS Infrastructure. Autonomous Agents and Multi-Agent Systems 7, 29–48 (1999)

[8] Fellifemine, F., Poggi, A., Rimassa, G.: JADE–A FIPA-compliant Agent Framework. In: Proceedings of PAAM, pp. 97–108 (1999)
[9] Quackenbush, J.: Computational analysis of microarray data. Nature Review Genetics 2(6), 418–427 (2001)
[10] Affymetrix. GeneChip® Human Genome U133 Arrays, http://www.affymetrix.com/support/technical/datasheets/hgu133arrays_datasheet.pdf
[11] Lipshutz, R.J., Fodor, S.P.A., Gingeras, T.R., Lockhart, D.H.: High density synthetic oligonucleotide arrays. Nature Genetics 21(1), 20–24 (1999)
[12] Foste, I., Zhao, Y., Raicu, I., Lu, S.: Cloud Computing and Grid Computing 360-Degree Compared. In: Grid Computing Environments Workshop, GCE 2008 (2008), doi:10.1109/GCE.2008.4738445
[13] Cerami, E.: Web Services Essentials: Distributed Applications with XML-RPC, SOAP, UDDI & WSDL. O'Reilly Media, Inc., Sebastopol (2002)
[14] Breiman, L., Friedman, J., Olshen, A., Stone, C.: Classification and regression trees. Wadsworth International Group, Belmont (1984)
[15] Quinlan, J.R.: C4.5: Programs for Machine Learning. Morgan Kaufmann Publishers Inc., San Francisco (1993)
[16] Glez-Bedia, M., Corchado, J.M.: A planning strategy based on variational calculus for deliberative agents. Computing and Information Systems Journal 10(1), 2–14 (2002)
[17] Vaquero, L.M., Rodero-Merino, L., Cáceres, J., Lindner, M.: A break in the clouds: towards a cloud definition. ACM SIGCOMM Computer Communication Review 39(1), 50–55 (2008)

A Support Vector Regression Approach to Predict Carbon Dioxide Exchange

Juan F. De Paz, Belén Pérez, Angélica González, Emilio Corchado, and Juan M. Corchado

Abstract. In this study, a new monitoring system for carbon dioxide exchange is presented. The mission of the intelligent environment presented in this work, is to globally monitor the interaction between the ocean's surface and the atmosphere, facilitating the work of oceanographers. This paper proposes a hybrid intelligent system integrates case-based reasoning (CBR) and support vector regression (SVR) characterised for their efficiency for data processing and knowledge extraction. Results have demonstrated that the system accurately predicts the evolution of the carbon dioxide exchange.

Keywords: Carbon dioxide, Support Vector Regression, Case-based Reasoning.

1 Introduction

One of the factors of greatest concern in climactic behaviour is the quantity of carbon dioxide (CO2) present in the atmosphere. Carbon dioxide is one of the greenhouse gases that helps to make the earth's temperature habitable, so long it maintains certain levels [6]. Traditionally, it has been considered that the main system regulating carbon dioxide in the atmosphere is the photosynthesis and respiration of plants. However, thanks to tele-detection techniques it has been shown that the ocean plays a highly important role in the regulation of carbon quantities, the full significance of which still needs to be determined [7]. Current technology allows us to obtain data and make calculations that were unimaginable some time ago. This data gives us an insight into carbon dioxide's original source, it's decrease and the causes for this decrease [1], which allow predictions on it's behaviour in the future.

This paper proposes a hybrid intelligent system that integrates case-based reasoning (CBR) and support vector regression (SVR) characterised for their efficiency for data processing and knowledge extraction. CBR is a type of reasoning

Juan F. De Paz, Belén Pérez, Angélica González, Emilio Corchado, and Juan M. Corchado
Departamento Informática y Automática
Universidad de Salamanca
Plaza de la Merced s/n, 37008, Salamanca, Spain
University of Salamanca, Spain
e-mail: {fcofds,lancho,angelica,escorchado,corchado}@usal.es

A.P. de Leon F. de Carvalho et al. (Eds.): Distrib. Computing & Artif. Intell., AISC 79, pp. 157–164.
springerlink.com

that uses past experiences to resolve new problems, and is very appropriate for use in scenarios where adaptation and learning abilities are necessary. In order to acquire intelligent behaviours, it is necessary to provide the systems with learning capabilities. One of the possibilities is learning from past experiences, which can facilitate cognitive knowledge. CBR systems are aimed at providing learning and adaptation capacities [3, 8, 9, 10]. The use of past experiences allows these systems to resolve new problems [8, 11]. SVR is a variation of support vector machines, able to provide regression models for non-linear datasets. The combination of CBR and SVR provides an added value to the prediction of the CO2 exchange. This proposal is a step in this direction and the first step toward the development of predictive models based on non-linear data. The model presented within this work provides great capacities for learning and adaptation to the characteristics of the problem in consideration by using novel algorithms in each of the stages of the CBR cycle that can be easily configured and combined. It also provides results that notably improve those provided by the existing methods for CO2 analysis.

Section 2 presents the problem that motivates this research. Then, in Section 3 the related work is presented. Section 4 describes the approach proposed in this research. Finally, in section 5 some preliminary results and the conclusions will be presented.

2 Carbon Dioxide Exchange

The oceans contain approximately 50 times more CO2 in dissolved forms than the atmosphere, while the land biosphere including the biota and soil carbon contains about 3 times as much carbon (in CO2 form) as the atmosphere [7]. The CO2 concentration in the atmosphere is governed primarily by the exchange of CO2 with these two dynamic reservoirs. Since the beginning of the industrial era, about 2000 billion tons of carbon have been released into the atmosphere as CO2 from various industrial sources including fossil fuel combustion and cement production. It is important, therefore, to fully understand the nature of the physical, chemical and biological processes, which govern the oceanic sink/source conditions for atmospheric CO2 [7, 4].

The need to quantify the carbon dioxide valence, and the exchange rate between the oceanic water surface and the atmosphere, has motivated us to develop the distributed system, presented here, that incorporates a CBR model capable of estimating such values using accumulated knowledge and updated information. The CBR model receives data from satellites, oceanographic databases and oceanic and commercial vessels. The case-based reasoning system incorporated is able to optimize tasks such as the interpretation of images using various strategies [5]. The information received is composed of satellite images of the ocean's surface, wind direction and strength, and other parameters such as water temperature, salinity and fluorescence. An improvement of the forecasting methods presented in [0, 1, 2] is incorporated in the CBR model presented in this paper.

It is possible to find different systems in literature aimed at predicting C02 exchange rates [15, 16, 1]. These works propose an approach based on obtaining regression models that are generated manually by experts. The works presented in [15, 16] focus on the variation of the exchange of CO2 produced during the day and during the night, while the work presented in [1] prioritizes the difference of pressures that exists between the ocean surface and the air. The regression models proposed in these works have, in general, a high level of complexity and sometimes require the incorporation of new variables once the model has been generated, which means recalculating the equations of the model. In this sense, the estimation of the CO2 exchange rate obtained by means of manual models presents deficiencies when working in dynamic environments, where the system needs to automatically adapt itself to the changes that occur in it's surroundings and evolve over time.

3 Support Vector Regression

SVR comes from Support Vector Machine (SVM) and is specialized in obtaining regression models by means of a change in the dimensionality of the data. SVM is a supervised learning technique that is applied to the classification and regression of different elements. SVM facilitates working with data that cannot be adjusted to linear models [12], initially conceived to obtain classifications in linear separable problems, by means of finding a hyperplan able to separate the elements of a set. One of the advantages of SVM is that it also allows separation of non-linear data. To obtain non-linear separation, SVM performs a mapping of the initial data into a high dimensionality space, where the data can be linearly separable using specific functions. Given that the dimensionality of the new space can be very high, most of the time it is not viable to use hyperplans to obtain linear separation. As a solution, non-linear functions called kernels are used. SVR is a variation of SVM to generate regressions [12, 13, 14]. The aim is to adjust the data. As in the case of SVM there is a mapping of the input data into a high dimensionality space. In this new space the regression can be carried out without the initial limitations. Equation (1) shows the linear regression obtained by means of $g_j(x)$ functions that transform the input vectors from their initial coordinates to a high dimensionality space.

$$f(\vec{x},\vec{w}) = \sum_{j=1}^{m} w_j g_j(\vec{x}) + b \tag{1}$$

4 System Description

The model proposed in this paper presents a case-based reasoning systems, which models the air-sea CO2 exchange rate. The CBR system has two aims. The first one is to generate models which are capable of predicting the atmospheric/oceanic interaction in a particular area of the ocean in advance. The second one is to permit the use of such models.

Moreover, the reasoning cycle is one of the activities carried out by the system. We can see how the reasoning cycle of a case-based reasoning system is included among the activities, composed of stages of retrieval, reuse, revise and retain. Also, an additional stage that introduces expert's knowledge is used.

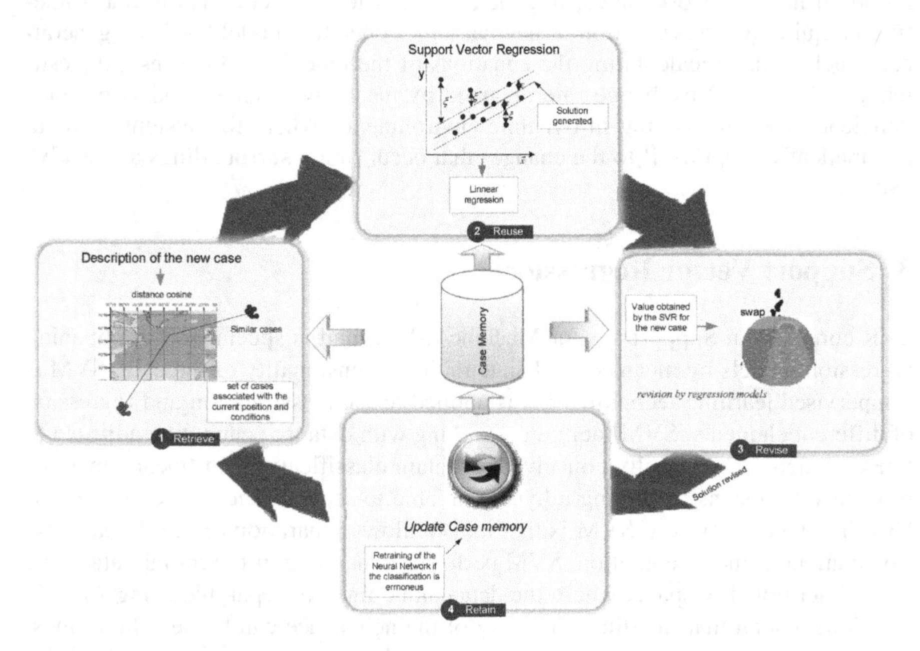

Fig. 1 Internal structure of CBR-System

Figure 1 shows the internal structure of the proposed CBR. Problem description (initial state) and solution (situation when final state is achieved) are represented as a set of values related to the oceanic and atmospheric status, the final state is the solution achieved for the problem (the predicted flux of CO2), and the sequences of actions are the steps carried out in each of the stages of the CBR cycle. The structure of a case for the CO2 exchange problem can be seen in Table 1. Table 1 shows the description of a case: DATE, LAT, LONG, SST, S, WS, WD, Fluo_calibrated, SW pCO2 and Air pCO2. Flux of CO2 is the value to be identified. DATE represents the date of the case, LAT represents the latitude of the location where the data has been obtained and LONG, the longitude in decimal degrees. SST represents the temperature of the ocean and S, the salinity. WS is the wind strength and WD is the wind direction. Fluo_calibrated represents the fluorescence calibrated with chlorophyll.

4.1 Retrieve

The prediction for the CO2 exchange rate is obtained from the parameters shown in Table 1. The prediction is carried out taking into consideration different regions

Table 1 Case Attributes.

Case Field	Measurement
DATE	Date (dd/mm/yyyy)
LAT	Latitude (decimal degrees)
LONG	Longitude (decimal degrees)
SST	Temperature (ºC)
S	Salinity (unitless)
WS	Wind strength (m/s)
WD	Wind direction (unitless)
Fluo_calibrated	Fluorescence calibrated with chlorophyll
SW pCO_2	Surface partial pressure of CO_2 (micro Atmospheres)
Air pCO_2	Air partial pressure of CO_2 (micro Atmospheres)
Flux of CO_2	CO_2 exchange flux (Moles/m^2)

of the Atlantic Ocean and, in order to obtain an effective prediction, the system needs to recover the appropriated past experiences. That is, those cases that contain problem descriptions for similar latitudes and longitudes. In order to establish this first filter in the retrieve stage, the oceanic region taken into consideration for this study was divided into grids of 10º for the latitudes and longitudes. The predictions and estimations are provided for the complete grid as a set. Once a region has been selected, the selection of the most similar case study is performed according to the cosine distance applied to the following set of variables SST, S, WS, WD, Fluo_calibrated, and Air pCO2. The cosine distance is used to avoid data normalization and corresponding problems with the data units.

4.2 Reuse

Once the most similar cases have been retrieved, the regression model is generated. As indicated in Section 4, the technique that will be used to create the regression model is Support Vector Regression (SVR). The input vector x represents a dataset with the structure presented in Table 1. The input vector can be represented as *x*=(DATE, LAT, LONG, SST, S, WS, WD, Fluo_calibrated, SW pCO2 and Air pCO2). The regression is obtained making use of all the vectors provided by the most similar cases retrieved in the previous stage of the CBR cycle, and the SVR is calculated following the algorithm presented in Section 4. The regression model is used to estimate the swap of the new case, which is used to generate the prediction value.

4.3 Revise

This phase is performed in an automatic fashion, and takes into account the error rate provided by the SVM. The error rate is calculated from the previous existing data using the coefficient of variation, in such a way that if the value obtained is minor than a pre-fixed value, then the prediction can be considered as successful. It is necessary to take into account that once the real data are obtained, the predicted exchange values are eliminated. The estimated values are only used to obtain prediction models under different conditions.

Moreover, during the revision stage an equation (F) is used to validate the proposed solution $p*$.

$$F = kso(pCO_2SW - pCO_2AIR) \tag{2}$$

Where: F: is the flux of CO_2 and k: is the gas transfer velocity. Then

$$k = (-5{,}204 Lat + 0{,}729 Long + 2562{,}765)/3600 \tag{3}$$

5 Results and Conclusions

In order to make evident the need to carry out a separation of the data in latitudes and longitudes, Figure 2 shows the results obtained after calculating the predictions using SVR with a dataset of 365 cases distributed in a homogeneous manner along the North Atlantic Ocean. The kernel function used for the experiments was polynomial and the loss function was ε-insensitive. The blue lines in Figure 2 represent the real value of the data and the red lines represent the predicted values. As can be seen in Figure 2, the error rate obtained in this experiment is very high compared the error rate obtained in Figure 3. The numerical values represent the millions of Tonnes of carbon dioxide that have been absorbed (negative values) or generated (positive values) by the ocean during each of the three months.

To evaluate the prediction capacities of the systems presented in this study, different tests were performed along the North Atlantic oceanic region with data obtained during 2009. In each of the tests, when a case containing the description of an oceanic area was introduced to the system, the most similar cases in the grid with the same latitude and longitude as the new case were taken into consideration

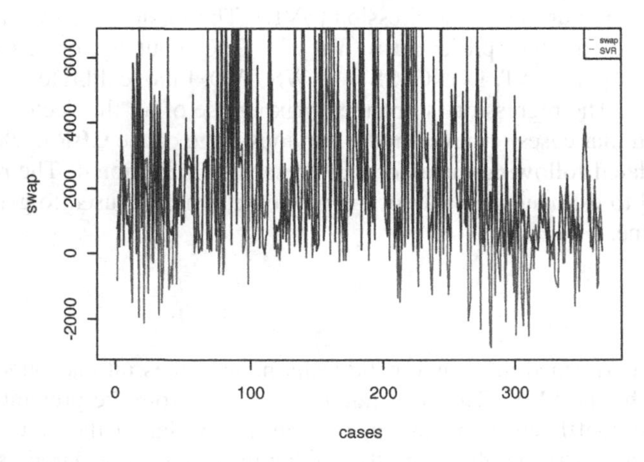

Fig. 2 Prediction if previous similar cases are selected

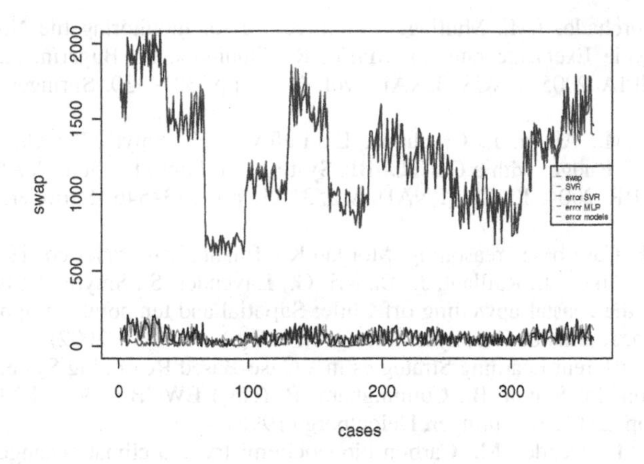

Fig. 3 Comparison between the real values and the prediction values for the CO2 exchange rate.

to obtain the prediction. Figure 3 shows the results obtained from the experiment. The blue line represents the real value and the red line represents the predicted value. Moreover, Figure 3 shows the absolute error rate obtained for the predicted value (red line) provided by the SVR. The absolute error rate obtained was 31.43, with an error deviation of 39.63. The error percentage obtained was 2.5%.

The absolute error rate obtained with the SVR has been compared to the error rate provided by alternative techniques, such as the multilayer perceptron and the oceanographers' manual models. Figure 3 shows the absolute error rate obtained for each of these predictions. The green line represents the error introduced in the system when the prediction is carried out using a multilayer perceptron. The multilayer perceptron used 27 neurons in the hidden layer and the final error percentage obtained was 5.1%. Finally, the error rate introduced in the system when the manual models are considered was 6.7%.

This study has presented a CBR intelligent system to predict and monitor the CO2 exchange rate in the North Atlantic Ocean. It applies a hybrid reasoning system specifically designed to analyze data from satellite images and vessels and predict potential CO2 fluxes in order to provide an innovative method for exploring the CO2 exchange prediction process and extract knowledge. This knowledge helps human experts to understand the prediction process and to obtain conclusions about the relevance of the situation of the oceanic environment.

Acknowledgements. This work has been supported by the MICINN TIN 2009-13839-C03-03 project.

References

1. Bajo, J., Corchado, J.M.: Evaluation and monitoring of the air-sea interaction using a CBR-Agents approach. In: Muñoz-Ávila, H., Ricci, F. (eds.) ICCBR 2005. LNCS (LNAI), vol. 3620, pp. 50–62. Springer, Heidelberg (2005)

2. Bajo, J., Corchado, J.M.: Multiagent architecture for monitoring the North-Atlantic carbon dioxide Exchange rate. In: Marín, R., Onaindía, E., Bugarín, A., Santos, J. (eds.) CAEPIA 2005. LNCS (LNAI), vol. 4177, pp. 321–330. Springer, Heidelberg (2006)
3. Corchado, J.M., Aiken, J., Corchado, E., Lefevre, N., Smyth, T.: Quantifying the Ocean's CO2 Budget with a CoHeL-IBR System. In: Funk, P., González Calero, P.A. (eds.) ECCBR 2004. LNCS (LNAI), vol. 3155, pp. 533–546. Springer, Heidelberg (2004)
4. Kolodner, J.: Case-based reasoning. Morgan Kaufmann, San Francisco (1993)
5. Lefevre, N., Aiken, J., Rutllant, J., Daneri, G., Lavender, S., Smyth, T.: Observations of pCO2 in the coastal upwelling off Chile: Sapatial and temporal extrapolation using satellite data. Journal of Geophysical research 107(6), 8.1–8.15 (2002)
6. Perner, P.: Different Learning Strategies in a Case-Based Reasoning System for Image Interpretation. In: Smyth, B., Cunningham, P. (eds.) EWCBR 1998. LNCS (LNAI), vol. 1488, pp. 251–261. Springer, Heidelberg (1998)
7. Sarmiento, J.L., Dender, M.: Carbon biogeochemistry and climate change. Photosynthesis Research 39, 209–234 (1994)
8. Takahashi, T., Olafsson, J., Goddard, J.G., Chipman, D.W., Sutherland, S.C.: Seasonal Variation of CO2 and nutrients in the High-latitude surface oceans: a comparative study. Global biochemical Cycles 7(4), 843–878 (1993)
9. Kolodner, J.: Maintaining organization in a dynamic long-term memory. Morgan Kaufmann, San Francisco (1993)
10. Kolodner, J.: Maintaining organization in a dynamic long-term memory. Cognitive Science 7, 243–280 (1983)
11. Kolodner, J.: Reconstructive memory, a computer model. Cognitive Science 7(4), 281–328 (1983)
12. Leake, D., Kendall-Morwick, J.: Towards Case-Based Support for e-Science Workflow Generation by Mining Provenance. In: Althoff, K.-D., Bergmann, R., Minor, M., Hanft, A. (eds.) ECCBR 2008. LNCS (LNAI), vol. 5239, pp. 269–283. Springer, Heidelberg (2008)
13. Vapnik, V.N.: An overview of statistical learning theory. IEEE Transactions on Neural Networks 10, 988–999 (1999)
14. Vapnik, V.: The Nature of Statistical Learning Theory. Springer, Heidelberg (1995)
15. Smola, A., Scolköpf, B.: A tutorial on support vector regression. Statistics and Computing (2003)
16. Jeffer, C.D., Woolf, D.K., Robinson, I.S., Donlon, C.J.: One-dimensional modelling of convective CO2 exchange in the Tropical Atlantic. Ocean Modelling 19(3-4), 161–182 (2007)
17. Jeffery, C.D., Robinson, I.S., Woolf, D.K., Donlon, C.J.: The response to phase-dependent wind stress and cloud fraction of the diurnal cycle of SST and air–sea CO2 exchange, vol. 23(1-2), pp. 33–48 (2008)
18. Rutgersson, A., Smedman, A.: Enhanced air–sea CO2 transfer due to water-side convection. Journal of Marine Systems 80(1-20), 125–134 (2010)

The Knowledge Modeling for the Simulation of Competition on Plant Community Growth

Fan Jing, Dong Tian-yang, Shen Ying, and Zhang Xin-pei

Abstract. Simulation of plant community growth has become a hotspot in virtual reality area. In order to improve the level of knowledge sharing and solve the problem that the simulation parameters are difficult to adjust, a novel approach is presented to simulate the plant community growth, which is knowledge modeling based on competition. After analysing the related botany and ecology knowledge, this paper presents two models: intra-species competition model and inter-species competition model. Ontology, as a formal description of the plant competition, is adopted to express these two models. In addition, these models are deployed to the knowledge query prototype system for plant growth and different tree species have been simulated to validate these models.

Keywords: Knowledge modeling, Inter-species competition, Intra-species competition, Simulation of plant growth.

1 Introduction

In recent years, with the development of technology in computer graphics and virtual reality, simulating the growth of plant community has become an attractive and challenging topic. Plant competition is an important factor to influence the composition and structure of plant community. Therefore, this paper combines the related knowledge of botany and ecology to make an abstract of plant competition, construct intra-species competition model and inter-species competition model and adopt ontology to express them.

The purpose of modeling plant competition is to construct an abstract description, which is simple but can reflect the real competition in the physical world. The competition model can be used to simulate the plant growth and dynamic succession of plant community, and also provide a useful approach to reveal the mechanism of plant growth, which is to discover the growth rhythm of plant and predict structure changes of plant community. By now, there are two typical

Fan Jing, Dong Tian-yang, Shen Ying, and Zhang Xin-pei
College of Computer
Zhejiang University of Technology
Hangzhou, 310014, China
e-mail: fangjing@zjut.edu.cn, dty@zjut.edu.cn,
shenying@zjut.edu.cn, eileen1495@sina.com

A.P. de Leon F. de Carvalho et al. (Eds.): Distrib. Computing & Artif. Intell., AISC 79, pp. 165–172.
springerlink.com

methods of modeling plant competition: shape simulation modeling and ecological simulation modeling [1]. The former focuses on the reality degree of the simulation. It mainly uses computer graphics techniques to show the plant's competition [2, 3, and 4]. This approach is widely applied in education, entertainment, computer-aided design and so on. The shortcoming of this method is involving a number of graphic rendering parameters, so the users need to debug these parameters repeatedly to obtain their desired results. The latter method focuses on the authenticity of botany and ecology knowledge. It primarily uses traditional mathematical methods (such as mathematical statistics, non-linear function, partial differential equations and so on) [5] and artificial intelligence methods (such as knowledge base, artificial neural networks, fuzzy systems, genetic algorithm and so on) [6, 7] to construct the plants competition model. The shortcoming of the second method is that users may have some difficulty in understanding the method since it involves a large number of professional terms, botany and ecology knowledge. This modeling method is mainly applied in agriculture and forestry research.

This paper focuses on exploring the competition laws of plant community, and then the knowledge modeling method is used to build the plant competition models. In this paper, the professional terms and models are abstracted as knowledge models that are transparent to the users. These knowledge models can reflect the competition between plants directly, predict the structure of plant community in a particular environment in a few years, and provide evidences for the succession of plant community. Knowledge modeling based on competition is a method to combine a lot of botany and ecology knowledge and lays the foundation for the simulation of plant competition.

2 Ontology Expression of Plant Community

In the past ten years, the research on ontology has become much more related to information technology, such as the development of knowledge engineering. The common definition of ontology was proposed in 1993 by Gruber. Ontology is defined as a formal description of the concepts within a domain and relationships between these concepts [7]. The aim of ontology is to capture knowledge of the relevant field, provide the common understanding of knowledge in the field and identify the common recognized vocabularies and relationships between them. The ontology has become a hotspot in the area of knowledge modeling.

This paper defines the formal description of the plant competition as follows:

$PO=\{PC, PA^C, PI, PX\}$, PC stands for the concept collection of plant community, PA^C represents the collection of attributes of the concepts in collection PC, PI is the instance collection of plant community, and PX is the axioms collection of the competition among plants. The four collections are illustrated in detail as follows.

- Conception collection: It explains botany and ecology terms and defines the parameters for simulation. This paper uses class to express the collection. Plants competition knowledge could be classified according to different conceptions. For example, plant competition class will be constituted of Intra-species competition class and Inter- species competition class.

- Attribute collection: It mainly describes the characteristic of the corresponding class. Class alone can't provide sufficient information to describe the details of plants competition knowledge, but using attributes can make the definition of knowledge more complete. Therefore, after the definition of class, this paper describes the internal structure of concept, that is, the attributes of class. Father class can be covered and inherited by subclass and subclass can also have their unique attributes.
- Instance collection: Class must be instantiated before used, since class is abstract. The definition of instance first needs to choose the classes, then fills the attributes value of the class. From the semantics, the instance is an object of class.
- Axioms collection: Each axiom in this collection is the restriction on attributes. In this paper, we extract the main rules of plants competition and express them using format: "IF…THEN". Axioms can represent the general plant competition in the real world.

This paper uses these four collections to describe the intra-species competition model and the inter-species competition model.

3 Knowledge Framework of Plant Competition

The structure of plant community is greatly affected by the plant competition. In fact, the plant community is an organization where different species adapt to each other and the environment. Because different plant species have different abilities of obtaining and using resources, the same plant species also have different sizes and distribution, the resources availability for different plants species and plants individuals have considerable differences. Therefore, plant competition is divided into the intra-species competition and inter-species competition. The knowledge framework of plants competition is shown in Figure 1.

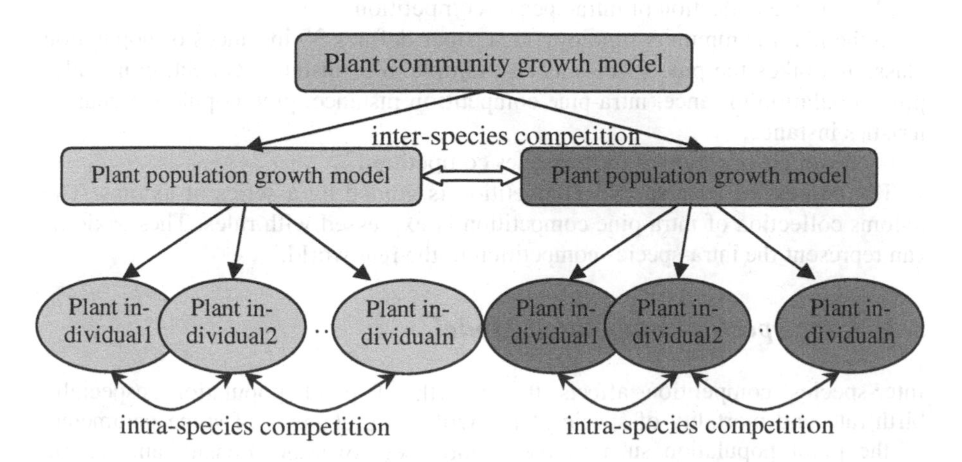

Fig. 1 The knowledge framework of plant competition.

3.1 Intra-species Competition Model

Intra-species competition refers to the same plant species competing resources in the same circumstances [8]. Intra-species competition mainly occurs among plants of different heights. The higher plant will receive more light and have more influence on the lower plant. Inversely, the lower plant will receive less light, because it is shaded by the higher plants. Intra-species competition has effects on the growth of the plant individual. In order to describe the process of intra-species competition, this paper introduces a parameter-influence circle. Each plant has its own influence circle, and it can be calculated by using the formula (1).

$$q=\pi r^2 \tag{1}$$

where r represents the max radius of plant crown. The intersection of influence circles represents the interaction among plants, while detached parts of influence circles represents that there are no impacts between plants. The process of intra-species competition can be simulated according to the three rules [9]. The intra-species competition model is illustrated as follows:

(1) Conception collection of intra-species competition

Conception collection of intra-species competition includes population class, intra-species competition class and population characteristics class. The population class is used to describe the collection of individual plants, which refers to the same species that occupy some space within a certain period of time. The intra-species competition class describes the intra-specific competition model. The population characteristics class mainly describes the characteristics of the population as a whole.

(2) Attribute collection of intra-species competition

Besides the shade-tolerant degree, the largest height, the maximum survival age and state of plants, the intra-species competition model also introduces other attributes as follows: area of influence circle, state of influence circle, position of plant individual, and distance between two individual plants.

(3) Instance collection of intra-species competition

In the plant community ontology, this paper defines 30 instances of population class, and takes the pine species as an example. The instance collection includes pine population instance, intra-pine competition instance, pine population characteristics instance.

(4) Axioms collection of intra-species competition

The process of intra-species competition is guided by a series of axioms. The axioms collection of intra-pine competition is expressed with rules. These axioms can represent the intra-species competition in the real world.

3.2 Inter-species Competition Model

Inter-species competition affects the growth of plant population, especially birth rate and mortality of species. That will cause changes of some parameters of the plant population such as the abundance, coverage, density and so on. Lotka-Volterra competition equation is specifically established for inter-species competition [8, 10]. Suppose that there are two plant populations named species1 and species2, N_1, N_2 refer to the amount of species1 and species2, K_1, K_2 represent the environment capacity of species1 and species2 (under the condition of without competition), species1's maximum instantaneous growth rate is r_1 and that of species2 is r_2. The following two differential equations can describe the growth situation of each species when they are competing with each other.

$$\frac{dN_1}{dt} = r_1 N_1 \left(\frac{K_1 - N_1 - \alpha_{12} N_2}{K_1}\right) \tag{2}$$

$$\frac{dN_2}{dt} = r_2 N_2 \left(\frac{K_2 - N_2 - \alpha_{21} N_1}{K_2}\right) \tag{3}$$

The inter-species competition model is illustrated as follows:

(1) Conception collection of inter-species competition

Conception collection of inter-species competition includes community class, inter-species competition class and community characteristics class. The community class is used to describe the collection of plant populations, which illustrates the relationship between plants and the relationship between plants and environment. The inter-species competition class mainly describes the competition for space and resources between populations. The community characteristics class mainly describes the features performed in community succession, such as dominant species and community area.

(2) Attribute collection of inter-species competition

By analysing the inter-species competition model, besides the attributes of environment capacity, plants amount, instantaneous growth rate and mortality rate, we also need to define following attributes: α_{12}, α_{21}, Inter-species competition results, k_1, k_2, etc.

(3) Instance collection of inter-species competition

Dinghushan National Nature Reserve[11, 12] is located in the Northeast of Zhaoqing City of Guangdong Province, with a total area of 1155 km2. It is a Pinus massoniana forest that is artificial cultivation. For the plant community of Dinghushan National Nature Reserve, the instance collection of Dinghushan community includes: Dinghushan community instance, Pinus massoniana-Schima superb competition instance, Dinghushan community characteristics instance.

(4) Axioms collection of inter-species competition

According to the inter-species competition model, the axioms collection of inter-species competition is also expressed with rules. These axioms can represent the inter-species competition in the real world.

4 Application

4.1 The Knowledge Query System for the Simulation of Plant Growth

Based on the intra-species competition model and inter-species competition model, as well as the ontology expression, we developed a knowledge query

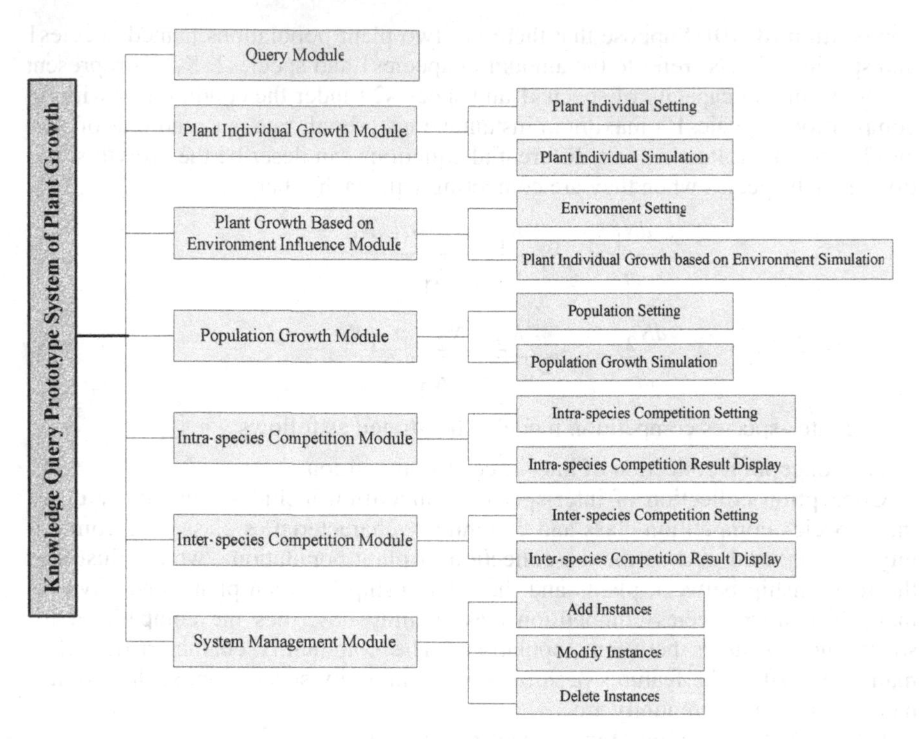

Fig. 2 The framework of knowledge query system.

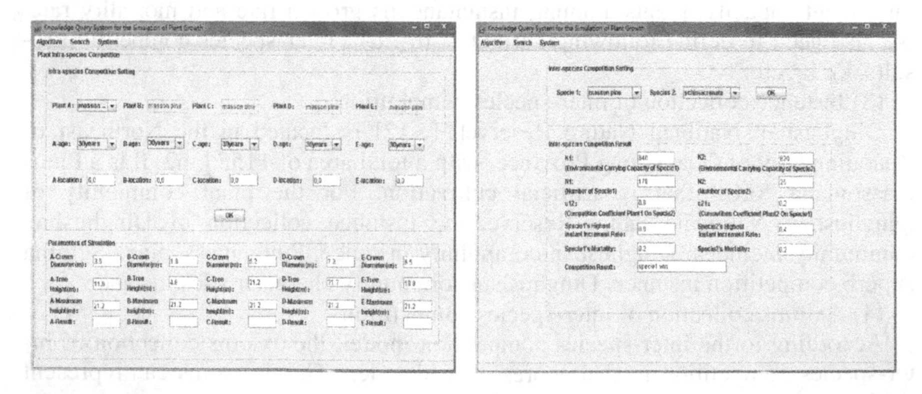

Fig. 3 Interface of intra-species competition. **Fig. 4** Interface of inter-species competition.

prototype system for the simulation of plant growth. It is mainly composed of seven modules, and the framework of knowledge query prototype system is shown in Figure 2.

Query module could enable users to inquire plant knowledge; individual plant growth module is designed to input the relevant parameters of the plant by users and conduct the simulation through calculation using use these parameters;

Intra-species competition module and inter-species competition module are provided to input competition conditions between plants, then it will carry out the transformation from condition to simulation parameters. In the system management module, users can add, modify, and delete instances and attributes.

The intra-species competition's interface is shown in Figure 3. Users can configure plant individual's age and location, which will be sent to the ontology to inquire and calculation, and then the results are displayed. The inter-species competition's interface is shown in Figure 4. The users first select species, then the system shows its environment capacity in the current ontology and mortality parameters. According to inter-species competition model, the inter-species competition result is displayed to the user.

4.2 Simulation of Plant Growth

In order to validate the models provided in the knowledge query prototype system for plant growth, we developed a plant growth simulation system using VC++ 6.0 and OpenGL. There is a plant morphological 3D model warehouse in this system. The plant models, built by Xfrog3.5, correspond to different growth stages of different plants. According to the calculation of the intra-species competition model and the inter-species competition model, this simulation system will use the corresponding morphological model and textures of plants. The plant growth simulation system is used to simulate the growth of several plants based on the competition model in the knowledge query prototype system. Figure 5 shows the growth of Cacti and Agave nearby some higher trees. The Cacti is a light-requiring plant, but the Agave is a shade-requiring plant. Therefore, the Cactis in the orange circle grows better than those Cactis in the purple circle and in the green circle, because the higher trees shield the sunlight. However, the Agaves in the orange circle grow more slowly than those agaves in the purple circle and in the green circle.

Fig. 5 Simulation result of plant growth.

5 Conclusions

Knowledge modeling has been considered as a very effective approach to research plant community, but its research and application are still in the initial stage. After analysing the plant competition models, this paper presents the intra-species competition model and the inter-species competition model by using four collections: conception collection, attribute collection, instance collection and axioms collection. Based on the plant competition models, a knowledge query prototype system for the simulation of plant growth is implemented. In addition, we developed the plant growth simulation system to validate the plant competition models.

Acknowledgements. The research work in this paper is sponsored by National Natural Science Foundation of China (60773116, 60403046), National 863 High Technology Planning of China (No.2008AA01Z302) and Zhejiang Natural Science Foundation of China (Z1090459, Y1080669).

References

[1] Hu, B.G., Zhao, X., Yan, H.: Plant growth modeling and visualizaition—review and perspective. Acta Automatica Sinica 27(6), 817–835 (2001)
[2] Prusinkiewicz, P., Lindenmayer, A.: The algorithmic beauty of plants. Springer, New York (1990)
[3] Runions, A., Fuhrer, M., Lane, B., Federl, P., Lagan, A., Prusinkiewicz, P.: Modeling and visualization of leaf venation patterns. ACM Transactions on Graphics 24(3), 702–711 (2005)
[4] Lintermann, B., Deussen, O.: A modelling method and user interface for creating plants. Computer Graphics Forum 17(1), 73–82 (1998)
[5] Zhan, Z., Wang, Y., Reffye, P., De, W.B., Xiong, Y.: Architectural modeling of wheat growth and validation study. In: 2000 ASAE Annual International Meeting, Milwaukee, Wisconsin (2000)
[6] Xiong, F.L.: Neural network based on intelligent systems in agriculture. In: Proceedings of IFAC Work shop on Expert Systems in Agriculture. International academia Publisher, Huangshang (1992)
[7] Gruber, T.R.: A translation approach to portable ontology specifications. Knowledge Acquisition 5(2), 199–220 (1993)
[8] Shang, Y.C.: General ecology, 2nd edn. Beijing University Press (2002) (in chinese)
[9] Lane, B., Prusinkiewicz, P.: Generating spatial distributions for multilevel models of plant communities. In: Proceedings of Graphics Interface 2002, Calgary, Alberta, May 27-29, 2002, pp. 69–80 (2002)
[10] Lane, B.: Models of plant communities for image synthesis [EB/OL], `http://algorithmicbotany.org/papers/laneb.th2002.html`
[11] Shi, J.H., Huang, Z.H., Zhou, X.Y., et al.: The regeneration strategies and spatial pattern of woody species in the mixed coniferous and broadleaf forest in Dinghu mountains. Journal of Nanjing Forestry University (Natural Sciences Edition) 12(4), 323–330 (2004) (in chinese)
[12] Zhang, Q.M., Chen, B.G.: Interspecific association of the dominant species in two typical communities in Dinghushan South China. Journal of South China Agricultural University 27(1), 79–83 (2006) (in chinese)

Pano UMECHIKA: A Crowded Underground City Panoramic View System

Ismail Arai, Maiya Hori, Norihiko Kawai, Yohei Abe, Masahiro Ichikawa, Yusuke Satonaka, Tatsuki Nitta, Tomoyuki Nitta, Harumitsu Fujii, Masaki Mukai, Soichiro Horimi, Koji Makita, Masayuki Kanbara, Nobuhiko Nishio, and Naokazu Yokoya

Abstract. Toward a really useful navigation system, utilizing spherical panoramic photos with maps like Google Street View is efficient. Users expect the system to be available in all areas they go. Conventional shooting methods obtain the shot position from GPS sensor. However, indoor areas are out of GPS range. Furthermore, most urban public indoor areas are crowded with pedestrians. Even if we blur the pedestrians in a photo, the photos with blurring are not useful for scenic information. Thus, we propose a method which simultaneously subtracts pedestrians based on background subtraction method and generates location metadata by manually input from maps. Using these methods, we achieved an underground panoramic view system which displays no pedestrians.

Keywords: spherical panorama, navigation system, background subtraction, Wi-Fi positioning.

1 Introduction

City panoramic view systems (e.g. Google Street View [1] and earthmine [2]) are expected to be highly usable navigation systems since their panorama photos offer

Ismail Arai, Yohei Abe, Masahiro Ichikawa, Yusuke Satonaka, Tatsuki Nitta, Tomoyuki Nitta, Harumitsu Fujii, Masaki Mukai, Soichiro Horimi, and Nobuhiko Nishio
Ubiquitous Computing and Networking Lab., Ritsumeikan University, 1-1-1, Noji-Higashi, Kusatsu, Shiga 525-8577, Japan

Maiya Hori, Norihiko Kawai, Koji Makita, Masayuki Kanbara, and Naokazu Yokoya
Vision and Media Computing Lab., Nara Institute of Science and Technology, 8916-5 Takayama, Ikoma, Nara 630-0192, Japan
e-mail: ismail@ubi.cs.ritsumei.ac.jp,
{maiya-h,norihi-k}@is.naist.jp,
{abebe,icchy,sacchin,tacky,nittan}@ubi.cs.ritsumei.ac.jp,
{dany,marshi,sow}@ubi.cs.ritsumei.ac.jp,
{koji-ma,kanbara}@is.naist.jp,
nishio@cs.ritsumei.ac.jp,yokoya@is.naist.jp
http://umechika.ubi.cs.ritsumei.ac.jp/

A.P. de Leon F. de Carvalho et al. (Eds.): Distrib. Computing & Artif. Intell., AISC 79, pp. 173–180.
springerlink.com

a person's eye view, as opposed to the bird's eye view of maps. The conventional navigation systems based on maps can guide a user to his/her destination, but he/she cannot know the details of the image, such as the shape and entrance position of a building. Also a user may want photos of turns in directions. Google Street View is the first web application which released world wide urban panoramic photos and contributed to the development of navigation systems. Also we have developed the Gooraffiti [3] which guides users various kind of urban information mashed up with Google Street View. However it published only outdoor panoramic photos shot from a car on a road. This conventional shooting method doesn't work in important areas for pedestrians, such as indoor and underground areas (out of GPS range) because the method automatically associates the photos with locations while driving the camera car. Therefore we are developing a method of generating indoor location metadata, and an indoor positioning method for a panoramic viewer.

Also, the privacy of people in the photos is also a serious issue. Google Street View blurs the faces of people automatically recognized, based on image processing. However most photos of public indoor areas crowded with pedestrians would end up full of blurring. Although we could shoot pedestrian-free photos while shops are closed, the atmosphere of the photo will not be informative for users.

To solve these problems, we developed a different shooting method for the crowded Umeda underground city (UMECHIKA) in Osaka Japan. We shoot a panoramic movie of 10 frames per minute, generally spanning one minute at the same position and apply a background subtraction technique to subtract pedestrians filmed in the frames. Then we manually input the correct locations since we have enough time while shooting.

2 Proposed Method

2.1 Pedestrian Subtraction and Inpainting Dead Space

To preserve privacy of pedestrians, their images are removed from the photos using a number of frames captured at a fixed point. In our assumption, frames which include a part of pedestrians are less than half of frames in capturing time. The pedestrian regions are interpolated by extracting median brightness value in timeseries. This can be found by arranging all the brightness values in each pixel. The pedestrian regions are interpolated by applying this technique to the whole panoramic image.

After pedestrians are removed, inpainting method [4] is applied to the photos to fill in the dead space below the camera, which exist in virtually every omnidirectional camera and decrease the realistic sensation of looking around. More concretely, first, based on the assumption that the ground is planar and the posture and height from the ground of the omnidirectional camera of each frame are invariant, omnidirectional images are projected onto the ground plane. In the images projected onto the plane, the appearance of textures is invariant between frames, unlike panoramic images. Next, a reference region having appropriate textures for the dead space of a target frame is determined in another frame. Then the dead space

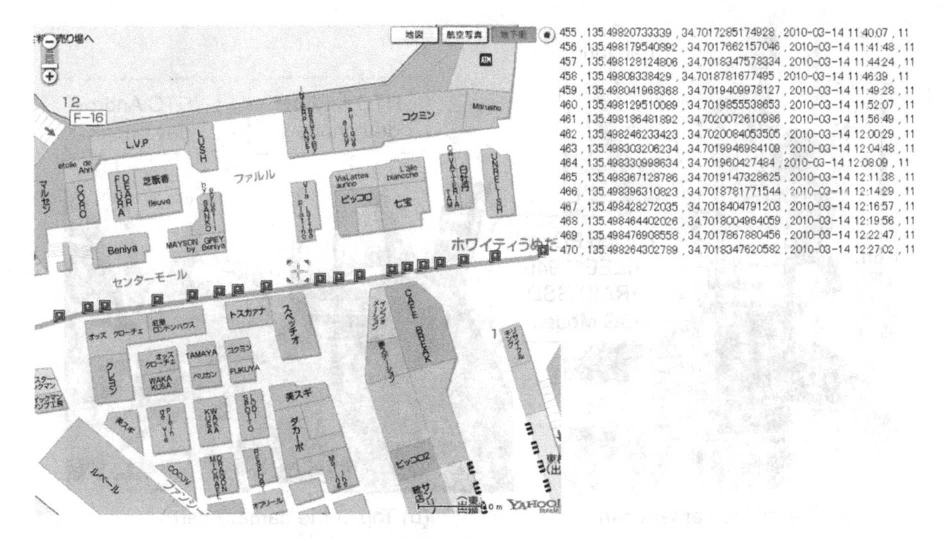

Fig. 1 A support tool for inputting shot location

is inpainted by minimizing an energy function based on the similarity of textures between the reference region and the dead space. Note that simple copying from the reference region to the dead space makes unnatural textures because of the difference in brightness and disconnection of edges. Minimizing energy-based evaluation is effective for these problems. Finally, textures in the images projected onto the ground plane are re-projected onto the original panoramic images.

2.2 *Tagging a Location and Sensor Data to a Panoramic Photo*

A panoramic view system should maintain links between adjacent panoramic photos and geographical relations between the photos and maps. If there is an accurate and feasible indoor positioning system, associating these information with the photos (geotagging) is trivial. Google has simultaneously stored urban panoramic photos and GPS sensor data to their server. GPS error range, which is around 10 meters, was the lowest possible error range to store panoramic photos at 10-meter intervals. Even if GPS were available in indoor areas, GPS is useless to store indoor panoramic photos because of the higher density of shops. In this project, we set the required shooting interval to 5 meters because the width of an indoor shop is generally around 5 meters.

Conventional indoor positioning methods such as a Wi-Fi positioning [5] and a beacon positioning [6] defeat the purpose of our project. Costs of Wi-Fi positioning are low because of the ubiquity of Wi-Fi AP (access points). However it is difficult to consistently achieve error ranges of under 5 meters, because densities of Wi-Fi APs vary widely. Beacon positioning achieves higher accuracy, but the costs of setting a large number of beacons in whole public indoor areas are high.

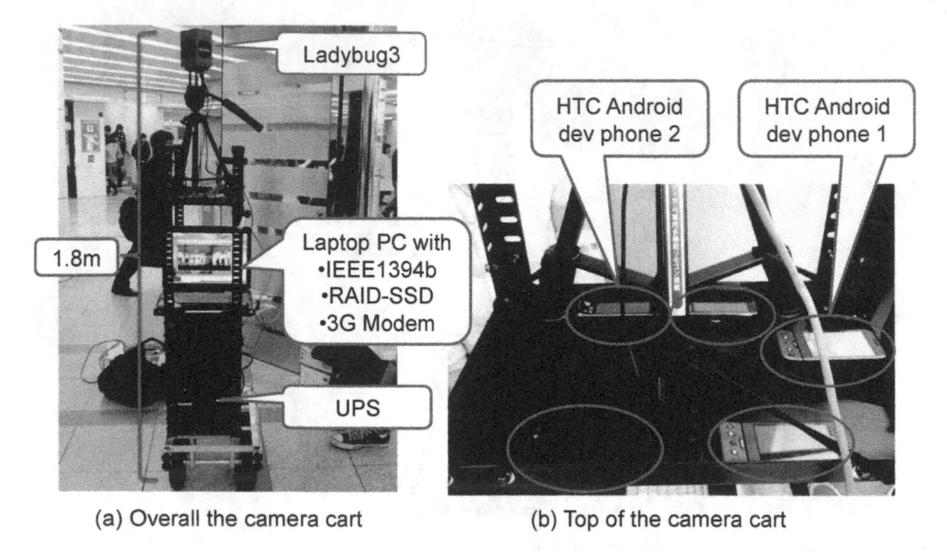

(a) Overall the camera cart (b) Top of the camera cart

Fig. 2 A Shooting equipment

Therefore we propose to store a shot location to a server by manually clicking on a web mapping application. The application uses a detailed underground map [7] and its API, released by Yahoo! Japan. As the map describes borderlines of each shop, skillful users can recognize the detailed location by looking at the map. We implemented a support tool for inputting shot location, as shown in Fig. 1. When a user clicks a point on the map, the current time and the coordinates (consisting of latitude and longitude) are displayed to the right of the map and the web application stores them to the server. The time stamp bonds the photo to its location. To stabilize directions of each trajectory, this application makes a preview of a trajectory from the last clicked point to the mouse cursor. A shot direction is calculated from the trajectory.

Although we stated Wi-Fi positioning is not suitable for our project, we validated the feasibility of Wi-Fi positioning as well as other sensor fusion techniques. Thus we installed five Android phones on the camera cart (Fig. 2) to collect as much sensor data as possible (GPS, Wi-Fi, Bluetooth devices, accelerometers and magnetic field).

3 Experimental Results

We pushed the camera cart shown in Fig. 2 through UMECHIKA for five days in March 2010, taking pictures at five-meter intervals for 2.2 kilometers to cover one third of total length of UMECHIKA. We plan to shoot the remaining area by the end of June 2010. We shoot panoramic movie spans from 20 to 120 seconds (depending on congestion) with 10 frames per minute at the same position at 5m interval.

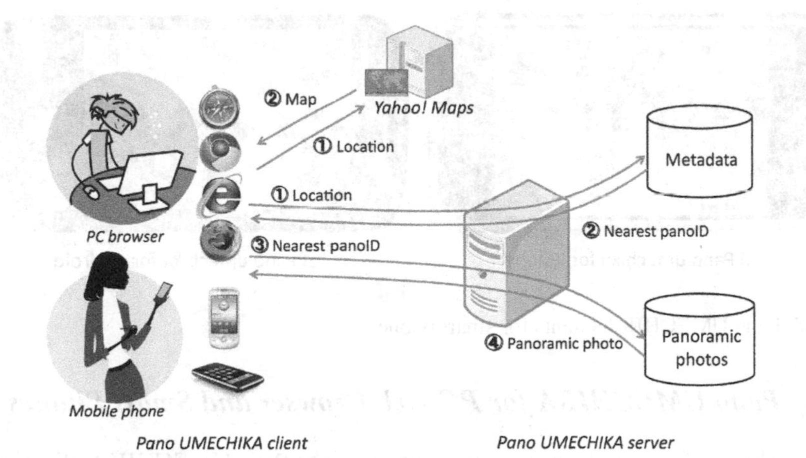

Fig. 3 Panorama viewer system overview

Fig. 4 Pano UMECHIKA client for PC web browser

This shot positions amount to 430. The camera cart carries Ladybug3, a spherical panoramic camera system with six camera units. The resolution of each camera is 1600(H) x 1200(V). The cart also carries a laptop PC featuring IEEE1394b to receive whole movies from the six high-resolution cameras, a RAID-SSD not to drop the huge movie, and a 3G Modem to communicate with the support tool for inputting location, as explained in section 2.2. A UPS is set at the bottom of the cart to supply enough energy to all the devices.

(a) Pano umechika for iPhone

(b) Pano umechika for Android

Fig. 5 Pano UMECHIKA clients for smart phones

3.1 Pano UMECHIKA for PC Web Browser and Smart Phones

Fig. 3. shows a panoramic view system consisting of Pano UMECHIKA clients and a Pano UMECHIKA server, which has the panoramic photos and their metadata. A Pano UMECHIKA client gets the nearest map from Yahoo! Maps, and the nearest panoramic photo from the server. If the client doesn't know its own location, the location is set to the center of the UMECHIKA area. The received panoramic photo is used as a texture of a sphere, which is drawn with a 3D graphics library. For PC browsers, we implemented a web application (Fig. 4) utilizing O3D [8], a 3D graphics JavaScript library released by Google. In the panoramic view, user can move to the next photo by clicking an arrow on a road. The user can also jump to another point by dropping a red mascot on the blue line drawn on the Yahoo! Maps.

As a result of generating metadata according to the proposal, we achieved the desired urban indoor panoramic viewer system. Along wide corridors, we shoot panoramic photos in two lines. Then ladder-shaped paths are calculated as shown at the left of the Yahoo! Maps in Fig. 4.

Furthermore, we implemented an OpenGL ES based Pano UMECHIKA client for iPhone and Android (Fig. 5). Due to the smaller main memory of a smart phone, the Pano UMECHIKA server makes the photos for them smaller (896(H) x 512(V)). Although the resolution is enough to make clear panoramic viewer on the smart phone.

3.2 The Results of Pedestrian Subtraction and Inpainting Dead Space

Fig. 6 (a) shows a panoramic image captured at a place crowded with pedestrians. As shown in this figure, many regions are full of blurring and are not useful for scenic information. Fig. 6 (b) shows the result of pedestrian subtraction based on background subtraction method. The panoramic image after pedestrian subtraction is informative for users.

Fig. 7 (a) shows a user's view when he/she looks down toward the ground. As shown in this figure, dead space appears in the view because Ladybug3 does not have a camera pointed downward. Fig. 7 (b) shows the result of inpainting dead

(a) Before pedestrian subtraction

(b) After pedestrian subtraction

Fig. 6 The result of pedestrian subtraction

space. The dead space was naturally filled in by the proposed method and the realistic sensation for users was drastically increased.

3.3 Reliability of the Sensor Data

We check the reliability of the stored Wi-Fi AP data (interesting values are BSSID and RSSI). The total number of scanned Wi-Fi APs is 17762. And the total number of unique Wi-Fi APs (BSSID) is 335. As a result of tagging panoId (an identification number of each panoramic photo) to the Wi-Fi AP data, only 219 of 430 photos has tagged with Wi-Fi APs. It means that the only around the half of the shot positions are available for Wi-Fi positioning. We should utilize another indoor positioning method such as template matching based on SIFT (Scale Invariant Feature Transform) [9] or SURF (Speeded Up Robust Features) [10] algorithm.

(a) without inpainting

(b) with inpainting

Fig. 7 The result of impainting

4 Conclusion

Toward a next generation navigation system, we focus a spherical panoramic viewer. In this research field, the main issues are how to care about the pedestrians' privacy and how to shoot and generate metadata of indoor environment as opposed to outdoor environment for which there are solutions. As a result of the experiment based on our proposed method, we achieve an urban underground panoramic viewer system. But the accuracy of Wi-Fi positioning for the Pano UMECHIKA client is still not sufficient. In the future we will combine a Wi-Fi positioning and a template matching method to achieve higher accurate indoor positioning.

References

1. Google Street View, `http://www.google.com/intl/en_us/help/maps/streetview/`
2. earthmine, `http://www.earthmine.com/index`
3. Nishio, N., Sakamoto, N., Arai, I.: Adjunct Proceedings of Pervasive 2009, pp. 269–272 (2009)
4. Kawai, N., Machikita, K., Sato, T., Yokoya, N.: Proc. Asian Conf. on Computer Vision (ACCV(2)), pp. 359–370 (2009)
5. Cheng, Y., Chawathe, Y., LaMarca, A., Krumm, J.: Proceeings of Mobisys 2005, pp. 233–245 (2005)
6. Harter, A., Hopper, A., Steggles, P., Ward, A., Webster, P.: The Fifth Annual ACM/IEEE International Conference on Mobile Computing and Networking, MOBICOM 1999, pp. 59–68 (1999)
7. Yahoo! Maps (JAPAN), `http://map.yahoo.co.jp/chika`
8. O3D, `http://code.google.com/apis/o3d/`
9. Lowe, D.G.: International Journal of Computer Vision 60(2), 91 (2004)
10. Bay, H., Ess, A., Tuytelaars, T., Gool, L.V.: Computer Vision and Image Understanding (CVIU) 110(3), 346 (2008)

Proposal of Smooth Switching Mechanism on P2P Streaming

Naomi Terada, Eiji Kominami, Atsuo Inomata, Eiji Kawai, Kazutoshi Fujikawa, and Hideki Sunahara

Abstract. In this paper we describe a smooth switching mechanism for enhancing the performance and low-waste traffic of distributed peer-to-peer video streaming. The mechanism was designed for when streaming topology set up a backup link and a predicted link to avoid being congested link each nodes. Furthermore this provides a *Dominant Keyword* procedure which enables to improve the performance of switching time for changing from a peer to another peer. Finally we shows an implementation design and discuss about an efficiency of this proposal.

Keywords: Streaming, Peer-to-peer networks, Zapping.

1 Introduction

During recent years, live streaming has gained much attention with the improvement of network infrastructure and client PC performance. Apparently, IP based streaming service has limitation of server and network capacity so that client cannot receive all channel simultaneously as TV broadcasting can. This means when a user

Naomi Terada, Eiji Kominami, Atsuo Inomata, Kazutoshi Fujikawa
Nara Institute of Science and Technology
e-mail: naomi-te@is.naist.jp, atsuo@is.naist.jp, fujikawa@itc.naist.jp

Eiji Kominami is now working in Asahi Broadcasting Corporation

Eiji Kawai
National Institute of Information and Communications Technology
e-mail: eiji-ka@nict.go.jp

Hideki Sunahara
Keio University
e-mail: suna@wide.ad.jp

A.P. de Leon F. de Carvalho et al. (Eds.): Distrib. Computing & Artif. Intell., AISC 79, pp. 181–184.
springerlink.com

changing the channel, he or she made to wait some seconds for retrieving content data. Those waiting period will cause a degrade QoS of live streaming.

Adding to this zapping time, start-up delay and playback continuity is the main issue on live streaming QoS. Different from server-client live streaming, P2P client node suffers from influence of leaving behavior of peers. Therefore, previous studies are mainly focused on maintaining distributed-tree, *e.g.*[2], or managing multi-source stream[1].

In this paper, we present our idea to the following challenge: 1) predict a probable next source and preload streaming data in advance to reduce zapping time 2) improve playback continuity with backup link.

2 Related Work

In the following, we show an overview of some work related to our system.

Peercast[4] adopts index server to manage multi-source streams. Figure 1 shows tree-based P2P distribution system for multi-source streaming. A peer willing to join a streaming S_1 send a query to a index server to find IP address of a root peer p_0. Next, peer joins the tree which is originating source S_1. Anysee[3] provides live media streaming to enlarge scale number of users. However, it doesn't focus on reducing waiting time.

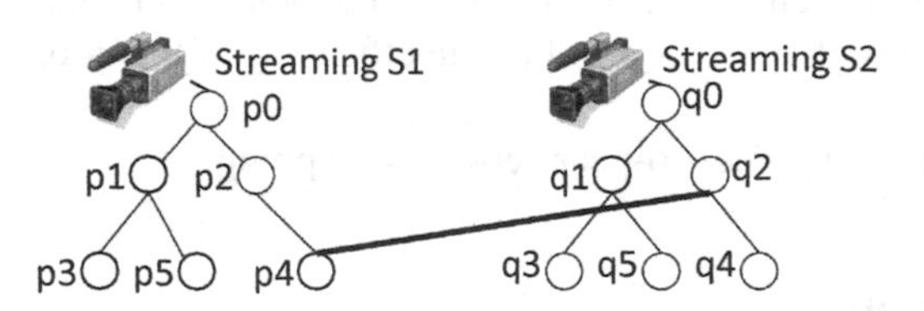

Fig. 1 Multi-source Live Streaming

3 System Overview and Prototype

We focused on reducing mean time to switch over multi-source streams. To minimize switching overhead, each peer should ideally have links to every stream(streaming S_1 to S_n), however, this approach is unrealistic because maintaining too much links generates excessive keep-alive traffic. Our basic idea is predicting a next stream likely to be switched using user's viewing history and prefetching the streaming data to enable switching smoothly. Our peer joining procedure is similar to Peercast. So it can use some peers to relay stream clips for other children peers. However this solution suitable for large-scale live streaming such a baseball stadium or concert hall that there are multiple cameras is simple to construct and easy to maintain but still weak in reliability.

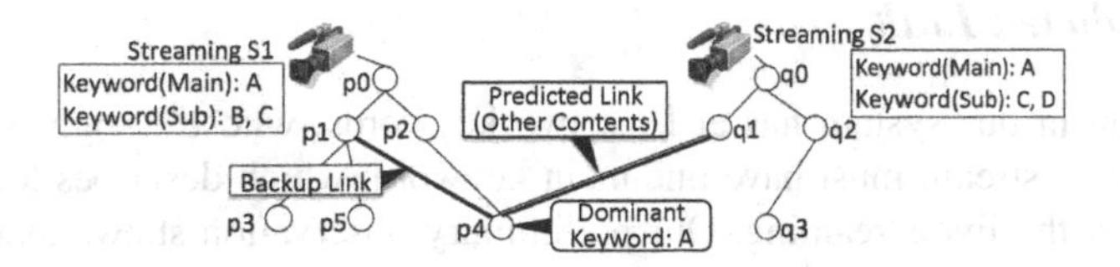

Fig. 2 Backup Link and Predicted Link

User's Viewing History

	stream
0 (now viewing)	Stream B
1	Stream A
2	Stream D

Distributed Streaming Keywords

	Keyword(Main)	Keyword(Sub)
Stream A	Stadium	Baseball, A-league
Stream B	Stadium	Baseball, B-league
Stream C	Concert Hall	Classical, Orchestra
Stream D	Stadium	Football
Stream E	Stadium	Baseball, C-league
Stream F	Theater	Movie, Action

Fig. 3 An example of selecting a Dominant Keyword process from a user's viewing list

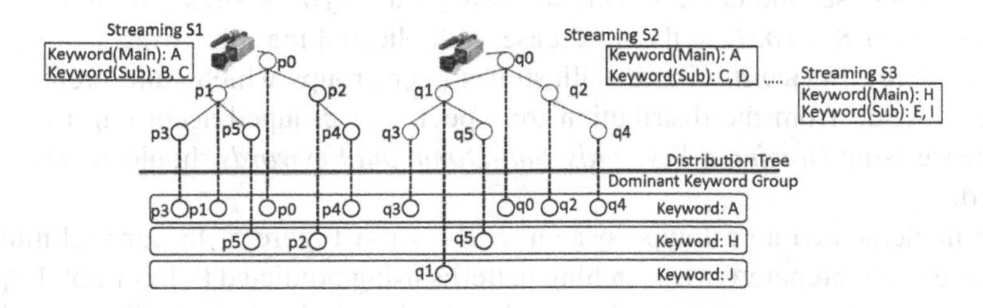

Fig. 4 Distribution tree and Dominant Keyword Groups Example

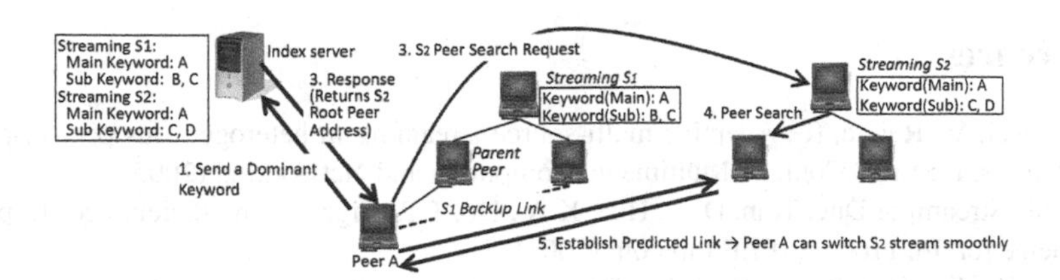

Fig. 5 Searching a Peer Mechanism

3.1 Backup Link

As illustrated Figure2, Every peer is linked to a peer(p_4-p_2) to receive stream data on P2P distribution system. Adding to this link, peer tries to link another backup link(p_4-p_1) in case original link failed.

3.2 Predicted Link

Each stream in our system has at least two keywords, whose design is depicted in Figure 2. A stream must have one main keyword, which describes location information of the live streaming. Our preliminary observation shows that location information is a key factor predicting a streaming which user switches next, while another additional information are described by sub keywords.

Some frequently-appearing keywords are extracted from a user's viewing history. Figure 3 shows how to determine the most frequently-appearing keyword(*Dominant Keyword*). In this example, "Stadium" is defined as a *Dominant Keyword* and "Baseball" is defined as a *sub-Dominant Keyword*. This means, a stream which has a keyword "Stadium" as a main keyword and "Baseball" as a sub keyword is most likely to be switched next.

We classified viewing history status into three cases as follows.

1. A peer has no user's viewing history
2. There is no corresponding keyword among user's viewing history
3. Some keywords are duplicated among user's viewing history

In the first and second case, a main keyword of a playback streaming is assigned to a *Dominant Keyword*. In the third case, a duplicated main keyword is assigned to a *Dominant Keyword*. Figure 4 illustrates peer groups which suite preferences of users. Aside from the distribution tree, peers are grouped according to user's preference using *Dominant Keywords*. *Sub-Dominant Keywords* should also be considered.

We implemented a prototype system as shown in Figure 5. In our preliminary experiment, we prepared two switching patterns using predicted links or not. Experimental result shows our proposed method can reduce switching time by up to 90%, even when a worst case, it can reduce switching time by 5%.

References

1. Agarwal, V., Rejaie, R.: Adaptive multi-source streaming in heterogeneous peer-to-peer networks. In: SPIE Conf. on Multimedia Computing and Networking (2005)
2. Media Streaming Duc, Tran, D.A., Hua, K.A., Do, T.T.: Zigzag: An efficient peer-to-peer scheme for. In: Proc. of IEEE Infocom (2003)
3. Liao, X., Jin, H., Liu, Y., Ni, L.M., Deng, D.: Anysee: Peer-to-peer live streaming. In: INFOCOM. IEEE, Los Alamitos (2006)
4. Zhang, J., Liu, L., Ramaswamy, L., Pu, C.: Peercast: Churn-resilient end system multicast on heterogeneous overlay networks. J. Network and Computer Applications 31(4), 821–850 (2008)

Low Cost Architecture for Remote Monitoring and Control of Small Scale Industrial Installations

Ignacio Angulo, Asier Perallos, Nekane Sainz, and Unai Hernandez-Jayo

Abstract. The high cost of existing industrial control systems prevents their implementation at small or medium size industrial plants. This paper describes a new architecture, based on low-power wireless sensor networks for remote monitoring and control of industrial equipment in small size plants. Through the use of very low consumption wireless devices, proposed architecture provides a low cost distributed control system easily deployable in small facilities.

Keywords: industrial control system, remote monitoring and control, low power wireless personal area network, wireless sensor network.

1 Introduction

The enormous economic damage caused by stopping the production line in industrial plants justifies the need to implement control systems. However, actual industrial control systems result cost-prohibitive to deploy in small or medium size industrial plants. Embedded control systems, as proposed in this paper, provide an affordable solution for small scale facilities.

The main objective of this work is to provide an industrial control system designed according to three fundamental characteristics:

- Reduce the final cost of the system using low cost devices
- Facilitate the deployment of the system through the use of wireless technology
- Minimize maintenance with low power devices self-powered by a battery

The proposed system has been initially designed to be used in intermediate goods and energy sectors. However, the challenge is to ensure that the solution can be easily adaptable to any production scope. Further, the process of adaptation and

---

Ignacio Angulo, Asier Perallos, Nekane Sainz, and Unai Hernandez-Jayo
DeustoTech - Deusto Institute of Technology, (Mobility Unit,)
University of Deusto, Avenida de las Universidades 24, 48007 Bilbao, Spain
e-mail: {ignacio.angulo,perallos,nekane.sainz}@deusto.es,
unai.hernandez@deusto.es

A.P. de Leon F. de Carvalho et al. (Eds.): Distrib. Computing & Artif. Intell., AISC 79, pp. 185–192.
springerlink.com

deployment of the developed technology to new sectors should require a low cost, as well as the cost of maintenance and use of the infrastructure. This feature is very important if our goal is that the proposed solution becomes an alternative to the current expensive SCADA systems. Moreover, this feature enables that the solution becomes a very interesting and competitive choice to be deployed in small plants. In fact, others sectors in which the new technology could be applied are the sectors of energy, environment, agriculture, capital goods, and so on.

2 Actual Scenario

There is a worldwide interest in the supervision and control industrial equipment in order to monitor their parameters every time they are needed to improve their efficiency. For example, this is the case of photovoltaic farms, where photovoltaic solar panels are physically distributed and they require technologically advanced solutions based on new technologies for positioning and communication. The recent evolution of these technologies helps to improve supervision systems, and there are new solutions arising with an acceptable success. These technologies are principally mesh sensor networks and ZigBee protocol for data capture and action signals distribution.

The ZigBee standard based upon the IEEE 802.15.4 specification has been specified by the ZigBee alliance with members of Freescale, Philips, Atmel, Siemens, Samsung, Analog Devices and Chipcon, and has the characteristics of large network capability (up to 65,000 nodes), long battery lifetime (up to a few years), short link establishment time (15-30 ms), but low data rate (up to 250 kbps) [1]. The 802.15.4 standard only specified physical and MAC layers, but for upper layers there are multiple different options to work with for improving features like battery life or fast network establishment. This is the case of the Beacon Networks [2].

ZigBee technology joined to mesh networking communications [3] is the last solution for monitoring the environment. The network is composed of nodes distributed around a wide area and a central base station that receives the information and controls the system. A simple medium access protocol and routing algorithm are proposed in [4] with the objective of reducing power consumption. The best solution is the one that achieves a good trade-off between power consumption and transmission delay improving communications algorithms and the architecture of the nodes.

However, although the industrial control is one of six application spaces for ZigBee identified by the ZigBee Alliance and many specialized companies are currently developing monitoring and control applications in industrial environments looking to wireless technologies like ZigBee to save the cost of wiring and installation and also to allow more flexible deployment of systems [5], there is job to do in technology research until a global solution for all the areas involved in supervision could be set up. This paper presents one new approximation based on the latest technologies for data acquisition, distribution and the remote monitoring and control.

3 Proposed Solution

The proposed solution consists of two distinct parts that will be evaluated with the recent implementation of a fully functional prototype.

- *Data acquisition system and communications*: wireless sensors network and the communications devices designed to be placed in the plants
- *Central server*: software infrastructure that enables remote monitoring and control through a graphical control panel.

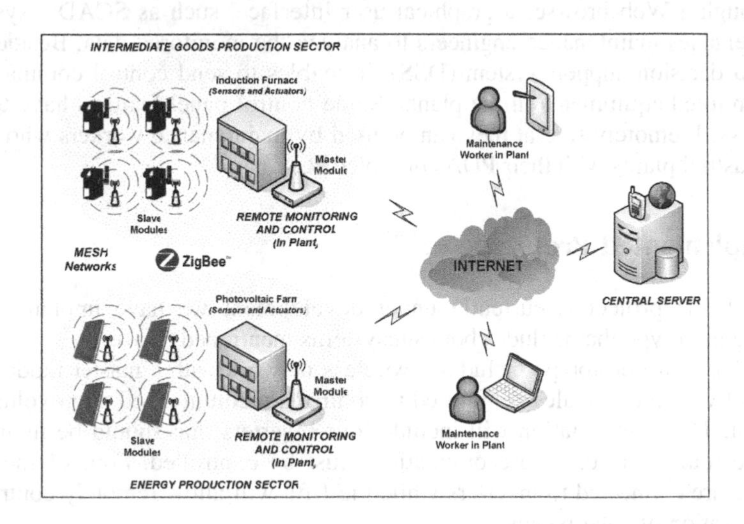

Fig. 1 Architecture of the proposed solution

Data Acquisition Systems and Communications

The infrastructure placed on industrial plants consists mainly of sensors, actuators and communication devices. In each plant there is a master module and several slave modules. The slave modules are very simple modules which control one or more machines of the plant using several sensors and actuators. Moreover they are making up a ZigBee communications network that connects them with the master module. Furthermore, each master module is able to manage the internal network of slave modules and is able to provide external communications using GPRS or HSPA technologies; thus it is performing as a network server. It communicates using the Internet infrastructure with the central server, which hosts a central repository with the data captured from all the monitored plants and provides a multichannel interface in order to access and exploit these data from other applications.

Moreover, the devices are not only units for reading and writing but they make up a network, they collaborate sharing the same communication channel to the outside, they filter captured data, or they manage the commands sent to all the equipment in a plant. All these features are highly demanded in the sectors in which we are working.

Central Server

Software infrastructure of one or multiple plants can be installed on an individual server that hosts a web application capable of managing communication with the controlled plant, maintain a data repository for all transactions and provide the user, through a Web browser a graphical user interface, such as SCADA systems, which enables maintenance engineers to analyze the monitored data. Besides acting as a decision support system (DSS), it enables to send control commands to the monitored equipments in the plants. Some control panel features have to also be accessed remotely so that they can be used by maintenance workers who are in the industrial plants with their PDAs or tablet-PCs.

4 Implemented Prototype

Although the project is currently under development we have implemented a working prototype that includes both subsystems mentioned above.

The hardware prototype includes a wireless network with a master module and a set of five slave modules configured to control a medium-sized photovoltaic solar plant. The power station will include four inverters that should be monitored and one rotating rotor whose orientation must be controlled. Four of the slave modules are connected to inverters while the fifth will allow remotely controlling the orientation of solar panels.

The developed software platform has been developed in collaboration with a company in the energy and waste management sector which has been actively involved in the specification of requirements providing some requisites that are usually not covered by standard industrial control systems not focused in mentioned sectors.

Although the first tests are currently being conducted in the laboratory, in the medium term, solution will be applied to a real plant in the province of Cantabria (Spain). These first tests in a controlled environment are being fundamental for power measurements and check the reliability of the packet transmission over the implemented beacon-enabled Medium Access Control (MAC) network taking into account different allowed configurations.

5 Architecture of the Data Acquisitions System and Communications

This subsystem is responsible for performing all tasks of control and data acquisition. It consists of all embedded controls arranged over the industrial plant and the central server that stores all the information captured and manages the communication between the parts.

Given that the proposed architecture is designed to be deployed on facilities where other industrial control systems are cost-prohibitive, the design of hardware modules has been carried out taking into account two important criteria: installation and maintenance of the proper solution must be cheap (1) and all hardware devices integrated in the modules must be low cost (2).

Low-power wireless sensor & control networks, like ZigBee, are a natural match with these requirements which provides inexpensive and low power usage nodes that allow communication between numerous embedded sensor and controls.

The infrastructure installed in industrial plant has a master module and several slave modules. Master acts as a gateway between the central server and the slaves which perform simple tasks of control and data acquisition.

In accordance with the requirements implementing the slave modules has been achieved using Texas Instruments (TI) ZigBee radio chip CC2430-F128. This System-on-Chip (SoC) solution includes the TI cc2420 RF transceiver and an 8 bit MPU based on high performance and low power industry-standard 8051 core. The integration into a single chip of ZigBee connectivity and processing power minimizes cost and size of the resulting nodes while reducing bill of materials and power consumption. The design of each slave module results in a small circuit board that contains the TI CC2430, an antenna, a crystal that allows the timing, a battery, very few external components and when it is necessary signal conditioning circuit that can receive information from each sensor or control a certain actuator. The peculiarities of industrial sensors prevent the design of a universal module while many sensors can be connected directly to the MPU through the integrated 12-bit ADC or via a serial interface as USART, I2C or SPI.

The master module requires connectivity with both ZigBee network and with the central server. To facilitate integration of the master module in the ZigBee network, it contains the same integrated circuit TI CC2430F128 included in the slave modules but performing now the tasks of the network coordinator. Connection to the central server is performed using GPRS / HSPA technology. This simplifies the installation of infrastructure in the industrial plant using a separate communications channel. Moreover, considering the future design of geographic information subsystem is desirable that the master module includes a GPS receiver. Wireless Module Cinterion XT75/65 provides both technologies: GPRS / HSPA and GPS including a high-performance ARM controller. This module will connect to the central server through a TCP socket and with the TI CC2430F128 via a serial port allowing a bidirectional communication between each slave module and the central server

Stringent battery-life requirements of the slave modules determine the use of a beacon-enabled Medium Access Control (MAC) network to be configured for optimal battery life and an adequate data rate. In this topology the coordinator, role played by the master module, sends beacon frames to other devices on the network indicating when to activate or deactivate their transceivers. Logically as long as they remain disabled minor will be the consumption increasing the battery-life. While the device is turned off coordinator cannot communicate with the device being impossible to access information that may be critical. For this reason the inter-beacon period depends on the nature of the elements monitored and controlled to determine maximum supported delays between the receipt of an instruction and its execution [6]. Two network configuration parameters are essential to determining the duty cycle of any device: the beacon order (BO) and the superframe order (SO). BO determines how often the beacon is broadcasted; SO determines the on-time duration within the superframe structure. Both parameters are easily configurable to meet the needs of each plant in which the application is installed.

The basic functionality of this system is to enable data exchange between the central server and each of the embedded sensor or control. The transfer of data can be two-way: upward to collect information from sensors, and downward to broadcast instructions to actuators or update configurations.

Each slave module can control up to 4 sensors or actuators. Each sensor has an associated timer that manages the regular sending of data. Each sensor also has a configurable alarm that causes the sending of sensor data at the next duty cycle available. All these tasks are performed by the MPU included in slave modules that continues working even when the transceiver is disabled.

The firmware of each slave module must be customized including a routine that allows access to information from each sensor as well as manipulate separate actuators. All sensor management functions should return a 64-bit unsigned integer value in the same way that all the routines that control the actuators receive a value of the same type which encodes the instruction to execute.

Information between Zigbee network coordinator (master module) and other nodes is performed through simple binary commands to minimize traffic and collisions. Each command is composed of an operation code (5 bits), a sensor/actuator identifier (3 bits) and a value (4 bytes). Available commands of the current version are "demand sensor reading", "enable/disable regular sensor reading", "set regular sensor reading period", "enable/disable alarm", "set alarm value" and "write value on actuator". Sensor readings (regular or on demand) are sent to the master module including sensor identifier (1 byte) and value (4 bytes).

Master module acts as a transparent gateway: receives requests from the central server that are translated and transferred to the involved sensor/actuator and responses to the server all data acquired from the slave modules.

6 Design of the Central Server

The central server is responsible for storing and managing the information exchanged with the plant and providing the resources to communicate with the elements that should be monitored or controlled by the system. An application that runs in the server manages constant communication with the plant, allowing the exchange of information via SOAP. All information exchanged between the plant and the central server is properly stored in a DBMS (currently SQL Server.)

Detailed information of the individual modules that are installed in the industrial plant as well as monitored signals and actuators is stored in an XML file that can be remotely managed from the control panel acceding with an administrator or installer profile user.

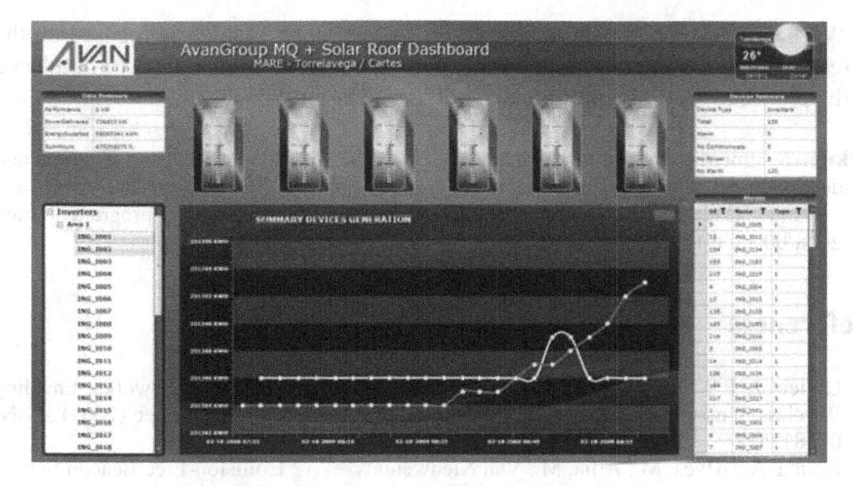

Fig. 2 Screenshot of the control panel.

Transmission between central server and infrastructure master module is performed using SOAP protocol.

All the services of monitoring and control systems installed in the plant are accessible through a web user interface (WUI) called control panel. This web application developed in ASP.NET provides a low cost control industrial system accessible from any Web browser over the Internet.

Web application framework Microsoft Silverlight has provided advanced graphical controls for the interface design that differentiate the application of traditional SCADA systems. Fig. 2 shows a screen shot of one of the user views of the control panel.

7 Conclusion and Future Work

In this paper we have presented low cost hardware and software technology for remote monitoring and control of industrial equipment. This technological infrastructure is made up of components placed inside the industrial plants and outside them. Inside the plants wireless sensors networks based on ZigBee technology are used in collaboration with actuators and communication devices based on GPRS to provide external connectivity. Outside the plants a central server to provide remote monitoring and control capabilities.

Proposed solution provides a technological infrastructure with very innovative contributions, such as the flexibility, adaptability and the low cost required to be deployed. Besides, the design of the devices network placed on the plants has been made taking into account the requirement of low energy consumption and cost of communications. In fact, these features enable that the proposed solution becomes a very interesting and competitive choice to be used in small productions plants.

We have shown a first prototype of this infrastructure which is focused on the wireless sensors network and the communications devices designed to be placed in the plants, as well as a preliminary web-based control panel

Acknowledgments. This work has been funded by the Ministry of Industry, Tourism and Trade of Spain under Avanza funding program (Grant TSI-020100-2009-719) and the Basque Country Government of Spain under Gaitek funding program (Grant IG-2008/0000099).

References

1. Gutierrez, J.A.: IEEE 802.15.4, Low-Rate Wireless Personal Area Networks: Enabling Wireless Sensor Networks. Institute of Electrical & Electronics Enginee (2003) ISBN: 0738135577
2. Kouba, A., Alves, M., Attia, M., Van Nieuwenhuyse, A.: Collision-Free Beacon Scheduling Mechanisms for IEEE 802.15.4/Zigbee Cluster-Tree Wireless Sensor Networks. In: 7th International Workshop on Applications and Services in Wireless Networks (ASWN 2007), Santander, Spain (May 2007)
3. Liang, N.C., Chen, P.C., Sun, T., Yang, G., Chen, L.J., Gerla, M.: Impact of Node Heterogeneity in ZigBee Mesh Network Routing. In: IEEE International Conference on Systems, Man, and Cybernetics (SMC 2006), Taipei, Taiwan (2006)
4. Jeon, J., Lee, J.W., Ha, J.Y., Kwon, W.H.: DCA: Duty-Cycle Adaptation Algorithm for IEEE 802.15.4 Beacon-Enabled Networks. In: IEEE 65th Vehicular Technology Conference (VTC2007-Spring), Dublin, pp. 110–113 (2007)
5. Egan, D.: The emergence of ZigBee in building automation and industrial controls. Comput. Control Eng. 16, 14 (2005), doi:10.1049/cce:20050203
6. Chen, A.F., Wang, N., German, R., Dressler, F.: Performance Evaluation of IEEE 802.15.4 LR-WPAN for Industrial Applications. In: Proceedings of 5th IEEE/IFIP Conference on Wireless On demand Network Systems and Services (IEEE/IFIP WONS 2008), Garmisch- Partenkirchen, Germany, January 2008, pp. 89–96 (2008)

Translators in Textual Entailment

Julio Javier Castillo

Abstract. This paper presents how the size of Textual Entailment Corpus could be increased by using Translators to generate additional $\langle t,h \rangle$ pairs. Also, we show the theoretical upper bound of a Corpus expanded by translators. Then, we propose an algorithm to expand the corpus size using Translator engines starting from a RTE Corpus, and finally we show the benefits that it could produce on RTE systems.

Keywords: Textual entailment, Translators, RTE datasets.

1 Introduction

The objective of the Recognizing Textual Entailment Challenge is the task of determining whether the meaning of the Hypothesis (H) can be inferred from a text (T).

In the past, RTEs Challenges machine learning algorithms were widely used for the task of recognizing textual entailment [1], [2], [3]. Some authors [4] showed how the accuracy increases when we add more training examples, and other authors holds the necessity of larger corpus [5]. In any case, a larger corpus enables a more detailed analysis of the problem domain and will let us build more accurate classifiers.

In this paper we show how a translator could increase the size of a RTE Corpus, and also suggest how a translator can be used as a tool to helps us with classification of new (unknowns) $\langle t,h \rangle$ pairs.

The remainder of the paper is organized as follows: Section 2 describes one approach driven by Translators and provides an analysis about the possible increasing size of the Corpus, whereas Section 3 shows some possible benefits of using Translators. Finally, Section 4 summarizes the conclusions and lines for future work.

Julio Javier Castillo
National University of Cordoba-FaMAF, Cordoba, Argentina
National Technological University-FRC, Cordoba, Argentina

A.P. de Leon F. de Carvalho et al. (Eds.): Distrib. Computing & Artif. Intell., AISC 79, pp. 193–196.
springerlink.com

2 Translation Driven Approach

In this section we propose to use Translators to expand the current RTE-Corpus sizes. First, we define the term "double translation" to describe the process of translating from Source (for example a string in English) to Target (another language) and back to the Source again; more commonly, this is referred as "round-tripping". Thus, our motivation is based on the fact that we can use a double translation process to produce equivalents Texts and Hypothesis, and so these new pairs can be taken as training set. Also, we suggest how a translator can be used as a tool to helps us with classification of new $\langle t,h\rangle$ pairs.

2.1 Corpus Sizes

Double translation process can be defined as the process of starting with a S (String in English), translate it to foreign language, for example Spanish, and back to the source (English) again. So, the observation of that double translation process can increase the Corpus size; can be generalized using N-Translator engines (Machine Translation systems).

It is important to note that the "quality" of the translation is given by the Translator engine, and we will suppose that the sense of the sentence should not be modified by the Translator. This, indeed, is the situation, almost for the majority of the cases in our first experiments (see Section 3). Bellow, we provide a theoretical justification of the increment of the corpus size with n-pairs using k-translators which is $O(n * k^2)$.

Notation: C is a RTE Corpus which consists of $\langle t,h\rangle$ pairs. C_q is the increased size of the Corpus C using q–Translators.

Notation: $\langle t,h\rangle \equiv t \rightarrow h$

Define:

$Tr : String \rightarrow String$

$$t \rightarrow Tr(t)$$

$Tr(t) = DoubleTranslationOfTheTrTranslator$

where: t and Tr(t) are in English.

Lemma. Given Tr_1, Tr_2,…,Tr_k translators and C a RTE Corpus with n-$\langle t,h\rangle$ pairs. If $Tr_i(t_p) \neq Tr_j(h_p) \quad \forall i,j \wedge i \neq j \wedge i,j \in \{1,...,k\} \wedge p \in \{1,..,n\}$ then $C_k = (k+1)^2 * n$.

Proof: By structural induction on K.

2.2 Using Double Translation over Textual Entailment Datasets

Translators can be used as a feature in machine learning algorithms [6]. Indeed, by using a translator it is possible to reduce the complexity of some of the sentences and by this way RTE task should be easier. Another use of Translator is to provide synonyms and expression with the same meaning. For example, the following pair 93 belongs to the RTE3 development and is given below.

Source pair:

T: *UN Secretary General Kofi Annan has noted that the Iraqi people turned out in large numbers to vote in the January 30 ballot.*

H: *Kofi Annan was elected in the January 30 ballot.*

Translated pair using Microsoft Bing Translator:

T: *Secretary General of the United Nations, Kofi Annan has pointed out that the Iraqi people resulting in large numbers to vote in the vote on 30 January.*

H: *Kofi Annan was elected in the vote on 30 January.*

In this example, we see how "*UN*" was translated to "*United Nations*" was translated (acronyms resolution). Additionally, we see that the expression "*UN Secretary General Kofi Annan*" is equivalent to "*Secretary General of the United Nations, Kofi Annan*". Also, we see some paraphrases "*has noted*" and "*has pointed out*", "*Iraqi people turned out in large numbers to vote*" and "*Iraqi people resulting in large numbers to vote*", and finally "*the January 30*" and "*on 30 January*". This example was taken randomly and seems to support our claim that using a translator engine it is possible to improve the semantic resources of RTE Systems.

3 Experimental Evaluation

In this section, we present an algorithm in order to obtain additional $\langle t,h\rangle$ pairs from a given Corpus:

1. Start with a RTE-x Corpus, with $|C|=n$
2. For each $\langle t_i,h_i\rangle \wedge i\in\{1,\ldots,n\}$
3. For each Translator Tr1, Tr2,...,Trk. If

$$Tr_j(t_i)\neq t_i \wedge Tr_j(h_i)\neq h_i \;\forall j\in\{1,..,k\}\rightarrow Add(Tr_j(t_i),Tr_j(h_i))\ to\ C_{new}$$

Where: C_{new} is the new Corpus obtained as the union between C and the new outputs pairs of the algorithm.

Based on the previous algorithm, we perform 3 experiments using the following Translator engines: Microsoft Bing Translator, Google Translator and Babel Fish Translator. We addressed our experiment over RTE3 development set only. Then, we assessed the output of the algorithm, by human judge who determinate whether a new $\langle t_{new},h_{new}\rangle$ pair produced change of the meaning from the original $\langle t,h\rangle$ pair. We tested by two human judge this approach over 50 randomly pairs of RTE3 development set using the translators mentioned before, and saw that approx. 90% of the cases the double translation process yielded by Bing does not affected the whole sense of that $\langle t, h \rangle$ pairs. Google Translator also produced an approx. 90% of the pairs, and Babel Fish Translator produced approx. 80% pairs that does not affected the original sense of the source pair. We concluded that both, Bing Translator and Google Translator are very useful to expand RTE Corpus size.

4 Conclusions and Future Work

In this work we first propose the use of Translators engines as a way to increase the corpus sizes. Then, we show the theoretical upper bound in which a corpus size (with n-pairs) could increase by using k-translators; and we also present an algorithm to increase training sets.

We concluded that for our algorithm, Microsoft and Google translators seem to be more useful than Yahoo translator.

However, further analysis is required to determine the impact of the Translators in Textual Entailment Systems and evaluation through ablation tests.

Future work is oriented to explore more deeply how Translation could improve the accuracy of the RTE Systems, and to test over different datasets.

Finally, our next steps will be testing the double translation process but passing through Spanish, Portuguese, Dutch, and Russian as intermediate language, and assessing the improvement that they can yield.

References

1. de Marneffe, M.-C., MacCartney, B., Grenager, T., et al.: Learning to distinguish valid textual entailments. RTE2 Challenge, Italy (2006)
2. Zanzotto, F., Pennacchiotti, M., Moschitti, A.: Shallow Semantics in Fast Textual Entailment Rule Learners, RTE3, Prague (2007)
3. Castillo, J.: A Study of Machine Learning Algorithms for Recognizing Textual Entailment. In: RANLP, Borovets, Bulgaria (2009)
4. Inkpen, D., Kipp, D., Nastase, V.: Machine Learning Experiments for Textual Entailment. RTE2 Challenge, Venice, Italy (2006)
5. Newman, E., Stokes, N., Dunnion, J., Carthy(2005), J.: UCD IIRG Approach to the Textual Entailment Challenge. In: PASCAL. Proc. of the First Challenge Workshop RTE (2005)
6. Agichtein, E., et al.: Combining Lexical, Syntactic, and Semantic Evidence for Textual Entailment Classification. In: TAC 2008, Gaithersburg, Maryland, USA (2008)
7. Bentivogli, L., Dagan, I., Dang, H.T., et al.: The Fifth PASCAL RTE. NIST, Maryland USA (2009)
8. Dolan, B., Quirk, C., Brockett, C.: Unsupervised construction of large paraphrase corpora: exploiting massively parallel news sources. In: COLING 2004, Morristown, NJ, USA, p. 350 (2004)

Strategies to Map Parallel Applications onto Meshes

Jose A. Pascual, Jose Miguel-Alonso, and Jose A. Lozano

Abstract. The optimal mapping of tasks of a parallel program onto nodes of a parallel computing system has a remarkable impact on application performance. We propose a new criterion to solve the mapping problem in 2D and 3D meshes that uses the communication matrix of the application and a cost matrix that depends on the system topology. We test via simulation the performance of optimization-based mappings, and compare it with consecutive and random trivial mappings using the NAS Parallel Benchmarks. We also compare application runtimes on both topologies. The final objective is to determine the best partitioning schema for large-scale systems, assigning to each application a partition with the best possible shape.

Keywords: mapping, parallel applications, QAP, k-ary n-mesh, scheduling.

1 Introduction

Message-passing parallel applications are composed of a collection of tasks that interchange information and synchronize among them using different communication patterns. These applications are often designed and developed with a parallel communications architecture in mind, and arrange the interchanges of data blocks using some scheme that tries to use efficiently the underlying network. The way tasks are arranged to perform communications is called the virtual topology. However, these applications, once programmed, can be executed in a wide variety of parallel architectures, with different interconnection networks. These architectures vary from clusters with simple LAN-based networks to supercomputers with custom-made networks organized as meshes, tori, trees, etc. Given this variety of interconnects, it is not uncommon to find a mismatch between the physical arrangement of the compute elements (that depends on the system's topology) and the virtual topology

---

Jose A. Pascual, Jose Miguel-Alonso, and Jose A. Lozano
Intelligent Systems Group, The University of the Basque Country UPV/EHU
e-mail: {joseantonio.pascual,j.miguel,ja.lozano}@ehu.es

A.P. de Leon F. de Carvalho et al. (Eds.): Distrib. Computing & Artif. Intell., AISC 79, pp. 197–204.
springerlink.com

(application dependent). This may result in an inefficient utilization of the network, that materializes in large delays and bandwidth bottlenecks that, in turn, results in a general performance loss. The way of mapping tasks onto network nodes has a remarkable impact on the overall system performance [10] [12] [1].

The most used network topology [5] in current supercomputers is the 3D torus. This topology provides desirable topological characteristics such as symmetry, low node degree and low diameter. However, sharing a 3D torus between multiple applications results in partitions with the form of 2D or 3D sub-meshes, maybe with some wrap-around links. Most of the research made in partitioning 3D topologies [6] focused on the search for 3D sub-topologies, assuming that all applications will benefit from them. However, this assumption does not take into account the virtual topologies of applications, that may not match with a 3D (sub-)mesh. Moreover, the search for just one shape for the partitions can affect the scheduling performance, due to an increase of external fragmentation [7].

In this paper, we adapt an optimization-based mapping strategy previously developed for 2D topologies [12] to work with 3D meshes. The procedure is expressed as an optimization problem, in which a target functions has to be minimized. Within this strategy, we can use a classic optimization criterion that tries to minimize the average distance traversed by messages. We also proposed an alternative criterion, called TD (Traffic Distribution) that tries to reduce contention for the use of network resources. Within this framework, a given mapping is better than another one if the value of the target function is smaller; still, we must check if a theoretically better mapping actually makes the application run faster. In order to carry out this validation we use simulation. The simulation workbench is INSEE [14] which, given a trace of an application, a target network, and a mapping, can provide an estimations of the execution time. As applications we use (traces of) a subset of the well-known NAS Parallel Benchmarks (NPB) [9]. For simplicity, we consider only configurations of 2D and 3D meshes in which the number of tasks equals the number of network nodes. Regarding mappings, in addition to those obtained using optimization, we have tested some trivial ones: consecutive and random.

All this work is necessary to answer the following question. If we have a very large-scale system, arranged in a 3D structure, in which we want to allocate a much smaller application, which option would be better, a 3D sub-structure (a sub-cube or sub-mesh) or a 2D structure (a plane, or a portion of it)? The way we deal with this issue will have a great impact on terms of application performance, and also on terms of scheduling costs. As we will see, 3D sub-structures perform better for many applications, but not for all. Meanwhile, it may be easier (less expensive, in terms of scheduling costs) to partition a cube into planes instead of sub-cubes.

2 Optimization-Based Mapping Framework

The mapping problem can be formally defined as follows: given a set of tasks belonging to a parallel job $T = \{t_1, ..., t_n\}$ and a set of processing nodes of a parallel computing system $P = \{p_1, ..., p_n\}$ find a mapping function $\pi : T \longrightarrow P$ that assigns a task t to a node p trying to optimize a given objective function. Note that we use (network) node and processor interchangeably.

The mapping problem can be expressed easily as a QAP (Quadratic Assignment Problem) [11] in the matrix form: given T and P two equal size sets representing parallel application tasks and processors respectively, matrix $W = [w_{i,j}]_{i,j \in T}$ representing the number of bytes interchanged between each pair of tasks, and matrix $C = [c_{i,j}]_{i,j \in P}$ representing some cost characteristic involving pairs of network nodes, find the bijection $\pi : T \longrightarrow P$ that minimizes:

$$\sum_{i,j \in P} w_{i,j} \cdot c_{\pi(i),\pi(j)} \tag{1}$$

The formulation of the problem as an instance of the QAP allows computing mappings using techniques developed for the generic QAP. In particular, we used a GRASP (Greedy Randomized and Adaptive Search Procedure) [13] solver, which is fast and provides the best results known when solving the generic QAP. This constructive, multi-step algorithm iterates over two steps: first, an initial solution is created by means of an adaptive greedy randomized algorithm; then a local search improves that solution. A detailed explanation of the algorithm can be found in [12].

3 Mapping Criteria on Meshes

The mapping problem has been stated often as a location problem [3], searching a mapping vector that minimizes the average distance traversed by application messages. This is what we call the *classic* (optimization-based) strategy. The result of using this strategy is a mapping that locates in neighbouring nodes those tasks that interchange large data volumes. Apparently, this is a good policy, because it exploits communication locality, reducing the utilization of network resources. However, experiments show that a reduction in average distance does not always result in a reduction of application running time, because it may create contention hot-spots inside the network [1] [2] [12]. For this reason, we propose a different cost matrix to be used instead of the distance matrix. The criterion of minimizing average distance is relaxed, in order to favour communication paths that distribute traffic through different network axes, reducing the risk of contention. We have called this the *Traffic Distribution* (TD) criterion.

The definition of both criteria (cost matrices) depends on the topological characteristics of the target network in which applications will run. In this work we only consider meshes with two and three dimensions. In these networks each node has an identifier i in the range $0 \dots N-1$. We will consider routing functions that make messages advance using shortest-path routes between source and a destination. The simplest form of routing is Dimension Order Routing (DOR) [4], in which messages traverse first all the necessary hops in each dimension, but we do not make any assumption regarding the routing algorithm, except that it must use minimal paths.

We will describe the different criteria for 3D meshes with $N = n_x \times n_y \times n_z$ nodes. The expressions for 2D meshes can be easily inferred, assuming that the number of nodes in the Z axis is 1. Given two nodes with identifiers i and j, the distance between them is given by Equation 2.

$$dist(i,j) = dist_x(i,j) + dist_y(i,j) + dist_z(i,j) \tag{2}$$

where each term of the sum is the number of hops through each axis a message must traverse when going from i towards j. Equation 3 shows terms $dist_x$, $dist_y$ and $dist_z$.

$$\begin{aligned} dist_x(i,j) &= |(j \bmod (n_x \times n_y)) \bmod n_x - (i \bmod (n_x \times n_y)) \bmod n_x| \\ dist_y(i,j) &= |(j \bmod (n_x \times n_y))/n_y - (i \bmod (n_x \times n_y))/n_y| \\ dist_z(i,j) &= |j/(n_x \times n_y) - i/(n_x \times n_y)| \end{aligned} \tag{3}$$

The classic criterion identifies C as the distance matrix, whose components can be filled using Equation 2 for each pair of source-destination nodes.

$$C = [c_{i,j}]_{i,j=1\ldots N} \text{ where } c_{i,j} = dist(i,j) \tag{4}$$

In this paper, we adapt the TD mapping criterion developed in [12] to 3D meshes. The new criterion is expressed as an alternative cost matrix that allows us to find mappings that, while not optimal in terms of distance, avoid network bottlenecks. In particular, we pay attention to the number of hops per dimension that messages have to travel. In a 3D mesh, a 6-hop route in which all the hops are in the X axis is worse than another 6-hop route with 2 hops per dimension, because the latter imposes less pressure in the X axis, spreading the utilization of network resources evenly among the three axes. Formalizing this, we penalize those routes in which the number of hops per dimension is not equal. The level of penalization depends on the level of asymmetry, as represented in Equation 5.

$$td(i,j) = dist + |dist_x - dist_y| + |dist_x - dist_z| + |dist_y - dist_z| \tag{5}$$

Hence, when using the TD criterion for the resolution of the mapping problem within our framework, the cost matrix is defined as:

$$C = [c_{i,j}]_{i,j=1\ldots N} \text{ where } c_{i,j} = td(i,j) \tag{6}$$

4 Experimental Set-up

In this section we detail the collection of experiments carried out to evaluate the effectiveness of the two approaches to the mapping problem described before. We use networks with 64 nodes: a 2D, 8x8 mesh and a 3D, 4x4x4 mesh. This way we can assess the influence of network topology on that effectiveness.

We use traces extracted from the NPB suite [9], a suite of small kernels and pseudo-applications which are derived from computational fluid dynamics (CFD) applications. For each application we need to generate the traffic matrix W. We have done so for the class A, size 64 instances of the benchmarks. These applications were run in a real parallel machine, in which traces of all the messages interchanged by tasks were captured to fill W.

The QAP Solver accepts as parameters matrix W (the communication matrix of the application trace to evaluate, expressed in bytes), matrix C (the cost matrix modeling a mapping criterion) and the number of iterations to be performed. The output is a mapping vector that obeys the criterion represented in the cost matrix.

Simulations have been carried out using INSEE [14] [8]. This tool simulates the execution of a message-passing application on a multicomputer connected via an interconnection network. It performs a detailed simulation of the interchange of the messages through the network, considering network characteristics (topology, routing algorithm) and application behaviour (causality among messages). The input includes the application's trace file and the mapping vector. The output is a prediction of the time that would be required to process all the application messages, in the right order, including causal relationships and resource contention. INSEE only measures the communication costs, assuming infinite-speed CPUs.

To perform the validation of mappings, we generated a set of 50 different vectors for each NPB application, using the classic and the TD expressions of the cost matrix. Each of the 50 tested vectors were selected after 50 GRASP iterations. In addition to the classic and the TD mapping strategies, we also evaluated the behaviour of the consecutive and random trivial mappings. Given a network, the consecutive mapping criterion allocates the application tasks onto the network nodes in order of identifiers, starting with task / node 0. Regarding the random criterion, application tasks are assigned to network nodes randomly; we used 50 different random assignments to evaluate this strategy.

5 Analysis of Results

We present the results of the experiments in three graphs. The first and the second allow us to assesses the quality of trivial vs. optimization-based mappings for the 64-node, 2D and 3D meshes. The third graph is designed to identify the topology that best matches each of the applications used in this study.

Figure 1 summarizes the results of experiments with 2D and 3D meshes. Due to limitations in space, we do not show average and standard deviation of the execution times as reported by the simulator for each application-topology-mapping combination. Instead, we discuss only average results, normalized to the consecutive case.

For the 2D mesh, it is clear that we can divide the applications into three groups. The first one contains applications BT, LU and SP. A property of these three applications is that the virtual topology matches the physical one; therefore, the consecutive mapping is the best performer. In a second group we include applications FT and IS; for these, optimization-based mappings are not better than the trivial ones – although the TD criterion provides almost the same results. These applications are implemented using non-optimized all-to-all communications, for which locality

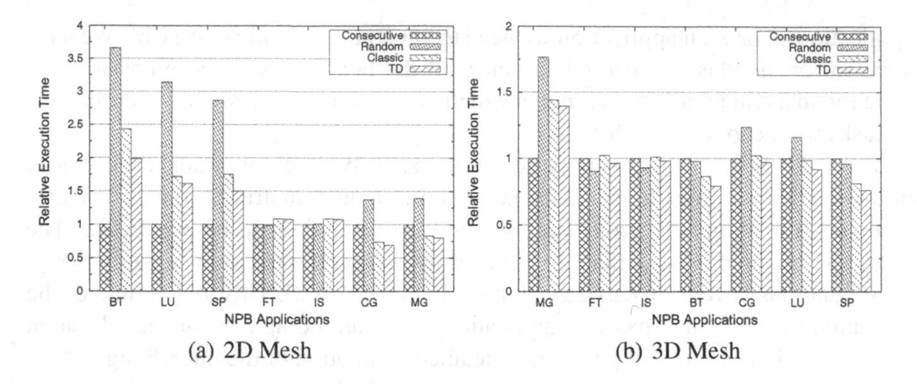

(a) 2D Mesh (b) 3D Mesh

Fig. 1 Execution times (simulation cycles) for the NPB applications (class A, size 64) running on a 2D mesh (a) and on a 3D mesh (b) for each mapping criterion. Values are normalized to the consecutive case.

cannot be efficiently exploited. In the third we put CG and SP. Is for these applications when the optimization-based strategies in general, and the TD in particular, shows their potential, providing results clearly better than those of the trivial criteria.

Regarding the 3D mesh, there is a case of perfect virtual-to-physical match of topology: the MG application – therefore the consecutive mapping is the best one. For FT and IS we observe again that most mappings provide similar results, being the random one slightly better. In the remaining cases TD is the best performer.

In summary, the TD procedure provides good mappings of applications onto mesh topologies. Trivial mappings are able to produce better results in those cases in which the virtual network matches the actual one (consecutive mapping) and when the communications pattern performs mainly all-to-all communications (random mapping). We have to remark that the TD criterion performs better than the classic for all the applications under test.

In the literature, it is generally assumed that the topological characteristics of 3D structures (that are better than those of same-size, 2D networks) should result in better execution times for applications. For this reason, partitions of 3D networks are generally done in terms of 3D sub-networks. The results of the experiments clearly state that applications perform best when there is a good virtual-to-physical matching of topologies. As not all applications have a 3D virtual topology, a mapping onto a 3D structure can be inadequate. In Figure 2 we have summarized the execution times of the NPB applications for 64-node, 2D and 3D meshes. As we can expect, for applications BT, LU and SP (2D-mesh virtual topology) a trivial, consecutive mapping onto a 2D network is faster than any mapping onto a 3D mesh. In contrast, for MG (3D-mesh virtual topology) the consecutive mapping onto the 3D network is clearly a better option. For the remaining applications, for which the structure of the communications does not match these topologies (or does not have any structure), the TD mapping onto a 3D partition provides excellent results.

We have to consider this in the context of a massively parallel computer managed by a topology-aware scheduler. Given an application, for which a system

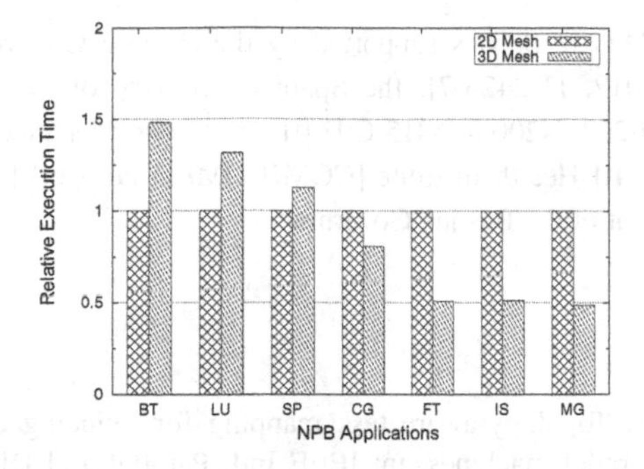

Fig. 2 Comparison of the execution time (simulation cycles) of the NPB applications in 2D and 3D meshes. Values are normalized to the 2D mesh case.

partition has to be assigned, it is desirable to locate a partition whose shape matches the most adequate one for that application; this can result in a search of a 2D sub-mesh (a plane, or a sub-plane) inside a 3D torus. If this match is impossible (or the application's virtual topology is unknown), then a 3D partition along with an optimization-based mapping is a good option. The search of a partition for an application should be done taking into account not only the effects on the performance of the application, but also the scheduling costs: the procedure should minimize system fragmentation.

6 Conclusions and Future Work

Current supercomputer centres share their resources between different users and applications. A system scheduler plays a crucial role, selecting the jobs to be executed and assigning some resources to it (nodes grouped as partitions). Often, the allocation/mapping procedures are carried out using simple mechanisms that ignore the communication patterns of the application and the topology of the network.

In this work we have focused on the mapping problem, describing a new criterion for mesh topologies to find good task-to-node assignments that takes into account the communication characteristics of the application, as well as the topology of the network. We have focused on meshes because, even though large-scale supercomputers are built around 3D torus, partitions do not always enjoy wrap-around links.

We have shown the importance of searching for the correct topology for a particular application; this will result in an improvement of applications run times. And we have also stated that the partitioning mechanism should be included in the scheduling process, selecting the best sub-network for a given job, but trying to keep fragmentation under control. A good mapping algorithm, such as the optimization-based TD strategy, is useful to keep good performance levels. Our main line of future work is precisely the integration of partitioning and mapping mechanisms into schedulers for large-scale systems.

Acknowledgements. This work was supported by the Basque Government [Saiotek, Research Groups 2007-2012, IT-242-07]; the Spanish Ministry of Science and Innovation [TIN2007-68023-C02-02, TIN2008-06815-C02-01 and Consolider Ingenio 2010 CSD2007-00018]; and the Carlos III Health Institute [COMBIOMED Network]. Mr. Pascual is supported by a doctoral grant of the Basque Government.

References

1. Agarwal, T., et al.: Topology-aware task mapping for reducing communication contention on large parallel machines. In: IEEE Intl. Parallel and Distributed Processing Symposium, Los Alamitos, CA, USA (2006)
2. Bhatele, A., Kalé, L.V.: An evaluation of the effect of interconnect topologies on message latencies in large supercomputers. In: Workshop on Large-Scale Parallel Processing (2009)
3. Bokhari, S.H.: On the mapping problem. IEEE Trans. on Computers 30(3), 207–214 (1981)
4. Dally, W., Towles, B.: Principles and practices of interconnection networks. Morgan Kaufmann Publishers Inc., San Francisco (2003)
5. Dongarra, J., Meuer, H., Strohmaier, E.: Top500 supercomputer sites, `http://www.top500.org`
6. Kang, M., et al.: Isomorphic strategy for processor allocation in k-ary n-cube systems. IEEE Trans. on Computers 52, 645–657 (2003)
7. Lo, V., et al.: Noncontiguous processor allocation algorithms for mesh-connected multicomputers. IEEE Trans. on Parallel and Distributed Systems 8, 712–726 (1997)
8. Miguel-Alonso, J., Navaridas, J., Ridruejo, F.J.: Interconnection network simulation using traces of MPI applications. Intl. Journal of Parallel Programming 37(2), 153–174 (2009)
9. NASA Advanced Supercomputer (NAS) division: NAS parallel benchmarks (2002), `http://www.nas.nasa.gov/Resources/Software/npb.html`
10. Navaridas, J., Pascual, J.A., Miguel-Alonso, J.: Effects of job and task placement on the performance of parallel scientific applications. In: 17th Euromicro Intl. Conf. on Parallel, Distributed, and Network-Based Processing, pp. 55–61 (2009)
11. Pardalos, P.M., Rendl, F., Wolkowicz, H.: The quadratic assignment problem: A survey and recent developments. In: DIMACS Workshop on Quadratic Assignment Problems, vol. 16, pp. 1–42 (1994)
12. Pascual, J.A., Miguel-Alonso, J., Lozano, J.A.: Optimization-based application framework for parallel applications. Tech. rep., The University of the Basque Country (2010)
13. Resende, M.G.C.: Greedy randomized adaptive search procedures. Journal of Global Optimization 6, 109–133 (1995)
14. Ridruejo, F.J., Miguel-Alonso, J.: INSEE: An interconnection network simulation and evaluation environment. In: 11th Intl. Euro-Par conf. on Parallel Processing, pp. 1014–1023. Springer, Berlin (2005)

A Location-Based Transactional Download Service of Contextualized Multimedia Content for Mobile Clients

Pablo Fernandez, Asier Perallos, Nekane Sainz, and Roberto Carballedo

Abstract. This paper explores the new opportunities offered by the emerging technologies of the last generation of mobile phones. Thanks to features like GPS facilities installed in the mobile terminals new value added services can be developed to offer the user the more suitable multimedia content depending on parameters like user's location and preferences. We describe the development of a contextualized and personalized multimedia content delivery platform using transactional communications for mobile terminals. Furthermore an actual test route has been made to proof the successful working of the platform.

Keywords: pervasive information system, context-aware, location-based, download service, transactional, multimedia contents, mobile device.

1 Introduction

Since the release of the first portable audio player in 1979, the Sony Walkman, the portable entertainment has evolved and we are in a moment of history where the intelligent mobile phone devices are the preferred portable entertainment devices, their prices have decreased and mobile Internet connection rates have became more accessible [1]. New terminals are equipped with multimedia capabilities and are easier to handle. This mobile device evolution motivates the change from voice-only communications towards mobile multimedia contents and the offer of new value added services in industries like entertainment and tourism.

Furthermore, GPS devices are being integrated in most of the modern mobile terminals [2], and thanks to this a lot of new intelligent value-added services that currently are not widely exploited can be offered. One of these services is the contextualized multimedia content delivery to mobility environment, attending to the client geographical position and personal preferences, as it is described in [3].

Pablo Fernandez, Asier Perallos, Nekane Sainz, and Roberto Carballedo
DeustoTech - University of Deusto, Avenida de las Universidades 24,
48007 Bilbao, Spain
e-mail: {pablo.fernandez,perallos,nekane.sainz}@deusto.es
roberto.carballedo@deusto.es

A.P. de Leon F. de Carvalho et al. (Eds.): Distrib. Computing & Artif. Intell., AISC 79, pp. 205–212.
springerlink.com

The aim of the work described in this paper is just to explore such possibilities through the development of a pervasive platform for providing contents on demand using mobile terminals in a mobility environment (transportation scenario) where the context of the user is considered (location, desired destination, preferences, etc.) in order to know which contents have to be provided in a proactive way.

2 Functionality

To review the system features and functionality we should first identify two main actors in the platform: *mobile clients* who discover new content and services when they are in motion; and *publishers* which contribute to the system with geo-referenced and subject-based contents. The platform is based in subscription model where a mobile client is offered with new contents when he is located in a specific location and he can subscribe to that content so the information is automatically downloaded to the mobile client.

Thanks to the functionality of the system, a tourist is able to arrive in a city he does not know and download the information he is interested in and it is near his location. Related work in tourism scenarios is described in [4] and [5]. It does not matter if he is driving or walking, the system has two different profiles not to bother the user with no necessary notifications. The application can be used in different mobile terminals because is compatible with a large number of devices and the system is able to personalize the multimedia content to the specific features of the user`s mobile phone. So if the tourist unfortunately loses his phone, he will be able to get a new one and start using the application as before. Furthermore, if the user receives a phone call in the middle of the downloading process of a multimedia content, the platform can resume that download after the phone call, thanks to the transactional downloading feature.

The technical functionalities of the system which we consider most innovative are briefly explained below:

- *Multimedia File Re-encoding for Personalized Mobile Content.* The platform is going to offer multimedia files to the clients, and they will usually be of a major size. In order to save data transmission traffic and resources, the platform is able to serve a personalized multimedia file to the client invoking its petition. So the platform recognizes the mobile phone specific features such as display size and codec playing ability, and sends him a re-encoded multimedia file with the display resolution and a suitable codec. This is done thanks to a large mobile phone database where the devices have associated a unique User-Agent with their specifications. When a client invokes a petition, their User-Agent is embedded in the HTTP request so the platform can compare it in its database and send the appropriate content file. If the User-Agent of the mobile phone is not found, the platform is able to read raw data of the display size of the device instead of looking in the database.
- *Large Compatibility with Mobile Devices.* There has been developed a code library in several programming languages like Java ME, Java, Windows Mobile,

so the client can be compatible with a large amount of devices, portables or not. However the library can be easily ported to the new generation operative systems of smart phones like Iphone's OS or Google's Android. Furthermore thanks to the personalized mobile content, the platform ensures that every mobile phone is going to receive a suitable media file to its specifications, so every terminal is compatible with the system.

- *Geo-Referenced and Subject-based Contents.* Using a user-friendly interface the content publisher is able to upload new content in an easy way. With an interactive map, the location coordinates and the activation range are automatically assigned, and then the publisher is able to insert keywords or subjects to describe the content. In the client-side the user is able to configure his favorite subjects and keywords, as well as the range within it is going to receive new content offers.
- *Load Balancing System and Distributed File Storage Architecture.* The platform architecture will be explained in further sections of this document but it is important to notice that multimedia files are stored in several servers so when a client invokes a download petition the platform can choose the proper server to download from, attending to parameters like, bandwidth, memory or CPU usage of the servers. Thanks to this feature, the platform can maintain a load balanced environment to try to always offer the best quality service.
- *Transactional Downloading.*

3 Platform Architecture

In the architecture schema we can identify four key components which are shown in Fig.1. Their responsibilities are briefly explained below.

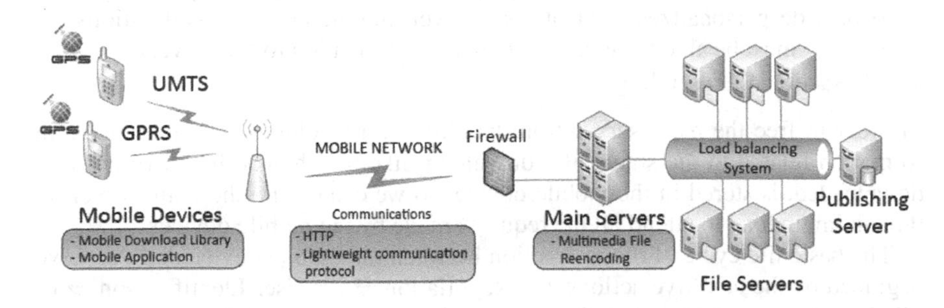

Fig. 1 Network architecture and key components

1. Main Server. It manages the user authentication, new content discovery services and business logic like enrooting the download request to a specific file server.
2. File Server. It contains all the content files of the system, it is a lightweight server that just has one task, replying requests from the Main Server.

3. Content Publishing Server. It runs in a web server where the publishing portal is hosted, when a new file is published, this is automatically replicated to all the file servers.
4. Mobile Devices. They are mobile terminals with GPS capabilities where the client application is running to retrieve the geo-referenced information.

The system is based on a centralized architecture. In this way the Main Server acts as a 'gateway' which receives all the requests made by the mobile phones. Thus, centralizing the connections grants the next features: (1) precise control of all the requests; (2) just one central point to ease the charging tasks; (3) security.

The mobile-server communications are established with a client-server topology, due to the fact that its transactional downloading and lightweight communications can be handled flawlessly by most of the mobile devices in the market.

4 System Design and Implementation

4.1 Communications Module

This module has the responsibility of implementing a lightweight communication protocol which has been designed in order to manage the interaction between the mobile device and the server in a secure and fast way. This lightweight protocol has several simple messages and can handle the next functionality:
With the porpoise to obtain a secure and fast communication, we have designed a lightweight protocol of simple messages that can handle the next functionality:

- Discover the available content for a specific user
- Send and receive GPS and location data
- Identify the mobile device and its specifications (because of the platform is able to provide personalized content for a given mobile device specifications, the protocol must be able to send the necessary information to the server).
- Transactional file sending

In order to free the main server from handling large volumes of data, all the information related to the status of a download, like the chunks that have not been downloaded, is stored in the mobile device, so we can ensure the main server has the minimum load to attend all the requests made by the mobile devices.

The basic life cycle of the interaction between the mobile device and the server is guided by the next five actions: (1) Negotiation → (2) User Identification → (3) File List Request → (4) File Information Request → (5) Chunk Request.

In addition to the previous cycle, there is a timer that is responsible of sending the location coordinates to the server using the GPS device installed in the mobile device. The time between each petition is configurable so the user can save battery life.

As we previously explained, one of the main features of the platform is the ability to offer personalized content according to the mobile device model and specifications. In order to make this identification, the mobile device sends its User-Agent. This is a unique string that identifies the mobile phone. But there may be situations where the mobile phone cannot be identified. In these cases the platform has an alternative mechanism to offer custom multimedia content: the mobile phone is able to send its screen resolution, so the platform can offer him the more suitable version of the file.

The transport protocol used for the communications is http because it is a standard, it has secure connections tools, and it can work through firewalls and proxies. Above the http protocol, the application layer protocol has been developed using custom Servlets and EJBs for attending the requests from the mobile client. As we have mentioned before, we have several messages that are part of our lightweight communication protocol. For attending these we have developed one Servlet per message that has to be sent. Working with Servlets is an easy task that involves using standard http connections that have implementations in all programming languages.

Web Services it is another good choice because they have many advantages like the standardization of the communications and the easy development process, but it has a major fault; it works with XML files, and mobile devices do not behave well with them because they take a lot of process time in the mobile terminal.

4.2 Mobile Device Module

4.2.1 Mobile Download Library for Mobile Devices

The library has been coded in three different programming languages, J2ME, J2SE and Windows. The functionality of this library is summarized in these four tasks:

1. Server Communication. As we mentioned in the previous section, in this library is where the lightweight communication protocol is implemented. Thanks to this, the client application can 'talk' with the server, authenticate, request new files and download chunks.
2. Download Management. This functionality is responsible of all the transactional downloading tasks. Including error handling and stopped downloads resuming.
3. GPS and Location Data Manager. The mobile module must be able to send periodically the GPS data. In this information are included the longitude and latitude coordinates as well as the direction towards the mobile device is moving. This periodical dispatch can be easily configured to save battery life. In this case the send rate will be lower and the accuracy of the position will be less precise.
4. File Storage. This module must first be able to store the chunks downloaded in a temporary folder, as well as the file information, and secondly it has to rebuild the file and store it in its final destination when all the chunks have been downloaded.

4.2.2 Mobile Application

Using the mobile download library, we have developed a fully functional J2ME and Windows Mobile client applications, which have been tested by real users. As we told in the previously the client has two preconfigured working profiles: Drive and Walk.

- In Drive mode, the application is continuously sending the GPS data to the server to ensure the high accuracy of the position. Due to this fact, is recommended to have the mobile device plugged in to the power line because this is a battery consuming profile. In this mode the user will not receive any notification; offers that are near his location and matches with his preferences will be automatically subscribed so the content is downloaded immediately.
- In Walk mode, the GPS data sending rate is decreased in order to save battery life. The position accuracy is lower but this is not a vital factor due to when a user is walking, his speed is not fast enough to lose any offer. When the user receives a notification, he can see a detailed view of the offer, including its description, keywords and position in a map. If he is interested in the offer he just has to subscribe to start downloading the multimedia content.

Besides this two working modes, with this application the final user can do all the tasks described in the section 2 of this paper. It means: (1) automatic transactional download of contents; (2) network traffic watch and download status; (3) downloaded contents management; (4) play downloaded contents: mp3, videos, and pictures; (5) user profile configuration: credentials, keyword interests, range offer activation.

4.3 Server Module

For all the three types of servers the technology chosen has been JBoss AS. This application server has already a web server included to support the EJBs and the Servlets needed to communicate with the mobile library. Moreover it is able to work in cluster mode in a very easy and secure way. This means one single application can be deployed and shared by several servers to always ensure the optimal performance of the application.

Main Server implements all the business logic to receive the petitions from the mobile devices; this logic is supported by EJBs and Servlets. File Servers have the only task to listen requests from the main server. These requests are made over JBoss Remoting, which is a substrate of JBoss AS and it is a lightweight communications framework that is able to use http connections. And finally Content Publishing Server is a web portal hosted in the JBoss Web Server responsible for the publishing of new content from a private or public provider. In this Web Portal the publisher can manage his offers and topics and upload multimedia content. To assign GPS coordinates and activation range of the offer we have designed an interactive application over Google Maps to manage the content in a very intuitive way. We just has to point in a map and he will receive a visual feedback of the location and activation range of his offer.

5 Testing and Results

For testing purposes we populated the platform with four different offers located in the city of Bilbao (Spain). Their GPS coordinates and information are presented in the Table 1. The test consisted in a route through the city with the mobile client application running in a *"HTC HD2"* mobile passing over all the activation ranges.

Table 1 Offer locations and information

Offer Name	Longitude	Latitude	Activation Range
Museo Guggenheim	-2.934250831604	43.2686437747697	400metres
Museo Arqueologico,	-2.92824268341064	43.2628937385262	450metres
Polideportivo Deusto	-2.94784426689148	43.2684719065613	500metres
Gran Hotel Domine	-2.93384313583374	43.2677375551133	200metres

We started our test route and put the application in Walk mode and we started walking. When we reached the first activation range, a notification appeared telling us there was an available offer in our position that matched our preferences. We subscribed to that offer and continue walking. The content of the first offer named *"Polideportivo de Deusto"* was automatically downloaded including three pictures in jpeg format and a pdf file with the prices of the establishment.

We continued our test route and put the application in Drive mode so it will not notify us of anything. We passed through the rest of the three activation ranges of the offers and the application automatically subscribed and started downloading the contents. In the middle of our test we stopped the application to test the transactional downloading process. When we restarted the mobile application, the download process resumed in the point it was before the restart. We got to the end of our test route and checked the application; it correctly had downloaded all the multimedia content of the offers, including two video files, one audio file, and several pdf files with information about the offers. This test was executed using the 3G mobile network for the download traffic and it successfully worked with an average download speed of 1832 kbps.

After finishing the experiment we can conclude we obtain successful and positive results in all the areas of the platform: the GPS data and location sending from the mobile client; new offer and content discovery in Walk mode; automatic subscribing and downloading in Drive mode, transactional downloading and personalized multimedia content based on mobile device specifications.

6 Conclusion and Future Work

In order to study some new possibilities that the mobile multimedia technologies are able to offer, we have successfully developed a platform capable of combine these ones with the advances in location solutions. The result of the work is a successful development of an innovative platform for the proactive provision (using on mobile devices) of multimedia contents which are contextualized to the user profile (location, desired destination, preferences, etc.). Therefore, it includes very advanced characteristics such as multimedia file re-encoding for personalize mobile contents, large compatibility of client application with mobile devices, distributed storage of contents and a load balancing system.

Future work is focused on the improvement of the content contextualization by using semantic webs and ontologies as described in [6], which can interpret the user data and preferences and discover which would be the best contents for his interests. Other goal is to allow the user to participate in the platform, allowing him to publish new contextualized content with his mobile phone, and integrating the application and content with the most important social networks.

References

1. Harroud, H., Karmouch, A.: A Policy Based Context-aware Agent Framework to Support Users Mobility. In: Proc of the IEEE Advanced Industrial Conference on Telecommunications / Service Assurance with Partial and Intermittent Resources Conference / ELearning on Tele-communications Workshop, Lisbon, Portugal, July 17-20 (2005)
2. Yu, Y.-H., Kim, J., Shin, K., Jo, G.-S.: Recommendation system using location-based ontolo-gy on wireless internet: An example of collective intelligence by using 'mashup' applications. Expert Systems with Applications (November 2009)
3. Vijayalakshmi, M., Kannan, A.: Proactive location-based context aware services using agents. International Journal of Mobile Communications (2009)
4. Martínez, D., Ruíz, N., Vera, P., Alcántara L.: Multimedia platform for mobile tourist guidance and services in the cities of Úbeda and Baeza (Spain). In: International Conference on Wireless Informatio Networks and Systems. Winsys 2009, Italy, Milan, July 7-10 (2009)
5. Kakaletris, G., Varoutas, D., Katsianis, D., Sphicopoulos, T., Kouvas, G.: Designing and Implementing an Open Infrastructure for Location-Based, Tourism-Related Content Delivery. Wireless Personal Communications (30), 153–165 (2004)
6. Bellavista, P., Corradi, A., Montanari, R., Toninelli, A.: Context-aware semantic discovery for next generation mobile systems. IEEE Communications Magazine (September 2006)

An Identification Method of Inquiry E-mails to the Matching FAQ for Automatic Question Answering

Kota Itakura, Masahiro Kenmotsu, Hironori Oka, and Masanori Akiyoshi

Abstract. This paper discusses how to match the inquiry e-mails to pre-defined FAQs(Frequently Asked Questions). Web-based interaction such as order and registration form on a Web page is usually provided with its FAQ page for helping a user, however, most users submit their inquiry e-mails without checking such a page. This causes a help desk operator to process lots of e-mails even if some contents correspond to FAQs. Automatic matching of inquiry e-mails to pre-described FAQs is proposed based on SVM(Support Vector Machine) and specific Jaccard coefficient. Some experimental results show its effectiveness. We also discuss future work to improve our method.

Keywords: FAQ, Support Vector Machine, Jaccard coefficient, Automatic Question Answering.

1 Introduction

Recently Web-based services, for instance, shopping and community management, are rapidly increasing, which usually provides basic interaction between a user and a service company by form-based input on Web pages. Such inquiry e-mails from a user are delivered to service company's operators and they send back e-mails by hand. Along with lots of inquiries, FAQ(Frequently Asked Question) page is introduced to reduce operators tasks. Many researches for helping with FAQ creating has been conducted.However, most users rush to submit their inquiry e-mails without a glance of FAQ page.

Kota Itakura, Masahiro Kenmotsu, and Masanori Akiyoshi
Osaka University, 2-1 Yamadaoka Suita Osaka, Japan
e-mail: itakura.kouta@ist.osaka-u.ac.jp,
akiyoshi@ist.osaka-u.ac.jp

Hironori Oka
Codetoys, 2-6-8 Nishitenma Kita Osaka, Japan

A.P. de Leon F. de Carvalho et al. (Eds.): Distrib. Computing & Artif. Intell., AISC 79, pp. 213–219.
springerlink.com

This situation motivates us to develop a new functionality that a service-side server can indicate a corresponding FAQ to a user's inquiry. Customers's inquiry e-mails tend to be contained within FAQs, which takes an amount of unnecessary time and effort of a help desk operator. In order to reduce the load of the operator, the automatic question answering system is required. The system outputs answers corresponding to questions by inputting inquiry e-mails as search query. There are a lot of researches of the automatic question answering system. [1, 2, 3, 4] Many of these research uses Jaccard coefficient [5] of an inquiry e-mails and FAQs for searching a corresponding FAQ. However, these researches have accuracy problems by reasons of that inquiry e-mails sentence is short, contains informal expression and spelled in several different ways. Moreover, FAQs resemble each other, which make it difficult to identify corresponding FAQ. On the other hand, another research uses the learning classifier method for determining whether or not inquiry e-mails correspond to each FAQ. The learning classifier judges inquiry e-mails based on the similarity between an inquiry and each past inquiry. However, because similarity determination between short sentences generally shows low accuracy, comparing one by one is inappropriate. The automatic question answering system needs another method to replace one to one comparing with an inquiry e-mail and a FAQ.

Therefore, we propose the automatic question answering system, which classifies an inquiry e-mail by comparing the similarity to a set of target FAQs.

2 An Identification Method of Inquiry E-mails to the Matching FAQ with SVM

2.1 Approach for Automatic Question Answering System

As for the similarity judgment, Jaccard coefficient, especially, can measure the similarity between inquiry e-mails and FAQs. Therefore, we conducted a preliminary experiment.

Fig.1 and Fig.2 shows Jaccard coefficient of some inquiry sentences and FAQs. The inquiries shown in solid lines in the graphs are the case of taking maximum Jaccard coefficient between inquiry sentence and correct FAQs. On the other hand, the inquiries shown in dashed lines are the case of NOT taking maximum Jaccard coefficient between them. These graphs show distinct patterns in shape, and expected to use these patterns in judging inquiries to answer automatically. However, the case of taking maximum Jaccard coefficient is not always useful in classifying, even if graphical form are similar. Despite another classifying method replaceing maximum value is required, it is difficult to determine judgment rules in advance.

Therefore, We focused on SVM (Support Vector Machine) [6, 7] , one of the learning classifier, for judging what inquiries can be answered automatically. By using SVM, we eliminate the need for determining judgment rules in advance. Suppose that operators replied as to questions, this enables learning classifiers to learn patterns and makes classifying possible. On the other hand, in the task of

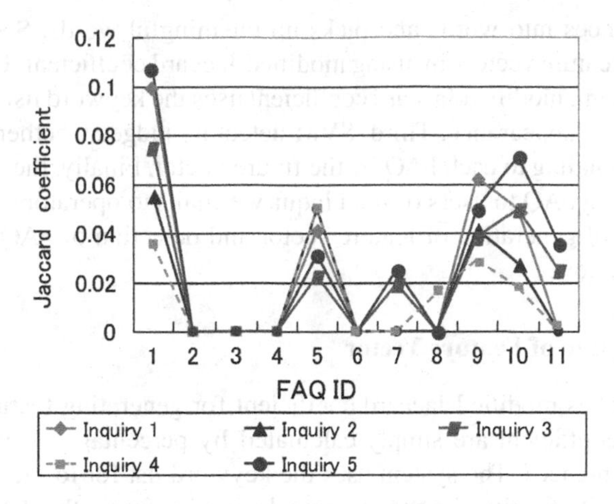

Fig. 1 Jaccard coefficients of the inquiry matching FAQ 1 to each FAQs

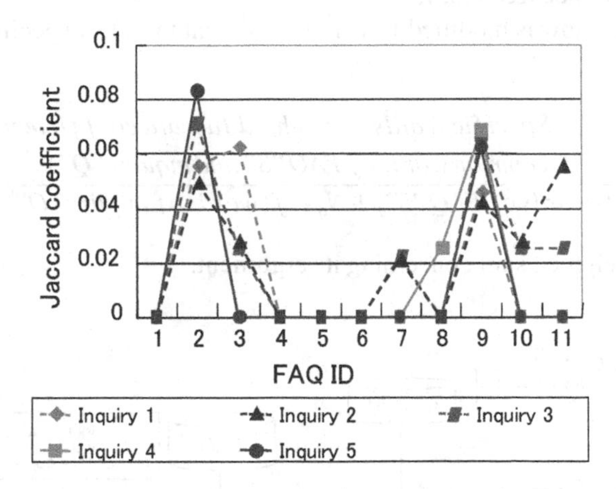

Fig. 2 Jaccard coefficients of the inquiry matching FAQ 2 to each FAQs

automatically answering, the system must have 100% precision rate. We consider this problem in our evaluation experiment.

2.2 An Identification Method of Inquiry E-mails with SVM

2.2.1 Outline of an Identification Method

Fig.3 shows an overview of our proposed system. When a user sends new inquiry e-mail, the system handle the inquiry as below.

First, the system separates inquiry sentences and FAQs into words. Japanese sentences do not have separator between words, so the morphological analyzer separates sentences into words and picks up meaningful words. Second, the system calculates feature vectors by using modified Jaccard coefficient. Unlike existing Jaccard coefficient, modified Jaccard coefficient uses the keyword list and weighting with frequency of appearance. Third, SVM detectors judges whether or not the inquiry is corresponding to each FAQ by the future vector. Finally, the system answer the corresponding FAQ to users or send inquiry e-mails to operators.

The method of generation of feature vector and detection of FAQ are discussed in more detail below.

2.2.2 Generation of Feature Vector

The system applies modified Jaccard coefficient for generating feature vector. Existing Jaccard coefficient are simply calculated by percentage of common words between two sentences. The system uses the keyword list for to the feature vector, because each FAQ has critical words making distinction from other FAQs. Such keywords span more than one FAQ, the system assigns significance of each keyword in inverse of appearance frequency.

The formula below is modified Jaccard Coefficient to reflect specificness in FAQ words.

$$Specific\,Words - weighted\,Jaccard\,coefficient$$
$$= \frac{f(common\,words\,of\,FAQ\,"S"\,and\,inquiry\,"Q")}{f(words\,of\,FAQ\,"S") + N_q - f(words\,of\,inquiry\,"Q")}$$

$f(\cdot)$ means a weighted sum concerning its argument.

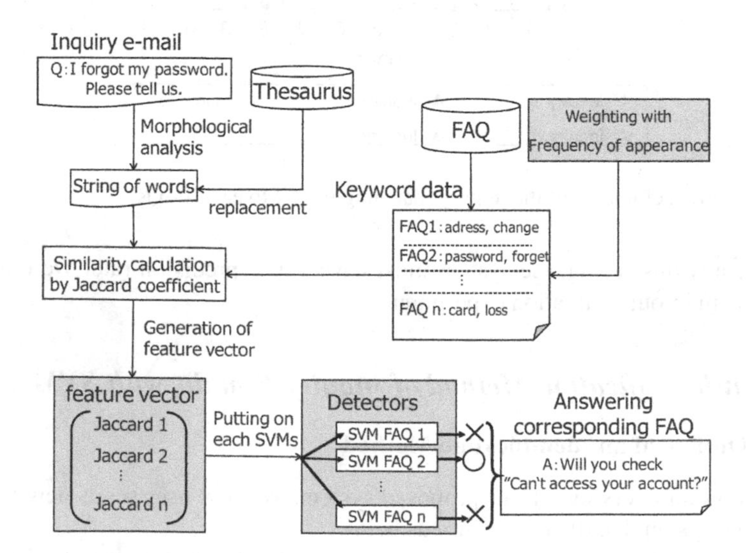

Fig. 3 Overview of identification methods

Each vector element is modified Jaccard coefficient of each FAQ. Thus, the number of dimensions of feature vectors are same as that of FAQs.

2.2.3 Detection of FAQ

The system conducts multi-class classification by using multiple SVMs. Each SVM learns supervised data and judges whether or not a inquiry corresponds to the FAQ sentences on one by one. If all SVMs judge that the inquiry e-mail does not correspond to FAQs, the system sends that inquiry e-mail to operators. The system requires 100 percent precision rate, so that the system fixes the margin parameters of SVMs to be able to judge in 100 percent precision rate by using supervised data.

3 Evaluation Experiment

We compared existing methods and our proposed method in precision and recall rates, by classifying inquiry e-mails sent to the member management system of the sports association. The number of e-mails are 1845, 545 of them corresponds to FAQs. As existing methods, we used word vector-based and maximum Jaccard coefficient judgment. In our method, feature vector is calculated by 4 FAQs, which has many past corresponding inquiry e-mails as supervised data. Table.1 shows the number of training data and classification target data. We extracted 22 keywords manually in these 4 FAQs and made synonyms data for 10 of these keywords. Fig.4 shows the experimental result of the precision and recall rates.

Table 1 The number of supervised data and classification target data

FAQ	The number of training data	The number of target data
Q3-3	100	150
Q7-2	23	30
Q9-8	20	21
Q4-7	18	20

Although proposal classification of Q4-7 is worse than existing method in precision rate, our method basically improves the precision and recall rate. Because Q4-7 has few data, 18 questions and answers, which may be reason of low accuracy. Our system aimes at the automatic question answering, it is also required to make precision rate 100 percent. However, 100 percent of recall rate of target classification data is not guaranteed as far as the system adjusts the threshold by using training data. Thus, we need another method which resetting threshold value with severity when the target classification data is severer than training data, shown in Fig.5. If we can adjust the threshold so that the target classification data is judged in 100 percent precision rate, the system classifies in 75 percent recall data, shown in Fig.6. Therefore, the future work is to set the threshold automatically.

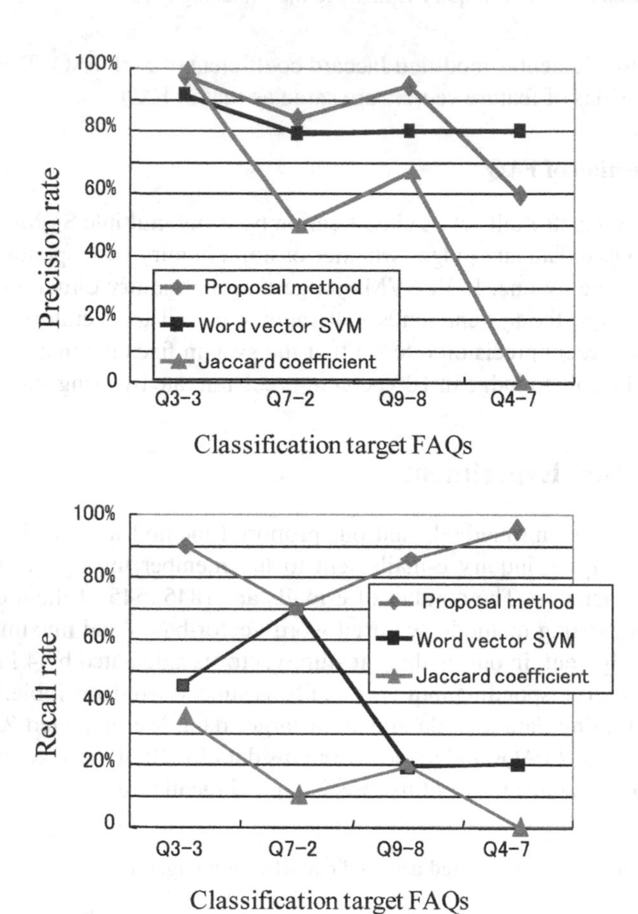

Fig. 4 Result of experiment

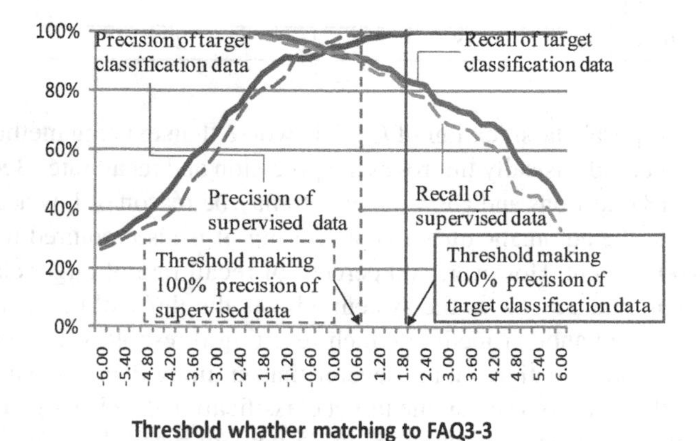

Fig. 5 Precision and Recall rate in changing threshold value

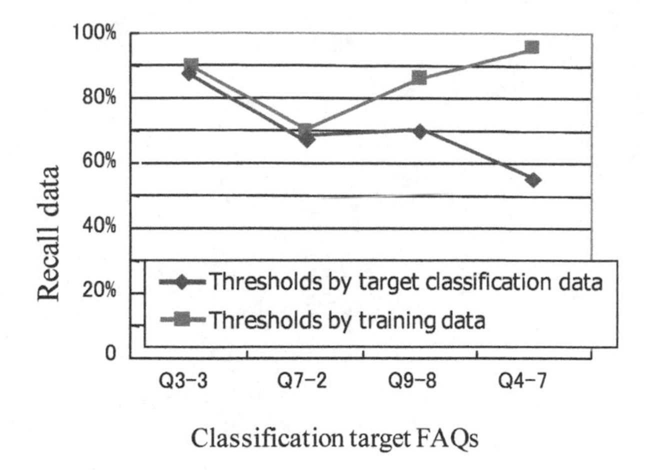

Fig. 6 Recall rate of proposal method under 100% precision rate by using classification target data

4 Conclusion

This paper addresses the automatic question answering system, which provides the function that lots of inquiry e-mails correspond to existing FAQs. Proposed system identifies inquiry e-mails to the matching FAQ by using SVMs. As feature vector of SVMs, we use Jaccard coefficient between inquiry e-mails and FAQ sentences, which is modified word weight by significance of each keyword.The experimental result shows its effectiveness. If we adjust the threshold by using target classifying data.

References

1. Sneiders, E.: Automated FAQ Answering with Question-specific Knowledge Representation for Web Self-service. In: 2nd Conference on Human System Interactions (HSI 2009), pp. 298–305 (2009)
2. Harksoo, K., Jungyun, S.: Cluster-Based FAQ Retrieval Using Latent Term Weights. IEEE Intelligent Systems 23(2), 58–65 (2008)
3. Sheng-Yuan, Y.: Developing an Ontological FAQ System with FAQ Processing and Ranking Techniques for Ubiquitous Services. In: First IEEE International Conference on Ubi-Media Computing, pp. 541–546 (2008)
4. Jibin, F., Jinzhong, X., Kekiang, J.: Domain Ontology Based Automatic Question Answering, vol. 2, pp. 346–349 (2009)
5. Jain, A.K., Dubes, R.C.: Algorithms for Clustering Data. Prentice Hall, Englewood Cliffs (1988)
6. Cortes, C., Vapnik, V.: Support-Vector Networks. Machine Learning 20(3), 273–297 (1995)
7. Joachims, T.: TextCategorizationwith SupportVector Machines: Learning with Many Relevant Features. In: Proceedings of the European Conference on Machine Learning. Springer, Heidelberg (1998)

Cryptanalysis of Hash Functions Using Advanced Multiprocessing

J. Gómez, F.G. Montoya, R. Benedicto, A. Jimenez, C. Gil, and A. Alcayde

Abstract. Every time it is more often to audit the communications in companies to verify their right operation and to check that there is no illegal activity. The main problem is that the tools of audit are inefficient when communications are encrypted.

There are hacking and cryptanalysis techniques that allow intercepting and auditing encrypted communications with a computational cost so high that it is not a viable application in real time.

Moreover, the recent use of Graphics Processing Unit (GPU) in high-performance servers is changing this trend.

This article presents obtained results from implementations of brute force attacks and rainbow table generation, sequentially, using threads, MPI and CUDA. As a result of this work, we designed a tool (myEchelon) that allows auditing encrypted communications based on the use of hash functions.

Keywords: CUDA, MPI, hash, audit tools, rainbow tables, brute force.

1 Introduction

From its origins computing has revolutionized the way in which companies communicate, coming to play an important role in their success. The constant emergence of new technologies and the ability to interconnect through networks gives significant improvements in productivity and business market. This has great benefits but also new challenges. The most important challenge is the computer security [1].

Given the importance of computer systems, laws and standards have been provided to regulate the use and transmission of information. For example, in Spain we can find the Data Protection Act [2], while at international level we can find the ISO 1799/BS [3] and ISO / IEC 27001 [4]. All these legal advances, together

J. Gómez, F.G. Montoya, R. Benedicto, A. Jimenez, C. Gil, and A. Alcayde
University of Almeria, Spain
e-mail: {jgomez,pagilm,rbenedicto,ajimenez,cgilm}@ual.es, aalcayde@ual.es

A.P. de Leon F. de Carvalho et al. (Eds.): Distrib. Computing & Artif. Intell., AISC 79, pp. 221–228.
springerlink.com

with the increased complexity of computer systems have caused an increment of computer security audits.

An audit of computer security [5] allows to check the security level of a computer system using all kinds of tools and techniques for finding the problems and weaknesses. To make a security audit is necessary to use a large set of tools to easily audit unencrypted communications. Therefore, the main challenge and objective of this work is to create an auditing tool that allows encrypted communications audit based on the use of hash functions.

For that, in section two, we analyze the attacks on hash functions. Section three explains how to use the hash functions attacks on MyEchelon to audit encrypted communications. In sections four and five we analized the results of implementing the brute force attacks and generation of rainbows table using different technologies multiprocessing to provide greater power to MyEchelon.

2 Criptoanalysis of Hash Functions

One of the great allies of computer security is cryptography. A clear example is the use of hash functions, which we can find in communications, to check the passwords, to verify the integrity of messages, etc.

A hash function allows to calculate the trace (the summary) that uniquely identifies a particular set of data. Table 1 shows the characteristics of the most important hash functions: MD5 [6], SHA-1 [7] and NTLM/MD4 [8].

The scientific community began to question about the security provided by the hash functions when Xiaoyun Wang [9] published the first results on the breaking of the MD5 hash function and Antoine Joux [10] demonstrated vulnerabilities in the SHA-0 hash function.

The classical way to break the hash functions is to use brute force attacks [11] that consists on generating all possibles solutions until you find the right one. This implies a high computational cost for large and complex passwords. For that, it becomes necessary to find an efficient solution to this problem. This solution was found in the use of tables Rainbow [12] that is an elegant solution from the known hash tables [13] created by Martin Hellman, avoiding as far as possible the large number of collisions inherited from his predecessor. The disadvantage of both techniques is that they have a very high computational cost, making unfeasible its use in conventional equipments.

Table 1 Comparative Hash functions

	MD5	SHA1	NTLM / MD4
Summary size	128 bits	160 bits	128 bits
Block size	512 bits	512 bits	512 bits
Number of steps	64	80	48
Message size	∞	∞	∞
Strength preimage	2^{128}	2^{160}	2^{128}

3 myEchelon

MyEchelon is created with the aim to automate the audit process of a system using a large set of tools. When myEchelon runs, it analyzes and takes control of the network where you are. Once the network scans, it carries out an attack Man In The Middle to the networks devices to audit communications (see Figure 1). Logically, unencrypted communications can be audited but the problem is the encrypted communications.

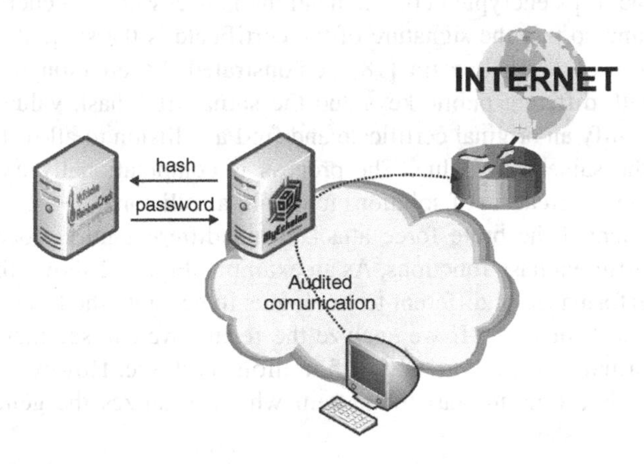

Fig. 1 Architecture of a network auditing myEchelon

To allow auditing the encrypted communications, myEchelon uses a high performance server which will be responsible for making the crypto-analysis (see Figure 1).

As we will see, the brute force attacks can audit https secure communications and the use of rainbow tables allows obtaining the passwords that are transmitted over the network.

To improve the performance of cryptanalysis it is going to be compared brute force attacks and generating rainbow tables sequentially, using threads [14], MPI [15] and using Graphics Processors Unit (GPU) using CUDA [16].

All results have been obtained on the server MX DUAL AZServer Xenon and NVIDIA TESLA 1070. The server has two processors Dual / Quad Core Intel Xenon 2.66 GHz, 8 2Gb SDRAM and two SATA hard drives of 1TB in RAID 1. Moreover, the TESLA team has four cards tesla 1070 T10, which makes a total of 960 cores of 1.44 GHz each.

4 Brute Force Using Advanced Multiprocessing

Basically, this type of cryptanalysis is based on the birthday attack [17]. This attack is that given a hash m, it has to find a text M´ whose value Hash (M´) is equal to the original hash (m).

For example, in the case of SHA1 (160 bits), this attack requires to generate 280 tests to obtain the solution.

One application of this cryptanalysis is the forgery of certificates https secure communications. An https security certificate is composed by a series of data that identifies the server (e.g. name, domain) that are signed by a certification authority. To ensure the integrity of the digital certificate is used a digital signature algorithm, which in the case of https certificate is the MD5 hash function.

To audit the https encrypted communications is necessary to generate a certificate in real time so that the signature of the certificate is the same than the original. In this sense, Arjen Lenstra [18] demonstrated the creation of two X.509 certificates with different public keys and the same MD5 hash value. Thus, it is possible to modify an original certificate and find a collision to allow the new certificate had the same hash value. The process is computationally expensive because it must test a great set of solutions to obtain a collision.

We implemented the brute force attack using different multiprocessing technologies for different hash functions. As an example, Figure 2 shows the comparative on the performance of different technologies to generate the brute force attack for the MD5 hash function. If we analyze the results we can see that CUDA has the best performance with more than 135 millions hash/sec. However, as happens to MPI, it needs a time to start the system which penalizes the generation of a small number of hashes.

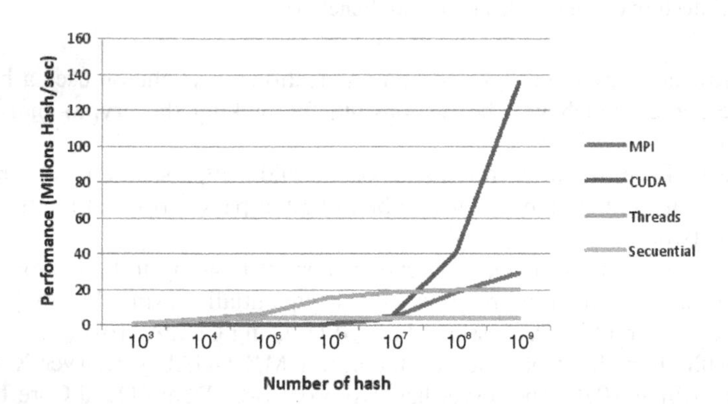

Fig. 2 Generation MD5 Hash

If we compare the results between the perfomance of different implementations with respect to the sequential implementation (see Figure 3 and Table 2), we see that the use of threads and MPI represents a better performance of 780% and CUDA provides the best results with a performance of 4410% over its sequential implementation.

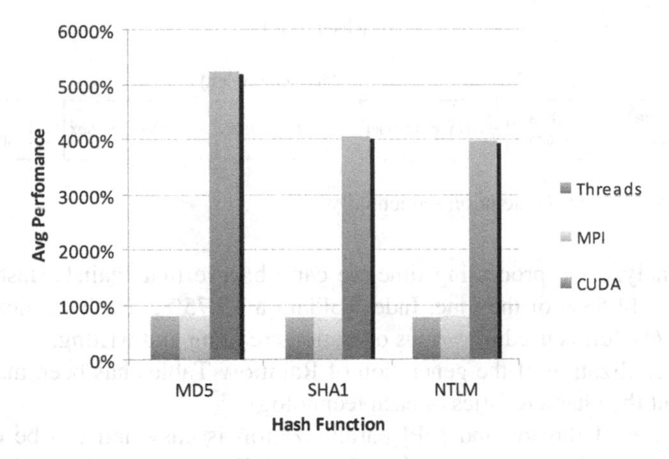

Fig. 3 Comparison of performance on the brute force attack

Table 2 Perfomance on the brute force attack

	MD5	SHA1	NTLM
Threads	806%	769%	764%
MPI	792%	766%	781%
CUDA	5222%	4046%	3962%

5 Rainbow Tables Using Advanced Multiprocessing

The use of Rainbow Tables is the fastest way to make brute force attacks. For the use of Rainbow Tables it has been used Rainbow-Crack project [19]. The process of breaking a hash consists of three phases: generation of tables (rigen), to put order in the tables (rtshort) and use of ordered tables to obtain the value of a given hash (rtcrack). The first two steps should be performed once, while the third step is repeated for each of the hash that we want to break.

The main problem is that the required size and the time for generating tables are related exponentially to the size of the password that you want to analyze. For example, to generate a rainbow table that enables to break an alphanumeric password of seven character is necessary to generate a table of 2Gb for what it would be necessary about three days. To decrypt passwords of nine characters we need 500 tables of 2Gb and it would take about three years.

Basically, a rainbow table is composed of a set of vector data. The generation of each vector can be done completely independently and for this, as it is shown in Figure 4, it is necessary to use mainly three functions: 1) IndexToPlain is responsible for converting a numerical value to a string belonging to the set of possible values 2) PlainToHash is the most important function since it is responsible for calculating the hash of the string above, 3) HashTolndex converts the hash value to a numeric value.

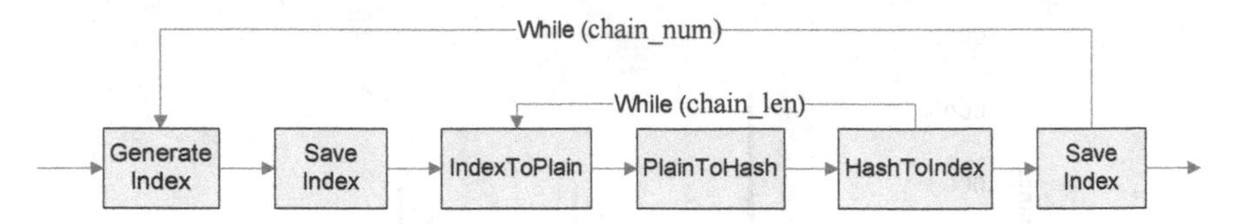

Fig. 4 Rainbow Tables Generation sequentially

If we analyze the processing time we can observe that PlainToHash function represents a 44.58% of the time, IndexToPlain a 15.75%, HashToIndex a 8.27% and the 31.6% left is used in various operations reading and writing.

The parallelization of the generation of Rainbow Tables has been made taking into account the characteristics of each technology.

In the case of threads and MPI parallelization is easy and can be calculated separately each of vector in the table. But in CUDA case parallelization is more complicated because CUDA does not allow operations W/R and therefore it generates a CPU thread for each vector. To complete the calculations on the GPU functions have been implemented PlainToHash, IndexToPlain and HashToIndex in CUDA. Thus, the CPU thread is responsible for carrying out the operations W/R and GPUs performs the calculations.

Figure 5 shows the results obtained by different technologies of multiprocessing. In this case, the use of threads and MPI represents a perfomance of 783% close to the ideal one while CUDA is much lower than the sequential implementation.

The CUDA implementation for the generation of tables Rainbows presents the worst results. It is produced as it makes a continuous movement of data between the CPU and the GPU system.

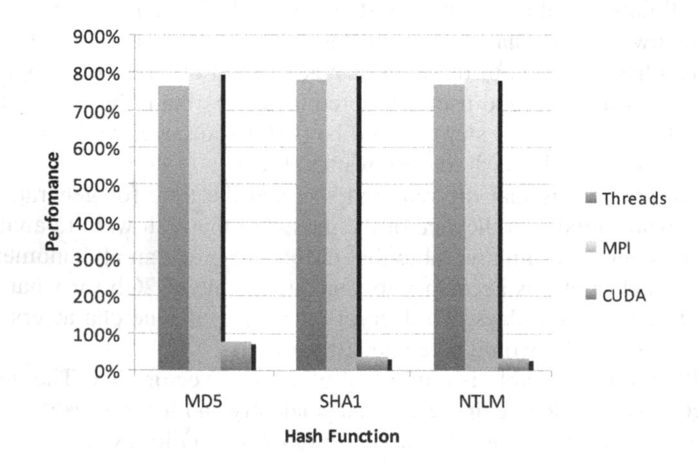

Fig. 5 Performance Comparison in Rainbow Tables generation

Table 3 Comparasion in Rainbow Tables generation

	MD5	SHA1	NTLM
Threads	765%	780%	770%
MPI	800%	797%	787%
CUDA	78%	36%	33%

6 Conclusions

This article presents the results of the implementation of brute force attacks and Rainbow Tables generation, so sequential, using threads, MPI, CUDA. Finally, we can say that CUDA has the best technology in brute force attacks while MPI presents the best results in the generation of tables Rainbow.

These results have been applied to myEchelon allowing intercepting and auditing encrypted communications using in each case the best multiprocessing technology.

Acknowledgements. This work has been financed by the Excellence Project of Junta de Andalucia (P07-TIC02988), in part financed by the European Regional Development Fund (ERDF).

References

[1] López, J.G., Navarro, R.B.: Seguridad en Sistemas Operativos Windows y Linux. Ra-Ma (2006)
[2] Ley orgánica 15/1999, de 13 de diciembre, de Protección de datos de Carácter Personal (1999)
[3] Kenning, M.: Security Management Standard — ISO 17799/BS 7799. Springer Netherlands (2001)
[4] Estándar Internacional ISO/IEC 27001 (2005)
[5] Mookhey, K.K., Burghate, N.: Linux - Security, Audit and Control Features. In: ISACA (2005)
[6] Rivest, R.: The MD5 Message-Digest Algorithm. Network Working Group (April 1992), http://www.ietf.org/rfc/rfc1321.txt
[7] Eastlake, D., Jone, P.: US Secure Hash Algorithm 1 (SHA1). Network Working Group (September 2001), http://www.ietf.org/rfc/rfc3174.txt
[8] Rivest, R.: The MD4 Message-Digest Algorithm. Network Working Group (April 1992) http://tools.ietf.org/html/rfc1320
[9] Wang, X., Yu, H.: How to Break MD5 and Other Hash Functions. Shandong University, Jinan 250100, China (2004)
[10] Joux, A., Chaveaud, F.: Differential Collisions in SHA-0. Centre d'Électronique de l'Armament. France (2004)
[11] Halevi, S., Krawczyk, H.: Strengthening Digital Signatures via Randomized Hashing. In: CFRG 2005 (May 2005)
[12] Oechslin, P.: Making a Faster Cryptanalytic Time-Memory Trade-Off. LASEC (2003)

[13] Hellman, M.E.: A Cryptanalytic Time-Memory Trade-Off. IEEE Tansactions on Information Theory (1980)
[14] Vargas, E.C.: Aplicaciones Multi Hebras. Arequipa – Perú (Octubre 2001)
[15] Alonso, J.M.: Programación de aplicaciones paralelas con MPI. Enero (1997)
[16] NVIDIA CUDA Compute Unified Device Architecture Programming Guide v2.0. (Septiembre 2008)
[17] Paul, C., et al.: Parallel Collision Search with Application to Hash Functions and Discrete Logarithms. In: Conference on Computer and Communications Security. ACM, New York (2004)
[18] Lenstra, A., Wang, X., de Weger, B.: Colliding X.509 Certificates, Cryptology ePrint Archive Report 2005/067 (March 1, 2005) (revised May 6, 2005), (retrieved July 27, 2008)
[19] `http://project-rainbowcrack.com/`, (accessed April 26, 2010)

Development of Transcoder in Conjunction with Video Watermarking to Deter Information Leak

Takaaki Yamada, Katsuhiko Takashima, and Hideki Yoshioka

Abstract. A transcoder incorporating video watermarking method has been developed. This paper reports the implementation and evaluation of the system. Visibility testing showed that degradation in the quality of the watermarked images was almost imperceptible. Robustness testing showed that the embedded watermarks were robust against re-encoding with scaling. Process-time testing showed that the total processing time is increasing by 17%. Use of this system should help in deterring information leak in viewed and distributed video content.

Keywords: transcoder, video watermark, information leak.

1 Introduction

The growing availability of video content online is exacerbating the problem of copyright violation. The copyright of video content can easily be violated because the Internet facilitates copying and redistribution of content. Moreover, video-sharing services, which provide a huge amount of the video content worldwide, enable illegal copies to be distributed anonymously [1]. An approach to solving this problem is to embed the copyright information into the content in the form of a digital watermark [2], but this requires creating video watermarks that are robust against image processing [3,4].

To deal with these problems, a previously developed method for embedding video watermarks creates watermarks robust against image processing and is suitable for content distribution using software on commodity personal computers (PCs) [5]. This method enables real-time processing including watermark

---

Takaaki Yamada and Katsuhiko Takashima
Hitachi., Ltd., {Systems Development Lab., Network Solution Div.}, Kanagawa, Japan

Hideki Yoshioka
Japan Broadcasting Corporation, Engineering Administration Dept., Tokyo, Japan

A.P. de Leon F. de Carvalho et al. (Eds.): Distrib. Computing & Artif. Intell., AISC 79, pp. 229–236.
springerlink.com

embedding, MPEG encoding, and hard disk drive recording. However, it can generate only MPEG-4 encoded watermarked files because the watermark embedding process is incorporated into the MPEG-4 encoding process. Moreover, it supports only the QVGA (320×240-pixel) format, which is converted from the VGA (640×480-pixel) format of the incoming video signal. Subsequent improvement enabled it to directly handle the SDTV format (equivalent to VGA) of the incoming video signal. It is separate from the encoding process and thus can be incorporated into various encoding and distribution systems. Moreover, the improved method can still run on commodity PCs. However, its input is restricted to SDTV format due to the limited ability of PCs to capture data. To enable our method to handle input formats larger than SDTV, we have now developed a transcoder for converting the HDTV format of the incoming signal to other formats.

In this paper, target application of the transcoder and its implementation are described. Results obtained in its experimental evaluation are then presented.

2 Digital Watermarking to Identify Source of Information Leak

2.1 Target Application

Consider a video archive that stores video files captured by surveillance cameras. These files may contain scenes showing personal information, copyrighted material, or trade secrets that need to be protected. When an authorized organization requests particular files, an official at the video archive packages the files and sends them to the recipient. The data size of the files might be reduced due to the limited capacity of portable media. A transcoder is a system that converts video files from one compression type to another compression type. It enables the official to re-compress video images for sending samples (e.g. by resizing from HDTV to SDTV).

If content from the files were found by chance on a video-sharing site, serious problems such as privacy violation, copyright piracy, and contract infringement would arise. It would be difficult, however, to determine whether the information leak occurred in the office of the organization to which the file was sent or in the video archive office. Quickly identifying the source of an information leak is therefore an important issue in cooperate governance to ensure accountability of the video archive.

One way to identify the site of information leakage is to use a digital watermarking technique that embeds the recipient's ID into the video file before it is sent. Watermark embedding process can be incorporated into transcoding process. If problematic video is found by chance, the embedded ID can be used to identify the party who redistributed it (see Fig. 1). It is, however, difficult to detect

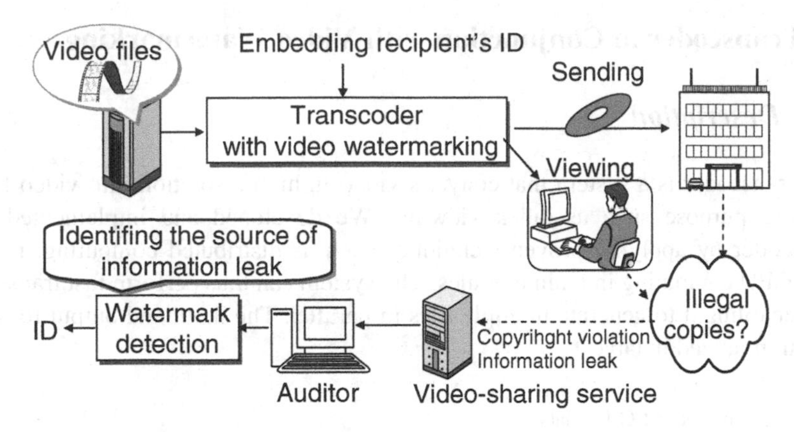

Fig. 1 Application image.

watermarks in content that has been severely degraded by the image processing used by video-sharing services, for instance, combinations of re-encoding and resizing.

2.2 Related Works

Many methods have been developed for general-purpose video watermarking for copyright protection [2–5]. For instance, spread spectrum communication schemes transmit a narrow-band signal via a wide-band channel by frequency spreading. Ideas from the spread spectrum communication are applicable to watermarking. Watermark signals can be seen as narrow-band signals and digital content can be seen as a wide-band channel [4]. In much of the early work, however, watermark robustness was evaluated against only encoding process such as MPEG. There has been little work on watermark robustness against the combination of image processing used in actual applications, such as re-encoding and resizing, except for some work on watermarking systems for HDTV [8].

A new approach to deterring information leakage is to use a transcoder in conjunction with video watermarking as described in section 2.1. The size of input image is often reduced by transcoder into smaller one. Watermarks should be embedded in such narrow area as VGA or less. The embedded watermarks, however, should survive the image processing of re-encoding and resizing. The system requirements when a transcoder is used differ from those of previous systems [2-5, 8]. This paper would contribute to security engineering by reporting the implementation and evaluation of such a system.

Moreover, with the publication of a second edition of the standard video set for advanced estimation of image processing [6], we can now evaluate changes caused by embedding watermarks in new types of scenes (for instance, a close-up image of a person's face (163) and scrolling tickers (116)). The results of evaluations using such samples would be valuable.

3 Transcoder in Conjunction with Video Watermarking

3.1 Description

Our transcoder is a system that converts video in high resolution into video for a specific purpose such as quick viewing. We developed and implemented the transcoder by applying proven techniques such as distributed computing, multi-core CPU computing in main modules. The system can thus perform fast transcoding, enabling it to generate multiple files in parallel. The input and output formats are summarized in table 1.

Table 1 Transcoder I/O formats

	Input	Output
System	MPEG2 TS, MXF	MPEG4
Video	MPEG2, I frame only, (100Mbps)	H.264 Baseline
Audio	PCM/BWF-J AAC-LC	AAC-LC

3.2 Encoding Process

The H.264/AVC encoding process used in our transcoder consists of two steps: (1) analysis of the complexity of and motion in images and (2) encoding based on the analysis results. The encoding step divides the encoding task into subtasks that process the entire image and slices of the image. It parallelizes the subtasks that are independent, so it encodes the data at a high degree of parallelism.

3.3 Video Watermarking Process

The watermarking module (WMM) maintains the image quality of watermarked images. It embeds watermarks into images by slightly changing the pixel values. It analyzes the original image by using a human visual model before embedding and extracts visually important areas such as edge shapes, where the changes are more perceptible than changes in other area. On the bases of its analysis, the module estimates the range of value changes for each pixel. These changes cannot be recognized by the human eye. Maximizing the watermark strength within these ranges results in watermarks that are imperceptible during playback but survive image processing.

It is generally possible to define a pattern of pixel values so that the probability of false positive errors occurring can be mathematically calculated. Our WMM embeds watermarks by using such patterns, i.e., by changing pixel values to meet these patterns. It can thus guarantee the avoidance of false positives by using patterns with a low probability.

Consider, for example, the embedding of a watermark by changing the pixel values of a one-frame image. Watermarked image $\mathbf{y}'\{y'_i \mid i = 1..n\}$ is obtained by adding the estimated watermark pattern $\mathbf{ep}$ to the original image $\mathbf{y}\{y_i \mid i = 1..n\}$. The estimated watermark pattern is calculated using the estimated strength and watermark pattern $\mathbf{m}\{m_i \mid i = 1..n\}$. That is,

$$\mathbf{y}' = \mathbf{y} + \mathbf{ep}. \quad (1)$$

Estimation value v for watermark detection is obtained by calculating the correlation between the watermark pattern and the watermarked image, as given by

$$v = \mathbf{y}' \bullet \mathbf{m}. \quad (2)$$

A watermark is detected by comparing the estimation value with a previously defined threshold value.

The original image is divided into regions and the embedding process is applied to each region one by one. The embedding process is applied to each frame image for the entire video file. The detection process is done in the same way.

3.4 Architecture

The transcoder architecture is shown in figure 2. The input video file is processed sequentially: de-multiplexing, decoding, scaling, marking if the marking option is used, encoding, and multiplexing into output video file. The mark-option module not only superimposes time codes and visible marks such as logos on video frames but also embeds watermarks into video frames. The watermark payload is 64-bits. The watermarking process in the mark-option module currently does not support parallel processing.

An example of the graphical user interface of the transcoder is shown in figure 3, and an image of the system installed in a server rack is shown in figure 4.

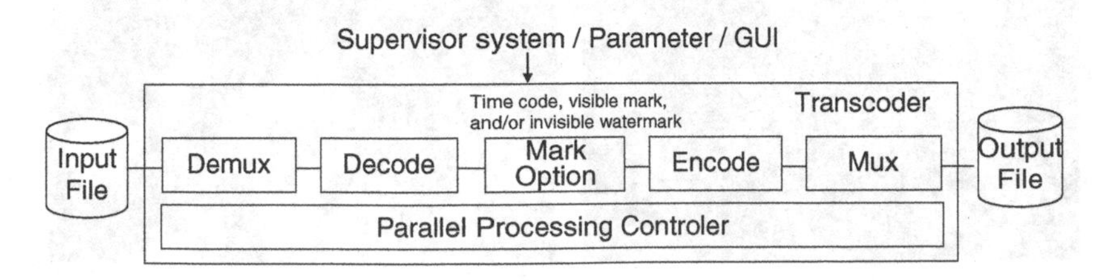

Fig. 2 Transcoder architecture.

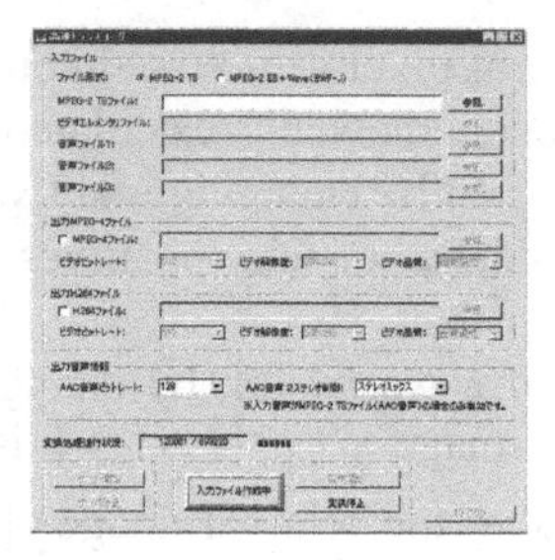

Fig. 3 Example of graphical user interface

Fig. 4 Transcoder image.

4 Evaluation

4.1 Experiment Conditions

Maintaining the image quality of images with embedded watermarks and providing watermark robustness against image processing are essential requirements for video watermarking. Video watermarks should survive not only video encoding process during legal viewing and distribution but also re-encoding process during illegal re-distribution to video sharing services. Re-encoding process also includes resizing.

Table 2 Requirement-A

Process	Size	Codec	Bit-rate
Encoding	640×480	H.264/AVC	2 Mbps
Re-encoding	480×360	H.264/AVC	512 kbps

Table 3 Requirement-B

Process	Size	Codec	Bit-rate
Encoding	352×240	H.264/AVC	512 kbps
Re-encoding	320×240	H.264/AVC	300 kbps

Fig. 5 Sample videos
left: night scene(fixed) (No.116), right: women and bouquet (No.163)

Two video samples (15-seconds-videos in the HDTV standard video set [6]), as shown in figure 5, were used for visibility and robustness testing. Moreover, three broadcast material samples (landscape, computer graphics, and music data) were used only for robustness test. We evaluated whether the watermark were robust against image processing for two technical requirements (shown in table 2 and 3) from the viewpoint of practical threats.

4.2 Visibility Test

The files containing the two video samples were encoded in accordance with requirement A (640x480) and were assumed to be distributed to users, as shown in figure 1.

The quality of images watermarked and encoded using our transcoder was evaluated subjectively using a procedure based on Recommendation ITU-R BT.500-11 [7]. H.264/AVC video images were displayed on a monitor and evaluated by five participants who rated watermark disturbance on a scale of 1 to 5 (5 for imperceptible, 4 for perceptible but not annoying, 3 for slightly annoying, 2 for annoying, and 1 for very annoying). The average score for both samples was 4.8, indicating that the evaluated images were practically useful.

4.3 Robustness Test

We considered watermark detection to be successful when 64 bits of information embedded in a video sample was correctly detected without any bit errors. The two samples used in the visibility test were re-encoded into 480x360 video images in accordance with requirement A using an encoder different from the transcoder. The watermarks in both re-encoded samples were detected successfully. Those in the three broadcast material samples were also successfully detected. These results show that the watermarking method satisfies both requirements A and B. That is, the watermarks in transcoded video images are robust against the image processing of video sharing services.

4.4 Processing Time

The software of our encoder was executed in a single thread to enable us to measure the processing time on a commodity PC. A 227-second video from the HDTV standard video set (from 101 to 117) was transcoded in accordance with requirement A. The total processing time for transcoding with video watermarking was 988.8 seconds. It was the average for three measurements including disc access time to input and output files. The processing time for transcoding without video watermarking was 842.3 seconds. The processing time spent on video watermarking was 146.5 seconds, calculated by subtraction. That is, transcoding time increases by 17% when the mark-option module (superimposing time codes and visible marks, embedding video watermarks) is fully used.

5 Conclusion

We have developed a transcoder in conjunction with video watermarking. Evaluation showed that the almost invisible embedded watermarks are sufficiently robust against the image processing of video sharing services. Use of this transcoder will help deter information leaks of relatively small video contents. Future work includes speed-up the processes used and further evaluation.

Acknowledgements. We thank Nippon Hoso Kyokai for lending us their video samples. We thank Hitachi Government & Public Corporation System Engineering, Ltd and Hitachi Consulting Co., Ltd. for implementing the system and supporting our experiments.

References

1. George, C., Scerri, J.: Web 2.0 and user-generated content: legal challenges in the new frontier. Journal of Information, Law and Technology 2 (2007)
2. Cox, I.J., et al.: Digital watermarking. Morgan Kaufmann Publishers, San Francisco (2001)
3. Bloom, et al.: Copy protection for DVD video. Proc. IEEE 87(7), 1267–1276 (1999)
4. Hartung, F., Girod, B.: Watermarking of uncompressed and compressed video. Signal Processing 66(3), 283–301 (1998)
5. Yamada, T., et al.: Evaluation of real-time video watermarking system on a commodity PC. In: Proc. of int'l workshop on long-term security (2006)
6. The second edition of HDTV standard video, The Institute of Image Information and Television Engineers (ITE) / Association of Radio Industries and Businesses, ARIB (2009)
7. Rec. ITU-R BT.500-7, Methodology for the subjective assessment of the quality of television pictures (1995)
8. Kim, K.S., et al.: Practical, real-time, and robust watermarking on the spatial domain for high-definition video contents. IEICE Transactions on Information and Systems E91-D(5), 1359–1368 (2008)

Novel Chatterbot System Utilizing BBS Information for Estimating User Interests

Miki Ueno, Naoki Mori, and Keinosuke Matsumoto

Abstract. Recently, the use of various chatterbots has been proposed to simulate conversation with human users. Several chatterbots can talk with users very well without a high-level contextual understanding. However, it may be difficult for chatterbots to reply to specific and interesting sentences because chatterbots lack intelligence. To solve this problem, we propose a novel chatterbot that can directly use Bulletin Board System (BBS) information in order to estimate user's interests.

Keywords: Chatterbot, BBS Information, Estimating User's Interests.

1 Introduction

Attempting to create an intelligent conversation system is one of the most important and interesting themes in computer engineering. However, approaches that use natural language processing and artificial intelligence techniques have not yet been able to create an enjoyable experience for users. On the other hand, chatterbots[1] have become a well-known method for simulating human conversation. The main objective of chatterbots is to produce interesting conversation, and many chatterbots do not care to simulate the actual human thought. Eliza[2], one of the very first chatterbots, could simulate a Rogerian psychotherapist. Many chatterbots have been proposed since Eliza[3, 4, 5]; however, the conversational level of the proposed chatterbots has not been sufficient to satisfy users. To solve this problem, we propose a novel chatterbot that directly uses Web information in order to estimate user's interests.

Miki Ueno, Naoki Mori, and Keinosuke Matsumoto
Department of Computer and Systems Sciences, College of Engineering,
Osaka Prefecture University, 1-1 Gakuencho, Nakaku, Sakai, Osaka 599-8531, Japan
e-mail: ueno@ss.cs.osakafu-u.ac.jp (M.U.)
mori@cs.osakafu-u.ac.jp (N.M.)
matsu@cs.osakafu-u.ac.jp (K.M.)

A.P. de Leon F. de Carvalho et al. (Eds.): Distrib. Computing & Artif. Intell., AISC 79, pp. 237–240.
springerlink.com

2 BBS Information for Chatterbot

2.1 *Target BBS*

We selected “2channel” (2ch)[6] as the target BBS because 2ch is the most comprehensive forum in Japan and covers diverse fields of interest. The top level unit is called “category”. Several boards belong to a “category”. It has more than 600 active boards including “Social News”, “Computers”, and “Cooking”. Each board usually has many active threads that have main topics for discussion.

2.2 *Utilizing BBS Information*

The following methods are adopted to use the information from the BBS.

1. If the chatterbot knows the user’s interests before the session, the chatterbot will try to use words that appear frequently on the board related to the user’s interests.
2. The chatterbot determines the user’s interests automatically by using statistical information from the user’s conversation log.
3. Make the internet community where people share interesting fields on the chatterbot.

2.3 *Distance between Boards*

2.3.1 Definition of Distance

In this study, we define the distance between two boards in the 2ch BBS as a simple Euclidean distance. If the similarity of two boards is high, the distance between those 2 boards will be low.

The words set throughout the entire BBS is defined as W, and the i-th word is denoted as w_i. We set $|W| = M$. Then the feature vectors of board x and y, $\hat{x}$ and $\hat{y}$ respectively, are defined as follows.

$$x = \begin{pmatrix} N^x_{w_1} \\ \vdots \\ N^x_{w_i} \\ \vdots \\ N^x_{w_M} \end{pmatrix}, \; y = \begin{pmatrix} N^y_{w_1} \\ \vdots \\ N^y_{w_i} \\ \vdots \\ N^y_{w_M} \end{pmatrix} \tag{1}$$

$N^x_{w_i}$: The number of words w_i in board x
$N^y_{w_i}$: The number of words w_i in board y

$$\hat{x} = \frac{x}{|x|}, \; \hat{y} = \frac{y}{|y|} \tag{2}$$

Then, the distance between boards x and y is defined as Euclidean distance.

2.4 Flow of Utilizing Sentences in the BBS

1. Save all sentences on board i from the Web to set S_i.
2. Let the number of all boards from which sentences are saved be n. Define $S = \bigcup_{i=1}^{n} S_i (\neq \phi)$ as the total set for n boards.
3. Set the probability of using the positive set p.
4. Let the first noun in the sentence be k. Set the positive set as $S^{\text{true}} = \{x | x \in S, x \text{ is the sentence which contains word } k\}$. Let $S^{\text{false}} = S \setminus S^{\text{true}}$.
5. Select a sentence randomly from S^{true} with probability p, or from S^{false} with probability of $1-p$, and output this sentence after the formatting procedure. If $S^{\text{true}} = \phi$ or $S^{\text{false}} = \phi$, a sentence is selected from S.

2.5 Flow of Estimating User Interests

1. The target user converses with the original chatterbot freely.
2. The chatterbot attempts to determine the board in which the target user is interested by using the conversational log obtained in Step 1. The distance between log and BBS board is utilized to find user's interests.
3. We adjust the chatterbot settings to use sentences from the board estimated in Step 2.

3 Experiment

To investigate the effects of applying BBS information to determine user's interests, we conducted the following questionnaire survey of users.

1. The target user freely converses with the chatterbot based on user's interests.
2. The chatterbot attempts to determine the board in which the target user is interested by using the conversational log obtained in Step 1.
3. The chatterbot show the result of the estimation of user's interests.
4. The target user evaluates the result of chatterbot estimation.

The results of the experiment is that the chatterbot could find the 3 different interests of Math, Fashion and TV Game.

4 Conclusion

In this paper, we proposed a novel chatterbot that directly uses Web information. Fig. 1 shows the look of our standalone chatterbot.

We confirmed that the proposed chatterbot is effective to find user's interests. Including a personality in the chatterbot will be studied in future research.

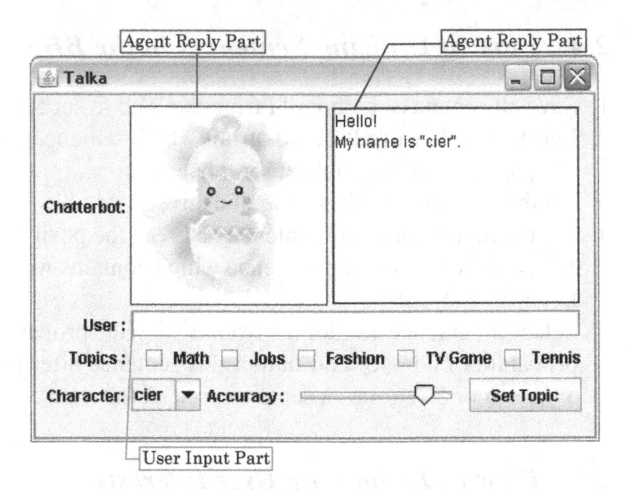

Fig. 1 GUI of standalone chatterbot

This research was supported in part by a Grant-in-Aid for Scientific Research (C), 22500208, 2010-2014 from the Ministry of Education, Culture, Sports, Science and Technology.

References

1. Chatterbot Is Thinking:
`http://www.ycf.nanet.co.jp/~skato/muno/index1.shtml`
2. Weizenbaum, J.: Eliza - a computer program for the study of natural language communication between man and machine. Communications of the ACM (9) (1966)
3. Chamberlain, W.: The Policeman's Beard is Half Constructed, Warner Books (1984)
4. Guzeldere, G., Franchi, S.: Dialogues with colorful personalities of early AI. Stanford Humanities Review 4(2) (1995)
5. Vrajitoru, D., Ratkiewicz, J.: Evolutionary Sentence Combination for Chatterbots. In: International Conference on Artificial Intelligence and Applications (AIA 2004), pp. 287–292. ACTA Press (2004)
6. 2ch: `http://www.2ch.net/`
7. Sen: `https://sen.dev.java.net/`

Do Engineers Use Convergence to a Vanishing Point when Sketching?

Raquel Plumed, Pedro Company, Ana Piquer, and Peter A.C. Varley

Abstract. We wish to determine whether design engineers commonly use central projections (convergence of parallel lines to a vanishing point) when sketching new shapes, rather than draw physically parallel lines as parallel. This paper describes a pilot experiment carried out to determine the presence and importance of central projections. Results suggest that designers rarely use vanishing points when sketching engineering shapes. Hence, convergence can safely be ignored when designing and implementing basic artificial intelligence systems which detect perceptual cues in engineering design sketches. Since we wish to develop an automated method for discriminating between central and parallel pictorial projections, the paper also presents a numerical analysis of our results which could be used to calibrate such a method. *Keywords: Sketch-based modeling reconstruction. Central projections. Parallel projections. Vanishing points.*

1 Introduction

Artificial intelligence aims to mimic human intelligence, and the most interesting artificial intelligence research is that which throws light on human intelligence (see, for example, [5]). *Visual perception* is a complex aspect of human intelligence, and its artificial equivalent, *machine vision*, has been much studied. There is a clear synergy between the studies of visual perception and machine vision: knowing as much as possible about human perception provides resources which help us to develop artificial perception methods which are intuitively correct [3], while creating artificial perception tools reminds us of what we still need to learn about human vision.

Our area of interest is creating computer-based tools for design engineers, using cue-based artificial perception processes which duplicate human vision processes. For this reason we wish to understand how designers make use of perceptual cues in engineering sketches.

Raquel Plumed, Pedro Company, Ana Piquer, and Peter A.C. Varley
Dept. of Mechanical Engineering and Construction,
Universitat Jaume I, 12071 Castellon – Spain
e-mail: {plumed,pcompany,Ana.Piquer,varley}@emc.uji.es

A.P. de Leon F. de Carvalho et al. (Eds.): Distrib. Computing & Artif. Intell., AISC 79, pp. 241–250.
springerlink.com

Parallel lines are a particularly important cue, and they have two common graphical representations in pictorial projections which have survived the test of time. One, *central projection* (used since the 15th Century and codified in Durer's *Four Books on Measurement* in 1522), is the convergence to one or more vanishing points of lines representing parallel lines in 3D space. The alternative, *axonometric projection* (at least as old in practice, but only codified in the 19th Century by Farish [4]), does not use vanishing points—convergence of parallel lines is deliberately absent. A third representation, *oblique projection*, also deliberately avoids the use of vanishing points, and in this paper oblique sketches are generally grouped with axonometric sketches.

Clearly, each representation has advantages and disadvantages, and engineers and designers must be trained to use both. However, which representation they prefer for any particular task has not been fully investigated. As far as we know, determining whether engineers and designers commonly use convergence of parallel lines while sketching new shapes is still an unresolved question. Such questions can only be answered by experiment. The work we present here describes a pilot experiment carried out to determine the presence and importance of vanishing points in sketches produced by engineers and designers. Section 2 discusses related work. Section 3 describes the hypotheses which we wish to test. Section 4 describes in detail our pilot experiment to determine whether central projections are preferred when making engineering sketches, and presents a visual analysis of the results, used to test our hypotheses. Section 5 presents a numerical analysis which could be used to calibrate an automated classification method, and Section 6 presents our preliminary conclusions.

2 Related Work

Physiologists have studied perceptual cues and how they guide the visual perception process (e.g. [1, 6, 9]), and some correspondences between perceptual organisation in biological and artificial vision have been established (see, for example, [13]). Some computer scientists have replicated cue-based perception processes in various approaches (e.g. [2, 8, 15, 16, 19]). Even so, the current situation is that we have no complete catalogue of perceptual cues, and the exact role even of known cues is not fully understood.

In particular, the importance of central projection as a perceptual cue specific to the interpretation of engineering sketches has not been investigated experimentally. Previous studies have in general either assumed or ignored convergence as one of their simplifying assumptions. For example, Parodi el al. [10] investigating the computational time complexity of labelling polyhedral scenes, assume the presence of convergence and show that calculating vanishing points is the rate-determining step. Sturm et al. [14] calculate vanishing points to calibrate single camera images of polyhedra, but require the user to specify manually which lines are intended to be parallel. Kanade [7] proposes a technique for recovering three dimensional shapes from a single image, based on mapping image regularities (in particular, parallelism of lines and skewed symmetry) into shape constraints.

Prats et al. [12] investigates the sketching process of designers and how they are able to obtain new shapes during sketching by the application of shape rules, but it does not consider the study of depth cues.

Wyeld [18] presents a psychology experiment directed at determining the range of variability in individuals' drawing ability and ability to read 3D images. The drawings used in the study included common perspective depth cues such shading, shadows and a ground plane. The population for this study was limited to first year undergraduate students. In contrast to Wyeld's study, we focus on one specific depth cue, but we consider a much wider range of technical drawing skill.

3 Hypothesis

The purpose of our experiment is to determine to what extent convergence to a vanishing point is used by designers while sketching engineering shapes. It is here that we must define *designer*. For the purposes of our experiment, a *product designer* is someone who has received specific training in 2D and 3D geometry and the commonly-used techniques for representing 3D objects in 2D as part of a technical education.

We contrast product designers with *graphic designers*, who have received training in the commonly-used techniques for representing 3D objects in 2D as part of a non-technical (often artistic) education.

Using these definitions, we hypothesise that the technical training received by product designers influences the way they represent three dimensional parts. In particular, we hypothesise that people who are trained to think in engineering terms will generally prefer parallel (axonometric or oblique) projection (which retains the important cues to functionality) rather than perspective projection.

By contrast, we might expect that graphic designers, after being exposed to more artistic training (where *how it looks* is more important than *what it does*), may show a trend towards using convergence when making design sketches.

Therefore, we should compare the behaviour of two distinct population groups: product designers whose background is in engineering, and graphic designers whose background is artistic.

Ideally, we should also consider a third group, those who have, as yet, received no design training either as engineers or as artists. In this case, we have no reason to hypothesise a preference either for parallel projections or convergence.

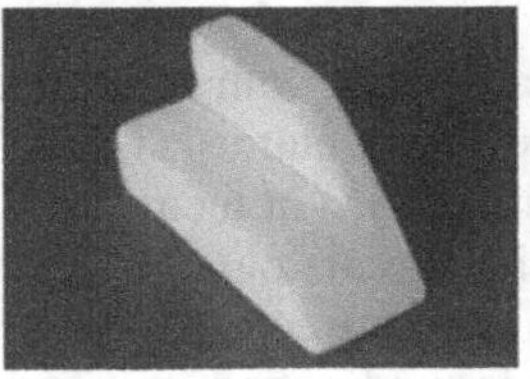

Fig. 1 Models used in the experiment, by rising order of difficulty from left to right: first, second and third model.

4 Design of the Experiment

The basis of our experiment is that we asked various people to draw pictures of three polyhedral solids. The solids and the accompanying instructions are described in detail in Section 4.1, and the participants in the study are described in detail in Section 4.2.

To determine the human perception of such drawings, the pictures were subjectively classified by a group of experts as central/axonometric projections and good/poor quality. The classification results are summarised in Section 4.3.

4.1 The Questionnaire

In designing our experiment, it is important to avoid any kind of implicit or explicit constraint or guidance on the way the task should be performed. In particular, we tried not to influence the participants either to use or not to use hidden lines, and left them free to choose the orientation of the model. For this purpose, we produced a minimal questionnaire which avoided as far as possible any unnecessary guidance to the participants. Expanded polystyrene was chosen for the model material as surface brightness helps the observer to recognise the object's faces and edges, thereby ensuring that all participants had a good mental model of the object they were to sketch.

To avoid any hint of how parallel edges should be represented, or whether or not hidden lines should be drawn, the physical object in the worked example was a tetrahedron oriented such that all of its edges were visible. Even this minimal task description contains one implicit hint: the participants are influenced to use pictorial projection, not multiview orthographic projections.

The questionnaire also included a 15 x 11 cm rectangular frame for the participants to draw their own objects. This helped to ensure that sketches were of a similar size.

We also collected personal data as studies level, studies field, sex and age.

4.2 Participants

The bulk of the population was drawn from several departments of the same university, and included industrial engineers, mechanical engineers, architects, designers and artists. The level of experience ranged from undergraduate students to professors. We also included a few participants with no technical drawing training.

A total of 147 questionnaires were returned. 16 (10.88%) were returned by participants with no university education. 2 of them (12.5%) were filled out by people with secondary education until 16 years old. 3 of them (18.75%) were solved by people with secondary education between 16 and 18 years old. And others 11 (68.75%) had received secondary education with professional orientation. Their ages ranged from 27 to 66 years old. 11 were male (68.75%) and 5 were female (31.25%).

73 questionnaires (49.66%) were returned by university students, of whom 58 (79.45%) studied an engineering speciality (mechanical, industrial), 7 (9.59%) studied architecture, and 8 (10.96%) studied other subjects. Their ages ranged from 18 and 43 years old. 53 were male (72.6%) and 20 were female (27.4%).

58 questionnaires (39.46%) were returned by participants with one or more university degree. 29 (50%) graduated in engineering, 14 (24.14%) graduated in architecture, 8 (13.79%) had artistic training via design studies or BBAA, and 7 (12.07%) graduated in other fields. Their ages ranged from 26 to 56 years old. 35 were male (60.34%) and 23 were female (39.66%).

4.3 Human Perception

In order to determine how humans would classify the sketches, each sketch was subjectively classified by six experts (four belonging to the research team and two external experts) as: clearly axonometric; clearly central; clearly non-pictorial orthographic; uncertain; and not classifiable because of poor quality.

Next, we compared the experts' classifications, and discarded those drawings where there was disagreement (*agreement* means here that four or more experts chose the same classification). As a result, 20 (13.6%) of the original 147 drawings were discarded. We also discarded 9 sketches which were considered by all six experts as so poor quality that trying to classify their contents as central or parallel projections was pointless. And finally, we discarded 7 sketches which were agreed as drawn using non-pictorial orthographic projections, as, showing only a single 2D view, these were not useful when determining how 3D objects are represented pictorially.

After discarding useless sketches, we were left with 111 valid sketches. Of these, 3 (2.54%) clearly used central projection and 108 (91.53%) clearly used axonometric projection. All 3 of those which used central projection depicted the first model (the cuboid).

In more detail:

71 of the valid sketches (63.96%) were created by product designers. All of these were classified as axonometric drawings by experts.

26 of the valid sketches (23.42%) were created by graphic designers. These included the 3 were classified by experts as perspective drawings.

14 of the valid sketches (12.61%) were produced by subjects without drawing training.

From these results, we find strong support for the hypothesis that designers (both product designers and graphic designers) prefer axonometric to perspective projection. There is only weak support for the hypothesis that graphic designers are more likely than product designers to use perspective projection, and there is also only weak support for the hypothesis that model complexity influences choice of representation, with perspective projection being more likely for simple objects than for complex objects.

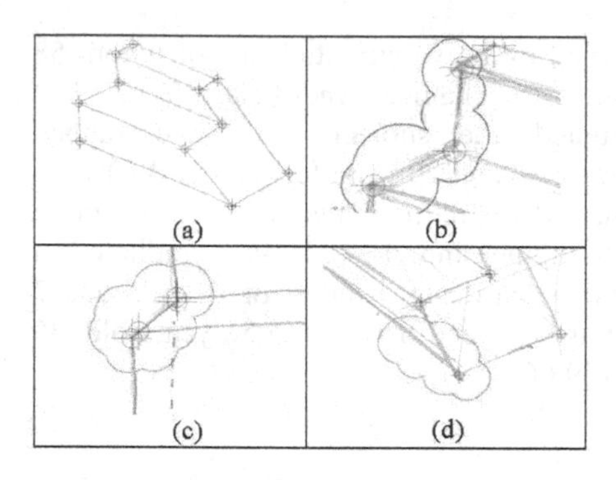

Fig. 2 Vertices defined by points intersection

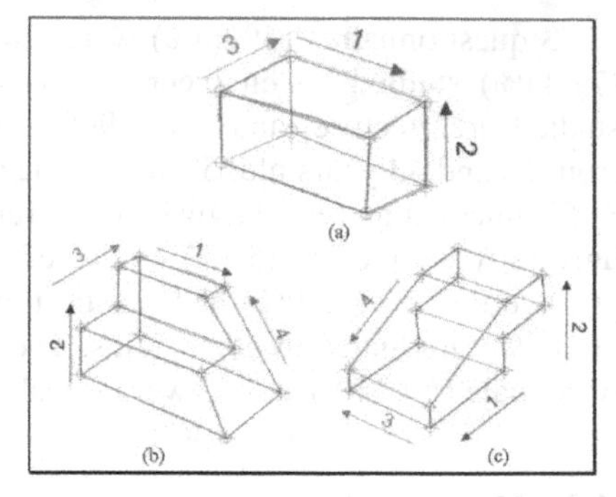

Fig. 3 Line drawings

5 Numerical Measurements

In order to preserve the questionnaires, the sketched images were scanned and saved as bitmaps. This has the additional advantage that it allows us to perform numerical analyses. This numerical data could be used to calibrate automatic classification of sketches into central/axonometric, as a step towards our aim of automating interpretation of engineering sketches. This data can be found in [11].

In order to produce this numerical data, we manually vectorised the scanned images into line drawings, by identifying vertex locations and tracing new lines from vertex to vertex, as described next. In most cases, vertex locations are clearly defined as junctions of two or more line segments, as shown in Figure 2 (a).

However, there were also cases in which vertex locations were not so well defined:

- Overtracing, as in Figure 2(b), results in several intersection points among several lines. In these cases, we defined the vertex location as the intersection of medial axes.
- Sometimes, as in Figure 2(c), junctions of lines which were intended to intersect at the same vertex were sufficiently separated to be considered as distinct vertices. In our processing, we merged any two vertex locations which differed by less than 3.5% of the length of the shortest line segment intersecting either vertex.
- Finally, some participants used scaffolding lines intermixed with pictorial lines, as in Figure 2(d) (*scaffolding* is any line or group of auxiliary lines in the sketch which is used to facilitate drawing and which does not correspond to any feature of the object). We assumed that thick lines are pictorial lines and thin lines are scaffolding, and vertex locations only occur at the intersections of pictorial lines.

Once all vertex locations were defined, we redrew pictorial lines in different colours, where each colour corresponds to a different direction (see Figure 3).

In order to facilitate later analysis, lines representing the same edges of each model were always drawn in the same colour.

Hidden lines were also vectorised, since they portray edges of the model, but other auxiliary lines (such as reference axes and scaffolding) were discarded.

From each vectorised line drawing we extracted: (a) geometrical information provided line slopes, line lengths and coordinates of every vertex; (b) general information, corresponding to geometric information sorted by direction; and (c) drawing information: on the existence or absence of hidden lines or auxiliary lines in drawings.

We analysed these numerical results with the objective of looking for criteria which could be used to discriminate between parallel and convergent sketches. This analysis is presented in Section 5.1. The analysis produced other interesting results, which are discussed briefly in Section 5.2.

5.1 Automatic Discrimination of Central Style Sketches

Obviously, not all people have the same ability to create accurate freehand sketches. We cannot simply look for parallel lines to discriminate parallel projections from central projections: in reality, no lines in a freehand sketch will be perfectly parallel.

Instead, we need a calculable function which can be used to distinguish deliberate parallelism and/or convergence from involuntary mistakes made during sketching. By analysing data from the 108 valid drawings classified as axonometric, we can then estimate a threshold value for this function.

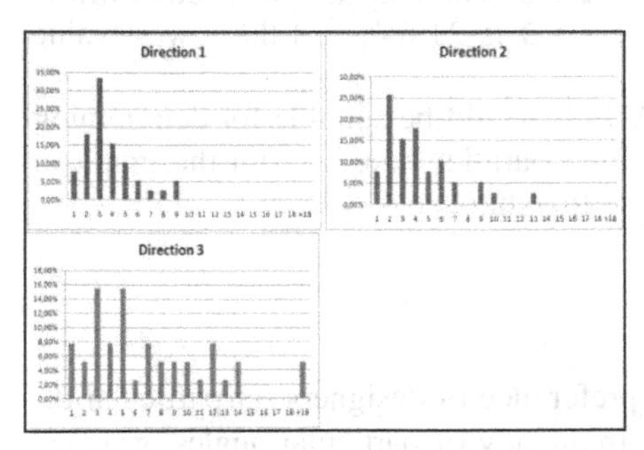

Fig. 4 Histograms of relative frequency vs intervals of angular dispersion for model 1.

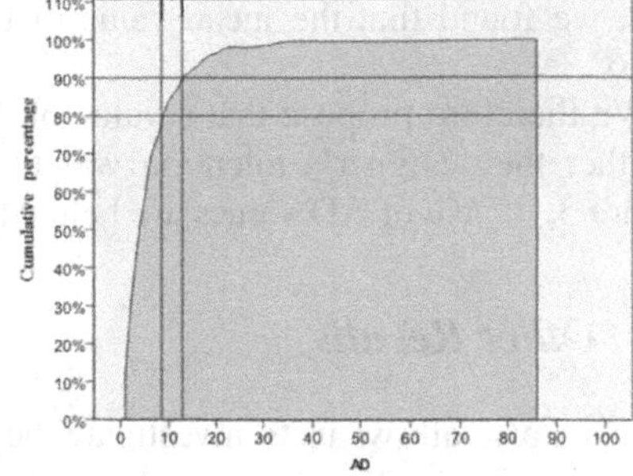

Fig. 5 Cumulative Frequency of overall AD

Our choice of function is *angular dispersion* (AD), the maximum angle value among the slope differences between each pair of lines representing parallel edges parallel of the original 3D model:

$$AD=\max |\alpha_i-\alpha_j| \;/\; i=1,2,..,n;\; j=1,2,..,n;\; i\neq j, \quad (1)$$

where *n* defines each of the set of lines belonging to a specific direction. It is intended to measure the maximum freehand sketching error which occurs when designers draw parallel lines.

Several aspects of the AD function were studied. First we created histograms which show the relative frequency (in 1° buckets) of AD for each model and *direction* (i.e. group of like-coloured lines). Figure 4 shows the histograms for model 1. From the original data, we also extracted the mean values of AD for each model and direction (see table 1).

Table 1 The mean values of AD for each model

	Model 1	Model 2	Model 3
Direction 1	2.98°	4.28°	7.25°
Direction 2	3.63°	4.77°	12.54°
Direction 3	6.19°	12.08°	8.65°
Direction 4		5.97°	2.96°

Secondly, we plotted the cumulative frequency of AD, as shown in Figure 5. We found that, owing to the part orientation chosen by some participants, in 22 cases there was only one edge in one or more of the directions, making it impossible to perform this calculation. These 22 cases were not included in this analysis.

We obtained a mean AD value of 6.5°. Fixing a threshold value of AD=9° would lead to 80.4% of directions being classified as parallel rather than converging. If we raise the threshold value to AD=13°, then 90.1% of directions would be classified as parallel.

Analysing the sketches which were classified manually as perspective drawings, we found that the mean value of their AD is 21.84°, and the lowest value is 8.6°.

We therefore propose that a value of AD=8° would be suitable for determining whether the designer's intention was to draw parallel lines. From the graph in Figure 5, 75.4% of AD values are below this threshold.

5.2 Other Results

Our data also allows us to investigate the preference of designers for particular direction angles in 2D. Figure 6 shows the frequency of particular angles, grouped in 4° buckets. As can be seen, there is a strong preference for 90° (i.e. vertical), followed by a weaker preference for 30°, 0° and -30°.

Earlier studies (e.g. [17]) have theorised that vertical edges of a 3D object will be drawn as vertical lines in 2D. The predominance of 90° angles, independent of the model drawn and the type of projection used, strongly supports this theory. Verticality is a dominant direction in pictorial representation, and both product and graphic designers are able to use it with accuracy.

The frequency of ±30° can be attributed to them being the required angles of the two horizontal axes in isometric projection.

Finally, in creating Table 1, we observed that the largest AD values corresponded to short lines. Taking this further, we looked for a possible correlation

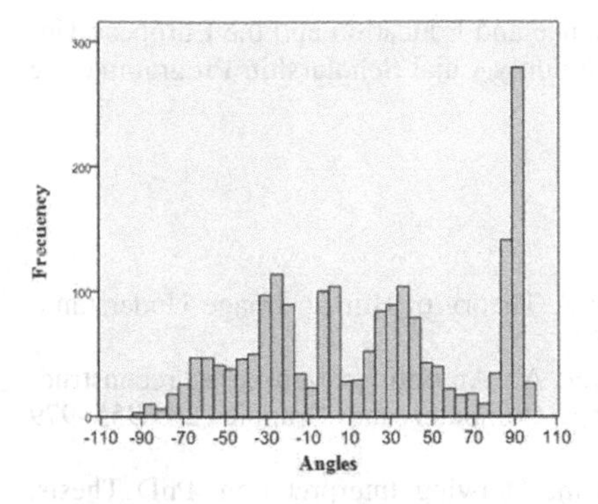

Fig. 6 Frequency of angles

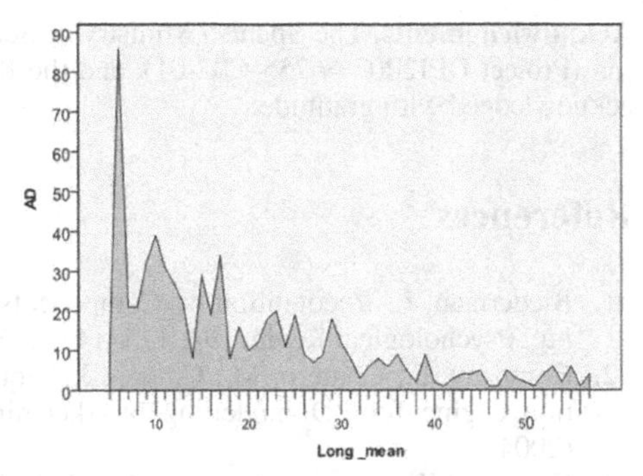

Fig. 7 Diagram of dispersion points between AD values and average lengths.

between AD values and average lengths of each set of lines. The results are plotted in Figure 7, which shows that there is indeed some correlation between the two parameters. This study falls outside the scope of our present paper, but is something which could profitably be investigated in the future.

6 Conclusions

From our experiment, we conclude that convergence is not an important cue for artificial perception of engineering sketches, as central projections are rarely used in engineering sketches. By contrast, parallelism is a very important cue, as parallel lines of three-dimensional shapes are usually represented as parallel lines in engineering sketches.

Both product and graphic designers tend to prefer axonometric projections when sketching simple polyhedral engineering parts. There is a suggestion that this preference becomes stronger with more complex objects, but the supporting evidence is weak and this remains a matter for future research.

This result is helpful when designing cue-based artificial perception processes. When implementing simple sketch-based modelling systems, perspective can safely be ignored. In more advanced systems, parallelism should be given more weight than convergence when, for example, competing processes must negotiate in order to identify the more sensitive cues.

Automatic discrimination between central and axonometric sketches remains problematic. The fact that line are or are not parallel (to within any particular threshold value) is not a sufficiently reliable indicator of the choice between central or axonometric projection. A threshold value (AD=8º) low enough to correctly identify all central projections would also lead to 24.6% of sketches which were identified manually as axonometric being misclassified as central projections.

What remains unclear is whether all of these problem sketches were intended by their creators to be axonometric. Is it possible that the machine is correct, that convergence is indeed present, but that some optical illusion prevents the human eye from detecting it. This too remains a matter for future research.

Acknowledgments. The Spanish Ministry of Science and Education and the European Union (Project DPI2007-66755-C02-01), and the Ramon y Cajal Scholarship Programme are acknowledged with gratitude.

References

1. Biederman, I.: Recognition-by-Components: A Theory of Human Image Understanding. Psychological Review 94, 115–147 (1987)
2. Company, P., Contero, M., Conesa, J., Piquer, A.: An optimisation-based reconstruction engine for 3D modelling by sketching. Computer and Graphics 28, 955–979 (2004)
3. Draper, S.W.: Reasoning about Depth in Line-Drawing Interpretation. PhD Thesis, Sussex University (1980)
4. Farish, W.: On Isometrical Perspective. Cambridge Philosophical Transactions 1 (1822)
5. Goel, V.: Sketches of Thought. The MIT Press, Cambridge (1995)
6. Hoffmann, D.: Visual Intelligence. How we create what we see. Norton Publishing, New York (1998)
7. Kanade, T.: Recovery of the Three-Dimensional Shape of an Object from a Single View. Artificial Intelligence 17, 409–460 (1981)
8. Lipson, H., Shpitalni, M.: Optimization-based reconstruction of a 3D object from a single freehand line drawing. Computer-Aided Design 28(8), 651–663 (1996)
9. Palmer, S.E.: Vision science. Photons to phenomenology. The MIT Press, Cambridge (1999)
10. Parodi, P., Torre, V.: On the Complexity of Labeling Perspective Projections of Polyhedral Scenes. Artificial Intelligence 70, 239–276 (1994)
11. Plumed, R., Company, P., Piquer, A., Varley, P.A.C.: Convergence measure of sketched engineering drawings. Technical Report Ref. 07/2010 Regeo. Geometric Reconstruction Group, `http://www.regeo.uji.es`
12. Prats, M., Lim, S., Jowers, I., Garner, S., Chase, S.: Transforming shape in design: observations from studies of sketching. Design Studies 30, 503–520 (2009)
13. Sarkar, S., Boyer, K.L.: Perceptual Organization in Computer Vision: A review and a Proposal for a Classificatory Structure. IEEE Trans. on Systems, Man and Cybernetics 23(2), 382–399 (1993)
14. Sturm, P.F., Maybank, S.J.: A Method for Interactive 3D Reconstruction of Piecewise Planar Objects from Single Images. In: Pridmore, A., Elliman, D. (eds.) Proc. British Machine Vision Conference, Nottingham, pp. 265–274 (1999)
15. Tian, C., Masry, M., Lipson, H.: Physical sketching: Reconstruction and analysis of 3D objects from freehand sketches. Computer aided design 41, 147–158 (2009)
16. Varley, P.A.C.: Automatic Creation of Boundary-Representation Models from Single Line Drawings PhD Thesis, University of Wales (2003)
17. Varley, P.A.C., Martin, R.R., Suzuki, H.: Frontal Geometry from Sketches of Engineering Objects: Is Line Labelling Necessary? Computer Aided Design 37(12), 1285–1307 (2005)
18. Wyeld, T.: The correlation between the ability to read and manually reproduce a 3D image: some implications for 3D information visualisation. In: 13th International Conference Information Visualisation, pp. 496–501 (2009)
19. Yuan, S., Tsui, L.Y., Jie, S.: Regularity selection for effective 3D objects reconstruction from a single line drawing. Pattern Recognition Letters 29(10), 1486–1495 (2008)

Fingerprinting Location Estimation and Tracking in Critical Wireless Environments Based on Accuracy Ray-Tracing Algorithms

Antonio del Corte, Oscar Gutierrez, and José M. Gómez

Abstract. This paper presents an alternative detection method in fingerprinting technique for indoor localization and trajectory estimation based on efficient ray-tracing techniques over Wireless Local Area Networks (WLAN). Firstly, the use of radio frequency (RF) power levels and relative time delay is compared as detection method to estimate the localization of a set of mobile stations using the fingerprinting technique. The localization algorithm computes the Euclidean distance between the samples of signals received from each unknown position and each fingerprint stored in the database or radio-map obtained by using the FASPRI simulation tool. Secondly, an indoor trajectory has been simulated and tested by means of ray-tracing techniques. Experimental results shows that more precision can be obtained in the trajectory estimation by means of relative ray delay instead of RF power detection method, enabling the deploy of new applications for critical environments such as airports, where security and safety requirements are strongly involved.

Keywords: Airport, Euclidean Distance, Localization, Ray-tracing, Safety, Tracking, Wireless.

1 Introduction

Indoor trajectory estimation and localization is an important aspect of GPS-alternative positioning systems and is of great importance for general communications areas. This paper aims to apply novel techniques to increase the airport

Antonio del Corte
Automatic Department
University of Alcalá
Alcalá de Henares. Madrid, Spain
e-mail: antonio.delcorte@uah.es

Oscar Gutierrez and José M. Gómez
Computer Science Department
University of Alcalá
Alcalá de Henares. Madrid, Spain
e-mail: {jose.gomez,oscar.gutierrez}@uah.es

A.P. de Leon F. de Carvalho et al. (Eds.): Distrib. Computing & Artif. Intell., AISC 79, pp. 251–258.
springerlink.com

operational security, ideal in areas where the satellite coverage is not good and clear implications on people and goods are involved. In this work, the problem of indoor localization based on the signals available from the wireless devices [1][2] that comprise the Wi-Fi standards is presented. The localization process is done by using the fingerprinting technique [3][4] that operate the relationship between the signal information by multipath reflections. In comparison with other techniques, such as angle of arrival (AOA) or time of arrival (TOA) that present several challenges due to multipath effects and non-line-of-sight (N-LOS)[2], the fingerprinting technique is relatively easy to implement. In traditional indoor localization systems based on Wi-Fi networks, the Euclidean distance is used as metric in the localization process and the fingerprinting technique is based on the power levels stored in the received signal strength indicator (RSSI) parameter available on the 802.11x standards. However, it is necessary to explore new techniques to improve the precision by using alternative detection methods. Firstly, the fingerprinting technique has been implemented by using relative ray delays as fingerprint in the radio-map. More precision was obtained compared to the power detection technique [8][9]. Secondly a simple indoor trajectory has been simulated based on the results previously obtained as direct application of this technique.

2 Ray-Tracing Model

Simulations have been performed by means of ray-tracing model and can be obtained with the FASPRI [4] simulation tool, that is able to make a 3D indoor propagation analysis by means of deterministic methods [5][6], ideal for the design of tracking and localization system. FASPRI is a ray-tracing code based on geometric optic (GO) and the uniform theory of diffraction (UTD). In order to optimize the program computing time, ray-tracing algorithms such as the angular zeta-buffer (AZB) or the space volumetric partitioning (SVP) [5][6] have been implemented. These algorithms make it possible to simulate a great number of case-studies in a reasonable amount of time. These results can be used to examine the effect of varying certain sensing parameters on the precision of the system such as the number of antennas, the position of the antennas and the number of tracks. The electric field levels can be obtained by the direct, reflected, transmitted, diffracted ray or combinations of these effects until fourth order. An advantage of using the ray-tracing techniques is that, besides obtaining the power level of a series of points, information can also be obtained about the multipath effects, such as the relative delays between rays and the directions of arrival. This information can be used as a fingerprint in the fingerprinting method with the purpose of improving the efficiency of the localization system and estimating trajectories.

3 Fingerprinting Technique

The fingerprinting technique can be divided in two phases [2]. In the first one, it obtains the radio map or fingerprints database. The radio-map of fingerprints is obtained by performing an analysis of the information available from devices and multiples access points over a defined grid. The vector of received signal of power and relative ray delay at a position on the grid is called the location fingerprint of that point. In the second phase, it analyzes the accuracy obtained in the localization process. For this purpose, the developed technique places a significant number of mobile stations into the area covered by the radio map and it obtains the vector of received samples from different APs [7]. The location estimation is made by an algorithm that computes the Euclidean distance between each measured mobile sample and all the fingerprints stored in the radio map. The X and Y coordinates associated with the fingerprint that results in the smallest Euclidean distance are returned as the position for the mobile. Vectors of signal power as well as relative delays are available from each access point to the mobile. In case of signal power (1) the samples are computed for each access point (N) but in case of relative delays (2) the number of ray-tracing effects (E) should also be considered.

$$Dp(x, y) = \sqrt{\sum_{i=1}^{N} \left[(Pm_i - Pf_i)^2 \right]} \tag{1}$$

$$Dr(x, y) = \sqrt{\sum_{i=1}^{N} \left[\sum_{j=1}^{E} \left[(Rm_{ij} - Rf_{ij})^2 \right] \right]} \tag{2}$$

4 Radio Frequency vs. Relative Delay for Localization

The use of radio frequency (RF) power levels and relative time delays based on ray-tracing is compared as detection method to estimate the localization of set mobile stations using the fingerprinting technique. The metric considered was the Euclidean distance. The geometry analyzed (Fig. 1) corresponds with a non-regulate section of the Polytechnic building. Fig.2. shows the relative delays between the rays detected in a fingerprint and their contribution to the total field due

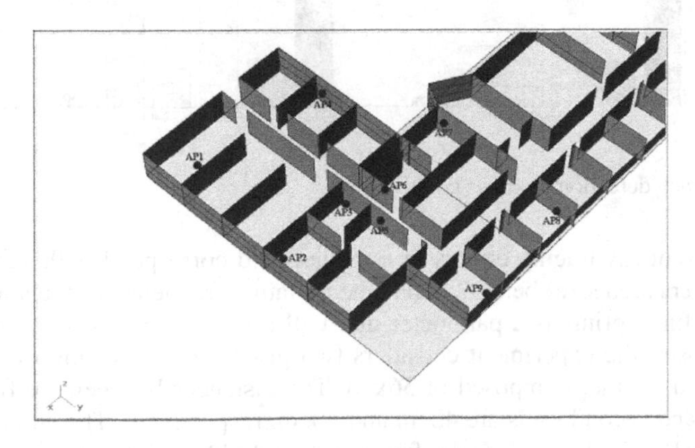

Fig. 1 Polytechnic building section

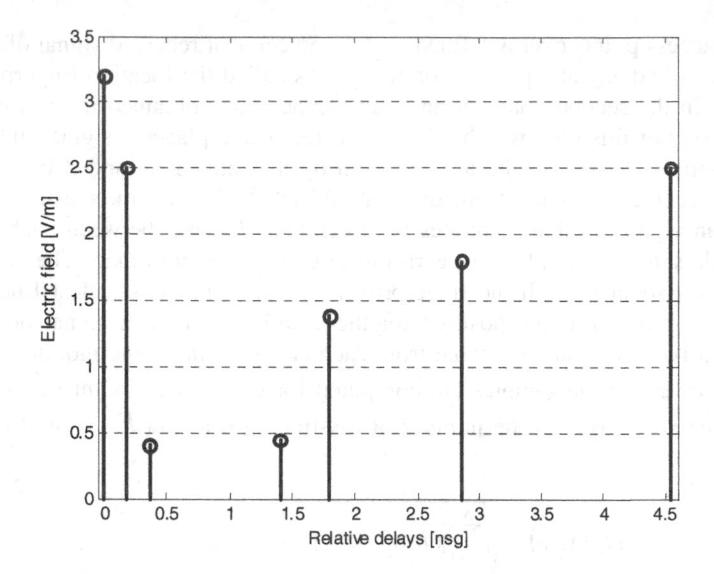

Fig. 2 Relative delays between rays in a fingerprint and their contribution to the total field

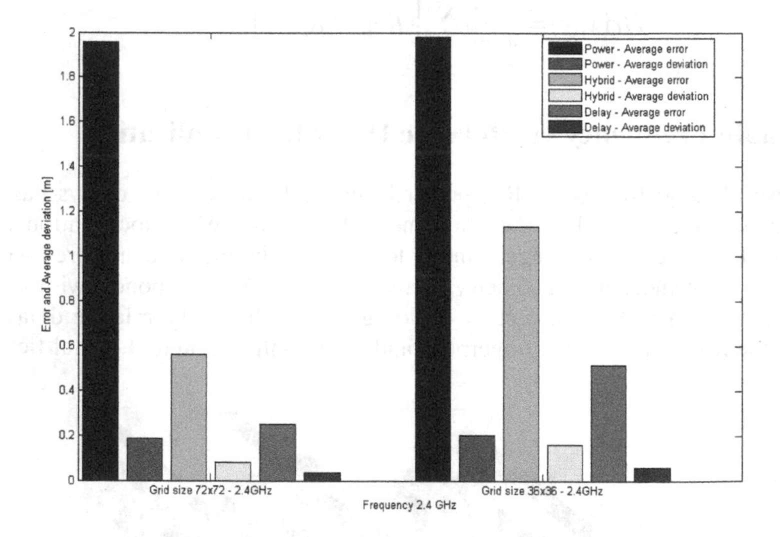

Fig. 3 Accuracy detection methods comparison

to the different ray-tracing effects in a regular grid correspond with 72x72 points (30x30 meters area size) being 2.4GHz the radiation frequency of the antenna. The number of fingerprints is a parameter that will affect the precision of the results. For this reason the experiment considers two grids: one consisting of 72x72 fingerprints and another composed of 36x36. The distances between the fingerprints in the first and second grids are 40cm and 80cm, respectively. The simulation also placed 9 AP's at the 2.4 GHz frequency and 99 mobile stations randomly distributed over the grids. The three detection methods analyzed are compared by means of statistical indicators (Fig. 3). With these results, we can confirm that the relative delay detection method provides better results than the power detection for any grid size. The hybrid detection, although it provides worse results than using delays, can reduce the detection ambiguity at the cost of increasing the mean error. However, it should be noted that the mean error is reduced when the number of fingerprints is increased, independent of the detection method used. Table 1 shows the error values obtained when comparing the three detection methods.

Table 1 Mean error and typical deviation detection methods comparison

	Power	Hybrid	Delay	Hybrid vs. Power	Delay vs. Power
Grid size	Mean error [m]			Mean error improvement (%)	
72x72	1.9554	0.5621	0.2504	71.28	87.17
36x36	1.9801	1.1337	0.5155	32.82	74.24

5 Indoor Trajectory Estimation

As application of the new localization technique based on ray-tracing described in this paper, an indoor trajectory estimation has been implemented. Over a defined

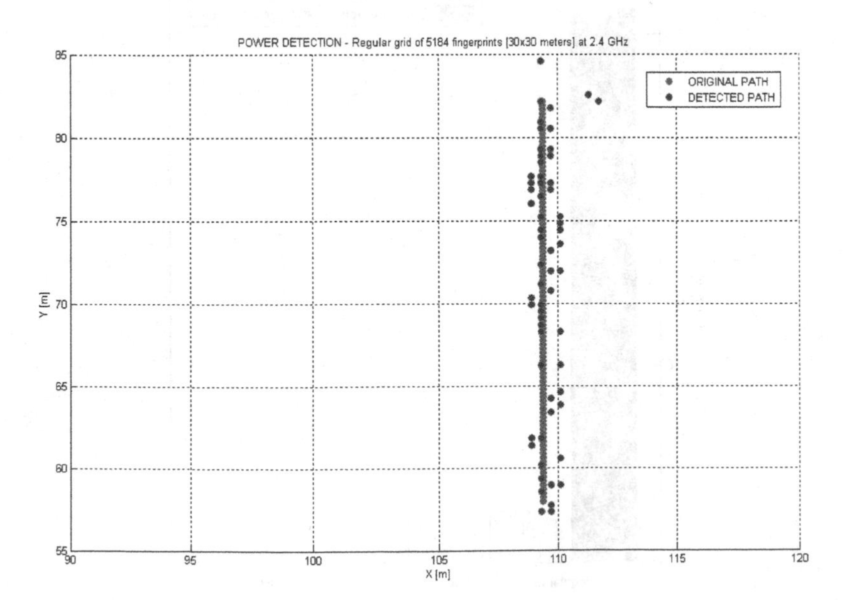

Fig. 4 Trajectory estimation by power detection

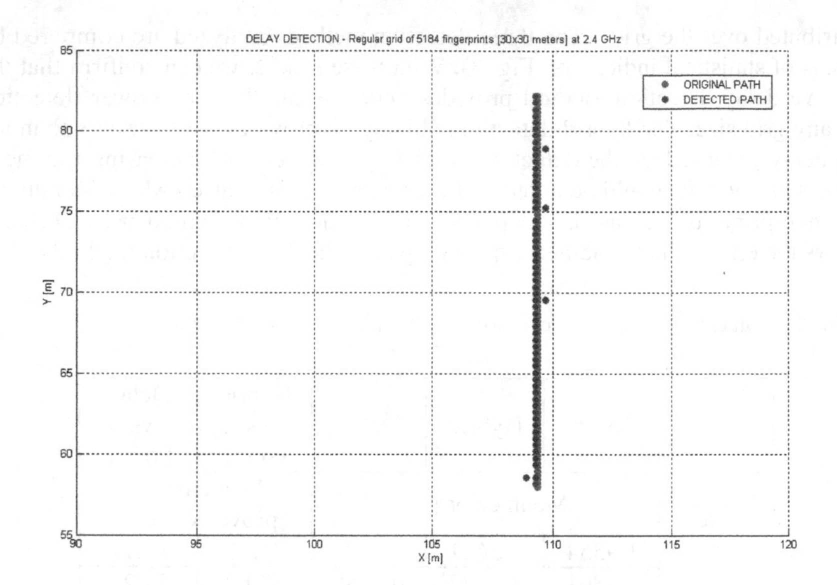

Fig. 5 Trajectory estimation by relative delay detection

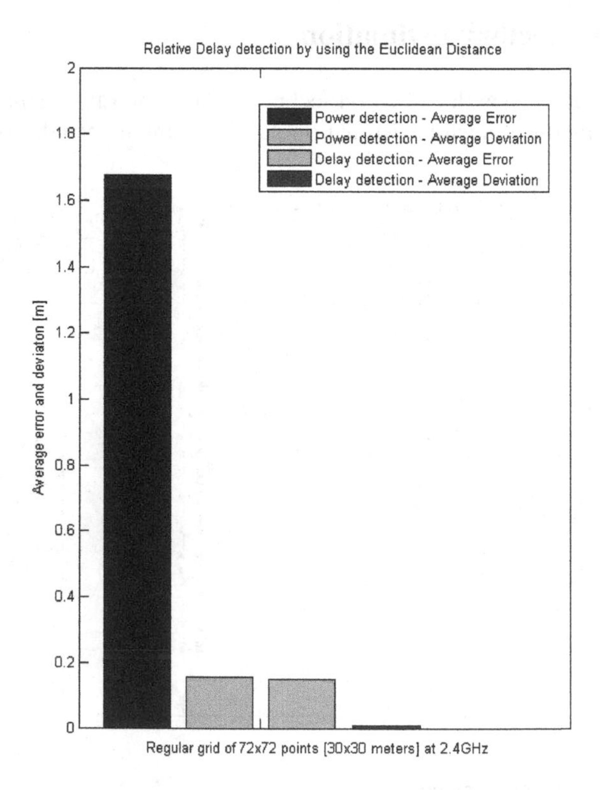

Fig. 6 Trajectory estimation: average error and deviation detection methods comparison

grid of 5184 fingerprints (radio-map) a straight line composed by 100 mobiles has been simulated. The fingerprints are separated 80 cm and the frequency of the 9 antennas was 2.4 GHz. The fingerprinting algorithm computes the Euclidean distance by means of both, power levels and relative time delays between rays, due to multipath ray-tracing effects. The results obtained in the trajectory estimation by means of power detection are showed in Fig. 4. As it can be seen, is obvious that power detection method does not produce the best results. As a conclusion, the trajectory will not be correctly estimated by this method due to great average error.

In the opposite, Fig. 5 shows the results obtained in the trajectory estimation by means of relative time delay detection method. Due to lower average error, less than 0.3 meters, the final trajectory can be correctly estimated.

Finally, Fig. 6 shows the average error and deviation obtained in the trajectory estimation comparing both power and relative delay detection methods.

6 Conclusions

In this work an alternative detection method / algorithm that can be used in the fingerprinting technique for indoor mobile localization and trajectory estimation has been presented. This technique makes possible the analysis and design of indoor services over WLAN networks. A comparative study between detection methods based on RF power and relative time delays has firstly been implemented concluding that relative delay detection technique presents better results than the power detection technique used in traditional Wi-Fi systems. Secondly, a simple trajectory has been estimated as practical application of the ray-tracing technique.

As final conclusion, this method enables deploy of new applications in critical environments such as airports, where security and safety requirements are strongly involved.

7 Future Work

We will simulate more complex trajectories and explore another metric instead of the Euclidean distance, such as Manhattan or Mahalanobis. Also a previous Clustering classification for improving real-time applications and an interpolation algorithm between the fingerprinting weighting in order to eliminate those fingerprints that do not contribute to the improvement in the accuracy will be implemented. Finally, we will introduce in our algorithm Bayesian estimation and Kalman filter to implement a proper tracking system.

Acknowledgements. This work has been financed by the Community of Madrid, project S2009/TIC-1485

References

[1] Cisco Wireless Location Appliance-Products. Datasheet (2006)
[2] Bahl, P., et al.: Enhancements of the Radar User Location and Tracking System. Microsoft Research Technology (February 2000)

[3] Kaemarungsi, K., Krishnamurthy, P.: Properties of Indoor Received Signal Strength for WLAN Location Fingerprinting. In: Proc. First Annual International Conf. on Mobile and Ubiquitous Systems: Networking and Services, August 2004, pp. 14–23 (2004)
[4] Sáez de Adana, F., et al.: Propagation model based on ray tracing for the design of personal communication systems in indoor environments. IEEE Trans. on Vehicular Technology 49, 2105–2112 (2000)
[5] Cátedra, M.F., Pérez-Arriaga, J.: Cell Planning for wireless communications. Artech House Publishers, Boston (1999)
[6] Kaemarungsi, K.: Distribution of WLAN Received Signal Strength Indication for Indoor Location Determination. In: Proc. First International Symposium on Wireless Pervasive Computing, January 2006, p. 6 (2006) CD-ROM
[7] del Corte-Valiente, A., Gómez-Pulido, J.M., Gutiérrez-Blanco, O., Cátedra-Pérez, M.F.: Efficient Techniques for Indoor Localization based on WLAN Networks. In: Second International Workshop on User-Centric Technologies and Applications Madrinet, Salamanca, Spain, pp. 5–15 (October 2008)
[8] del Corte-Valiente, A., Gómez-Pulido, J.M., Gutiérrez-Blanco, O.: High Precision for Indoor Localization Applications based on Relative Delay and the Fingerprinting Technique. In: Third International Workshop on User-Centric Technologies and Applications, Madrinet, Salamanca, Spain, pp. 25–32 (June 2009)
[9] del Corte-Valiente, A., Gómez-Pulido, J.M., Gutiérrez-Blanco, O.: Efficient Techniques for Improving Indoor Localization Precision on WLAN Networks Applications. In: IEEE International Symposium on Antennas & Propagation and USNC/URSI National Radio Science, Charleston, South Caroline, USA (June 2009)

FASANT: A Versatile Tool to Analyze Radio Localization System at Indoor or Outdoor Environments

Lorena Lozano, Mª Jesús Algar, Iván González, and Felipe Cátedra

1 Introduction

Due to the increase of the technological area related to the radio localization, coverage study, both indoor and outdoor environments, the role of simulation tools rahter than traditional measurement techniques, are becoming more important. These tools are based on High Frequencies.

These are applied to analyze and optimize radio localization systems at indoor or outdoor scenarios, to characterize the channel propagation in these environments. A theoretical deterministic model based on the Uniform Theory of Diffraction is considered, in which the electrical field in an observation point is the sum of the contribution of all the rays that arrive to such point due the several propagation mechanisms like reflections, diffractions, transmissions and combinations of them.

2 Fasant Tool

FASANT is a versatile tool to analyze the radio localization on 3D complex structures electrically large such as buildings, airports, cities, etc. using the GTD/UTD and PO approach. These structures can be defined by planar and/or curved surfaces that can be modelled by any kind of material: perfect electrically conductor or dielectric materials.

In previous versions the number of effects to take into account in the simulations was limited up two bounces and not all combinations between the different effects were possible. At present, the FASANT tool allows the multiple interaction of different kind of effects with no limit in the number of bounces. Figure 2 shows

Lorena Lozano, Mª Jesús Algar, Iván González, and Felipe Cátedra
Electromagnetic Computing Group, Computer Sciences Department
University of Alcala. 28871. Alcala de Henares. Madrid. Spain
e-mail: felipe.catedra@uah.es

A.P. de Leon F. de Carvalho et al. (Eds.): Distrib. Computing & Artif. Intell., AISC 79, pp. 259–266.
springerlink.com

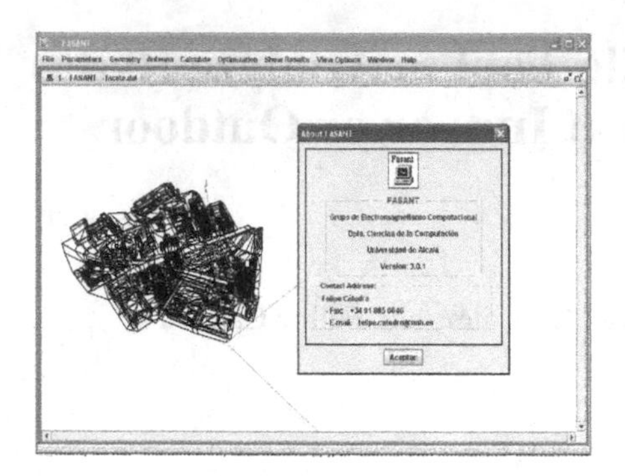

Fig. 1 FASANT simulation tool.

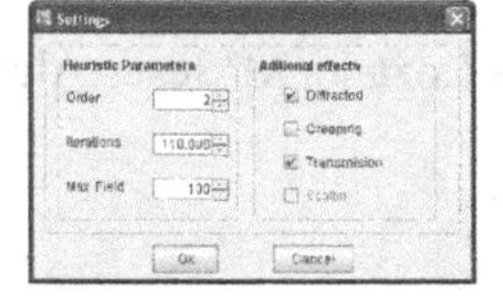

Fig. 2 Settings parameters.

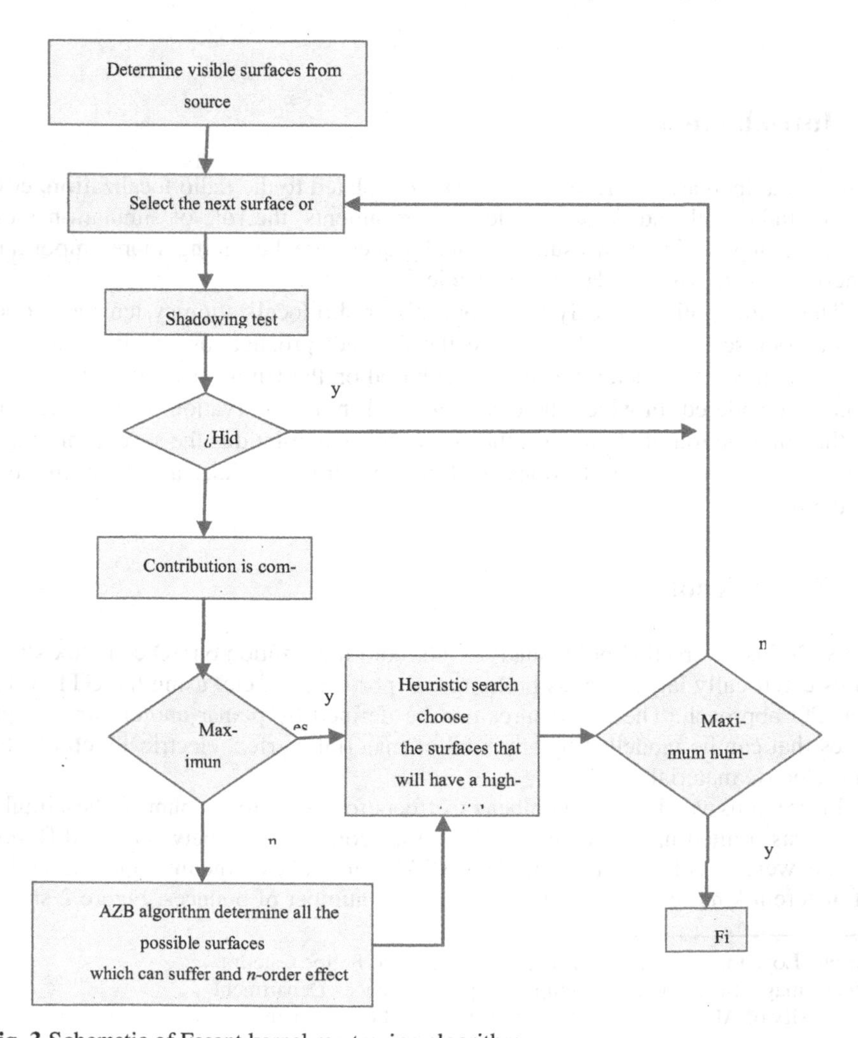

Fig. 3 Schematic of Fasant kernel ray tracing algorithm

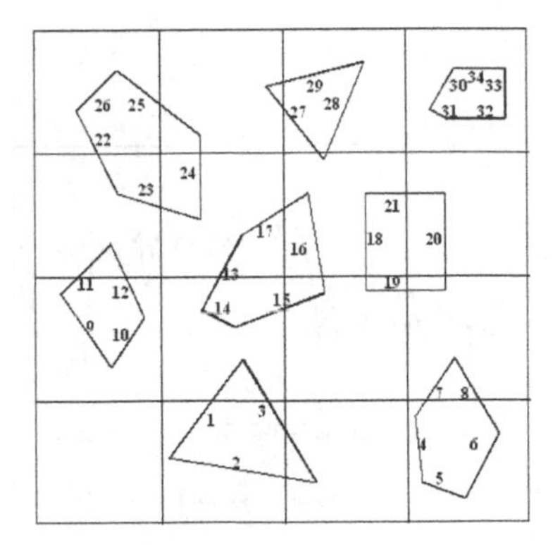

Fig. 4 Example of AZB scenario.

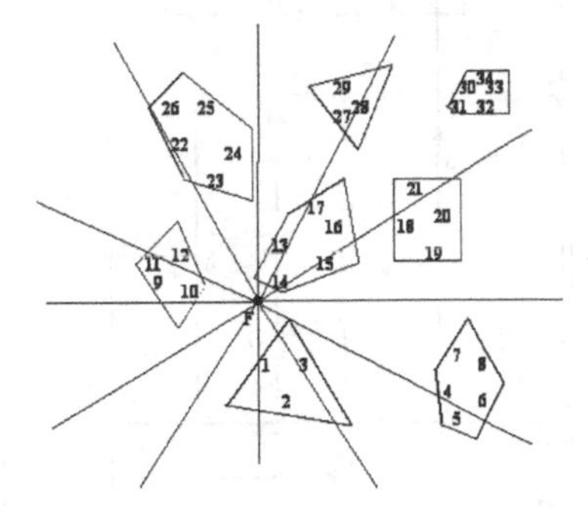

Fig. 5 Example of SVP scenario.

the settings parameters of the code, where the user can set the order of the simulation, the maximum number of iterations and the maximum number of contributions to be obtained in every observation point.

To accomplish this, kernel of FASANT has been improved with the new ray-tracing algorithm based on a recursive application of the Angular Z-Buffer (AZB) and the Space Volumetric Partitioning (SVP) [2] together with the A* heuristic method [3-4], to compute efficiently interactions with flat and curved surfaces, requiring less CPU time and memory resources than previous versions.

In figure 3 is shown the schematic of the kernel process to calculate the electrical field in an observation point. The AZB technique is used to determine all the possible surfaces that can contribute to a N order effect, with the A* heuristic search the surfaces that can contribute potentially to electrical field is used, using a cost function that depends on the length path from the source point to the surfaces and the losses due to a transmission effect. The SVP is used to obtain the ray path segments hidden. This is made in an iterative process until the number of contributions for every point is reached or the maximum number of iterations is exceeded.

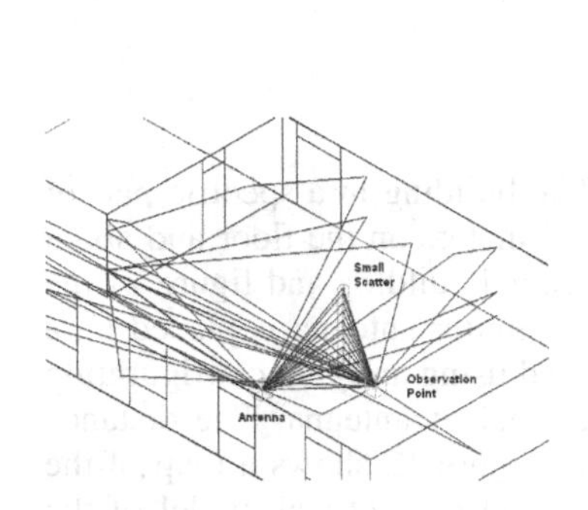

Fig. 6 Example of ray tracing wit small scatter.

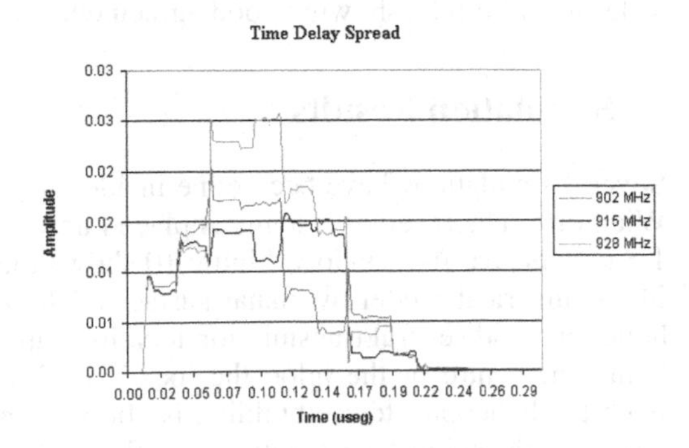

Fig. 7 Time delay spread for indoor propagation.

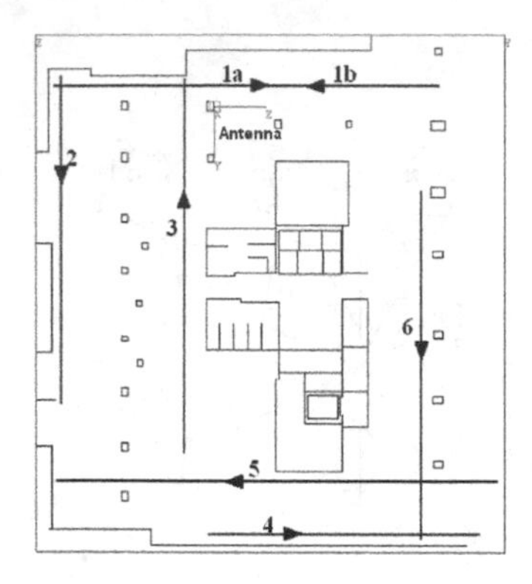

Fig. 8 Sota building image and

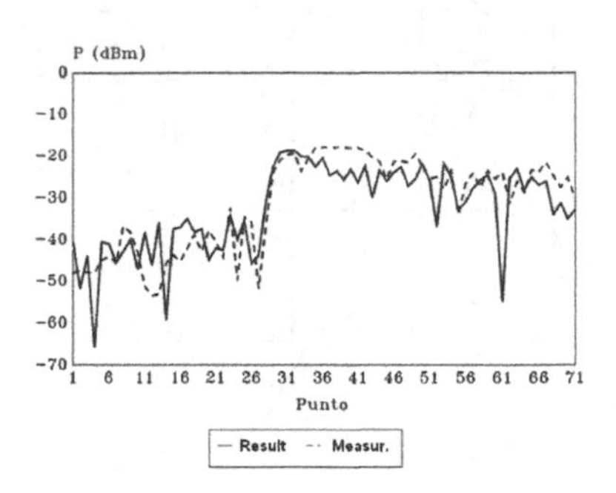

Fig. 9 Comparison between simulations and measurement paths measurements for path 4.

So, using this technique, the M stronger rays (higher field intensities) are obtained from the emitting antenna to the observation point considering all possible rays.

When there are small obstacles that not satisfy the requirements of the GTD/UTD, a heuristic approach based on PO can be applied. The scatter response is modelled by a set of equivalent currents that depends on the magnitude and phase of the incoming waves that reach the scatter. These incoming waves can be any order combination of direct and/or reflected, transmitted and diffracted ray.

When the kernel has finished the calculation, the user's interface could be used to visualize the rays from the antenna according with the kind of interaction selected on the scenario. Also, an individual ray can be select with the mouse to obtain all the information about it: path of the ray, length, propagation time and the electrical field contribution for every frequency of the ray.

The code has been applied successfully to the analysis of propagation in indoor/urban environments. Figure 9 show a comparison between measurements and simulations of the received power inside the Sota building in Bilbao city made by Telefonica Mobile, showing good agreements.

3 Simulation Results

Several simulations have been done inside an office building in a specific part of Wien city. The receiver antenna is placed above 1.5 meters on the floor and about 1.8 meters near the window. Figure 10 shows the actual building and figure 11 the 3D geometrical model by planar surfaces. The transmitter antenna is placed on a helicopter and several transmission locations are used using an Igloo configuration being the centre of the igloo the location of the receiver antenna. The distance from the helicopter to the building is about 1 km. Figure 12 shows a map of the city with the igloo configuration and figure 13 the 3D geometrical model of the Wien part analyzed. The transmission frequency is 2.57 GHz and the power transmitted is 40 dBm. A horn antenna has be considered in the simulations for the transmitted antenna.

Fig. 10 Real building with the location of receiver antenna

Fig. 11 3D geometrical model of the building.

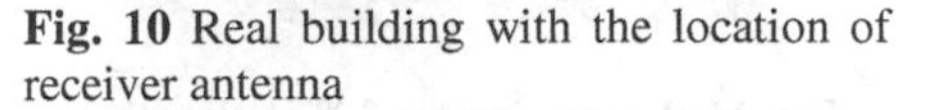

Fig. 12 Locations of the transmitter antenna.

Fig. 13 3D geometrical model of the part of the city considered in the abalysis.

Three elevation angles from the ground have been consider in this first study 30º, 45 º and 60º. Figure 14 shows the ray tracing for the propagation simulations of Wien showing the parts of the city that are contributing to the electrical field inside the building. Most of the contributions are transmitted, diffractions and diffraction-transmissions effects. The simulations have been done considering the most important fields contributors considering effects with a maximum of until 5 bounces and a maximum or 10 bounces. No significant differences in the results of the field values are obtained, so the field computation considering only ray-paths with a number of bounces of 5 at maximum is enough. More detailed views of the ray tracing are shown in figures 15-17 for a zoom-in near the building where the surrounding buildings and the terrain that acts as obstacles disturb the propagation.

The ray tracing obtained from our heuristic approach removing the surrounding buildings are shown in figures 18-20. The effects that contribute mainly to the electrical field are the transmitted, diffractions and diffractions-transmission fields. The heuristic approach obtains only the strongest effects that contribute to the total electrical field, and in this case there are simple diffractions that are in the frame of the window because the window was open. There is not facet in the model representing the glass of the window.

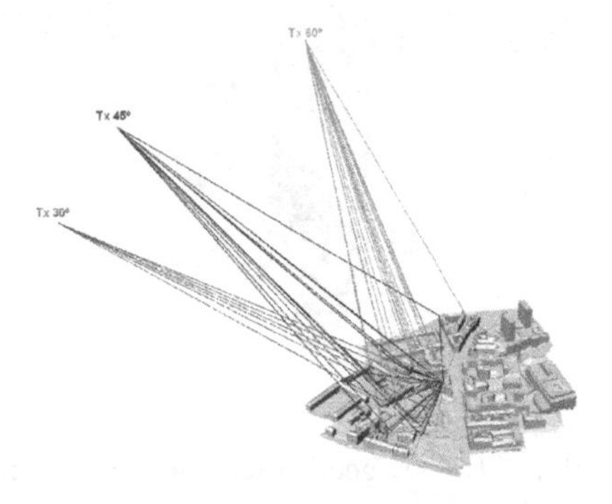

Fig. 14 Locations of the transmitter antenna.

Fig. 15 Ray-tracing for 30° of elevation.

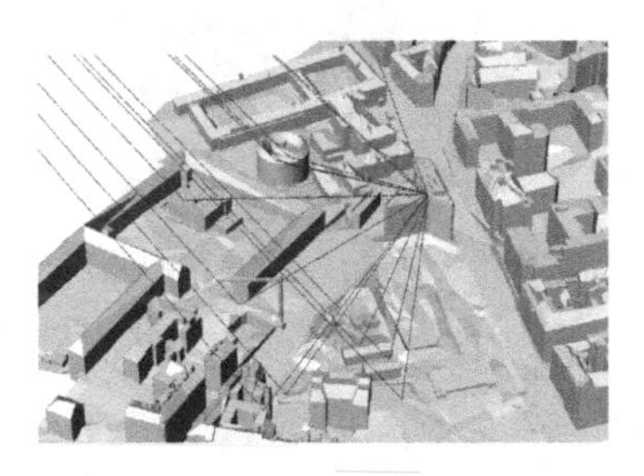

Fig. 16 Ray-tracing for 45° of elevation

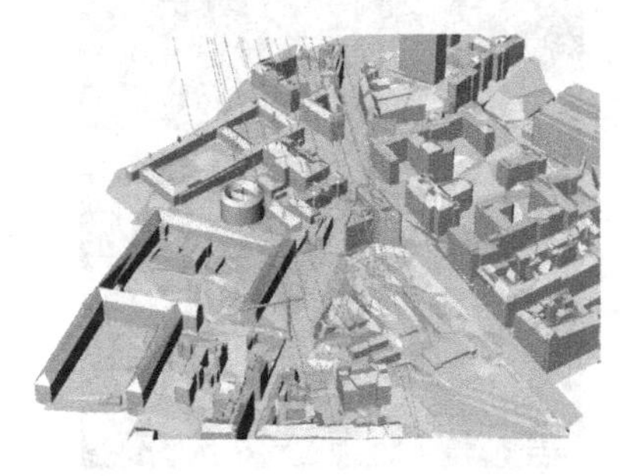

Fig. 17 Ray-tracing for 60° of elevation.

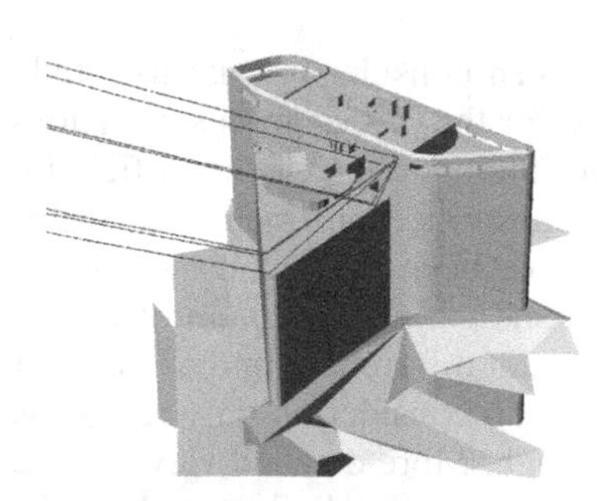

Fig. 18 Ray-tracing for 45° of elevation building alone.

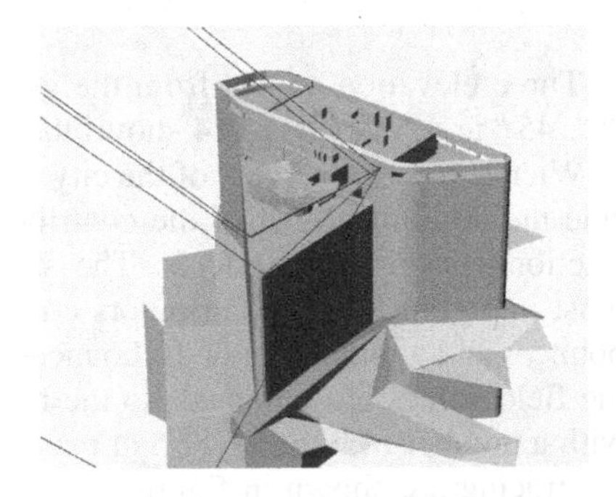

Fig. 19 Ray-tracing for 60° of elevation building alone.

The area of Wien considered in the study has about 6.600 facets and the simulation CPU-time to process until 10000 contributions (10000 ray-paths considering a maximum of 5 bounces) was 45 minutes. The CPU-time for processing the building alone was 35 minutes considering the same number of contributions that in the

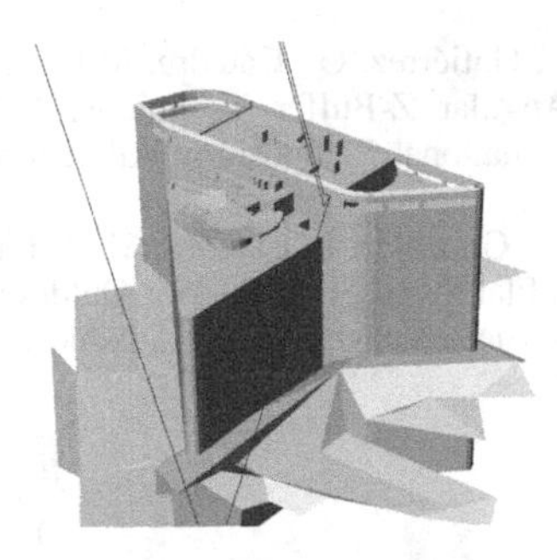

Fig. 20 Ray-tracing for 45° of elevation for the building alone case.

Table I Received power in dBm

Tx location	Building Alone	Wien
30°	-57.32	-80.86
45°	-85.60	-84.79
60°	-80.15	-76-72

previous case. Both cases were run in a single Opteron machine at 2.4 GHz using only 320 MB of RAM; Table I shows the received power in the observation point where the received antenna is located. At the moment of writing this paper measurements were not available and therefore it has not been possible to compare simulations with measurements.

4 Conclusions

This paper shows a numerical method based on GTD/UTD suitable for studying the propagation to indoor/outdoor complex environments. With this approximation the electrical field in an observation point is obtained as the contribution of all rays that arrive the observation point. The most consuming time computation is to obtain the path from every ray from the emitting antenna to the receiver antenna, so it is necessary to use ray tracing acceleration techniques to accelerate this process.

Acknowledgments. This work has been financed by the Community of Madrid, "CONcepts and Technology Enablers for conteXt-aware Service (CONTEXTS)", project S2009/TIC-1485

References

[1] Cátedra, M.F., Pérez, J.: Cell Planning for Wireless Communications. Artech House Publishers, Boston (1999)

[2] Pérez, J., Saez de Adana, F., Gutiérrez, O., González, I., Guzmán, J.: FASANT: Fast Computer Tool for the Analysis of on Board Antennas. IEEE Antennas & Propagation Magazine, 94–98 (April 1999)

[3] Delgado, C., Lozano, L., Gutiérrez, O., Cátedra, M.F.: Iterative PO Method based on Currents Modes and Angular Z-Buffer Technique. In: 2006 IEEE Antennas and Propagation Society International Symposium, Albuquerque, NM, EEUU, pp. 1853–1856 (July 2006)

[4] Lozano, L., Gutiérrez, O., González, I., Cátedra, M.F.: Fast Ray-Tracing For Computing N-Bounces Between Flat Surfaces At In-door/Outdoor Propagation. In: 2007 International Conference on Electromagnetics in Advanced Applications, Torino, Italy, September 17-21 (2007)

Piecewise Linear Representation Segmentation as a Multiobjective Optimization Problem

José Luis Guerrero, Antonio Berlanga, Jesús García, and José Manuel Molina

Abstract. Actual time series exhibit huge amounts of data which require an unaffordable computational load to be processed, leading to approximate representations to aid these processes. Segmentation processes deal with this issue dividing time series into a certain number of segments and approximating those segments with a basic function. Among the most extended segmentation approaches, piecewise linear representation is highlighted due to its simplicity. This work presents an approach based on the formalization of the segmentation process as a multiobjetive optimization problem and the resolution of that problem with an evolutionary algorithm.

1 Introduction

Time series (sequences of data having, among other components, a timestamp for each of their points) are of great importance for a wide variety of domains, such as financial [1], medicine [13] of manufacturing applications [7]. In recent years, the fast development of storage and collection technologies has lead to an increasing role of time series in the industry. A clear example can be found in the tracking of stock prices [8], being constantly updated in the different markets all over the world.

The required amount processing for these huge volumes of data is unaffordable, and thus the need for an approximate representation emerges. Time series segmentation is a tool designed to deal with this issue, by means of dimensionality reduction. A segmentation technique basically divides a certain time series into a number of segments and approximates those segments with a basic function. According to the different choices for this function, several segmentation techniques can be defined: Fourier Transforms [1], Wavelets [4], Symbolic Mappings [2], etc.

Among the different segmentation techniques, probably the most extended one is Piecewise Linear Representation (PLR, also named Piecewise Linear Approximation, PLA), [10] [15]. This segmentation technique is highlighted by its ease of use, since the basic function used to approximate the different segments is a linear

José Luis Guerrero, Antonio Berlanga, Jesús García, and José Manuel Molina
Group of Applied Artificial Intelligence (GIAA), Computer Science Department
University Carlos III of Madrid. Colmenarejo, Spain
e-mail: {joseluis.guerrero,antonio.berlanga}@uc3m.es,
{jesus.garcia,josemanuel.molina}@uc3m.es

A.P. de Leon F. de Carvalho et al. (Eds.): Distrib. Computing & Artif. Intell., AISC 79, pp. 267–274.
springerlink.com

function. Due to its wide usage, several processes have been designed regarding the result of this segmentation technique, such as fast similarity search [12] or the definition of data mining approaches [11].

Traditionally, the results of the different proposed segmentation techniques were compared according to the final error obtained, regardless of the number of segments required to obtain that error. Recently, this fact has been pointed out [15] and new approaches at least consider this number of segments as a quality metric over the final results. Considering this from an optimization point of view, this means that originally only one objective function was considered (the measured error), but the introduction of the number of segments as an additional objective function has turned this problem into a multi-objective optimization problem (MOOP) [6].

MOOPs are complex problems which require that a set of objective functions (usually in conflict) are optimized (maximized or minimized) jointly. In the PLR segmentation problem defined, the two objective functions are approximation error and number of segments. These objectives are in conflict, since a greater number of segments implies a finer approximation, and thus a lower error value. Both objective functions have to be minimized.

Evolutionary algorithms (EAs) have obtained remarkable results applied to MOOPs, being classified as Multi-objective Evolutionary Algorithms (MOEAs) [5]. The objective of this work is to define the proper problem dependent items for the application of a MOEA to the PLR segmentation issue, define the required configuration for the algorithms and compare the obtained results with one of the techniques specifically designed for this purpose: bottom up segmentation. The test set used will be a set of trajectories coming from the Air Traffic Control domain.

The organization of the paper will introduce the segmentation techniques in general and the bottom up technique used in the second section, followed by the required MOEA definition and configuration, performed in the third section. After the definition of the two techniques, their results and comparison are presented, along with the conclusions which these results point to.

2 Piecewise Linear Representation Segmentation Techniques

Segmentation techniques can be defined as the process which divides a given trajectory into a series of segments and afterwards approximates each of these segments with a given basic function. PLR segmentation techniques in particular, use a piecewise linear model as its basic function.

Several classifications can be made over PLR segmentation techniques, being probably the two most important ones their online or offline nature and the use of linear interpolation or regression functions. Online segmentation techniques perform their time series division processing the data as it is received, being the sliding window one of its most known examples [15]. Offline segmentation techniques, on the other hand, require the trajectory to be complete prior to the application of the technique, allowing them to use global information about its behavior. This characteristic allows them, in general, to obtain better segmentation results, but it also makes their complexity order higher, especially according to the trajectory size. The most extended approaches are top down and bottom up techniques [10].

The use of linear interpolation or regression functions usually depends on the need to obtain continuous piecewise lines. Linear interpolation uses a model defined by the first and last point of the segment, so that contiguous segments will always have one point in common. This characteristic may be a requirement in some domains. Linear regression, on the other hand, obtains the regression line considering all the different points belonging to the segment, obtaining, thus, discontinuous piecewise lines. The overall error of the regression functions is always less than or equal to the one obtained with linear interpolation [10], leading to its usual choice as the approximation function (it must be mentioned, though, that the required complexity for its calculation is also considerably higher). Both techniques can be applied along with any of the previously mentioned segmentation algorithms.

The traditional criteria to determine the quality of a segmentation process [10], [15] are the following:

1. Minimizing the overall representation error (*total_error*)
2. Minimizing the number of segments such that the representation error is less than a certain value (*max_segment_error*)
3. Minimizing the number of segments so that the total representation error does not exceed *total_error*

where *total_error* and *max_segment_error* are user defined parameters for the algorithm. The segmentation problem, seen as a multiobjective optimization problem, can be defined with (1)

$$T=\{\vec{x}_k\}\rightarrow S(T)=\{B_m\}\rightarrow B_m=\{\vec{x}_j\}j\in[k_{\min}...k_{\max}]\rightarrow \underset{\max}{\min} f_{quality}(\{B_m\}) \quad (1)$$

where T is the original trajectory, $\vec{x}_k$ are the points belonging to it, S(T) is the segmentation process, B_m is a given resultant segment from that process and $f_{quality}(\{B_m\})$ are the quality metrics used. Particularizing this general formulation to the criteria presented, the segmentation problem is defined in (2)

$$S(T)=\{B_m\}\rightarrow\{\vec{x}_j\}j\in[k_{\min}...k_{\max}],m\in[1...seg_{num}]\rightarrow\min\{d(S(T),T),seg_{num}\}$$
$$\begin{cases} d(S(T),T)\le total_{error} \\ \forall m, d(f_{ap}(B_m),B_m)\le \max\_segment\_serror \end{cases} \quad (2)$$

where $d(x, y)$ is a distance error function between segments x and y, $f_{ap}(x)$ is the approximation function result over segment x (in PLR the resulting line which approximates the data in segment x), and seg_{num} is the number of segments obtained by the applied segmentation algorithm.

Among offline algorithms, the bottom up technique is usually reported to produce the best results [10], so this will be the chosen technique to compare its results with the multi-objective approach presented. The heuristic applied by this technique consists in an initial division of the trajectory into its finest possible set of segments, followed by an iterative merge of these segments until no pair of segments can be merged without obtaining a segment with an error above the *max_segment_error* boundary. An overview of this process is shown in figure 1:

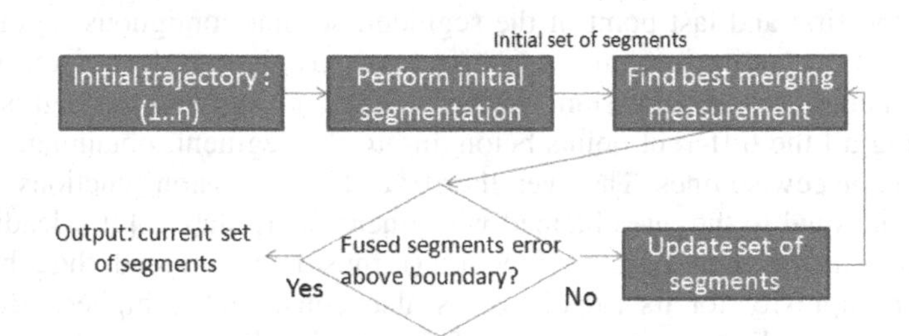

Fig. 1 Overview of the different operations in the bottom up segmentation algorithm

3 Multi Objective Evolutionary Algorithms Configuration for the PLR Segmentation Problem

Multi-objective evolutionary algorithms have reached an enormous expansion in their use in the recent years, helped by their implementation in tools such as PISA [3] or the general metaheuristics environment PARADISEO [14], which allow the user to focus on his particular problem. This section will cover the problem dependent issues which must be implemented under these tools in order to resolve the PLR problem

The PLR segmentation problem requires a codification which allows expressing a variable number of segments (with values ranging from one to the series length) represented by the position of their boundaries in the time series. According to these boundaries the chosen codification was a vector of integer values (which represent the number of a measurement in the time series) with a length of the number of measurements in the time series, *n,* minus two. These values represent the intermediate segmentation points in the trajectory (the first point of the first segment is always one, not included in the representation, and the last one is always *n*, neither included in the codification). This representation must be sorted, repeated gene values are ignored. Figure 2 shows an example.

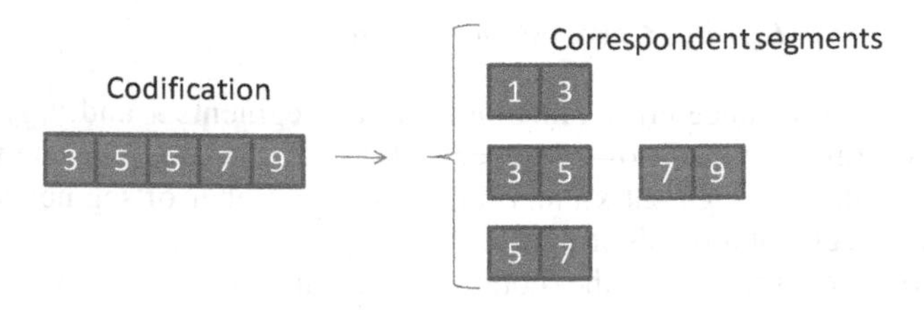

Fig. 2 Codification of an example solution for a time series of length nine

The distance function used to measure the error is the Euclidean Distance calculating the sum of squares over the least squares regression line. To evaluate a given solution, the algorithm analyzes the codification components sequentially, adding one to the number of segments and the calculated Euclidean distance value to the error whenever it finds a different value, until the maximum possible value is found (n) or the vector ends. This sequential evaluation of the chromosome requires it to be sorted (in order to obtain the output segments of the solution and be able to calculate the objective functions values). This leads to the application of a sorting procedure after chromosome modifications.

The initialization function seeks to introduce the highest possible diversity into the random initial population of the MOEA. According to the genotype, that was approached by means of a random choice of an integer value for every gene. However, as shown in figure 3, this lead to a very poor diversity in the phenotype values (especially regarding the number of segments) which lead to poor final results.

To resolve this issue, the following alternative initialization was designed: a certain number of segments are randomly chosen, followed by the random choice of the extreme values for those segments, duplicating the values where necessary to fill the codification vector completely. The results obtained were better, but were highly dependant on the initial boundaries values, leading to a final initialization function which uses one or the other alternative randomly.

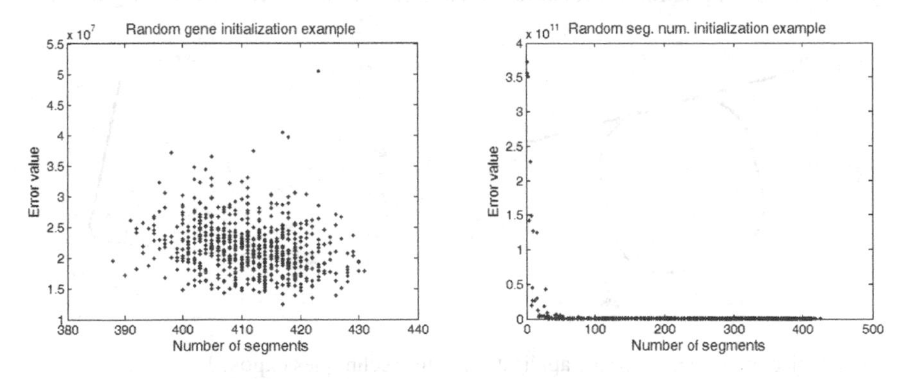

Fig. 3 Comparison of the objective function values obtained for an initial population of size 650 with the proposed initialization methods individually applied

The crossover transformation function is a standard crossover with two split points. The mutation function however, presented similar difficulties to the ones introduced in the initialization. The initial choice was to mutate a certain number of genes according to a *gene_mutation_probability* to a random integer value defined by an *epsilon* percentage (referred to the trajectory length value). This mutation biased the evolution towards those solutions with the highest number of segments. A complementary method was introduced: whenever a gene was chosen to mutate its value, it could either change according to the random mutation exposed or change its value to one of its surrounding genes. This mechanism was used to allow the mutation operator to increase or decrease the number of segments in the mutated chromosome. In practice, this approach obtained more disperse final Pareto fronts than the random mutation, but the evolution was not satisfactory (the results obtained were worse than those of the bottom up technique). The final implemented mutation operation applies one or the other mechanism randomly to the whole chromosome (instead of a random application to every individual gene).

Several MOEAs from the PARADISEO framework have been used to test the proposed operations, (their individual results cannot be presented due to space restrictions), obtaining SPEA2 [16] the best results (probably due to its archive use). This will be the chosen algorithm for the final results, with an archive size equal to the time series length minus one (to be able to store, ideally, one solution for every possible number of segments).

4 Experimental Validation

The proposed MOEA configuration has been tested on an Air Traffic Control test set similar to the one used in [9]. This test set includes the measurements recorded by sensor devices of different trajectories performed by aircrafts (with an added measuring error). Due to space restrictions, the results for only two of these trajectories can be shown, being the chosen ones a racetrack (the trajectory performed by aircrafts during landing procedures) and a turn trajectory, shown in figure 4:

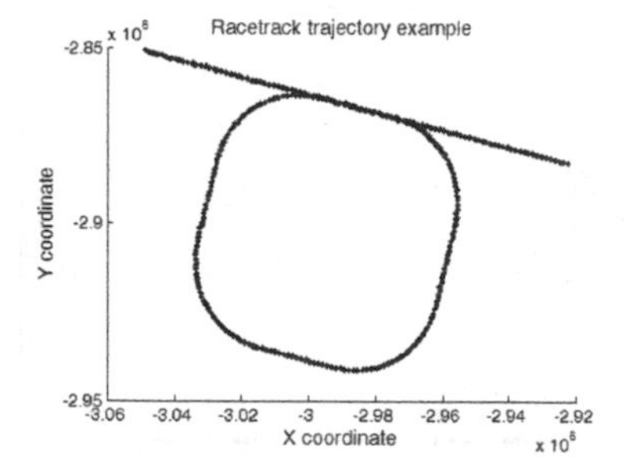

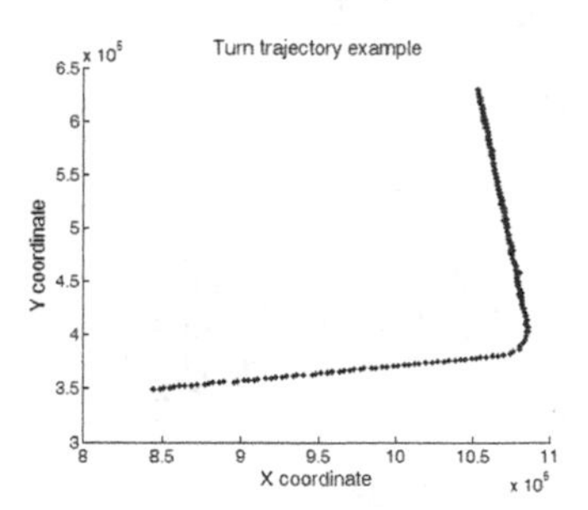

Fig. 4 Trajectories chosen for the application of the techniques exposed

Along with the introduced codification vector, a different one with size *n/2* was also tested, in order to focus in solutions with a smaller number of segments. Table 1 shows the configuration parameters:

Table 1 Parameter configuration for the MOEA algorithm

Parameter	Value	Parameter	Value
initial population	3000	mutation epsilon	0.2
mut. probability: chrom. / gene	0.3 / 0.01	Crossover probability	0.5
reduced codif. iterations	500	complete codif. iterations	1000

Figure 5 presents the obtained Pareto Fronts for the presented solutions. For the bottom up algorithm, the presented front was obtained by a trial process with different *max_segment_error* values, focusing in the search space zone corresponding to a number of segments around 50% of the number of measurements in the time series. The MOEA solution is composed of the non-dominated solutions obtained both in the whole codification and the reduced one. This approach exhibits better results than the bottom up alternative in the whole Pareto front

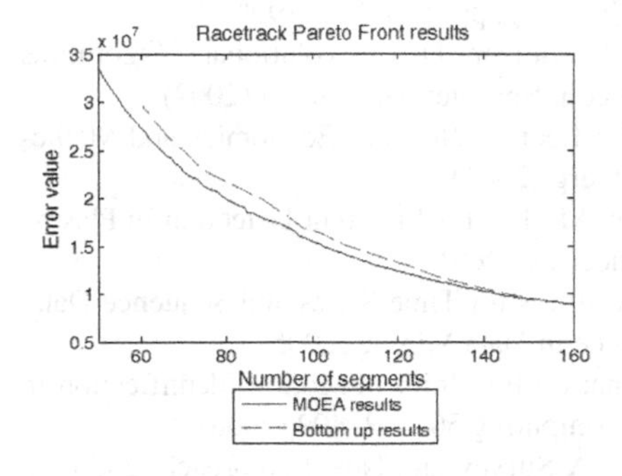

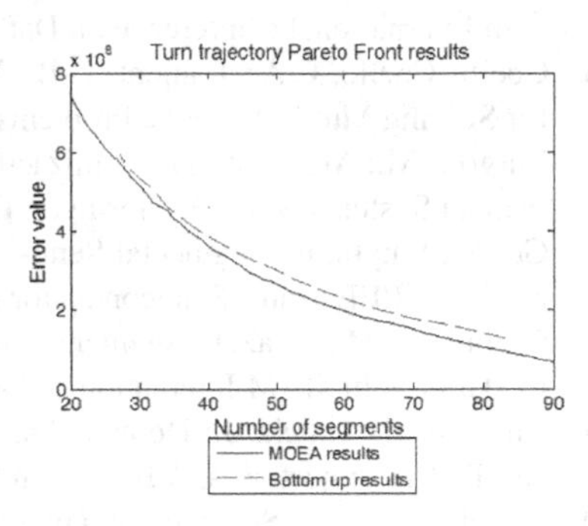

Fig. 5 Pareto Front results comparison for the two selected trajectories

5 Conclusions

Segmentation is a requirement to process the huge amount of data in actual time series. Among the variety of techniques which can be applied for this process, Piecewise Linear Representation is the most extended approach, probably due to its easy implementation. The results presented in this work show that this process can be faced with a Multi objective evolutionary algorithm obtaining better results than a classical offline technique reported to be very accurate: bottom up segmentation. Obviously, due to its computational complexity, these approaches cannot be used as a general segmentation technique, but their results can be useful for the development of new heuristics or as a tool for quality assessment. Future lines include the analysis of the results over a wider set of test problems, the comparison with curve approximation algorithms and the optimization of the configuration (for example with the inclusion of a global stopping criteria) to improve the time required to obtain the solutions.

Acknowledgments. This work was supported in part by Projects CICYT TIN2008-06742-C02-02/TSI, CICYT TEC2008-06732-C02-02/TEC, CAM CONTEXTS (S2009/TIC-1485) and DPS2008-07029-C02-02.

References

1. Agrawal, R., Faloutsos, C., Swami, A.: Efficient similarity search in sequence databases. In: Proceeding of the 4th Conference of Data Organization and Algorithms, pp. 69–84 (1993)

2. Agrawal, R., Lin, K.I., Sawhey, H.S., Shim, K.: Fast similarity search in the presence of noise, scaling and translation in time-series databases. In: Proceedings of 21st International Conference on Very Large Databases, pp. 490–501 (1995)
3. Bleuler, S., Laumanns, M., Thiele, L., Zitzler, E.: PISA - A Platform and Programming Language Independent Interface for Search Algorithms. In: Fonseca, C.M., Fleming, P.J., Zitzler, E., Deb, K., Thiele, L. (eds.) EMO 2003. LNCS, vol. 2632, pp. 494–508. Springer, Heidelberg (2003)
4. Chan, K., Fu, W.: Efficient time series matching by wavelets. In: Proceedings of the 15th International Conference on Data Engineering, pp. 126–133 (1999)
5. Coello Coello, C.A., Lamont, G.B., Van Veldhuizen, D.A.: Evolutionary Algorithms for Solving Multi-Objective Problems, 2nd edn. Springer, Heidelberg (2007)
6. Ehrgott, M.: Multicriteria optimizaction. In: Lecture Notes in Economics and Mathematical Systems, vol. 491, Springer, Heidelberg (2005)
7. Ge, X., Smyth, P.: Segmental Semi-Markov Models for Endpoint Detection in Plasma Etching. IEEE Trans. Semiconductor Engineering (2001)
8. Gionis, A., Mannila, H.: Segmentation algorithms for Time Series and Sequence Data. In: Tutorial in SIAM International Conference in Data Mining (2005)
9. Guerrero, J., Garcia, J.: Domain Transformation for Uniform Motion Identification in Air Traffic Trajectories. Advances in Soft Computing 50, 403–409 (2008)
10. Keogh, E., et al.: Segmenting Time Series: A Survey and Novel Approach, 2nd edn. Data Mining in Time Series Databases, pp. 1–21. World Scientific, Singapore (2003)
11. Keogh, E., Pazzani, M.: An enhanced representation of time series which allows fast and accurate classification, clustering and relevance feedback. In: Proceedings of the 4th International Conference of Knowledge Discovery and Data Mining (1998)
12. Keogh, E., Chakrabarti, K., Pazzani, M., Mehrotra, S.: Dimensionality Reduction for Fast Similarity Search in Large Time Series Databases. Journal of Knowledge and Information Systems 3(3), 263–286 (2001)
13. Koski, A., Juhola, M., Meriste, M.: Syntactic Recognition of ECG Signals by Attributed Finite Automata. Pattern Recognition 28(12), 1927–1940 (1995)
14. Liefooghe, A., Jourdan, L., Talbi, E.-G.: A Unified Model for Evolutionary Multiobjective Optimization and its Implementation in a General Purpose Software Framework: ParadisEO-MOEO. Research Report RR-6906, INRIA (2009)
15. Liu, X., Lin, Z., Wang, H.: Novel Online Methods for Time Series Segmentation. IEEE Trans. on Knowledge and Data Engineering 20(12) (2008)
16. Zitzler, E., Laumanns, M., Thiele, L.: SPEA2: Improving the Strength Pareto Evolutionary Approach. In: EUROGEN 2001. Evolutionary Methods for Design, Optimization and Control with Applications to Industrial Problems, pp. 95–100 (2002)

An Architecture to Provide Context-Aware Services by Means of Conversational Agents

David Griol, Nayat Sánchez-Pi, Javier Carbó, and José M. Molina

Abstract. In human-human interaction, a great deal of information is conveyed without explicit communication. This context information characterizes the situation of the different entities involved in the communication process (users, place, environment and computational objects). In this paper, we present an agent-based architecture that incorporates this valuable information to provide the most adapted service to the user. One of the main characteristics of our proposal is the incorporation of conversational agents handling different domains and adapted taking into account the different users requirements and preferences by means of a context manager. This way, we ensure a natural communication between the user and the system to provide a personalized service. The implementation of our proposed architecture to develop and evaluate a context-aware railway information system is also described.

1 Introduction

Ambient Intelligence (AmI) emphasizes on greater user-friendliness, more efficient services support, user-empowerment, and support for human interactions.To achieve this goal it is necessary to provide an effective, easy, safe and transparent interaction between the user and the system. To do so, in the last years there has been an increasing interest in simulating human-to-human communication, employing the so-called conversational agents [5]. Conversational agents, which enhance agents with computational linguistics, have became a strong alternative to provide computers with intelligent communicative capabilities.

In the literature, there are several approaches developing platforms, frameworks and agents applications for offering context-aware services [4, 1, 6]. The processing of context is essential in conversational agents to achieve an adapted behaviour and also cope with the ambiguities derived from the use of natural language. For

David Griol, Nayat Sánchez-Pi, Javier Carbó, and José M. Molina
Group of Applied Artificial Intelligence (GIAA), Computer Science Department,
Carlos III University of Madrid
e-mail: {david.griol,nayat.sanchez,javier.carbo}@uc3m.es
josemanuel.molina@uc3m.es

A.P. de Leon F. de Carvalho et al. (Eds.): Distrib. Computing & Artif. Intell., AISC 79, pp. 275–282.
springerlink.com

instance, context information can be used to resolve anaphoric references, to take into account the current user position as a data to be used by the system or to decide the strategy to be used by the dialog management module by taking into account the specific user preferences [8]. For this reason, the result of the interaction can be completely different depending on the environment conditions (e.g. people speaking near the system, noise generated by other devices) and user skills. This information usually describes the user state (e.g. communication context and preferences) and the environment state (e.g. location and temporal context). One of the most important goals of our work is to combine both sources of information to carry out the system adaptation.

For this reason, in this paper we describe an agent-based architecture that provides context awareness and situation sensitivity to discover, process and provide context aware services adapted to users' location, preferences and needs. In order to ensure a natural and intelligent interaction between the user and the system, advanced conversational agents have been developed including a context manager module to facilitate the provisioning of personalized services similar to human-human communication. We also provide a complete implementation of our architecture for the provisioning of personalized services to users in a railway information system and evaluate the influence of context information in the operation of the dialog model.

2 Our Proposed Multi-agent architecture

The proposed agent-based architecture manages context information to provide personalized services by means of users interactions with conversational agents. As it can be observed in Figure 1, it consists of five different types of agents that cooperate to provide an adapted service. *User agents* are configured into mobile devices or PDAs. *Provider Agents* are implemented by means of *Conversational Agents* that provide the specific services. A *Facilitator Agent* links the different positions to the providers and services defined in the system. A *Positioning Agent* communicates with the ARUBA positioning system [7] to extract and transmit positioning information to other agents in the system. Finally, a *Log Analyzer Agent* generates user profiles that are used by Conversational Agents to adapt their behaviour taking into account the preferences detected in the users' previous dialogs.

Eight concepts have been defined for the ontology of the system. The definition is: *Location* (*XCoordinate* int, *YCoordinate* int), *Place* (*Building* int, *Floor* int), *Service* (*Name* String), *Product* (*Name* String, *Characteristics* List of Features), *Feature* (*Name* String, *Value* String), *Context* (*Name* String, *Characteristics* List of Features), *Profile* (*Name* String, *Characteristics* List of Features), *DialogLog* (*Log* List of Strings).

Our ontology also includes six predicates with the following arguments: *HasLocation* (*Place*, *Position*, and *AgentID*), *HasServices* (*Place*, *Position*, and *List of Services*), *isProvider* (Place, Position, AgentID, Service), *HasContext* (*What*, *Who*),

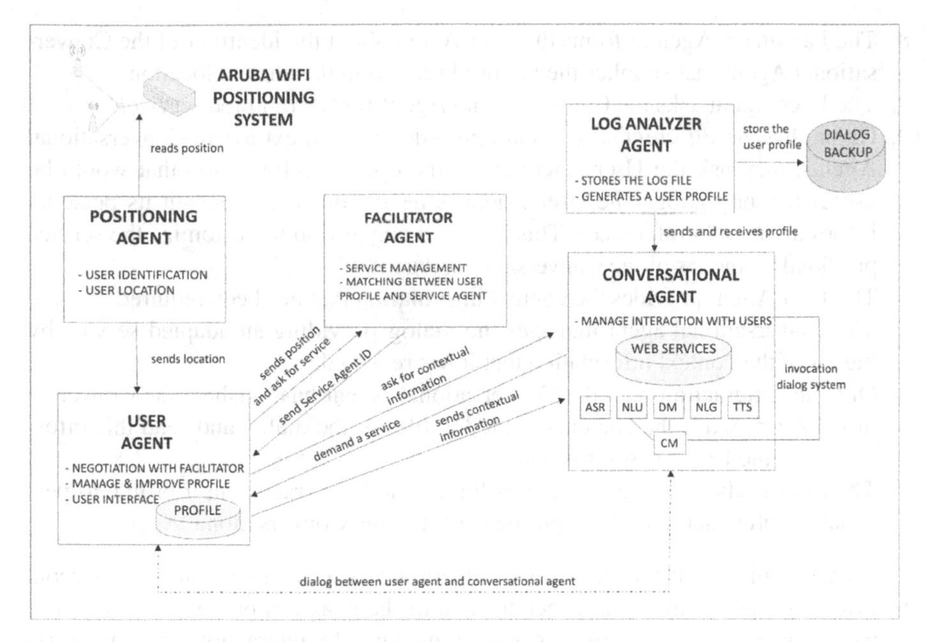

Fig. 1 Schema of the proposed multi-agent architecture

HasDialog (*DialogLog* and *AgentID*), *HasProfile* (*Profile* and *AgentID*), and *Provide* (*Product* and *AgentID*).

The interaction with the different agents follows a process which consists of the following phases:

1. The ARUBA positioning system is used to extract information about the positions of the different agents in the system. This way, it is possible to know the positions of the different User Agents and thus extract information about the Conversational Agents that are available in the current location.
2. The Positioning Agent reads the information about position (coordinates x and y) and place (*Building* and *Floor*) provided by the ARUBA Positioning Agent by reading it from a file, or by processing manually introduced data.
3. The Positioning Agent communicates the position and place information to the User Agent.
4. Once a User Agent is aware of its own location, it communicates this information to the Facilitator Agent in order to find out the different services available in that location.
5. The Facilitator Agent informs the User Agent about the services available in this position .
6. The User Agent decides the services in which it is interested.
7. Once the User Agent has selected a specific service, it communicates its decision to the Facilitator Agent and queries it about the service providers that are available.

8. The Facilitator Agent informs the User Agent about the identifier of the Conversational Agent that supplies the required service in the current location.
9. The User Agent asks the Conversational Agent for the required service.
10. Given that the different services are provided by context-aware Conversational Agents, they ask the User Agent about the context information that would be useful for the dialog. The User Agent is never forced to transmit its personal information and preferences. This is only a suggestion to customize the service provided by means of the Conversational Agent.
11. The User Agent provides the context information that has been required.
12. The conversational agent manages the dialog providing an adapted service by means of the context information that it has received.
13. Once the interaction with the Conversational Agent has finished, the Conversational Agent reads the contents of the log file for the dialog and send this information to the Log Analyzer Agent.
14. The Log Analyzer Agent stores this log file and generates a user profile to personalize future services. This profile is sent to the Conversational Agent.

As stated in the introduction, a conversational agent is a software that accepts natural language as input and generates natural language as output, engaging in a conversation with the user. To successfully manage the interaction with the users, conversational agents usually carry out five main tasks: automatic speech recognition (ASR), natural language understanding (NLU), dialog management (DM), natural language generation (NLG) and text-to-speech synthesis (TTS). These tasks are usually implemented in different modules.

In our architecture, we incorporate a Context Manager in the architecture of the designed conversational agents, This module deals with loading the context information provided by the User and Positioning Agents, and communicates it to the different modules of the Conversational Agent during the interaction.

To manage context information we have defined a data structure called *user profile*. Context information in our user profile can be classified into three different groups. *General user information* stores user's name and machine identifier, gender, prefered language, pathologies or speech disorders, age, *Users Skill level* is estimated by taking into account variables like the number of previous sessions, dialogs and dialog turns, their durations, time that was necessary to access a specific web service, the date of the last interaction with the system, etc. Using these measures a low, medium, normal, high or expert level is assigned. *Usage statistics and preferences* are automatically evaluated taking into account the set of services most required by the user during the previous dialogs, date and hour of the previous interactions and preferred output modality.

The free software JADE (Java Agent DEvelopment Framework)[1] has been used for the implementation of our architecture. It was the most convenient option as it simplifies the implementation of multi-agent systems through a middle-ware that complies with the FIPA specifications and through a set of graphical tools that

[1] http://jade.tilab.com/

supports the debugging and deployment phases. The agent platform can be distributed across machines and the configuration can be controlled via a remote GUI.

3 Case Study: A Context-Aware Railway Information System

We have applied our context aware methodology to design an adaptive system that provides information in natural language about train services, types, schedules, and fares. The compiled code of this application can be executed and is freely available through our website. The requirements for the task have been specified by taking into account the ontologies defined for this task in the DIHANA project [3]. Users can ask for information about *Hour*, *Price*, *Train-Type*, *Trip-Time*, and *Services*. They also can provide task-independent information like *Affirmation*, *Negation*, and *Not-Understood*. The attributes needed by the system to answer to the different user queries are *Origin*, *Destination*, *Departure-Date*, *Arrival-Date*, *Ticket-Class*, *Departure-Hour*, *Arrival-Hour*, *Train-Type*, *Order-Number*, and *Services*. The system responses can be classified into the following categories: *Opening*, *Closing*, *Not-Understood*, *Waiting*, *New-Query*, *Acceptance*, *Confirmation*, *Rejection*, *Question*, and *Answer*. A total of 51 different system actions were defined taking into account the information that the system provides, asks or confirms to the user.

A set of scenarios were manually defined to cover the different queries to perform to the system including different user requirements and defining one or several objectives (e.g., to obtain timetables and prices given a specific origin, destination and date). An example of these scenarios is as follows:

```
User name: John Smith
Location: Atocha Station
Date and Time: 2009-05-01, 9:00am
Device: PDAQ 00-18-41-32-0B-59
Objective: To know timetables and prices to Valencia
```

The first step is to execute the different agents using the JADE platform. Then the information about the position is sent from the Positioning Agent to the User Agent as shown in Figure 2. In the figure, the Positioning Agent is called Sensor Network and the name of the User Agent is John Smith. This figure shows the message that is sent by the Positioning Agent and received by the User Agent and how it sets the new position values.

In the next phase, the User Agent John Smith looks for the Facilitator Agent and ask it about the services that are available in the user's location. The following phase consists of selecting the service. From the list of two services that are available in the current location (TrainTicket and TrainInfo), one of them has to be selected. The service TrainTicket is selected and the Facilitator Agent answers the user's query by informing about the AgentID of the Provider Agent which provides the required service. Once the User Agent knows how to contact with the Conversational Agent that supplies the TrainTicket service, it sends a query to be provided with this service. Figure 3 shows this query to the Conversational Agent, called in the figure "Dialog System" Agent.

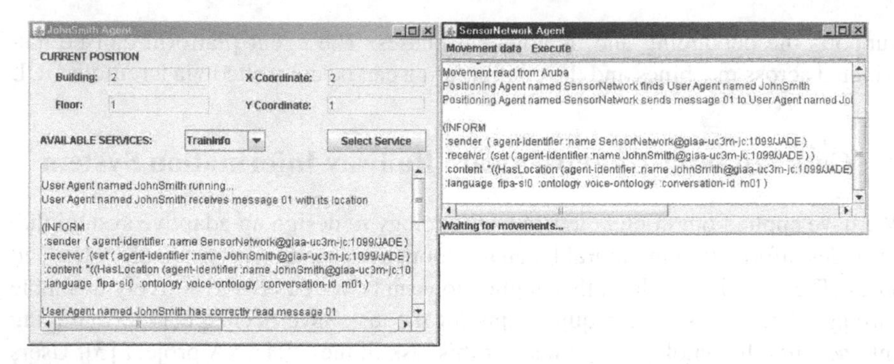

Fig. 2 Positioning information sent from the Positioning Agent to the User Agent

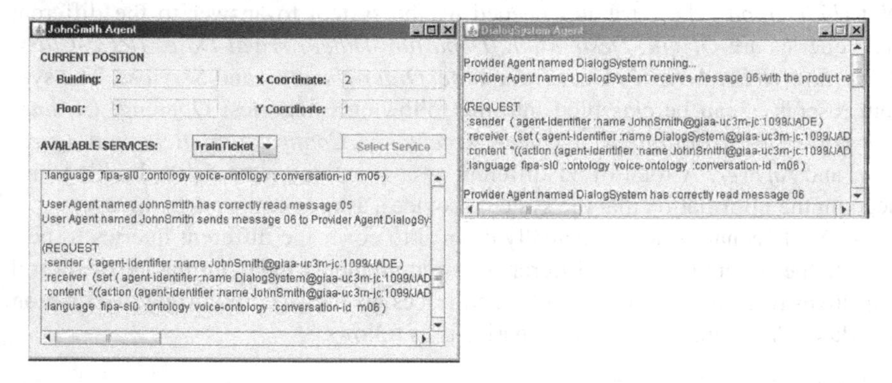

Fig. 3 Query to the Conversational Agent that provides the service

Then, the Provider Agent asks the User Agent about the context information that can be useful for the Conversational Agent to provide an adapted service. The User Agent decides which context information to send as a result of the previous query and transmits this information to the Provider Agent. Once the Conversational Agent has received the context information, it transmits this information to the Context Manager that is included in its architecture. The rest of modules of the Conversational Agent requires the Context Manager for the specific information that they require to adapt its behaviour using the user profile.

Once the interaction with the Conversational Agent has finished, this agent communicates the result of the dialog to the LogAnalyzer Agent. Once the results of this dialog have been interpreted, the LogAnalyzer Agent adds this information to the set of previous dialogs that have been stored regarding the interactions of the User Agent John Smith with the Conversational Agent DialogSystem. This way, this Provider Agent can provide the most adapted service taking into account the specific user's preferences detected in the previous dialogs. This process is automatically carried out, reducing the participation of the user in this process.

To evaluate our architecture, we acquired ten different dialogs for each one of the scenarios considering or not the inclusion of the Context Manager in our

Table 1 Results of the high-level dialog features for the comparison of the two kinds of dialogs

	Without Context Inform.	**Using Context Inform.**
Percentage of successful dialogs	61.6%	78.4%
Average number of turns per dialog	12.6	6.2
Percentage of different dialogs	92.9%	71.9%
Number of repetitions of the most seen dialog	5	12
Number of turns of the most seen dialog	9	5
Number of turns of the shortest dialog	5	5
Number of turns of the longest dialog	25	17

architecture. A dialog simulation technique has been used to acquire the total of 1000 successful dialogs whether introducing the modules that manage context information in our architecture or not [2]. In this technique we automatically acquire a dialog corpus by means of the introduction in our architecture of a dialog manager simulator and a user simulator. A user request for closing the dialog is selected once the system has provided the information defined in the objective(s) of the dialog. The dialogs that fulfill this condition before a maximum number of turns are considered successful. Using this technique it is possible to carry out a detailed exploration of the dialog space and reduce the effort that would be necessary with real users.

We defined seven measures for the comparison of the dialogs acquired using or not context information: the percentage of successful dialogs, the average number of turns per dialog, the percentage of different dialogs, the number of repetitions of the most seen dialog, the number of turns of the most seen dialog, the number of turns of the shortest dialog, and the number of turns of the longest dialog. Using these measures, we tried to evaluate the success of our approach as well as its efficiency with regard to the different objectives specified in the scenarios. Table 1 shows the comparison of these measures. As it can be observed, the first advantage of our approach is regarding the number of dialogs that was necessary to simulate in order to obtain the 1000 successful dialogs of each kind. While, only a 61.6% of successful dialogs is obtained using context information, this percentage increases to 78.4% when the context manager is introduced. The second improvement is the reduction in the number of turns. Using the context manager it is possible to obtain a reduction greater than 50% in the average number of turns per dialog. This is due to the context-aware system requires less information from the user, therefore the possibility of the ASR module introducing errors and the number of data confirmations are reduced. This reduction can also be observed in the number of turns of the longest, shortest and most seen dialogs. Finally, the number of different dialogs is lower using the context information due to the reduction in the number of turns, as can be observed in the number of repetitions of the most seen dialog.

4 Conclusions

In this paper, we have presented a multi-agent architecture for developing context aware adaptable services. This allow us to deal with the increasing complexity that the design of this kind of systems requires, adapting the services that are provided by taking into account the user requirements and preferences by means of a context manager. We have described the different agents that are included and the modules that make it up. One of the main characteristics of our architecture is to ensure a natural and intelligent interaction by means of a conversational agent, whose modules are adapted by means of the context information contained in specific user profiles. The results of the application of our methodology to design a railway information system shows how the main characteristics of the dialogs can be improved by taking into account context information. As a future work, we want to evaluate our methodology with real users and also carry out a detailed study of the user rejections of system-hypothesized actions. Finally, we also want to apply our technique to carry out more complex tasks in which conversational agents no only provide information but also carry out additional functionalities.

Acknowledgements. Funded by projects CICYT TIN2008-06742-C02-02/TSI, CICYT TEC2008-06732-C02-02/TEC, SINPROB, CAM MADRINET S-0505/TIC/0255, and DPS2008-07029-C02-02.

References

1. Bajo, J., Julian, V., Corchado, J.M., Carrascosa, C., De Paz, Y., Botti, V., De Paz, J.F.: An execution time planner for the artis agent architecture. Journal of Engineering Applications of Artificial Intelligence 21(8), 769–784 (2008)
2. Griol, D., Hurtado, L., Sanchis, E., Segarra, E.: Acquiring and Evaluating a Dialog Corpus through a Dialog Simulation Technique. In: Proc. of the 8th SIGdial, pp. 39–42 (2007)
3. Griol, D., Hurtado, L.F., Segarra, E., Sanchis, E.: A Statistical Approach to Spoken Dialog Systems Design and Evaluation. Speech Communication 50(8-9), 666–682 (2008)
4. Jong-yi, H., Eui-ho, S., Sung-Jin, K.: Context-Aware Systems: A Literature Review and Classification. Expert Systems with Applications 36(4), 8509–8522 (2009)
5. McTear, M.F.: Spoken dialogue technology. Springer, Heidelberg (2004)
6. Nieto-Carvajal, I., Botía, J., Ruiz, P., G'omez-Skarmeta, A.: Implementation and evaluation of a location-aware wireless multi-agent system. In: Proc. of EUC 2004, pp. 528–537 (2004)
7. Sánchez-Pi, N., Fuentes, V., Carbó, J., Molina, J.: Knowledge-based system to define context in commercial applications. In: Proc. of SNPD 2007, pp. 694–699 (2007)
8. Seneff, S., Adler, M., Glass, J., Sherry, B., Hazen, T., Wang, C., Wu, T.: Exploiting Context Information in Spoken Dialogue Interaction with Mobile Devices. In: Proc. of Int. Workshop on Improved Mobile User Experience, pp. 1–11 (2007)

A Conversational Academic Assistant for the Interaction in Virtual Worlds

D. Griol, E. Rojo, Á. Arroyo, M.A. Patricio, and J.M. Molina

Abstract. The current interest and extension of social networking are rapidly introducing a large number of applications that originate new communication and interaction forms among their users. Social networks and virtual worlds, thus represent a perfect environment for interacting with applications that use multimodal information and are able to adapt to the specific characteristics and preferences of each user. As an example of this application, in this paper we present an example of the integration of conversational agents in social networks, describing the development of a conversational avatar that provides academic information in the virtual world of Second Life. For its implementation techniques from Speech Technologies and Natural Language Processing have been used to allow a more natural interaction with the system using voice.

1 Introduction

Social Networking has been a global consumer phenomenon during the last few years [10]. The staggering increase in the amount of time people are spending on these sites is changing the way people spend their time online and influence on how people behave, share and interact within their normal daily lives. The development of so-called Web 2.0 has also made possible the introduction of a number of applications into many users' lives, which are profoundly changing the roots of society by creating new ways of communication and cooperation.

The advance of social networking has entailed a considerable progress in the development of virtual worlds [8, 1]. Virtual robots (*metabots*), with the same appearance and capabilities that the avatars for human users, thus intensify the perception of the virtual world, providing gestures, glances, facial expressions and movements

D. Griol, E. Rojo, M.A. Patricio, and J.M. Molina
Computer Science Department. Carlos III University of Madrid
e-mail: {david.griol,eduardo.rojo,miguelangel.patricio}@uc3m.es,
josemanuel.molina@uc3m.es

Á. Arroyo
Applied Intelligent Systems Department. Technical University of Madrid
e-mail: aarroyo@eui.upm.es

A.P. de Leon F. de Carvalho et al. (Eds.): Distrib. Computing & Artif. Intell., AISC 79, pp. 283–290.
springerlink.com

necessary for the communication process. Therefore, these virtual environments are very useful to enhance human-machine interaction.

This way, virtual worlds have become real social networks useful for the interaction between people from different places who can socialize, learn, be entertained, etc. Thanks to the social potential of virtual worlds, they have also become an attraction for institutions, companies and researchers with the purpose of developing virtual robots with the same look and capabilities of avatars for human users. However, social interaction in virtual worlds are usually carried out using only text communication by means of chat-type services. In order to enhance communication in these environments, we propose the integration of conversational agents to develop intelligent metabots with the ability of oral communication and, at the same time, which benefit from the visual modalities provided by these virtual worlds.

Conversational agents [9, 7, 6] can be defined as automatic systems that are able of emulating a human being in a dialog with another person, in order to complete a specific task (usually providing information or perform a particular task.) Two main objectives are fulfilled thanks to its use. The first objective is to facilitate a more natural human-machine interaction using the voice. The second one allows the accessibility for users with motor disabilities, so that the interface avoids the use of traditional interfaces, such as keyboard and mouse.

Our work focuses on three key points. Firstly, since it is very difficult to find studies in the literature that describe the integration of Speech Technologies and Natural Language Processing in virtual worlds, to show that this integration is possible. Secondly, to show a practical application of this integration through the development of a conversational metabot that provides academic information in the virtual world Second Life. Finally, we promote the use of open source applications and tools for the creation and interaction in virtual worlds, such as OpenSim[1] and OsGrid[2].

2 Conversational Agents

As stated in the introduction, a conversational agent is a software that accepts natural language as input and generates natural language as output, engaging in a conversation with the user. To successfully manage the interaction with the users, conversational agents usually carry out five main tasks: automatic speech recognition (ASR), natural language understanding (NLU), dialog management (DM), natural language generation (NLG) and text-to-speech synthesis (TTS). These tasks are usually implemented in different modules.

Speech recognition is the process of obtaining the text string corresponding to an acoustic input. It is a very complex task as there is much variability in the input characteristics, which can differ depending on the linguistics of the utterance, the speaker, the interaction context and the transmission channel. Different applications demand different complexity of the speech recognizer. [4] identify eight parameters that allow an optimal tailoring of the speech recognizer: speech mode, speech style,

[1] http://opensimulator.org
[2] http://www.osgrid.org

dependency, vocabulary, language model, perplexity, SNR and transductor. Regarding the speech mode, speech recognizers can be classified into isolated-word or continuous-speech recognizers. Regarding the speech style, discourse can be read or spontaneous, the latter has peculiarities such as hesitations and repetitions that make it more complex to recognize.

Once the conversational agent has recognized what the user uttered, it is necessary to understand what he said. Natural language processing is the process of obtaining the semantic of a text string. It generally involves morphological, lexical, syntactical, semantic, discourse and pragmatical knowledge. In a first stage lexical and morphological knowledge allow dividing the words in their constituents distinguishing lexemes and morphemes. Syntactic analysis yields a hierarchical structure of the sentences, however in spoken language frequently phrases are affected by the difficulties that are associated to the so-called disfluency phenomena: filled pauses, repetitions, syntactic incompleteness and repairs [5].

Semantic analysis extracts the meaning of a complex syntactic structure from the meaning of its constituents. In the pragmatic and discourse processing stage, the sentences are interpreted in the context of the whole dialog, the main complexity of the stage is the resolution of anaphora, and ambiguities derived from phenomena such as irony, sarcasm or double entendre.

There is not a universally agreed upon definition of the tasks that the dialog management module has to carry. [11] state that dialog managing involves four main tasks: i) updating the dialog context, ii) providing a context for interpretations, iii) coordinating other modules and iv) deciding the information to convey and when to do it. Thus, the dialog manager has to deal with different sources of information such as the NLU results, database queries results, application domain knowledge, knowledge about the users and the previous dialog history. Its complexity depends on the task and the dialog flexibility and initiative. When it is necessary to execute and monitor operations in a dynamically changing application domain, an agent-based approach can be employed to develop the dialog management module. The modular agent-based approach to dialog management makes it possible to combine the benefits of different dialog control models, such as finite-state based dialog control and frame-based dialog managing Similarly, it can benefit from alternative dialog management strategies, such as the system-initiative approach and the mixed-initiative approach Furthermore, it makes it possible to combine rule-based and machine learning approaches.

Natural language generation is the process of obtaining texts in natural language from a non-linguistic representation. It is usually carried out in five steps: content organization, content distribution in sentences, lexicalization, generation of referential expressions and linguistic realization. It is important to obtain legible messages, optimizing the text using referring expressions and linking words and adapting the vocabulary and the complexity of the syntactic structures to the user's linguistic expertise. The simplest approach consists in using predefined text messages (e.g. error messages and warnings). Although intuitive, this approach completely lacks from any flexibility. The next level of sophistication is template-based generation, in which the same message structure is produced with slight alterations. Using this approach, it is possible to provide adapted system prompts that take into account context information.

Finally, text-to-speech synthesizers transform a text into an acoustic signal. A text-to-speech system is composed of two parts: a front-end and a back-end. The front-end carries out two major tasks. Firstly, it converts raw text containing symbols such as numbers and abbreviations into their equivalent words. Secondly, it assigns a phonetic transcriptions to each word, and divides and marks the text into prosodic units, i.e. phrases, clauses, and sentences. The back-end (often referred to as the synthesizer) converts the symbolic linguistic representation into sound. On the one hand, speech synthesis can be based in human speech production, this is the case of parametric synthesis which simulates the physiological parameters of the vocal tract, and formant-based synthesis which models the vibration of vocal chords.

3 SecondLife and OpenSim

Second Life[3] (SL) is a three dimensional virtual world developed by Linden Lab in 2003 and accessible via the Internet. A free client program called the Second Life Viewer enables its users, called *Residents*, to interact with each other through avatars. Residents can explore, meet other residents, socialize, participate in individual and group activities, and create and trade virtual property and services with one another, or travel throughout the world, which residents refer to as the grid. Resident population is nowadays of millions of real people from around the world. Each person is represented by an avatar that represents their chosen digital persona. Second Life is currently used as a platform for education by many institutions, such as colleges, universities, libraries and government entities (e.g. Ohio University, Royal Opera House, Universidad Pública de Navarra, Instituto Cervantes, Universidad Carlos III de Madrid, etc.). We own an island in Second Life called TESIS, in which we built its Virtual facilities in which numerous educational activities are performed. Figure 1 shows an image of the TESIS island.

We decided to use SL as an experimental laboratory of our research for several reasons. Firstly, because it is one of the most popular social virtual worlds, and its population is now of millions of residents worldwide. Secondly, because it uses very advanced technologies for the development of realistic simulations, making avatars and the environments more credible and similar to real world users. Thirdly, because the possibility of customizing SL is extensive and supports innovation and user participation, which increases the naturalness of the interactions in the virtual world.

OpenSim is an open source simulator that uses the same standard as Second Life to communicate with their users, and emulates virtual environments independently from the world of Second Life, using its own infrastructure. OsGrid is a network that allows linking free virtual worlds developed using simulators such as OpenSim.

[3] http://secondlife.com/

Fig. 1 An image of the TESIS island in Second Life

4 Practical Implementation: Metabot That Provides Academical Information

We have developed a conversational metabot that facilitates academic information (courses, professors, doctoral studies and enrollment) based on the functionalities provided by a previously developed dialog system [2, 3]. The information provided by the metabot can be classified into four categories: subjects, teachers, doctoral studies, and registration. The system has been developed using the typical architecture of current spoken conversational agents, including a module for automatic speech recognition, a dialog manager, a module for access to databases, data storage, and the generation of an oral response through a language generator and a text to speech synthesizer, as described in the previous section.

Figure 2 shows the architecture developed for the integration of conversational metabot both in the Second Life and OpenSim virtual worlds. The conversational agent that governs the metabot is outside the virtual world, using external servers that provide both data access and speech recognition and synthesis functionalities. The speech signal provided by the text to speech synthesizer is captured and transmitted to the voice server module in Second Life (SLVoice) using code developed in Visual C #. NET and the SpeechLib library. This module is external to the client program used to display the virtual world and is based on the Vivox technology, which uses the RTP, SIP, OpenAL, TinyXPath, OpenSSL and libcurl protocols to transmit voice data. We also use the utility provided by Second Life lipsynch to synchronize the voice signal with the lip movements of the avatar.

In addition, we have integrated a keyboard emulator that allows the transmission of the text transcription generated by the conversational avatar directly to the chat in Second Life. The system connection with the virtual world is carried out by using the libOpenMetaverse library. This .Net library, based on the Client /Server

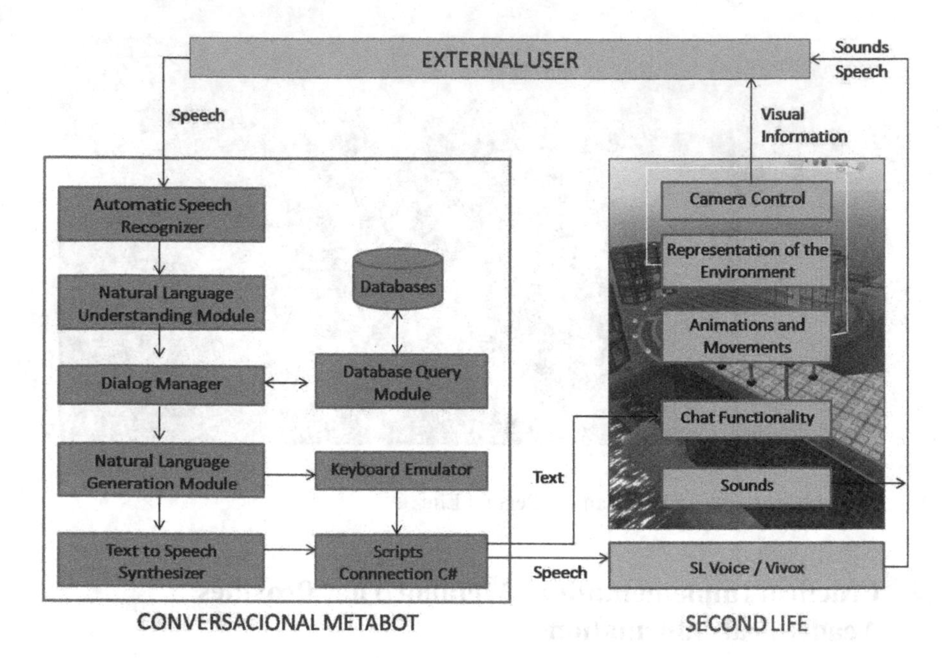

Fig. 2 Architecture defined for the development of the conversational metabot

paradigm, allows to access and create three-dimensional virtual worlds, and it is used to communicate with servers that control the virtual world of Second Life.

Speech recognition and synthesis are performed using the Microsoft Speech Application Programming Interface (SAPI), integrated into the Windows Vista operating system. To enable the interaction with the conversational in Spanish using the chat in Second Life, we have integrated synthetic voices developed by Loquendo.

Using this architecture user's utterances can be easily recognized, the transcription of these utterances can be transcribed in the chat in Second Life, and the result of the user's query can be communicated using both text and speech modalities. To do this, we have integrated modules for the semantic understanding and dialog management implemented for the original dialog system, which are based on grammars and VXML files. Using OsGrid we have also developed our own free virtual world and integrated our conversational metabot with OpenSim.

Through the participation of students and professors of our university we have acquired a set of dialogs using the same scenarios defined for a previous acquisition using the conversational agent that governs the avatar with real users outside the virtual world of Second Life. Table 2 shows the statistics of the acquisition of 50 dialogs. The main conclusion that can be extracted from this preliminary study is the absence of differences in these statistics among the dialogs acquired using only the conversational agent and the dialogs acquired by means of the interaction with the conversational metabot in Second Life. Figure 3 shows the developed metabot providing information about tutoring hours of a specific professor.

Fig. 3 Conversational metabot developed to interact in virtual worlds

Table 1 Statistics of the acquired dialogs

Average number of turns per dialog	4.99
Percentage of confirmations from the metabot	13.51%
Questions from the metabot to request information	18.44%
Prompts generated by the metabot after a database query	68.05%

5 Conclusions

The development of social networks and virtual worlds brings a wide set of opportunities and new communication channels that can be incorporated to traditional interfaces like conversational agents. In this work, we propose a methodology for creating conversational metabots which are able to interact in virtual worlds. Using our the proposal we have implemented a conversational metabot that provides academic information in Second Life. This virtual world offers a number of possibilities for the development of educational applications, given the possibility for users to socialize, explore and access a large number of educational and cultural resources. As future work we want to evaluate new features to be included in the conversational metabot to improve the communication process, and carry out a detailed analysis of the integration of different modalities for the presentation of information provided by SL in addition to the use of voice.

Acknowledgements. Funded by projects CICYT TIN2008-06742-C02-02/TSI, CICYT TEC2008-06732-C02-02/TEC, SINPROB, CAM MADRINET S-0505/TIC/0255, and DPS2008-07029-C02-02.

References

1. Arroyo, A., Serradilla, F., Calvo, O.: Multimodal agents in second life and the new agents of virtual 3d environments. In: Mira, J., Ferrández, J.M., Álvarez, J.R., de la Paz, F., Toledo, F.J. (eds.) IWINAC 2009. LNCS, vol. 5601, pp. 506–516. Springer, Heidelberg (2009)
2. Callejas, Z., López-Cózar, R.: Implementing modular dialogue systems: a case study. In: Proc. of Applied Spoken Language Interaction in Distributed Environments (ASIDE 2005), Aalborg, Denmark (2005)
3. Callejas, Z., López-Cózar, R.: Relations between de-facto criteria in the evaluation of a spoken dialogue system. Speech Communication 50(8-9), 646–665 (2008)
4. Cole, R., Zue, V.: Survey of the State of the Art in Human Language Technology, pp. 1–49. Cambridge University Press, Cambridge (1997)
5. Gibbon, D., Mertins, I., Moore, R.: Resources, Terminology and Product Evaluation. In: Handbook of Multimodal and Spoken Dialogue Systems. Kluwer International Series in Engineering and Computer Science, vol. 565. Kluwer Academic Publishers, Dordrecht (2000)
6. Griol, D., Hurtado, L., Segarra, E., Sanchis, E.: A Statistical Approach to Spoken Dialog Systems Design and Evaluation. Speech Communication 50(8-9), 666–682 (2008)
7. López-Cózar, R., Araki, M.: Spoken, Multilingual and Multimodal Dialogue Systems. In: Development and Assessment. John Wiley & Sons, Chichester (2005)
8. Lucia, A.D., Francese, R., Passero, I., Tortora, G.: Development and evaluation of a virtual campus on second life: The case of seconddmi. Computers & Education 52(1), 220–233 (2009)
9. McTear, M.F.: Spoken Dialogue Technology: Towards the Conversational User Interface. Springer, Heidelberg (2004)
10. Nielsen: Global Faces and Networked Places: A Nielsen Report on Social Networking's New Global Footprint. Nielsen Online (2009)
11. Traum, D., Larsson, S.: The Information State Approach to Dialogue Management. In: Current and New Directions in Discourse and Dialogue, pp. 325–354. Kluwer Academic Publishers, Dordrecht (2003)

Using Context-Awareness to Foster Active Lifestyles

Ana M. Bernardos, Eva Madrazo, Henar Martín, and José R. Casar

Abstract. This paper describes a context-aware mobile application which aims at adaptively motivating its users to assume active lifestyles. The application is built on a model which combines 'motion patterns' with 'activity profiles', in order to evaluate the user's real level of activity and decide which actions to take to give advice or provide feedback. In particular, a 'move-to-uncover' wallpaper puzzle interface is employed as motivating interface; at the same time, context-aware notifications are triggered when low activity levels are detected. In order to accelerate the application's design and development cycle, a mobile service oriented framework – CASanDRA Mobile - has been used and improved. CASanDRA Mobile provides standard features to facilitate context acquisition, fusion and reasoning in mobile devices, making easier access to sensors and context-aware applications cohabitation.

1 Introduction

Current mobile technologies may be especially efficient to support preventive-proactive healthcare protocols [1], as mobile devices are quickly augmenting their processing, communication, interface and embedded sensing capabilities. In the boost of mobile applications, personal healthcare has captured the attention of mobile application developers: according to a recent report [2], there are more than 5000 commercial mobile health applications available for general users, patients and healthcare professionals. The offer is wide, covering from cardio and sport training control to sleep or pregnancy monitoring or fulfillment of diet or smoking cessation programmes. For the most part, these applications include basic features such as information, monitoring, calendar, reminders, calories calculators, etc. and some of them are prepared to use location data and connect to online web 2.0 services. However they often require permanent feedback from the user, lacking automation and transparency and therefore, usability.

But mobile devices provide powerful elements to design next-generation healthcare applications, which are to be 'context-aware and persuasive' ones. The concept of 'persuasive computing' (or captology) was coined in the late nineties [3] to define the capability of technology to 'shape, reinforce or change behaviors,

Ana M. Bernardos, Eva Madrazo, Henar Martín, and José R. Casar
Universidad Politécnica de Madrid, Av. Complutense 30, 28040 Madrid (Spain)
e-mail: {abernardos,eva.madrazo,hmartín}@grpss.ssr.upm.es,
jramon@grpss.ssr.upm.es

A.P. de Leon F. de Carvalho et al. (Eds.): Distrib. Computing & Artif. Intell., AISC 79, pp. 291–299.
springerlink.com

feelings or thoughts about an issue, object or action'. Almost a decade earlier, pioneer context-aware applications began to show how the use of sensors and information processing could make possible to deliver applications capable of dynamically adapting their performance to the user's situation [4].

In this paper we address the design and development of one of these persuasive context-aware applications, in particular designed to prevent sedentary behavior. It has been demonstrated that insufficient physical activity is a health risk, which is associated to other factors - such as overweight, stress or sleep problems – that worsen the quality of life and may contribute to develop serious diseases - such as cardiovascular problems or type II diabetes mellitus. As it may be difficult to adhere to healthy activity patterns in modern lifestyle, the application aims at persuading the user to reasonably move, taking into account his personal situation.

The application is developed on top of our embeddable framework to provide Context Acquisition Services anD Reasoning Algorithms, *CASanDRA Mobile* [5]. As described below, CASanDRA Mobile relies on a light data fusion service-oriented architecture (mOSGi) and offers a set of standard off-the-shelf features to accelerate the application's design and development life cycle.

The paper is structured as follows. Section II reviews the state-of-the-art of context-aware mobile healthcare applications which use activity detection as input. Section III addresses application's design aspects. Next Section IV explains how context information is managed to achieve the application's persuasive objectives. Section V explains the implementation details on top of the CASanDRA Mobile framework. Finally, section VII concludes the work.

2 Related Work

Up to now, a good number of mobile applications dealing with activity monitoring can be found in literature. Their target users include healthy people who want to keep fit or to adopt a healthier lifestyle [6-9], but also patients suffering different types of chronic diseases [10]. Following there is a short review of some applications which combine activity monitoring with social networks [6] [8], take into account past activity personal history [6], adapt their output to real-time biometric performance [7] or aims at providing fun interfaces [8] to guarantee user's adherence.

Walkabout [6] is a mobile application designed to propose motivating walking alternatives to ordinary routes. Apart from route planning and performance monitoring, the application includes a social component, which allows inviting people to walks and receiving invitations to walks from others. Similarly, Footpaths [7] aims at suggesting outdoors walking routes taking into account the user's cardio-respiratory fitness level (which is calculated by using the Rockport 1-Mile Walk Test). A network of body sensors (two accelerometers attached at both user's legs and one ECG sensor) and a GPS-equipped mobile phone are assumed to be worn.

The UbiFit system [8] encourages regular physical activity through a glanceable display which uses the metaphor of a garden that blooms as the individual performs activities. UbiFit is connected to the Mobile Sensing Platform; MSP transmits a list of activities (walking, running, cycling, using elliptical trainer and using a stair machine) and their predicted likelihoods to the mobile phone. UbiFit's authors state that the application capability to adapt to normal life breaks (due to multiple reasons, such as colds, work changes, etc.) is important not to discourage the user, and that social networking may be a two-edged sword when dealing with self-motivation.

Mattila et al. [9] presents two 3-months user studies on the Wellness Diary, a mobile application based on Cognitive-Behavioral Therapy (CBT), which tries to foster continue self-monitoring to make the patient aware of his health goals. The correct use of the application is very demanding from the user's point of view, as requires entering data manually each time he weights himself, exercises or eats or drinks. As a result, the number of entries decreases with the time of use of the application.

Finally, [10] presents a wearable assistant for Parkinson's disease patients with the freezing of gait symptom (a sudden and transient inability to move). It uses on-body acceleration sensors to measure the patients' movements, and generates a rhythmic auditory signal to help the patient to resume walking when the symptom is detected. The work underlines to which extent the system is sensitive to the diversity of gait patterns, requiring personal calibration and adaptation.

As the reader will notice, most of the applications are prepared to work outdoors, not giving advice when the user is working or performing daily tasks at home. [6] is useful to plan daily transportation events while [8] is also prepared to monitor sport activities. With respect to their sensing needs, [6] relies on the GPS sensor embedded in the mobile device, while others require wearable accelerometers [7] [8] [10], pedometers [9], biometric sensors [8] or data annotation delivered by different devices without a wireless interface (e.g. scales) [9]. *Motion state* estimation is considered in [7] and [10].

Every application above aims at informing the users to help them to make decisions, but varies in their data gathering strategy and level of adaptation when providing feedback to the user. For example, [9] claims that automation in data acquisition may not be effective from the therapeutic point of view, as the user loses awareness of his state. But very demanding applications in terms of user interaction (both for data acquisition and feedback) may result tiring and discouraging. [8] gives feedback in an attractive way, although a nice interface is not enough if individuals are not attracted to it in decision points.

From this analysis, in the next Section we gather the design principles of our persuasive application to control sedentary behavior.

3 Design Principles

Our objective is to build a persuasive context-aware mobile application to induce individuals to holistically modify their daily life activity habits, in order to make them internalize healthier motion behaviors when at home, at work, commuting or practicing sports. Basically, the application will process data coming from different sensing sources which will give sufficient information to infer (and store) the user's movements (*motion states*) which, combined with location and time data will deliver his *activity profile*. With this information, the application's logic will control the user's activity level. Each hour, the application will evaluate if the activity level is enough to show progress. If so, a block in the puzzle interface will be uncovered. The user will be able to see the complete image if he has maintained a satisfactory activity level. During the day, the application will deliver context-aware notifications in order to encourage the user to increase its activity when low levels are detected.

Fogg [11] states that there are three elements which guarantee that a person will perform a target behavior: ability, motivation and 'effective triggers' which remind and initiate the action. In our application:

- The user is assumed to have the necessary 'ability' to perform the proposed activities (walk, run, stand up, climb the stairs, etc.). For better adaptation, a configuration panel will get some input about the user's habits (e.g. no. of working hours, no. of weekly sport sessions, etc.) when starting the application the first time.
- The 'motivation' aspect will be mainly driven through a visual interface capable to feedback the user at a glance: a wallpaper puzzle (Fig. 1) hiding an attractive image will be completed according to the periodic evaluation of motion levels. Additionally, alerts giving advice when low motion is detected may include quiz questions to increase the user will to uncover the whole wallpaper puzzle.

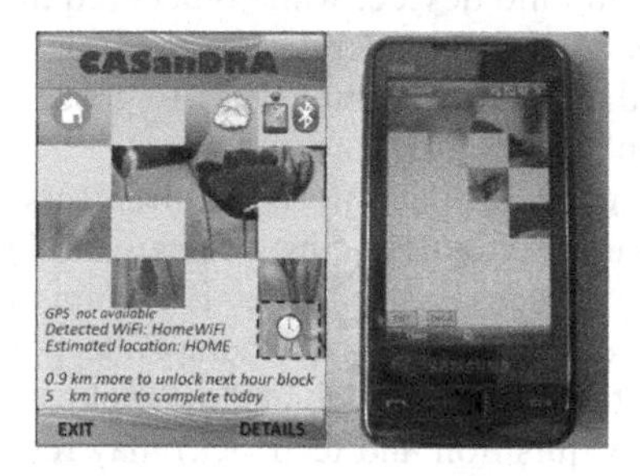

Fig. 1 Snapshot of the user interface. Prototype and implementation.

- With respect to 'effective triggers', aforementioned context-aware alerts will be generated when convenient to attract the attention of the user towards the interface. It is important to note that the delivering period of alerts is not previously set, but handled in an adaptive way depending on the user situation. We aim at providing the user with adequate information at point of decision.

From user studies such as [9], it is possible to understand the convenience to reduce to a minimum the interaction episodes with the user, as very demanding interaction schemes usually have discouraging effects. For this reason, our application will automate data gathering as much as possible; the user will need to provide data for configuration just when starting the application for first time.

4 Managing the User's Context: Sensing Needs and Reasoning Patterns

In order to infer *activity profiles*, it is necessary to handle a set of sensors delivering raw data which will be fused to extract context features. Most of sensors will be available in the mobile device, but in order to have better activity estimates, a external Shimmer mote [12] (equipped with a 3-axis accelerometer, gyroscope and ZigBee and Bluetooth interfaces) is assumed to be permanently attached to the user's instep. Following there is a summary of the context parameters needed and its relationship with sensors:

a) *Motion state:* Still-walking-running states will be detected by using both the accelerometer available in the mobile device and the Shimmer mote; an algorithm using thresholds on variance data will be used for this purpose. This redundancy of sources for motion estimates will help us to have better quality of context and failure tolerance. Transitions between motion states will generate detectable events.

b) *Indoors location:* As GPS is not available indoors, two indoor location systems are featured in the application. One of them connects to an infrastructure-based location system [13] which combines Bluetooth and WiFi received signal strengths (RSS) to calculate the user's coordinates (which will be translated into zones). The second one bases on a continuous scan of the WiFi and Bluetooth environment, in order to recognize common scenarios which may be identified by networks and devices (e.g. when the user arrives at work, his mobile device will detect his colleagues' mobile phones). Both of these algorithms require previous knowledge of the environment or additional infrastructure. Input information for the second method is to be added in the configuration interface.

c) *Outdoors location:* GPS will be used to locate the user when outdoors. In order to enhance the battery use, roaming between indoors and outdoors will be managed by a software component (the Location Fusion Enabler), which will be in charge of powering the GPS and the communication interfaces on and off in indoors to outdoors transitions and the other way round.

d) *Walked distance:* The walked distance will serve as primary input for activity inference. It may be calculated from location systems output (more accurately with GPS than with indoors positioning technology). Nevertheless, the most reliable way to have permanent feedback on the walked distance is to implement a step counter (pedometer). The number of steps is calculated by using a threshold on the envelope signal which combines the 3-axis acceleration signals. In order to calculate the covered distance in each step, some user information (gender and weight) is gathered in the configuration interface.

e) *Date & time:* All the gathered data need to be date and time stamped, as time is an essential input for context inference.

Using several sensors which may infer the same context parameter can cause conflicts when inferring the user activity; for example, the accelerometer sensor can inform of user movement while the GPS sensor is reporting stillness. It is necessary to have this aspect in mind when implementing the application logic, as the system will need to have a conflict strategy resolution and to determine which the most reliable estimates are.

The location information will be fused with date and time in order to infer the most probable *activity profile* for the user at a given time of the day. Initial *activity profiles* include AT WORK, SLEEPING, TRANSFERRING, AT HOME and PRACTICING SPORTS, although the list will grow to detect other sedentary activities (e.g. watching TV at home). Every *activity profile* has an associated *motion pattern* that user is expected to fulfill (e.g. one move every hour when the user is working, or 1 km walked when user commutes from his place to work). Conjoint processing of *motion state and location* data will be used to infer the user's *motion pattern*, which will be stored during an hour. At the end of the hour, the application will evaluate how the user is doing (by comparing the stored *motion pattern* to the predefined one) and if the evaluation is positive (>75% of the expected motion level), the interface will be conveniently modified. Depending on the evaluation result, context-aware alerts will be queued to be delivered at the right time. Each *activity profile* will have predefined rules to handle notifications (*alerts pattern*), to avoid interaction overload.

Table 1 Overview of activity and motion patterns

Profile Name	**Profile Properties**	
	Activity level	**Motion pattern**
WORK	Low	One move every hour
SLEEP	Very Low	10 hours máx.
TRANSFER	High	Walk 1km at least
HOME	Low	One move every hour
SPORTS	Very High	Run 4km at least

5 Description of the Application Components and Development Issues

The application has been built using the CASanDRA Mobile framework [5], which architecture (Fig. 2) is composed by three building blocks - Acquisition Layer, Context Inference Layer and Core System. The Acquisition Layer decouples the access to embedded and external sensors from upper processing levels by using software 'Sensors', which deal with low-level hardware information retrieval. The Context Inference Layer gathers a number of 'Enablers' - modules that process data coming from 'Sensors', fuse them, and infer complex context parameters. Finally, the Core System provides several features to integrate these components in the middleware, such as discovery and registry management of new elements and some common utility libraries. Both 'sensors' and 'enablers' publish their output data in the middleware through an event manager. Applications run on top of CASanDRA Mobile middleware, consuming context information provided by Enablers and Sensors and using its standard features.

The first step when building an application on top of CASanDRA Mobile is to define and separate every sensor or enabler module in an independent OSGi bundle. Fig. 1 shows all the bundles needed for the application to work. Using top-down design, firstly we define the application bundle including the application logic and also the user interface module. This application will program the rules that define the different *activity profiles* in the reasoning tool, will subscribe to that activity profile context parameter, and will perform an evaluation every hour according to this profile and to user activity stored during that hour. It will also show progress and alerts to user through the designed user interface.

Then, given the rules, four enablers need to be developed to infer context parameters. A *Step Counter Enabler* will process the external accelerometer measures using appropriate algorithms for detecting and storing every step user takes. The *Indoors Location Enabler (ILE)* will use an infrastructure positioning service [13] for providing location. The *Nearby Resources Location Enabler* (NRLE) will be used when the infrastructure service is not available, it will use a visible device and networks fingerprint to estimate the user's position. The *Location Fusion Enabler* will combine the indoors location and the GPS available data, in order to handle roaming and resources. With respect to persistence requirements, besides the context history, the database also will store other configuration parameters, e.g. the map of known devices and networks needed to make the NRLE work when the ILE is not available.

Finally, it is necessary to identify the sensor bundles. The Step Counter Enabler needs an external accelerometer bundle to access the Shimmer device using Bluetooth; an internal accelerometer bundle providing data about mobile internal inertials will be used in some rules; a GPS bundle will access internal GPS data when available; a Wi-Fi bundle will detect visible Wi-Fi networks and also provide RSS data to the RSS Indoors Location Enabler; a Bluetooth bundle will provide a list with close Bluetooth devices used in the other NRLE. GPS, internal accelerometer

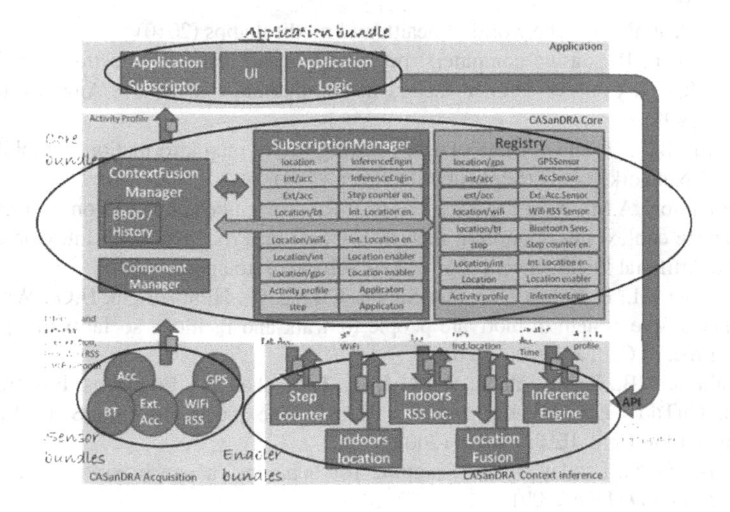

Fig. 2 Bundle deployment over CASanDRA Mobile.

and Bluetooth bundles have been already developed as reusable bundles and are available to use in CASanDRA Mobile.

6 Concluding Remarks

From the design and development of our persuasive context-aware application, it has been possible to validate the CASanDRA Mobile framework and detect missing features which need to be incorporated to enhance the middleware. For example, the use of different context sources inferring the same context parameter may cause logic conflicts which have had to be directly handled by the application. CASanDRA Mobile will be improved to offer transparent management of Quality of Context in a probabilistic way, allowing the comparison of different conflictive measures in order to select one or combine both when possible.

To evolve the application, apart from adding new sensors and features working on them, it is necessary to study how persuasion may be modeled and translated into performance. For this reason, the next step is to proceed with a user study which, besides evaluating the application, will focus on shedding some light about how the system should learn from real user interaction in order to dynamically modify the application's persuasion strategies.

Acknowledgments. This work has been supported by the Government of Madrid under grant S-2009/TIC-1485 and by the Spanish Ministry of Science and Innovation under grant TIN2008-06742-C02-01.

References

[1] IPTS, eHealth in 2010: Realising a Knowledge-based Approach to Healthcare in the EU (2004)
[2] MobiHeatlhNews, The world of health and medical apps (2010)
[3] Fogg, B.J.: Persuasive computers: perspectives and research directions. In: Proc. of the SIGCHI Conf. on Human factors in computing systems, Los Angeles, pp. 225–232 (1998)
[4] Schilit, B.N., Theimer, M.M.: Disseminating active map information to mobile hosts. IEEE Network, 22–32 (September/October 1994)
[5] Bernardos, A.M., Madrazo, E., Casar, J.R.: An embeddable fusion framework to manage context information in mobile devices. In: Proc. of the 5th Int. Conf. on Hybrid Artificial Intelligence Systems, San Sebastián (to appear, 2010)
[6] Brehmer, M., El-Zohairy, M., Jih-Shiang Chang, G., Himmetoglu, H.G.: Walkabout: a persuasive system to motivate people to walk and facilitate social walks planning. In: Proc. of CHI, Atlanta (2010)
[7] Waluyo, A.B., Pek, I., Yeoh, W.-S., Kok, T.S., Chen, X.: Footpaths: Fusion of mObile OuTdoor Perosnal Advisor for walking rouTe and Health fitnesS. In: Proc. 31st Annual Int. Conf. IEEE EMBS (2009)
[8] Consolvo, S., Landay, J.A.: Designing for behavior change in everyday life. Computer, 86–89 (June 2009)

[9] Mattila, E., et al.: Mobile diary for wellness management – Results on usage and usability in two user studies. IEEE Trans. on Information Tech. in Biomedicine 12(4) (2008)
[10] Bächlin, M., Plotnik, M., Roggen, D., Maidan, I., Hausdorff, J.M., Giladi, N., Tröster, G.: Wearable assistant for Parkinson's disease patients with the freezing of gait symptom. IEEE Trans. on Information Technology in Biomedicine 14(2) (2010)
[11] Fogg, B.J.: A Behavior Model for Persuasive Design. In: Proc. of Persuasive 2009, Claremont (2009)
[12] Shimmer motes: http://www.eecs.harvard.edu/~konrad/projects/shimmer
[13] Aparicio, S., Pérez, J., Bernardos, A.M., Casar, J.R.: A fusion method based on Bluetooth and WLAN Technologies for Indoor Location. In: IEEE Int. Conf. in Multisensor Fusion and Integration for Intelligent Systems, Seoul (2008)
[14] 3APLm, http://www.cs.uu.nl/3apl-m/

[9] Mattila, E., et al.: Mobile diary for wellness management – Results on usage and usability in two user studies. IEEE Transactions on Information Technology in Biomedicine 12(4) (2008)
[10] Mattila, M., [illegible] IEEE Transactions on [illegible] Technology in Biomedicine [illegible]
[11] [illegible] (2009)
[12] [illegible]
[13] Amoretti, S., Perez, [illegible] Fusion and [illegible] Intelligent Systems [illegible]
[14] [illegible]

Multi-camera and Multi-modal Sensor Fusion, an Architecture Overview

Alvaro Luis Bustamante, José M. Molina, and Miguel A. Patricio

Abstract. This paper outlines an architecture for multi-camera and multi-modal sensor fusion. We define a high-level architecture in which image sensors like standard color, thermal, and time of flight cameras can be fused with high accuracy location systems based on UWB, Wifi, Bluetooth or RFID technologies. This architecture is specially well-suited for indoor environments, where such heterogeneous sensors usually coexists. The main advantage of such a system is that a combined non-redundant output is provided for all the detected targets. The fused output includes in its simplest form the location of each target, including additional features depending of the sensors involved in the target detection, e.g., location plus thermal information. This way, a surveillance or context-aware system obtains more accurate and complete information than only using one kind of technology.

1 Introduction

Video surveillance has been the most popular security tool for years. Banks, retail stores, and countless other end-users depend on the protection provided by video surveillance. And thanks to the new breakthroughs in this evolving technology, security cameras are more effective, cheaper, and easy to deploy than even before.

This advances has issued the increase of image sensors, thanks in part also to the IP-based video emerging technology, opening a new research field in the last decade. The huge amount of video sensors installed in some scenarios makes unaffordable use humans operators for real-time monitoring. This way, new automated tracking systems are proposed in order to solve this problem [13]. These systems addresses the task of multiple people tracking in multi-camera environments. So, many effort put in this area consists in perform fusion information provided by the different

Alvaro Luis Bustamante, José M. Molina, and Miguel A. Patricio
Applied Artificial Intelligence Group, Universidad Carlos III de Madrid,
Avd. de la Universidad Carlos III, 22, 28270, Colmenarejo, Madrid, Spain
e-mail: {alvaro.luis,miguelangel.patricio}@uc3m.es,
josemanuel.molina@uc3m.es
http://www.giaa.inf.uc3m.es

A.P. de Leon F. de Carvalho et al. (Eds.): Distrib. Computing & Artif. Intell., AISC 79, pp. 301–308.
springerlink.com

optical sensors, solving problems such as background and foreground detection, object tracking, tracking occlusion, track continuity, and so on [18].

Meanwhile, with the technical advances in ubiquitous computing [1], wireless networking and the proliferation of mobile computing devices, there has been an increasing need to capture the context information for context-aware systems and services. Context-aware computing is a mobile computing paradigm in which applications can discover and take advantage of contextual information (such as user location, time of day, nearby people and devices, and user activity) [3]. Therefore, much research has focused on developing services architectures for location-aware systems [16], and also many attention has been paid to the fundamental and challenging problem of locating and tracking mobile users, especially in in-building environments, since, as discussed in [9], context-aware systems are based fundamentally in the user location. Hence, new systems for indoor location have emerged using different wireless technologies such as Wifi [14], Ultra Wide Band (UWB) [5], Radio Frequency IDentification (RFID) [6], etc.

Such deploy of multi-camera environments, indoor-localization systems, and automated specific processes and services, has allowed the coexistence of both video surveillance and indoor location systems in the same environment, but usually with different scopes. Video surveillance and automated tracking systems are fundamentally used for security purposes such as intrusion and event detection, activity recognition, or simply as *a posteriori* forensic tool. Meantime, indoor location is used especially for context-aware services, access control, personnel monitoring, augmented reality, etc.

The differenced use of both kind of technologies coexisting in the same environment is feasible, but could be improved if both techniques complements each other. For example, consider a location platform which provides a rich information of each target, such as, high-accuracy location, trajectory, speed, shape, color, size, thermal information, real-time video, and so on. One single sensor cannot provide all this information, so a fusion platform is needed to process all the independent sensor data and provide a single fused source of information. All this features could be used in complex systems like event recognition [12], behavioural profiling [2], action recognition [17], or simply, advanced surveillance and context-aware systems.

This way, our proposal consists in a hybrid fusion architecture which enables the fusion information of image and location sensors. Fuse these different sources is not a trivial task, therefore we define a first approach clarifying the different aspects, processes, and design decisions involved in such fusion architecture. In this topic there are not a sizeable literature, since most effort in the fusion field has been placed in fuse sensors of similar characteristics [8]. Some works, as described in [4, 11] deals with vision and location sensors fusion, but appears to be *ad-hoc* solutions to specific problems.

The rest of the paper is organized as follows: Section 2 overviews the fusion architecture, defining their basic inputs and outputs. Section 3 describes in more detail the architecture, paying special attention to the more relevant parts of the system.

Section 4 concludes with some reflexions about the architecture and describing the future work.

2 Architecture Overview

The main idea of the architecture is to be able to process multiple location and image sensors and provide a single fused, non-redundant output. The sensors for location could be Wifi, UWB, RFID, Bluetooth, etc. In fact, any wireless technology which can provide a location of a target with their associated identifier (i.e, the MAC address of the location device). In the vision field can be used standard image color sensors, thermal cameras, infrared sensors, time of flight cameras, and so on. Figure 1 represents the high-level input and output of the desired architecture.

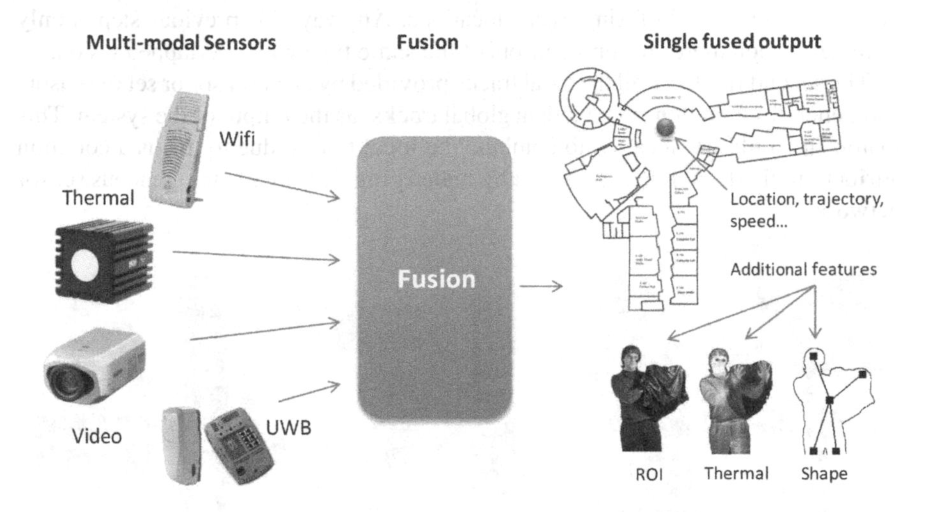

Fig. 1 High-level fusion architecture input/output

The architecture should provide all the information available of each target for each client subscribed to the fusion system, not only the fused location. That is, if a fusion is performed between a UWB location system and a color image sensor, the system should provide a single fused location and also the additional information offered by a image sensor, like color, shape, the image of the target, etc. So *a posteriori* processing could be done if required.

Regarding the final users or systems accessing the fusion architecture, the goal is that final output were accessible from any subscribed client, and therefore, be accessible both by surveillance and context aware systems at the same time. So, in some way the fusion system should also acts as a location server.

The full specification of the proposed architecture, with all their algorithms, protocols, etc, would exceed the length of the paper. Instead, we provide a brief overview of the different parts.

3 Detailed Fusion Architecture

Processing algorithms can be organized in different fusion architectures. The proposed solution is organized in different distributed tiers, each one processing inputs from the bottom tiers and feeding the upper one. The bottom tier is associated with the local sensor processing, so each sensor is the responsible of process it own information and provide a list of local tracks. This is generally achieved by the local track processing module as shown in figure 2.

Depending on the set of sensors used in the system, may be required an intermediate fusion step. As shown in figure 2, color and thermal image sensors are previously fused before the general location fusion. This particular sensor fusion node can achieve advantage in the fusion step, since sensors of the same type can obtain similar target features in order to improve fusion, i.e., in the color sensor fusion node, we can use attributes like location, shape, size, color, etc, to perform a better fusion than only fusing tracks locations. Anyway, this previous step is only necessary when there are some sensors of the same type with overlapped vision.

The second tier fuses all the local tracks provided by each sensor or set of sensors and generates a set of non-redundant global tracks, as the output of the system. This fusion step only use location to combine the local tracks, due to this is a common attribute in the local tracks provided by underlying tiers in a heterogeneous sensor network.

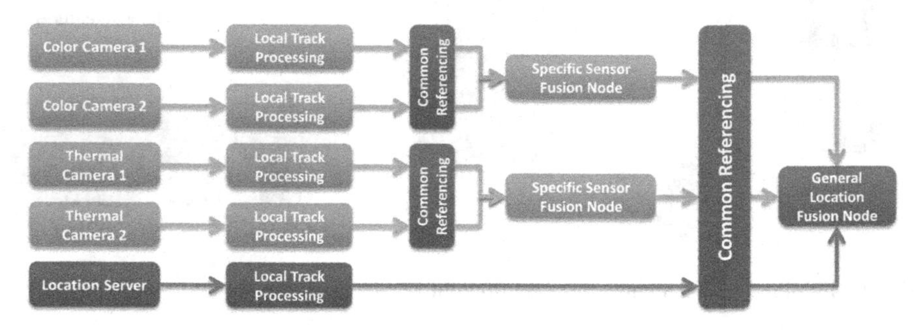

Fig. 2 Detailed fusion architecture

The main advantage of this architecture resides in the distributed data processing which lets each processing adjusts to the particularities of each sensor. This advantage is highlighted in this particular system, where imaging and location sensors are used. In this way particular processing algorithms could be defined for each sensor type. Another advantage of a decentralized architecture could be that the computational load could be balanced across different processors executing each one of the tasks on the system.

Each system module and some considerations are explained in more detail in the following sections, overviewing the different aspects needed to be taking into account when developing such fusion architecture.

3.1 Local Track Processing

The local track processing is the part of the architecture which deals directly with the data streams provided by sensors. The input to this module is the data provided by their associated sensor (an image in image sensors and locations in location sensors). In their turn, the output should be a set of local tracks, that is, all the targets detected by the sensor, with all their available features. This system should keep updated the track list in various ways, i.e, updating existing tracks with new associated plots, creating new tracks, and deleting tracks after some lack of updates. This module can be outlined in figure 3 where a simple local track processor is presented. In such processor, classical gating, association and filtering processes are performed [8].

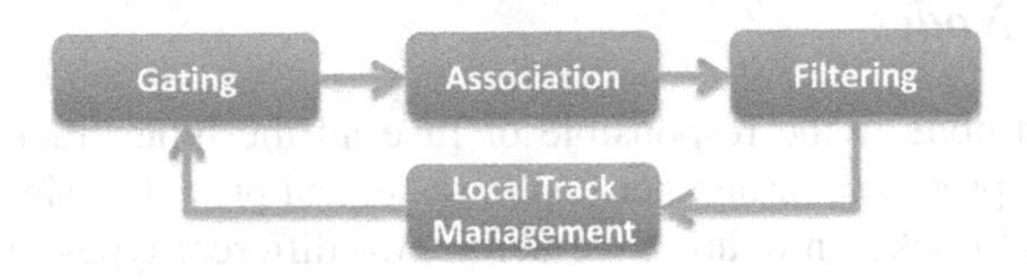

Fig. 3 Local sensor processing overview

Image sensors addresses the problem that cannot provide a location (plots) of the moving targets present in the scene, as they only provide an image. So is needed in this case apply a real-time object tracking based on image analysis [18]. In the other hand, location sensors usually provides some kind of identification with each target location, so normally is not needed a preprocessing step.

3.2 Common Referencing

Location-based fusion using sensors of diversity nature also introduces a handicap when trying to represent all locations in a common coordinate system. Usually, camera tracking is achieved directly over the image, that is, in the camera perspective of the scene, and this is a 2D representation with X and Y pixels coordinates. Location sensors usually lets the user establish the coordinate system and its location, so in this case the main problem arise with image sensors.

Fortunately, there are many approaches in the multi-camera fusion literature in order to provide a common referencing between multiple views. The most used is the based in the concept of homography [10]. In the computer vision field, any two images of the same planar surface in space are related by a homography (assuming a pinhole camera model). This has many practical applications, such as image rectification or image registration. For example is possible to change the perspective view of a camera and then process a synthesized image plane, as shown in figure 4.

In any case, this module must be able to transform the location and speed of each local track, in a common coordinate system. This way, fusion nodes can perform, at least, the location-based fusion.

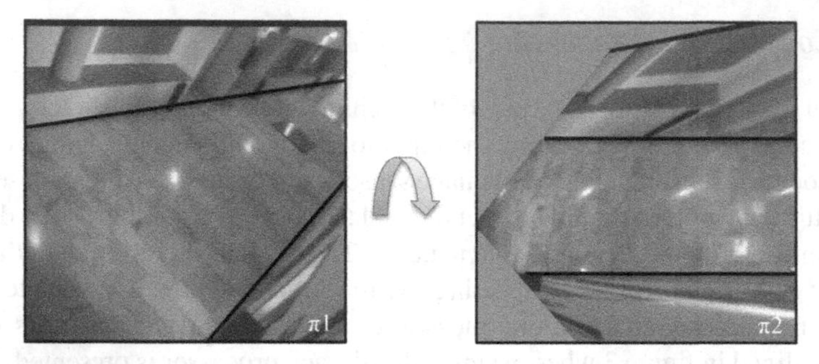

Fig. 4 Projective plane transformation

3.3 Fusion Nodes

A single fusion node is the responsible of fuse all the input tracks provided by each local track processor or another fusion node, and provide a single fused, non-redundant, set of tracks. In figure 2 we define two different types of fusion nodes, one sensor-specific, and other more general based only in location. This differentiation is due to the sensor-specific implementation must take into account the extra features provided by a set of common image sensors, like, color, shape, temperature, etc, achieving a location and feature-based fusion. The feature-based fusion has been widely used and tested in many fields with successful [7], so is useful distinguish both kind of fusion nodes. The general location-based fusion achieved in the second level only should consider the location of the tracks to perform the fusion, since there are not common features between, i.e., a image sensor and a location sensor that provide X, Y, Z coordinates.

The fusion nodes, are also the responsible, as the local track processors, of maintain updated the set of output tracks, attending to the input tracks. So, it should manage the creation, update and deletion when required. Classical processes of gating, association and filtering are also performed over the tracks, as described in the local track processors.

Would be also useful that fusion nodes could provide for each output track, the input tracks identifiers that has contributed to generate them. This way, the client regarding the output of the second level fusion node could know all the local tracks contributing for a final global track, and then, know all the independent features of a global track.

3.4 Infrastructure Considerations

There is an inherent problem in the fusion architecture proposed, and is the transmission of all the information from video and location sensors over the different local track processors, fusion nodes, and the final client when required. The decentralized architecture enables a distributed sensor processing, so is needed to enable some infrastructure allowing multiple video and general data transmission. For both image and location sensors is required to be enabled independent servers which could broadcast all the information over the same network. For image sensors we propose the digital video streaming system described in [15], implementing a JPEG2000 RTP streaming service allowing broadcast transmissions. Some similar location server must be enabled for each location system attached to the architecture.

4 Conclusions and Future Work

In this paper, a overview of an architecture for multi-camera and multi-modal sensor fusion has been given. The architecture is scalable in the way that many heterogeneous image and location sensors can be attached to the system. This provides an improved location service, taking the advantages of the different sensors used. The architecture is the responsible of processing efficiently all the data sensors, so a distributed sensor processing is proposed, with all their inherently benefits.

We have noticed along the description of this architecture that such system can be very complex to design, so it is needed to define well all the aspects and algorithms involved. In future works the different processing algorithms, communication protocols, and other important aspects of the architecture will be further described. Working prototype of the architecture is being developed using the infrastructure of VISLAB, with some color image sensors, one thermal camera, and a UWB indoor localization system.

Acknowledgements. This work was supported in part by Projects CICYT TIN2008-06742-C02-02/TSI, CICYT TEC2008-06732-C02-02/TEC, SINPROB, CAM CONTEXTS S2009/TIC-1485 and DPS2008-07029-C02-02

References

1. Abowd, G., Mynatt, E.: Charting past, present, and future research in ubiquitous computing. ACM Transactions on Computer-Human Interaction (TOCHI) 7(1), 58 (2000)
2. Atallah, L., ElHelw, M., Pansiot, J., Stoyanov, D., Wang, L., Lo, B., Yang, G.: Behaviour profiling with ambient and wearable sensing. In: 4th International Workshop on Wearable and Implantable Body Sensor Networks (BSN 2007), pp. 133–138. Springer, Heidelberg (2007)
3. Chen, G., Kotz, D.: A survey of context-aware mobile computing research. Technical report, Citeseer (2000)
4. Germa, T., Lerasle, F., Ouadah, N., Cadenat, V.: Vision and RFID data fusion for tracking people in crowds by a mobile robot. Computer Vision and Image Understanding (2010)
5. Gezici, S., Tian, Z., Giannakis, G., Kobayashi, H., Molisch, A., Poor, H., Sahinoglu, Z.: Localization via ultra-wideband radios: a look at positioning aspects for future sensor networks. IEEE Signal Processing Magazine 22(4), 70–84 (2005)
6. Hahnel, D., Burgard, W., Fox, D., Fishkin, K., Philipose, M.: Mapping and localization with RFID technology. In: Proceedings of IEEE International Conference on Robotics and Automation, 2004. ICRA 2004, vol. 1 (2004)
7. Hall, D., Llinas, J.: An introduction to multisensor data fusion. Proceedings of the IEEE 85(1), 6–23 (1997)
8. Hall, D., Llinas, J.: Handbook of multisensor data fusion. CRC, Boca Raton (2001)

9. Harter, A., Hopper, A., Steggles, P., Ward, A., Webster, P.: The anatomy of a context-aware application. Wireless Networks 8(2), 187–197 (2002)
10. Hartley, R., Zisserman, A.: Multiple view geometry in computer vision. Cambridge Univ. Pr., Cambridge (2003)
11. Hofmann, U., Rieder, A., Dickmanns, E.: Radar and vision data fusion for hybrid adaptive cruise control on highways. Machine Vision and Applications 14(1), 42–49 (2003)
12. Hongeng, S., Nevatia, R., Bremond, F.: Video-based event recognition: activity representation and probabilistic recognition methods. Computer Vision and Image Understanding 96(2), 129–162 (2004)
13. Kanade, T., Collins, R., Lipton, A., Burt, P., Wixson, L.: Advances in cooperative multi-sensor video surveillance. In: Proceedings of DARPA Image Understanding Workshop, Citeseer, vol. 1, p. 2 (1998)
14. Lim, H., Kung, L., Hou, J., Luo, H.: Zero-configuration, robust indoor localization: theory and experimentation. In: Proceedings of IEEE Infocom, Citeseer, pp. 123–125 (2006)
15. Luis, A., Patricio, M.: Scalable Streaming of JPEG 2000 Live Video Using RTP over UDP. In: International Symposium on Distributed Computing and Artificial Intelligence 2008 (DCAI 2008), pp. 574–581. Springer, Heidelberg (2008)
16. Maass, H.: Location-aware mobile applications based on directory services. Mobile Networks and Applications 3(2), 157–173 (1998)
17. Yamato, J., Ohya, J., Ishii, K.: Recognizing human action in time-sequential images using hidden Markov model. In: Proc. Comp. Vis. and Pattern Rec., pp. 379–385 (1992)
18. Yilmaz, A., Javed, O., Shah, M.: Object tracking: A survey. ACM Computing Surveys (CSUR) 38(4), 13 (2006)

Multi-sensor and Multi Agents Architecture for Indoor Location

Gonzalo Blázquez Gil, Antonio Berlanga de Jesús,
and José M. Molina Lopéz

Abstract. This paper aims to present a new architecture to provide location services using multiple communication technologies such as Wifi, UWB, RFID and so on. Firstly, it will explain the advantages of multi sensor architecture against to use unique indoor location system and the reasons which led us to take this solution. Besides, this paper discusses the suitability of using ontologies for modeling message structure to locate in context-aware services platforms. This message will be described based on the concept of Asterix format used in aerospace multi-sensor communications.

1 Introduction

Context-aware Systems allow to develop a new kind of location-aware mobile applications in different sectors such as healthcare, military, emergencies, and recently retail and agriculture [2, 5]. These applications could be represented as a context based scenario where there are individuals who require a satisfaction of their needs and there are providers who can solve these lacks. In context-aware computing, context is any information that can be used to characterize the situation of an entity (i.e. a person, computing device, or other). In this case, the architecture proposed focused on improving the location performance.

Positioning (also called Location aware) has been a main factor in the development of context applications. Location Awareness in general describes applications, which change their behavior according to the position of the user [3]. Location-aware enables new kinds of services and applications [4].

In outdoor environments GPS provides an effective solution to determine the location of mobile devices. However, in GPS-denied areas such as urban,

Gonzalo Blázquez Gil, Antonio Berlanga de Jesús, and José M. Molina Lopéz
Applied Artificial Intelligence Group, Universidad Carlos III de Madrid,
Avd. de la Universidad Carlos III, 22, 28270, Colmenarejo, Madrid, Spain
e-mail: {gonzalo.blazquez,antonio.berlanga}@uc3m.es,
josemanuel.molina@uc3m.es
http://www.giaa.inf.uc3m.es

A.P. de Leon F. de Carvalho et al. (Eds.): Distrib. Computing & Artif. Intell., AISC 79, pp. 309–316.
springerlink.com

indoor, and subterranean environments, unfortunately, an effective location technique does not exist. In this paper, it is presented an architecture which use multiple wireless technologies (UWB, Wifi, RFID, Bluetooth) [7] to provide location services. Hence, it is difficult to develop a location-aware application without making assumptions about the indoor location technology. So, it is possible to combine multiple indoor location systems to meet applications requirements.

Multi-sensor integration is becoming an essential aspect of indoor location systems [8]. Merge data from multiple indoor location systems and gather that information in order to achieve inferences, which will be more efficient and more accurate than if data were obtained of a single source. Applying Multi-sensor concepts could be improved the system performance, whether it is in coverage, availability or accuracy [1].

In [9] is described a new Ambient Intelligent Platform based on Multi Agent Systems (MAS) where location is included in the Locator Agent. Based on this architecture, it will be proposed that the Locator as Fusion architecture (Fusion Agent) and each Indoor Location System (ILS) is an agent which sends information about the location of each user to Fusion agent. To make easy the fusion process, fusion agent needs to know the message structure of each ILS, since this structure is different in each ones, it is necessary to standardize the structure message to ILS message structure.

Researchers in agent communication languages cite three important elements in MAS interaction [6]: (i) a common agent communication language and protocol; (ii) a common format for the content of communication and (iii) a shared ontology. The first two points are solved with declarative agent languages such as KQML or FIPA. Nevertheless, we will focus on creating an ontology which represents the conceptualization of positioning based on Asterix (All Purpose Structured Eurocontrol Surveillance Information Exchange) format, used in air navigation. In this paper, it is presented GONZ, Global Ontology for Indoor Localization, which aims to standardize the communication between ILS agent and Fusion Agent for indoor environments.

In this paper we propose an architecture which processed the measurement data received from indoor location system and besides, enhances the association process using the attributes of GONZ format. This paper is organized as follows: Section 2 describes the new fusion architecture for indoor applications. Section 3 presents a new ontology for indoor localization based on Asterix. Section 4 provides some concluding comments and future researches lines.

2 Multi-agent Fusion Architecture

In this section, we will describe the Fusion architecture based on Multi Agent system. Figure 1 represents a context-aware Multi Agent system architecture for heterogeneous domain used by Venturini et al [9] to describe Ambient

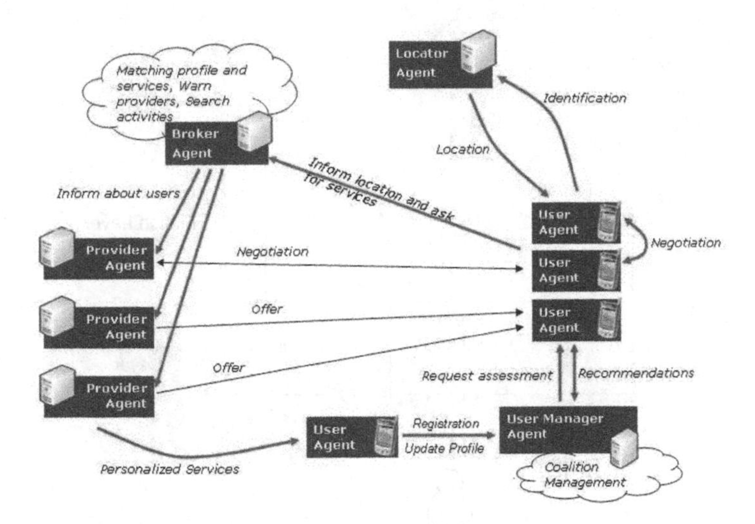

Fig. 1 Context-aware Multi Agent system architecture for heterogeneous domain.

Intelligent Platform based on Multi Agent Systems. Locator Agent, composed of a single indoor location system, is responsible to locate each user agent.

The structure of this architecture is composed of service provider agents (including Locator agent) and User agents. Provider agent offers services (Location, advertisement, weather information and whatever you want) to User Agent. This architecture presents several disadvantages in indoor environments due to indoor location issues: Multipath, reflexions, refraction, etc.

Multi-sensor architecture improves the precision, correcting systematic errors of some technologies with those ones they have not. Besides, if the Context aware architecture is able to obtain user position for more than one technology (Locator Agent) the system availability is improved. Finally, user availability is improved if multiple technologies could offer services to them. So, users need not to carry on a different mobile device in each environment. For all these reasons, it is interesting to design a new concept of Locator agent based on Multi-sensor fusion architectures.

Normally, indoor location systems run in their own server, therefore, a distributed architecture is needed. Multi agent system offers good features (Modularity, scalability and robustness) to solve distributed problems against other architectures like server-client or ad-hoc systems which present important lacks. The modularity of MAS permits make easy to develop new functions. For example, if new Indoor location Systems is added to the architecture, it is only necessary to create an agent which carry out the message format proposed in the last section.

The architecture proposed for the Locator agent is based on a layered architecture for fusion process as illustrated in figure 2, with separate layers for raw sensor data (ILS API), for features extracted from ILS (Locator Layer), and for context derived from ILS (Fusion Layer). Each component is

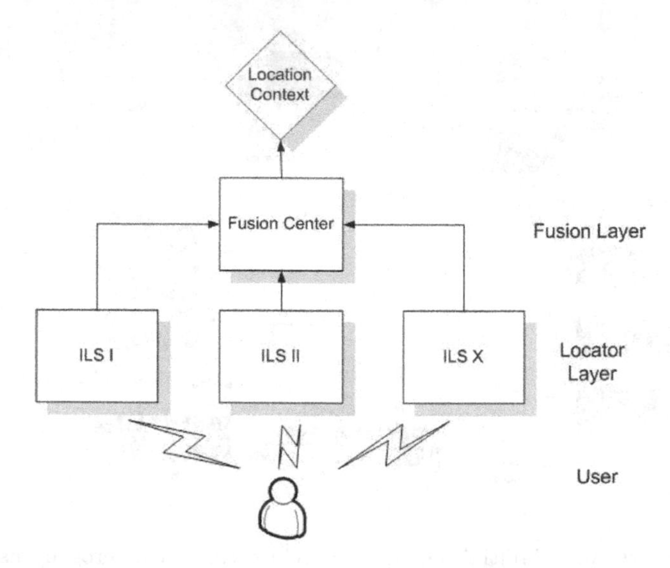

Fig. 2 Locator agent architecture

developed as an agent the first one is the (i) Fusion Center, which integrates data from all technologies and provides a location context for each user. The second one, called (ii) Locator layer, is responsible for providing location measurements to Fusion Center (Locator Agent in Figure 1).

2.1 Fusion Layer

The core function is collect and fuse information from available indoor location system (ILS agent) to provide user location context of each user. The fusion center received periodically (each T seconds) the detected data from ILS. Fusion layer could be split in other three layers (also see Figure 3).

2.1.1 Processing Layer

Processing layer is an agent which aims to receive measurement information of each ILS agent and starts the fusion process. Firstly, the new user position is stored in a vector structure which contains the last valid positions of this target. Then, the new measurement is transformed to unify coordinates

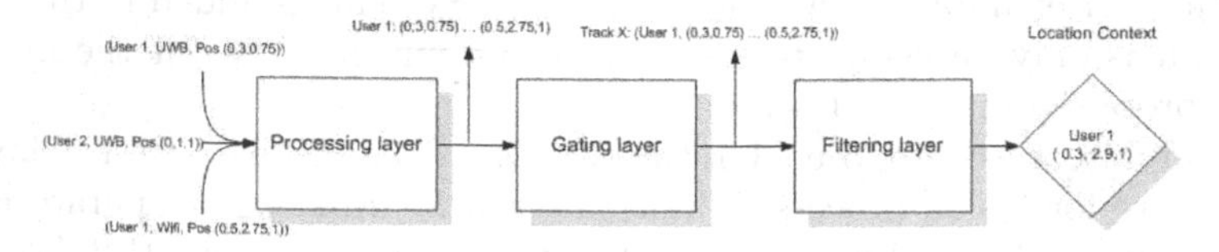

Fig. 3 Architecture of Fusion layer.

values with respect to the same global position (it is defined in the next section Global Ontology for Indoor location). Each measurement is preprocessed depending on the position of the ILS which generates it. To conclude, the structure will pass to the next layer and the fusion process continues.

2.1.2 Gating Layer

Gating layer aims to evaluate the chance of a new measurement will become a track and besides, search for associations between measurements and tracks, managing the structure of new measurements and the non-assigned plots of previous periods.

2.1.3 Filtering Layer

The filtering layer correlates the tracks belonging to the same located-object (user device) from various sources (ILS) and it aims to manage of tracks: Create, delete or fuse tracks with similar features.

2.2 Locator Layer

Since indoor location systems tend to be proprietary and only provide us a communication protocol or API in C♯ , java, C, etc to access to measurement, it is necessary to create a wrapper which aims to obtain, process and send to fusion layer measurements in a common language. Therefore, Locator layer is an agent which allows Fusion center works together that normally could not because of each ILS provide a different message structure. Hence, the Locator layer is responsible for transforming data into appropriate form. For instance, if a concrete ILS send position values in inches but other one send position in meters, the Locator layer would be responsible for unit conversion.

2.3 Multi Agent System Interaction

Interaction diagram (AUML Interaction Protocol Diagrams) describes better the Multi Agent Architecture proposed, with the advantage to identify clearly several agents roles. The figure below shows three agents, two of them belonging to the Location layer (ILS I and ILS II) and the other one belonging to Fusion Layer (Processing Layer).

ILS agents receive location information from each ILS, process this information and send location information to the Fusion layer, These actions are performed periodically, according to the sampling frequency each Indoor Location System (In this case, ILS II sends measurement faster than ILS I). Processing Agent also works periodically but the frequency does no have to be the same as the ILS Agent. In this case, Processing agent receive user

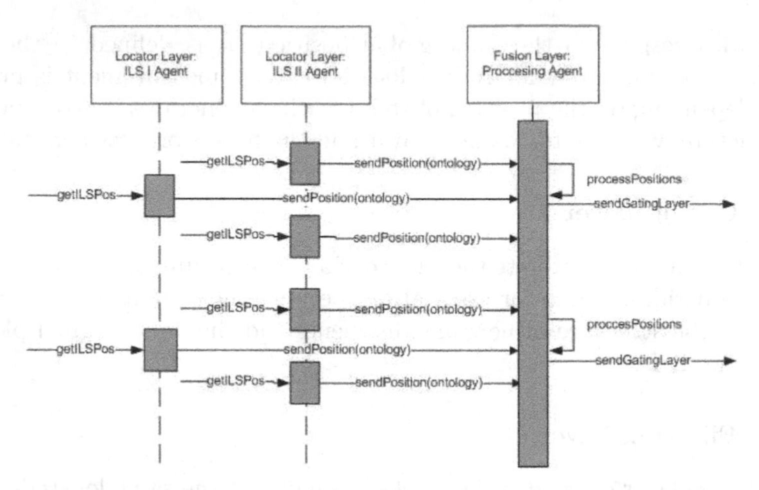

Fig. 4 Interaction Diagram of Location.

position measurement, store it and send it to the next layer of the of the Fusion Center (Gating layer).

3 Global Ontology for Indoor Location

To allow interoperability between Fusion Layer and Locator layer, it is necessary that the location context terminology is understood by all participating devices. In this section we propose an adaptable and extensible location context ontology (Global ontology for indoor location, GONZ) for which describes the knowledge in the communication process between agents. GONZ format presents the same structure message for all kind of indoor location system since the differences between every sensor are not considerable.

Other important point is that this message structure contains high level parameters since several ILS does not provide raw measurement (TDoA, azimuth, RSSI, etc). In this work we only have considered parameters in the format message for systems that provide the height and those ones that do not. Later, if new technologies were developed and provides remarkable features to fuse, it could create a new message format to GONZ standard.

Then, we will specify the principal parameters of the Global Ontology for Indoor Location format:

- Identification: Value which identified a track unambiguously, this identifier is composed for two strings: $id_{track} - id_{devicelocator}$.
- Time measurement (UTC): The time in UTC format of the last measurement taken.
- Sampling frequency (ms): Defines the time between two consecutive measurements.

Height known
Identification
Time measurement
Sampling frequency
[Origin of coordinates]
Pos X
Pos Y
Pos Z
Error

Height unknown
Identification
Time measurement
Sampling frequency
[Origin of coordinates]
Pos X
Pos Y
Error

Fig. 5 GONZ format message for height known and unknown.

- Origin of coordinates: In this attribute is stored (If it is possible) in Geographic coordinates or a common representation the origin of coordinate's of the indoor location system.
- Position X, Y, Z coordinates (m): Vector which contains the x, y, z position in meters of the user.
- Error: Estimated error of the measurement given.

An ontology represents a hierarchy structure of the environment knowledge where now the root node is GONZ and its child nodes are a concrete message (Height know message and Height unknown message). If new kind of message was created, a new leaf node could be added in the actual structure of the ontology. E.g. A new message structure could be created for a GPS devices or GSM devices since none of them are valid in the current standard. Then, it is necessary to create a new child node whose parents (GONZ, $Height_{known}$ or $Height_{unknown}$) depending on the features that new technology has.

4 Conclusions

This paper presents a Multi agent architecture to enhance the location in indoor environments with different indoor location technologies. The proposed architecture is responsible to obtain, process and fuse the measurements of each indoor location system and provide the result of fusion as location context.Besides, a structure message (GONZ) is defined for describing the communication between each ILS with the Fusion Center.

The principal advantages which offers this architecture is a highly interchangeable and adaptable system. If a new indoor technology will be developed, it will not involve major changes in its architecture. Only it will affect the inclusion of a new kind of message in the Global Ontology for Indoor Location.

Acknowledgements. This work was supported in part by Projects CICYT TIN2008-06742-C02-02/TSI, CICYT TEC2008-06732-C02-02/TEC, CAM CONTEXTS (S2009/TIC-1485) and DPS2008-07029-C02-02.

References

1. Aparicio, S., Pérez, J., Tarrío, P., Bernardos, A., Casar, J.: An indoor location method based on a fusion map using Bluetooth and WLAN technologies. In: International Symposium on Distributed Computing and Artificial Intelligence 2008 (DCAI 2008), pp. 702–710. Springer, Heidelberg (2008)
2. Becker, C., Dúrr, F.: On location models for ubiquitous computing. Personal and Ubiquitous Computing 9(1), 20–31 (2005)
3. Chen, G., Kotz, D.: A survey of context-aware mobile computing research. Technical report, Citeseer (2000)
4. Cheverst, K., Davies, N., Mitchell, K., Friday, A.: Experiences of developing and deploying a context-aware tourist guide: the GUIDE project. In: Proceedings of the 6th annual international conference on Mobile computing and networking, pp. 20–31. ACM, New York (2000)
5. Clarke, S.: Context-aware trails (2004)
6. Gruber, T., et al.: A translation approach to portable ontology specifications. Knowledge acquisition 5, 199–199 (1993)
7. Hii, P., Zaslavsky, A.: Improving location accuracy by combining WLAN positioning and sensor technology. Monash University, AU
8. Leonhardt, U., Magee, J.: Multi-sensor location tracking. In: Proceedings of the 4th annual ACM/IEEE international conference on Mobile computing and networking, p. 214. ACM, New York (1998)
9. Verónica Venturini, J.C., Molina, J.M.: An Ambient Intelligent Platform based on Multi-Agent System, pp. 1–9 (2008)

Multi-agent Based Distributed Semi-automatic Sensors Surveillance System Architecture

Jesús Tejedor, Miguel A. Patricio, and Jose M. Molina

Abstract. In the present paper, we describes a semi-automated and decision support sensor surveillance architecture used to develop an intelligent sensor surveillance system. The proposed architecture is grouped in three agents layers: the sensors agents layer, sensor processing agents layer and finally, the support assistant agents layers. The sensor agents layer is formed by sensor managing agents and sensor data flow agents that they control the sensor devices and retransmit data streams to upper layer respectively. In sensor processing agents layer is an agents collection that process data flows produced by sensors, allowing elements tracking. The last layer is formed by special agents for helping and supporting the user monitoring and user choice. This architecture proposes a fully decentralized multi-agent system using FIPA Agent Communication Language.

Keywords: Multi-Agent System, Sensor Surveillance, Monitoring and Control, Agent, System Architectures, User Decision Support.

1 Introduction

This proposed architecture is a surveillance system that integrates sensor analysis and agent technology. The architecture is projected for outdoor conditions where is necessary tracking and processing elements (like land vehicles, ships) through many sensor (especially video cameras). The main goal is to coordinate video sensor and other sensor types (like radar, global positioning system, thermal cameras, etc) to tracking these desired targets and facilitates user operations, for example to maintain a constant vision frontier with minimum overlap. Valera Espina and Velastin had written a concise review of totally automated visual surveillance systems [1].

Human monitoring in surveillance functions is expensive and quite ineffective because each sensor provides a huge quantity of information [2]. Even trained users would lose concentration and miss a great amount of critical events in minimum lapses. Therefore, surveillance operators should be helping by support automatic assistant, filtering data to provide only most relevant information to user. Even, replace them exclusively by software systems [1].

Jesús Tejedor, Miguel A. Patricio, and Jose M. Molina
Applied Artificial Intelligence Group, Universidad Carlos III de Madrid,
Avd. de la Universidad 22., 28270 Colmenarejo, Spain
e-mail: {jtejedor,mpatrici}@inf.uc3m.es, molina@ia.uc3m.es

A.P. de Leon F. de Carvalho et al. (Eds.): Distrib. Computing & Artif. Intell., AISC 79, pp. 317–324.
springerlink.com

Some authors like Henry Detmold propose surveillance middleware architecture, based on service-oriented architecture (SOA) [3] for all computational tasks. The SOA inconvenience lies in data re-encapsulation by simple object access protocol, not allowing a fully and correct communications between agents and data flows.

In this paper, we proposed a multi-agent architecture with two flows: The first flow, to maintain sensor controlling and managing (such as movement, focus or zoom in cameras, change parameters in radars, etc) through coordination between agents, and another flow to maintain a visual stream for the system users. The interaction between agents is controlled by a BDI - like architecture [8].

The coordination between agents would be based in FIPA ACL message [4] for keep flexibility and adaptability that they are desired conditions to system. At the same time ACL is a communication language based on the speech-act theory [5] and used in communications between last layers.

In section 2 we explain optimal conditions that system requires and mains goals that the system complains. Section 3 is focused on the architecture itself and we describe the different agents and their functionalities. Section 4 presents our conclusions.

2 System Characteristics

The perfect desired architecture must group not only essential surveillance requirements, but also complain other requests such as integration, net security, scalability, availability, flexibility and intelligent image processing.

This architecture is:

- Open, any agent can be adding to system thanks to standard language to communicate between agents.
- Flexible, the system can incorporate new functionalities and layers without changing previous work
- Scalable, increase the number of sensors or agents has a linear growth in computational complexity and an agent does not often communicate with all other agents.

The system should compute a huge data from different sensors types. Also can manage all sensors to coordinate them avoiding overlap, allowing a correct tracking and showing most important sensor signals to user. These are system main goals.

3 System Architecture

We have designed an open and generic multi-layer architecture for semi-automatic surveillance systems.

The first layer of our architecture called Devices Layer, it is formed by a set of agents that interact with the sensors and monitors. Each agent (especially sensor device agents) delivers output and receive requests for changes their parameters its

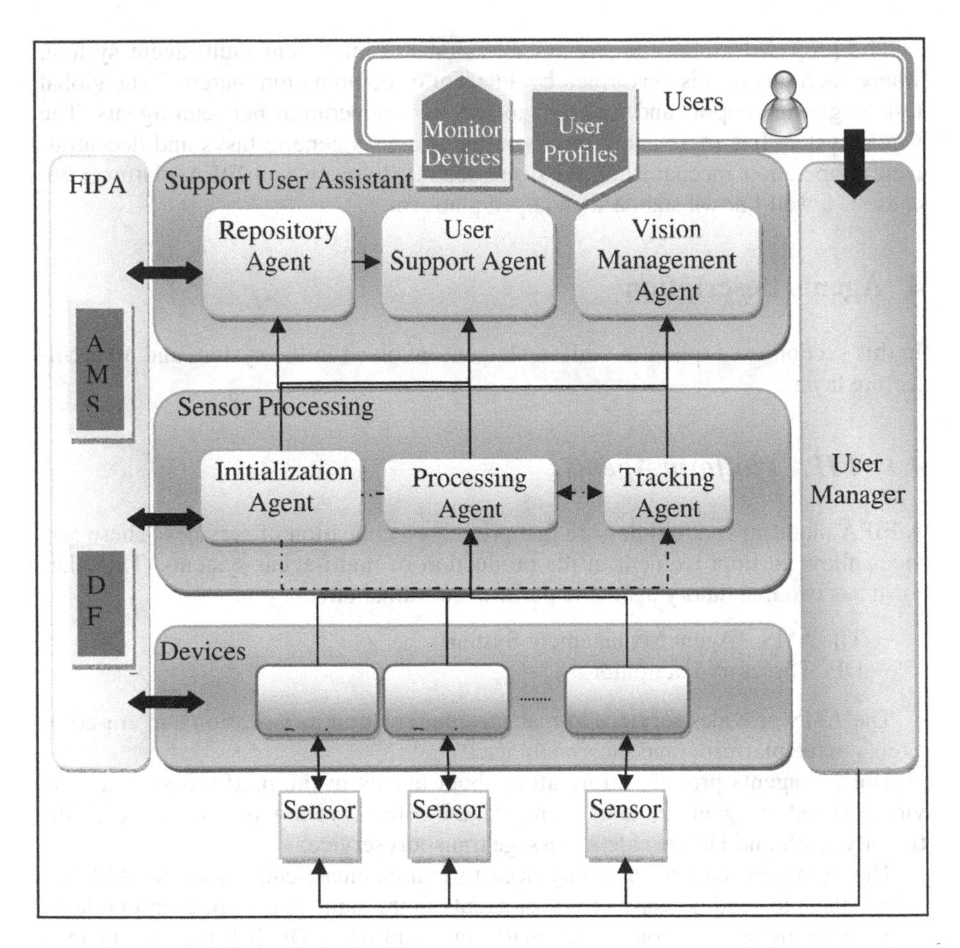

Fig. 1 Multi-Agent Based Distributed Semi-Automatic Surveillance System Architecture. This image shows the three different layers which compounds the whole system relying in FIPA platform.

associated device. Each monitor, camera or other sensor has associated in this layer only one agent.

The tasks of second layer, Sensor Processing Layer, are the construction of marked target, introduces different sensor data to maintain tracking in the proper way to facilitate the users visualization. This layer refines the sensor data producing semantic description to interpret for upper-layer and coordinates three agents, initialization agent, processing agent and tracking agent.

The third layer, support user assistant, analyses bottom-layer data in order to produce actions and suggests to user different ways to perform a correct management. Generally, these actions could result in generate movement in sensors or change their parameters for example, avoiding possible hole in global line vision, keep a tracking or show user most relevant image in monitor system.

The proposed architecture is a semi-automatic intelligent multi-agent system, where each function is performed by an agent or combination thereof. The global task is giving support and assistant to user by cooperation between agents. The whole system lies in an agent platform that provides generic tasks and decentralized cooperation mechanisms. As discussed we have chosen FIPA platform because is a well-known standard and open platform.

4 Agents Description

In this section we explain the different agent involved in the system and all architecture layers.

4.1 FIPA Platform Agents

A FIPA platform is a middleware that provides a collection of services. These services allow an improvement in the production of multi-agent systems. This platform has two mandatory agents to perform coordination:

- The AMS - Agent Management System
- DF - Directory Facilitator.

The AMS provides services to management creation, registration and erased of agents in the platform (and the system itself).

The DF agents provide information about agents in the platform such as services offered by agent, name of agent, etc. DF offers a yellow pages service. Additionally AMS and DF provide a message transport service.

The agents in the platform only need to communicate call a specific API that allows them to send messages to other agents in the same platform, even in others. A possible implementation of this platform could be JADE [6], perform in Java language. JADE has the property of portability and it does not depend on native computer architecture.

4.2 Device Agent

This agent perform controlling and management of sensor. It provides the image to next layer for corresponding processing. Definitely the agent has two tasks: provides sensor data to system and receives the possible data that modifies their attributes.

The agent should have negotiation capacities for communicate with other devices agent in the same layers.

This agent should offer the same interface for any sensor to facilitate inclusion of new sensor devices with minimum impact in the architecture. However, when communicating with the top layer should be informed the services it can offer. The different sensor types provides different data types.

4.3 Initialization Agent

The main goal of this agent is initializing tracking through targets detection. This function starts when the user notifies to system about some element to track or sensor data processing like radars.

When the target is located, the information is transmitted to tracking agent and processing agent. This agent also classifies the element to track such as suspicious target or not suspicious target with a confidence factor (high when the user requests this tracking).

4.4 Tracking Agent

This agent is a multi-process agent, with the purpose of tracking the targets trajectory and speed, in order to predict their position for commutate between sensors to keep target tracking when actual sensor that it performs tracking cannot follow this functionality.

It is possible to use a large number of algorithms to perform this tracking, one example may be the work done in [7] by Luis Botehlo. This prediction process is formed by two algorithms: prediction algorithm thanks a model and learning model to readjust the model factors.

4.5 Processing Agent

The vision frontier must be constant, avoiding overlap and hole in its vision. Therefore it is necessary managing the movement and other sensor characteristic to maintain this frontier.

This agent manages the movement and sensor range, and should keep the maximum distance in its vision range trough the sensor intervals.

The final functionality that this agent has is to provide sensor data to others agents in the same or upper layer. It could manipulate the sensor data to visualize them better.

4.6 Repository Agent

The principal aim of this agent is to store content offered by sensor and accumulate the performed reports about the system activity (tracking, changes in vision line…).

This agent offer a great historical data that the user can consult, even is possible discard target thanks to recognize in this repository a possible target that has no impact on tracking process because it is not suspicious.

This data could be accessed by the user through agents in upper-layer. Even this agent in upper-layer could be accessed this information to perform a possible choices.

4.7 User Support Agent

To facilitate the user work for inspecting unusual or suspicious target in a concrete environment, this rational agent compiles and processes the information in repository agent and tracking agent.

Thanks to initialization agent, it could suggest more suspicious target to track with sensor devices, perform different report types such as last target tracking or sensor incidents or state and even it suggest stop tracking a concrete target because it is not suspicious.

This agent needs learning algorithms to adapt different user profiles and promote a proper sync with the user.

4.8 Vision Management Agent

In large security rooms where there are many monitor devices is difficult appreciate all events that produce in vision line. Also many sensor devices are most interesting than others. For this reason, it is necessary selecting most important sensor device to view in monitor devices.

This agent provides this capability due concise analysis about user preferences and relevant data extracts to repository. Monitor characteristic are different between them, even if all monitor devices have same characteristics, their locations changed and some specific location offer a better view than others. In consequence, the monitor characteristics and location are useful for users because they could watch a limited number of screens.

For example, tracking an objective is most important with a sensor device than others where nothing unusual is reporting. The target data that sensor is tracking thus view in a screen.

Also it can perceive the shutdown or occupation in other activities of sensors to avoid holes in the frontier. Also it can perceive an excessive overlap between sensors and order to separate the vision sensor ranges

The agent should integrate proper algorithms that provides to user these capacities and must be in this agent.

5 Agent Management Sample

When this agent perceives any discontinuity, it begins a communication between device agents that stimulates a negotiation between involved agents and continues until the hole or overlap is avoided. Also the users modify the vision range or charge other profiles that specifies a concrete vision range, this agent evaluates and notify to devices agents the new ranges leaving the negotiations between them.

The figure represents this data flow.

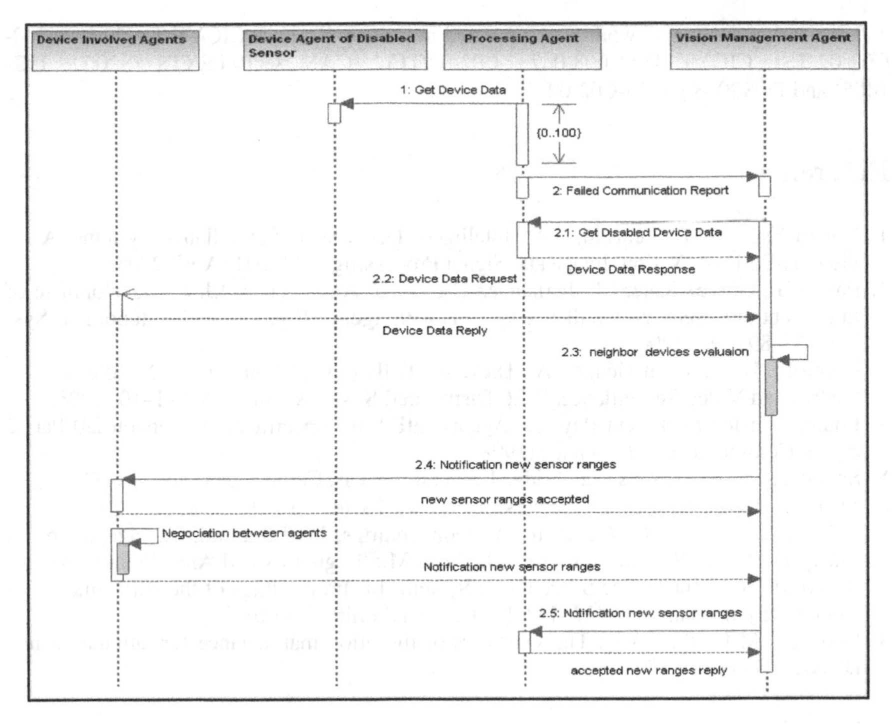

Fig. 2 This image shows interaction diagrams that it represents when the system detects a discontinuity

6 Conclusion

We have proposed an intelligent distributed semi-automated system based on technology multi-agent, the architecture avoids a reliable management of different sensor, tracks targets and advises users to facilitate their work.

The main advantage of using agents over video algorithms is:

- The tasks performed by each agent could be incrementally sophisticated without impact on system.
- Autonomous programs could be tested and developed independently because each service and device control is encapsulated in an agent by using middleware technology, which can improve intelligent and automation degree of information processing, which ensure that all raw data could be processed efficiently.

The information that could be processed by users is minor, promoting the user efficiency because the user can focus in most priority events and cases.

This generic architecture provides a middleware to build a concrete implementation in concrete environments depending local circumstances and characteristics.

Acknowledgments. This work was supported in part by Projects CICYT TIN2008-06742-C02-02/TSI, CICYT TEC2008-06732-C02-02/TEC, CAM CONTEXTS (S2009/ TIC-1485) and DPS2008-07029-C02-02

References

1. Valera Espina, M., Velastin, S.A.: Intelligent Distributed Surveillance Systems: A review. IEEE Proc. Vision, Image and Signal Processing, 192–204 (April 2005)
2. Pavón, J., Gómez-Sanz, J.J., Fernández-Caballero, A., Jiménez, J.J.V.: Development of intelligent multisensor surveillance systems with agents. Robotics and Autonomous Systems 55, 892–903 (2007)
3. Detmold, H., van den Hengel, A., Dick, A., Falker, K., Munro, D.S.: Middleware for Distributed Video Surveillance. IEEE Distributed Systems Online 9(2), 1–10 (2008)
4. Foundation for Intelligent Physical Agents, FIPA 97 Specification, Version 2.0 Part 2 Agent Communication Language (1998)
5. Searle, J.R.: Speech Acts. Cambridge University Press, Cambridge (1969)
6. Java Agent Development Framework, `http://jade.tilab.com/`
7. Luis, B.A., Botelho, L., Cavallaro, A., Douxchamps, D., Ebrahimi, T., Figueiredo, P., Macq, B., Mory, B., Nunes, L., Orri, J., José, M.: Trigueiros and Ana Violante, Video-Based Multi-Agent Traffic Surveillance System. In: Proceedings of the IEEE Intelligent Vehicles Symposium, pp. 457–462. IEEE, Los Alamitos (2000)
8. Georgeff, M.P., Rao, A.S.: The semantics of intention maintenance for rational agents. IJCAI, 704–710 (1995)

Interactive Video Annotation Tool

Miguel A. Serrano, Jesús Gracía, Miguel A. Patricio, and José M. Molina

Abstract. Increasingly computer vision discipline needs annotated video databases to realize assessment tasks. Manually providing ground truth data to multimedia resources is a very expensive work in terms of effort, time and economic resources. Automatic and semi-automatic video annotation and labeling is the faster and more economic way to get ground truth for quite large video collections. In this paper, we describe a new automatic and supervised video annotation tool. Annotation tool is a modified version of ViPER-GT tool. ViPER-GT standard version allows manually editing and reviewing video metadata to generate assessment data. Automatic annotation capability is possible thanks to an incorporated tracking system which can deal the visual data association problem in real time. The research aim is offer a system which enables spends less time doing valid assessment models.

Keywords: Ground Truth, Tracking, Automatic Annotation.

1 Introduction

Over the last years, the amount of multimedia resources has grown due to the popularization of Web 2.0. This information is unstructured and poorly organized, being very hard to browse and retrieve it. In order to overcome this limitation, it is necessary to semantically label and organize all this data.

Annotation is a process which provides visual metadata superimposed over resources without modifying the analyzed element. Manual annotation is an unfeasible task for large video collections however automatic video annotation systems can automatically add metadata through computer vision techniques.

Normally, new automatic systems based on computer vision techniques try to improve their benefits comparing with the current state of art. To demonstrate these improvements in a scientific way it is only possible through assessment models.

Therefore annotation and label systems need tools to assess the performance of their techniques. Such evaluation is often carried out by comparing results obtained from a given algorithm against ground truth - a set of results determined a

Miguel A. Serrano, Jesús Gracía, Miguel A. Patricio, and José M. Molina
GIAA, Carlos III University, Spain
e-mail: {miguel.serrano,jesus.garcia}@uc3m.es,
{miguelangel.patricio,josemanuel.molina}@uc3m.es

A.P. de Leon F. de Carvalho et al. (Eds.): Distrib. Computing & Artif. Intell., AISC 79, pp. 325–332.
springerlink.com

priori to be correct [4]. More specifically, in the video evaluation scope, generate ground truth annotations for large scale video collections has involved huge amount of effort. Traditional techniques don't work because a temporal dimension generated from video frame sequence is added to the images spatial dimensions.

This paper presents a new supervised automatic annotation tool. The basic infrastructure is a tracking system integrated to an extended version of the ViPER-GT annotation tool.

Perform tracking tasks during the video analysis, facilitates the automatic annotation feature. In tracking low level tasks, such as segmentation or trajectory analysis, the tool detects, label and annotates tracks. In addition, user may manually create tracks or modify the location, size and trajectory if an error occurs.

Moreover, with this tool high level semantic tasks at scene and object level can be developed. Semantic annotations are done manually trough the ViPER-GT tool interface. All these actions can be done in real time because the system is adaptable to the changes.

The paper is organized as follows. In Section 2 annotation and assess fields are studied briefly; Section 3 the annotation tool overall architecture is presented; Section 4 shows the experimental results; Section 5 explains the conclusions obtained and the future work.

2 Brief Summary about Annotation

For years researchers in annotation have been worked in two different ways. Initial approaches focused on low level visual descriptors such as texture, shape... After, researches turned to knowledge approaches, which try to extract semantic descriptions with the goal of save the semantic gap.

Low level descriptors approaches imitate the way users assess visual similarity [5] and don't try to extract directly semantic assertions from visual content. The main feature of this kind of methods is the capability to find patterns from frame/image features. These techniques are based on machine learning methods. Most used techniques in this field are Hidden Markov Models and Neural Networks.

On the other hand, knowledge-based approaches uses a higher abstraction level when annotate content. To that end, make use of "a priori" knowledge such as models, rules... These approaches, normally allow realizing inference operations between the elements and the spatial relationships. New hidden domain knowledge results of these operations.

The actual trend is to blend both approaches, extracting relevant semantic elements from videos by combining several low-level descriptors [7]. In this context, one of the keys is the MPEG-7 standard. MPEG-7 represents audiovisual information and allows content descriptions. For instance, MPEG-7 Visual Part support low level features such as color, texture, shape or motion. There are also MPEG-7 Multimedia Description Schemes which support spatial relations between detected segments.

Nowadays general purpose approaches don't exist due to the knowledge dependency with the specific context domain. High level semantic applications are based on specific context such as sports, movies, security…

Inside evaluation field there are some resources such as databases and semi-automatic annotation tools. Large scale general purpose databases are scarce. Examples of publicly available sets of databases are: NIST TRECVID databases which contain several hours of publicly available *ground truth*, from over 1000 visual concept categories, PETS datasets [18] which actually include outdoor people and vehicle tracking, indoor people tracking, annotated hand posture classification data..., the Surveillance Performance Evaluation Initiative (SPEVI) [12] which includes an audiovisual people dataset, a single face dataset and a multiple face dataset all of them made with the Video Performance Evaluation Resource (ViPER), on the other hand as single dataset we can list, cVSG [13], OTCBVS [14], VISOR [15], ETISEO [16], CANDELA [17], etc.

We can also find notable tools to create new *ground truths*, the IBM MPEG-7 Annotation Tool, for example, provides a rich user interface that displays the video, semi-automatically detects shot boundaries and selects key frames, automatically propagates prior shot labels, presents a hierarchy of 133 suggested visual concept labels but also accepts new user-created visual concept labels [8]. ViPER Ground Truth (ViPER-GT) is another interesting video annotation tool, which may be used as a viewer of algorithmically generated markup, a tool for assisting performance evaluation of such markup and more. ViPER Performance Evaluation tool (ViPER-PE) complete the capabilities of ViPER-GT providing the ability to compare result data with ground truth tools for solving the evaluation problem [11].

3 Overall Architecture

Annotation system presented in this paper is based in two fundamental elements, an annotation system which is a modified version of ViPER Ground Truth tool and a tracking system optimized to perform video analysis in real time. The user supervisor monitors the automatic annotation process between these components through the ViPER-GT interface. The overall architecture of the proposed framework is called MViPER-GT and is illustrated in Fig. 1.

System works as follow. MViPER-GT tool sends raw frames from the video selected to the tracker when the analysis starts. Supervisor may update the track annotations during the analysis. These changes consist on create and delete tracks and adjust their size and position. Update track annotations can be done in real time if an error is detected, for instance, if a track is not created automatically by the tracking system or if a location prediction of a track is not properly done. MViPER-GT sends changes done by the supervisor to the tracking system. Information sent between subsystems is performed through a communication layer. This layer interacts in a bidirectional manner, transforming information to operate on each subsystem and enabling the communication. Bidirectional communication between VIPER-GT and the tracking system is performed one time per frame.

Tracking system is responsible for carrying out the establishment, updating and deleting tracks automatically. Once the tracker has received information concerning to MViPER-GT (a raw image and in some cases, also, user annotation updates), it realizes a complete tracking analysis, which includes segmentation, association, trajectory prediction, etc. User updates are taken into account during the analysis. Feature modifications may cause reaction in tracking system, changing the size or the location of the tracks. For instance, if a track trajectory is modified in the annotation tool, this will perform modifications in the calculated trajectory of the tracker, normally, trajectories followed at this time, will probably suffer alterations in the same sense of the new positioning. Thanks to the trajectory prediction algorithms housed in the tracking system, MViPER-GT receives non linear trajectories of each track or object in each frame. When the analysis is completed, tracking system predictions (updated feature tracks, size and position) are sent to the annotation tool. The annotation tool receives these predictions annotates them and starts a new cycle with the next frame.

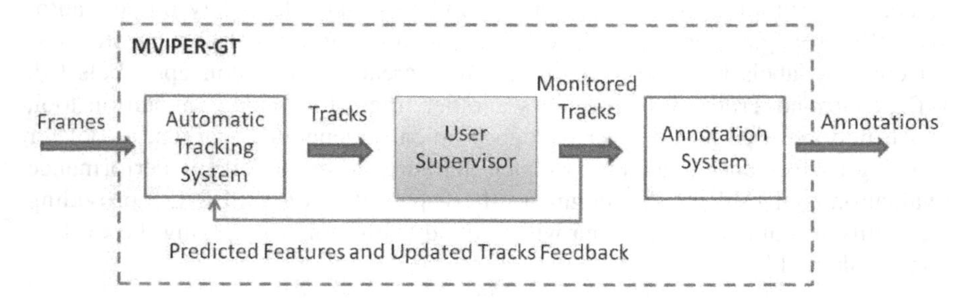

Fig. 1 Overall Architecture.

3.1 Annotation System: ViPER-GT and MViPER-GT

Create an annotation model was a tedious task, especially in the video domain, because it was required review sequences of frames with the similar content from frame to frame. Annotation system based on ViPER-GT makes it easier. ViPER-GT is a Java open source development project, supported by the ViPER API. When ViPER was thought, the first goal was the creation of a flexible ground truth format and second goal was to provide tools to easily create and share ground truth data [4]. Developed GUI could be used to record the requisite information in a single scan of the video content. For a given frame, users could select a cell representing a spatial attribute (point, bbox, obox or circle) [4].

The modified version of ViPER-GT presented in this paper, allows user to be merely a supervisor of the annotation task. The tool allows users to configure data generation and evaluation. Descriptors represent the data generation structure of each video. Structures of the objects are defined by the descriptors which can contain different kinds of attributes. Annotations describe the object state through its attributes. Attributes contain the feature values which can represent, for instance, the name or the location of an object. MViPER-GT starts with two predefined descriptors, a static metadata structure which represents video information like the number of frames or the frame rate and a dynamic metadata structure which represents the object information. This structure has two attributes, the track identification number and the bounding box which represent the position and the size of each track. New descriptors may be created to this basic configuration, in order to separate different knowledge levels. Descriptors based, for cxample, on semantic information may be created before or during the analysis; however the annotations should be done manually.

ViPER offer some annotation possibilities like creation, deletion of tracks and modifications on the features of each track. As we seen before MViPER tool includes trajectory prediction algorithms. ViPER standard version includes a default linear interpolation utility which can also be used with MViPER. This linear interpolation utility fills in new intermediate values of spatial attributes between two separated frames.

Sometimes tracking system detects systematically a new track which the user does not want to annotate. ViPER propagation utility is the best way to treat this kind of situations. Propagation copies the current frame's value of selected object descriptors to all frames in the range of propagations [11]. This is especially helpful for spatial attributes that do not change much across frames. All attribute values of the chosen object descriptors are overridden with the values in the current frame [11]. Unwished labelling may be avoided, enabling propagation and disabling track annotations.

3.2 Tracking System

Architecture is based on a video chain with different modules that run in sequence, which correspond to the successive phases of the tracking process. The tracking system is composed by four modules: Foreground/Background Detection module shows when a pixel has moved and group them in blobs. Association module predicts the blobs positions, assign sets of blobs to tracks and finally update the tracks positions. Initialize or Delete module, create and delete tracks when have not assigned to any blob. Trajectory Generator module detects anomalous behaviours studying tracks trajectories. Algorithms belonging to each module are interchangeable in each run. Each module has a specific task to be implemented by all the algorithms that correspond to a certain module [10]. The input data for the pipeline is the image of current frame and the output data is the tracks position and size.

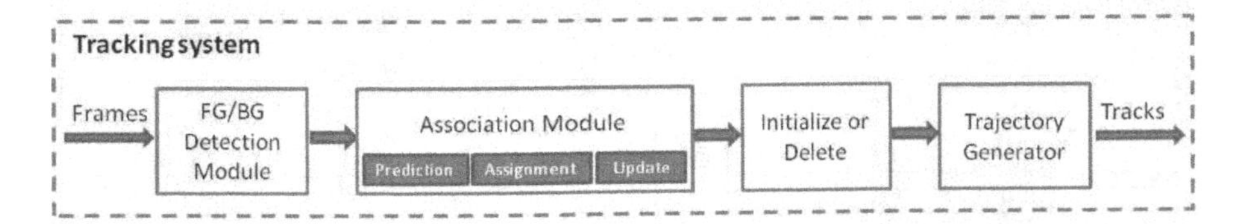

Fig. 2 Tracking system.

4 Experimental Results

We have implemented a prototype to get some experimental result. A JNI communication layer has been developed to achieve bidirectional interaction between ViPER-GT annotation system and OpenCV tracking system.

This system has been tested with the Computer Vision Based Analysis in Sport Environments (CVBASE) dataset [19]. Video features are 25 frames per second, 384x576 pixels of resolution and M-JPEG compression.

Selected video is a zenithal record of two players playing squash. They are in close proximity to each other, they are dressed similarly and are moving quickly, and there are constant crossings and occlusions between players, which make the video an interesting challenge to the quality measure of the system.

As we can see in this first image, the operation of the tracking system for this video is correct. Tracks are detected with quite accurately and the annotations are automatically done. However there is not still a complicated situation where the user has to intervene.

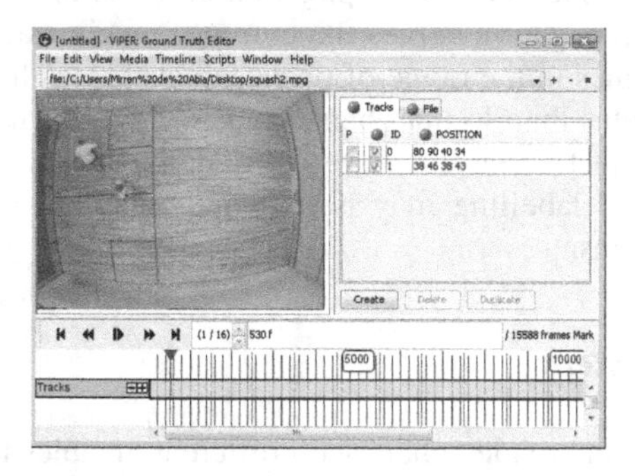

Fig. 3 Video analysis. Frame 530.

The two images below show the performance under critical circumstances of occlusion between tracks. The system has also an optimal behavior in such cases, therefore, it is not necessary, in this case, the intervention of the supervisor to make changes in the annotations.

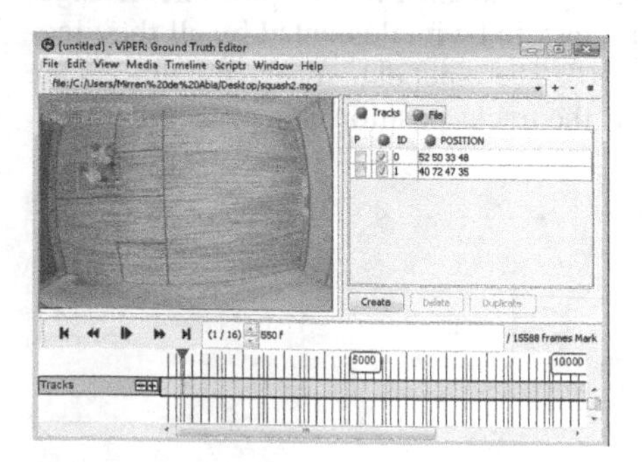

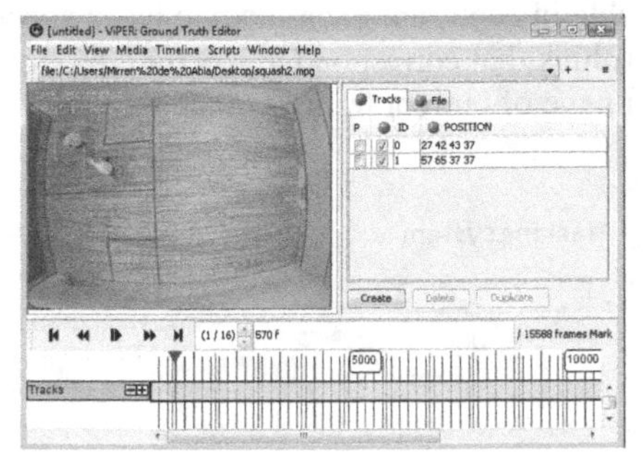

Fig. 4 Video analysis. **4A** Frame 550. **4B** Frame 570.

In general this modified version of ViPER improves notably annotation times. In many cases it is not necessary to realize changes from one frame to other. In some cases only it is necessary to do small and simple modifications in a fully annotated frame.

Ground truths are stored in XML files as sets of descriptor records. Each descriptor annotates an associated range of frames by instantiating a set of attributes for that range [4]. This is a sample code which represents the track position in a range of frames. Positions are denoted by bounding boxes.

```
<object framespan="-2147483648:2147483646" id="2" name="Tracks">
    <attribute name="POSITION">
        <data:bbox framespan="146:148" height="11"
            width="16" x="26" y="72"/>
        <data:bbox framespan="301:303" height="10"
            width="15" x="26" y="72"/>
    </attribute>
</object>
```

Fig. 5 Sample code.

Manually semantic annotation it is possible, thanks to the ViPER Schema Editor utility. Adding a new descriptor and attributes it is feasible to carry out a semantic description about what is happening in the video. This images show how users can annotate the name of the players and when it is producing a cross between them.

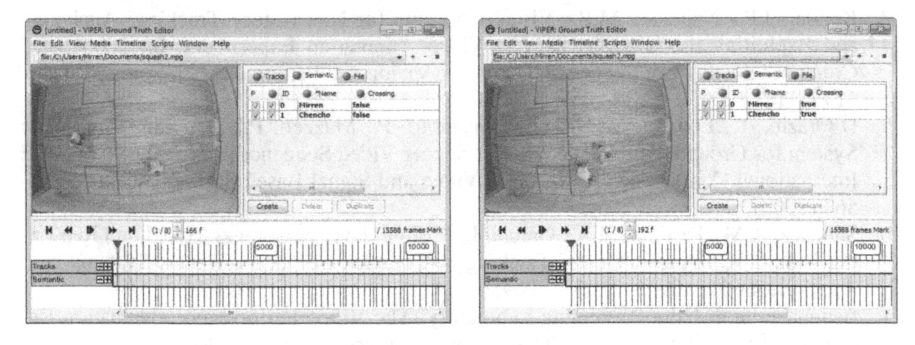

Fig. 6 Semantic analysis. **6A** Frame 166. **6B** Frame 192.

5 Conclusion

We have presented a new annotation tool for interactive *ground-tuth* generation. This system integrates a tracking module for a multi-level automatic and supervised labeling. In general MViPER-GT improves notably annotation times. In many cases it is not necessary to realize changes from one frame to other, and only in some cases it is necessary to do small and simple modifications in a fully annotated frame.

Future works will be addressed to the configuration of the tracking system through a XML file, the integration of a context-based module to introduce semi-automatic annotation at scene and object level and the capability to develop automatic annotations at the semantic level based on the behavior of the tracked elements.

References

1. Snoek, C.G.M., Worring, M.: Multimodal Video Indexing: A Review of the State-of-the-Art. Multimedia Tools and Applications 25(1), 5–35 (2004)
2. Bloehdorn, S., Petridis, K., Saathoff, K., Simou, N., Tzouvaras, V., Avrithis, Y., Handschuh, S., Kompatsiaris, Y., Staab, S., Strintzis, M.G.: Semantic Annotation of Images and Videos for Multimedia Analysis. In: Gómez-Pérez, A., Euzenat, J. (eds.) ESWC 2005. LNCS, vol. 3532, pp. 592–607. Springer, Heidelberg (2005)
3. Butler, M., Zapart, T., Li, R.: Video Annotation – Improving Assessment of Transient Educational Events. In: Proceedings of the 2006 Informing Science and IT Education Joint Conference (2006)
4. Doermann, D., Mihalcik, D.: Tools and Techniques for Video Performance Evaluation. In: 15th International Conference on Pattern Recognition, vol. 4, p. 4167 (2000)
5. Panagi, P., Dasiopoulou, S., Papadopoulos, G.T., Kompatsiaris, I., Strintzis, M.G.: A Genetic Algorithm Approach Ontology-Driven Semantic Image Analysis. In: IET International Conference on Visual Information Engineering, pp. 132–137 (2006)
6. Black, J., Ellis, T., Rosin, P.: A Novel Method for Video Tracking Performance Evaluation. In: Joint IEEE International Workshop on Visual Surveillance and Performance Evaluation of Tracking and Surveillance (2003)
7. Assfalg, J., Bertini, M., Colombo, C., Del Bimbo, A.: Semantic Annotation of Sports Videos. IEEE Multimedia Magazine 9(2), 52–60 (2002)
8. Kender, J.R., Naphade, M.R.: Visual Concepts for News Story Tracking: Analyzing and Exploiting the NIST TRECVID Video Annotation Experiment. In: 2005 IEEE Computer Society Conference on Computer Vision and Pattern Recognition, vol. 1, pp. 1174–1181 (2005)
9. D'Orazio, T., Leo, M., Mosca, N., Spagnolo, P., Mazzeo, P.L.: A Semi-Automatic System for Ground Truth Generation of Soccer Video Sequences. In: 2009 Sixth IEEE International Conference on Advanced Video and Signal Based Surveillance, pp. 559–564 (2009)
10. Sánchez, A.M., Patricio, M.A., García, J., Molina, J.M.: A Context Model and Reasoning System to Improve Object Tracking in Complex Scenarios. Expert Systems with Applications 36, 10995–11005 (2009)
11. Language and Media Processing Laboratory. The Video Performance Evaluation Resource, `http://viper-toolkit.sourceforge.net`
12. Surveillance Performance EValuation Initiative (SPEVI), `http://www.elec.qmul.ac.uk/staffinfo/andrea/spevi.html`
13. A chroma-based Video Segmentation Ground-truth, `http://www-vpu.ii.uam.es/CVSG/`
14. OTCBVS Benchmark Dataset Collection, `http://www.cse.ohio-state.edu/otcbvs-bench/`
15. Video Surveillance Online Repository (VISOR), `http://imagelab.ing.unimore.it/visor/video_categories.asp`
16. ETISEO Video undestanding Evaluation, `http://www-sop.inria.fr/orion/ETISEO/`
17. CANDELA project, `http://www.multitel.be/~va/candela/`
18. Computational Vision Group, `http://www.cvg.rdg.ac.uk/`
19. CVBASE dataset, `http://vision.fe.uni-lj.si/cvbase06/downloads.html`

Data Modeling for Ambient Home Care Systems

Ana M. Bernardos, M. del Socorro Bernardos, Josué Iglesias, and José R. Casar

Abstract. Ambient assisted living (AAL) services are usually designed to work on the assumption that real-time context information about the user and his environment is available. Systems handling acquisition and context inference need to use a versatile data model, expressive and scalable enough to handle complex context and heterogeneous data sources. In this paper, we describe an ontology to be used in a system providing AAL services. The ontology reuses previous ontologies and models the partners in the value chain and their service offering. With our proposal, we aim at having an effective AAL data model, easily adaptable to specific domain needs and services.

1 Introduction

The population pyramid in many developed countries shows a constrictive profile, due to a lower percentage of young people than some decades ago and an expansive number of elderly and mature citizens, with higher life expectancy but also with more chronic diseases to cope with. This situation is a serious challenge for social (health)care systems, which need to offer high quality services while guaranteeing their economic sustainability. For this reason, technologies to promote independent living are in the research strategic agenda in many countries [1]. In particular, the development of Ambient Care Systems (ACS) are attracting a lot of interest in the research community: enhancing the elder's care network, detecting and handling emergency situations and helping the users to accomplish activities of daily living (eg. fulfilling medication intake schedules) are some of their functional objectives.

Ambient Care Systems are based on an opportunistic and intelligent acquisition of information from heterogeneous sources (mobile phones, personal biometric sensors or infrastructure devices), its subsequent aggregation with static data (such as profile information, personal calendars or electronic health records) and an accurate and real-time inference of information about users' contexts [2]. ACSs are supposed to take decisions on the user's 'context images', triggering events and actions, and offering support to decision processes. At the same time, ACSs need to be configured to serve as channels for third parties to integrate external services, so they need to provide interoperable communications and authentication

Ana M. Bernardos, M. del Socorro Bernardos, Josué Iglesias, and José R. Casar
Universidad Politécnica de Madrid, Madrid (Spain)
e-mail: {abernardos,josue,jrcasar}@grpss.ssr.upm.es,
sbernardos@fi.upm.es

A.P. de Leon F. de Carvalho et al. (Eds.): Distrib. Computing & Artif. Intell., AISC 79, pp. 333–340.
springerlink.com

and security features. Then, ACSs are context-aware systems needing to cope with a great diversity of data, which should be sharable among different types of consumer applications with different informational needs. How to represent context parameters in an expressive and extendable way is still an open issue [3], as general data models are not adapted to the ACSs information needs.

A trend in context modeling is to use ontologies to represent context data; due to their versatility in terms of distribution, validation, formalization, ambiguity control and completeness. In this paper we propose an application-oriented ontology (ACS-Ont), specifically designed to cover the representation needs of an ambient care system. ACS-Ont reuses previous ontologies and expands some others, in order to generate a domain data model.

Our proposal differs from previous ones in its holistic approach: ACS-Ont covers the deployment of services both in residential environments and outdoors and considers the integration of different stakeholders in the service provision value chain. ACS-Ont is scalable in terms of sensor and service diversity, and has been built following the reusability paradigm for data modeling, by focusing on the exploitation of existent ontologies.

The paper is structured as follows. Section II reviews the state-of-the-art in ontology-based context modeling, with a special focus on remote healthcare systems. Section III explains the service scenario and its requirements. Section IV covers the ACS-Ont design process and our ontology models reuse strategy. Section V underlines the open issues.

2 State of the Art

An ontology is a formal explicit specification of a shared abstract model representing some phenomenon [5]. In practice, an ontology defines a common vocabulary based on the main concepts defining a discourse universe or domain, their properties and the restrictions on the relationships among concepts [6]. Ontologies are powerful tools to specify concepts and relationships, and they simplify interpretation and managing of non-complete data. For these reasons, ontologies have been chosen to model context information (see e.g. [7] [8]). Some examples are: SOUPA [9], CONON [10], the GAIA's context model [12] and CoDAMoS ontology [13]. SOUPA combines a number of useful vocabularies from different ontologies (FOAF, Rei, etc.) to model *Person*, *Policy&Action*, *Time* or *Event*. CONON provides an upper ontology to model general concepts and supports extensibility for adding domain ontologies. CONON defines 14 core classes which model *Person*, *Location*, *Activity* and *ComputationalEntities*. The GAIA architecture uses a context model based on predicates which are afterwards formalized into a DAML+OIL ontology. The ontology facilitates the predicates' consistency checking: e.g. its (XML) description is validated with the ontology when a new entity is included. The particularity of GAIA's proposal is that it aims at modeling uncertainty by attaching a confidence value to predicates. CoDAMoS ontology considers the provision of ambient services through a platform, determining four main entities: *User*, *Environment*, *Platform* (hardware and software description of a device) and *Service* (specific functionalities to the user).

Regarding the needs of ambient care systems, several works have addressed partial aspects of the problem. The K4CARE project [14] proposes a healthcare model for home services which is composed by two ontologies: the Actor Profile Ontology (APO) and the Case Profile Ontology (CPO). This data model neither considers the representation of the infrastructure nor personal sensing devices, nor the service provision. The SOPRANO project [15] is built on an OWL-Lite ontology which only aims at describing the person's state, taking into account past states. Nugent et al. [16] present a XML-based schema representation of information in smart homes. The hierarchical representation describes 1) the home physical structure and the devices in it; 2) the contact details of the institutional caregiver in charge and 3) the inhabitant details (name, ambulatory devices and care plan). In this case, the data model does not consider how to deal with real-time context information. Ko et al. [17] propose the use of a context ontology for U-HealthCare. The data model classifies context according to the *Person context ontology, Device context ontology* and *Environment context ontology*. In [18], the context model is composed by several OWL ontologies to detect alarms in home environments. The ontology describes the patients, the home domain, the alarm management system and the social care network. Other proposals are focused on the messages exchange system. For example, [19] present a data model for a telemedicine system, which aims at supporting the management of messages exchange between different actors in the telemedicine system. There are also some attempts to model the singularities of special users. It is the case of [20], which proposes a context model to monitor and handle agitation behavior in persons suffering dementia. A great challenge in ambient care systems design is how to make them emotion-aware. With respect to data modeling, ontologies have been also used to model feelings and emotional states (e.g. [17]).

The design of the ACS-Ont has a special focus on reusing existing ontologies while considering the particularities of the ACS domain and its value chain. To the best of our knowledge, none of the previous initiatives completely cover the ACS provision problem.

3 Model Requirements

Ambient Care Systems aim at providing different type of assistive services in different scenarios, always taking into account the user's real-time needs. In this Section, our approach to ACS service offering is addressed. From it, we derive the requirements of the data model which will be afterwards implemented in ACS-Ont ontology. Fig. 1 represents our understanding of the service paradigm for ACS. The ACS will be working on:

1. A *Personal Network* (PN), which will include all the devices that a person must wear or use. For instance, localization may be supported by a mobile device with GPS capabilities (for outdoor positioning) or with communication capabilities such as WiFi, Bluetooth and/or ZigBee (for outdoor positioning), or by

other means (e.g. RFID-based user-object interactions). Additionally, depending on the user profile and the services to be configured, the PN may include continuous health monitoring sensors (ECG, oxymeters, etc.).
2. A *Home Network*, which will include home infrastructure sensors, actuators and appliances capable of notifying their status. The Home Network Unit will be able to communicate with the Personal Network by using ad hoc networking capabilities. It will include local intelligent features to dispatch events and orders depending on the situation. These processing capabilities will be part of a home gateway which will connect the home environment with the Core Care Network.
3. A *Core Care Network*, serving as a bridge of communication between the home infrastructure and third parties and service or context providers. Services may be enabled through the Core Care Network; it will authorize the connection of external service providers (External Care Network) and External Context Data Providers, centralize system monitoring and guarantee the security of personal data.

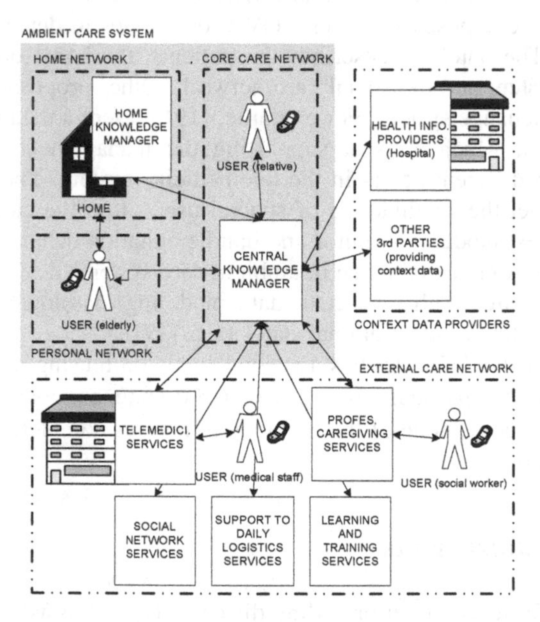

Fig. 1 The service concept behind the Ambient Care System

On top of this structure, the ACS will build its own basic services, such as: a) context-aware notifications/reminders dispatching, b) emergency detection or c) activity logging. The first set of services is provided to the elder or to his caregiver, and can be configured to dispatch reminders of medication intake, appointment schedule or general context-aware notifications. The second set of services relies on the processing of biomedical data and environmental information checking to detect risky states. Emergencies are treated both as dangerous situations in the present and long-term deviations from typical behavior that can trigger a complicated situation in the future. Connecting with this case, last group of services stores all the information about activity, allowing an off-line pattern analysis which may feedback the system intelligence.

The system using the ontology will need to:

- Dynamically check consistency when including new concepts.
- Deal with information retrieval and data fusion at different levels of abstraction.
- Support probabilistic inference mechanisms, by handling uncertainty and quality of context.
- Offer a coarse-grained design for general service support, in order to facilitate service scalability.
- Provide data structures to support security: authentication, traceability and data logging.

This operative scenario and the requirements it imposes to a context management system have been taken as the starting point to design ACS-Ont. Next step in the development of the ontology has been the specification of both class representing concepts and relations among them, as explained in the following section.

4 ACS-Ont Structure: Methodology, Main Concepts and Sub-ontologies

We started the development of the ontology by defining its domain and scope. Then we conducted a knowledge acquisition task based on the analysis of different documentation related to the field and on the background of some members of the team. As a result of this process, we could establish the relevant terms of the domain and started to define and (hierarchically) organize the classes of the ontology. We selected the terms that describe objects having independent existence rather than terms that describe these objects.

Due to the large scale of the proposed ACS domain (ranging from specific device and sensor characteristics to concepts related to the users' health status), the domain of interest has been divided into smaller ones: a general ontology regarding the most abstract concepts and relations (those from Fig. 1) has been created and these concepts and relations have been extended generating other ontologies ('sub-ontologies'). The process followed for developing the taxonomies of sub-ontologies has been a combination of top-down and bottom-up approaches: first the more salient classes have been defined and then they have been generalized and specialized appropriately.

ACS-Ont has been built taking into account that an advantage of ontology models is their capability to be reused. Thus, groups of experts in specific domains might put all their effort into developing solid ontological models of a specific domain that other developers may integrate (as a whole or, more often, just in part) in their own works. This 'share-ability' capability is especially important for

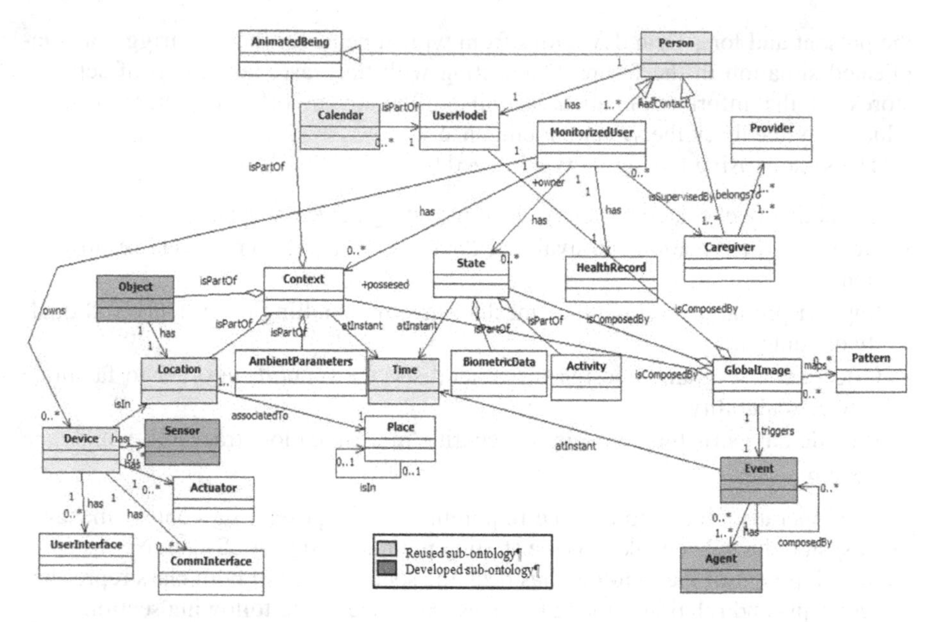

Fig. 2 ACS-Ont structure.

highly heterogeneous domains as the present one (with a high diversity of sensor and devices, but also of services and actors). ACS-Ont incorporates different ontologies to describe some common aspects of Ambient Intelligent systems (See Table 1). These models have been directly integrated (either completely or partially) –specially when they are available in the language used for ACS-Ont: OWL– or taken just as reference models –in other cases. Color-coded schema of Fig. 2 shows the relations among concepts of general ontology and sub-ontologies, also highlighting reused models.

Some other well-known ontologies have been indentified in order to be reused in future extensions of ACS-Ont (mainly in the development of sub-ontologies). Regarding healthcare operability, an OpenEHR [21] ontology may extend the concepts below HealthRecord class, where SNOMED-CT model [22] can be also integrated (when modeling, for instance, medicines or electronic prescriptions);

Table 1 Overview of reused ontologies*

Entity/Concept	From ontology	Ontology URI
Time	OWL-Time	http://www.isi.edu/~pan/damltime/time-entry.owl
Location	Geo-OWL	http://www.opengis.net/gml/
Person	FOAF	http://xmlns.com/foaf/0.1/
Calendar	RDF Calendar	http://www.w3.org/2002/12/cal/icaltzd
Device	Delivery Context Ont.	http://www.w3.org/2007/uwa/context/

*General structure of SOUPA, CONON, CODAMOS and X73 models has also been considered.

nevertheless, OWL-based ontologies associated to these models are still an open issue. Finally, from a more general view, the ACS security support may be modeled by means of REI Policy Ontology [23] that can be used to formally describe permissions and obligations.

ACS-Ont has been implemented in OWL, fulfilling the OWL-DL expressivity and therefore allowing using any Description Logics reasoner to exploit it.

5 Conclusions/Discussion and Further Work

Although this ontology is in a stage of development (the appropriate subontologies have been identified but some of them have not been completely developed yet), it constitutes a headway towards an optimal structuring of the AAL domain. If properly completed, it could be considered as a generic semantic frame for use within this field, since it contributes to the shaping of a common, shared knowledge based in reusability and that can be reused.

A future phase of this work will deal with: a) the enrichment of the ontology by adding possible rules, axioms and constants; b) extending the model to cover specific service needs (e.g. modeling activities of daily living to support behavioral analysis) and c) offering uncertainty control and Quality of Context.

In addition, the use of the application will lead to the introduction of instances that will populate the ontology and allow the application to make inferences or other types of reasoning.

Acknowledgments. This work has been supported by the Government of Madrid under grant S-2009/TIC-1485 and by the Spanish Ministry of Science and Innovation under grant TIN2008-06742-C02-01. The authors also acknowledge related discussions with partners within the AmIVital Cenit project, financed by the Centre for the Development of Industrial Technology (CDTI).

References

[1] The Ambient Assisted Living (AAL) Joint Programme, http://www.aal-europe.eu/
[2] Dey, A.K., Abowd, G.: Towards a Better Understanding of Context and Context-Awareness. In: Gellersen, H.-W. (ed.) HUC 1999. LNCS, vol. 1707, pp. 304–307. Springer, Heidelberg (1999)
[3] Bolchini, C., et al.: A Data-oriented Survey of Context Models. SIGMOD Rec. 36(4), 19–26 (2007)
[4] Boury-Brisset, A.-C.: Ontology-based Approach for Information Fusion. In: Proc. of the 6th Int. Conference on Information Fusion (2003)
[5] Gruber, T.: Toward Principles for the Design of Ontologies Used for Knowledge Sharing. Int. Journal Human-Computer Studies 43(5-6), 907–928 (1995)
[6] Noy, N.F., McGuinness, D.L. (2005) Desarrollo de Ontologías, http://protege.stanford.edu/publications/ontology_development/ontology101-es.pdf (accessed June 2008)

[7] Baumgartner, N., Retschitzegger, W.: A Survey of Upper Ontologies for Situation Awareness. In: Proc. of the Int. Conf. on Knowledge Sharing and Collaborative Engineering (2006)
[8] Ye, J., et al.: Ontology-based models in pervasive computing systems. The Knowledge Engineering Review 22(4), 315–347 (2007)
[9] Chen, H., et al.: SOUPA: Standard Ontology for Ubiquitous and Pervasive Applications. In: Proc. First Annual International Conference on Mobile and Ubiquitous Systems (MobiQuitous 2004), pp. 258–267 (2004)
[10] Wang, X.H., Zhang, D.Q., Gu, T., Pung, H.K.: Ontology Based Context Modeling and Reasoning using OWL. In: Proc. of the 2nd IEEE Annual Conf. on Pervasive Computing and Communications Workshops (2004)
[11] Khedr, M., Karmouch, A.: Negotiating Context Information in Context-Aware Systems. IEEE Computer 19(6), 21–29 (2004)
[12] Ranganathan, A., et al.: Ontologies in a Pervasive Computing Environment. In: Proc. of the Workshop on Ontologies and Distributed Systems (2003)
[13] Preuveneers, D., et al.: Towards and Extensible Context Ontology for Ambient Intelligence. In: Markopoulos, P., Eggen, B., Aarts, E., Crowley, J.L., et al. (eds.) EUSAI 2004. LNCS, vol. 3295, pp. 148–159. Springer, Heidelberg (2004)
[14] K4CARE: Knowledge-based HomeCare eServices for an Ageing Europe. Deliverable 01: The K4Care Model (2006)
[15] Klein, M., et al.: Ontology-Centred Design of an Ambient Middleware for Assisted Living: The Case of SOPRANO. In: 30th Annual German Conf. on Artificial Intelligence (2007)
[16] Nugent, C.D., et al.: homeML-An Open Standard for the Exchange of Data Within Smart Environments. LNCS 4551, pp. 121-129 (2007)
[17] Ko, E.J., et al.: Ontology-Based Context Modeling and Reasoning for U-HealthCare. IEICE Trans. Inf.&Syst. E90-D(8) (2007)
[18] Paganelli, F., Giuli, D.: An ontology-based context model for home health monitoring and alerting in chronic patient care networks. In: Proc. of the 21st Conf. on Advanced Information Networking and Applications Workshops (2007)
[19] Nageba, E., et al.: A model driven ontology-based architecture for supporting the quality of services in pervasive telemedice applications. In: Proc. of Pervasive Health (2009)
[20] Fook, V.F.S., Tay, S.C., Jayachandran, M., Biswas, J., Zhang, D.: An ontology-based context model in monitoring and handling agitation behaviour for persons with dementia. In: Proc. IEEE Int. Conf. on Pervasive Computing and Communications Workshops (2006)
[21] openEHR Foundation, http://www.openehr.org/
[22] College of American Pathologists. Snomed Clinical Terms® Technical Reference Guide, release (July 2003)
[23] Kagal, L., et al.: A Policy Based Approach to Security for the Semantic Web. In: Fensel, D., Sycara, K., Mylopoulos, J. (eds.) ISWC 2003. LNCS, vol. 2870, pp. 402–418. Springer, Heidelberg (2003)

Face Recognition at a Distance: Scenario Analysis and Applications

R. Vera-Rodriguez, J. Fierrez, P. Tome, and J. Ortega-Garcia

Abstract. Face recognition is the most popular biometric used in applications at a distance, which range from high security scenarios such as border control to others such as video games. This is a very challenging task since there are many varying factors (illumination, pose, expression, etc.) This paper reports an experimental analysis of three acquisition scenarios for face recognition at a distance, namely: close, medium, and far distance between camera and query face, the three of them considering templates enrolled in controlled conditions. These three representative scenarios are studied using data from the NIST Multiple Biometric Grand Challenge, as the first step in order to understand the main variability factors that affect face recognition at a distance based on realistic yet workable and widely available data. The scenario analysis is conducted quantitatively in two ways. First, an analysis of the information content in segmented faces in the different scenarios. Second, an analysis of the performance across scenarios of three matchers, one commercial, and two other standard approaches using popular features (PCA and DCT) and matchers (SVM and GMM). The results show to what extent the acquisition setup impacts on the verification performance of face recognition at a distance.

1 Introduction

The growth of biometrics has been very significant in the last few years. A new research line growing in popularity is focused on using biometrics in less constrained scenarios in a non-intrusive way, including acquisition "On the Move" and "At a Distance" [7], which are user-friendly, and often do not need user cooperation.

The most common biometric modes used for recognition at a distance are face, iris and gait, being face the most popular of them. Face recognition is a challenging problem in the field of computer vision which has been the subject of active

R. Vera-Rodriguez, J. Fierrez, P. Tome, and J. Ortega-Garcia
ATVS, Escuela Politecnica Superior - Universidad Autonoma de Madrid,
Avda. Francisco Tomas y Valiente, 11 - 28049 Madrid, Spain
e-mail: {ruben.vera, julian.fierrez, pedro.tome}@uam.es,
javier.ortega@uam.es

A.P. de Leon F. de Carvalho et al. (Eds.): Distrib. Computing & Artif. Intell., AISC 79, pp. 341–348.
springerlink.com

research for the past decades because of its many applications in domains such as surveillance, covert security and context-aware environments. Face recognition is very appealing as a biometric as it offers several advantages in terms of being non-intrusive, non-invasive, cost-effective, easily accessible (i.e., face data can be conveniently acquired with a few inexpensive cameras) and relatively acceptable to the general public. However, employing the face for recognition also presents some difficulties since the appearance of the face can be altered by intrinsic factors such as age, expression, facial hair, glasses, make up, etc., as well as extrinsic ones such as scale, lighting, focus, resolution, or pose amongst others [13].

This paper is focused on the study of the effects of acquisition distance variation on the performance of automatic face recognition systems. This is motivated by the analysis of the results from the recent NIST Multiple Biometric Grand Challenge (MBGC 2009) [8] and the Face Recognition Vendor Test (FRVT 2006) [9], which show that a lot of research is still needed to overcome these problems. In this sense, three different scenarios have been defined from the NIST MBGC depending on the acquisition distance between the subject and the camera, namely "close", "medium" and "far" distance. We use a subset of this benchmark dataset consisting of images of a total of 112 subjects acquired at different distances and varying conditions regarding illumination, pose/angle of head, and facial expression. This analysis is conducted quantitatively at two levels for the considered scenarios: 1) main data statistics such as information content, and 2) performance of recognition systems: one commercial, and two other based on popular features (PCA and DCT) and matchers (SVM and GMM).

Depending on the distance to the camera, face recognition could be applied in two different applications [1, 7]:

- Requiring cooperative users (near distance), such as in border control (e-passport) or security access (for example access to stadium in 2008 Olympic Games). In these cases a verification (one to one) of the identity is carried out.
- Not requiring cooperative users (medium and far distances), such as face surveillance (for example subway watch-list) or in large database search (such as national registration data or black-list data). In these cases an identification (one to many) is normally carried out.

Other applications could be on social network webs for automatic face tagging and finding people[1]. Apart from the person recognition applications, there are other applications in which face recognition technology can be useful such as activity detection (for smart homes [14], ambient assisting living [3] or video games [6, 12]), or in pedestrian detection to avoid accidents. In this last case a possible fusion between face and gait would be of interest [5]. Figure 1 shows some examples of applications of face recognition.

The paper is structured as follows. Sect. 2 describes the dataset and scenarios under study. Sect. 3 analyzes the main data statistics of the scenarios. Sect. 4 studies the performance of the three considered recognition systems on the different scenarios. Sect. 5 finally discusses the experimental findings and outlines future research.

[1] For example http://picasaweb.google.com

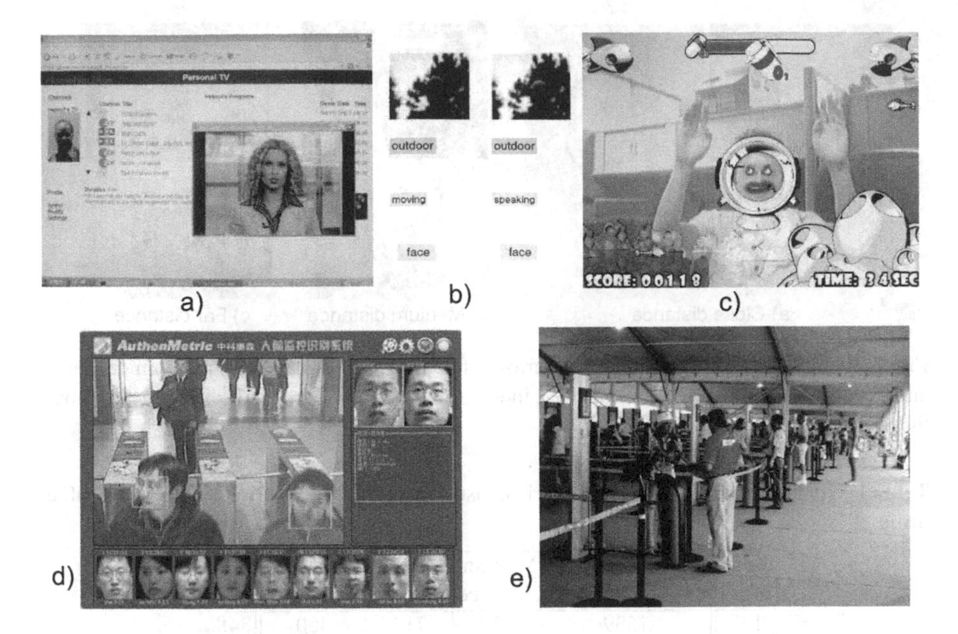

Fig. 1 Example images of different applications of face recognition: a) Web interface for smart TV program selection by face recognition [14]. b) Classification results of activity detection [2]. c) Example of video game using face and activity detection [12]. d) Example of a watch-list surveillance and identification system [7]. e) Face verification system used in Beijing 2008 Olympic Games [1].

2 Scenario Definition

The three scenarios considered are: 1) "close" distance, in which the shoulders may be present; 2) "medium" distance, including the upper body; and 3) "far" distance, including the full body. Using these three general definitions, the 3482 face images from the 147 subjects present in the dataset NIST MBGC v2.0 Face Stills [8] were manually tagged. Some sample images are depicted in Fig. 2. A portion of the dataset was discarded (360 images from 89 subjects), because the face was occluded or the illumination completely degraded the face. Furthermore, although this information is not used in the present paper, all the images were marked as indoor or outdoor.

Finally, in order to enable verification experiments considering enrollment at close distance and testing at close, medium, and far distance scenarios, only the subjects with at least 2 images in close and at least 1 image in both of the two other scenarios were kept. The data selection process is summarized in Table 1, which shows that the three considered scenarios result in 112 subjects and 2964 face images.

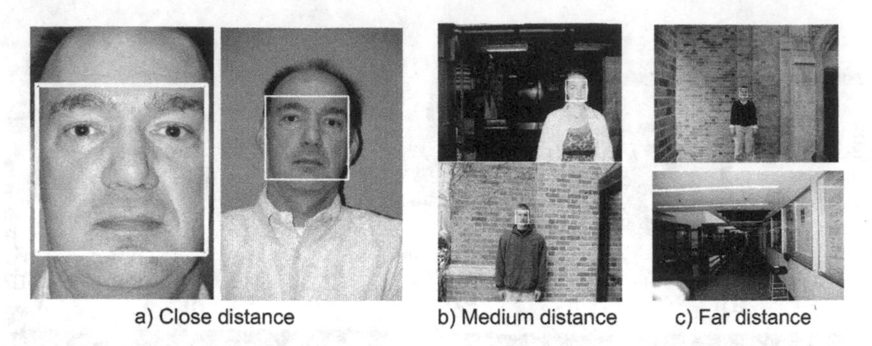

Fig. 2 Example images of the three scenarios defined: a) close distance, b) medium distance, and c) far distance. Images are collected indoors and outdoors and with different illuminations.

Table 1 Number of images of each scenario constructed from NIST MBGC v2.0 Face Visible Stills.

Num. users	Close distance	Medium distance	Far distance	Discarded images	Total
147	1539	870	713	360	**3482**
	*At least **2** images per user*	*At least **1** images per user*			
112	1468	836	660		**2964**

Table 2 Segmentation results based on errors produced by face Extractor of VeriLook SDK.

	Close distance	Medium distance	Far distance	Discarded	Total
Num. Images	1468	836	660	360	**3324**
Errors	21	151	545		**848**
Errors(%)	1.43%	18.06%	**82.57%**		

3 Scenario Analysis: Data Statistics

First of all, faces were localized and segmented (square areas) in the three acquisition scenarios using the VeriLook SDK discussed in Sect. 4.1. Segmentation results are shown in Table 2, which shows that segmentation errors increase significantly across scenarios, from only 1.43% in close distance to 82.57% in far distance. Segmentation errors here mean that the VeriLook software could not find a face in the image. For all the faces detected by VeriLook we conducted a visual check, where we observed 3 and 10 segmentation errors for medium and far distance respectively. All the segmentation errors were then manually corrected by manually marking the eyes. The face area was estimated based on the marked distance between eyes.

As a result of the defined scenarios, we observe that the sizes of the segmented faces decrease with the acquisition distance. In particular, the average face size in pixels for each scenario is: 988×988 for close, 261×261 for medium, and

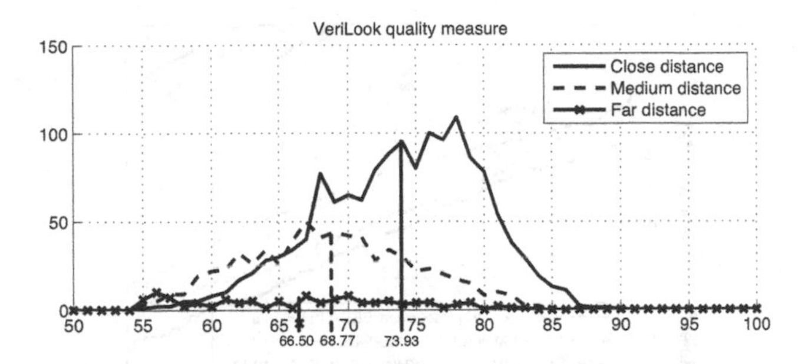

Fig. 3 Histogram of face quality measures produced by VeriLook SDK.

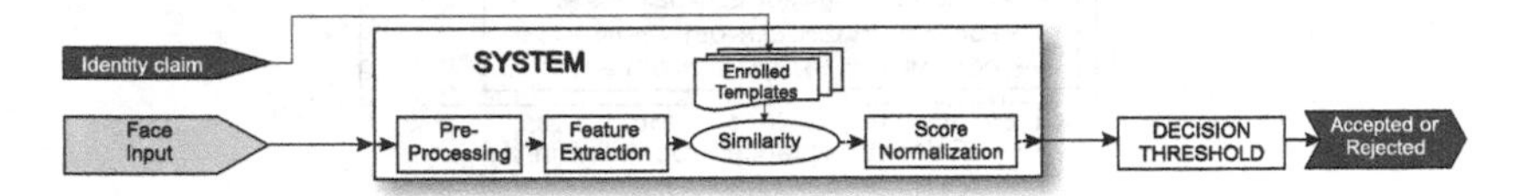

Fig. 4 Diagram of face recognition system used for VeriLook SDK, DCT-GMM and PCA-SVM.

78×78 for far distance. For the experimental work, the face size is normalized to 64×80 pixels.

Another data statistic that was computed for the three scenarios was the average face quality index provided by VeriLook ($0 =$ lowest, $100 =$ highest): 73.93 for close, 68.77 for medium, and 66.50 for far distance (see Fig. 3, computed only for the faces correctly segmented by VeriLook). As stated by VeriLook providers, this quality index considers factors such as lightning, pose, and expression.

4 Scenario Analysis: Verification Performance Evaluation

4.1 Face Verification Systems

The architecture of the face recognition system used is shown in Fig. 4. In a similar way as in previous work [10], three approaches are used for face verification:

- **VeriLook SDK.** Commercial face recognition system developed by Neurotechnology[2].
- **PCA-SVM system.** This verification system uses Principal Component Analysis (PCA). The evaluated system uses normalized and cropped face images of size 64×80 (width $\times$ height), to train a PCA vector space where 96% of the variance is retained. This leads to a system where the original image space of 5120 dimensions is reduced to 249 dimensions. Similarity scores are computed in this PCA vector space using a SVM classifier with linear kernel.

[2] http://www.neurotechnology.com/

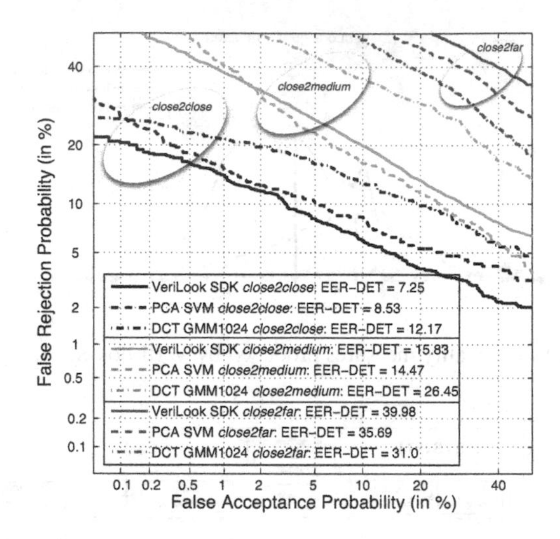

Fig. 5 Verification performance results for the three scenarios and three systems considered.

- **DCT-GMM system.** This verification system also uses face images of size 64 × 80 divided into 8 × 8 blocks with horizontal and vertical overlap of 4 pixels. This process results in 285 blocks per segmented face. From each block a feature vector is obtained by applying the Discrete Cosine Transform (DCT); from which only the first 15 coefficients ($N = 15$) are retained. The blocks are used to derive a world GMM Ω_w and a client GMM Ω_c [4]. From previous experiments we obtained that using $M = 1024$ mixture components per GMM gave the best results. The DCT feature vector from each block is matched to both Ω_w and Ω_c to produce a log-likelihood score [4].

4.2 Experimental Protocol

Three main experiments are defined for the verification performance assessment across scenarios:

- *Close2close.* This experiment gives an idea about the performance of the systems in ideal conditions (both enrollment and testing using close distance images). About half of the close distance subcorpus (754 images) is used for development (training the PCA subspace, SVM, etc.), and the rest (714 images) is used for testing the performance.
- *Close2medium*, and *close2far* protocol. These two other experiments use as training set the whole close distance dataset (1468 face images). For testing the performance of the systems the two other datasets are used: 836 medium distance images for *close2medium*, and 660 far distance images for *close2far*.

4.3 Results

Fig. 5 shows the verification performance for the three considered scenarios: *close2close*, *close2medium*, and *close2far*. As can be seen, VeriLook is the best of the three systems in *close2close* with an EER of around 7%. At the same time, this commercial system is the most degraded in uncontrolled conditions, with an EER close to 40% in *close2far*, much worse than the other two much simpler systems. This result corroborates the importance of analyzing and properly dealing with variability factors arising in biometrics at a distance.

Fig. 5 also shows that the GMM-based system works better in far distance conditions than the other systems, although being the less accurate in *close2close* and *close2medium*. This result demonstrates the greater generalization power of this simple recognition approach, and its robustness against uncontrolled acquisition conditions.

5 Discussion and Future Work

An experimental approach towards understanding the variability factors in face recognition at a distance has been reported. In particular, a data-driven analysis of three realistic acquisition scenarios at different distances (close, medium, and far) has been carried out as a first step towards devising adequate recognition methods capable of working in less constrained scenarios.

This analysis has been focused on: 1) data statistics (segmented face sizes and quality measures), and 2) verification performance of three systems. The results showed that the considered systems degrade significantly in the far distance scenario, being more robust to uncontrolled conditions the simplest approach.

Noteworthy, the scenarios considered in the present paper differ not only in the distance factor, but also in illumination and pose (being the illumination variability much higher in far distance than in close distance). Based on the data statistics obtained and the performance evaluation results, a study of the effects of such individual factors is source for future research.

Also, depending on the application, fusion with other biometrics would be of interest, such as in the case of pedestrian detection in order to avoid car crashings it would be very useful a fusion with gait, or also with footsteps [11] in scenarios like walking through an identification bow. This also could be used in ambient intelligence applications such as monitoring the behavior of elderly people [3].

Acknowledgements. This work has been supported by project Contexts (S2009/TIC-1485). P. Tome is supported by a FPU Fellowship from Univ. Autonoma de Madrid.

References

1. Ao, M., Yi, D., Lei, Z., Li, S.Z.: Face Recognition at a Distance: System Issues. In: Handbook of Remote Biometrics for Remote Biometrics, pp. 155–167. Springer, Heidelberg (2009)

2. Casale, P., Pujol, O., Radeva, P.: Face-to-Face Social Activity Detection Using Data Collected with a Wearable Device. In: Pattern Recognition and Image Analysis, pp. 56–63 (2009)
3. CAVIAR: Context Aware Vision using Image-based Active Recognition, `http://homepages.inf.ed.ac.uk/rbf/CAVIAR/`
4. Galbally, J., McCool, C., Fierrez, J., Marcel, S., Ortega-Garcia, J.: On the vulnerability of face verification systems to hill-climbing attacks. Pattern Recognition 43(3), 1027–1038 (2010)
5. Jafri, R., Arabnia, H.R.: Fusion of face and gait for automatic human recognition. In: Third International Conference on Information Technology: New Generations, pp. 167–173 (2008)
6. Lee, Y.J., Lee, D.H.: Research on detecting face and hands for motion-based game using web camera. In: Proc. International Conference on Security Technology (SECTECH 2008), pp. 7–12 (2008)
7. Li, S.Z., Schouten, B., Tistarelli, M.: Biometrics at a Distance: Issues, Challenges, and Prospects. In: Handbook of Remote Biometrics for Surveillance and Security, pp. 3–21. Springer, Heidelberg (2009)
8. MBGC: Multiple biometric grand challenge. NIST - National Institute of Standard and Technology, `http://face.nist.gov/mbgc/`
9. Phillips, P.J., Scruggs, W.T., O'Toole, A.J., Flynn, P.J., Bowyer, K.W., Schott, C.L., Sharpe, M.: Frvt 2006 and ice 2006 large-scale experimental results. IEEE Transactions on Pattern Analysis and Machine Intelligence 32, 831–846 (2010)
10. Tome, P., Fierrez, J., Alonso-Fernandez, F., Ortega-Garcia, J.: Scenario-based score fusion for face recognition at a distance. In: Proc. IEEE Confererence on Computer Vision and Pattern Recognition (2010)
11. Vera-Rodriguez, R., Mason, J., Evans, N.: Assessment of a Footstep Biometric Verification System. In: Advances in Pattern Recognition. Handbook of Remote Biometrics. Springer, Heidelberg (2009)
12. Wang, S., Xiong, X., Xu, Y., Wang, C., Zhang, W., Dai, X., Zhang, D.: Face-tracking as an augmented input in video games: enhancing presence, role-playing and control. In: CHI 2006: Proceedings of the SIGCHI conference on Human Factors in computing systems, pp. 1097–1106. ACM, New York (2006)
13. Zhou, S.K., Chellappa, R., Zhao, W.: Unconstrained Face Recognition. Springer, Heidelberg (2006)
14. Zuo, F., de With, P.: Real-time embedded face recognition for smart home. IEEE Transactions on Consumer Electronics 51(1), 183–190 (2005)

Distributed Subcarrier and Power Allocation for Cellular OFDMA Systems

Ruxiu Zhong, Fei Ji, Fangjiong Chen*, Shangkun Xiong, and Xiaodong Chen

Abstract. Dynamic resource allocation according to user's link quality is cirical in OFDMA system to improve network capacity. In this paper we consider joint subcarrier and power allocation of the downlink communication of multi-cell OFDMA system. The allocation problem is formulated with the goal of minimizing the transmitted power subject to individual rate constraint of the users. We propose a suboptimal distributed algorithm which consists of two stages. In the first stage each cell ignore the inter-cell interference and perform single-cell resource allocation. In the second stage the cells iteratively exchange the allocation result and update resource allocation until users' rate requirements are met. The proposed algorithm is evaluated with computer simulations and compared with existing centralized algorithm. It is shown that the proposed algorithm obtain satisfactory tradeoff between quality of solution and complexity.

Keywords: Distributed Resource Allocation; OFDMA; Cellular System.

1 Introduction

Orthogonal-frequency-division-multiple-access (OFDMA) has emerged as one of the prime multiple access scheme in broadband wireless system including 3G LTE and WIMAX. In an OFDMA system the total bandwidth is divided into non-overlapped traffic channels (one or a cluster of subcarriers) such that each user occupies a subset of traffic channels for transmission. In cellular OFDMA system, each cell can apply the whole bandwidth for transmission, which may cause severe inter-cell interference (ICI) if adjacent cells apply same traffic channels. Radio resource allocation is an important technique to coordinate ICI and optimize network

Ruxiu Zhong, Fei Ji, and Fangjiong Chen
School of Electronics and Information Engineering, South China University of Technology
e-mail: zhongruxiu@sina.com, {eefeiji, eefjchen}@scut.edu.cn

Shangkun Xiong and Xiaodong Chen
Department of Mobile Communication, Guangzhou R&D Center, China Telecom
e-mail: {xiongsk, chenxiaod}@gsta.com

* Corresponding author.

A.P. de Leon F. de Carvalho et al. (Eds.): Distrib. Computing & Artif. Intell., AISC 79, pp. 349–356.
springerlink.com

performance. Radio resource allocation in OFDMA system usually refer to the technique that assigns to each user a subset of the available radio resources (mainly power and sbcarrier) according to a certain optimality criterion and some practical constraints. However optimal resource allocation is still an NP-hard problem that is fundamentally difficult to tackle. In practice, additional constraints, e.g. individual users' rate requirements, further complicate the problem. Hence, suboptimal solution with acceptable complexity is the focus of research in the literature.

Various heuristic algorithms have been proposed for the resource allocation of OFDMA systems. Two important criteria are commonly applied. The first one is to maximize network throughput subject to the constraint on transmission power[2][4][8]. The second one is to minimize transmission power subject to user's individual rate requirement[1][6]. These algorithms are centralized. They need a central network unit to collect the channel information of all users, perform allocation algorithm and notify all cells the allocation results. However, such network unit may not be available in practice. Distributed algorithms, which is performed at each cell based on information exchange between neighbor cells, may be more suitable in multi-cell systems or ad hoc systems[3]. Distributed algorithms based on maximizing throughput were proposed in [5][10]. The algorithm in [5] first ignores ICI and performs single-cell allocation as initialization. Then iteratively deactivates some users' subcarrier if their cause too much ICI. The algorithm is computationally efficient. But due to subcarrier deactivation, it may not be able to fulfil user's rate requirement. The algorithm in [10] also consider bit loading, consequently it is much more complicated. Distributed algorithms based on game theory[3][9] instead maximize network utility mapped from throughput. Applying utility is more efficient to describe human perception of throughput. However the nonlinear mapping between throughput and utility may complicate the problem. The distributed algorithm in [7] is based on minimizing the transmission power. The authors proved that the algorithm is asymptotically optimal. However only 2-cell system is considered.

In this paper we consider distributed algorithm for joint subcarrier and power allocation of the downlink communication of cellular OFDMA system. The applied criterion is transmission power minimization subject to individual rate constraint of the users. The proposed algorithm consists of two stages. In the first stage each cell ignore the ICI and perform single-cell resource allocation. In the second stage the cells iteratively exchange the allocation result and update resource allocation until users' rate requirements are satisfied. The algorithm result in very simple computation. Moreover, except for some special cases, the algorithm guarantee to satisfy users' rate requirement. Computer simulations shows that the proposed algorithm obtain better throughput than the centralized heuristic algorithm in [1].

2 System Model

We consider an OFDMA system with K cells. The cells apply same frequency bandwidth for transmisison and the frequency bandwidth is divided into M subcarriers. In this paper, the problem of subcarrier allocation is formulated to minimize the overall power consumed by the OFDMA system on a radio TTI (Transmission Time Interval) given a certain constraint on the rate of each user. However, physical layer performance depends on the chosen coding scheme and modulation format, while in general traffic requests from the scheduler are in terms of informative bit rates. For simplification, we assume the coding scheme and modulation format on each channel are the same.

Given a set of subcarriers $D = \{1, \cdots, M\}$, a set of cells $C = \{1, \cdots, K\}$, and for each cell k a set of users $U_k = \{1, \cdots, N_K\}$. Let $U = \bigcup U_k$ be the set of all users in the system. We assume each user has a fixed target spectral efficiency formulated as $\eta_i = log_2(1 + SINR_i)$ (in bit/s/Hz)[1]. Hence the spectral efficiency can be interpreted as the required signal-to-interference-and-noise-ratio (SINR) level of the users. The rate requirements for a given user i correspond to a certain number of subcarriers $r_i = R_i/\eta_i$, where η_i is set in such a way that r_i is an integer.

For each user $i \in U$, we denote by $b(i) = k$ the cell of user i. Let $U(j) \subseteq U$ be the set of users (belonging to different cells) which are assigned the same sub-carrier j. the measured SINR for user i on subcarrier j is

$$SINR_i(j) = \frac{G_i^{b(i)}(j)P_i(j)}{\sum_{h\in U(j),h\neq i} G_h^{b(i)}(j)p_h(j) + BN_0} \tag{1}$$

where $G_i^k(j)$ is the channel gain of user i to the kth basestation(BS) on subcarrier j. $P_i(j)$ is the transmission power of user i on subcarrier j. B is the band width of subcarrier and N_0 is the power density of the white channel noise. We assume the users can estimate the channel gains and feedback to the BS. Hence $G_i^k(j)$ is known to the transmitter.

Note $SINR_i$ is the SINR level corresponding to the target spectral efficiency of user i. From (1) we note that the minimum power assigned to user i on subcarrier j is

$$P_i^{MIN}(j) = SINR_i \frac{\sum_{h\in U(j),h\neq i} G_h^{b(i)}(j)p_h(j) + BN_0}{G_i^{b(i)}(j)} \tag{2}$$

The applied criterion in this paper is to minimize the total transmission power defined as $\sum_{i\in U,j\in D} P_i(j)$. Taking into account the rate constraints and other practical constraints, the problem of joint subcarrier and power allocation can be formulated as follows.

$$\min \sum_{i\in U,j\in D} P_i(j) \tag{3}$$

$$s.t.\;\; p_i(j) \geq x_{ij} SINR_i \frac{\sum_{h\in U(j),h\neq i} G_h^{b(i)}(j)p_h(j) + BN_0}{G_i^{b(i)}(j)} \tag{4}$$

$$p_i(j) \leq Q x_{ij}, \forall i, j \tag{5}$$

$$\sum_{i \in U_k, j \in M} p_i(j) \leq pmax_k, \forall k \tag{6}$$

$$\sum_j x_{ij} = r_i, \forall i \tag{7}$$

$$p_i(j) \geq 0, \forall i, j \tag{8}$$

$$x_{ij} \in \{0, 1\}, \forall i, j \tag{9}$$

Where x_{ij} is a binary variable equal to 1 if user i is assigned subcarrier j (and 0 otherwise). Constraint (4) imposes that if subcarrier j is assigned to user i, then power $p_i(j)$ is not less than the minimum power defined in (2). Q in (5) is a suitable large positive number. Hence constraint (5) will force $p_i(j)$ to be 0 if subcarrier j is not assigned to user i. Constraint (6) impose that the transmission power of each cell cannot exceed a predefined value. Constraints (7) requires that r_i subcarriers must be assigned to user i to satisfy its rate requirement.

3 Distributed Allocation Algorithm

In this section, we develop a distributed subcarrier and power allocation algorithm based on the solution of single-cell allocation. The algorithm involve single-cell allocations as initialization and an iterative procedure to reduce the ICI and guarantee users' rate requirement.

The proposed algorithm is based on the observation that the minimum power required to achieve the target spectral efficiency can be divided into two parts.

$$P_i^{MIN}(j) = SINR_i \frac{\sum_{h \in U(j), h \neq i} G_h^{b(i)}(j) p_h(j)}{G_i^{b(i)}(j)} + SINR_i \frac{BN_0}{G_i^{b(i)}(j)} \tag{10}$$

It can be observed that the second item, denoted as $p_i^0(j)$, is the minimum transmission power when there is no ICI. Its value can be determined by each cell which distributively performs single-cell resource allocation. The first item, denoted as $\Delta p_i(j)$, is the required transmission power due to ICI. Its value can be calculated based on the transmission powers of other users assigned the same subcarrier. Note $p_i(j)$ also introduce ICI to other users. The calculation of $\Delta p_i(j)$ for different users cannot be decoupled. We therefore propose to iteratively calculate $p_i(j)$ based on

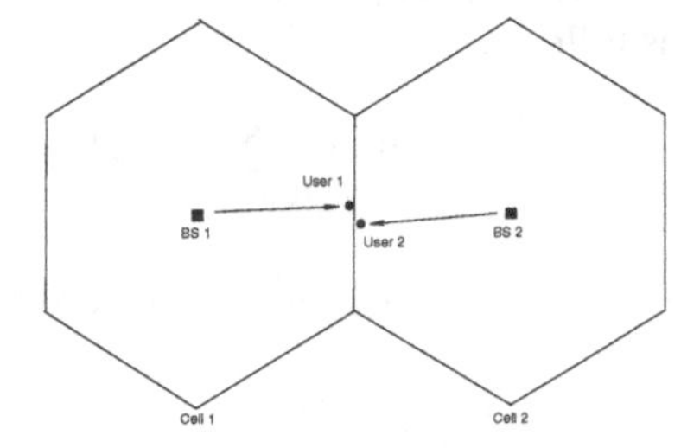

Fig. 1 A special situation where the users are at the cell egde and have similar channel gains to both basestations. In this case it is impossible to increase the SINR levels for both users.

the initial values of $p_i^0(j)$. That is, we first assume no ICI and calculate $p_i^0(j)$ using single-cell resource allocation method. Then we iteratively apply (10) until the target SINR levels are achieved.

It is important to note that increasing user i's transmission power will cause ICI to the users that apply the same subcarrier, which will force the BS to increase the transmission power of the interfered users. As a result, the ICI of user i increase and force the transmission power of user i further increase. It is possible that the transmission power of all users keep increasing but never achieve the target SINR level. It is not difficult to find such a situation as shown in fig.1, where user 1 and user 2 are at the cell edge and both apply subcarrier j. We assume the users' channel gains to BS1, BS2 are extremely close such that $G_1^1(j) \approx G_1^2(j) \approx G_2^1(j) \approx G_2^2(j)$. The transmission power calculated from (10) will have close values, i.e. $p_1(j) \approx p_2(j)$. The SINR of user 1, calculated as follows,

$$SINR_1(j) = \frac{G_1^1(j)P_1(j)}{G_2^1(j)p_2(j) + BN_0} \tag{11}$$

never larger than 0dB no matter how large transmission power is applied. We argue that in this scenario the two users cannot be assigned the same subcarrier. Hence, we proposed to switch or deactivate user's assigned subcarrier if its SINR level is not significantly improved during iterations. Following is the detailed algorithm flow.

- **Initialization.** Ignoring the ICI, each cell perform single-cell resource allocation and then transmit the allocation results, i.e. x_{ij} and $p_i^0(j)$ $(i \in U,\ j \in M)$ to the adjacent cells. Set $p_i(j) = p_i^0(j)$.
- **Iteration.** The iteration is separately applied to every subcarrier in each cell. Each iteration consist of the following steps.
  - step1: Based on the allocation information from other cells, Each cell forms the user set for each subcarrier, denoted as $\tilde{U}(j)$. Note that $\tilde{U}(j)$ may be a subset of $U(j)$. Because in a large network it is not practical to obtain the allocation information of all cells. Based on $\tilde{U}(j)$, the BS calculate the power need to be added on every subcarrier.

$$\Delta P_i(j) = SINR_i \frac{\sum_{h \in \tilde{U}(j), h \neq i} G_h^{b(i)}(j) p_h(j)}{G_i^{b(i)}(j)} \tag{12}$$

  - step2: Set $p_i(j) = p_i(j) + \Delta p_i(j)$, Calculate the SINR for every subcarrier as follows.

$$SINR_i(j) = \frac{G_i^{b(i)}(j)P_i(j)}{\sum_{h \in \tilde{U}(j), h \neq i} G_h^{b(i)}(j) p_h(j) + BN_0} \tag{13}$$

  - step3: If $SINR_i(j)$ is smaller than the required SINR level, i.e. $SINR_i(j) < SINR_i$, we further increase the transmission power by a smaller step.

$$p_i(j) = p_i(j) + \beta \Delta p_i(j) \quad (14)$$

where β is a small positive number.

- step4: Calculate the difference of $SINR_i(j)$ between two adjacent iterations. Find out the user and subcarrier corresponding to the lease difference of SINR. Switch the subcarrier to an unoccupied subcarrier if available, otherwise deactivate the subcarrier from the user with probability γ.
- step5: If all $SINR_i(j)$ meet the SINR requirement, stop the iteration, else notify the adjacent cells the new allocation result and go to step 1.

4 Computer Simulation

In the simulation we consider a system with 19 cells as shown in fig.1.2. The cell radius is 1000 meters. The number of users of each cell is 16, which are uniformly distributed in the cell. The channel is frequency selective Rayleigh fading with an exponential power delay profile. The RMS delay spread is =0.5s, typical of an urban environment. The spectral efficiency need to be satisfied are 2, 2.5, 3, 3.5, 4, in other words, the SINR need to be satisfied for each user are 4.77dB, 6.68dB, 8.45dB, 10.5dB, 11.7dB. Other parameters in simulation are shown in Table1.1.

Table 1 System parameters in the simulation

Parameter	Value
Subcarrier frequency	2G Hz
Noise power spectrum density	-174dBm/Hz
Path loss model	L = 128.1+37.6log10(d) (dB)
Channel bandwidth	15MHz
Subcarrier bandwidth	180KHz
BS antenna gain	17dB
User antenna gain	0dB
Cable loss	2dB
Penetration loss	10dB
Noise figure	5dB

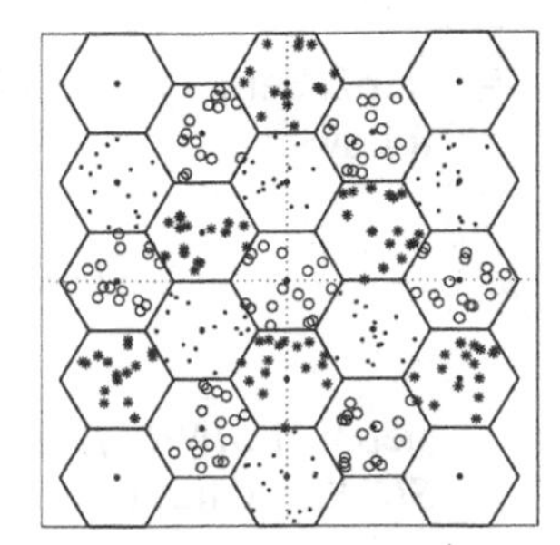

Fig. 2 The scenario of simulation: 19 cells, 16 users in each cells. The users are randomly generated and uniformly distributed in the cell.

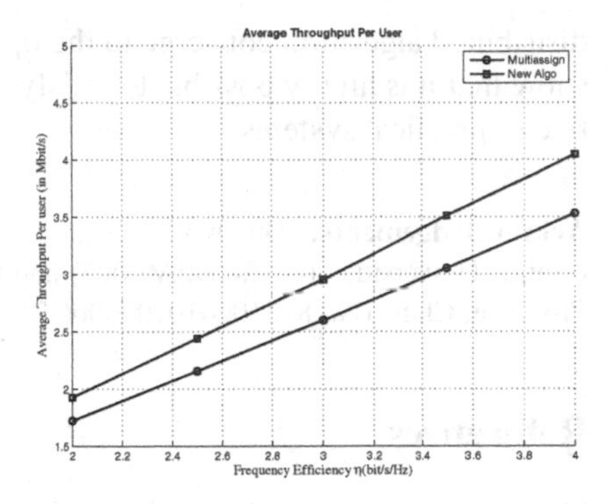

Fig. 3 The experienced average throughput of all users under difference simulation scenario, i.e., different values of η. "Multiassing" represents the result of the *Multiassing* algorithm in [1]. "New Algo" represents the proposed algorithm.

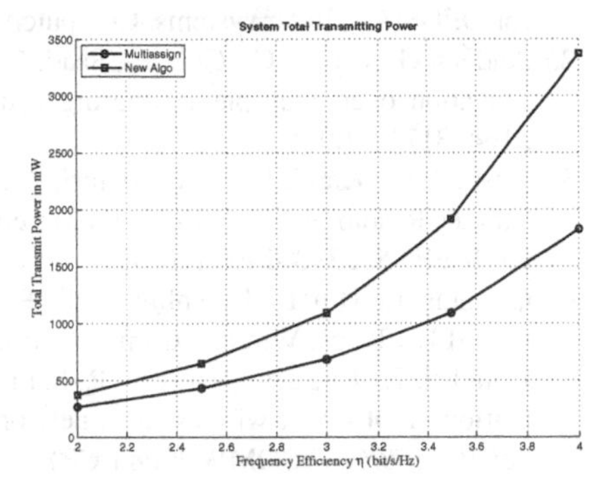

Fig. 4 The total transmission power of the system under difference simulation scenario, i.e., different values of η. "Multiassing" represents the result of the *Multiassing* algorithm in [1]. "New Algo" represents the proposed algorithm.

The proposed algorithm is compared with the centralized heuristic algorithm in [1] (denoted as "Multiassign"). The average throughput of the users and total transmission power under different scenario are plotted in fig.3 and fig.4, respectively.

It can be observed that the proposed algorithm obtains much better average throughput, with the cost of higher transmission power. This is because the Multiassign algorithm has higher probability to deactivate subcarrier. as a result, user's rate requirement may not be satisfied. We argue that in practical system it is more important to satisfy user's rate requirement than to save the transmission power. Because rate requirement is one of the key factors in user's perception of Quality-of-Service. In the scenario of large spectrum efficiency, the extra transmission power of the proposed algorithm is much more significant. This is because for a given SINR level, the required power increase exponentially with the spectrum efficiency (we remark that $SINR_i = 2^{\eta_i} + 1$). Hence, $\Delta P_i(j)$ calculated from (12) increase almost exponentially with η_i, which implies in the proposed algorithm extra transmission power may be allocated to user, and consequently user's experienced transmission rate may exceed its rate requirement.

5 Conclusion

A distributed algorithm with simple computation has been proposed for subcarrier and power allocation in OFDMA system. Currently we cannot prove that the distributed algorithm converge to the optimal solution. However simulation results show that it is highly possible to satisfy user's rate requirement and hence is attractive to practical systems.

Acknowledgements. This work is supported by the Fundamental Research Funds for the Central Universities (No.2009ZM0005) and the Natural Science Foundation of Guangdong Province, China (No.8151064101000066, 9351064101000003).

References

1. Andrea, A., Alessandro, A., Paolo, D., Marco, M.: Radio resource allocation problems for OFDMA cellular systems. Computers & Operations Research 36, 1572–1581 (2009)
2. Anders, G., David, G., Geir, E., Saad, G.K.: Binary Power Control for Sum Rate Maximization over Multiple Interfering Links. IEEE Trans. Wireless Communication 7, 3164–3173 (2008)
3. Gesbert, D., Kiani, S.G., Gjendemsj, A., Ien, G.E.: Adaptation, Coordination, and Distributed Resource Allocation in Interference-Limited Wireless Networks. Proceedings of the IEEE 95, 2393–2409 (2007)
4. Guoqing, L., Hui, L.: Downlink Radio Resource Allocation for Multi-Cell OFDMA System. IEEE Trans. Wireless Commnication 5, 3451–3459 (2006)
5. Honghai, Z., Luca, V., Narayan, P., Sampath, R.: Distributed inter-cell interference mitigation in OFDMA wireless data network. In: IEEE GLOBECOM Workshops (2008), doi:10.1109/GLOCOMW.2008.ECP.81
6. Ksairi, N., Bianchi, P., Ciblat, P., Hachem, W.: Resource Allocation for Downlink Cellular OFDMA SystemsłPart I: Optimal Allocation. IEEE Trans. on Signal Processing 58, 720–734 (2010)
7. Nassar, K., Pascal, B., Philippe, C., Walid, H.: Allocation for Downlink Cellular OFDMA SystemsłPart II: Practical Algorithms and Optimal Reuse Factor. IEEE Trans. Signal Processing 58, 735–749 (2010)
8. Ronald, Y.C., Zhifeng, T., Jinyun, Z., Jay, K.: 3494 Multicell OFDMA Downlink Resource Allocation Using a Graphic Framework. IEEE Trans. Vehicular Tech. 58, 3494–3507 (2009)
9. Schmidt, D., Shi, C., Berry, R., Honig, M., Utschick, W.: Distributed resource allocation schemes. IEEE Signal Processing Magazine 26, 53–63 (2009)
10. Yonghong, Z., Leung, C.: A Distributed Algorithm for Resource Allocation in OFDM Cognitive Radio Systems. In: IEEE Vehicular Technology Conference (2008), doi:10.1109/VETECF.2008.262

Distributed Genetic Programming for Obtaining Formulas: Application to Concrete Strength

Alba Catoira, Juan Luis Pérez, and Juan R. Rabuñal

Abstract. This paper presents a Genetic Programming algorithm which applies a clustering algorithm. The method evolves a population of trees for a fixed number of rounds or generations and applies a clustering algorithm to the population, in a way that in the selection process of trees their structure is taken into account. The proposed method, named DistClustGP, runs in a parallel environment, according to the model master-slave, so that it can evolve simultaneously different populations, and evolve together the best individuals from each cluster. DistClustGP favors the analysis of the parameters involved in the genetic process, decreases the number of generations necessary to obtain satisfactory results through evolution of different populations, due to its parallel nature, and allows the evolution of the best individuals taking into account their structure.

Keywords: Genetic Programming, concrete mixture, civil engineering, clustering, k-means, Evolutionary Computation.

1 Introduction

Genetic Programming (GP) [1] is an inherently parallel search technique and it requires a lot of complex computations when used in symbolic regression tasks on a set of experimental data, so it is interesting to reduce this cost by the distribution of work. In this way it is usual to obtain evolutionary processes that remain in local minima when the search space is large and complex. One purpose of this work

Alba Catoira
RNASA-IMEDIR group, Information and Communication Tecnologies of Information and Communications (TIC) Department, School of Computer Engineering,
University of A Coruña, Campus de Elviña, A Coruña, Spain

Juan Luis Pérez
Technologies Research Centre (CITIC), University of A Coruña,
Campus de Elviña, A Coruña, Spain

Juan R. Rabuñal
Centre of Technological Innovations in Construction and Civil Engineering (CITEEC),
University of A Coruña, Campus de Elviña, A Coruña, Spain

A.P. de Leon F. de Carvalho et al. (Eds.): Distrib. Computing & Artif. Intell., AISC 79, pp. 357–364.
springerlink.com

is to allow parallel execution of the algorithm using several populations and provide the necessary mechanisms to communicate and exchange individuals between them. One of the most important features of DistClustGP is the appliance of a clustering algorithm [2,3] in populations to group individuals in clusters, according to a similarity measure, appointing the best in the cluster as their representatives.

2 Parallel Genetic Programming

GP has been parallelized frequently in recent years [4,5,6,7,8,9]. As in the rest of evolutionary algorithms [1,10,11,12] it is always saving time by using a high number of processors. One of the aspects which underlies this paper, is the good results that the distributed cellular GP system *dCAGE* [13] (an extension of *CAGE* [9]) provides for pattern classification. This system is a distributed environment to run genetic programs by a hybrid variation of the classic island model, which is a combination of island model and cellular model [14,15] and propose a distributed boosting cellular GP classifier to build the ensemble of predictors. To take advantage of the cellular model of GP, the islands are evolved independently using the *CGPC* algorithm [16], and the outermost individuals are asynchronously exchanged. In each generation, a clustering algorithm is applied to the population of trees and the individuals who have the best fitness [1] are selected. DistclustGP (the algorithm proposed in this paper) is based on the island model [17,18,19] and *dCAGE* uses a hybrid model [20]. On the other hand, *dCAGE* uses a clustering algorithm in every generation for not getting a large number of predictors (the size of a population is not small generally). However, DistClustGP is oriented to different types of problems, especially symbolic regression tasks (in which experiments are based) and applies the clustering method after a fixed number or generations and a migration process between slaves and master nodes to bring greater diversity to the populations of nodes, taking into account the structure of individuals. The system is developed for a distributed environment. The distributed architecture provides significant advantages in flexibility and extensibility.

3 DistClustGP

The system is based on the parallel execution of the application on different population or portions thereof. Each population evolves locally applying the GP algorithm independently, and contributes to the entire system through the periodic migration of the best solutions generated to a master node. After a number of generations, the node applies the clustering algorithm to the population, and if there are individuals from the master node, adds them to its population. One of the most important features of DistClustPG is that it is oriented to all types of problems, not only to classification problems (like *dCAGE*).

3.1 DistClustGP Architecture

The basic environment in which the system will be used is an Ethernet network of computers. DistClustGP consists of a *master* node and other nodes named *slaves*, according to the scheme of the *master-slave* architecture. The master node is responsible for sending orders to slaves and providing the necessary data for execution (parameters, patterns, etc.). Slaves contain the GP algorithm and the clustering algorithm needed to evolve the population. Once they have received from the master node the configuration parameters which to run the algorithm and data patterns, are waiting for the master to send them the start order. If such an order is received the slave nodes runs the GP algorithm until it reaches the desired number of gencrations or it reaches a minimum error. The clustering algorithm is executed after a predetermined number of generations, grouping the current population in groups or clusters according to a similarity measure between individuals. Fig.1. shows a diagram of the DistClustGP system. The master indicates the problem to be solved and the parameters, and slaves running the algorithm, sending the best individuals to the master after a fixed number of generations. The master receives the individuals and incorporates them to his own population.

All communications between master and slaves are performed using the MPI (Message Passed Interface) library [25], to allow cooperation among islands.

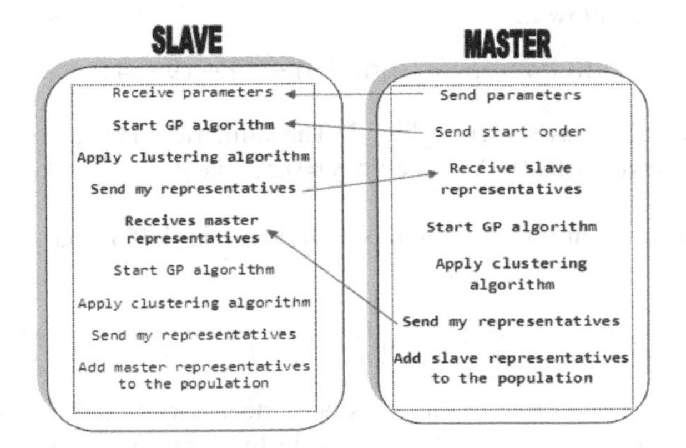

Fig. 1 DistClustGP architecture

3.2 Clustering Algorithm of DistClustGP

In the field of GP, it can occur that the best individual obtained after the evolution of a population in a number of generations, have a minor fitness compared with other individuals, but the structure is more complex,there are individuals with the same fitness and, however, are structurally very different. The goal of the clustering algorithm is to measure the structural differences between individuals, providing the expert the possibility to redirect the process according to the structural difference. The clustering algorithm in the system developed is based on the main idea of the classical k-means algorithm and partitional algorithms based on similarities [21]. Every possible key in a tree node (operator, constant, variable or function) is assigned a numeric value. The distance to empty tree ϕ (considered as the origin tree) is calculated for each individual, represented as a tree. The metric adopted to measure the structural distance between two genetic trees is that introduced by Ekárt and Németh [22]. The distance between two trees h_1 and h_2 is calculated in three steps: 1) h_1 and h_2 are filled with empty nodes to have the same structure. 2) For each pair of nodes al matching positions, the difference of their keys is computed. 3) The differences computed in the previous steps are combined in a weighted sum. Formally, the distance of two trees h_1 and h_2 with roots R_1 and R_2 is defined as

$$\mathrm{dist}(h_1,h_2) = \mathrm{d}(R_1,R_2) + \frac{1}{K}\sum_{i=1}^{m}\mathrm{dist}(\mathrm{child}_i(R_1),\mathrm{child}_i(R_2)) \tag{1}$$

where $\mathrm{d}(R_1, R_2) = |\mathrm{key}(R_1) - \mathrm{key}(R_2)|$, where key is a coding function that assigns a numeric code to each node of the tree, $\mathrm{child}_i(Y)$ is the ith of the m possible children of a node Y, if $i \leq m$, or the empty tree, otherwise. The constant K is used to give different weights to nodes belonging to different depth levels. In the case of calculating the distance between a tree and the empty tree, the result is the sum of their keys weighted by the depth level of the node in the tree. Below is an example. Suppose the following

$$\mathrm{sqrt}((\,day * 5) + (\mathrm{sqrt}(\,9))) + (4 + (9 + 4)) \tag{2}$$

represented as a tree as shown in Fig. 2. The numeric codes associated with the nodes are key(constant)=0.3, key(day)=0.5, key(+)=1.1, key(*)=1.3 and key(sqrt)=1.7.

The distance to empty tree for this individual is 4.453 and is computed as follow:

$$(1.1) + \left(\frac{1.7}{2}\right) + \left(\frac{1.1}{2}\right) + \left(\frac{1.1}{3}\right) + \left(\frac{0.3}{3}\right) + \left(\frac{1.1}{3}\right) + \left(\frac{1.3}{4}\right) + \left(\frac{1.7}{4}\right) + \left(\frac{0.3}{4}\right) + \left(\frac{0.3}{4}\right) + \left(\frac{0.5}{5}\right) + \left(\frac{0.3}{5}\right) + \left(\frac{0.3}{5}\right) = 4.453 \tag{3}$$

To group individuals into clusters has been necessary to define a sort function under this new criterion of structural distance to empty tree. Once sorted, individuals are distributed in clusters according to a threshold value, to be recalculated during the execution.

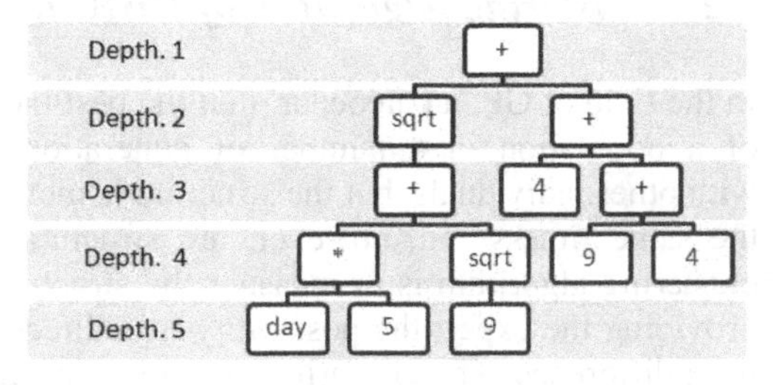

Fig. 2 An example of a tree

The expert sets the maximum percentage of clusters he wants to achieve in relation to population size and the threshold is obtained as follows:

$$\frac{d_{\max} - d_{\min}}{\max number\_of\_clusters} = threshold \quad (4)$$

where d_{max} and d_{min} are the maximum and minimum distance to the empty tree present in the current population and *max number of clusters* is the result of:

$$\frac{percentage\_of\_clusters * size\_population}{100} \quad (5)$$

To obtain the clusters, the first individual in the sorted population is taken as reference, and will be in the same cluster those individuals whose empty tree distance value does not exceed this threshold. If this condition is not fulfilled, the first individual to exceed the threshold will constitute the next point of reference for a new cluster. Once individuals are classified into different groups according to their structure, find the best fitness individual in each cluster is necessary. If the result of the clustering algorithm is the existence of *k* clusters, there will be *k* representatives of the clusters, being the representative the individual with the best fitness within it. These *k* representatives will be sent to the master node to be part of a new population that will evolve by the master GP algorithm. Just a slave, the master will apply the clustering algorithm to every *x* generations, it will send representatives of their clusters to all slaves and will add individuals from the slaves to its population. Unlike *dCAGE*, the number of cluster changes dynamically therefore also changes the number of individuals sent.

4 Experimental Results

High-performance concrete (HPC) is rather new terminology used in the concrete construction industry after well-known high strength concrete (HSC) in recent years. Although there are various definitions of HPC in many countries, the essence of HPC is emphasized on such characteristics as high strength, high workability with good consistency, dimensional stability and durability [23,24]. Table 1 shows the seven variables involved in the problem.

Table 1 Ranges of components of data sets.

Component	Minimum (kg/m^3)	Maximum (kg/m^3)	Average (kg/m^3)
Cement	71	600	232.2
Fly ash	0	175	46.4
Blast furnace slag	0	359	79.2
Water	120	228	186.4
Superplasticizer	0	20.8	3.5
Coarse aggregate	730	1322	943.5
Fine aggregate	486	968	819.9

After several experiments with classical GP, the best results have been obtained with the parameter configuration: maximum height=9, initial height=6, mutation height=6, parsimony=0.01, percentage of crossover=80% and percentage of mutation=20%. This is the reason that this configuration is used for testing DistClustGP.

Table 2 Average results obtained with classic GP

Parsimony	0.0001		0.001		0.01	
% crossover - % mutation	80-20	95-5	80-20	95-5	80-20	95-5
EM	5.25	5.25	5.28	5.29	5.17	5.42
ECM	46.69	46.06	49.17	47.73	45.89	50.77

The evolution of mean square error in nodes during 1000 generations is shown in Fig. 3.

The co-evolutionary strategy avoids local minima. Fig. 3 shows that during the 300, 400 and 500 generation the *slave 1* value is stationary. In generation 500 the *slave 1* receives individuals from master and reduces the MSE. In contrast, classic GP does not obtain a lower error compared with the co-evolutionary strategy. In Table 3 shows the coefficients of determination R^2, the mean square error (MSE) and medium error (ME) of the experiments.

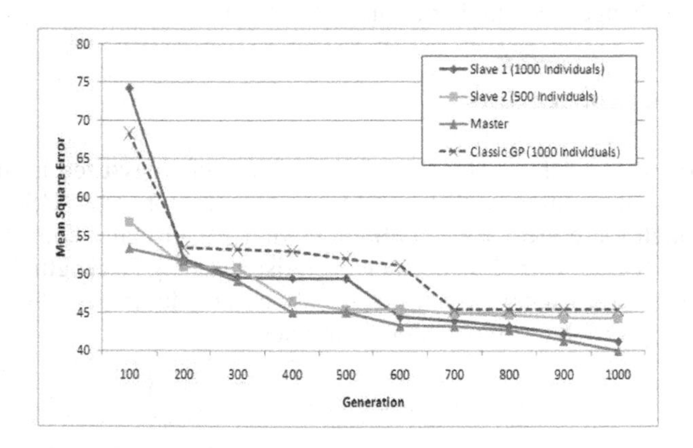

Fig. 3 Mean square error evolution in nodes.

Table 3 DistClustGP vs Classic GP (best result of algorithm).

	Slave 1	Slave 2	Master	Classic GP
R^2	0.83	0.81	0.85	0.81
MSE	41.20	44.20	40.00	45.30
ME	5.03	5.17	5.01	5.03

The best results have been obtained in the master node, which contains the best individuals from the slave nodes so it evolves faster. The slave nodes evolve faster classic GP too, because they receive the best individuals from master.

5 Conclusion

The system DistClustGP has been presented. The method evolves a population of trees for a fixed number of generations in several nodes in a network, and applies a clustering algorithm to bring greater diversity to the population. The populations generated using this method contains better individuals due to the exchange of the best individuals between nodes. The co-evolutionary strategy avoids local minima (static populations) in search process. A main advantage of the distributed architecture is that it enables for flexibility, extensibility, and efficiency.

Acknowledgments. This work was partially supported by the General Directorate of Research, Development and Innovation of the Xunta de Galicia (Ref.07TMT011CT and Ref.08TMT005CT). The work of Juan L. Pérez is supported by an FPI grant (Ref. BES-2006-13535) from the Spanish Ministry of Education and Science.

References

[1] Koza, J.R.: Genetic Programming: On the Programming of Computers by Means of Natural Selection. MIT Press, Cambridge (1992)
[2] Dubes, R.C., Jain, A.K.: Algorithms for Clustering Data. MIT Press, Cambridge (1988)
[3] Gan, G., Ma, C., Wu, J.: Data Clustering: Theory, Algorithms and Applications. ASA-SIAM, Pennsylvania (1979)
[4] Andre, D., Koza, J.R.: Parallel genetic programming: A scalable implementation using the transputer network architecture. In: Angeline (ed.) Advances in Genetic Programming, vol. 2, The MIT Press, Cambridge (1996)
[5] Weinbrenner, T.: Genetic Programming Kernel Version 0.5.2 (1997), `http://thor.emk.e-technik.thdarmstadt.de/~thomasw/gpkernel1.html`
[6] Zongker, D., Punch, W.F.: Lilgp, a C system for genetic programming. Michigan State University, Ann Arbor (1998)
[7] Fernández, F., Sánchez, J.M., Tomassini, M., et al.: A parallel genetic programming tool based on PVM. In: Advances in parallel virtual machine & message passing interface, Springer, Heidelberg (1999)
[8] Fernández Tomassini, M., Punch, W.F., et al.: Experimental Study of Multipopulation Parallel Genetic Programming. In: Proceedings of Euro Gp 2000, pp. 283–293 (2000)
[9] Folino, G., Pizzuti, C., Spezzano, G.: CAGE: A Tool for Parallel Genetic Programming Applications. In: Euro GP 2001, pp. 64–73 (2001)
[10] Rechenberg, I.: Evolutions strategia: Optimierung technischer systeme nach prinzipien der biolgischen evolution. Stutgart (1973)
[11] Fogel, L.J.: Autonomous automata. Industrial Research 4, 14–19 (1962)

[12] Holland, J.H.: Adaptation in Natural and Artificial Systems. Univ. of Michigan Press, Ann Arbor (1975)
[13] Folino, G., Pizzuti, C., Spezzano, G.: Training Distributed GP Ensemble With a Selective Algorithm Based on Clustering and Pruning for Pattern Classification. IEEETrans. Evol. Comput. 12(4), 458–468 (2008)
[14] Tomassini, M.: The Parallel genetic cellular automata: Application to global function optimization. In: Proceedings of the International Conference on Artificial Neural Networks and Genetic Algorithms, pp. 385–391 (1993)
[15] Whitley, C.: Cellular Genetic Algorithms. In: Proceedings of the Fifth International Conference on Genetic Algorithms, vol. 658 (1993)
[16] Folino, G., Pizzuti, C., Spezzano, G.: A cellular genetic programming approach to classification. In: Proc. Genetic Evol. Comput. Conf. (GECCO 1999), FL, pp. 1015–1020 (1999)
[17] Whitley, D., Rana, S., Heckendorn, R.B.: Island Model Genetic Algorithms and Linearly Separable Problems. In: Proc. of the AISB Workshop on Evolutionary Computation, pp. 109–125 (1999)
[18] Koza, J.R., Bennet, F.H., Andre, D., et al.: Automated Synthesis of analog electrical circuits by means of GP. IEEE Transactions on evolutionary computation 1, 109–128 (1999)
[19] Koza, J.R., Bennet, F.H., Andre, D., et al.: Genetic Programming III. In: Darwinian invention and problem solving, Morgan Kaufman Publishers, San Francisco (1999)
[20] Folino, G., Pizzuti, C., Spezzano, G.: A scalable cellular implementation of parallel genetic programming. IEEE Trans. Evol. Comput. 7(1), 37–53 (2003)
[21] ANL Mathematics and Computer Science MPI: A Message Passing Interface Standard (2010), http://www.mcs.anl.gov/research/projects/mpi/
[22] Ekárt, A., Németh, S.Z.: Maintaining the diversity of genetic programs. In: Foster, J.A., Lutton, E., Miller, J., Ryan, C., Tettamanzi, A.G.B. (eds.) EuroGP 2002. LNCS, vol. 2278, pp. 162–171. Springer, Heidelberg (2002)
[23] Ta-Peng, C., Chung, F., Lin, H.: A mix proportioning methodology for high-performance concrete. Journal of the Chinese Institute of Engineers 19(6), 645–655 (1996)
[24] Hwang, C.L., Lee, L.S., Lin, F.Y.: Densified mixture design algorithm and early properties of high performance concrete. Journal of the Chinese Institute of Civil Engineering and Hydraulic Engineering 8(2), 217–229 (1996)
[25] Yeh, I.: A Mix proportioning methodology for fly ash and slag concrete using artificial neural networks. Journal of Science and Engineering 1(1), 77–84 (2003)

A Distributed Clinical Decision Support System Applied to Prostate Cancer Diagnosis

Oscar Marín, Irene Pérez, Daniel Ruiz, and Antonio Soriano

Abstract. Currently, the best way to reduce the mortality of cancer is to detect it and treat it in the earliest stages. Automatic decision support systems are very helpful in this task but their performance is constrained by different factors and sometimes it is difficult to find a method with high sensitivity and specificity rates. One solution to this problem can be the collaboration between independent decision support systems. This article presents a proposal for a distributed and collaborative prostate cancer automatic diagnosis system based on artificial neural networks, which pretends to increase the accuracy of the decision support system combining the independent contributions of different artificial diagnosis entities.

Keywords: Automatic decision support systems, distributed systems, artificial neural networks, prostate cancer.

1 Introduction

Cancer is a major public health concern in the developed countries. A total of 1,479,350 new cancer cases and 562,340 deaths from cancer were expected to occur in the United States in 2009 [1]. From those, approximately 192,000 men were diagnosed with prostate cancer, and 27,000 men were expected to die from this disease what makes prostate cancer the second most common cause of cancer death among men aged 80 years and older [2].

Like the case of many other kinds of cancer, early detection of prostate cancer symptoms is the best way to treat the disease at its first stages reducing the morbidity and mortality [3]. The survival rate of prostate cancer soars from 34% when the cancer is detected at the advanced stage to nearly 100% at the early stage [4].

In prostate cancer case early detection is mainly based on a biomarker, a protein called prostate-specific antigen (PSA) [2]. There are also other factors that a doctor should take into account as digital rectal examination (DRE) results, free and total PSA, patient age, PSA velocity, PSA density, family history, ethnicity etc.

A clinical decision support system can be useful to help specialists in the difficult task of diagnosis [5]. A second expert opinion, even if it is from an artificial entity or software acting as a human expert, can support the decision of the doctor.

Oscar Marín, Irene Pérez, Daniel Ruiz, and Antonio Soriano
Departamento de Tecnología Informática y Computación,
University of Alicante, Ctra. San Vicente s/n. C.P:03690, San Vicente del Raspeig, Spain
e-mail: {omarin, iperez, druiz, soriano}@dtic.ua.es

A.P. de Leon F. de Carvalho et al. (Eds.): Distrib. Computing & Artif. Intell., AISC 79, pp. 365–372.
springerlink.com

In other cases, when the decision from the artificial entity does not agree with the doctor's opinion, the clinical decision support system can suggest alternative tests to increase the degree of certainty in a specific diagnosis. In the medical problem we are working with, the prostate cancer, a clinical decision support system can help the specialist to improve the certainty in the diagnosis and, for example, to avoid useless biopsies.

2 Decision Support Systems and Diagnosis

Since the aim of a decision support system (DSS) is to support or even be an alternative of a certain subject expert's opinion, it has to acquire the basic knowledge and skills needed to simulate the expert's work. That is, the automatic system has to learn and become an expert using different computing techniques from the machine learning theory.

From its rising, machine learning theory has had as one of its main goals its applying, with several purposes, to health and clinical field [5]. This paradigm describes algorithms to solve automatic learning and classification problems, which are the basis to implement systems for clinical decision support (CDSS). Nowadays CDSS are used in healthcare programs, including cancer screening and diagnosis [6] [7].

Within machine learning, ANNS are not the only, but an extensively used tool to perform automatic classification tasks [8].

An ANN is a mathematical model consisting of a number of highly interconnected processing elements, neurons, organized into layers, the geometry and functionality of which have been inspired by that of the human brain. As a consequence of its parallel distribution, an ANN is generally robust, tolerant of faults, able to generalize well and capable of solving nonlinear problems.

In general, ANNs are able to model complex biological systems by revealing relationships among the input data that cannot always be recognized by conventional analyses [9].

To have a set of examples representing previous experience is essential to construct an ANN-based classifier that assures good rates on learning and generalization processes. These examples would be the inputs to the designed ANN. After applying to these values a collection of mathematical functions in different stages, an output is obtained. The output value has a useful meaning for classify or express the probability of being a member of a target class.

A classification of artificial neural networks mainly divides them into two types depending on the sort of learning process used: supervised or unsupervised learning. Besides these two groups there are many more that are different in their performance but related to them or inspired by these two.

The training process of the ANNs designed following a supervised learning paradigm implies a supervision of its outputs where it is specified explicitly what output value corresponds to an input. Since the learning algorithm estimates the error of each output comparing them with the desired value, it is possible to adjust the net parameters in order to obtain a better performance.

Our distributed system contains four nodes that consist on different classifiers developed using this learning technique. There are two different implementations of an ANN called multilayer perceptron (MLP) [10] and the other two are implementations of radial basis functions (RBSs), another supervised ANN.

MLPs and RBSs have been widely used in researches that imply the use of automated methods on clinical environment, i.e. support the cancer diagnosis [11] [12] [13].

Finally, in the unsupervised learning it is expected that the neural network classify the inputs in different groups without knowing explicitly the relation between input and output. The net would find is there are any patterns within the input data set and then associates each pattern to a target class to finally divide the samples in the input data into the different classes. We have implemented a self organized map (SOM) which is an unsupervised neural network algorithm, which has been used with great level of success in the clinical field [14].

3 Distributed and Collaborative System

The proposed distributed diagnosis system consists of six independent nodes, which could be geographically distributed, divided into two groups depending on their assigned task: system core and diagnostic entities (Figure 1).

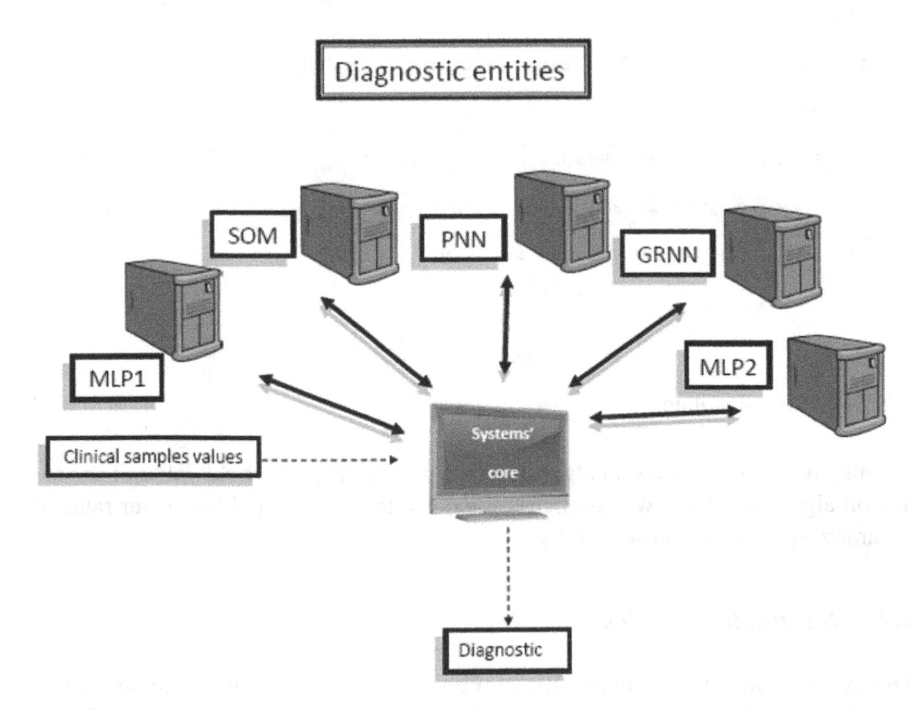

Fig. 1 Distributed proposed system.

3.1 System Core

It is the node that is in charge of the system's coordination. It has as input values from patient's clinical samples that are sent to each diagnostic entity and receives the individual diagnostic value that they send back as a result of their classification task. Once it has all the diagnosis values, it applies to them a selection algorithm to obtain the diagnostic values that the system will give to the user as an output. The selection algorithm pays attention to the accuracy, specificity and sensitivity rates of each classifier in order to choose a final diagnostic with the highest accuracy possible (Figure 2). Bearing this in mind we have as a main objective to increase the sensitivity rate because it is the parameter related to positive detection of cancer.

```
Result = 0
For all networks n = 1 to n = 5
        if Diagnostic(n) == 1 then Result = Result + Sensitivity(n)
        else Result = Result - Specificity(n)
        endif
endfor
if  Result <= 2.75 then Result = 0 endif
if  Result >= 2.75 then Result = 1 endif
if  Result > 2.75 and Result < 2.75 then
        if Diagnostic(PNN)  == 1 then Result = 1
        else Result = Diagnostic(GRNN)
        endif
endif
```

Fig. 2 Selection algorithm.

This node also includes an algorithm that updates the threshold value of the selection algorithm if a new classifier is joined to the system and has better rates of accuracy, specificity and sensitivity.

3.2 Diagnostic Entities

The system contains five diagnostic entities and each one of them implements a classifier based in one of the previously explained ANN designs. Specifically there are two MLPs (with a different parameters configuration), one SOM and two

RBF: one probabilistic neural network (PNN) and a generalized regression neural network (GRNN). Each identity behaves in the same way. At the beginning, the kernel node sends to them a set o patients' samples to classify. Next, they classify each sample and finally send back the results of this process and their accuracy, specificity and sensitivity rates to the core node.

3.3 Communication between Nodes

There is a bidirectional communication between the system's elements. It is based on sending and receiving XML messages which makes this communication process fast, easy to understand, and scalable. Each message from the core node to the classifiers contains a set of patients' samples to be classified. In response to this message each diagnostic entity sends a message that includes three numerical fields in the header (accuracy, sensitivity and specificity rates of the classifier) and a diagnostic for each classified sample in the body of the message.

4 Experimentation

Our clinical database contains 950 samples from patients who have been tested by an expert urologist to check if they suffer from prostate cancer, in 381cases the result was affirmative. Besides the diagnosis results for all of the samples, it also has values for 14 characteristics more for each patient. These characteristics are commonly used by urology experts for prostate cancer diagnosis: age, PSA in blood level, PSA density, prostate volume, rectal examination results, transitional zone flow, peripheral zone transitional, intralesional IR, intraprostatic IR, periprostatic IR, state of the prostate capsule, state of the seminal vesicles, quotient, and prostateseminal angle.

Not all the fields are numerical, 5 of them are filled using a subset of medical terms. In order to use these text fields, we have related each term with a number (e.g. adenoma, LD nodule, LI nodule and bilateral nodule, which are values of "rectal test results" fields, are translated to 1, 2, 3 and 4 respectively). On the other hand, the diagnosis value has two possible values: 'yes' or 'no' that we have identified with 1 and 0 respectively.

In this work we have divided the input data in three non-overlapping sets to carry out the training (60% of the input samples), validation (20%), and testing (20%) processes.

There are a set of customizable parameters whose best values should be obtained by trial. For this reason, we wrote an executable script to test in a batch way several parameters for each MLP, SOM, PNN and GRNN network, and compiling metrics after the execution of each one.

This process allow us to choose the proper parameters configurations that gives us the best performance rates of each ANN. Firstly, we try to find the number of hidden layers and the suitable size of each net's layer. We have tested designs that

Table 1 Results (in percentage) of the testing executions

	SOM	MLP1	MLP2	GRNN	PNN	Collaborative system
Accuracy	45,71	64,74	71,05	81,05	85,26	84,21
Sensitivity	42,42	42,42	46,96	57,57	74,24	81,81
Specificity	67,76	76,61	83,87	93,54	91,12	85,48

contain 1 and 2 hidden layers with a range from 5 to 20 neurons in each layer. Secondly, for the MLPs we look for the transfer function that will control the input data through the net. We used the tan-sigmoidal, log-sigmoidal, and lineal transfer functions. We have tried different combinations of parameters for the different networks, and finally we have used for the SOM network, a grid topology, Euclidean distance as a function of distance and 1000 training epochs. A spread of 0.1 for PNN and GRNN, and finally, we need to know the ideal algorithm that should be used during the training process.

After the previous task, we know the parameters' configuration that obtains the best results for each of the five networks.

Finally we test each classifier separately and then the whole distributed system. The obtained results from each execution can be seen on the Table 1.

5 Conclusions

We have designed a distributed and collaborating clinical decision support system in order to achieve better rates of sensitivity (and reduce the rate of false negative diagnostics). The system is focused on the diagnostic of prostate cancer and uses a clinical database of patients who have been through tests to be diagnosed from this disease.

The distributed system contains nodes that implement different neural networks (two MLPs, a SOM, a PNN, and a GRNN). We have trained, validated and tested them separately, obtaining a set of metrics about their performance. These metrics are used to choose a final diagnostic as an output of the system.

Finally we have tested the whole system and compared its performance with the most accurate independent classifiers one. We have seen that, although the distributed system is almost as accurate as the most accurate classifying node (1% less) it has a best sensitivity rate than the rest of the classifiers. This means that the rate of false negatives (patients who suffer from prostate cancer but haven't been correctly diagnosed) is much lower using the proposed system than using each classifier separately. This was one of our main goals at the beginning of the project since the accuracy rate is limited by the consistency of the database.

The specificity decreases slightly with respect the results of some of the independent classifiers, but it still remains fairly high and continues to exceed by far the best specificity rate of the independent classifiers.

As future works we will try to repeat this job procedure adding other automatic classifiers based on ANNs or other machine learning theories like genetic algorithms, decision trees or support vector machines and different classifier-combining techniques like bagging and boosting.

Acknowledgments. This work has been granted by the Ministerio de Ciencia e Innovación of the Spanish government (Ref. TIN2009-10855) and co-financed by FEDER.

References

1. Jemal, A., Siegel, R., Ward, E., Hao, Y., Xu, J.: USA Cancer Statistics. CA Cancer Journal for Clinicians (2009), doi:10.3322/caac.20006
2. Wolf, A.M.D., Wender, R.C., Etzioni, R.B., Thompson, I.M., D'Amico, A.D.: American Cancer Society Guideline for Early Detection of Prostate Cancer: Update. CA Cancer Journal for Clinicians (2010), doi: 10.3322/caac.20066
3. Brawley, O.W., Ankerst, D.P., Thompson, I.M.: Screening for Prostate Cancer. CA Cancer Journal for Clinicians (2009), doi:10.3322/caac.20026
4. Brawer, M.K.: Prostate-specific antigen: Current status. CA Cancer Journal for Clinicians (1999), doi:10.3322/canjclin.49.5.264
5. Ruiz, D., Soriano, A.: A distributed Approach of a Clinical Support System Based on Co-operation. In: IGI Global (ed) Mobile Health Solutions for Biomedical Applications. Hersey, New York (2009)
6. Lisboa, P.J., Taktak, A.F.G.: The use or artificial neural networks in decision support in cancer: A systematic review. Neural Networks (2006), doi:10.1016/j.neunet.2005.10.007
7. Abbod, M.F., Catto, J.W.F., Linkens, D.A., Hamdy, F.C.: Application of Artificial Intelligence to the Management of Urological Cancer. The Journal of Urology (2007), doi:10.1016/j.juro.2007.05.122
8. Andina, D.: Computational Intelligence for Engineering and Manufacturing. Dordrecht, The Netherlands (2007)
9. Shin, H., Markey, M.K.: A machine learning perspective on the development of clinical decision support systems utilizing mass spectra of blood samples. Journal of Biomedical Informatics (2006), doi:10.1016/j.jbi.2005.04.002
10. Haykin, S.: Neural Networks and Learning Machines. Upper Saddle River, New Jersey (2008)
11. Anagnostou, T., Remzi, M., Lykourinas, M., Djavan, B.: Artificial Neural Networks for Decision-Making in Urologic Oncology (2003), doi:10.1016/S0302-2838(03)00133-7
12. Coiera, E.: Guide to Health Informatics, London, UK (2003)

13. Mehrabi, S., Maghsoudloo, M.: Application of multilayer perceptron and radial basis function neural networks in differentiating between chronic obstructive pulmonary and congestive heart failure diseases. Expert Systems with Applications (2008), doi:10.1016/j.eswa.2008.08.039
14. Hautaniemi, S., Yli-Hara, O., Astola, J.: Analysis and Visualization of Gene Expression Microarray Data in Human Cancer Using Self-Organizing Maps. Kluwer Academic Publishers, Dordrecht (2003), http://www.springerlink.com/content/m5583x8n810p4315/fulltext.pdf

A Survey on Indoor Positioning Systems: Foreseeing a Quality Design

Tomás Ruiz-López, José Luis Garrido, Kawtar Benghazi, and Lawrence Chung

Abstract. The plethora of current positioning technologies, each one with very different features, together with the variety of environments wherein they are to be implanted, force system architects to thoroughly consider the choice for one of them in an isolated way, without combinining several options. Additionally, what makes a technology very appropriate in a certain constraints, may be the result of failing to fulfill others. Thus, trade-off solutions are usually to be made. In this paper, we provide a survey on different positioning techniques in relation to the satisfaction of certain non-functional requirements such as accuracy, responsiveness, complexity, scalability, etc, so that it can serve as guide to system designers in their ultimate decisions. The survey serves as an analysis and intends to highlight the need to undertake a new design capable of adapting this kind of distributed systems to specific characteristics of those technologies and environments; this objective could be achieved on the basis of a design considering non-functional such as requirements.

Keywords: distributed systems, mobile and wireless systems, location-based systems, indoor positioning, wireless technology, quality attributes.

1 Introduction

Over recent years, there has been a growing interest in Context-aware Systems, and more precisely, Location-aware or Location-based Systems (LBS). LBS are information services accessible with mobile devices through a communication network and employing the ability to make use of its location. The rapid development of mobile technology, together with other technologies like the Global Positioning System (GPS), has contributed to the appearance of such distributed systems [1, 3].

Tomás Ruiz-López, José Luis Garrido, and Kawtar Benghazi
Software Engineering Department, University of Granada, Spain
e-mail: tomruiz@correo.ugr.es, jgarrido@ugr.es, benghazi@ugr.es

Lawrence Chung
Department of Computer Science, The University of Texas at Dallas, USA
e-mail: chung@utdallas.edu

A.P. de Leon F. de Carvalho et al. (Eds.): Distrib. Computing & Artif. Intell., AISC 79, pp. 373–380.
springerlink.com

However, GPS does not achieve the accuracy needed for indoor areas; therefore, new positioning techniques must be sought out in order to obtain finer grained accuracy. In addition to that, a lot of technologies have been used in the literature to build positioning systems. Knowing the features of different technologies and how they are used to compute a position helps to choose among them.

Moreover, positioning components are key constituents of LBS, and often they require common functionality, more than just a location. Sesigning the positioning component as a service, would allow designers to build their LBS over a common, reusable grounding.

In addition, from the software engineering perspective, some non-functional requirements are expected from a positioning service. These requirements, such as, the just-mentioned accuracy, robustness or scalability, often force the designer to adopt trade-off solutions which have to be studied thoroughly too.

In this paper, we try to analyze and review what it has been done up to date in positioning systems, as well as what should be considered for upcoming designs. First, we need to study adopted distributed architectures and the usual functional needs of LBS from a positioning service; that is, *what* a positioning service should provide them in order to make their design easier, and to be a reusable base (section 2). Secondly, we have to review which methods, techniques and associated algorithms (section 3) exist and have been proved to work indoors, together with the technology (section 4) that supports them; namely, *how* the positioning is done. Last, but not least, we have to compile and analyze the non-functional requirements (section 5) for these systems. Then, we expose the conclusions obtained from this study (section 6).

2 Location-Based Systems

A LBS is usually composed of several physical components: one or more *mobile devices*, usually carried around by the users; a *communication network*, to support user-to-service communication; a *content and data provider*, which maintains transverse information to different LBS; a *service and application provider*, which processes the users' requests; and finally, a *positioning component*, which provides the user's location. The user's location can be either entered manually or computed. In this paper, we will focus on the latter.

As for the positioning component, there are usually two kinds of entity: *base stations* and *mobile devices*. Base stations are fixed in a known location, and they can either transmit a signal, which is measured at the mobile devices, or they wait to receive the transmitted signal by the mobile device. Base stations help to compute mobile devices' location respect to their position.

Depending on *where* the location computation is performed, we can establish a taxonomy of positioning systems. In a *Network-based* approach the base station network has to either detect the mobile device or receive a signal transmitted from it. The network computes the location of the user. In a *Terminal-based* architecture, the base stations transmit a signal, which is received at the mobile device. It performs some measurements and computes its own location. Finally, mixing both alternatives, in the *Terminal-assisted* approach the mobile device takes some measurements and a server computes its location.

Examining the tasks performed by various LBS [8, 9], we can extract some common needs grouped into different categories [2]. The most obvious one is *orientation* and *localization*, which needs to make use of operations such as geocoding and geodecoding, and, of course, positioning. We will go through this one in section 3 since it is the key aspect o LBS. Other operations fall into different categories, each one of them trying to solve a necessity from the user. Categories are *search* and *navigation*, to find something in the surroundings and move to it; *identification*; or *checking*, to know what is happening at a certain place.

Gratsias et al. [4] present a taxonomy of LBS based on the entities involved in a system. In an LBS there are some entities which are querying and some others answering. This classification considers if those entities are static or mobile. The authors present several algorithms to be used depending on the kind of system we are dealing with. Further details of routines needed and implementation details can be found in the authors' paper.

As we can note, there are some needs, as well as algorithms to solve them, which are common in a wide range of LBS. Furthermore, we can solve some of them by composing simpler operations, e.g., combining positioning and routing, we can obtain guiding. Therefore, some discussion on which functionality has to be provided by the positioning service, and which should be in a higher-level service using it, is to be done.

3 Positioning Methods and Techniques

In this section, we are going to analyze the existing alternatives for positioning, together with their advantages and disadvantages [5]. The usual approach of almost every positioning method consists of performing some measurements on one or more signals, then optionally transform those measurements, and finally estimate the position. Depending on how those transformations and estimations are done, we can set 4 groups of techniques: Proximity-based techniques, Dead Reckoning, Triangulation and Scene Analysis. In Proximity-based techniques, user's location is taken as the position of the closest base station. For Dead Reckoning [6], inertial devices are used to obtain acceleration, velocity and direction, among others, and numerically integrating them, we can obtain the user's location.

Methods based on *Triangulation* estimate positions on the basis of geometrical properties of triangles. Inside this category, there are two sub-categories: *Lateration*, which employs distance estimations from the mobile device to the base stations, and *Angulation*, which uses the angles with which the signal is received.

In techniques based on lateration, we have to measure some characteristics from the signal which is proportional to the distance it has travelled from the transmitter to the receiver. Most common feature which is measured to estimate distance is *Time of Arrival* (TOA). TOA can be transformed to distances, and then, geometrical computations or least-squares approximations can be done to obtain the location. However, there exists several problems: signals suffer form multipath effect in indoor environments, which affects to distance estimation; all clocks have to be precisely synchronized; and the signal must carry a timestamp with the time it was transmitted. Other approaches dealing with time are *Time Difference of Arrival* (TDOA) and *Roundtrip time of flight* (RTOF), which both aim to reduce the problems TOA problems regarding clock synchronization.

Other feature which varies with distance is the *Received Signal Strength* (RSS). There are several empirical and theoretical methods to estimate a distance given the RSS value from a signal. Finally, we can take the *Received Signal Phase* (RSP) of the signal, assuming we are dealing with sinusoidal waves. This technique is usually combined with others, such as the already seen ones, in order to improve their accuracy.

As we said before, angulation based techniques employ the angle with which the signals are received from, referring to this as *Angle of Arrival* (AOA), or the direction it comes from, *Direction of Arrival* (DOA). With these measurements we can compute the intersection of several straight lines and obtain a position. We have to remark that in this case less measurements are needed to obtain a location (2 for a plane, 3 for space). Unfortunately, it also has some disadvantages. It requires additional complex hardware to obtain the measurements. Also, it suffers degradation as the mobile device moves further from the base stations.

Scene Analysis, also known as *Location Fingerprinting*, refers to a series of methods which are based on the extraction of features (most commonly RSS) from the received signal in different reference points, obtaining what is known as *fingerprints*, which are tagged with the location they have been taken. Once a set of fingerprints has been collected in all the locations of interest, positioning is calculated by taking the same features and comparing it with the reference set using a given algorithm. It will return the location of the reference fingerprint which is most similar, given a similarity criterion. As we can see, there are two different phases in Scene Analysis. One consists of the *offline stage*, in which we cover the area of interest, take measurements and tag them. In the *online stage*, the system is working and making positioning.

The offline stage, also known as the training phase, requires some important decisions to be made, which may subsequently affect the overall performance of the system. An important decision is the *granularity* of the grid of *reference points*. The usual approach is to take a homogeneous grid, in which reference points are equally spaced. Reducing granularity leads to more accurate results, although reducing it too much may degrade the accuracy of the system [12] or not imply an improvement in the same order of magnitude [13].

This technique has an important handicap. The offline stage is time-consuming, and changes in the location of B.S. implies retraining of the system. To leverage this fact, Krigin interpolation [12] can be done to estimate fingerprints in some locations, or setting heterogeneous grids with different resolutions at different places. Also, there's a correlation between user's orientation and RSS. Thus, at each point, several measurements should be taken for each direction.

In the online stage, we have to choose one algorithm to give us the closest match in the reference set, given the observed fingerprint. Any Pattern Matching algorithm might be applied at this point. In the following we will study some of the most habitual ones among the wide variety. A very simple algorithm is *k-Nearest Neighbors* (kNN), or *k-Weighted Nearest Neighbors* (kWNN). This algorithm obtains the *k* closest matches in the reference fingerprint set according to a similarity criteria. Examples of this criterion used in the literature are Euclidean Distance and Manhattan Distance. This is a simple yet powerful algorithm which usually gives good results. Other methods, based on mathematical classifiers are *Neural Networks* [14] and *Support Vector Machines* [18], solving the latter the problems that the former has.

Finally, we can consider *probabilistic methods*. At each location, estimation of probability of being at a certain place given the observed measurements needs to be done. Moreover, connections between different places could be considered, since someone cannot walk through a wall. Thus, a *Markov Model* can be built to estimate the position, picking the one which maximizes the probability. This approach is more complex and requires more computation, but presents better results. Again, we have to find a tradeoff between accuracy and performance.

4 Technologies Used in Positioning Systems

Although in previous sections we have already mentioned some technologies which are used in positioning systems, we are going to present more details about them here. Technologies for positioning can be grouped into four categories: *infrared*, *radio frequency*, *ultrasound* and *inertial*. Among these four, radio frequency signals are the most popular ones. The Global Positioning System (GPS) is probably the most famous technology to obtain a device position by triangulation. However, GPS cannot give us the required accuracy we need indoors. Nevertheless, it is still interesting its consideration in order to combine it to design hybrid systems for indoors and outdoors.

IEEE 802.11, the set of protocols for Wireless Local Area Networks, popularly known as Wi-Fi, is another technology which has been widely used in several pieces of work [8, 15]. This technology is used in conjunction with fingerprinting methods. Received signal strength is measured to obtain the fingerprints, since it is usually available and it gives good results. For Personal Area Networks, Bluetooth and ZigBee are possible alternatives. Bluetooth can be used for positioning, but its latency to take the measurements may not be desirable for the system [16]. ZigBee is a new technology intended to be used for home control and automation. Some work has been successfully done using this technology, applied to Ambient Assisted Living [14]. Finally, Radio Frequency Identification (RFID) is a good technology when proximity-based techniques are being applied, since its range is reduced. Navigational and tracking systems have been successfully developed using this technology [17].

5 Non-functional Requirements for Positioning Systems

Non-functional characteristics are of paramount importance; however, positioning systems by and large have been centered around functional characteristics in the past. In this section, from the software engineering perspective, we present a summary of some of the non-functional requirements that have appeared frequently and commonly in various literature on positioning services - such non-functional requirements can be used as a benchmark [10, 5] in comparing and analysing different positioning systems in a rational way.

Usually, there is a tradeoff between some of the non-functional requirements, and we have to study them in order to be able to determine what is preferred for the applications that will be built on top of our system. Designing and building a positioning service with *interoperable* components can address this fact. It gives the capability of creating hybrid, distributed systems by combining different technologies and methods. Moreover, those components may be replaced by new or better ones.

Accuracy and *precision* are two desirable properties in our system. Accuracy tries to measure how exact the estimations of the system are, compared to the actual locations. Precision considers the distribution of errors. We are interested in keeping the error low after several computations. Otherwise, if the error grows after every positioning request, the applications using the positioning service would offer wrong results. This relates to *robustness* of the system, which implies it should work under adverse conditions.

Complexity of the system can refer to different aspects: hardware complexity, software complexity or operation factors, among others. Since it is difficult to derive an analytical formula over systems complexity, we may take computation time as a metric to quantify complexity of the system. As we said before, we have to find a compromise solution between accuracy and complexity; more accurate results may need more computations, but time taken to calculate them must not be too long, especially if real-time requirements are desirable for the system. Complexity is an important factor to deal with; thus, to release the mobile device of computational tasks, we may send the data to be computed in a dedicated machine, but it would increase the communication time.

At some point, systems must *scale*. Positioning systems can scale *geographically* (extending the scope of the system) or in *density* (number of mobile units per time and space unit). Finally, *cost* restrictions may arise. Sometimes, the positioning system must be built over existing infraestructure used for another purpose. Deployment and maintenance time, space needed by the hardware, weight or energy consumptions are other requirements which have to be considered.

6 Conclusions

This paper provides a survey of various techniques and technologies for positioning services, with their relative advantages and disadvantages, while identifying some key decisions to be made during the design of such a system. Many choices are possible, and must be studied carefully depending on the environment they have to be deployed in.

Regarding functional requirements, we have seen several common needs in LBS. Some of the operations can be obtained by combining lower level ones. Therefore, we can design them as services inside a Service-oriented Architecture. The positioning service has to provide the just-mentioned low level operations, while other services are designed and built on top of it, ensuring reusability. Concreting which minimum functionality must be provided by the positioning service on a certain environment is left as future work.

We have seen that there exist many positioning methods. Some of them can work standalone, while others are conceived to help the firsts to improve their accuracy. Therefore, a design for a positioning system should take this fact into account, i.e., it should be designed in such a way that allows several positioning methods working. Moreover, it can help to satisfy some of the non-functional requirements.

As for the technology, several choices are also possible, and the election may fulfill some non-functional requirements too. Cost restrictions may push us to use existing infraestructure, such as WiFi routers used in Internet connections. A design for a positioning service must also take into account the possibility of combining several technologies. This allows to obtain different resolution levels for positioning to be obtained. Furthermore, technologies for both indoors and outdoors may be integrated to develop more efficient and robust systems.

Non-functional requirements can be addressed from concrete design decisions, as we have already mentioned, but also serve as a benchmark to compare different positioning systems, and know which one has better performance for the environment we are interested in.

Finally, a design for positioning services based on interoperable components would be needed in order to be able to combine the different positioning techniques available, as well as different technologies which support them. This would allow to build hybrid systems in which those components can be replaced easily and switch between them. It should be done at runtime in order to satisfy quality requirements successfully.

Acknowledgements. This research has been funded by the Spanish Government's Ministry of Science and Innovation, via the projects TIN2008- 05995/TSI and TIN2007-60199.

References

1. Marmasse, N.: comMotion: A Context-Aware Communication System. In: Proceedings of CHI 1999 (1999)
2. Steiniger, S., Neun, M., Edwardes, A.: Foundations of Location Based Services
3. Abowd, G.D., Atkeson, C.G., Hong, J., Long, S., Kooper, R., Pinkerton, M.: Cyberguide: A mobile context-aware tour guide. Wireless Networks 3, 421–433 (1997)
4. Gratsias, K., Frentzos, E., Delis, V., Theodoridis, Y.: Towards a Taxonomy of Location Based Services. In: Li, K.-J., Vangenot, C. (eds.) W2GIS 2005. LNCS, vol. 3833, pp. 19–30. Springer, Heidelberg (2005)

5. Liu, H., Darabi, H., Banerjee, P., Liu, J.: Survey of Wireless Indoor Positioning Techniques and Systems. IEEE Transactions on Systems, Man, and Cybernetics – Part C: Applications and Reviews 37(6) (November 2007)
6. Randell, C., Djiallis, C., Muller, H.: Personal Position Measurement Using Dead Reckoning
7. Kjaergaard, M.B.: A Taxonomy for Radio Location Fingerprinting. In: Hightower, J., Schiele, B., Strang, T. (eds.) LoCA 2007. LNCS, vol. 4718, pp. 139–156. Springer, Heidelberg (2007)
8. Kawaguchi, N.: WiFi Location Information System for Both Indoors and Outdoors. In: Omatu, S., Rocha, M.P., Bravo, J., Fernández, F., Corchado, E., Bustillo, A., Corchado, J.M. (eds.) IWANN 2009. LNCS, Part II, vol. 5518, pp. 638–645. Springer, Heidelberg (2009)
9. Mateos, M., Berjon, R., Sanchez, M.A., Beato, E., Fermoso, A.: A Case Study of a Pull WAP Location-Based Service Incorporating Maps Services. In: IWANN 2009. LNCS, Part I, vol. 5517, pp. 1240–1247. Springer, Heidelberg (2009)
10. Lin, T.N., Lin, P.C.: Performance Comparison of Indoor Positioning Techniques based on Location Fingerprinting in Wireless Networks (2005)
11. Li, B., Kam, J., Lui, J., Dempster, A.G.: Use of Directional Information in Wireless LAN based Indoor Positioning. In: IGNSS Symposium (2007)
12. Li, B., Salter, J., Dempster, A.G., Rizos, C.: Indoor Positioning Techniques based on Wireless LANs
13. Prasithsangaree, P., Krishnamurthy, P., Chrysanthis, P.K.: On Indoor Position Location with Wireless LANs
14. Blasco, R., Marco, A., Casas, R., Ibarz, A., Coarasa, V., Asensio, A.: Indoor Localization Based on Neural Networks for Non-dedicated ZigBee Networks in AAL. In: Cabestany, J., Sandoval, F., Prieto, A., Corchado, J.M. (eds.) IWANN 2009. LNCS, Part I, vol. 5517, pp. 1113–1120. Springer, Heidelberg (2009)
15. Ekahau: `http://www.ekahau.com/`
16. Aparicio, S., Pérez, J., Tarrío, P., Bernardos, A.M., Casar, J.R.: An Indoor Location Method Based on a Fusion Map Using Bluetooth and WLAN Technologies. In: DCAI 2008, ASC, vol. 50, pp. 702–710 (2008)
17. Chon, H.D., Jun, S., Jung, H., An, S.W.: Using RFID for Accurate Positioning. Journal of Global Positioning Systems (2005)
18. Brunato, M., Battiti, R.: Statistical Learning Theory for Location Fingerprinting in Wireless LANs

Distributed and Asynchronous Bees Algorithm: An Efficient Model for Large Scale Problems Optimizations

Antonio Gómez-Iglesias, Miguel A. Vega-Rodríguez, Francisco Castejón, and Miguel Cárdenas-Montes

Abstract. There are several different algorithms based on the ideas of collective behaviour of decentralized systems. Some of these algorithms try to imitate the distributed and self-organized systems that can be found in nature. Algorithms based on the mechanisms of distributed evidence gathering and processing of bee swarms are recent optimisation techniques. The distributed schema makes these algorithms suitable for a distributed implementation using the distributed computational infrastructures (DCIs) available. With these DCIs, large scale scientific problems can be optimized in a feasible time. However, the distributed paradigm of these infrastructures introduces several challenges in the design and development of any optimization technique. A distributed and asynchronous bees (DAB) algorithm running in a DCI is here presented with the aim to optimize any large scale problem.

Keywords: distributed, asynchronous, bees, optimization, grid.

1 Introduction

In nature, we can find different types of swarm. The efficiency they apply to their own challenges becomes interesting when we think of optimization problems. Examples of these swarm intelligence algorithms are those algorithms based on ants, bees, wasps or flocks. In the case of bees, we have a swarm continuously foraging for nectar in a changing environment. These changes depend on the discovery or the abandonment of a food source. The discovery of new sources is done by the scout bees, while the exploitation of a food source is carried out by bees recruited by these scouts. The exchange of information among bees in the colony is called

Antonio Gómez-Iglesias, Francisco Castejón, and Miguel Cárdenas-Montes
CIEMAT, Avda. Complutense 22, Madrid, Spain
e-mail: antonio.gomez@ciemat.es, francisco.castejon@ciemat.es, miguel.cardenas@ciemat.es

Miguel A. Vega-Rodríguez
University of Extremadura, Escuela Politécnica - Campus Universitario S/N, Cáceres, Spain
e-mail: mavega@unex.es

A.P. de Leon F. de Carvalho et al. (Eds.): Distrib. Computing & Artif. Intell., AISC 79, pp. 381–388.
springerlink.com

waggle dance. Only a few individuals have relevant information, like the location of the food, the quality of the source or the route. Those individuals without this information reach a consensus considering the leadership within the group [6].The decision making process about which food source is going to be explored is not only taken in this waggle dance: on the route from the colony to the source, some interactions take place among the bees going for foraging. Some bees can go to different sources due to these interactions.

Each day the colony will know a few potential food sources, each one with a different level of nectar and at different distance from the hive. Translating the discovery and foraging process into computer-aided optimisation problems means some processes are looking for new solutions for a given problem; every time an optimised solution is found, more processes are allocated to look for approximated solutions using it as base element to explore.

DCIs are useful for different problems where the computational requirements are high. If the communication among the different processes is a critical element, then high performance computers are the best option. However grid computing fits better into the requirements of the problem, when the computation is much more important than communication. Nevertheless, the grid introduces more complexity in the system as a result of the heterogeneous resources shared in the infrastructure. Our goal is to design and develop a distributed system based on bees foraging behaviour that, running on the grid, is able to optimize large scale computational problems in a reasonable time with an optimal exploitation of the resources. The result will be a totally distributed bio-inspired algorithm that will efficiently use the existing computational resources.

The rest of this paper is organised as follows: section 2 summarizes the related work and previous efforts done. Section 3 details the implementation of the DAB algorithm while section 4 introduces an example of an optimization process using the algorithm. Finally, in section 5, we conclude the paper and summarise a variety of perspectives of the explained work.

2 Related Work

In the last years, different works have been focused on the optimization and solving of problems by means of metaheuristics with DCIs [2, 5, 14, 15]. However, there are not many relate works where the computational requirements of the problem to be solved are so expensive that only large DCIs can be used. Moreover, these works solve problems with some level of homogeneity and where the communication and computation requirements are similar.

There are also different frameworks allowing implementing and deploying parallel and distributed metaheuristics. Most of them are restricted to only parallel and distributed evolutionary algorithms [3, 7] or they use several different metaheuristics although they are not running on the grid [1].

Honey bees based algorithms can be classified following the characteristics of the behaviour they are trying to imitate:

- Marriage behaviour: in nature, the sperm from different drones will be deposited in the queen to form the genetic pool for the hive.
- Foraging behaviour: it is related to the feeding process of bees. The waggle dance process is used to search for food sources and to exchange information about the food sources among the bees. Our system follows this model.
- Queen bee: this technique consists of improvements of GAs (Genetic Algorithms), especially during the crossover.

We have not found any implementation or design similar to what we propose here. Also, the application described in this article has not been previously optimized by means of metaheuristics.

3 Distributed and Asynchronous Bees Algorithm

The main reason to develop a distributed and asynchronous algorithm was to avoid, as much as possible, bottlenecks. In previous algorithms [10, 11, 9, 12] we always found that they could find optimized solutions, but the time required was very long. Usually, just a few processes, when not only one, where running. Also, the continuous submission of jobs introduced delays in the execution.

3.1 Avoiding Bottlenecks

Previous algorithms followed an iterative model which led to these bottlenecks. Therefore, a key point is to remove iterative model and create a paradigm where the information and the decision making process could be distributed among all the processes involved in the optimization process. The result was an algorithm that can be classified as a bee algorithm.

In the system, there is not any process that has to wait for other processes to finish. New processes are created, although they do not introduce dependencies in terms of delays for other tasks. Each process has the information required to perform its computation at any time. Also, each process will produce information that can be used by new processes. If a new optimized solution is found, this information is distributed among the elements in the system when they finish the operation they were performing. This follows the behaviour of bees in nature, where they share the information in the waggle dance.

3.2 Minimizing the Creation of New Tasks

In nature, when bees return to the colony after foraging, they do not die: they get more information and go back for foraging or scouting. In our system, when a task finishes its evaluation, it sends this information to the colony, and waits for more input to perform new evaluations. It stays idle while this communication takes place, and once new data has been received, automatically restarts the computations. As real bees - they keep on foraging until they die - a process will evaluate solutions until it finishes (usually due to problems in the infrastructure).

In the colony, there are bees waiting for information. After the waggle dance, they go foraging, and return to the colony with new information. A new waggle dance starts. But not all the dances are successful, so when the bee does not find any other follower, it remains in the colony looking other dances. In our system, there are processes that imitate this behaviour: they remain idle until optimized configurations are found. These processes are sent just when needed, they will perform their calculation and finish, without waiting for more input.

3.3 Description of the Algorithm

The system consists of the exploration of the solution space and it devotes more computational resources to the best solutions as these solutions are found. Fig. 1 shows the different processes of the system in the form of bees.

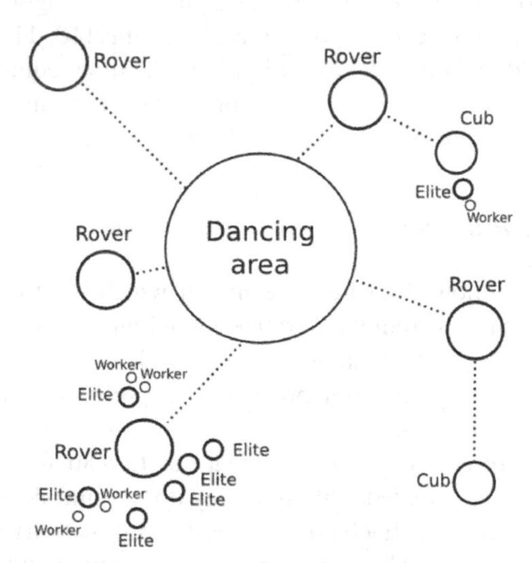

Fig. 1 System overview.

This figure shows the processes with their names in the system. Based on the original concept of two different bees in the system - bees looking for new food sources and forager bees, going to explored sources - we have two different main processes in the system. Furthermore, considering the concept of leadership, we have four different types of bees:

- Two levels of *employed*:
  1. *Elites*: perform a wide search using an approximated configuration. This approximated configuration has been previously found by a *scout*, so they have

the information about the food source. They allocate a computational resource that will be used as long as possible.
2. *Workers* (associated to an *elite* bee): by using a local search procedure, they explore deeply the best configuration found by *elite* bees. Once they finish their computations, the resource is released.

- Two levels of *scouts*:

1. *Rovers*: they use diversification methods to explore the solution space considering the paths previously explored. They also allocate the resources for the longer allowed time.
2. *Cubs*: they just follow a rover and, at some point, bring a modification in their paths. The resource is released as soon as the computations have finished.

3.4 Exploitation Mechanism

Bees, like ants and other social foragers, search for food sources trying to maximize the ratio E/T, where E is the energy obtained, the amount of nectar, and T is the time required for foraging. The selection of a food source by a bee is done taking into account the amount of nectar of a food source, compared to the total amount of nectar of all food sources. For this study, we also consider the number of computational resources previously used by a candidate solution. So, the probability p_i of selecting a food source i is determined by equation 1.

$$p_i = \frac{fit_i}{r_i \sum_{j=1}^{n} fit_j} \tag{1}$$

where fit_i if the fitness of the solution i, r_i is the number of computational resources previously devoted to the exploration of the solution i and n is the number of food sources. The main target of this study is to get a reasonable distribution of the computational resources among the different candidate resources. If we do not introduce r_i, the best solution would always receive much more resources than other candidates solution. This would lead to a wrong scenario where the convergence would be extremely high for a single solution.

4 Application

In nuclear fusion, the neoclassical transport consists of the movement of energy and particles towards the plasma boundary. This is due to the gradients of temperature, potential and density, as well as the collisions and the inhomogeneities of magnetic field [4]. Improving the neoclassical transport, we improve the plasma confinement within the fusion device, so the probabilities of getting fusion reactions are higher, increasing the efficiency of the device.

4.1 The Fitness Function

After many expressions involving concepts of magnetic fields and plasma physics [4], we get the target function given by

$$F_{target function} = \sum_{i=1}^{N} \left\langle \left| \frac{\vec{B} \times \vec{\nabla} |B|}{B^3} \right| \right\rangle_i \quad (2)$$

In this equation, i represents the different magnetic surfaces inside a fusion device, whereas B represents the intensity of the magnetic field in each magnetic surface. Our target is to minimise the value given by this function. To get all of the values involved in this function, we need to execute a workflow application to calculate the magnetic surfaces of plasma inside a fusion device. This workflow can be found in the related work [8]. The number of parameters we can modify goes up to 300, making this a very large optimization problem. For optimised configurations, this workflow has more than 40,000,000 Millions of Instructions (MI).

4.2 Results

On previous efforts [13] we had seen how this kind of algorithm was optimal for optimization processes where large computational capabilities were required. However, there was some tendency to get a greedy system devoting a large number of resources to the best candidate solution found so far. Here we try to avoid this issue by introducing a factor to modify the probability to select a given solution. Fig. 2 shows the evolution of the number of computational resources devoted to the exploration of three different candidate solutions. Depending on the quality of the solutions and the number of computational resources previously assigned to each solution, the probabilities of each of them to receive more resources will vary.

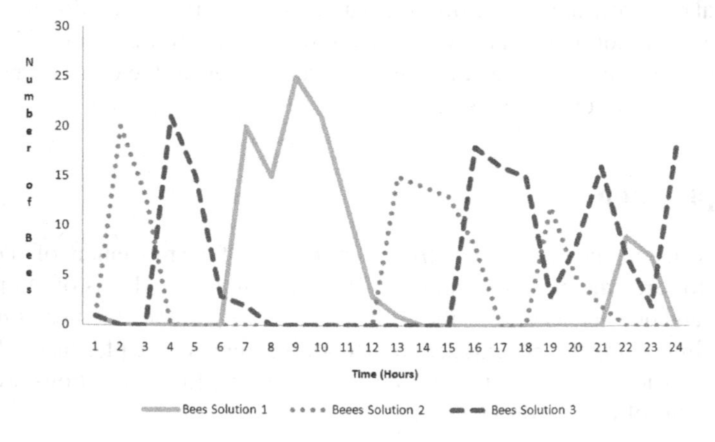

Fig. 2 Computational resources assigned to different solutions

Table 1 shows the required wall clock (WC) time and the overall execution time to get comparable results with different metaheuristics. These results represent optimized configurations with optimal values for the fitness function. As it can be seen, DAB algorithm offers results faster due to its distributed paradigm and the lack of bottlenecks. Compared to other metaheuristics, it clearly shows how this model fits better the requirements of the problems we are facing.

Table 1 Comparison to previous implementations (GA = Genetic Algorithm, SS = Scatter Search, DAB = Distributed and Asynchronous Bees)

Algorithm	Maximum Number of Jobs	WC Time	CPU Time
GA	4,000	251:49:25	31,871:26:32
SS	100	207:12:49	9,955:49:52
DAB	745	19:02:37	11,120:28:20

5 Conclusions and Future Work

Our objective for an optimal use of the resources has been accomplished. In this algorithm, there are not bottlenecks or dependencies among different resources. This helps to obtain good results in a reasonable time.

The measure introduced to avoid a single candidate solution hoarding a vast majority of the computational resources provides an optimal distribution of the resources amongst the different approximated solutions.

The next step is to introduce more target functions to be optimized, beyond the neoclassical transport. Mercier and Ballooning stability criteria are the following optimisation functions that will be considered in our algorithm. With these two new functions we will have three different criteria to optimise a fusion device. The development of an implementation devoted to be used with HPC (High Performance Computing) is also being considered for some special cases.

Acknowledgements. This work was partially funded by the Spanish Ministry of Science and Innovation and ERDF (the European Regional Development Fund), under the contract TIN2008-06491-C04-04 (the MSTAR project).

References

1. Alba, E., Almeida, F., Blesa, M., Cabeza, J., Cotta, C., Díaz, M., Dorta, I., Gabarró, J., León, C., Luna, J., Moreno, L., Petit, J., Roas, A., Xhafa, F.: MALLBA: a Library of Skeletons for Combinatorial Optimization. In: Monien, B., Feldmann, R.L. (eds.) Euro-Par 2002. LNCS, vol. 2400, pp. 927–932. Springer, Heidelberg (2002)
2. Alba, E., Nebro, A.J., Troya, J.M.: Heterogeneous Computing and Parallel Genetic Algorithms. J. Parallel Distrib. Comput. (2002) doi:10.1006/jpdc.2002.1851

3. Arenas, M.G., Collet, P., Eiben, A.E., Jelasity, M., Merelo, J.J., Paechter, B., Preuß, M., Schoenauer, M.: A Framework for Distributed Evolutionary Algorithms. In: Guervós, J.J.M., Adamidis, P.A., Beyer, H.-G., Fernández-Villacañas, J.-L., Schwefel, H.-P. (eds.) PPSN 2002. LNCS, vol. 2439, pp. 665–675. Springer, Heidelberg (2002)
4. Bellan, P.M.: Fundamentals of Plasma Physics. Cambridge University Press, Cambridge (2006)
5. Cahon, S., Talbi, E., Melab, N.: ParadisEO: A Framework for the Reusable Design of Parallel and Distributed Metaheuristics. Journal of Heuristics (2004) doi: 10.1023/B:HEUR.0000026900.92269.ec
6. Couzin, D., Krause, J., Franks, R., Levin, S.: Effective leadership and decision-making in animal groups on the move. Nature 7025, 513–516 (2005)
7. Gagné, C., Parizeau, M., Dubreuil, M.: Distributed BEAGLE: an Environment for Parallel and Distributed Evolutionary Computations. In: Proceedings of the 17th Annual International Symposium on High Performance Computing Systems and Applications, vol. 1, pp. 201–208 (2003)
8. Gómez-Iglesias, A., Vega-Rodríguez, M.A., Castejón-Magaña, F., Rubio-del-Solar, M., Cárdenas-Montes, M.: Grid Computing in Order to Implement a Three-Dimensional Magnetohydrodynamic Equilibrium Solver for Plasma Confinement. In: 16th Euromicro International Conference on Parallel, Distributed and network-based Processing (2008), doi: 10.1109/PDP.2008.61
9. Gómez-Iglesias, A., Vega-Rodríguez, M.A., Cárdenas-Montes, M., Morales-Ramos, E., Castejón-Magaña, F.: Grid-Oriented Scatter Search Algorithm. Adaptive and Natural Computing Algorithms (2009), doi:10.1007/978-3-642-04921-7_20
10. Gómez-Iglesias, A., Vega-Rodríguez, M.A., Castejón-Magaña, F., Cárdenas-Montes, M., Morales-Ramos, E.: Using a Genetic Algorithm and the Grid to Improve Transport Levels in the TJ-II Stellarator. In: 7th International Symposium on Parallel and Distributed Computing (2008), doi:10.1109/ISPDC.2008.30
11. Gómez-Iglesias, A., Vega-Rodríguez, M.A., Castejón-Magaña, F., Cárdenas-Montes, M., Morales-Ramos, E.: Grid-Enabled Mutation-Based Genetic Algorithm to Optimise Nuclear Fusion Devices. Computer Aided Systems Theory (2009), doi:10.1007/978-3-642-04772-5_104
12. Gómez-Iglesias, A., Vega-Rodríguez, M.A., Castejón-Magaña, F., Cárdenas-Montes, M., Morales-Ramos, E.: Evolutionary computation and grid computing to optimise nuclear fusion devices. Cluster Computing (2009), doi:10.1007/s10586-009-0101-3
13. Gómez-Iglesias, A., Vega-Rodríguez, M.A., Castejón-Magaña, F., Cárdenas-Montes, M., Morales-Ramos, E.: Artificial Bee Colony Inspired Algorithm Applied to Fusion Research in a Grid Computing Environment. In: 18th Euromicro International Conference on Parallel, Distributed and network-based Processing (2010), doi:10.1109/PDP.2010.50
14. Lim, D., Ong, Y.S., Jin, Y., Sendhoff, B., Lee, B.S.: Efficient Hierarchical Parallel Genetic Algorithms Using Grid Computing. Future Generation Computer Systems (2007), doi: 10.1016/j.future.2006.10.008
15. Melab, N., Cahon, S., Talbi, E.: Grid Computing for Parallel Bioinspired Algorithms. Journal of Parallel and Distributed Computing (2006), doi:10.1016/j.jpdc.2005.11.006

Integrating Data Mining Models from Distributed Data Sources

Ingrid Wilford-Rivera, Daniel Ruiz-Fernández, Alejandro Rosete-Suárez, and Oscar Marín-Alonso

Abstract. Data mining has been widely applied to analyze data for decision makers. However, traditional data mining techniques are insufficient for analysis of multiple data sources. To mine multiple data sources, one possible way is reusing local data mining models discovered from each data source and searching for valid patterns that are useful at the global level. This paper presents a Knowledge Integration Model for integrating data mining models discovered from different data sources. This proposal is especially helpful for organizations which distributed data sources have been mined locally, and don't share their original databases.

Keywords: Multiple data sources, data mining models, distributed knowledge, integration.

1 Introduction

Data mining aims at the discovery of useful information from large databases and has been widely applied to analyze data for decision makers. Most of the current data mining researches focus on mining a single database, using traditional data mining techniques [1-4]. However, there are many information systems where data is distributed among several nodes located in distant places. Advances in computer communication networks have favored the development of distributed applications. Because of data privacy issues, it is possible that some data sources of an organization may share their patterns but not their original databases. On the other hand, due to the massive data volume, it may be inconvenient to collect data from different data sources for centralized processing. Therefore traditional data mining techniques may be insufficient for large organizations with multiple data sources.

Ingrid Wilford-Rivera and Alejandro Rosete-Suárez
Center of Engineering and Systems, José Antonio Echeverría Institute of Technology, Havana City, Cuba
e-mail: {iwilford,rosete}@ceis.cujae.edu.cu

Daniel Ruiz-Fernández and Oscar Marín-Alonso
Department of Computer Technology, University of Alicante, AP.99-03080, Alicante, Spain
e-mail: {druiz,omarin}@dtic.ua.es

A.P. de Leon F. de Carvalho et al. (Eds.): Distrib. Computing & Artif. Intell., AISC 79, pp. 389–396.
springerlink.com

Few works have been reported on integrating data mining models from different data sources. The process of gathering, analyzing, and synthesizing data mining models or patterns discovered from different data sources, is named also postmining [5]. In this paper, we propose a Knowledge Integration Model for integrating data mining models (patterns set) discovered from different data sources. The rest of this paper is organized as follows. In Section 2 our research problem is formulated and other related works are described. Section 3 presents the Knowledge Integration Model proposed. Section 4 illustrates experiments performed. At the end, in Section 5, we conclude this paper.

2 Problem Statement and Related Works

There are three possible ways to mine multiple data sources. One of them consists on putting all distributed data together to create a single data set for centralized processing, using traditional or parallel data mining techniques [1; 2; 4; 6]. The second way consists on using distributed mining techniques [7-9]. Nevertheless, putting all distributed data together for centralized processing or using distributed data mining techniques has important limitations. Due to data privacy issues, most organizations share their data mining models discovered, but not their original databases. Because of the massive data volume it may be inconvenient to collect data from different data sources for centralized processing. Some data mining algorithms are sequential in nature and cannot make use of parallel hardware. Also, parallel and distributed data mining algorithms do not make use of local models at different data sources; however, in real-world applications these local models are useful for the local decision makers, and would have to be generated.

The third way to mine multiple data sources consists on reusing local data mining models discovered from each distributed data sources and searching for valid patterns that are useful at the global level (postmining) [10-13]. Our research problem is directly related with this way of mining.

We have *n* local data mining models obtained from *n* data sets respectively, so we know general information about these data sets (for example: size of data sets, attributes and their domains, etc), but not the originally analyzed records. It is our goal to integrate these models to find a global data mining model that contains valid patterns for the global level, which would have been discovered from the union of these *n* different data sets. That is, to discover global patterns for a set of distributed data sources, which have been mined locally, and don't share their original databases but general information about these mined data sets.

Formally, the problem above formulated can be defined as follows:

Let $M = \{\mu_1, \mu_2, \dots \mu_n\}$ be a set of local data mining models discovered from different data sets. Each local data mining model $\mu_i \in M$ can be defined as a set of patterns $\mu_i = \{\rho_1, \rho_2, \dots \rho_n\}$, and each pattern $\rho_i \in P$ has the form $\rho_i = \langle id_j, me_j\rangle$, where id_j represents the characteristics that identify the pattern ρ_i, and me_j represents the measures of ρ_i. For example, if ρ_i is an association rule, id_j would be the antecedent and the consequent of the rule and me_j would be its support and confidence. Also, let $\Phi = \{\varphi_1, \varphi_2, \dots \varphi_n\}$ be a set of files, where each file $\varphi_i \in \Phi$

contains general information and technical specifications about a different local data mining model and its original data set analyzed. Then, the problem of integrating n local data mining models $\mu_i \in M$, using its files $\varphi_i \in \Phi$, to obtain a global data mining model μ_G can be formalized as follows:

$$MI = \prod_{i=1}^{n} (\mu_i, \varphi_i) \Big| \mu_i \in \mathrm{M} \wedge \varphi_i \in \Phi \quad (1)$$

Few research efforts reported on postmining: integrate or synthesize local patterns from different data sources. Some of these researches are now briefly described.

In [13] authors propose a method to integrate association rules sets mining from distributed XML data sets, based on the mathematical formulations defined in [11]. These mathematical formulations are appropriate to find global exceptional patterns, but not to identify global patterns valid for all the data sets. This can be considered a limitation of the method proposed in [13]. Authors in [5; 10-12] advocated an approach for mining association rules in multi-database by weighting. They defined a new process for multi-database mining that performs three steps: (1) search for a good classification of distributed databases, (2) identify two types of new patterns from local patterns: high-vote patterns and exceptional patterns, and (3) synthesize local patterns (association rules) by weighting. This approach has some limitations such as:

1. Authors assume that each database from each data source contains about the same amount of data. If the data sources are of different sizes, they suggest make them of similar size by splitting the large ones and merging the smaller ones. However, when accessing the local databases for preprocessing is not possible, this approach is not viable.
2. The same minimum support and minimum confidence for all databases analyzed is assumed. Nevertheless, in real-world applications, local association mining models may be generated with different thresholds of support and confidence for each local database.

In addition, all proposals reported on postmining lack of generality, since they are specific for integration of only one type of data mining model: association rules.

3 Knowledge Integration Model

The general model that supports our proposal (Knowledge Integration Model - KIM), will be described by means of using two different views or sub-models: Conceptual Model (CMKIM) and Functional Model (FMKIM).

The CMKIM is basically the view of the processes (P) defined by the KIM. So, this sub-model focuses on the description of processes and sub-processes, as well as on the description of the main tasks identified as part of them. On the other hand, FMKIM, is the view of the agents (A) involved in the performance of these processes.

3.1 Conceptual Model

In this section, Conceptual Model of Knowledge Integration Model (CMKIM) is described, by defining its main processes graphically based on the Eriksson-Penker notation [14]. CMKIM consists of three processes: Translation (P_T), Synthesis (P_S), and Representation (P_R). Figure 1 illustrates the workflow of CMKIM.

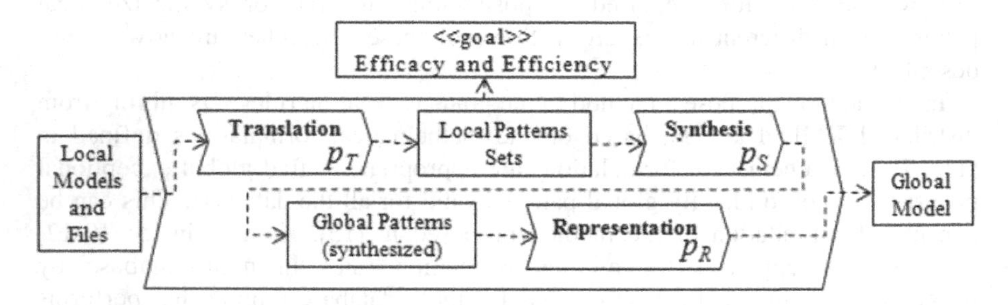

Fig. 1 Knowledge Integration Process workflow.

The first process (P_T) examines the files and selects the local models that will be integrated. Then, each selected model is represented as a set of patterns, using a homogeneous format: different types of patterns (association rules, clusters, classification rules) are codified in the same form. The last process (P_R), builds the global model by representing the global patterns synthesized, using a standard format like PMML (Predictive Model Markup Language). These processes (P_T and P_R) are the simplest. Then, Synthesis (P_T) is the most complex and important processes of our Knowledge Integration Model (KIM). For that reason, this process is explained in detail.

In figure 2 the workflow of Synthesis process is shown. This (P_S) consists of three sub-processes: *Initial Solution Construction* (P_{ISC}), *Solution Improvement* (P_{SI}), and *Measures Assignment* (P_{MA}). The input of P_S is the local models and its files, and the output is a set of synthesized global patterns.

Process P_{ISC} is responsible for the construction of an initial solution, using a random method or a deterministic method. In our model (KIM), a solution correspond to a set of global patterns that lack of measures. Then, the workflow of P_{ISC} consists of three tasks: *Select Solutions Codification*, *Construct Initial Solution* and *Evaluate Solution*.

There are two types of solutions codification, which correspond to the two integration level supported in KIM: basic level and advanced level. In the basic level all patterns of the global model must be included in at least one of the local models integrated. This means that is not possible to discover new patterns. On the other hand, in advanced level, the global model may consist of not only patterns included on local models integrated, but also new patterns discovered.

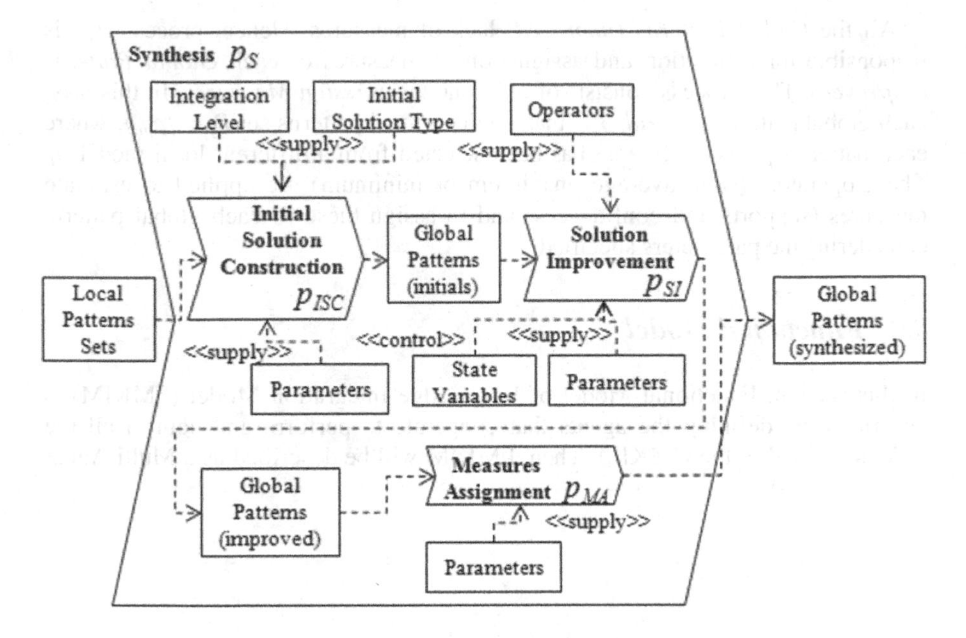

Fig. 2 Synthesis Process workflow.

The *Evaluate Solution* task, must evaluate the quality of the candidate solution received. The quality of the given candidate solution can be defined as the similarity between corresponding global model and all the local models that must be integrated. Then, the goal of the KIM is minimize the distance between given global model (candidate solution) and all the local models. Subsequently, the objective function used to evaluate each candidate solution, may be defined as follows:

$$f(S_l) = \sum_{i=1}^{m} w_{\mu_i} * d_M(MI_l, \mu_i) \tag{2}$$

where $w\mu_i$ is a weigh corresponding to local model μ_i and $d_M(MI_l, \mu_i)$ is a function of distance between a given candidate global model and a local model. The range of values of the objective function defined is [0..1].

Otherwise, P_{SI} is an iterative process, responsible for the improvement of a candidate solution (global model). Each iteration consists on generating several solutions (neighbors) from current candidate solution. Then, these solutions must be evaluated to select the next candidate solution, using the objective function defined above. Therefore, the workflow of P_{SI} consists of three tasks: *Generate Neighbors Solutions*, *Evaluate Neighbors Solutions* and *Select Candidate Solution*. The *Generate Neighbors Solutions* task must apply different neighborhood operators, until generate the count of solutions specified as parameter. Finally, the *Select Candidate Solution* task must update the variables that control process P_{SI}. The input of this process is the initial solution constructed in P_{ISC} (*Global Patterns initials*) and the output is a final solution (*Global Patterns improved*).

All the *Global Patterns (improved)* lack of measures. Hence, process P_{MA} is responsible for estimation and assignation of measures of each *Global Patterns (improved)*. This process consists of only one task: *Assign Measures*. In this task, each global pattern $\rho_{Gj} = \langle id_{Gj} \rangle$ is matched to a local patterns set $P_j = \{\rho_{j\mu i}\}$, where each pattern $\rho_{j\mu i} = \langle id_{j\mu i}, me_{j\mu i} \rangle$ has been selected from a different local model μ_i. Then, operators (sum, average, maximum or minimum) are applied to estimate measures (supports and confidences) and to assign these to each global pattern, considering the parameters specified.

3.2 Functional Model

In this section, Functional Model of Knowledge Integration Model (FMKIM) is described, by defining the agents that cooperate to perform and control all the tasks identified in the CMKIM. Then, FMKIM will be described as a Multi Agent System (MAS).

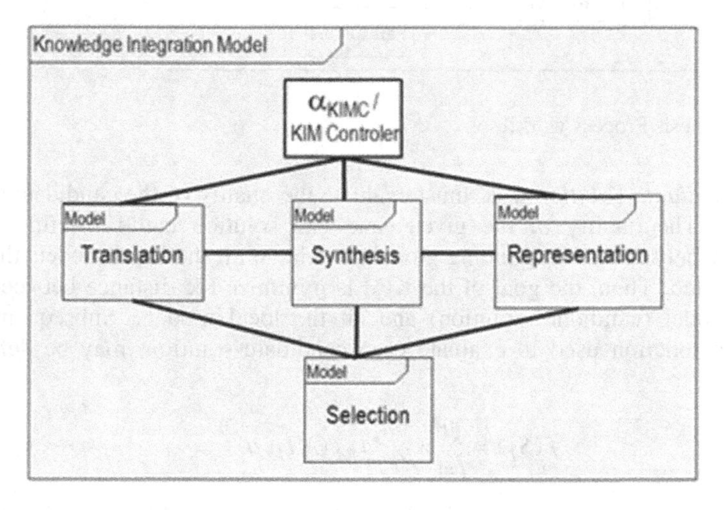

Fig. 3 Knowledge Integration Model.

Agents, according to its role in the MAS, may be classified as: General Coordinator Agent, Model Coordinator Agent or Operator Agent. *KIM Controller agent* (α_{KIMC}) is the General Coordinator Agent of the MAS, responsible to control the performance of this system, interacting with the Model Coordinator Agents: *Translation Coordinator agent* (α_{TC}), *Synthesis Coordinator agent* (α_{SC}) and *Representation Coordinator agent* (α_{RC}). Also, there are some direct relations between Model Coordinator Agents, as shows the AUML (Agent Unified Modeling Language) collaboration diagram illustrated in figure 3. On the other hand, Operator Agents perform all the tasks defined in CMKIM. Following table resumes the Operator Agents of the Synthesis Model and the tasks performed by these.

Table 1 Operator Agents of the Synthesis Model

Agent	Tasks
Initial Solutions Constructer (α_{ISC})	Construct Initial Solution, Start initial solution evaluation, Send initial solution
Solutions Improver (α_{SI})	Start neighbors solutions generation, Start neighbors solutions evaluation, Select Candidate Solution, Send improved solution
Measures Estimator (α_{ME})	Assign Measures, Send global patterns with its measures
Solutions Evaluator (α_{SE})	Evaluate Solution, Evaluate Neighbors Solutions, Send evaluation of initial solution, Send evaluation of Neighbors solutions
Neighborhood Generator (α_{NG})	Generate Neighbors Solutions, Send generated Neighbors solutions

4 Experiments

This section describes the performed experimental study, whose objective was to test the feasibility of our method for integrating association rules models. This consisted of following phases: preparing data, creating association rules models, integrating local models and analyzing results. To perform the experiment we used a database that stores clinical information for 8624 patients. From this database four training sets or data sets were created: D, D1, D2 and D3. The data set D (centralized data) had a total of 8624 records; while the disjoint sets: D1, D2 and D3, generated randomly from D, were composed of a total of 2587, 3449 and 2588 records respectively.

Then, to create association rules models discovered from each data set the Apriori algorithm implemented in Weka [15] was used. The minimum confidence factor specified was 0.846 for data set D and 0.85 for D1, D2 and D3. The count of rules resulting from D, D1, D2 and D3 was 7, 8, 7 and 4 respectively.

Once created association rules models, the three local models were integrated by means of using our proposal. A total of 20 executions of our integration method were performed, obtaining 20 proposals for global models. In each execution, a random initial solution (process P_{ISC}) was built and then 1000 iterations of the process P_{SI} were performed. The mean of the optimal values of the objective function was equal to 0.1156, where the minimum value and the maximum value were equal to 0.1105 and 0.1270, respectively. All optimal values were close to zero. This means that all the integrated models generated were similar to the local models set. Finally, we compared the centralized model, discovered from D, with the synthesized global model. Both models had five identical association rules. Moreover, the distance between these models, calculated by the distance function between models defined to evaluate the objective function was equal to 0.0453. So, we can conclude that these models were very similar. These results demonstrated the feasibility of our proposal for the integration of association rules models.

5 Conclusions

In this paper, we propose a Knowledge Integration Model useful for integrating local data mining models from distributed data sources and for searching valid global patterns. Our approach is suitable when each data source has been mined locally, and there is no access to their original databases. In contrast with related works reported, our proposal is applicable, not only to association rules models, but also to other types of data mining models. On the other hand, our approach is appropriate for mining data sources of different sizes. Furthermore, each local model can be generated with different measures thresholds, such as minimum support and minimum confidence.

The developed experiments demonstrate the feasibility of the proposal for the integration of models of association rules. As future work, we are designing and performing other experiments to validate our proposal for the integration of clustering models and classification rules models.

References

1. Wang, J.: Encyclopedia of Data Warehousing and Mining. Idea Group Reference, United States of America (2006)
2. Witten, I.H., Frank, E.: Data Mining. Practical Learning Tools and Techniques. Morgan Kaufmann, San Francisco (2005)
3. Han, J., Kamber, M.: Data Mining: Concepts and Techniques. Morgan Kaufmann, Oxford (2006)
4. Little, B.: Data Mining: Method, Theory and Practice. WIT Press (2009)
5. Wu, X., Zhang, S.: Synthesizing High-Frequency Rules from Different Data Sources. IEEE Transactions on Knowledge and Data Engineering 15(2), 353–367 (2003)
6. Cios, K.J., Pedrycz, W., Swiniarsky, R.W., Kurgan, L.A.: Data Mining. A Knowledge Discovery Approach, Springer Science Business Media, LLC (2007)
7. Cannataro, M., Congiusta, A., Pugliese, A., Talia, D., Trunfio, P.: Distributed Data Mining on Grids: Services, Tools, and Applications. IEEE Transactions on Systems, Man, and Cybernetics-Part B: Cybernetics 34(6) (2004)
8. Miller, H.J., Han, J.: Geographic Data Mining and Knowledge Discovery. Chapman &Hall/CRC, Taylor & Francis Group, LLC (2009)
9. Kargupta, H., Han, J., Yu, P. S., Motwani, R., Kumar, V.: Next Generation of Data Mining. Chapman &Hall/CRC, Taylor & Francis Group, LLC (2009)
10. Zhang, S., Wu, X., Zhang, C.: Multi-Database Mining. IEEE Computational Intelligence Bulletin 2(1), 5–13 (2003)
11. Zhang, C., Liu, M., Nie, W., Zhang, S.: Identifying Global Exceptional Patterns in Multi-database Mining. IEEE Computational Intelligence Bulletin 3(1), 19–24 (2004)
12. Zhang, S., Zhang, C., Wu, X.: Knowledge Discovery in Multiple Databases. Springer, Heidelberg (2004)
13. Paul, S., Saravanan, V.: Knowledge integration in a parallel and distributed environment with association rule mining using XML data. International Journal of Computer Science and Network Security (IJCSNS) 8(5), 334–339 (2008)
14. Eriksson, H., Penker, M.: Business Modeling with UML: Business Patterns at work. Wiley & Sons, Chichester (1999)
15. Hall, M., Frank, E., Holmes, G., Pfahringer, B., Reutemann, P., Witten, I.: The WEKA Data Mining Software: An Update. SIGKDD Explorations 11(1) (2009)

Using a Self Organizing Map Neural Network for Short-Term Load Forecasting, Analysis of Different Input Data Patterns

C. Senabre, S. Valero, and J. Aparicio

Abstract. This research uses a Self-Organizing Map neural network model (SOM) as a short-term forecasting method. The objective is to obtain the demand curve of certain hours of the next day. In order to validate the model, an error index is assigned through the comparison of the results with the real known curves. This index is the Mean Absolute Percentage Error (MAPE), which measures the accuracy of fitted time series and forecasts. The pattern of input data and training parameters are being chosen in order to get the best results. The investigation is still in course and the authors are proving different patterns of input data to analyze the different results that they will be obtained with each one. Summing up, this research tries to establish a tool that helps the decision making process, forecasting the short-term global electric load demand curve.

Keywords: Self-Organizing Maps and Short-Term Load Forecasting.

1 Introduction

Industrialized countries have experienced a global electricity demand growth this decade. With the power system growth and their complexity increase, many factors have become influential to the electric power generation and consumption. Therefore, the forecasting process has become even more complex and accurate forecasts are needed. The supply industry requires forecasts with lead times that range from the short term (a few minutes, hours, or days ahead) to the long term [1]. But the relationship between the load and its exogenous factors is complex and nonlinear, making it quite difficult to demonstrate it through conventional

C. Senabre and S. Valero
Department of Industrial Systems Engineering.
E.P.S.E., Universidad Miguel Hernández de Elche
Campus of Elche (Quorum-V Building). Avd. de la Universidad s/n, 03202 Elche. Spain
Tel.:+34 96 6658969
e-mail: svalero@umh.es, csenabre@umh.es

J. Aparicio
Operational Research Center. Universidad Miguel Hernández de Elche
Avd. de la Universidad s/n, 03202 Elche. Spain

A.P. de Leon F. de Carvalho et al. (Eds.): Distrib. Computing & Artif. Intell., AISC 79, pp. 397–400.
springerlink.com

techniques, such as time series and linear regression analysis. Short-term forecasting techniques are useful tools in the decision making to keep the generation/consumption balance [2, 3 & 4].

2 Case of Study and Objective

Historical data of real global load demand is being used for the research. This data is presented to the SOM as daily load value, in hours, from the workdays of the year 2002 and 2003. The load curves are obtained from the Spanish Electrical System Operator [5]. For testing purposes, 2004 data are also collected. The main objective of the research is to use the capacity of SOM maps to classify historical data and, in a following step, to take advantage of the memorization of this classification to identify similarities between the trained map (year 2002 and 2003) and the first demand hours of a new day corresponding to year 2004. Certain preprocessing of the input data is needed in order to obtain good results as the demand evolves over the years. It is necessary to apply a small increase to the input data with the aim of expressing estimated demand growth in 2004. Vectors are also normalized using the maximum value of demand.

3 Self Organizing Maps and Others Neural Models

The SOM is an algorithm used to visualize and interpret large high-dimensional data sets. This methodology was introduced by Kohonen two decades ago [6]. These networks are a kind of unsupervised ANN that performs a transform from the original input space (n dimensional data vector) to a reduced output space (bidimensional). The advantage of SOM is that the relationship between the original vectors is to some extent preserved in the output space, providing a visual format where a human operator can "easily" discover clusters, relations, and structures in the usually complex input space database. On the other hand, other models of neuronal networks also have been applied to load forecasting, for example certain models of Multilayer Perceptron Neural Network (MLP) and Support Vector Machine (SVM).

4 Methodology and First Results

Once understood the operation of the SOM and their capacity, the authors work with different parameters to face the process. In order to express the accuracy of the tool, a measurement index is defined. This index is the Mean Absolute Percentage Error (MAPE), which measures the accuracy of fitted time series and forecasts.

$$MAPE = \frac{1}{N}\sum_{i=1}^{N}\left[\frac{\left|L_{Ri} - L_{Pi}\right|}{L_{Ri}}\right]x100\%$$

Where: N is the number of forecasted hours (in this case is 24 values), L_{Ri} is the real load value of the i hour and L_{Pi} is the forecasted load of the i hour. For example, in some cases the average value of the MAPE index in a 22 week testing period for Tuesdays in 2004 was 1.71, which is a good index. A first methodology is based on the training of the network with daily load demand curves of the years 2002 and 2003. Afterwards, the maps are tested network bombing with some load curves of the year 2004, but simply using the first hours of the day to be forecasted. Different simulations were carried out using 8 and 10 hours as input data, and 12 hours in the future research. The objective is to associate the 2004 input data to the most similar days of the network, formed by 2002 and 2003 days, and then obtain the pattern for the most suitable day evolution. Input data is labelled as "mmddyy", i.e. month-day-year. After testing the network with the first hours of the day to be forecasted, the winner cell is chosen as the most similar curve shape by the software. This allows estimating the evolution of the following hours. One of the pros of simulating two or three different size vectors (8 and 10 hours) is to check if the map assigns the same or different winning cells. If the assignation remains always the same, the success is almost guaranteed. Figure 1 shows an example of the trained map with the labels in each neuron.

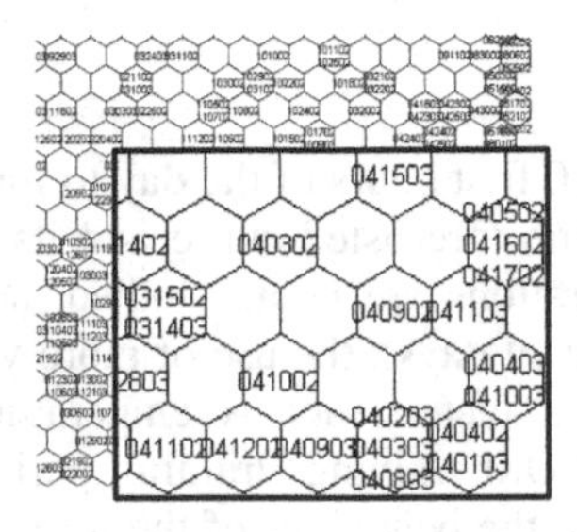

Fig. 1 Trained SOM for the load curves of the years 2002 and 2003.

Good results were obtained with training SOM parameters identificated in the Table I. This table also shows the MAPE indexes for 22 days of July 2004, when the first 8 and 10 hours are used, respectively:

Table 1 The best Training Parameters and Average MAPE

Input data	Randinit	MAPE	8 hours	2.8 %
Training algorithm	Sequent			
Network size	25 X 25		10 hours	2.1 %
Neighbour function	Bubble			
Iteration number	5000			

Figure 2 shows the estimate load curve versus the real curve of the 13th of July 2004. We have used the first 8 hours of this day for testing in the trained map. The winning neurons contain the labels of the 29th of May, 4th and 06th of June 2003.

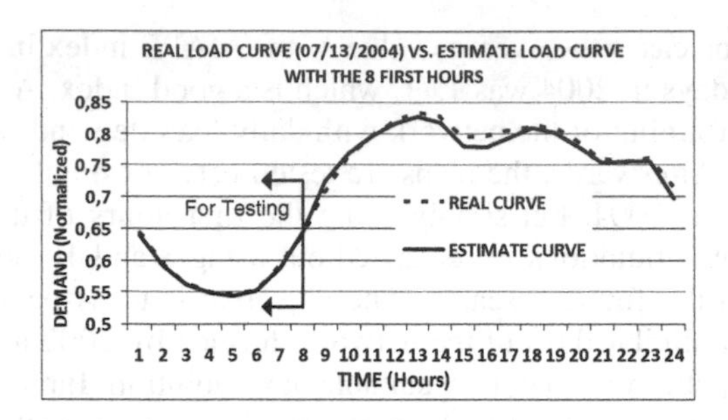

Fig. 2 Real load demand curve vs. estimated curve using the first 8 hours as input data.

After we make the average of these three associate curves by obtain the final estimate load curve, that is to say the following 16 hours. The average MAPE index, for 22 days of July 2004, is 2.8 with the 8 first hours and 2.1 with the 10 first hours.

5 Conclusions

The results with the 8 and 10 first hours of the day to forecast show a low error index in the comparison of the forecasted curve and its real one. At the moment, several input data configurations are being studied, such as longer vectors (12 hours or with more consequent days), the use of more years for the historical data or the introduction of weather factors such as temperature. More different parameter configurations for the SOM training (training periods, algorithms, etc...) are also being tested to improve the behaviour of the forecast. This methodology is a tool that could assist Companies, Utilities and Independent System Operators (ISO) in predicting the short-term demand of energy.

Acknowledgments. The work described in this paper is being supported by the "Generalitat Valenciana, Conselleria d´Educació" under Research Project GV/2010/080.

References

1. Makarov, Y.V., Reshetov, V.I., Stroev, A., Voropai, I.: Blackout Prevention in the United States, Europe, and Russia. Proceedings of the IEEE 93, 1942–1955 (2005)
2. Mohd Hafez, H.H., Muhammad, M.O., Ismail, M.: Short Term Load Forecasting (STLF) Using Artificial Neural Network Based Multiple Lags of Time Series. In: Köppen, M., Kasabov, N., Coghill, G. (eds.) ICONIP 2008 Part II, LNCS, vol. 5507, pp. 445–452. Springer, Heidelberg (2009)
3. Fan, S., Chen, L.: Short-term load forecasting based on an adaptive hybrid method. IEEE Transactions on Power Systems 21(1), 392–401 (2006)
4. Tafreshi, S.M.M., Farhadi, M.: Improved SOM based method for short-term load forecast of Iran power network In: Power Engineering Conference, IPEC (2007)
5. REE, Red Eléctrica de España, `http://www.ree.es`
6. Kohonen, T.: Self-organisation and associative memory. Springer, Berlin (1989)

SOM for Getting the Brake Formula of a Vehicle on a Brake Tester and on Flat Ground

C. Senabre, E. Velasco, and S. Valero

Abstract. The objective of the research is to prove the capability of Self-Organizing Map (SOM) to classify brake formula of a vehicle on a bank of roller tester from the MOT (Ministry of transport) and on flat ground. The neural network demonstrated good generation of the brake-slide relationship when presented with data not used in network training. This tool will easily find brake-slide equation of each experience and we will compare the brake on two different experimental tests. This article demonstrates that the MOT brake testing do not check the car brake in its usual way of driving. We will provide data and graphs to prove that tyre pressure is a determining factor when assessing the condition of brakes.

Keywords: Self-Organizing Maps and brake-slide relationship.

1 Introduction

When a vehicle is taken to the (Ministry of Transport) MOT testing facilities, this includes a brake test made on a roller bed to check the brake circuit. Several questions need to be answered over the efficiency of MOT testing facilities: Does braking on a roller tester faithfully reproduce braking on flat ground?, How far does tyre pressure affect the measurements taken on the roller bed?, Is this a safe test to assess the condition of brakes?, Is the brake test 100% efficient?.

The aim of the study is to find out a vehicle's braking capacity by measuring slippage on a brake roller tester at MOT centres, compare the measurements with other similar ones taken on flat ground, and use the result to assess the reliability of the machine in testing brake systems.

The effective contribution to these programmes require detailed knowledge of each brake through definition of "The Pacejka96 formula" [1] of experimental data, and the characterisation of these brake clusters.

C. Senabre, E. Velasco, and S. Valero
Department of Industrial Systems Engineering.
E.P.S.E., Universidad Miguel Hernández de Elche
Campus of Elche (Quorum-V Building). Avd. de la Universidad s/n, 03202 Elche. Spain
Tel.:+34 96 6658907;
e-mail: csenabre@umh.es, emilio.velasco@umh.es, svalero@umh.es

A.P. de Leon F. de Carvalho et al. (Eds.): Distrib. Computing & Artif. Intell., AISC 79, pp. 401–408.
springerlink.com

This high dimensional data set cannot be modelled easily, and advanced tools to synthesise structures from such information are needed, i.e. data mining research and applications [2]. The evaluation of Self-Organizing Map (SOM) [3] to get new Magic formula of each brake on bank of roller are the aims of this work.

The main objective of this paper is to report the results of training a neural network using field data to predict the brake-slide relationship. These predictions would then be compared with the results of a regression-based model.

2 Testing Methods

The vehicle used in the test was a used Renault 21, Model: Nevada, 7-seater, diesel. The front brakes are disc with sliding clamps. The rear wheels have drum brakes. Measurements were obtained from 2 encoders fitted to the brake roller tester and wheel of the vehicle, and a pressure sensor fitted to the vehicle's brake circuit. These instruments were used to measure the pressure in the hydraulic circuit after pressing the brake pedals, and to relate the data to the speed of the wheel. The two tests carried out are described below:

2.1 Test 1

Measuring braking on the brake roller tester at the MOT centre is carried out by placing the vehicle on rollers rotating at 5 km/h, which the wheel tries to stop by braking. The torque on the rotation axis of the rollers is measured by a strain gauge. The pressure in the brake hydraulic circuit on the vehicle and the slippage value are also measured, using known data on the angular velocity of the rollers and vehicle wheels, see Figure 1.

2.2 Test 2

In this test, the vehicle runs on flat ground, with the same signals recorded as the ones that had been taken on the brake roller tester as in test 1.A fifth wheel has been driven by the vehicle to measure slippage, as you see at Figure 2.

Fig. 1 Test 1 measuring on the brake roller tester at the MOT centre.

Fig. 2 Fifth wheel, or drive wheel.

3 Results

Tyres have been inflated to 1, 1.5 and 2 bars. These three values clearly show how events develop, although the range was increased in later studies. The measurements obtained from both tests were: the brake pressure on the vehicle and the slippage on the braking wheel. A comparative analysis was made of the braking and slippage measurements for the test carried out with different tyre pressures [4] and [5], to see how the test evolved.

4 The Pacejka96 Formula

After an important study we choose the Pacejka formula because its curve is the most similar to the experimental data [6].

The official Pacejka-96 longitudinal formula goes like this [1] & [7]:

$$F(x) = D \sin[C \arctan \{ B(1- E)x + E^* \arctan(Bx)\}] \quad (3)$$

The B, C, D and E variables are functions of the wheel load, slip angle, slip ratio and camber that we consider constant.We will need to know F_x= longitudinal brake, and x= longitudinal slip. We have used "The Pacejka96 formula" databases with many possible variables of to get different equations of brake and slide. We introduce in a data base many variations of value of each parameter and we get the graphic of group, and we label it with a number. So with the help of Self-Organizing Maps (SOM) and modelling methodologies as support tools we can compare experimental and mathematic formulas. We get an experimental curve of brake and slide on a MOT tester and on flat ground. We introduce the experimental array in the SOM toolbox and we get the value of each parameter of the Formula so we compare both curves and we can quantify how different are.

5 Introduction to Self-Organizing Maps (SOM)

The SOM is an algorithm used to visualize and interpret large high-dimensional data sets. These networks are a kind of unsupervised ANN that performs a transform from the original input space (n dimensional data vector) to a reduced output space (bidimensional). The advantage of SOM is that the relationship between the original vectors is to some extent preserved in the output space, providing a visual format where a human operator can "easily" discover clusters, relations, and structures in the usually complex input space database, see Figure 4 and 5. Once trained the SOM map, it memorizes the topographic configuration and it is a possible to test a new curve that has not been used in the training. The aim is to analyze the capacity of the network to identify and associate a new experimental curve with the theoretical curves trained in the network. We have realized successive simulations using the network SOM. Diverse parameters of training have been changed to obtain the best results with the spectrum of information of entry. The parameters of SOM used are: Map grid size: 10 x 10 neurons, Epochs of training: 1000 for rough training and 2000 for fine-tuning training, Algorithm for initialization of the input data: Randinit, Algorithm for training the input data: Sequent.

It is important to indicate that for the present study it has been considered to be a spectrum of 180 theoretical curves (Figure 3) corresponding to a few certain ranges of data of entry for each of the variables. The range of 180 curves considered in this study is sufficient to identify the brake curve of the vehicle on rollers, and to analyze the capacity of the maps SOM to classify and group the types of theoretical curves. This consideration is owed to which in none of the real curves used for testing has excelled of the maximum values of the theoretical curves used as information of entry to the SOM. This "The Pacejka96 formula" databases have many possible variables to get different equations of brake and slide:

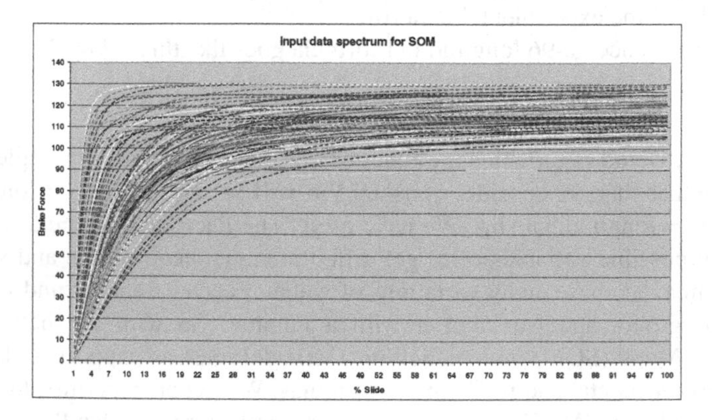

Fig. 3 "The Pacejka96 formula" databases.

We obtain the average of the values of variables of the "Pacejka formula" and we observe that the theoretical resultant curve is like the experimental curve.

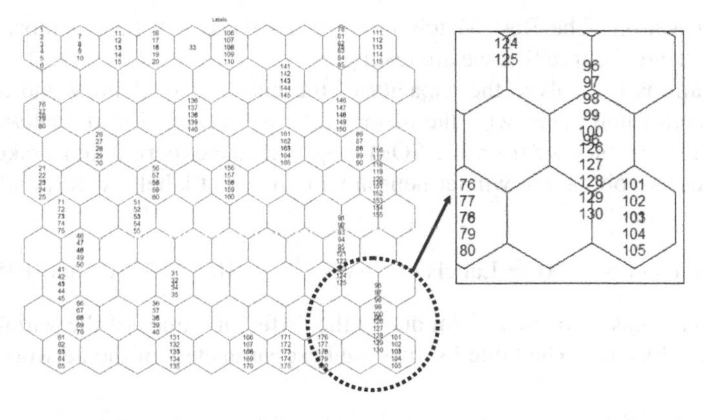

Fig. 4 Databases neurons of SOM.

The Figures 4&5 show the map trained with 180 curves used for the training. The first map shows the numerical label that we have assigned to every curve placed in every winning neuron.

The following map shows in colours the 5 clusters found for the network once trained. In every cluster the most similar curves are grouped and classified.

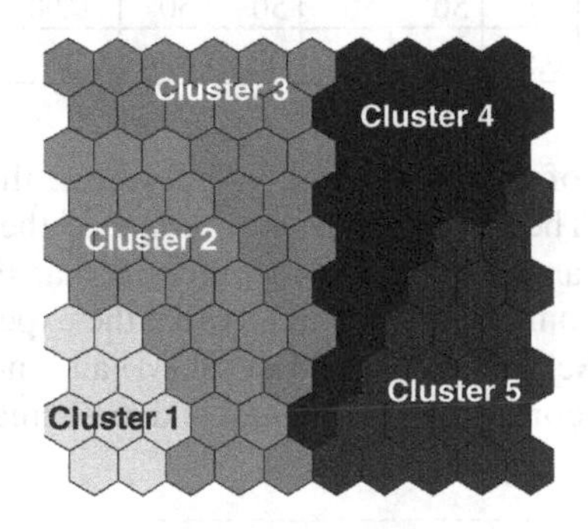

Fig. 5 Results of the map trained with labels and map of clusters

Labels are grouped in every cluster, so we can determine the set of curves and its associated parameters that define them.

Each cluster has different ranges of values of the parameters of the "The Pacejka96 formula" that define the behaviour of the curves classified under every cluster. Once trained the SOM map, it memorizes the topographic configuration and it is possible to test a new curve that has not been used in the training.

"The Best Matching Unit" will contain the theoretical curve or theoretical curves that are like the real one. In case of the "The Best Matching Unit" contained several curves, we have verified that the average of all these curves provides to us a curve very near to the real curve that we have tested. Therefore, we have to identify the parameters associated with the theoretical curves of the winning neuron (The Best Matching Unit), so these will be the most similar to those of the real curve that we are testing.

The aim is to analyze the capacity of the network to identify and associate a new experimental curve with the theoretical curves trained in the network. In the following example, we train the SOM map with the experimental brake curve on MOT, and we obtain the winner neuron with different labels corresponding to different curves:

Neuron number 100 → Labels of curves = 101, 102, 103, 104 and 105

Every curve has a concrete form due to the different values of the variables of the Pacejka96 formula. The table I shows the best parameters of the network SOM:

Table I Label of the curves of "Pacejka Formula" of the winner neuron.

	Labels of curves of the winning neuron					
	101	102	103	104	105	Media
	110	115	120	125	130	121,67
	1	1	1	1	1	1,00
	50	50	50	50	50	50,00
	0,5	0,5	0,5	0,5	0,5	0,46

We represent curves of the winnin neuron and we see that the curves are not so similar between them. Then, we obtain the average of the values of variables of the "Pacejka formula" and we observe in Figure 6 that the theoretical resultant curve (represented with outlines) is practically like the experimental curve P1.

So with the help of Self-Organizing Map (SOM) and modelling methodologies as support tools we can compare experimental and mathematic formulas.

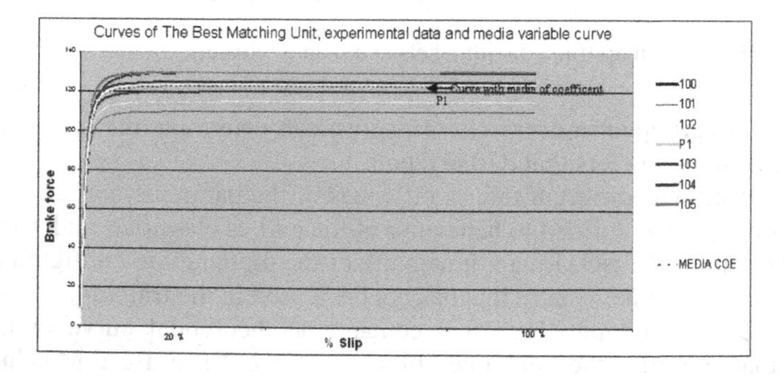

Fig. 6 Comparison of curves of the winning neuron, experimental data (P1) and curve with media of coefficient of the winning neuron.

On the other hand we get an experimental curve of brake-slide on a MOT tester and on flat ground. We introduce the experimental array in the SOM toolbox and we get the value of each parameter of "Magic Formula" so we compare both curves and we can quantify how different are.

For the example: This is the " The Pacejka96 formula " using the average of the coefficients of table I of the " winning neuron":

F(x) for 1 bar of pressure tire inflated on MOT tester:

$$F(x) = D \sin[C \arctan \{ B(1- E)x + E^* \arctan(Bx)\}]$$

$$\mathbf{F(x)_{media\ ceof} = 121,67 \sin[1^* \arctan \{ 50^* \arctan(0,46x) - 22,54\, x \}]}$$

Also, we have used "The Pacejka96 formula" databases with all possible formula of brake and slide and to find the data for each variable D, C, B and E of the curve obtained on flat ground. So we can know how different are the curve of experimental data of brake-slip on flat ground and on the MOT tester, and how different are de parameters of the pacejka96 formula for both experiments, you see Figure 7.

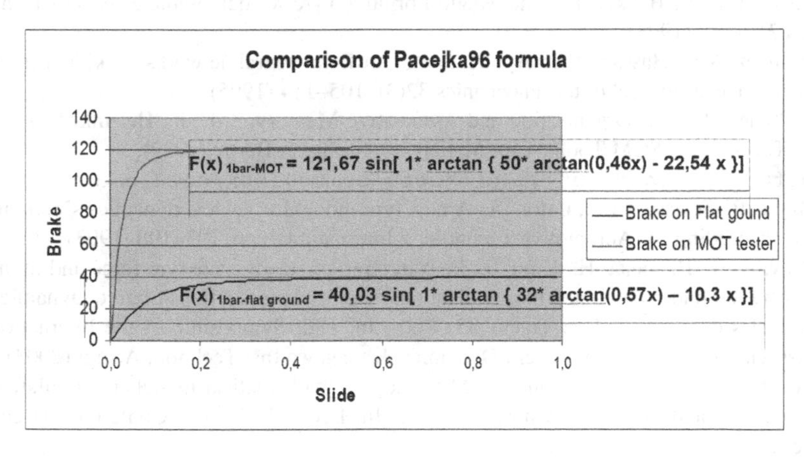

Fig. 7 Comparison of curves of Pacejka formula of brake-slide on Flat ground and MOT tester, when tire inflated with 1bar pressure.

6 Conclusions

The maps SOM have been used in numerous scientific applications and of research for the versatility and capacity of the same ones. Nowadays and thanks to environments like Matlab it is possible to use and to adapt predefined bookshops of neural networks as the case of these maps to apply them to different areas of the knowledge. The principal characteristic of these maps is that in view of a very big set of curves independently of that they are very similar or not (spectrum of information of entry) they allow to identify, to classify and to group the same ones in clusters or segments with very similar bosses. Through the proposed integration of tools such as SOM and PBLM methodologies, with the options selected by the SOM, we can know the parameter value of the "The Pacejka96 formula" of any experimental data. We get an experimental curve of brake and slide on a tester and on flat ground. We introduce the experimental array in the SOM program and we get the value of each parameter of "The Pacejka96 formula" so we compare both curves and we can quantify how different are. We must stand out that SOM methodology identify the most similar curve from the data bank to the new experimental array. But if there is not any curve equal SOM gives as result a group of curves. So if you get the media of that group of curves you get the equal curve to the experimental data. The next step would be to be changing some characteristics of the tire as pressure of inflated, and angles of drift and fall to be able to observe different behaviours and classifications different from the networks. So we will find standards of behaviour and to compare them between slides in roller and on flat ground.

References

1. Pacejka, H.B., Bakker, E.: The Magic Formula Tyre Model. Vehicle System Dynamics 21, 1–18 (1993)
2. Tohmaz, A.S., Hassan, A.E.: Application of artificial neural networks to skidder traction performance. Journal of terramechanics 32(3), 105–114 (1995)
3. Kohonen, T.: Self-Organisation and Associative Memory, 3rd edn. Helsinki University of Technology, SOM Toolbox for Matlab 5.0. Springer, Berlin (1989), `http://www.cis.hut.fi/projects/somtoolbox/download`
4. Bayle, P., Forissier, J.F., Lafon, S.: A new tyre model for vehicle dynamics simulations. In: Proceedings of Automotive Technology International, pp. 193–198 (1993)
5. Mavros, G., Rahnejat, H., King, P.: Investigation of steady-state tyre force and moment generation under combined longitudinal and lateral slip conditions Source Dynamics of Vehicles on roads and on Tracks 351-360 . In: 18th Symposium of the International-Association-for-Vehicle-System-Dynamics, Kanagawa Inst Technol., Atsugi (2003)
6. Senabre, C., Velasco, E., Sanchez, M.: Study of mathematical models to simulate tyre movement and interaction with the surface. In: León Mechanical Conference (December 2004)
7. Pacejka, H.B., Besselink, I.J.M.: Magic Formula Tyre Model with Transient Properties. Vehicle System Dynamics Supplement 27, 234–249 (1997)

Towards an Effective Knowledge Translation of Clinical Guidelines and Complementary Information

J.M. Pikatza, A. Iruetaguena, D. Buenestado, U. Segundo, J.J. García,
L. Aldamiz-Echevarria, J. Elorz, R. Barrena, and P. Sanjurjo

Abstract. Clinical guidelines enable best medical evidence transfer to where best practice is needed. Although technology is considered the best way to reach this goal, the desired results have not been achieved yet.

In this work, we introduce a technological platform that allows the definition of guidelines including complementary information required by users. It is also capable of generating platform-independent executable versions, thus improving the profitability of the undertaken effort. The systematisation of development using Model-Driven Development methods facilitates adaptation to changes and continuous improvement of quality in both guidelines and infrastructure.

Developed guidelines, together with their browsable graphical representation, are made available to health professionals through our Web Portal e-GuidesMed.

After evaluating our guideline implementations on Rare and respiratory diseases, independent experts have emphasised their usefulness in daily practice and how valuable the technology is for supporting the development of new guidelines.

Keywords: Computerised clinical guidelines, Model-driven software development, Archetypes, Clinical Decision Support Systems.

1 Introduction

Clinical guidelines enable best medical evidence transfer to where best practice needed. A well-developed guideline [13] improves quality in health care by

J.M. Pikatza, A. Iruetaguena, D. Buenestado, U. Segundo, J.J. García,
R. Barrena, and P. Sanjurjo
Languages and Computer Systems Department, Computer Science Faculty,
Universidad del País Vasco/Euskal Herriko Unibertsitatea (UPV/EHU)

L. Aldamiz-Echevarria and P. Sanjurjo
Paediatrics Service, Cruces Hospital, Osakidetza – Servicio Vasco de Salud

J. Elorz
Neumology Unit, Paediatrics Service, Basurto Hospital,
Osakidetza – Servicio Vasco de Salud

A.P. de Leon F. de Carvalho et al. (Eds.): Distrib. Computing & Artif. Intell., AISC 79, pp. 409–416.
springerlink.com

reducing variability and improving accuracy of diagnoses and therapies, while discouraging ineffective and dangerous surgical procedures [9]. However, although a great number of guidelines in text format (CG) have been developed, these have not managed to reach daily practice [19]. The methodology followed in their elaboration, often poorly defined, varies widely within and between organisations [16]. It is estimated that technology could help to improve the development, update and implementation of CG [19].

Given the failure of the traditional method of documentary transmission, Clinical Decision Support Systems (CDSS) based on executable clinical guidelines (e-CG) are considered appropriate to transfer knowledge and information [2] on drugs, bibliography, evidence, etc. A large number of randomised controlled trials and systematic reviews have examined the cost-effectiveness relationship of different strategies for implementation of guidelines [9]. Research in areas such as effective strategies for implementation of CG on CDSS [9], e-CG verification, search for up-to-date information, etc. is still missing.

In summary, multiple problems need to be solved: creation of quality CG using defined methodologies, development and systematic conversion of the abundant existing knowledge into e-CG, integration of such knowledge in CDSS, and maintenance and management of knowledge and infrastructure.

Using our experience acquired by implementing e-CG, we have developed a CG translation platform. Our guidelines on Rare and respiratory diseases and their integration in our Web-based CDSS e-GuidesMed have been positively assessed by independent experts. They have remarked the usefulness of our technology in supporting the creation of CG by medical experts. Model Driven Development methods (MDDM) [17] provide the systematisation, version compatibility, adaptability and interoperability necessary to solve the problems before us.

The next section offers some background information on these problems. Section 3 explains the methods taken into account, while Section 4 shows the results obtained. Section 5 contains our conclusions.

2 Background

The conversion from CG to e-CG has traditionally focused on the representation and execution of processes contained in the CG but not in its complete life cycle. In [10] we can find a recent comparative of CG development tools: *Arezzo^{TM}*, *Degel*, *GLARE*, *GLEE*, *HeCaSe2*, *NewGuide*, *SAGE* and *SPEM*, each one of them including its own language and execution engine. These use similar, but differently related, elements of representation. The revised features (CG repository, editor, coordination, process execution systems, CDSS architecture, EHR access and clinical management systems) have a similar importance in most of them. Also the majority of them are sparsely deployed and do not make extensive use of standards. Given these commonalities, standardisation is considered possible and sensible. This study highlights the achievable benefits, the importance of medical vocabularies, the need of task coordination during the execution of e-CG, and the influence of knowledge acquisition and verification in its quality. Additionally, the compliance to standards and the two-level modelling (archetypes + reference models [4]) approach proposed by *openEHR* have meant a great advance.

Aiming at an effective knowledge translation, the *American Medical Informatics Association* (AMIA) published a roadmap [15] proposing a collaborative process on the basis of standardisation, thus overcoming barriers while obtaining continuous improvement.

In other domains there exist modelling languages with a higher level of standardization. For instance, *Business Process Modelling* (BPM) is based on workflows and features standards such as *Business Process Modelling Notation* (BPMN) [3], which offers a higher expressivity and mappings to standard executable languages such as *Business Process Execution Language* (BPEL). BPMN 2.0 will include a set of metamodels to make tools implementation easier.

Having mentioned the concept of metamodel, we must point out the benefits brought by the use of MDDM, due to their impact on quality through systematisation introduced by automatic code generation.

To bring the knowledge to its recipients, hence improving clinical practice is necessary to build a CDSS available at the time and place where decisions have to be made while providing applicable recommendations [11]. Looking ahead is necessary to extract detailed information about the way users and CDSS interact [11].

3 Methods and Resources

Taking into account AMIA's proposal [15] to achieve an effective knowledge translation, we intend to find a technological solution that:

1. provides the best available knowledge in the best format on a CDSS,
2. overcomes barriers providing CDSS useful for translation and dissemination of knowledge,
3. supports continuous improvement of knowledge and development methodology.

In order to facilitate the capture knowledge from official CGs, we have used the terminology available in UMLSKS [20] and, alternatively, a local repository created with *LexBIG* [12]. We have also defined a structure of archetypes based on the reference and archetype models [1] defined by *openEHR*. The activities prescribed by the CG are represented as instances of these archetypes.

Due to the large number of archetype instances to be defined for each e-CG, we have developed an editor to collect the information from the CG plus any complementary content. This editor is automatically generated from a metamodel (Fig 1.) using tools provided by EMF [6]. Using *OpenArchitectureWare* (OAW) [17], we perform *Model-to-Text* transformations to generate the executable e-CG.

Some e-CG nodes may include classification systems for the implementation of diagnostic tasks. So far, we have used Fuzzy Cognitive Maps [18] and Fuzzy Inference Systems. The execution is performed by Mairi: a Web-Service aware version of our EHSIS inference engine [6].

Our second action line focuses on the creation of a Web-based CDSS called *e-GuidesMed*, which manages multiple e-CG and supports communication between users and experts. The e-CG execution is achieved using BPM tools. Currently, we use the JPDL language and the jBPM engine [10].

Our third action line is feasible due to the advantages provided by the methods employed while attaining the two previous lines.

4 Results

In this section we will present the tools we have developed to edit and execute e-CGs. In the first place, we have developed a UMLS medical terminology search application, which includes a graphical browser of the relationship between medical concepts. This tool is useful to identify medical concepts during multidisciplinary work with experts. We have also developed an e-CG editor based on the metamodel shown in Fig 1. For our archetype-based e-CG representation, we have created the A-POM reference model (*Archetype-based Process Object Model*) (Fig. 2b) based on the *openEHR Archetype Object Model* (Fig. 2a).

The nodes of an e-CG behave similar to the ones outlined in [9]: *Question*, *Decision*, *Recommendation*, *Action*, *Calculation* and *Final*. Any classification systems that could be required by an e-CG is executed in *Calculation* nodes using patient data. Since different e-CG may have common sections, the editing tool offers

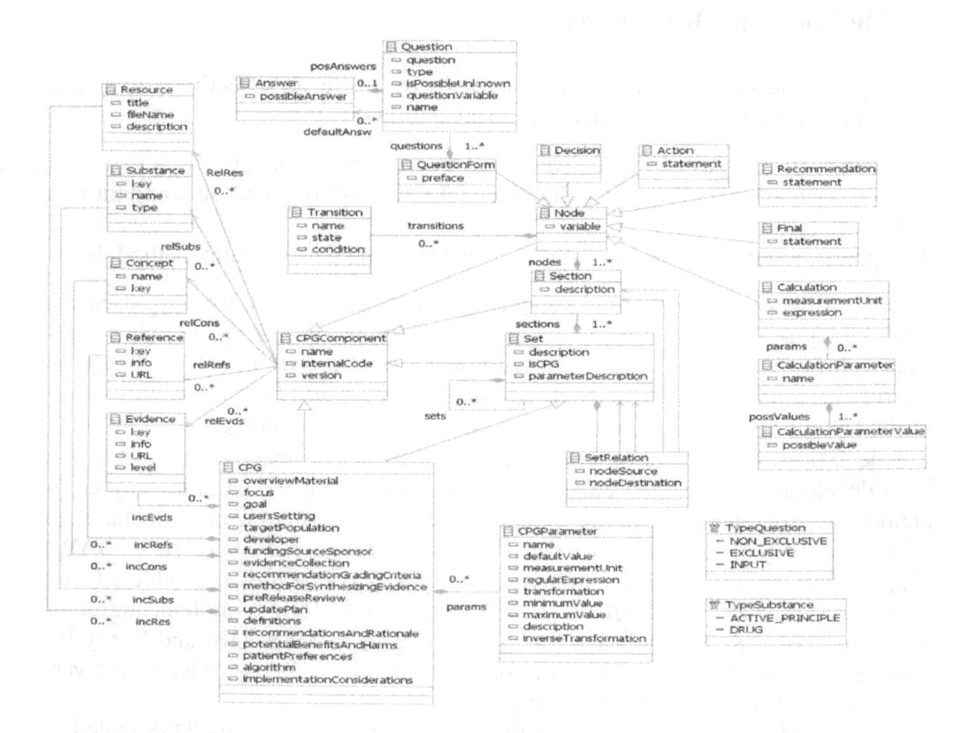

Fig. 1 Metamodel of an e-CG

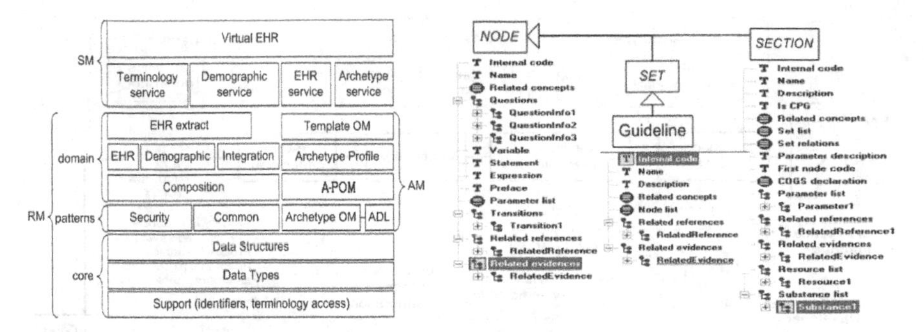

Fig. 2a A-POM in the Archetipe Model of the openEHR architecture.

Fig. 2b Metamodel elements and related archetypes in A-POM

the possibility to define abstraction levels higher than *Nodes* using *Sections* and *Sets*. A *Section* is formed only by *Nodes*, while *Sets* can contain both *Sections* and *Sets*. The whole guideline is represented as a *Set*. For each one of these elements we have defined an archetype using the results of the *openEHR* initiative (Fig 3b). The CG models created using the editing tools are transformed to instances of these archetypes by means of templates (Fig. 3a) to execute *Model-to-Text* transformations.

These transformations allow generating instances of archetypes that will store all the information of the CG in a platform-independent format. The result is a compressed file including all these instances (Fig. 3b). This design allows the re-use of knowledge across different platforms. In consequence, the development and standardisation efforts are required only once throughout the creation of an e-CG. A general overview of the e-CG creation and execution can be seen in Figure 4.

Regarding the execution of e-CG, *e-GuidesMed Portal* is a Web-based CDSS (see Fig 5.) that aims to distribute consensus knowledge along and complementary information to non-specialist users. In e-GuidesMed the information inherited in the instances of A-POM archetypes is extracted and stored in a data base. Additionally a JPDL executable process is obtained through a Model-to-Text transformation. This process controls the tasks to be performed during the execution of the e-CG.

```
DEFINE EvaluationNode FOR Node-»
 archetype (adl_version=1.4)
    openEHR-EHR-EVALUATION.Node-«internalCode».v«version»
 concept
    [at0000]
 description
    author = <"unknown">
    lifecycle_state = <"0">
 definition
    EVALUATION[at0000] matches {
         data matches {
         ITEM_TREE[at0001] matches {
             items cardinality matches {0..*; unordered} matches {
         ELEMENT[at0002] occurrences matches {0..1} matches {
                   value matches {
                        DV_TEXT matches {
                                  value matches {"«internalCode»"}
                        }
                   }
         }
```

Fig. 3a Archetype template extract

Fig. 3b Archetype structure of a e-CG

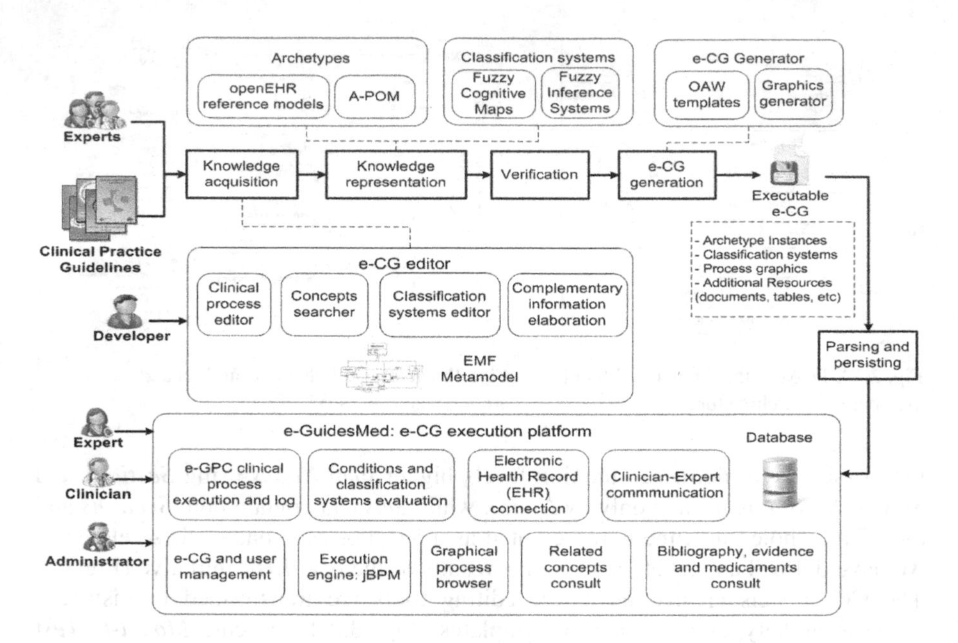

Fig. 4 e-CG edition and execution overview

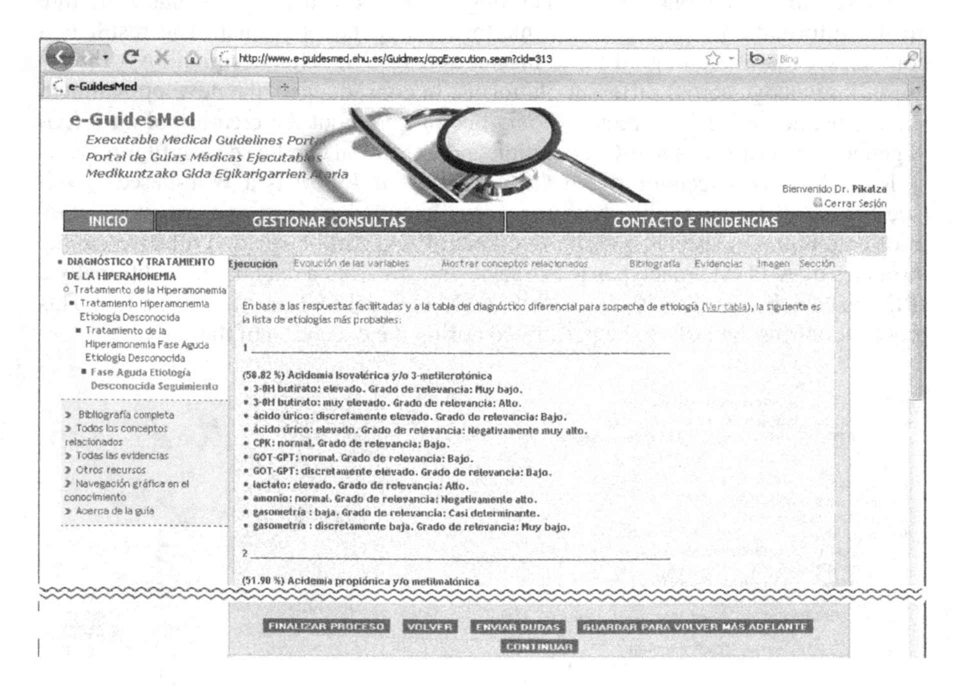

Fig. 5 Execution of an e-CG in e-GuidesMed.

In order to evaluate our *e-GuidesMed Portal*, we have developed an e-CG for Rare diseases [7] and two e-CG on respiratory diseases [15] with direct participation of the group of medical experts. These guidelines describe the processes to be followed by the healthcare professional and their execution offers information on medical terminology, bibliography, active principles and pharmaceutical products, management of communication with experts, monitoring of data variables, graphical navigation of knowledge, etc.

Both the hyperammonaemia and asthma e-CGs have passed the stage of evaluation by independent experts and will be tested in real situations through the *e-GuidesMed Portal*. We are currently extending the use of e-GuidesMed in collaboration with *Sección Española de los Errores Innatos del Metabolismo de la AEP* (SEEIM) and *Swedish Orphan International* company.

e-GuidesMed and the hyperammonemia e-CG have won the prize to the best oral communication in *VIII Congreso Nacional de Errores Congenitos del Metabolismo, 2009*.

5 Conclusions

In this paper, we have presented the capabilities of our current technology for the translation of knowledge contained in CG together with any complementary information a user may need to apply that knowledge in daily practice. It also offers the possibilities of browsing graphically said knowledge and keeping a record of its application to a particular patient.

This technology has been used to build the *e-GuidesMed Portal* and e-CG on Rare Diseases and respiratory diseases. The knowledge about the former is somewhat superficial and vague, while in the latter case, the available knowledge offers a higher level of detail.

The level of systematisation of e-GC development offers the possibility to: reuse of knowledge represented using *Set* or *Section* archetypes, rapid e-CG development from the information collected by the specific editor (generated from the metamodel) and the generation of execution platform-independent CG. We also have the advantages of developing using MDDM: fast development, quality, maintainability, abstraction, interoperability, portability, etc.

The performance evaluation by independent experts concluded that it is useful as a tool for non-specialist professionals and satisfactory supports the creation of new CG and knowledge management.

Acknowledgments. This work has been developed with funding received from the Department of Innovation and Knowledge Society of the Gipuzkoa Provincial Council [OF-94/2008], UPV/EHU [GIU08/27], Department of Industry, Commerce and Tourism - Basque Government [S-PE08UN79] [S-PE09UN60], and the Spanish Ministry of Science and Innovation [TIN2009-14 159-C05-03].

References

[1] Beale, T.: Archetypes: Constraint-based domain models for future-proof information systems. In: Baclawski, K., Kilov, H. (eds.) Eleventh OOPSLA Workshop on Behavioral Semantics: Serving the Customer, pp. 16–32. Northeastern University, Boston (2002)

[2] Bouaud, J., Séroussi, B., Falcoff, H., et al.: Design Factors for Success or Failure of Guide-line-Based Decision Support Systems: an Hypothesis Involving Case Complexity. In: AMIA Annual Symposium Proc., pp. 71–75 (2006)

[3] BPMN, http://www.omg.org/spec/BPMN/2.0/Beta1/PDF/ (accessed April 27, 2010)

[4] Chen, R., Klein, G.: The openEHR Java reference implementation project. Stud Health Technol. Inform. 129(1), 62–68 (2007)

[5] Eclipse Modeling Framework, http://www.eclipse.org/modeling/emf (accessed April 27, 2010)

[6] EHSIS, http://erabaki.ehu.es/ehsis (accessed April 27, 2010)

[7] Grupo de Consenso Hispano-Portugués para las Hiperamoniemias, Protocolo Hispano-Luso de diagnóstico y tratamiento de las hiperamoniemias en pacientes neonatos y de más de 30 días de vida. 2ª edición. Ergon (2009), ISBN: 978-84-8473-781-0

[8] Heselmans, A., Van de Velde, S., Donceel, P., et al.: Effectiveness of electronic guideline-based implementation systems in ambulatory care settings – a systematic review. Implementation Sci. 4(82) (2009), doi:10.1186/1748-5908-4-82

[9] Isern, D., Moreno, A.: Computer-based execution of clinical guidelines: A review. I. J. Medical Informatics 77(12), 787–808 (2008)

[10] Jboos Community. jBPM, http://docs.jboss.org/jbpm/v3/userguide (accessed April 27, 2010)

[11] Kawamoto, K., Houlihan, C.A., Balas, E.A., et al.: Improving clinical practice using clinical decision support systems: a systematic review of trials to identify features critical to success. BMJ 330, 765 (2005)

[12] LexBIG, https://cabig-kc.nci.nih.gov/Vocab/KC/index.php/LexBig_and_LexEVS (accessed April 27, 2010)

[13] NICE, The guidelines manual. London: National Institute for Health and Clinical Excellence (2009), http://www.nice.org.uk

[14] Osakidetza, Guía de Práctica Clínica sobre Asma. Osakidetza / Servicio Vasco de Salud (2006), http://www.respirar.org/pdf/gpcpv.pdf (accessed April 27, 2010)

[15] Osheroff, J.A., Teich, J.M., Middleton, B.F., et al.: A Roadmap for National Action on Clini-cal Decision Support. J. Am. Med. Inform Assoc. 14(2), 141–145 (2007)

[16] Rosenfeld, R.M., Shiffman, R.N.: Clinical practice guideline development manual: A quality-driven approach for translating evidence into action. Otolaryngol. Head Neck Surg. 41, S1–S43 (2009)

[17] Stahl, T., Voelter, M., Czarnecki, K.: Model-Driven Software Development: Technology, Engineering, Management. John Wiley & Sons, Chichester (2006)

[18] Stylios, C.D., Georgopoulos, V.C., et al.: Fuzzy cognitive map architectures for medical decision support systems. Appl. Soft. Comput. 8, 1243–1251 (2008)

[19] Tu, S.W., Campbell, J.R., Glasgow, J., et al.: The SAGE Guideline Model: Achievements and Overview. JAMIA 14, 589–598 (2007)

[20] UMLSKS, http://umlsks.nlm.nih.gov/ (accessed April 27, 2010)

Decision Making Based on Quality-of-Information a Clinical Guideline for Chronic Obstructive Pulmonary Disease Scenario

Luís Lima, Paulo Novais, Ricardo Costa, José Bulas Cruz, and José Neves

Abstract. In this work we intend to advance towards a computational model to hold up a Group Decision Support System for VirtualECare, a system aimed at sustaining online healthcare services, where Extended Logic Programs (ELP) will be used for knowledge representation and reasoning. Under this scenario it is possible to evaluate the ELPs making in terms of the *Quality-of-Information* (*QoI*) that is assigned to them, along the several stages of the decision making process, which is given as a truth value in the interval 0...1, i.e., it is possible to provide a measure of the value of the *QoI* that supports the decision making process, an end in itself. It will be also considered the problem of *QoI* evaluation in a multi-criteria decision setting, being the criteria to be fulfilled that of a Clinical Guideline (CG) for Chronic Obstructive Pulmonary Disease.

Keywords: quality of information, clinical guidelines, artificial intelligence.

1 Introduction

In general terms, Decision Theory (DT) is a means of analyzing which set of alternatives should be chosen when there is uncertainty about the results, in order to make an option. DT focuses its attention on identifying the "best" choice. The

Luís Lima and Ricardo Costa
CIICESI
Escola Superior de Tecnologia e Gestão Felgueiras do Instituto Politécnico do Porto, Felgueiras, Portugal
e-mail: {lcl,rcosta}@estgf.ipp.pt

Paulo Novais and José Neves
Departamento de Informática
Universidade do Minho, Braga, Portugal
e-mail: {pjon,jneves}@di.uminho.pt

José Bulas Cruz
University of Trás-os-Montes e Alto Douro, Vila Real, Portugal
e-mail: jcruz@utad.pt

A.P. de Leon F. de Carvalho et al. (Eds.): Distrib. Computing & Artif. Intell., AISC 79, pp. 417–424.
springerlink.com

notion of "best" has different meanings, being the most common the one that maximizes the expected utility for the decision maker. On the other hand, Utility Theory (UT) attempts to infer subjective value (utility) from choices in three traditional ways, namely the descriptive, the normative and the prescriptive ones. The descriptive approach tries to describe people's utility functions. The normative approach attempts to use utility in a rational model for decision making. The third approach, the prescriptive one, tries to reduce the differences between the former two by considering the limitations people usually have with the normative one [1].

Indeed, any entity operating in a complex environment is naturally cautious about the state of the world. It does not have complete information about the state nor how it will evolve. Also, the purpose to maximize the expected utility for the decision maker is not always practical, once there are bounds on computational resources which prevent the search for the optimal solution. This situation call for decision models under bounded rationality, which aims to be rational in the sense of recommending the option with maximum expected utility, but which admit bounds on their resources, and so relax some premises of the optimal approach.

The Carnegie Decision Making Model (CMDM) also known as Cyert-March-Simon model [2] is an example of decision models that emphasizes bounded rationality. Decisions are made to *satisface* rather than to optimize the solution. In group decision making it will be accepted a solution perceived as satisfactory to all members, contrary to the rational approach, which assumes that every reasonably alternative is analyzed, a quick short-run solution is looked around and typically the first satisfactory one that emerges is adopted. In contexts of high uncertainty, when the information available is incomplete, and the outcome can be hazardous, the first solution that emerges cannot be adopted irrespective of the Quality-of-Information (QoI) available. We propose a method to evaluate the quality of information and a computational model that extends the CMDM, incorporating as threshold its QoI.

In this paper it is shown how this skeleton can be applied to problem solving on the health sector, where clinical guidelines set the criteria to be followed, and are given in terms of Extended Logic Programs (ELPs). We start by summarizing previous work on the evaluation of QoI, in section two. In section three we elaborate on a computational model for decision making that extends and subsumes CMDM. In section four we present some results of the combination of these methods and techniques with the Analytic Hierarchy Process (AHP) [3], used to support the decision process. Finally, in section five, we draw some conclusions.

2 Evaluation of the Quality of Information

Clinical guidelines (CG) have been developed for more than fifty years. More recently the emphasis has been centred on the development of evidence-based guidelines and their evaluation, and ease-of-use in daily practice. CG have drawn the attention of the Artificial Intelligence (AI) community, leading to the development of specific models, tools and languages to support their practical design and implementation, in what may be called Computer-interpretable Guidelines (CIG) [4] [5]. Guideline-based Decision Support Systems (GbDSS) are also emerging as a promising way to apply AI to healthcare practice [6] [7].

One of the critical issues to implement computer-based CG is the depiction model. Several approaches using different description models are in use, namely Arden Syntax [8], Guideline Interchange Format (GLIF) [9], Asbru [10] and Pro*forma* [11], among others. Like the Proforma-based systems, we are particularly interested in approaches using logic programming and in combining general models of human decision-making with formal ones. On the other hand we draw on ELPs in order to handle positive and negative information in an explicit way, making possible the use of null values (Program 1). With this kind of construction it is possible to compute the QoI, with respect to an extension of a generic predicate P of an ELP, based on the cardinality of the exception set for that predicate. Combining the QoI for all predicate extensions, a global measure of the QoI in the decision process is made available at any time [12].

3 Decision Making

The background for decision making is set in terms of the VirtualECare project [13]; indeed, Group Decision Support Systems (GDSSs) in VirtualECare must address multi-criteria problems, layed down as incomplete ELPs.

The GDSS that supports VirtualECare is based on the limited or empiric rationality of Hebert Simon [14]. The propensity phase come about persistently, as a consequence of the natural interaction of GDSS with other components that make the VirtualECare framework. The identification of a problem triggers the formation of a decision group. The group assembling occurs in the pre-meeting phase, and a facilitator is entitled to choose the participants. The activities associated to the conception and choice occurred already at the meeting phase.

The process matures alongside a time line, centered on the description of the problem, until a suggested solution is reached; in the meantime it goes through consecutive stages (Fig. 1), namely that of options description (Generation), value judgments (Structuration), and operative rules (Evaluation), under a cycling mode.

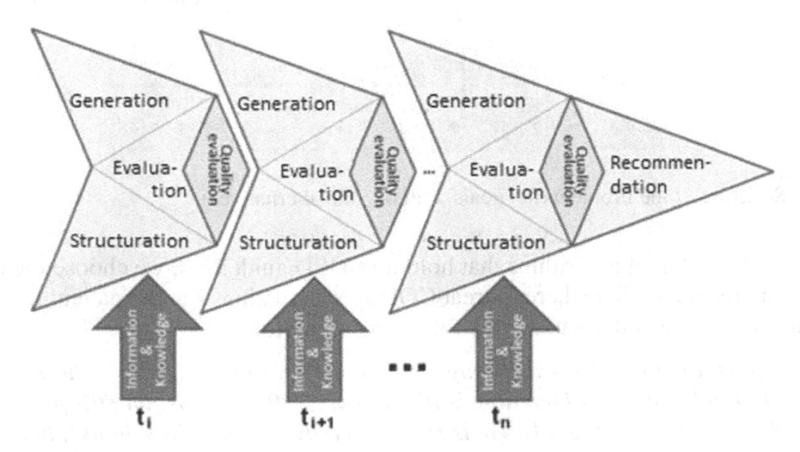

Fig. 1 Evaluation of the QoI along with the Decision Process

As a result, we finished with a meta model, with a neighborhood built around four layers, which can be instantiated with methods, tasks and tools – that is to say, instantiated with a specific model – and an area that borders all the previous ones, the Evaluation of the Quality of Information, as it is stated below :

Generation – it is a departure zone, of simultaneous exploration of potential alternative paths, information research, and problematic questions evaluation;
Structuration - it is a discussion zone, of understanding of other people's perspectives, of clarifying criteria, of revising the conjectures and restrictions, of creating a context that can be shared, that is of structuration of the decision process;
Evaluation – it is a convergence zone, of risk and consequence evaluation, hypothesis reduction, and voting; and
Recommendation – this is the end of the process, voting or final preference aggregation, following the selected decision method.

4 A Case Study

As an example we select a CG for the Chronic Obstructive Pulmonary Disease (COPD) from the National Guideline Clearinghouse (NGC) [15]. According to the World Health Organization (WHO), COPD is already responsible for 3 (three) million deaths a year, and will be the third world death cause by 2030. As demands for more patient care will continue to grow and the shortcomings of medical services are more and more recognized, systems like VirtualECare will be needed to help in the treatment and prevention of diseases like COPD, at the patient natural habitat.

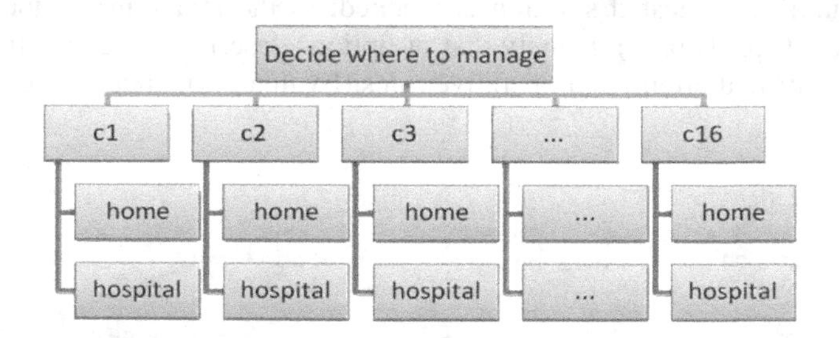

Fig. 2 Structure of the Problem (i.e. goals, criteria, and alternatives).

From the different algorithms that hold up COPD guidelines, we choose the one that supports the verdict where to treat COPD exacerbations - at the patient natural habitat or in the hospital – under the following scenario:

John is a patient that was brought to the hospital by neighbors, who he asked help, and to which now is diagnosed COPD. It was not still possible to contact the family. He is retired and lives at relatives' house, but there is not the certainty to be a structured family. Seemingly he has enough mobility to accomplish his tasks of personal hygiene, but the neighbors don't see him outdoors very often. He doesn't have recent clinical analyses.

In such a scenario, the COPD guideline suggests the evaluation of sixteen criteria (Table 1). The CG does not define a minimum set of criteria or possible combinations of criteria in order to make a final decision. We use the AHP method to structure the problem and compute a recommendation (Figure 2). We also presume that there is not complete information about all the sixteen criteria.

The weight of each criterion was evaluated using pairwise comparison, and is shown in Table 1.

Table 1 Exacerbations Treatment Criteria

	Criteria	Favours treatment at home	Favours treatment in hospital	Weight (wi)
c1	Able to cope at home	Yes	No	0,0101
c2	Breathlessness	Mild	Severe	0,0151
c3	General condition	Good	Poor/deteriorating	0,0075
c4	Level of activity	Good	Poor/confined to bed	0,0079
c5	Cyanosis	No	Yes	0,0228
c6	Worsening peripheral edema	No	Yes	0,0525
c7	Level of consciousness	Normal	Impaired	0,1463
c8	Already receiving LTOT	No	Yes	0,0129
c9	Social circumstances	Good	Living alone/not coping	0,0336
c10	Acute confusion	No	Yes	0,2022
c11	Rapid rate of onset	No	Yes	0,0595
c12	Significant comorbidity (particularly cardiac and insulin dependent diabetes)	No	Yes	0,0747
c13	Sao2 < 90%	No	Yes	0,0325
c14	Changes on the chest radiograph	No	Present	0,0480
c15	Arterial pH level	≥ 7.35	< 7.35	0,1412
c16	Arterial Pao2	≥ 7 KPa	< 7 KPa	0,1332
	Total			1.0000

As it may be observed, at a first glance, the information available leads to a knowledgeable representation as the one depicted by Program 1. With this (incomplete) data, the local values for each alternative, using Saaty scale [3], is the one given in Table 2.

```
¬   ableToCopeAtHome(E,V) ← not   ableToCopeAtHome(E,V),
        not exception(ableToCopeAtHome(E,V))
exception(ableToCopeAtHome(E,V)) ← ableToCopeAtHome(E,⊥)
ableToCopeAtHome (john, ⊥)
exception(dyspnoea(john, moderate))
exception(dyspnoea(john, bad))
generalCondition(john, bad)
exception(levelOfActivity(john, sedentary))
exception(levelOfActivity(john, moderate))
cyanosis(john, yes)
arterial_pH_level(john, ⊥)
arterial_PaO2(john, ⊥))
```

Program 1 Initial state of Knowledge (excerpt), where the symbol ⊥ stands for a null value of the type unknown.

Table 2 Local values of the alternatives for each criterion

	c1	c2	c3	c4	c5	c6	c7	c8	c9	c10	c11	c12	c13	c14	c15	c16
home	0	0	.1	0	.1	.9	.9	.9	0	.9	0	.9	0	0	0	0
hospital	0	0	.9	0	.9	.1	.1	.1	0	.1	0	.1	0	0	0	0

Table 3 Some QoI values for the Predicate Extensions in Program 1 (partial table)

$V_{ableToCopeAtHome}(john) = 0$	$V_{socialCircumstances}(john) = 0$
$V_{dyspnoea}(john) = .5$	$V_{acuteConfusion}(john) = 1$
doLTOT	arterial_PaO2

Now we can compute the global weighing of each alternative using the values from Table 3 and the weights from Table 1, which leads us to: patient natural habitat = 0,44; hospital = 0,07. This means that the information available favours the treatment at the patient natural habitat (home). Let us analyze the QoI that supports this recommendation. Table 3 shows the scoring values for each predicate`s extension, according to the available information.

Now we can compute the global QoI (also graphically depicted in Figure 3 in terms of the dashed area), using the expression (1):

$$V_{COPD}(john) = \sum_{j=1}^{16} w_j * V_j(john) = 0.5304 \tag{1}$$

As it can be seen the QoI is very low. A minimum threshold of 0.8 was defined, so no decision is made in the meantime. We need more information in order to

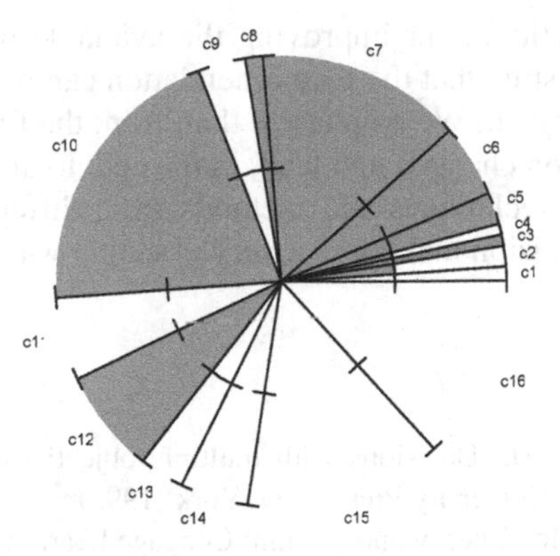

Fig. 3 QoI for Program 1

ableToCopeAtHome (john, no)
dyspnoea(john, severe)
levelOfActivity(john, sedentary)
socialCircumstances(john, no)
rapidRateOfOnset(john, yes)
lowSaO2(john, no))
arterial_pH_level(john, 7.32)
arterial_PaO2(john, 6.8))

Program 2 New knowledge (excerpt)

reduce uncertainty. The patient family is contacted and some clinical exams and analysis are made, so that in a second moment we have more information. Program 2 shows the corresponding changes at the knowledge representation level.

Computing again the values for the two recommendations, new values are obtained: patient natural habitat = 0.39; hospital = 0.55. As we can see, the recommendation has changed! Let's compute the value of the QoI now, to validate the premises for a decision. The value is now 0.95, fairly above the 0.8 threshold. So the system may deliver a recommendation for the decision.

5 Conclusions

In this paper we present an example of the evaluation of QoI to a multi-criteria decision process. A CG for COPD was used to define the criteria and conditions for the decision in a simulated clinical scenario. An ELP was used for the knowledge representation and reasoning and to support the QoI evaluation. AHP method was used to compute the preferences for each alternative.

In the beginning, the system was able to issue a recommendation but with a very low value for QoI, indeed bellow the predefined threshold. This value of QoI discouraged any immediate action based on the recommendation. In a posterior moment, a second iteration, after improving the available information, leads to a much better QoI, suggesting that the recommendation can be accepted and the corresponding actions executed. We emphasize that, from the first to the second iteration, the recommendation changed and lead, in the end, to an opposite alternative.

The combination of techniques and methods from different areas, namely AI and DT, can support decision making in a very effective way.

References

1. Keeney, R.L., Raiffa, H.: Decisions with multiple objectives: Preferences and value tradeoffs. Cambridge University Press, New York (1993)
2. Daft, R.L.: Organization Theory and Design: Cengage Learning (2009)
3. Liberatore, M.J., Nydick, R.L.: The analytic hierarchy process in medical and health care decision making: A literature review. EJOR 189(1), 194–207 (2008)
4. Colombet, I., et al.: Electronic implementation of guidelines in the EsPeR system: A knowledge specification method. IJMI 74(7-8), 597–604 (2005)
5. Hommersom, A., Lucas, P., Balser, M.: Meta-level verification of the quality of medical guidelines using interactive theorem proving. LNCS, pp. 654–666 (2004)
6. Patel, V.L., et al.: The coming of age of artificial intelligence in medicine. Artificial Intelligence in Medicine 46(1), 5–17 (2009)
7. Goud, R., Hasman, A., Peek, N.: Development of a guideline-based decision support system with explanation facilities for outpatient therapy. Computer Methods and Programs in Biomedicine 91(2), 145–153 (2008)
8. Downs, S.M., Biondich, P.G., Anand, V., Zore, M., Carroll, A.E.: Using Arden Syntax and Adaptive Turnaround Documents to Evaluate Clinical Guidelines. In: AMIA Annu. Symp. Proc., pp. 214–218 (2006)
9. Boxwala, A.A., et al.: GLIF3: a representation format for sharable computer-interpretable clinical practice guidelines. JBI 37(3), 147–161 (2004)
10. Miksch, S., Shahar, Y., Johnson, P.: Asbru: A Task-Specific, Intention-Based, and Time-Oriented Language for Representing Skeletal Plans, p. 9–1. Open University, UK (1997)
11. Sutton, D.R., Fox, J.: The syntax and semantics of the PROforma guideline modeling language. JAMIA 10(5), 433–443 (2003)
12. Lima, L., Novais, P., Cruz, J.B.: A Process Model for Group Decision Making with Quality Evaluation. In: Distributed Computing, Artificial Intelligence, Bioinformatics, Soft Computing, and Ambient Assisted Living, Proceedings, Salamanca, Spain, pp. 566–573 (2009)
13. Costa, R., Novais, P., Lima, L., Bulas-Cruz, J., Neves, J.: VirtualECare: Group Support in Collaborative Networks Organizations for Digital Homecare. In: Yogesan, K., et al. (eds.) Handbook of Digital Homecare. Springer, Heidelberg (2009)
14. Simon, H.A.: Models of Bounded Rationality: Empirically Grounded Economic Reason, vol. 3. MIT Press, Cambridge (1982)
15. Nice: Chronic obstructive pulmonary disease (2009), http://guidance.nice.org.uk/CG12 (06.12.2009)

Decision Support System for the Diagnosis of Urological Dysfunctions Based on Fuzzy Logic

David Gil, Magnus Johnsson, Juan Manuel García Chamizo, Antonio Soriano Payá, and Daniel Ruiz Fernández

Abstract. In this article a fuzzy system with capabilities for urological diagnosing is proposed. This system is specialized towards the diagnosis of urological dysfunctions with neurological etiology. For this reason the system specifies all the neural centres involved in both the urological phases, voiding and micturition. The fuzzy system allows to classify every dysfunction of all patients by means of their membership functions. The results of the experiments show that the fuzzy approach allows the diagnosis of urological dysfunctions from the relationship between neural centres and their associated neurological dysfunction.

Keywords: fuzzy logic systems, urology, artificial intelligence in medicine.

1 Introduction

Urinary incontinence is one of the problems of the urinary system that can affect persons of any age, but it occurs particularly frequently in the geriatric population, among children, paraplegics and postpartum [1].

In the Lower Urinary Tract (LUT) elements intervene which are not linear and of difficult characterization and it is one of the systems solely controlled by the sympathetic, parasympathetic and somatic nervous systems [2]. Furthermore, etiological studies of the LUT have demonstrated a great

David Gil, Juan Manuel García Chamizo, Antonio Soriano Payá, and Daniel Ruiz Fernández
Computing Technology and Data Processing, University of Alicante, Spain

Magnus Johnsson
Lund University Cognitive Science, Sweden

A.P. de Leon F. de Carvalho et al. (Eds.): Distrib. Computing & Artif. Intell., AISC 79, pp. 425–433.
springerlink.com

pathological diversity [3]: the same dysfunction can have neurogenic, anatomical, infectious and inflammatory causes etc.

Consequently, articles displaying manifest contradictions at various levels have been published: with regard to the morphology of the musculature with regard to the connections of the nervous peripheral system, with regard to reflexes and with regard to pontine and supra-pontine connections and their role in the control [4].

The use of classifier systems in medical diagnosis is increasing gradually. There is no doubt that evaluation of data taken from patients and decisions of experts are the most important factors in diagnosis. Classification systems can help in increasing accuracy and reliability of diagnoses and minimizing possible errors, as well as making the diagnoses more time efficient [5].

Some of the related work in the field of the urological diagnosis has been developed basically by means of Artificial Neural Networks (ANNs) [6] [7].

The work carried out in this paper explores however other systems much less used in diagnosis in medicine: fuzzy logic systems. Besides, it allows to know the degree of relationship between a damaged part of the neural regulator of the LUT and the type of dysfunction. This correlation is performed by means of a fuzzy logic system and its membership functions since the classic logic lacks biomedical sense and its approach to treat medical information. This work, as a part of two research projects in the field of urology ("Cooperative diagnosis system applied to the urinary dysfunction" between 2005), is described in the following section. The remaining part of this paper is organized as follows: first, a description is given of the approximation of the artificial model of the biological system, then the experiments, which have been implemented by means of a fuzzy approach, will be described. Later, the subsequent testing carried out in order to analyze the results is described. Finally, the relevant conclusions are drawn.

2 Background to the Artificial Model

The presented model of the LUT based on the agent paradigm is shown in figure 1 [8]. The LUT is divided into two parts: the mechanical system (MLUT) and the neuronal regulator (RLUT). The agents which constitute the multiagent system correspond to the neuronal centres of the RLUT. These agents collect information generated by the MLUT and process/transmit it back towards the mechanical part.

Each agent makes a contribution to the system, called influence, in such a way that the total number of the different influences will determine the overall state of the system and the activation or non activation of the different signals involved. In the process, the model of the LUT is defined as:

$$LUT = \langle MLUT, RLUT, {}^{MLUT}I_{RLUT} \rangle \tag{1}$$

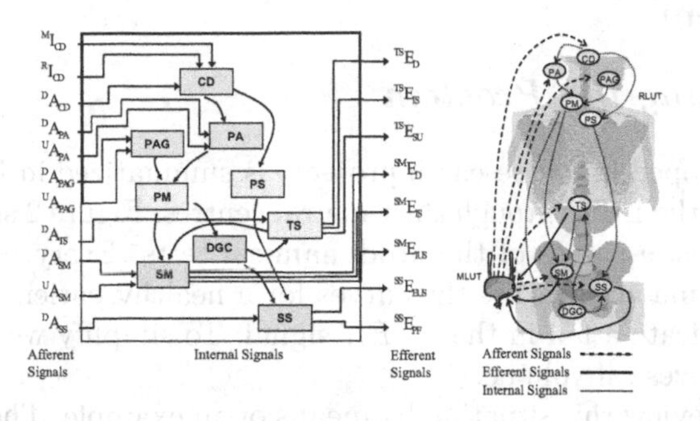

Fig. 1 Approximation of the artificial model to the biological system. The nine neural centres are represented in the artificial model with the same types of signals (afferent, efferent and internal) as the biological model in order to display the highest possible degree of similarity.

where the MLUT models the mechanical part of the lower urinary tract, the RLUT models the neuronal regulator of the lower urinary tract and the $^{MLUT}I_{RLUT}$ approaches the relation between both parts. Since the interface regards the LUT as a system of actions and reactions, it is defined as:

$$^{MLUT}I_{RLUT} = \langle \Sigma, \Gamma, P \rangle \tag{2}$$

In the above equation Σ is the set of possible states in which the system can stay. Γ is the set of the possible intentions of actions (an action proposed by an agent is represented as an intention of modification) in the system. P is the set of the actions (plans) that the different agents can execute with the objective of modifying their states.

The original model of the LUT implemented by means of multiagent systems displayed a high degree of similarity with the biological model [8] with flexibility regarding the capacity to add or change the functions and the components. This flexibility was also a good reason to carry out the implementation of the model.

The nomenclature used in figure 1 is $^{D}A_{CD}$ and indicates an afferent signal from the mechanical part (detrusor) towards the CD neural centre for the afferent signals. For the efferent signals $^{SM}E_{D}$ indicates an efferent signal from the SM neural centre towards the mechanical part (detrusor).

Some studies show a detailed description of this model [8]. The model allows to reproduce exactly the studies made by the urological analysis. For this reason, the model has been tested with real data to verify the correct functioning in both normal and dysfunctional situations due to neurological causes.

3 Fuzzy System

3.1 Approaching the Problem

Previous work developed in our research projects is summarized in Figure 1. This is the model of the LUT through their neural centres. Figure 2 shows the four most characteristic curves of the urodynamical tests. Every figure has two curves, the continuous ones are the curves for a healthy patient whereas the dashed ones indicate a fail in the ${}^{SM}E_D$ signal. To simplify we will just explain the curve of vesical volume.

We are going to review this situation by means of an example. The dashed curve of figure 3a corresponds to urodynamic measures carried out with the highest degree of injury in the SM centre (signal ${}^{SM}E_D$ which was already referenced in figure 1). The continues curve from a healthy patient is overwritten to the dashed ones from an illness patient. The goal is to highlight the major differences as well as the most critical points between both of them. We are obviously dealing with theoretical models in which the differences between the normal models and those which present some failure in a nervous signal can easily be quantified. In real life this situation is not always as obvious. Some signals, although they work in an incorrect manner, do not produce a visible output that can be explicitly reflected in the values of the urodynamical curves. The circumstance that a great number of cases exist with different outputs leads to the need for methods of learning regarding these curves for the prediction and diagnosis of new curves to be evaluated. These methods of learning as well as the explanatory rules by means of fuzzy logic with their membership functions allow a more real explanation which

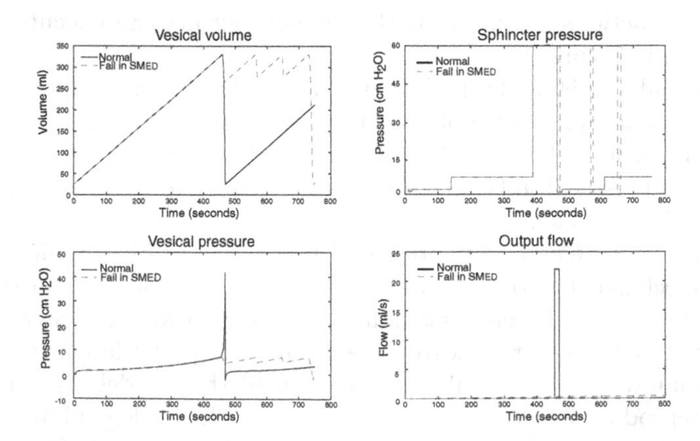

Fig. 2 These are the four most characteristic curves of the urodynamical tests. Every figure has two curves, the continuous ones are the curves for healthy patients whereas the dashed ones indicate a fail in the ${}^{SM}E_D$ signal.

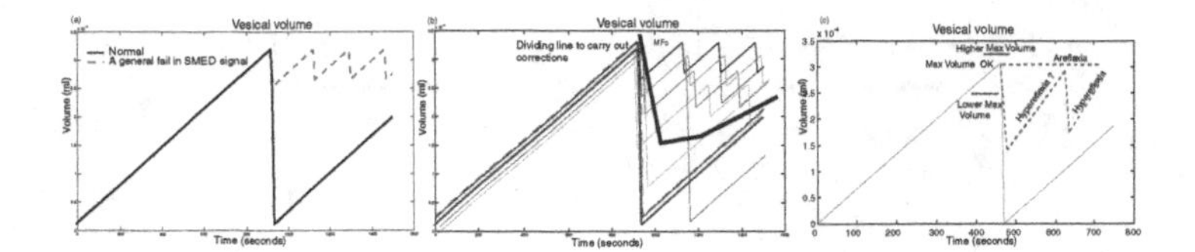

Fig. 3 a. Curves with an entire fail in the centre SM for the signal ${}^{SM}E_D$. b. High variety of curves with different degrees of fail in the ${}^{SM}E_D$ signal. c. Indications of curves with and without dysfunctions. The continuous line shows the curve of urine volume of a healthy patient, whereas the dashed lines indicate the points which determine neurological dysfunctions. This is implemented by means of the membership functions.

is closer to the biological models unlike the classic logic that this would only be so if all the cases were equal.

This real situation with the curves processing for every curve will indicate a certain degree of error in the ${}^{SM}E_D$ signal. Figure 3b tries to represent this situation with a good quantity of curves, all of them with degrees of lesion in ${}^{SM}E_D$ signal. It indicates the degree of dysfunction according to the degree of membership.

3.2 Construction of the Fuzzy System

Figure 3b with its variety of curves suggests a fuzzy approach for this problem. Then, figure 3c, as a starting point, represents a general vision which is essential to deal with many similar graphs. The continuous curve shows the curve of a healthy patient in the storage and voiding phases. The dashed curves overwritten indicate the points which determine neurological dysfunctions.

The curves represented in Figure 3c were obtained through the urodynamical model presented in the previous paragraph. Through a close collaboration with the urologists they have been simulated graphs of their patients. In other words, it has developed an empirical study with the urologists to simulate urodynamical curves of different behaviours: healthy patients and patients with dysfunctions due to neurological causes.

When a curve is overwritten on the healthy patient (continuous line) patients have no dysfunction. However, when a new curve deviates from this, the closeness or distance to each of these dashed curves indicate the degree of dysfunction associated with each curve. The same curve may have various degrees of dysfunction. This could be the case of a patient with several possible diagnoses (or mixture of them).

Fuzzy tool of matlab is the tool chosen to implement the system since it has a high variety of elements and it has been used in a broad range of areas for solving fuzzy logic problems [9]. The implementation consists of three input variables and one output variable according to Figure 3c (it is constructed taking that figure as a starting point). This implementation has

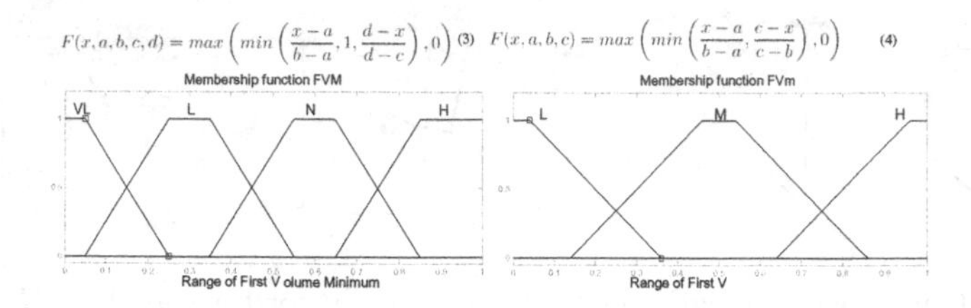

Fig. 4 a. Fuzzy membership values for First Volume Maximum (FVM). According to Figure 3c, this variable measures the maximum value of the curve reaches in his first peak and it compares with normal and dysfunctions curves, measuring their degree of approximation. b. Fuzzy membership values for First Volume minimum (FVm). Also according to Figure 3c, this variable now measures the minimum value which also reaches its first peak curve and it compares with normal and dysfunctions curves, measuring their degree of approximation.

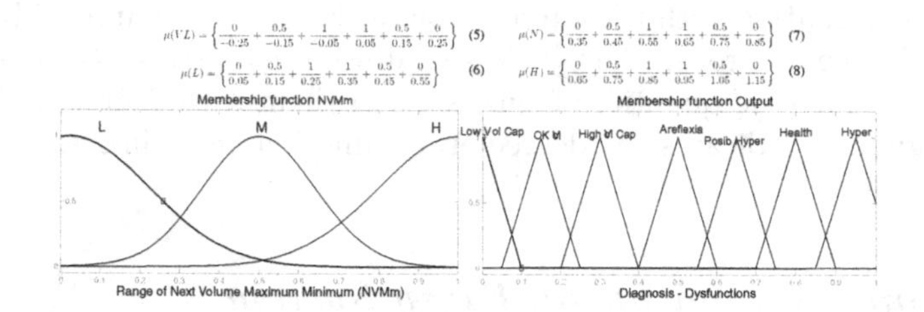

Fig. 5 a. Fuzzy membership values for Next Volume Maximum Minimum (NVMm). Also according to Figure 3c, the last variable input measures the maximum and minimum values reaching the curve but now not in its first rise but in the next one and it compares with normal and dysfunctions curves, measuring their degree of approximation. Figure 3c shows the dashed curves that simulate dysfunction compared with healthy patients. This is what is represented by the membership functions of this variable. b. Fuzzy membership values for kind and level of Dysfunction (GD) The output variable relates the vesical volume presented on the input variables and the diagnosis curves. There are many curves, some for healthy patients, some for dysfunctions, as areflexia and hyperreflexia, but also some of them in between which indicates a possible dysfunction as "PosibHyper" membership function which indicates a possible hyperreflexia or "HighVolCap" or "LowVolCap" which indicate possible problems with the vesical volume that may appear and they should be clarified with more urological tests.

been performed in a close cooperation with the urologists. It has been decided with urologists using 4 variables (3 input variables and output) to explain the functioning of the different curves (healthy patients and patients with urological dysfunctions). Each of the input variables measured basic parameters of both the storage and voiding phases. The membership functions within each of these variables gives an indication of the degree of approximation of these curves with the real ones of a patient. The output variable will help to make a diagnosis approaching the membership functions of output variable to the output of a simulation of patient real data. These input variables are "FirstVolMax", "FirstVolMin", "NextVolMaxMin" and the output variable is "DisfDegree".

These zones or areas that entail a dysfunction or lack of it are represented by fuzzy membership values using an intuition technique which involves contextual and semantic knowledge [10]. The two membership functions used are trapezoidal and triangular. The trapezoidal and triangular membership functions of a vector x depends on four(a, b, c and d) and three (a, b and c) scalar parameters in equations 3 and 4 respectively, which corresponds to all the problem areas as mentioned.

The components of the input vector consist of membership values to the linguistic properties such as very low volume, low volume, normal volume and high volume. Here, it is explained the first input "First Volume Maximum" (FVM). It can be represented as FVM = VL (very low), L (low), N (normal), H (high). The membership value of VL, L, N and H can be written as in Eqs. 5-8 and can be depicted in the figure 4a. The remain variables (two other inputs and the output) have similar membership and they are summarized in the figures 4b 5a and 5b.

3.3 Experiments

The experimentation of the system determinate the degree of exactness of the fuzzy system. The urological model presented in the section 2 has been tested

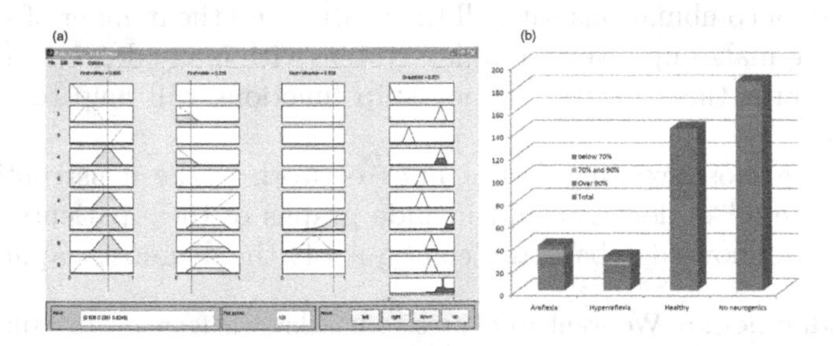

Fig. 6 (a) Experimentation with the fuzzy logic system. The output value lies between three different membership functions, possible hyperreflexia, healthy patient and hyperreflexia (two dysfunctions and healthy patients) which means that it is a good example for applying fuzzy logic. (b) Total patients 300. 20 with areflexia. 10 are over 90% of the fourth membership function of areflexia; 6 between 70% and 90%; 4 below 70%. 15 with hyperreflexia; 10 are over 90% of the seventh membership function of hyperreflexia and many of them in the fifth one(possible hyperreflexia); 2 between 70% and 90%; 5 below the 70%. 80 healthy patients; 63 above 90% of the sixth membership function and also most of them in the second one. Rest of the not neurogenic dysfunctions, are not covered by the fuzzy system that considers only neurogenic dysfunction.

for the urologists. An example of the experimentation of the system is showed in figure 6a. In particular it shows a test which points out exactly the clearest definition of fuzzy logic. The output value lies between three different membership functions (possible hyperreflexia on the left, healthy patient in the middle and hyperreflexia on the right). In this situation the degree of each of the membership functions indicate a complex diagnosis. i.e. not a particular dysfunction but a situation with three different health states. All these types of experiments could be compared to the urodynamical tests carried out at the hospital. These tests offer approximations to several kinds of dysfunctions or health situations. The total number of patients is 300. Among all of them figure 6b presents the results with every neurogenic dysfunction.

4 Conclusions and Future Work

In this paper the functioning of an urological model is evaluated. The model consist of two parts, the mechanical one (MLUT) and the neural regulator (RLUT).

The ability to carry out simulations not only for healthy patients but also for the ones who present dysfunctions has allowed the recreation of any situation which in the real world presents manifest inconveniences. The results of the experimentation identify curves for patients with neurological dysfunctions and healthy patients.

However, it is not always possible to establish with 100% accuracy which are the neurological dysfunctions in every individual case. The fuzzy logic approach will make it possible to deal with this imperfection. Furthermore, the quantity of combinations with all the centres and the number of signals of each of them, makes up a very complex system with many degrees of approximation towards the diseases. Membership functions will help to approach this model.

One of the most exciting future lines of work is the construction of a website where all urologists could include graphs of their patients, thereby improving the knowledge base and consequently the diagnosis capability.

Acknowledgements. We want to express our acknowledgement to Mikael Skylv for very helpful comments on the manuscript. The collaboration with urologists of the Hospital of San Juan (Alicante-Spain) has made it possible to reach a better understanding of the complex neurological parts of the LUT. This work has been developed in the context of the project funded by the Ministry of Science and Innovation and co-financed by FEDER (Ref.: TIN2009-10855).

References

1. Harris, M.R., Chute, C.G., Harvell, J., White, A., Moore, T.: Toward a National Health Information Infrastructure: A Key Strategy for Improving Quality in Long-Term Care. DHHS, Office of the Assistant Secretary for Planning and Evaluation (2003)

2. Yoshimura, N., Groat, W.C.: Neural Control of the Lower Urinary Tract. International Journal of Urology 4, 111–125 (1997)
3. Blaivas, J.G., Chancellor, M.: Atlas of urodynamics. Williams & Wilkins (1996)
4. Blok, B.F.M., Holstege, G.: The central control of micturition and continence: implications for urology. BJU Int 83, 1–6 (1999)
5. Akay, M.F.: Support vector machines combined with feature selection for breast cancer diagnosis. Expert Systems With Applications (2008)
6. Gil, D., Soriano, A., Ruiz, D., Montejo, C.: Embedded system for diagnosing dysfunctions in the lower urinary tract. In: Proceedings of the 2007 ACM symposium on Applied computing, pp. 1695–1699. ACM Press, New York (2007)
7. Gil, D., Johnsson, M., Chamizo, J.M.G., Paya, A.S., Fernandez, D.R.: Application of artificial neural networks in the diagnosis of urological dysfunctions. Expert Systems with Applications 36, 5754–5760 (2009)
8. Maciá, F., Chamizo, J.M.G., Payá, A.S., Fernández, D.R.: A robust model of the neuronal regulator of the lower urinary tract based on artificial neural networks. Neurocomputing 71, 743–754 (2008)
9. Jiménez, L., Angulo, V., Caparrós, S., Pérez, A., Ferrer, J.L.: Neural fuzzy modeling of ethanolamine pulping of vine shoots. Biochemical Engineering Journal 34, 62–68 (2007)
10. Zadeh, L.: Fuzzy sets. World Scientific Publishing Co., Inc, River Edge (1996)

10. [illegible]

Market Stock Decisions Based on Morphological Filtering

Pere Marti-Puig, R. Reig-Bolaño, J. Bajo, and S. Rodriguez

Abstract. In this paper we use a nonlinear processing technique based on mathematical morphology to develop a simple day trading system that automatically decides the timing to commute the marked strategy in terms of sort/long positions. In this short paper we show preliminary results.

Keywords: Mathematical Morphology, Nonlinear Processing, Algorithmic trading.

1 Introduction

The financial markets are supported by electronic platforms that provide real time efficient services. It is known as algorithmic trading the use of computer software to generate trading orders. By means of algorithms it is obtained support to decisions in aspects such as the timing, the price or the volume of the operation, managing risk and the market impact. Furthermore, in most cases the computer algorithms introduce orders in the electronic market without human intervention. Hedge founds, pension funds, mutual funds or institutional traders are some of the big users of these techniques. According to Boston-based financial services industry research and the consulting firm Aite Group, in 2006 a third of all EU and US stock trades were driven automatically. In 2009 the trading firms account for 73% of all US equity trading volume [1][2]. In this work we explore the mathematical morphology (MM) to develop a simple day trading system that automatically decides the timing to commute the marked strategy in terms of sort/long positions.

2 Mathematical Morphology: An Introduction

Mathematical morphology was first proposed by J.Serra and G. Matheron in 1966, was theorized in the mid-seventies and matured from the beginning of 80's. It can

Pere Marti-Puig and R. Reig-Bolaño
Grup de Tecnologies Digitals, University of Vic, C/ de la Laura, 13, 08500, Vic, España

J. Bajo and S. Rodriguez
BISITE Plaza de la Merced s/n, 37008, Salamanca, España
e-mail: pere.marti@uvic.cat, ramon.reig@uvic.cat

A.P. de Leon F. de Carvalho et al. (Eds.): Distrib. Computing & Artif. Intell., AISC 79, pp. 435–439.
springerlink.com

process binary signals, originally, and it was fast extended to gray level signals. This technique is proved to be very useful in digital image processing. Mathematical morphology is based on two fundamental operators: dilation and erosion. These two basic operations are done by means of a structuring element. The structuring element is a set in the Euclidean space and it can takes different shapes as circles, squares, or lines. Using different structuring elements it will achieve different results; therefore, the selection of a suitable structuring element is fundamental. A binary signal can be considered a set and dilation and erosion are Minkowski addition and subtraction with the structuring element [3]. In the context of stock prices we can work with series of data that can be modeled as gray level signals. In this context, the addition and subtraction operations that are applied in binary morphology are replaced by suprermum and infimum operations. Moreover, on the digital signal processing framework, supremum and infimum can be changed by maximum and minimum operations. In this context, the erosion can be seen as the minimum value of the part of the function inside a mobile window that is defined by the structuring element. Then, given a one-dimensional signal, the function f containing the stock prices and a flat structuring element Y, the erosion is defined as:

$$\varepsilon_Y(f)(x) = \min_{s \in Y} f(x+s) \tag{1}$$

As the erosion computes the minimum gray level inside the mobile window function it decreases the peaks and accentuates the valleys of the original function f. On the other hand the dilation (for gray level signals) is defined as:

$$\delta_Y(f)(x) = \max_{s \in Y} f(x-s) \tag{2}$$

The dilation gives the maximum gray level value of the part of the function included inside the mobile template defined by the structuring element, accentuating peaks and minimizing valleys. By combining dilation and erosion we can form other morphological operations. The opening and the closing are basic morphological filters. The morphological opening of a signal f by the structuring element Y is denoted by $\gamma_Y(f)$ and is defined as the erosion of f by Y followed of a dilation by the same structuring element Y. This is:

$$\gamma_Y(f) = \delta_Y(\varepsilon_Y(f)) \tag{3}$$

The morphological closing of a signal f by the structuring element Y is denoted by $\varphi_Y(f)$ and it is defined as the dilation of f by Y followed of the erosion by the same structuring element:

$$\varphi_Y(f) = \varepsilon_Y(\delta_Y(f)) \tag{4}$$

Opening and closing are dual operators. Closing is an extensive transform and opening is an anti-extensive transform. Both operations keep the ordering relation between two images (or functions, in our case) and are idempotent transforms [3]. In the 1-D context these operations create a more simple function than the original. By combining an opening and a closing, both of them with the same structuring element, we can only create four different morphological filters. Considering the operations γY and φY, the four filters we could obtain are γYφY, φYγY, γYφYγY and φYγYφY. No other different filter can be produced as a consequence of idempotency property. To derive different families of morphological filters we need to combine openings and closings with different structuring elements. New filters, alternating sequential filters, can be obtained by alternating appropriately theses operators [3].

3 The Proposed Trading System

The system is very simple and has been developed using MATLAB. Initially was thought to operate in the day tracking context but it could be modified to intraday operations. Once the markets of interest are closed, the system updates the stock information from any public Internet source, then it processes the data and, it is the case, it generates the orders to next day. The only stock information required is the open, close, high and low prices that the stocks, or any financial asset, reached along every day session. As we initially look for a simple system, we establish that the system will take the decisions from a reduced set of signals. Basically it computes the opening and the closing of the prices and uses a linear mean that is computed from the close day prices. Based on the cross of these three signals the system generates the decisions. To generate the market orders we have developed a finite state machine that governs the switching between the short and long scenarios. In fig.1 we can see de opening and closing signals (in blue) computed using a flat structuring element of length 8. The red signal is the mobile mean taken from 34 elements.

Next, in fig. 2 we have used the Santander stock prices along a period of 10 years. We maintain the same parameters of the system. In fig. 2 (a) we show the opening and closing signals (in blue) with L=8 and the mobile mean of 34 samples (in red). In fig. 2 (b) there are represented the long (blue) and short (red) operations.

The system parameters can be optimized for different kind of data. Using the same data of fig.2, the Santander stock prices, we have searched the structuring element and the median filter that maximize the profits. The system performance is summarized in fig. 3.

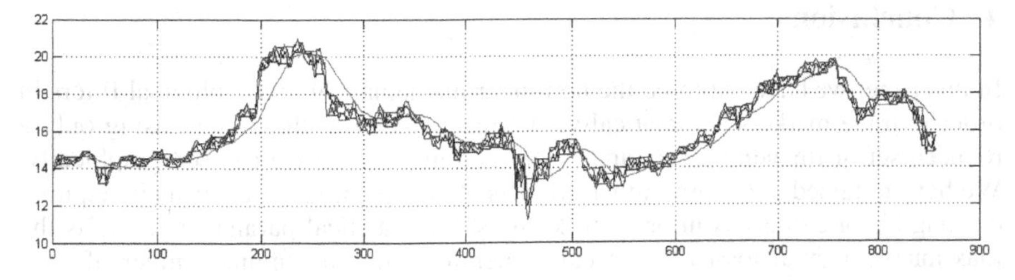

Fig. 1 Telefonica stock prices. Opening and closing signals (blue) computed with a flat structuring element of length 8 and the mobile mean of 34 elements (red).

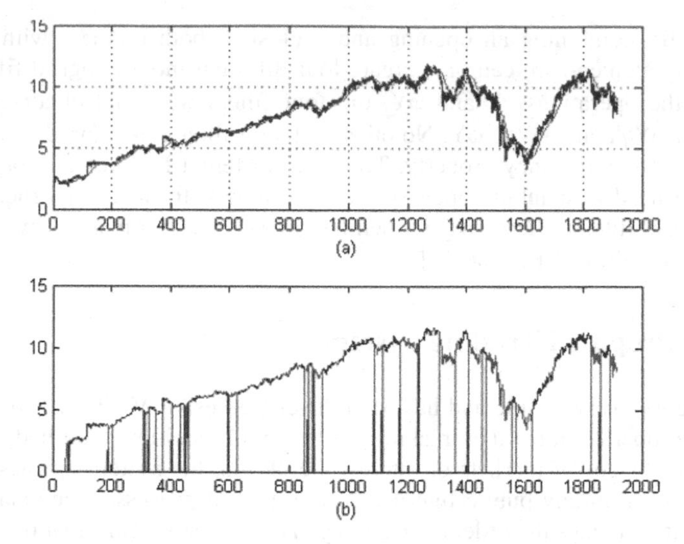

Fig. 2 Santander stock prices representation (1917 days). (a) Opening and closing signals with L=8 (in blue) and the mobile mean with T=34 (in red). (b) System performance.

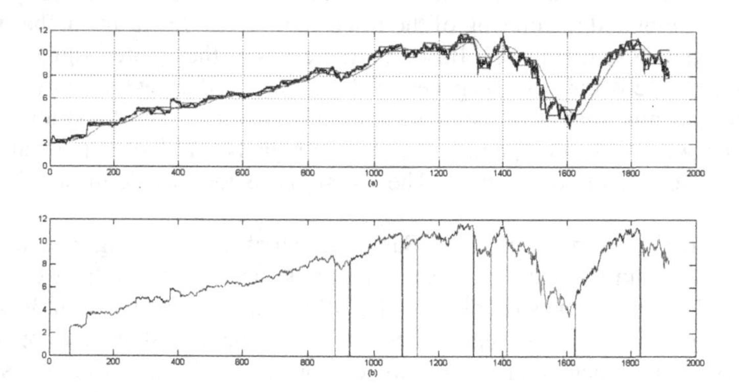

Fig. 3 Santander stock prices representation (1917 days). (a) Opening and closing signals with L=34 (in blue) and the mobile mean with T=55 (in red). (b) System performance.

4 Conclusions

In this paper we have explored the nonlinear mathematical morphological filters in order to trade markets automatically. We have evaluated the system using only a reduced set of financial assets and it works quite well as we can see graphically. We have obtained some preliminary results. More quantitative analysis is required on long historic datasets in order to know some statistical parameters such as the maximum potential losses that it can generate or the maximum number of days that the system can keep in losses. It is desirable that the system can follow the great market movements as well as generates small number of false signals and, if it is the case, that those signals do not incur in large losses. Some optimizations have to be done. New structuring elements and more sophisticated morphological filters must be evaluated. In the context of financial data processing, the mathematical morphology can be explored in a lot of different ways showing that could be a good tool to include among other well-known techniques.

Acknowledgments. This work has been supported by UVIC.

References

1. Timmons, H.: A London Hedge Fund That Opts for Engineers, Not M.B.A.'s (August 18, 2006)
2. Aldridge, I.: High-Frequency Trading: A Practical Guide to Algorithmic Strategies and Trading Systems. Wiley & Sons, Chichester (2009)
3. Serra, J.: Image Analysis and Mathematical Morphology. Academic, New York (1982)

Swarm Intelligence, Scatter Search and Genetic Algorithm to Tackle a Realistic Frequency Assignment Problem

José M. Chaves-González, Miguel A. Vega-Rodríguez,
Juan A. Gómez-Pulido, and Juan M. Sánchez-Pérez

Abstract. This paper describes three different approaches based on complex heuristic searches to deal with a relevant telecommunication problem. Specifically, we have tackled a real-world version of the FAP *–Frequency Assignment Problem* by using three very relevant and efficient metaheuristics. Realistic versions of the FAP are NP-hard problems because the number of available frequencies to cover the entire network communications is always much reduced. On the other hand, it is well known that heuristic algorithms are very appropriate methods when tackling this sort of complex optimization problems. Therefore, we have chosen three different strategies to compare their results. These methods are: a very novel metaheuristic based on swarm intelligence (ABC *–Artificial Bee Colony*) which has not ever been used previously to tackle the FAP; a very efficient Genetic Algorithm (GA) which is a classical and effective algorithm tackling optimization problems; and one of the approaches that provides better results solving our problem: Scatter Search (SS). After a detailed experimental evaluation and comparison with other approaches, we can conclude that all methodologies studied here provide very competitive frequency plans when they work with real-world FAP, although the best results are provided by the SS and the GA strategies.

Keywords: FAP, Frequency Planning, SS, ABC, GA, real-world GSM network.

1 Introduction

The frequency assignment problem (FAP) is one of the most relevant optimization problems in the Telecommunications domain. In fact, it is considered a key task for current, and future, real-world GSM (Global System for Mobile) operators

José M. Chaves-González, Miguel A. Vega-Rodríguez,
Juan A. Gómez-Pulido, and Juan M. Sánchez-Pérez
Dept. Technologies of Computers and Communications. University of Extremadura,
Escuela Politécnica, Cáceres, Spain
e-mail: {jm,mavega,jangomez,sanperez}@unex.es

A.P. de Leon F. de Carvalho et al. (Eds.): Distrib. Computing & Artif. Intell., AISC 79, pp. 441–448.
springerlink.com

because only with an optimum frequency plan, which makes the most of the scarce range of available frequencies, is possible to perform a communication of quality between the cell phones of a realistic network. Moreover, the GSM technology is, even nowadays, the most used mobile communication system around the world. Indeed, by mid 2009 GSM services were in use by around 3.5 billion subscribers [1] across more than 220 countries, representing approximately the 80% of the world cellular market. Therefore, GSM is expected to play an important role as a dominating technology for many years because, it is widely accepted that the third and fourth generations of mobile telecommunication will coexist with the enhanced releases of the GSM standard, at least, in the medium term.

Furthermore, the FAP is an NP-hard problem [2], therefore, using exact algorithms to solve real-sized instances of the problem is not a practical option. On the contrary, metaheuristics [3] are, if not compulsory, the best choice to obtain competitive frequency plans when working with real-world FAPs. For this very reason, we decided to choose three relevant heuristic algorithms: the Artificial Bee Colony (ABC) algorithm [4], which is a very novel approach (never used previously to tackle our problem) based on swarm intelligence; the Genetic Algorithm (GA, [5]) and the Scatter Search (SS) metaheuristic [6], which are very reliable and efficient strategies which have been very successfully used in many works of the literature as optimization solvers.

We have performed complete sets of experiments with all the metaheuristics, and after a detailed statistical study we can conclude that all the methods can obtain very competitive frequency plans which give optimal solutions to real-world instances of the frequency assignment problem.

The rest of the paper is organized as follows: In section 2 we present very briefly the FAP. Sections 3, 4 and 5 describe the algorithms we have used to solve the problem, with the changes performed to adapt them to our realistic version of the FAP. After that, the experiments performed and the results obtained are summarized in Section 6. Finally, we show the main conclusions and future lines of work in Section 7.

2 The Frequency Assignment Problem

The two most relevant components which refer to frequency planning in GSM systems are the *antennas* or, as they are more known, base transceiver stations (BTSs) and the *TRX*. The TRXs of a network are installed in the BTSs where they are grouped in sectors, oriented to different points to cover different areas. The instance we use in our experiments is quite large (it covers the city of Denver, USA, with more than 500,000 inhabitants) and the GSM network includes 2612 TRXs, grouped in 711 sectors, distributed in 334 BTSs. We are not going to extend the explanation of the GSM system here. More information can be found in [7].

FAP lies in the assignment of a channel (or a frequency) to every TRX in the network. The optimization problem arises because the usable radio spectrum is very scarce and frequencies have to be reused for many TRXs in the network (for example, the instance we have used for this study, includes 2612 TRXs and only

18 available frequencies). However, the multiple use of a same frequency may cause interferences that can reduce the quality of service down to unsatisfactory levels. In fact, significant interferences will occur if the same or adjacent channels are used in near overlapping areas [8].

Although there are several ways of quantifying the interferences produced in a telecommunication network, the most extended one is by using what is called the *interference matrix* [9], denoted by M. Each element $M(i,j)$ of this matrix contains two types of interferences: the *co-channel interference*, which represents the degradation of the network quality if the cells i and j operate on the same frequency; and the *adjacent-channel interference*, which occurs when two TRXs operate on adjacent channels (e.g., one TRX operates on channel f and the other on channel $f+1$ or $f-1$). An accurate interference matrix is an essential requirement for frequency planning because the final goal of any frequency assignment algorithm will be to minimize the sum of all the interferences. In addition to the requirements described above, frequency planning includes more complicating factors which occur in real life situations (see [7] for a detailed explanation).

Finally, in the following subsection we give a brief description of the mathematical model we use (for more information, consult references [10], [8]).

2.1 Mathematical Description

We can establish that a solution to the problem is obtained by assigning to each TRX t_i (or u_i)$\in T = \{t_1, t_2,\ldots, t_n\}$ one of the frequencies from $F_i. = \{f_{i1},\ldots, f_{ik}\} \subset N$. We will denote a solution (or frequency plan) by $p \in F_1 \times F_2 \times \ldots \times F_n$, where $p(t_i) \in F_i$ is the frequency assigned to the transceiver t_i. The objective, or the plan solution, will be to find a solution p that minimizes the cost function (C):

$$C(p) = \sum_{t\in T} \sum_{u\in T, u\neq t} C_{sig}(p,t,u) \tag{1}$$

The smaller the value of C is, the lower the interference will be, and thus the better the communication quality. In order to define the function $C_{sig}(p,t,u)$, let s_t and s_u be the sectors (from $S = \{s_1, s_2,\ldots, s_m\}$) in which the transceivers t and u are installed, which are $s_t=s(t)$ and $s_u=s(u)$ respectively. Moreover, let $\mu_{s_ts_u}$ and $\sigma_{s_ts_u}$ be the two elements of the corresponding matrix entry $M(s_t,s_u)$ of the interference matrix with respect to sectors s_t and s_u. Then, $Csig(p,t,u) =$

$$\begin{cases} K & \text{if } s_t = s_u, |p(t)-p(u)| < 2 \\ C_{co}(\mu_{s_ts_u}, \sigma_{s_ts_u}) & \text{if } s_t \neq s_u, \mu_{s_ts_u} > 0, |p(t)-p(u)| = 0 \\ C_{adj}(\mu_{s_ts_u}, \sigma_{s_ts_u}) & \text{if } s_t \neq s_u, \mu_{s_ts_u} > 0, |p(t)-p(u)| = 1 \\ 0 & \text{otherwise} \end{cases} \tag{2}$$

K>>0 is a very large constant defined by the network designer to make undesirable allocating adjacent frequencies to TRXs serving the same area (e.g., placed in the same sector). $C_{co}(\mu,\sigma)$ is the cost due to *co-channel interferences*, whereas $C_{adj}(\mu,\sigma)$ represents the cost in the case of *adjacent-channel interferences* [8].

3 The ABC Algorithm

Artificial Bee Colony (ABC) [4] is one of the most recently defined metaheuristics used to solve optimization problems (created in 2005). ABC is motivated by the intelligent behavior of honey bees. The colony consists of three groups of bees: employed, onlookers and scouts. In ABC heuristic, the position of a food source represents a possible solution to the optimization problem (in our case, a valid frequency plan) and the nectar amount of a food source corresponds to the quality (fitness) of the associated solution. The outline of the algorithm adaptation to deal with our problem is shown in Algorithm 1.

Algorithm 1 – Pseudo-code for ABC

```
1: initialize (population)
2: population ← localSearch (population)
3: while (not time-limit) do
4:     population ← mutationMethod (employedBees, population)
5:     probVector ← generateSolutionProbability (population)
6:     population ← generateSolutions (onlookerBees, probVector, population)
7:     population ← replacePoorerSolutions (scoutBees, population)
8: endwhile
9: return bestIndividual (population)
```

The algorithm starts with the random generation of the population (line 1) and the improvement of the solutions (or frequency plans) within the population using a local search method which improves the quality of the solution previously generated (line 2). Then, each generation will be divided into the following 4 stages: firstly, the employed bees perform a random mutation in which the frequencies of a set of TRX chosen randomly from a solution are reassigned with a valid frequency chosen also randomly. After that, the same local search applied in line 2 is used to improve the mutated solutions. Secondly, a probability vector is generated according to the fitness of each solution. This vector contains the probability that each solution in the population has of being explored by the onlooker bees. The best solutions have more probability of being chosen by this kind of bees. The onlooker bees will generate then new individuals by taking solutions from the population according to the probability vector previously created. The mutation method applied in line 6 to create new frequency plans will take the solutions according to the probability vector created in line 5. Finally, the scout bee replaces the worst solution in the population by another one randomly generated (line 7). At the end of the process, when the time limit arises, the best solution is returned (line 9).

4 The GA Algorithm

The Genetic Algorithm (GA) [5] is probably the metaheuristic which has been most widely used in the bibliography. It provides very good results in a great variety of optimization problems. A brief description of the algorithm is shown in Algorithm 2. As we can see, our GA starts with the random generation of the population (or frequency plans), so that all the TRXs of each individual are randomly assigned with one of their valid frequencies (line 1). After that, a local search method specially adapted to improve the random frequency plans generated is applied over each single solution of the population (line 2).

Algorithm 2 – Pseudo-code for GA

```
1: initialize (population)
2: population ← localSearch (population)
3: while (not time-limit) do
4:     parents ← binaryTournament (population)
5:     offspring ← uniformCrossover (parents)
6:     offspring ← randomMutation (offspring)
7:     offspring ← localSearch (offspring)
8:     population ← updatePopulation (offspring)
9: endwhile
10: return bestIndividual (population)
```

Our GA uses binary tournament as selection scheme (line 4) to choose the parents from which the offspring will be generated. This offspring will contain a certain number of solutions which will be used in the next generation if they improve the worst individuals within the current population. The algorithm applies then uniform crossover to each pair of parents in which every frequency of each TRX from the solution is chosen randomly from one of the two parents (line 5). The mutation operator used is the random mutation in which the frequencies of a set of randomly chosen TRXs of the solution are reassigned with a random valid frequency. After the mutation, the same local search applied in line 2 is used to improve the offspring fitness (line 7). Finally, with the new offspring, the population is updated (line 8) by replacing the worst individuals if the newly generated ones are better (lower FAP cost). This process will be repeated until the stop condition (a time limit in our case) of the algorithm is reached (line 3). Then, the best individual in the population will be returned as final solution (line 10).

5 The SS Algorithm

We have published recently a study in which this metaheuristic has been thoroughly studied and adjusted to solve in the same conditions the same FAP as the proposed here. Please, consult reference [11] to extend the information given here.

6 Experiments and Results

We have performed a complete set of experiments with all the metaheuristics with the aim of optimizing the results obtained by each one when tackling a realistic FAP. For this reason, all tests have been performed using the real world instance described in section 2 (2612 TRX and only 18 available frequencies). Moreover, in order to provide the results with statistical confidence and study the behaviour of the algorithms within short and long periods of time, we have performed 30 independent executions for each experiment taking in consideration the results every

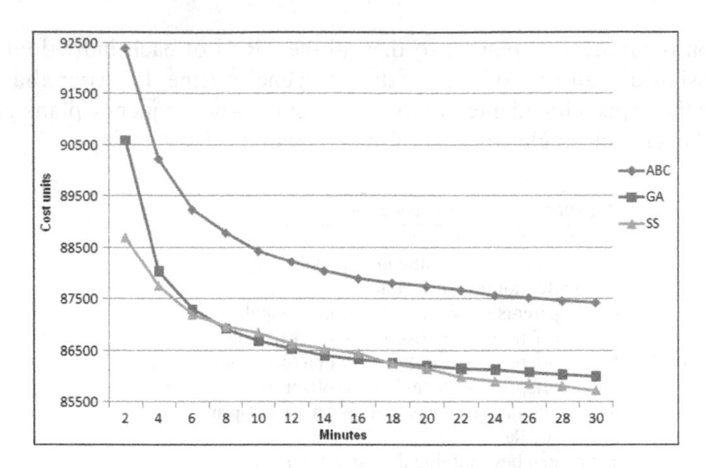

Fig. 1 Results evolution from 2 to 30 minutes for the 3 metaheuristics used

two minutes (as we can see the evolution of both algorithms in Fig. 1). The results are given in function of the cost that a frequency plan is able to obtain (see section 2.1). Thus, the smaller the value of the cost is, the better the frequency plan will be. Moreover, all the experiments have been limited to 30 minutes in order to fairly compare the results with other studies and to make possible to perform a great amount of tests in a reasonable period of time (some of them would take a lot of hours without time limitation).

We will start with the explanation of the parameter adjustment of the GA. We have performed a great number of experiments in order to adjust this algorithm (section 4). We summarize here the most relevant ones, which are relative to the parameter adjustment. Concretely, we have done experiments to fix the population size (10, 20, 30, 40, 50, 100), the parents selection scheme (randomly, best parents, best-random parents and binary tournament), the crossover operations per generation (1, 2, 3, 4, 6 and 8), and the mutation probability (from 0.01 to 0.4). Due to the limitation in the length of this contribution, we are forced to reduce the discussion about the parameter adjustment to the conclusions obtained after the complete statistical study performed. Therefore, the best set of parameter values, which provides the results shown in Fig. 1 and in Table 1 are the following: population size = 30, uniform crossover with 6 crossover operations per generation, binary tournament as parent selection, mutation probability = 0.02.

As occurred with the GA, the number of experiments which were run to adjust the ABC metaheuristics was also very large, but due to the page limit of this paper we are forced to summarize here the most representative experiments performed, which were related to the colony size (10, 20, 30, 40, 50, 100), the proportion of employed/onlookers bees (from 50% each to 10%-90% and 90%-10%), the number of scouts bees (from 0 to 20% of the colony size), and the mutation probability (from 0.01 to 0.8 –here the mutation operator is more important than in the GA, because in the ABC there is not crossover operator). After a complete statistical analysis of the results, we can conclude that the best configuration for the ABC method (section 3) to solve our realistic FAP (section 2) is: colony size = 30,

Table 1 Empirical results (in cost units) for different metaheuristics. The first 3 rows of the table correspond with the results obtained by the methods described in this paper. It is shown the best, average and standard deviation of 30 independent executions.

	120 seconds			600 seconds			1800 seconds		
	Best	Avg.	Std.	Best	Avg.	Std.	Best	Avg.	Std
SS	86169.4	88692.7	1124.9	84570.6	86843.8	950.5	84234.5	85767.6	686.3
GA	88529.7	90586.8	1023.2	85253.6	86680.1	754.2	84994.9	85994.9	542.6
ABC	90361.1	92383.8	1182.1	86463.5	88428.7	912.4	86045.8	87432.5	612.4
ACO	90736.3	93439.5	1318.9	89946.3	92325.4	1092.8	89305.9	90649.9	727.5
DE	92145.8	95414.2	1080.4	89386.4	90587.2	682.3	87845.9	89116.8	563.8
LSHR	88543.0	92061.7	585.3	88031.0	89430.9	704.2	87743.0	88550.3	497.0
GRASP	88857.4	91225.7	1197.2	87368.4	89369.6	1185.1	86908.4	88850.6	1075.2

proportion of employed/onlooker bees = 20%-80%, number of scouts = 1, and mutation probability = 0.2. This parameter setting provides very good evolution in the results (such as we can see in Fig. 1) and nice frequency plans with reduce cost values (Table 1). Results are not the best ones, but they are very competitive, as will be discussed in the next section.

Finally, the SS metaheuristic was also adjusted thoroughly to the FAP with a wide set of experiments. We encourage the interested reader to consult reference [11] to obtain a complete explanation about the study performed with SS algorithm when it is applied to solve the same real version of the FAP tackled here.

7 Conclusions and Future Work

In this work we have studied three different meta-heuristics (ABC, GA and SS) to solve a real-world version of the frequency assignment problem. All the experiments have been performed using data from a real GSM network composed of 2612 transceivers, only 18 available frequencies and the typical requirements and constrains of these kind of networks (section 2). The ABC algorithm was chosen because of its novelty and because it represented a different approach for solving our problem (based on swarm intelligence). On the other hand, we selected two effective metaheuristics (GA and SS) to compare the results with. In any case, after a complete parameter adjustment we can conclude that all the strategies give very good results. However, it is true that GA and SS algorithms obtain frequency plans which have more quality than the ABC metaheuristic. In fact, SS is the sequential strategy which obtains the best results published so far. Table 1 summarizes the results obtained by our approaches and by other relevant methods that tackle the same problem as ours. As we can see, our results are very good, in fact, they beat other recent results obtained with other very different approaches (Ant Colony Optimization –ACO, Differential Evolution –DE, Local Search with Heuristic Restarts –LSHR, and Greedy randomized adaptive search procedure –GRASP) using the same problem instance and measurements as ours [10-14]. Future work includes the evaluation of the algorithms using additional real-world instances and the use of parallel approaches to improve the quality of frequency plans obtained here.

Acknowledgments. This work has been partially funded by the Spanish Ministry of Education and Science and ERDF (the European Regional Development Fund), under contract TIN2008-06491-C04-04 (the M* project).

References

1. GSM World (2009), http://www.gsmworld.com/news/statistics/index.shtml
2. Hale, W.K.: Frequency assignment: Theory and applications. Proceedings of the IEEE 68(12), 1497–1514 (1980)
3. Blum, C., Roli, A.: Metaheuristics in Combinatorial Optimization: Overview and Conceptual Comparison. ACM Computing Surveys 35, 268–308 (2003)
4. Karaboga, D., Basturk, B.: A powerful and efficient algorithm for numerical function optimization: artificial bee colony (ABC) algorithm. Journal of Global Optimization 39, 459–471 (2007)
5. Goldberg, D.E.: Genetic Algorithms in Search, Optimization, and Machine Learning. Addison-Wesley Longman Publishing, Amsterdam (1989)
6. Glover, F., Laguna, M., Martí, R.: Scatter search. In: Advances in Evolutionary Computing: Theory and Applications, pp. 519–537. Springer, Heidelberg (2003)
7. Eisenblätter, A.: Frequency Assignment in GSM Networks: Models, Heuristics, and Lower Bounds. PhD thesis, Technische Universität Berlin (2001)
8. Mishra, A.R.: Fundamentals of Cellular Network Planning and Optimisation: 2G/2.5G/3G... Evolution to 4G, pp. 21–54. Wiley, Chichester (2004)
9. Kuurne, A.M.J.: On GSM mobile measurement based interference matrix generation. In: 55th Vehicular Technology Conference, VTC Spring 2002, pp. 1965–1969 (2002)
10. Luna, F., Blum, C., et al.: ACO vs EAs for Solving a Real-World Frequency Assignment Problem in GSM Networks. In: GECCO 2007, London, UK, pp. 94–101 (2007)
11. Chaves-González, J.M., Vega-Rodríguez, M.A., et al.: Solving a Real–World FAP Using the Scatter Search Metaheuristic. In: Moreno-Díaz, R., Pichler, F., Quesada-Arencibia, A. (eds.) Computer Aided Systems Theory - EUROCAST 2009. LNCS, vol. 5717, pp. 785–792. Springer, Heidelberg (2009)
12. da Silva Maximiano, M., et al.: A Hybrid Differential Evolution Algorithm to Solve a Real-World Frequency Assignment Problem. In: International Multiconference on Computer Science and Information Technology, Wisła, Poland, pp. 201–205 (2008)
13. Luna, F., Estébanez, C., et al.: Metaheuristics for solving a real-world frequency assignment problem in GSM networks. In: GECCO 2008, Atlanta, GE, USA, pp. 1579–1586 (2008)
14. Chaves-González, J.M., Vega-Rodríguez, M.A., et al.: Solving a Realistic FAP Using GRASP and Grid Computing. In: Advances in Grid and Pervasive Computing. LNCS, vol. 5529, pp. 79–90. Springer, Heidelberg (2009)

Multi-Join Query Optimization Using the Bees Algorithm

Mohammad Alamery, Ahmad Faraahi, H. Haj Seyyed Javadi,
Sadegh Nourossana, and Hossein Erfani

Abstract. Multi-join query optimization is an important technique for designing and implementing database management system. It is a crucial factor that affects the capability of database. This paper proposes a Bees algorithm that simulates the foraging behavior of honey bee swarm to solve Multi-join query optimization problem. The performance of the Bees algorithm and Ant Colony Optimization algorithm are compared with respect to computational time and the simulation result indicates that Bees algorithm is more effective and efficient.

Keywords: Bees algorithm, Database Management System, Multi-join, optimization.

1 Introduction

One of difficulties in relational database management system (RDBMS) which has been solved faultily is multi-join query optimization (MJQO). In traditional applications of RDBMS, the number of join N involved by a single query is relatively small, Usually, N<10. With the expansion of the database application areas, the traditional query optimization technology cannot support some of the latest database applications. Such as, applications of decision support system (DSS), OLAP and data mining (DM), which may produce a query including more than 100 joins. In this condition, the shortfall of the traditional query optimization technology is exposed gradually. Therefore, it is necessary to explore new technology to solve MJQO problem.

Mohammad Alamery and Ahmad Faraahi
Department of Information Technology, Payame Noor University, Tehran, Iran

H. Haj Seyyed Javadi
Department of Mathematics and Computer Science, Shahed University, Tehran, Iran

Sadegh Nourossana and Hossein Erfani
Computer Engineering Department, Science and Research Branch,
Islamic Azad University, Tehran, Iran

A.P. de Leon F. de Carvalho et al. (Eds.): Distrib. Computing & Artif. Intell., AISC 79, pp. 449–457.
springerlink.com

MJQO is an NP hard problem [1]. With the increase of join number, the number of query execution plan (QEP) corresponding to a query grows exponentially, which lead to computational complexity of MJQO problem is very large. Recently, solving the problem with heuristic algorithm becomes a hotspot. Such as, ACO [1], Greedy Algorithm [2], GA [3], AB [4], etc. Several approaches have been proposed to model the specific intelligent behaviors of honey bee swarms and applied for solving combinatorial type problems [5–9]. In this paper, Bees algorithm was adopted to solve the problem MJQO.

2 Description for Multi-Join Query Optimization Problem

The process of RDBMS managing user query is as follows: After receiving query submitted by users, query parser checks syntax, verifies relations, translates the query into its internal form. It is usually translated into relational algebra expression, which can be denoted as query syntax tree. A relational algebra expression may have many equivalent expressions, so it also corresponds to many equivalent query syntax trees.

Then, query optimizer selects appropriate physical method to implement each relational algebra operation and finally generate query execution plan (QEP). The QEP consists of the order in which the operations in a query are to be processed, and the physical method to be used to process each operation. Amongst all equivalents QEP, query optimizer chooses the one with lowest cost output to the query-execution engine, then, the query-execution engine takes the QEP, executes that plan, and returns the answers to user. The process is depicted in Fig 1. This paper is to study how to make query optimizer select a QEP with lower cost in shorter time.

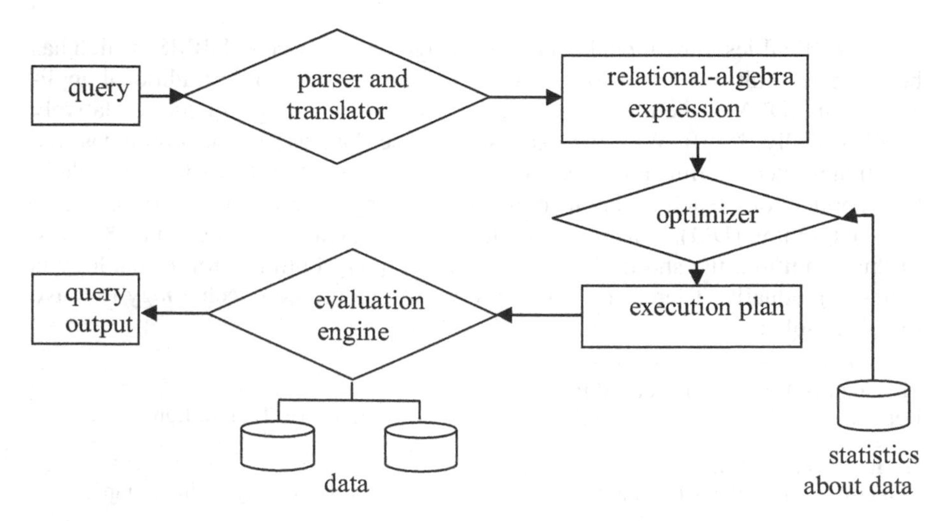

Fig. 1 Process of query execution

After being implemented optimizing operation of pushing down the select operation and project operation, one query that include project, select and join operations will transform to relational algebra expression constituted by N join operations which can be denoted as join tree. An example multiple query Q include A, B, C, D, E five relations, which can be denoted as three kinds of join tree shown in Fig 2: (a) left-deep tree (b) bushy tree (c) right-deep tree. The leaf nodes are relations constituting query Q and the internal nodes express join operation and intermediate results. Executive order is bottom-up execution. The different order of N join and the different physical methods selected to implement join operation lead to the cost of join trees have great differences. Assume that each join operations are implemented by the same physical method; Multi-join query optimization problem is simplified as setting a good join order, making the join tree has the lowest cost. Hence, tree in the left linear space can take full advantage of the index, and often contain the best strategy or the strategy whose cost is similar to the best strategy at least, therefore, consider left linear space as a search space.

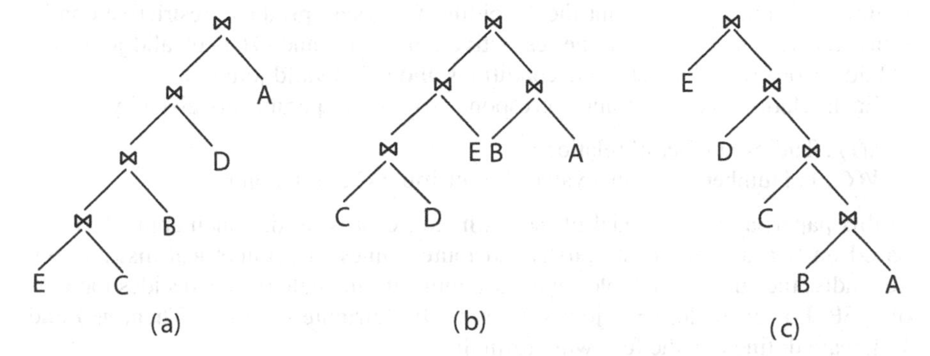

Fig. 2 Three Kinds of join tree

In order to reduce the search space furthermore, avoiding the emergence of Cartesian product often is considered as constraint of the issue. An example multiple join query Q include A, B, C, D, E five relations. The attributes associating between five relations which are founded from statistical information of the database catalog, could be denoted as a query graph G= (V, E), shown as Figure 3. Nodes in query graph are relations and an edge connecting two relations, indicates attributes associating between two relations.

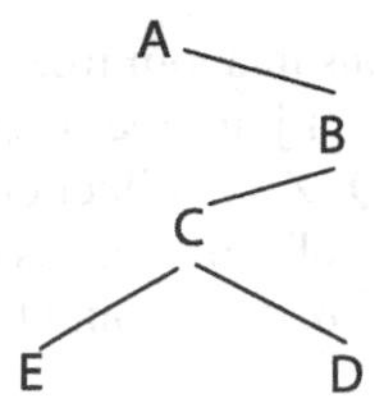

Fig. 3 Query graph

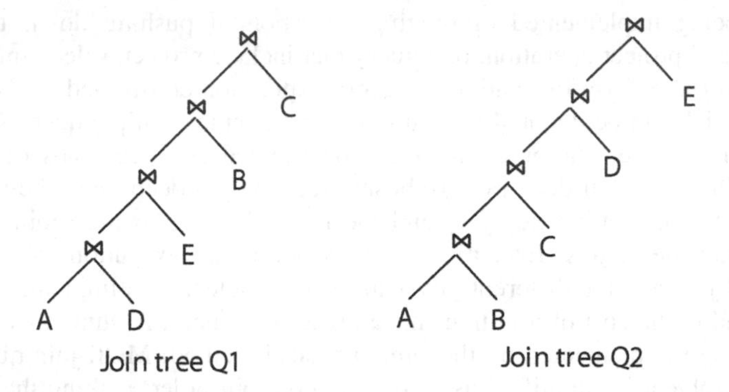

Fig. 4 Two join trees

Q1 and Q2 are two join trees in the left linear space of query Q, depicted in Figure 4. Taking into account the "avoiding Cartesian product" restrictive conditions, Q1 do not accord with the restrictive conditions and Q1 is invalid join tree; Q2 do accord with the restrictive conditions and Q2 is valid join tree.

Each relation in query graph corresponds to a set of parameters given by:

n(r) : Tuples number of relation r
V(C,r) : Number of distinct values for attribute C in relation r

In this paper, a simple model of the estimated cost is used, which applied in [1], based on two assumptions: Firstly, attribute values in symmetrical distribution. Secondly, the sum of the tuples number about intermediate results decides the cost of QEP. For example, t = r join s, C is public attribute over r, s. Then, *n(t)* and *V(A,t)* are defined by the following formulas:

$$n(t) = \frac{n(r) \times n(s)}{\prod_{C_i \in C} \max(V(C_i, r), V(C_i, s))} \tag{2.1}$$

$$V(A,t) = \begin{cases} V(A,r) & A \in r - s \\ V(A,s) & A \in s - r \\ \min(V(A,r), V(A,s)) & A \in r, A \in s \end{cases} \tag{2.2}$$

Assume that there are N relations in a join tree; the cost of QEP is the sum of the tuples number of internal nodes t_i in join tree. $n(t_i)$is number of tuples about intermediate result t_i . For a query Q, Z is collection of all the possible QEP corresponding to Q. Each member z in collection Z has query execution cost --- *Cost*(*z*) , then, Z_0 meeting $Cost(Z_0) \approx \min_{z \in Z} Cost(z)$ should be found.

3 Bees Algorithm for MJQO Problem

3.1 Bees Algorithm

The foraging bees are classified into three categories; employed bees, onlookers and scout bees [10]. All bees that are currently exploiting a food source are known as employed. The employed bees exploit the food source and they carry the information about food source back to the hive and share this information with onlooker bees. Onlookers bees are waiting in the hive for the information to be shared by the employed bees about their discovered food sources and scouts bees will always be searching for new food sources near the hive. Employed bees share information about food sources by dancing in the designated dance area inside the hive. The nature of dance is proportional to the nectar content of food source just exploited by the dancing bee. Onlooker bees watch the dance and choose a food source according to the probability proportional to the quality of that food source. Therefore, good food sources attract more onlooker bees compared to bad ones. Whenever a food source is exploited fully, all the employed bees associated with it abandon the food source, and become scout. Scout bees can be visualized as performing the job of exploration, whereas employed and onlooker bees can be visualized as performing the job of exploitation. In the Bees algorithm [11], each food source is a possible solution for the problem under consideration and the nectar amount of a food source represents the quality of the solution represented by the fitness value. The number of food sources is same as the number of employed bees and there is exactly one employed bee for every food source. This algorithm starts by associating all employed bees with randomly generated food sources (solution). In each iteration, every employed bee determines a food source in the neighborhood of its current food source and evaluates its nectar amount (fitness). The ith food source position is represented as $X_i = (x_{i1}, x_{i2}, \ldots, x_{id})$. $F(X_i)$ refers to the nectar amount of the food source located at X_i. After watching the dancing of employed bees, an onlooker bee goes to the region of food source at X_i by the probability p_i defined as

$$p_i = \frac{F(X_i)}{\sum_{k=1}^{S} F(X_k)} \tag{3.1}$$

where S is total number of food sources. The onlooker finds a neighborhood food source in the vicinity of X_i by using

$$X_i(t+1) = X_i(t) + \delta_{ij} * u \tag{3.2}$$

where δ_{ij} is the neighborhood patch size for j^{th} dimension of ith food source defined as

$$\delta_{ij} = x_{ij} - x_{kj} \tag{3.3}$$

where k is a random number $\in(1, 2, \ldots; S)$ and $k \neq i$, u is random uniform variate $\in[-1, 1]$. If its new fitness value is better than the best fitness value achieved so far, then the bee moves to this new food source abandoning the old one, otherwise it remains in it sold food source. When all employed bees have finished this process, they share the fitness information with the onlookers, each of which selects a food source according to probability given in Eq.(3.1). With this scheme, good food sources will get more onlookers than the bad ones. Each bee will search for better food source around neighborhood patch for a certain number of cycle(limit), and if the fitness value will not improve then that bee becomes scout bee.

3.2 Pseudo Code for Bees Algorithm

1: Initialize
2: REPEAT.
3: Move the employed bees onto their food source and evaluate the fitness
4: Move the onlookers onto the food source and evaluate their fitness
5: Move the scouts for searching new food source
6: Memorize the best food source found so far
7: UNTIL (termination criteria satisfied)

3.3 Foraging (Neighborhood Search)

Each preferred path which a bee takes is a complete QEP which passes contains all relations. So each relation is connected to other relations who are called nearby neighborhoods of this relation.

The decision of each bee for changing these nearby neighborhoods results in the invention of new QEP which are considered as neighborhood QEP of the preferred path.

In the suggestive model of the neighborhood search, each bee tries to follow its own preferred path with the probability ω, and with the probability $(1-\omega)$ tries to make better paths by changing the nearby neighborhoods of its preferred path relation. The value of ω is calculated by Eq.(3.4).

$$\omega = \frac{Problem\ \ Size - Search\ \ range}{Problem\ \ Size} \tag{3.4}$$

Where *Problem size* is the number of all relations of the problem, and *Search range* is a positive parameter which identifies the extension of the neighborhood searching area.

This way, each bee begins to make a new QEP. It will be randomly located in a relation and selects the next relation by following the below rules:

(a) When a bee has decided to follow its preferred path, and none of the nearby neighborhoods have been visited. In this case it will choose one of them randomly and moves to it.
(b) When a bee has decided to follow its preferred path, but there is only one nearby neighborhood unvisited. So it will move to this unvisited relation.
(c) When a bee has decided to follow its preferred path, but all of the nearby neighborhoods have been already visited. In this case the bee will select the next relation based on the probability function (3.5).

$$I(i,j)=\begin{cases}0 & j\in l\\ \dfrac{[m(i,j)][1/\eta(i,j)]^{\beta}}{\sum_{s=1,s\notin l}^{n}[m(i,j)][1/\eta(i,s)]\beta} & j\notin l\end{cases} \tag{3.5}$$

Where $I(i, j)$ is the probability with which the bee moves from relation i to j, $h(i, j)$ the distance between i and j relation, b positive parameter, whose values determine the relative importance of memory versus heuristic information, n the number of relations, and l a list of all the visited relations so far.

(d) When a bee has decided not to follow its preferred path and choose a new nearby neighborhood, in this case it will do the same as in rule c.

4 Experimental Results

In order to illustrate the effect of Bees-MJQO in solving this problem, experiments have been implemented on computer with Pentium4 2.93G + 1024

Table 1 Algorithmic parameters.

Bees	No. of bees = 16 No. of iterations= (No. of Relation)^2
ACO	No. of Ants = 10 α=1, β=3, q0=0.2, ρ = 0.9

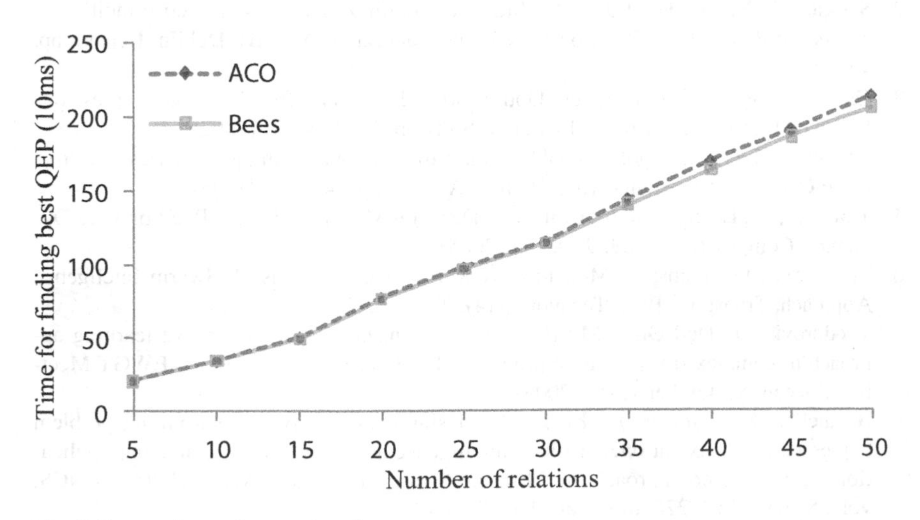

Fig. 5 Comparison of execution time

RAM + Windows XP Pro. A database including 50 relations that have attributes association with each other has been used as test data. ACO [1] and Bees are used to solve this problem respectively.

As is shown in Fig. 5, number of relations corresponding to query Q is taken as X-axis and time for generating optimal solution—query execution plan is taken as Y-axis. The simulation results show that Bees finds optimum solutions more effectively in time than ACO. The figure indicates ACO algorithm spends more time than Bees algorithms on finding optimal solution especially with the incensement of relation number.

5 Conclusions

MJQO problem is hotspot in database research field. A good optimization algorithm not only can improve the efficiency of queries but also reduce query execution costs. In this paper, the Bees algorithm, which is a new, simple and robust optimization algorithm, was proposed to solve the problem of MJQO. The performance of the proposed algorithm is compared with the ACO algorithm. The results reveal that Bees algorithms converge faster compared to ACO algorithm for this problem.

The simulation results show that Bees Algorithm finds optimum solutions more effectively in time than ACO especially with the incensement of relation number.

References

1. Li, N., Liu, Y., Dong, Y., Gu, J.: Application of Ant Colony Optimization Algorithm to Multi Join Query Optimization. Springer, Heidelberg (2008)
2. Shekita, E., Young, H., Tan, K.L.: Multi-join optimization for sym-metric multiprocessors. In: Proc. Of the Conf. on Very Large Data Bases (VLDB), Dublin, Ireland, pp. 479–492 (1993)
3. Cao, Y., Fang, Q.: Parallel Query Optimization Techniques for Multi-Join Expressions Based on Genetic Algorithms. Journal of Software 13, 250–256 (2002)
4. Swami, A., Iyer, B.: A polynomial time algorithm for optimizing join queries. In: Proc. IEEE Conf. on Data Engineering, Vienna, Austria, pp. 345–354 (1993)
5. Tereshko, V., Loengarov, A.: Collective Decision-Making in Honey Bee Foraging Dynamics. Comput. Inf. Sys. J. 9(3), 1–7 (2005)
6. Teodorovi , D.: Transport Modeling By Multi-Agent Systems: A Swarm Intellgence Approach. Transport. Plan. Technol. 26(4), 289–312 (2003)
7. Teodorovi , D., Dell'Orco, M.: Bee colony optimization—a cooperative learning approach to complex transportation problems. In: Proceedings of the 10th EWGT Meeting, Poznan, September 13-16 (2005)
8. Benatchba, K., Admane, L., Koudil, M.: Using bees to solve a data-mining problem expressed as a max-sat one, artificial intelligence and knowledge engineering applications: a bioinspired approach. In: Mira, J., Álvarez, J.R. (eds.) IWINAC 2005. LNCS, vol. 3562, pp. 212–220. Springer, Heidelberg (2005)

9. Wedde, H.F., Farooq, M., Zhang, Y.: Bee Hive: an efficient fault-tolerant routing algorithm inspired by honey bee behavior, ant colony, optimization and swarm intelligence. In: Proceedings of the 4th International Workshop, ANTS 2004 (2004)
10. Sabat, S.L., et al.: Artificial bee colony algorithm for small signal model parameter extraction of MESFET. Engineering Applications of Artificial Intelligence (2010)
11. Karaboga, D., Basturk, B.: On the performance of artificial bee colony (ABC) algorithm. Appl. Soft Comput. 8(3), 687–697 (2008)

Towards an Adaptive Integration Trigger

Vicente García-Díaz, B. Cristina Pelayo G-Bustelo,
Oscar Sanjuán-Martínez, and Juan Manuel Cueva Lovelle

Abstract. Continuous integration in software development is a practice recommended by the most important development methodologies. It promises many advantages such as early detection of bugs. An important element of continuous integration, although largely forgotten by the scientific literature, is the trigger, which initiates the process of building software from development sources. This paper discusses the possibility of improving this software component and opens the way for research that could be applied to other computer-related fields. To this end, we have implemented a prototype that shows for a case study, the results obtained when using existing triggers.

Keywords: Continuous Integration, Adaptive, Trigger, Optimize.

1 Introduction

Continuous Integration (CI) is a practice recommended by many software methodologies. So, it was picked as one of the 12 original practices of the Extreme Programming (XP) [1] and it is part of the recommendations of the Unified Process (UP) [2]. There are many advantages of its use, among which can be highlighted: risk reduction, bugs deletion, more accurate estimates or lower costs. Works such as Fleischer [3] reveal the importance as well as the improvement achieved by those who use it. Moreover, CI serves as a member of the development team which is responsible for monitoring the source code, compiling each change, testing construction and notifying the responsible any problems that occurred during the process [4].

The construction of artifacts step begins when one of the triggers, configured with the CI tool, determines it under a certain condition. However, the motivation

Vicente García-Díaz, B. Cristina Pelayo G-Bustelo, Oscar Sanjuán-Martínez,
and Juan Manuel Cueva Lovelle
University of Oviedo, Department of Computer Science, Sciences Building,
C/ Calvo Sotelo s/n 33007, Oviedo, Asturias, Spain
e-mail: garciavicente@uniovi.es, crispelayo@uniovi.es,
osanjuan@uniovi.es, cueva@uniovi.es

A.P. de Leon F. de Carvalho et al. (Eds.): Distrib. Computing & Artif. Intell., AISC 79, pp. 459–462.
springerlink.com

for this work is the lack of automatic triggers to reduce the workload of developers and at the same time to optimize the software integration.

The aim of this paper is to show the need for a trigger that minimizes the number of times that launches the construction of software without any changes in the software repository while minimizing the time that software in the repositories is not built. In fulfilling this objective, we achieve reducing the load on source control systems caused by the polling for modifications performed by other triggers and integrating software frequently, that is one of the practices recommended by Fowler [5] to reduce the risk as much as possible in software development.

There are several features that the proposed adaptive software should has: 1- machine learning, without requiring training; 2- evolution of knowledge throughout the entire process, i.e. what the algorithm knows as true may become false and vice versa; 3- ability to know if it makes a mistake, making a waste of resources, which are limited.

The remainder of this paper is structured as follows: In Section 2 we presented a brief related state of the art. In Section 3 we discussed the case study. Finally, in Section 4 we indicated our conclusions and future work to be done.

2 State of the Art

It should be noted that there are still few scientific studies related to CI such as Holck and Jorgensen [6]. Therefore, works focused on the triggers of CI tools are practically nonexistent. But even so, there are a variety of triggers that do not require manual intervention to be executed. Some of the best known are:

- Dependency trigger. It is used to run each build after another on which it depends.
- Interval trigger. It is used to run a build each a specified time interval after the last cycle of integration.
- Multiple trigger. It is used to simultaneously run multiple builds.
- Schedule trigger. It is used to specify days of the week and time in which it will run the build phase.
- Startup trigger. It is used to run the build whenever the CI tool starts.
- Url trigger. It is used to run the build whenever a Url changes. To this end, a check is made each time interval. Thus avoiding any increase in load, which on version control systems, can cause interval triggers.

3 Case Study

To show the problem to be solved, a study has been carried out for a month in a course project carried out by 4 people who work Monday through Friday from 8:30 to 15:30 (they also work Tuesday and Wednesday from 18:00 to 20:00). Keep in mind that during this month there are 4 weekends and a holiday Wednesday. To obtain the results shown below, we have built a prototype used to simulate the process. The study reveals that the development team has made 60 changes to the

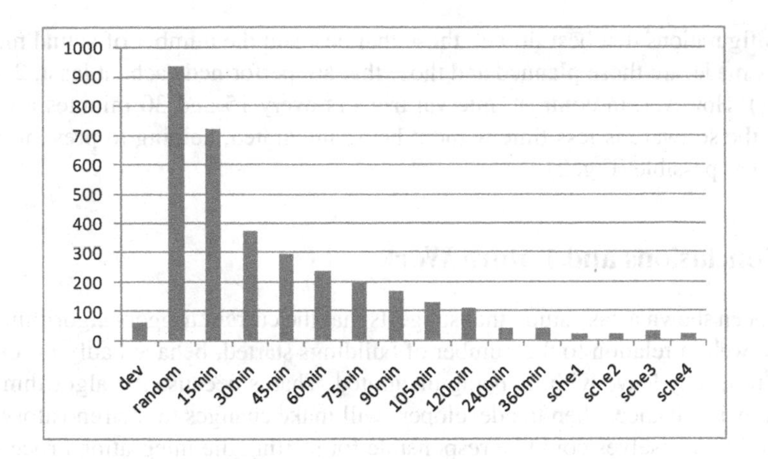

Fig. 1 Number of integrations started with each configuration

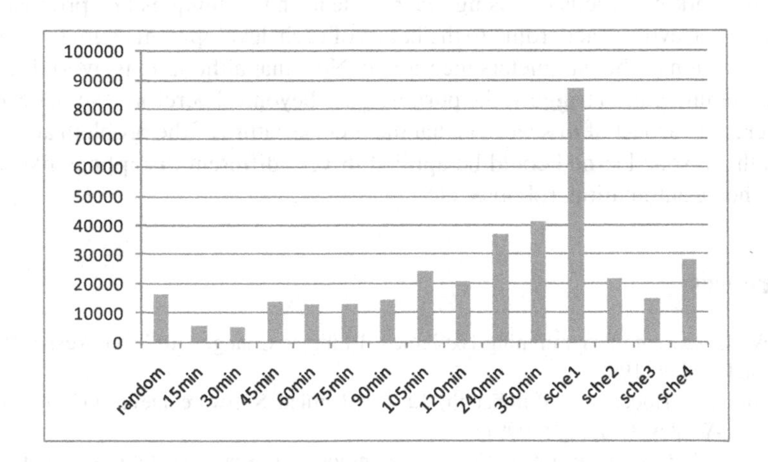

Fig. 2 Time without performing an integration with each configuration

repository governed by the version control system. Thus, a hypothetical optimal and automatic trigger would only have to make 60 attempts to build software sources, doing it instantly, after changes made by developers. That way, it would save CPU cycles and software would be integrated as early as possible.

Using currently available technologies in continuous integration tools, we have used interval triggers with a filter that allows work only in business hours (each 5, 15, 30, 45, 60, 75, 90, 105, 120, 240, 360 minutes). We have done the same with schedule triggers used in working hours (at 8:30, 15:30, 15:30, 20:00 hours)[1].

[1] Some people have worked overtime for several days, making changes in the repository after the time when the trigger runs.

Configurations that best fit, i.e., those that best suit the number of actual modifications made, are those planned and those that are performed each, at least, 2 hours (Fig. 1). However, in contrast, interval triggers every 15 and 30 minutes are with which the software is less time without being integrated, helping to prevent errors as soon as possible (Fig. 2).

4 Conclusions and Future Work

It has been shown a case study that suggests that the current triggers algorithms that behave well in relation to the number of buildings started, behave badly in relation to the time they leave without being integrated. That is because the algorithms can not know in advance when the developers will make changes to the repository. The developers themselves could be responsible for starting the integration process but this would require a manual or an ad-hoc adaptation of each version control system, which is not desired.

Future work will focus on, using machine learning techniques [7], proposing an algorithm that evolves according to the habits of each development team, looking for an optimization of the parameters mentioned. Note that although this work focus on continuous integration triggers, the purpose goes beyond. Therefore, we believe that the interesting aspect of this work is that the precise nature of the problem addressed makes the lessons learned could be applied in very different disciplines like video games, home automation, robotics, etc.

References

1. Beck, K.: Extreme Programming Explained: Embrace Change. Addison-Wesley Professional, Reading (1999)
2. Jacobson, I., Booch, G., Rumbaugh, J.: The Unified Software Development Process. Addison-Wesley, Reading (1999)
3. Fleischer, G.: Continuous integration. what componies expect and solutions provide. Tech. rep., Fontys University of Applied Sciencies (2009)
4. Richardson, J., Gwaltney, W.: Ship it! A Practical Guide to Successful Software Projects, Pragmatic Bookshelf (2005)
5. Fowler, M.: Continuous integration. Web (2009), `http://martinfowler.com/articles/continuousIntegration.html`
6. Holck, J., Jorgensen, N.: Australian Journal of Information Systems, vol. 11(1), p. 40 (2004)
7. Michalski, R.S., Carbonell, J.G., Mitchell, T.M. (eds.): Machine Learning: An Artificial Intelligence Approach, pp. 3–23. Springer, Heidelberg (1984)

Multi-Objective Evolutionary Algorithms Used in Greenhouse Planning for Recycling Biomass into Energy

A.L. Márquez, C. Gil, F. Manzano-Agugliaro, F.G. Montoya,
A. Fernández, and R. Baños

Abstract. Advanced parallel Multi-Objective Evolutionary Algorithms (MOEA) have been used in order to solve a wide array of problems, including the planning of greenhouse crops. This paper shows the application of MOEA using the Island Parallel Model to solve a problem involving greenhouse crop planning in order to maximize profits and the production of biomass while reducing economic risks. The interest in maximizing biomass waste lies in the possibility of recycling it into heat and energy.

Keywords: multi-objective, optimization, island parallel model, greenhouse, crop planning.

1 Introduction

The planning of greenhouse crops constitutes an interesting problem involving several objectives is the planning of greenhouse crop surfaces. The greenhouse-covered area in southeast Spain is mainly used to grow vegetables, such as tomato, pepper, melon... This production system has undergone intensive development and is therefore highly profitable [8], but at the same time it produces a negative impact on the area, due to the residues generated and waste byproducts such as plant biomass,

A.L. Márquez, C. Gil, R. Baños, and A. Fernández
Dpt. of Computer Architecture and Electronics
University of Almería
e-mail: almarquez@ual.es, cgil@ual.es,
rbanos@ual.es, afernandezmolina@gmail.com

F. Manzano-Agugliaro and F.G. Montoya
Dpt. of Rural Engineering
University of Almería
e-mail: fmanzano@ual.es, pagilm@ual.es

A.P. de Leon F. de Carvalho et al. (Eds.): Distrib. Computing & Artif. Intell., AISC 79, pp. 463–470.
springerlink.com

which needs to be removed at the end of its life cycle. Leaving the biomass untreated is a risk factor that increases not only the likelihood of developing diseases for the crops, but also the contamination of surface waters and aquifers as it decomposes.

One desirable way to get rid of these waste byproducts is the gasification of the organic matter, a partial oxidation of the biomass at high temperatures, which provides heat and/or electricity. Its main advantages are that it needs a lower amount of biomass to work, it produces less contamination, and it can be easily connected to electrical networks.

While recycling biomass into energy is very interesting, it is important to not forget the real interests of greenhouse owners, namely increasing profits (Gross Margin) while reducing market risks, which could lead to reduce profit or even economical losses.

This paper aims to maximize benefits by means of careful planning and reducing market risks, while also selecting crops that produce a higher amount of biomass to transform into energy.

2 Concepts in Multi-Objective Optimization

The use of Multi-Objective Optimization as a tool to solve Multi-Objective Problems (MOP) implies explaining some key concepts that are of invaluable importance. Without them it would be inaccurate to describe what a good approximation to the Pareto Front is, in terms of criteria such as closeness to the Pareto set, diversity, etc [5, 3, 13].

Multi-Objective Optimization is the exploration of one or more decision variables belonging to the function space, which simultaneously satisfy all constraints to optimize an objective function vector that maps the decision variables to two or more objectives.

$$minimize/maximize(f_k(s)), \forall k \in [1,K] \quad (1)$$

Each decision vector $s=\{(s_1, s_2, .., s_m)\}$ represents accurate numerical qualities for a MOP. The set of all decision vectors constitutes the *decision space*. The set of decision vectors that simultaneously satisfies all the constraints is called *feasible set (F)*. The objective function vector (f) maps the decision vectors from the decision space into a K-dimensional objective space $Z \in \Re^K$, $z=f(s)$, $f(s)=\{f_1(s), f_2(s),..., f_K(s)\}$, $z \in Z$, $s \in F$.

In order to be able to compare the solutions of a given MOP with $K \geq 2$ objectives, instead of giving a scalar value to each solution, a partial order is defined according to Pareto-dominance relations, as detailed below.

Order relation between decision vectors: Let s and s' be two decision vectors. The dominance relations in a minimization problem are:

$$\begin{cases} s \; dominates \; s' \; (s \prec s') \quad iff \\ \quad f_k(s) < f_k(s') \wedge f'_k(s) \not> f'_k(s'), \; \forall k' \neq k \in [1,K] \end{cases} \tag{2}$$

$$\begin{cases} s, \; s' \; are \; incomparable \; (s \sim s') \quad iff \\ \quad f_k(s) < f_k(s') \wedge f'_k(s) > f'_k(s'), \; k' \neq k \in [1,K] \end{cases} \tag{3}$$

Pareto-optimal solution: A solution s is called *Pareto-optimal* if there is no other $s' \in F$, such that $f(s') < f(s)$. All the Pareto-optimal solutions define the *Pareto-optimal set*, also called *Pareto Front.*

Non-dominated solution: A solution $s \in F$ is *non-dominated* with respect to a set $S' \in F$ if and only if $\nexists s' \in S'$, verifying that $s' \prec s$.

To summarize these definitions, figure 1 shows the Pareto-dominance concept for a MOP with two objectives (maximizing f_1 and minimizing f_2). The filled circles represent non-dominated solutions, while the non-filled ones are dominated or indifferent solutions. Figure 1(a) shows the location of several solutions in regard to the Pareto Front, while figure 1(b) shows the relative distribution of the solutions in reference to a certain solution s. There exist solutions that are *worse* (in both objectives) than s, *better* (in both objectives) than s, and *incomparable* (better in one objective and worse in the other).

Obtaining a wide and evenly distributed Pareto Front is also of key importance because such a set of solutions is more useful for the decision making process.

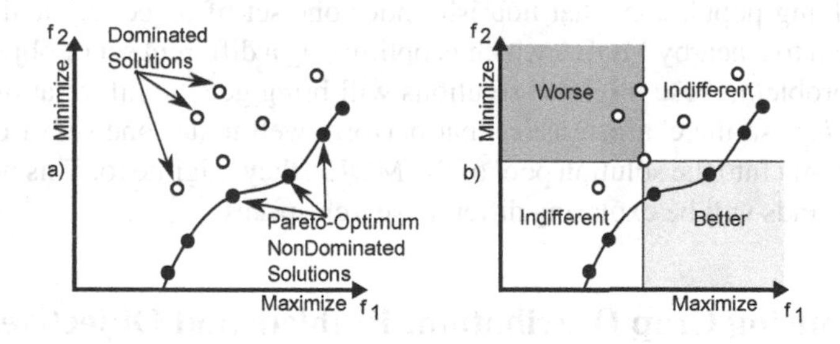

Fig. 1 Pareto-dominance relations in a two-objective problem

3 Multi-Objective Evolutionary Algorithms

The following MOEAs have been used to perform the experiments needed to gather the data used in this paper:

- *NSGA-II*, Non-dominated Sorting Genetic Algorithm II [6]
- *SPEA2*, Strength Pareto Evolutionary Algorithm [15]
- *PESA*, Pareto Envelope-based Selection Algorithm [4]
- *msPESA*, Mixed Spreading PESA [7]

4 Island Parallel Model

Parallelization of single/multi-objective optimization algorithms has long been of interest to researchers [12, 1]. The Island Parallel Model is highly effective in parallelizing the problem in order to obtain better results without a significant penalization in run time.

Island paradigm parallel MOEAs are based on the phenomenon of natural populations evolving in relative isolation, such as those that might occur within some ocean island chain with limited migration. The population is divided into a few subpopulations or islands, and each one evolves separately on different processors. On each island the population is free to converge towards a different optimum. The exchange of information between islands is possible via a migration operation. The island parallel model has often been reported to display better search performance [14], in terms of solution quality, than serial population models.

The topology is often based on logical or physical geometric structures such as rings, meshes, etc [12]. The chosen implementation of the geometry for the experiments performed on this paper is a ring, where 5 solutions of the N_{th} island travel to the $(N-1)_{th}$ island, while the other 5 migrate to the $(N+1)_{th}$ island.

The migration operation allows that otherwise unrelated and isolated MOEAs can share their most promising subjects with other MOEAs in their whereabouts. This is usually done after a certain number of generations or subject evaluations has been performed. The aim of this operation is to avoid one or more of the independent MOEAs getting stuck on local minima.

By evolving populations that flourish under one set of objectives and then migrating them to a nearby MOEA which is optimizing a different set of objectives for the same problem, these migrated solutions will bring genetic information that has been *evolving*, so the characteristics that perform well under one set of objectives will be brought into the solution pool of the MOEA they migrate to. This means that different islands will be exploring different search spaces.

5 Optimizing Crop Distribution: Problem and Objectives

The province of Almería (south-eastern Spain) is home to approximately 30,000ha of intensive crops, with an estimated production of $3 \cdot 10^9$ kg of produce at an approximate value of $1,384 \cdot 10^6$€ Pepper (Pe), tomato (T), green beans (GB), cucumber (Cu), courgette (Co), watermelon (W) and melon (M) account for 80% of total produce [8, 9]. The distribution of greenhouse crop surfaces changes every year, with the seven main crop varieties mentioned above combined in sixteen vegetable crop alternatives that are the object of this study.

The data fed to the solver has been obtained from an accountancy tracking of 46 and 49 greenhouses in two recent years. With this information the Gross Margin (GM) for each crop option and each year can be obtained. Therefore, the problem to solve has sixteen variables, represented as an array of floating point values, each one of them corresponding to the surface for every one of the sixteen crop alternatives[10]. There are three main objectives to optimize:

Maximizing Profit: To do this, the gross margin (GM) is maximized for the various options proposed:

$$GM = \sum_{i=1}^{n} (GM_i \cdot X_i) \tag{4}$$

Where GM_i is the gross margin of option i per surface area unit (€$/m^2$) and X_i is the surface area that the crop alternative covers.

Minimizing Risk: To calculate market risks, the variance and covariance matrix is used for the gross margins of the different crop options, based on market data [10].

$$R = \overline{X_i}[cov]\overline{X_i} \tag{5}$$

Maximizing Biomass production: The total biomass (Bm) is the volume of organic matter produced by the individual average biomass waste Bm_i that each one of the crops generates per surface area (m^3/m^2).

$$Bm = \sum_{i=1}^{n} (Bm_i \cdot X_i) \tag{6}$$

There are also two constraints to be taken into account:

- The total surface area is limited to 2.5ha, which is the average greenhouse surface, so in the simulation the sixteen crop alternatives would spread over it.

$$\sum_{i=1}^{n} X_i = 25,000m^2 \tag{7}$$

- The maximum surface area for a certain crop should never be higher than 40% of the total surface, because it would flood the market, leading to a major drop in prices.

6 Evolutionary Operations

A floating point numerical representation has been chosen for each of the sixteen variables, because it is the most natural representation for this problem.

Mutation. Changes up to ±25% of the initial value of every problem variable.
Crossover. The procedures use a multipoint chromosome crossover.
Chromosome repair procedure. It normalizes the total surface area represented by each of the variables to a fixed area of 2.5ha by calculating the total of each of the seven crops, to check that none accounts for over 40% of the total.

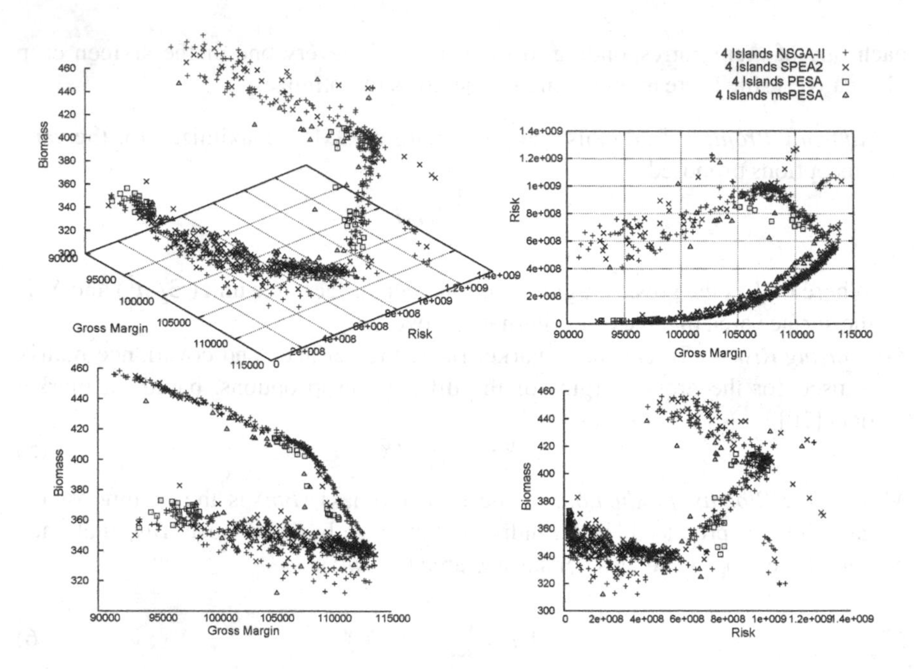

Fig. 2 Three simultaneous objectives; Gross Margin, Risk and Biomass.

7 Experimental Results

To determine which Pareto surface is the best one, the following metric has been used to compare their quality:

Coverage of two sets (C):[16] Let X, X' be two solution sets. Function C maps the sorted pair X, X' to the interval $[0,1]$ (see equation 8). The value $C(X,X') = 1$ means that all the points (solutions) in X' are *dominated* or *indifferent* to the points of X. The opposite value, $C(X,X') = 0$ means that no point in X' is covered by any other point in set X. It is important to note that both $C(X,X')$ and $C(X',X)$ have to be considered because they are not complementary values, nor are they necessarily equal. For instance, if X dominates X' then $C(X,X') = 1$ and $C(X',X) = 0$, but if X is indifferent to X', and X' is also indifferent to X, then the Coverage would be $C(X,X') = 1$ and $C(X',X) = 1$.

$$C(X,X') := \frac{|a' \in X'; \exists a \in X : a \prec a'|}{|X'|} \tag{8}$$

These are the chosen experimental parameters:

Number of Evaluations. 30,000 evaluations on each island.
Migration Rate. Migrate solutions between islands each 1000 evaluations.

Table 1 Coverage Metric for the experiments performed (Algorithm, number of islands)

	msPESA, 8	**msPESA, 16**	**NSGAII, 8**	**NSGAII, 16**	**PESA, 8**	**PESA, 16**	**SPEA2, 8**	**SPEA2, 16**
msPESA, 8		0.03	0.00	0.00	0.63	0.51	0.00	0.01
msPESA, 16	0.52		0.00	0.00	0.72	0.61	0.00	0.01
NSGAII, 8	0.54	0.52		0.08	0.60	0.60	0.14	0.16
NSGAII, 16	0.55	0.52	0.17		0.62	0.61	0.13	0.18
PESA, 8	0.28	0.09	0.00	0.00		0.04	0.01	0.14
PESA, 16	0.44	0.21	0.01	0.00	0.27		0.03	0.19
SPEA2, 8	0.48	0.46	0.25	0.17	0.55	0.54		0.16
SPEA2, 16	0.55	0.52	0.28	0.20	0.63	0.61	0.18	

Number of Islands. 4, 8 and 16 Islands, the odd islands optimizing Risk and GM, while even islands run Biomass production and GM as target objectives.

Crossover/Mutation probabilities. There is a 90% probability for the crossover operation, while the mutation operation has a probability of 10%.

In figure 2, the economic risks are never 0 the graphs only show that the risk is much smaller [10] compared to the high-risk crop configurations.

8 Conclusions

Table 1 shows that there are improvements in coverage as the number of islands increases. This shows that the algorithms have been able to find a good approximation to the Pareto Front of this problem. Due to the high amount of indifferent solutions, the experiments do not show a clear winner between the different options, though SPEA2 and NSGA-II show a better coverage.

With the data obtained from the simulations, and as figure 2 shows, the crop configurations that show a greater increase in biomass production are also the ones with higher economic risk, while the gross margin they produce is also reduced in comparison with other configurations that produce a lower amount of biomass.

The present paper present a novel use of MOEAs to solve the crop planning problem, which constitutes a step forward since weighted goal programming using utility functions has been clasically used as a methodology for the analysis and simulation of agricultural systems [11, 2], usually ignoring the multi-objective nature of the problem.

Acknowledgements. This work has been financed by the Spanish Ministry of Innovation and Science (TIN2008-01117), Ministry of Science and Technology (AGL2002-04251-C03-03) and the Excellence Project of Junta de Andalucía (P07-TIC02988), in part financed by the European Regional Development Fund (ERDF).

References

1. Alba, E., Tomassini, M.: Parallelism and evolutionary algorithms. IEEE Transactions on Evolutionary Computation 6(5), 443–462 (2002)
2. Amador, F., Sumpsi, J.M., Romero, C.: A non-interactive methodology to asses farmers utility functions: An application to large farms in andalusia, spain. European Review of Agricultural Economics 25, 92–109 (1998)
3. Coello, C., van Veldhuizen, D.A., Lamont, G.: Evolutionary Algorithms for solving Multi-Objective Problems. Kluwer Academic Publishers, Dordrecht (2002)
4. Corne, D.W., Knowles, J.D., Oates, M.J.: The pareto envelope-based selection algorithm for multiobjective optimization. In: Proceedings of the Parallel Problem Solving from Nature VI Conference, pp. 839–848. Springer, Heidelberg (2000)
5. Deb, K.: Multi-Objective Optimization using Evolutionary Algorithms. John Wiley & Sons, Chichester (2002)
6. Deb, K., Agrawa, S., Pratap, A., Meyarivan, T.: A fast elitist non-dominated sorting genetic algoritm for multiobjective optimization: Nsga-ii. In: Deb, K., Rudolph, G., Lutton, E., Merelo, J.J., Schoenauer, M., Schwefel, H.-P., Yao, X. (eds.) PPSN 2000. LNCS, vol. 1917, pp. 849–858. Springer, Heidelberg (2000)
7. Gil, C., Márquez, A., Baños, R., Montoya, M.G., Gómez, J.: A hybrid method for solving multi-objective global optimization problems. Journal of Global Optimization 38(2), 265–281 (2007)
8. IEC: Análisis de la campaña hortofrutícola de Almería, campaña 2007-2008 (greenhouse crop analisys in Almeria for the 2007/2008 season) (2009), `http://fundacioncajamar.es/instituto.htm`
9. Manzano Agugliaro, F.: Gasification of greenhouse residues for obtaining electrical energy in the South of Spain: Localization by gis. Interciencia 32(2), 131–136 (2007)
10. Márquez, A., Manzano-Agugliaro, A., Gil, C., Cañero-León, R., Montoya, F., Baños, R.: Multiobjective evolutionary optimization of greenhouse vegetable crop distributions. In: INSTICC (ed.) Proceedings of International Joint Conference on Computational Intelligence 2009 (2009)
11. Sumpsi, J.M., Amador, F., Romero, C.: On farmers' objectives: A multi-criteria approach. European Journal of Operation Research 96, 64–71 (1997)
12. Veldhuizen, D., Zydallys, J., Lamont, G.: Considerations in engineering parallel multiobjective evolutionary algorithms. IEEE Transactions on Evolutionary Computation 7(2), 144–173 (2003)
13. Voorneveld, M.: Characterization of pareto dominance. Operations Research Letters 31(1), 7–11 (2003)
14. Whitley, D., Rana, S., Heckendorn, R.B.: The island model genetic algorithm: On separability, population size and convergence. Journal of Computing and Information Technology 7, 33–47 (1998)
15. Zitzler, E., Laumanns, M., Thiele, L.: Spea2: Improving the strength pareto evolutionary algorithm. Tech. rep. (2001)
16. Zitzler, E., Thiele, L.: Multiobjective evolutionary algorithms: a comparative case study and the strength pareto approach. IEEE Trans. Evolutionary Computation 3(4), 257–271 (1999)

Parallel Hyperheuristics for the Antenna Positioning Problem

Carlos Segura, Yanira González, Gara Miranda, and Coromoto León

Abstract. Antenna Positioning Problem (APP) is an NP-Complete Optimisation Problem which arises in the telecommunication field. It consists in identifying the infrastructures required to establish a wireless network. Several objectives must be considered when tackling APP and multi-objective evolutionary algorithms have been successfully applied to solve it. However, they required a deep analysis, and a correct parameterisation in order to obtain high quality solutions. In this work, a parallel hyperheuristic island-based model approach is presented. Several hyper-heuristic scoring strategies are tested. Results show the advantages of the parallel hyperheuristic. On one hand, the testing of each sequential configuration can be avoided. On the other hand, it speeds up the attainment of high-quality solutions even when compared with the best sequential approaches.

Keywords: Parallel Hyperheuristics, Antenna Positioning Problem, Multi-Objective Evolutionary Algorithms.

1 Introduction

The Antenna Positioning Problem (APP) is one of the main problems which arises in the engineering of mobile telecommunication networks [8]. APP solves the positioning of Base Stations (BS) or antennas on potential sites, in order to fulfil some objectives and constraints. It plays a major role in various engineering, industrial, and scientific applications because its outcome usually affects cost, profit, and other business performance metrics. APP is also referred in the literature as Radio Network Design (RND) and Base Station Transmitters Location Problem (BST-L). It has been shown to be an NP-complete problem [6].

Carlos Segura, Yanira González, Gara Miranda, and Coromoto León
Dpto. Estadística, I. O. y Computación. Universidad de La Laguna
La Laguna, 38271, Santa Cruz de Tenerife, Spain
e-mail: csegura@ull.es, ygonzale@ull.es, gmiranda@ull.es, cleon@ull.es

A.P. de Leon F. de Carvalho et al. (Eds.): Distrib. Computing & Artif. Intell., AISC 79, pp. 471–479.
springerlink.com

Several objectives can be considered when tackling APP. Most typical considered objectives are: minimise the number of antennas, maximise the quality of service, and/or maximise the covered area. Most of the proposals for APP in the literature are mono-objective [7]. In such cases, several objectives are integrated into a fitness function. In [11] some objectives are translated into constraints. In [9, 10] several objectives are considered simultaneously and multi-objective strategies are applied. The main advantage of multi-objective approaches is the improvement in the diversity of the obtained solutions. In this paper we address a multi-objective definition of APP [9].

Many strategies have been applied to the mono-objective and multi-objective APP. In [11] ad-hoc heuristics were used. In [12] ad-hoc heuristics were integrated into a Tabu Search based algorithm. The aforementioned strategies use problem-dependent information, so it is difficult to adapt them to other definitions of the problem. Metaheuristics can be considered as high-level strategies that guide a set of simpler heuristic techniques in the search of an optimum. They are more general than ad-hoc heuristics. In [6] several metaheuristics were applied to the mono-objective canonical formulation of APP and they were extensively compared. In [1] the same definition of APP is solved by means of genetic algorithms. APP has also been solved by incorporating problem-dependent mutation operators [14] inside an evolutionary approach. In [9] several *Multi-Objective Evolutionary Algorithms* (MOEAs) and operators were successfully applied to the here considered APP formulation. MOEAs were able to achieve high quality solutions. However, two main-drawbacks were identified. On one hand, comparisons with the best mono-objective approaches show that there exists some room for improvement. On the other hand, it was necessary to perform a deep analysis of the evolutionary operators and its parameterisations, in order to obtain high quality solutions.

Several studies have been performed in order to reduce the execution time and the resource expenditure when using evolutionary approaches. These studies naturally lead to its parallelisation.Several models of Parallel Evolutionary Algorithms (PEAs) have been designed. PEAs can be classified [2] in four major computational paradigms: master-slave, island-based, diffusion or cellular, and hybrid paradigm. These evolutionary approaches are proved effectively solving problems, but they are often time and domain knowledge intensive. The heavy dependence on problem specific knowledge affects their reusability. In order to provide a reusable and robust approach, applicable to a wide range of problems and instances, authors proposed hyperheuristics island based-models [5]. This model combines the operation of an island-based scheme with the hyperheuristic approach to manage the choice of which lower-level metaheuristics should be applied at any given time, depending upon the characteristics of the algorithm, problem, and instance itself.

The proposal presented here lies on the application of the parallel hyperheuristic island-based model using the algorithms proposed in [9] as low-level metaheuristics. Our present work has three main aims:

- Avoid the requeriment of manually testing several evolutionary approaches.
- Improve the quality of the solutions achieved for a real-world instance.
- Analyse the benefits and drawbacks of several hyperheuristic scoring strategies.

The remaining content is structured in the following way: the mathematical formation of the multi-objective APP is given in Section 2. Section 3 is devoted to describe the hyperheurstic island-based model and the applied low-level metaheuristics. The computational study is presented in section 4. Finally, the conclusions and some lines of future work are given in section 5.

2 Mathematical Formulation

APP is defined as the problem of identifying the infrastructures required to establish a wireless network into a geographical area G. The geographical area is discretised into a finite number of locations. Tam_x and Tam_y are the number of vertical and horizontal considered subdivisions, respectively. They are selected by communications experts, depending on several characteristics of the region and transmitters.

The considered mathematical formulation [9] derives from the mono-objective one proposed in [1, 14]. Specifically, it comprises the maximisation of the coverage of a given geographical area while minimising the BS deployment. Thus, it is an intrinsically multiobjective problem. In our definition of APP, BS can only be located in a set of potential locations. U is the set of locations where BS can be deployed: $U = \{(x_1,y_1),(x_2,y_2),...,(x_n,y_n)\}$. Location i is referred using the notation $U[i]$. The x and y coordinates of location i are named $U[i]_x$ and $U[i]_y$, respectively. The region irradiated by a BS is called a cell. When a BS is located in position i its corresponding cell ($C[i]$) is covered. In our definition we use the canonical APP problem formulation, i.e., an isotropic radiating model is considered. The set P determines the locations covered by a BS: $P = \{(\Delta x_1,\Delta y_1),(\Delta x_2,\Delta y_2),...,(\Delta x_m,\Delta y_m)\}$. Thus, when BS i is deployed, the covered locations are given by the next set: $C[i] = \{(U[i]_x + \Delta x_1, U[i]_y + \Delta y_1),(U[i]_x + \Delta x_2, U[i]_y + \Delta y_2),...,(U[i]_x + \Delta x_m, U[i]_y + \Delta y_m)\}$.

Being $B = [b_0,b_1,...,b_n]$ the binary vector which determines the deployed BS, APP is defined as the MOP given by the next two objectives:

$Minimize\ f_1 = \sum_{i=0}^{n} b_i$

$Maximize\ f_2 = \sum_{i=0}^{Tam_x} \sum_{j=0}^{Tam_y} covered(i,j)$

where:

$$covered(x,y) = \begin{cases} 1 \text{ If } \exists\ i/\{(b_i = 1) \wedge ((x,y) \in C[i])\} \\ 0 \text{ Otherwise} \end{cases}$$

3 Hyperheuristics Island-Based Model

Island-based models divide the population into a number of independent subpopulations. Each subpopulation is associated to an island and a heuristic configuration is executed over it. A *configuration* is constituted by an optimisation algorithm and its parameterisation. Usually, each available processor constitutes an island which evolves in isolation for the majority of the parallel run. Occasionally, some solutions can be transferred among islands following a migration scheme. Usually, it is difficult to know a priori which configurations are suitable to solve a problem. Therefore, a previous analysis must be performed in order to successfully apply island-based schemes. Hyperheuristics have been integrated with island-based models in order to avoid such a step [5]. The hyperheuristic manages the choice of which lower-level configuration is executed on each island at each optimisation stage with the aim of granting more resources to the most promising configurations.

The hybrid model is constituted by a set of *slave islands* that evolve in isolation. A tunable migration scheme allows the exchange of solutions among neighbour islands. Moreover, a new special island is introduced into the scheme. That island is in charge of applying the hyperheuristic principles. In the proposed model local stop criteria are fixed for the island executions. When an island local stop criterion is reached, such an island is stopped. Based on the achieved results,a score is assigned to the corresponding configuration. Then, a selection strategy is applied in order to select the next configuration that should be executed on the idle island.

Two different hyperheuristics, following the same scheme, have been considered. First, a scoring strategy is applied in order to evaluate the suitability of each configuration. Then, a probability-based selection scheme [13] is applied. The parameter β represents the minimum selection probability of each configuration. Thus, being n_h the number of low level heuristics, a random selection is performed in $\beta * n_h$ percentage of the cases. In the first hyperheuristic, *HH_Imp*, the score assigned to each configuration is an estimation of the hypervolume improvement that each configuration can achieve, when breaking from the current solution. In order to perform the estimation the previous hypervolume improvements are used. They are calculated when a configuration reaches its stop criterion. Improvements obtained during the migration stage of the algorithm are discarded. Considering a configuration c, which has been executed j times, the score is calculated as a weighted average of the last k improvements. The weighted average assigns greater importance to the last executions. The score - $s(c)$ - and selection probability - $p(c)$ - are given by:

$$s(c) = \frac{\sum_{i=1}^{k} i * imp[c][j-i]}{\sum_{i=1}^{k} i} \qquad p(c) = \beta + (1 - \beta * n_h) * \left[\frac{s(c)}{\sum_{i=0}^{n_h} s(i)} \right]$$

The second hyperheuristic, *HH_Syn*, tries to detect synergies between pairs of configurations. The hyperheuristic estimates how well a (meta)-heuristic operates in parallel with another (meta)-heuristic. *HH_Syn* assigns two different scores to each configuration. The first one, is called the visibility - *vis* - and represents the independent performance of each configuration. It is calculated as s in *HH_Imp*. The second one, is called the cooperation between pairs (c_p) and represents the performance of a (meta)-heuristic in the presence of other metaheuristics. The improvements achieved along the execution by a metaheuristic m_1, in the presence of $m2$, is referred as $imp[m_1][m_2]$. Given two metaheuristics m_1 and m_2, which have been executed in parallel j times, the cooperation $c_p(m_1, m_2)$ is calculated as a weighted average of the last k hypervolume improvements achieved by m_1, in the presence of m_2. Given a metaheuristic m_1 and the set of currently assigned metaheuristics

$m\_set = \{h_1, h_2, ..., h_n\}$, the cooperation $c_s(m_1)$ is calculated as the maximum c_p of any of its components, i.e., $c_s(m_1) = max\{c_p(m_1, h_1), c_p(m_1, h_2), ..., c_p(m_n, h_n)\}$.

When every island gets idle, the hyperheuristic must grant the resources among the available configurations. The first assignment is performed as in *HH_Imp*, but substituting the score s by the visibility *vis*. For the remaining assignments c_s is used with a probability γ, and *vis* is used with a probability $(1 - \beta * n_h - \gamma)$.

As has been stated in the previous paragraphs the model requires a set of low-level approaches. Several MOEAs and evolutionary operators for APP were analysed in [9]. Although any of the MOEAs were suitable to obtain high quality solutions, they require a correct parameterisation in order to attain them in reasonable times. The set of low level approaches managed by the hyperheuristics derives from the ones presented in it. Since the aim is to avoid the detailed analysis of each APP approach, the low-level algorithms were selected in a blind-manner, i.e., the comparison among algorithms performed in [9] was not considered.

The MOEA selected for performing the analysis is SPEA2 [15]. Tentative solutions are represented as binary strings with n elements. Each gene determines whether the corresponding BS is deployed. The mutation operator applied was a random operator: Bit-Inversion Mutation(BIM). Each gene is inverted with a probability p_m. Two different crossover operators - one random, and one directed - are applied. The crossover operators are the following: One-Point Crossover (OPC) [4], and a Geographic Crossover(GC) [10] which exchanges the BS that are located within a given radius, r, around a randomly chosen BS.

4 Experimental Evaluation

In this section the experiments performed with the optimisation schemes described in the paper are presented. The analysis is focused in detecting the advantages achieved by the incorporation of the hyperheuristics principles inside the parallel model. Parallel executions have been performed using 4 slave islands. Tests have been run on a Debian GNU/Linux cluster of 8 HP nodes, each one consisting of two Intel(R) Xeon(TM) at 3.20GHz and 1Gb RAM. A real world-world-sized problem instance [3] was used. It is defined by the geographical layout of the city of Malaga.

Since stochastic algorithms are applied, each execution was repeated 30 times. In order to provide the results with confidence, comparisons have been performed following the next statistical analysis. First, a Kolmogorov-Smirnov test is performed in order to check whether the values follow a normal distribution or not. If so, the Levene test checks for the homogeneity of the variances. If samples have equal variance, an ANOVA test is done; otherwise a Welch test is performed. For non-gaussian distributions, the non-parametric Kruskal-Wallis test is used. A confidence level of 95% is considered. The analysis is performed using the hypervolume metric [16].

First experiment performs a comparison of the hypervolume obtained by a set of sequential configurations and by the proposed parallel approaches. The configurations were constituted by combining different parameterisations of mutation and crossover operators. The BIM operator was used with $p_m = \{0.001, 0.002, 0.004\}$.

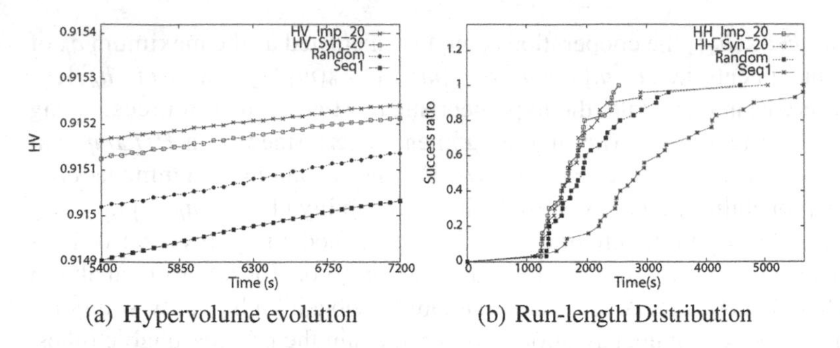

(a) Hypervolume evolution (b) Run-length Distribution

Fig. 1 Comparison of parallel and sequential approaches

The set of crossover operators was the following: OPC with $p_c = \{0.5, 1\}$, and GC with $(p_c, r) = \{(0.5, 15), (0.5, 30), (0.5, 45), (0.5, 60), (1, 15), (1, 30), (1, 45), (1, 60)\}$. The population and archive sizes were fixed to 100. Each considered scheme was executed fixing a stop criterion of 2 hours. Considering the obtained results, sequential algorithms were ordered based on the mean hypervolume achieved at the end of the executions. An index based on such an order is assigned to each configuration, being "Seq1" the one obtaining the highest hypervolume.

The parallel model was executed using both defined hyperheuristics, with the 24 described configurations. Two different configurations were defined for each hyperheuristic. In every hyperheuristic β was fixed to 0.2. In *HH_Imp_20* k was fixed to 20, while in *HH_Imp_40*, it was fixed to 40. In the *HH_Syn* hyperheuristic schemes γ was fixed to 40, i.e. the visibility and cooperation scores were used with the same probability. In *HH_Syn_20* k was fixed to 20, while in *HH_Syn_40* it was fixed to 40. Also, a random parallel scheme was used. In such a scheme, every configuration is scored with the same value. For the parallel executions the local stop criteria was fixed to 1 minute. Migration was performed following an asynchronous scheme with a migration probability of 1, but it only takes place when new non-dominated individuals have been generated. The topology consisted in an all to all connected structure. Migrated individuals are selected following an elitist scheme. Replacements were performed also following an elitist scheme. In order to preserve a good diversity the subpopulation sizes were fixed to 100.

Fig. 1-(a) shows the hypervolume evolution for "Seq1", "random", and for the best and worst hyperheuristic parallel models. It shows that every parallel model is better than the best sequential configuration. Therefore, the parallel models are useful, not only to avoid the testing of each sequential configuration, but also, to speed up the attainment of high-quality solutions. The worst and best hyperheuristics achieve a similar quality, showing the robustness of the approach. Moreover, its results clearly improve the ones achieved by the "random" model. A statistical comparison was performed among the models. It shows that the parallel models are better than the sequential configuration. Also, they show that the parallel models which make use of hyperheuristics, are better than the random scheme. However, the analysis shows that differences among hyperheuristics are not significant.

Table 1 Speedup of parallel models versus some selected sequential configurations

	HH_Imp_20 Sp.	HH_Imp_40 Sp.	HH_Syn_20 Sp.	HH_Syn_40 Sp.	Random Sp.
Seq1	1.71	1.60	1.69	1.71	1.49
Seq5	5.15	4.81	5.09	5.12	4.49

The previous experiment has compared the schemes, mainly focused in terms of the achieved quality. However, since the parallel executions use more computational resources than the sequential ones, the improvement must be quantified. In order to measure the improvement of the parallel approach, the second experiment analyses the run-time behaviour of the models. Each model was executed using as finalisation condition the achievement of a fixed level of hypervolume. Fig. 1-(b) shows, the run length distribution - success ratio vs. time - for the best and worst hyperheuristic-based model, for the random parallel model and for the best behaved sequential configuration. It shows the similarity among the hyperheuristics, and the superiority of such models when compared with the best sequential configuration and with the random scheme. Table 1 shows the speedup of the parallel models versus a set of selected sequential configurations. It has been calculated considering the median of each set of executions. Although linear speedup is not achieved when comparing with the best configuration, it must be taken into account that when solving a problem, the best configuration is not known a priori, so, the time saving is much greater than the speedup calculated versus the best configuration. In fact, the speedup highly increases when comparing to other configurations.

5 Conclusions and Future Work

This paper presents a set of approaches used to deal with APP. Previously, MOEAS had been applied to the here tackled APP. However, in order to attain high quality solutions it was necessary to perform a deep analysis of the evolutionary operators and its parameterisations. In order to avoid it, a parallel approach, based on hybridising hyperheuristics and island-based models is applied. The proposal adds an adaptive property to the island model by applying the operation principles of hyperheuristics. The model combines a set of low-level metaheuristics in an intelligent way, granting more computational resources to the most promising configurations. Two different scoring methods - based on the hypervolume metric - are tested. *HH_Imp* estimates the hypervolume improvement that each configuration can achieve, when breaking from the currently achieved solutions. *HH_Syn* tries to detect synergies between pairs of configurations. Results demonstrate the validity of the scheme. The parallel approach is useful, not only to avoid the testing of each sequential configuration, but also, to speed up the attainment of high-quality solutions. The analysis of the scoring approaches shows that they are very similar. Therefore, at least for APP, the added complexity of *HH_Syn* is not well-grounded.

In order to improve the quality of the results, several modifications can be performed. On one hand, more problem-dependent information can be included into the strategy. For instance, by incorporating a local search scheme, or by applying some directed mutation operators. On the other hand, it would be very useful to perform an scalability analysis of the model, when applied to APP.

Acknowledgements. This work was partially supported by the EC (FEDER) and the Spanish Ministry of Science and Innovation as part of the 'Plan Nacional de I+D+i', with contract number TIN2008-06491-C04-02 and by Canary Government project number PI2007/015. The work of Carlos Segura was funded by grant FPU-AP2008-03213.

References

1. Alba, E.: Evolutionary algorithms for optimal placement of antennae in radio network design. In: Parallel and Distributed Processing Symposium, International, vol. 7, p. 168 (2004), http://doi.ieeecomputersociety.org/10.1109/IPDPS.2004.1303166
2. Cantú-Paz, E.: A survey of parallel genetic algorithms. Calculateurs Paralleles 10 (1998)
3. Gómez-Pulido, J.: Web site of net-centric optimization, http://oplink.unex.es/rnd
4. Holland, J.H.: Adaptation in natural and artificial systems. MIT Press, Cambridge (1992)
5. León, C., Miranda, G., Segura, C.: Hyperheuristics for a Dynamic-Mapped Multi-Objective Island-Based Model. In: Omatu, S., Rocha, M.P., Bravo, J., Fernández, F., Corchado, E., Bustillo, A., Corchado, J.M. (eds.) IWANN 2009. LNCS, vol. 5518, pp. 41–49. Springer, Heidelberg (2009)
6. Mendes, S.P., Molina, G., Vega-Rodríguez, M.A., Gomez-Pulido, J.A., Sáez, Y., Miranda, G., Segura, C., Alba, E., Isasi, P., León, C., Sánchez-Pérez, J.M.: Benchmarking a Wide Spectrum of Meta-Heuristic Techniques for the Radio Network Design Problem. IEEE Transactions on Evolutionary Computation, 1133–1150 (2009)
7. Mendes, S.P., Pulido, J.A.G., Rodriguez, M.A.V., Simon, M.D.J., Perez, J.M.S.: A differential evolution based algorithm to optimize the radio network design problem. In: E-SCIENCE 2006: Proceedings of the Second IEEE International Conference on e-Science and Grid Computing, p. 119. IEEE Computer Society, Washington (2006), http://dx.doi.org/10.1109/E-SCIENCE.2006.3
8. Meunier, H., Talbi, E.G., Reininger, P.: A multiobjective genetic algorithm for radio network optimization. In: Proceedings of the 2000 Congress on Evolutionary Computation, pp. 317–324. IEEE Press, Los Alamitos (2000)
9. Segura, C., González, Y., Miranda, G., León, C.: A Multi-Objective Evolutionary Approach for the Antenna Positioning Problem. In: 14th International Conference on Knowledge-Based and Intelligent Information & Engineering Systems. LNCS (LNAI), Springer, Heidelberg (to appear, 2010)
10. Talbi, E.G., Meunier, H.: Hierarchical parallel approach for gsm mobile network design. J. Parallel Distrib. Comput. 66(2), 274–290 (2006)
11. Tcha, D.w., Myung, Y.S., Kwon, J.h.: Base station location in a cellular CDMA system. Telecommunication Systems 14(1-4), 163–173 (2000)
12. Vasquez, M., Hao, J.K.: A heuristic approach for antenna positioning in cellular networks. Journal of Heuristics 7(5), 443–472 (2001), http://dx.doi.org/10.1023/A:1011373828276
13. Vinkó, T., Izzo, D.: Learning the best combination of solvers in a distributed global optimization environment. In: Proceedings of Advances in Global Optimization: Methods and Applications (AGO), Mykonos, Greece, pp. 13–17 (2007)

14. Weicker, N., Szabo, G., Weicker, K., Widmayer, P.: Evolutionary multiobjective optimization for base station transmitter placement with frequency assignment. IEEE Transactions on Evolutionary Computation 7(2), 189–203 (2003)
15. Zitzler, E., Laumanns, M., Thiele, L.: SPEA2: Improving the Strength Pareto Evolutionary Algorithm for Multiobjective Optimization. Evolutionary Methods for Design, Optimization and Control, 19–26 (2002)
16. Zitzler, E., Thiele, L.: Multiobjective Optimization Using Evolutionary Algorithms - A Comparative Case Study. In: Eiben, A.E., Bäck, T., Schoenauer, M., Schwefel, H.-P. (eds.) PPSN 1998. LNCS, vol. 1498, pp. 292–301. Springer, Heidelberg (1998), `citeseer.ist.psu.edu/zitzler98multiobjective.html`

An Improved AntTree Algorithm for Document Clustering

M.L. Pérez-Delgado, J. Escuadra, and N. Antón

Abstract. The AntTree algorithm is a clustering method based on artificial ants which has been applied to document clustering, reaching good results. In this paper an improvement to the basic algorithm is proposed, based on the use of the information provided by the silhouette statistic. Computational results show that the improvement generates better results than the basic method.

Keywords: artificial-ants, clustering.

1 Introduction

The aim of cluster analysis is to find groupings or structures within data. The partitions found should result in similar data being assigned to the same cluster and dissimilar data assigned to different clusters.

Recently, a clustering algorithm based on the use of artificial ants was proposed by Azzag, [1]. This algorithm, called AntTree, is inspired from the self-assembly behavior observed in some species of real ants. These ants are able to build structures by connecting themselves to each others. They can form drops constituted of ants, [12], or chains to link leaves together, [9]. It has been observed that the structures disaggregate after a given time.

The behavior observed in the ants can be used to build a hierarchical tree-structured partitioning of a set of elements, according to the similarities between those elements. Several variants of the basic AntTree algorithm are proposed in [2].

The AntTree algorithm has been applied to document clustering, on-line mining of web sites usage, and automatic construction of portal sites, [2]; brain images clustering, [5]; image segmentation, [14]; texture segmentation, [7]; graphical

María-Luisa Pérez-Delgado, J. Escuadra, and N. Antón
Escuela Politécnica Superior de Zamora. Universidad de Salamanca, Av. Requejo, 33, C.P. 49022, Zamora, Spain
e-mail: {mlperez, jeb, nanton}@usal.es

A.P. de Leon F. de Carvalho et al. (Eds.): Distrib. Computing & Artif. Intell., AISC 79, pp. 481–488.
springerlink.com

symbol recognition in architectural plans, [13]; and hierarchical summarisation of video, [10].

Many clustering algorithms require the definition of the number of clusters beforehand, [8]. To overcome this problem, various cluster validity indexes have been proposed to assess the quality of a clustering partition, [6]. The silhouette validity index has shown to be an useful value for the prediction of optimal clustering partitions, [11]. Although the AntTree algorithm does not require to determine the number of clusters beforehand, the silhouette value is used in this paper to improve the result obtained by the AntTree algorithm.

This paper is organized as follows. Section 2 presents the AntTree algorithm. Section 3 describes the silhouette statistic, considered to improve the AntTree. The improvement is presented in Section 4. The next section includes computational results. The last section presents the conclusions of the work.

2 The AntTree Algorithm

Let us consider a set of n documents to be clustered. A function $Sim(i, j)$ is used to measure the similarity between two documents, i and j. If two documents are completely different, its similarity is cero; when they are equal, the similarity is one. In this paper, the cosine measure has been considered to compute similarities.

To cluster n documents with AntTree, a set of n ants is considered; the ant a_i represents the document i, with $i \in [1,n]$. The ants will be connected in a tree structure to cluster the associated documents. The root of the tree is represented by a nodo a_0, called the support. Initially all the ants are on the support. The ants will be incrementally connected either to the support or to other already connected ants. The proccess continues until all the ants are connected to the tree; at this moment all the documents have been clustered. The tree can be interpreted as a data partition by considering each subtree connected to a_0 as a cluster.

An ant can be in two states: it is connected to the structure or it can move over the structure. The algorithm is applied until there is no moving ants. At every iteration of the algorithm, a moving ant a_i is selected. Let us consider that this ant is on the node a_{pos}, which can be the support or another node in the tree (an ant connected to the structure). Three situations can be presented when a_i is selected, each with a different treatment:

- if no ant, or only one ant is connected to a_{pos}, a_i is connected to a_{pos}.
- if two ants are connected to a_{pos} for the first time, the sub-tree with root in the second ant is disconnected, and all the dropped ants are moved back to the support. Next, a_i is connected to a_{pos}.
- if two ants are connected to a_{pos}, but not for the first time, or if more than two ants are connected, the lowest dissimilarity value which can be observed among the daughters of a_{pos} is computed:
  $T_{Dissim}(a_{pos}) = Min\{Sim(a_j, a_k) | a_j, a_k \in \{ \text{ ants connected to } a_{pos}\}\}$.
  Let a^+ be the ant connected to a_{pos} which is most simmilar to a_i. If a_i is dissimilar enough to a^+, a_i connects to a_{pos}; otherwise, a_i moves towards a^+.

The two first ants are automatically connected to a_{pos} without any test, since the value $T_{Dissim}(a_{pos})$ can not be calculated if a_{pos} has less than two daughters. This may result in a bad connection for the second ant; therefore this ant is disconnected as soon as the third ant is connected to a_{pos}.

Table 1 summarizes the steps of the algorithm.

Table 1 AntTree algorithm

```
PROCEDURE ANT_TREE (Sim, n, tree)
  While not all the ants are connected to the tree
    Take a moving ant, a_i
    If no ant or only one ant connected to a_pos then
      Connect a_i to a_pos
    else-If 2 ants connected to a_pos, and for the fist time then
        Disconnect from a_pos the subtree with root in the second ant
        Place all the ants of the subtree back onto the support
        Connect a_i to a_pos
    else
      Find a+ : ant connected to a_pos most simmilar to a_i
      Compute T_Dissim(a_pos)
      If Sim(a_i, a+) < T_Dissim(a_pos) then
          Connect a_i to a_pos
      else
          Move a_i towards a+
      end-if
    end-if
  end-while
END
```

3 Silhouettes

The silhouette is a validity index which can be calculated to assess the quality of a clustering partition. It has shown to be a robust strategy for the prediction of optimal clustering partitions, [3], [11].

When a partition has been obtained, silhouettes can be calculated. The silhouette function estimates the membership degree of an arbitrary ant with respect to the cluster under consideration.

For each document i in the data set, a silhouette value $s(i)$ is computed, according to Eq. 1, where $a(i)$ is the average similarity between i and all other objects of the cluster to which i belongs, and $b(i)$ is the maximum average similarity of i to all objects of other clusters. From the expression it follows that $-1 \leq s(i) \leq 1$.

$$s(i) = \frac{a(i) - b(i)}{max\{a(i), b(i)\}} \tag{1}$$

The value $s(i)$ measures how well the object i has been classified. If it is close to 1, i is well clustered (it was assigned to an appropriate cluster); if it is close to 0, i could also be assigned to the nearest neighbouring cluster; if it is close to -1, i has been misclassified.

For each cluster j, the cluster silhouette, S_j, can be calculated by applying Eq. 2, where m is the number of documents in cluster j.

$$S_j = \frac{1}{m} \sum_{i=1}^{m} s(i) \qquad (2)$$

The global silhouette value for a partition U of c clusters is given by Eq. 3.

$$S_U = \frac{1}{c} \sum_{i=1}^{c} S_j \qquad (3)$$

The global silhouette can be used to estimate the most appropiate number of clusters for U: the partition with the maximum global silhouette is taken as the optimal partition.

4 AntTree Combined with the Improvement Process

In this paper the AntTree algorithm is modified, including an improvement applied to the solution generated by AntTree.

Table 2 shows the steps of the improvement procedure.

Table 2 Algorithm to perform the improvement

```
PROCEDURE IMPROVE (Sim, n, tree)
Do
   Compute_silhouette( Sim, tree, s, second )
   mov = 0
   For i = 1 to n
      If(s(i) < TH) then
         Move ant i to the root of cluster second(i)
         Apply AntTree to connect i to subtree second(i)
         mov = mov + 1
      end-if
   end-for
While (mov > 0)
END
```

The AntTree algorithm generates a tree containing all the ants (*tree*). The silhouette of all the documents is calculated, according to Eq. 1, and stored in the vector s. Moreover, the second-best cluster for each document is stored in the vector *second*.

If the silhouette of a document is smaller than a threshold, $TH < 0$, it has been misclassified. Therefore, it is moved from its cluster to the second-best cluster. The ant is dropped from the tree and it goes to the root of the second-best cluster. The AntTree algoritm is then applied to move this ant until it connects to the subtree in an adequate place.

The test is performed for all the documents in the collection, and the ones which have a silhouette smaller than TH are moved. Experimental tests show that good results are reached for negative values of TH close to cero. The process is applied iteratively until no movement is performed during an iteration.

5 Computational Results

The algorithm has been coded in C language. The tests have been performed on a personal computer with Intel Centrino Core 2 Duo processor, 2.2 GHz, with 2G RAM memory and working on Linux Operating System.

Table 3 Databases used to check the algorithm

Name	Instances	Keys	Clusters	Instances per cluster
Iris	150	4	3	{50, 50, 50}
Wine	178	13	3	{59, 70, 49}
Thyroid	215	5	3	{150, 35, 30}
Glass	214	9	7	{70, 70, 17, 0, 13 ,9, 29}

The proposed solution has been applied to some real databases taken from the Machine Learning Repository, [4]. Due to the limited extent of the paper, only the results for 4 databases are showed. Table 3 summarizes the main features of the selected databases: the name, the number of documents, the number of attributes (Keys), the number of clusters and the distribution of the instances over each cluster. Since it is known for each instance the cluster it belongs to, the algorithm can be evaluated with respect to the classification error measure proposed in [1], given by Eq. 4, where ε_{ij} is given by Eq. 5. k_i represents the real cluster of the document i, and k'_i represents the cluster computed by AntTree.

$$E_c = \frac{2}{n(n-1)} \sum_{(i,j)\in 1,\ldots,n^2, i<j} \varepsilon_{ij} \quad (4)$$

$$\varepsilon_{ij} = \begin{cases} 0 \text{ if } (k_i = k_j \text{ and } k'_i = k'_j) \text{ OR } (k_i \neq k_j \text{ and } k'_i \neq k'_j). \\ 1 \quad \text{otherwise} \end{cases} \quad (5)$$

Table 4 shows the results obtained by the basic and the improved AntTree algorithm. It shows the name of the problem, the number of clusters found (K'), the time in seconds to reach a solution (T), the global silhouette (S_U), and the classification error measure (E_c). Bold typeface is used to highlight the best value for E_c and S_U. The AntTree algoritm must know the maximum number of daugthers any node can have, L; this value limits the maximum number of clusters that the algorithm

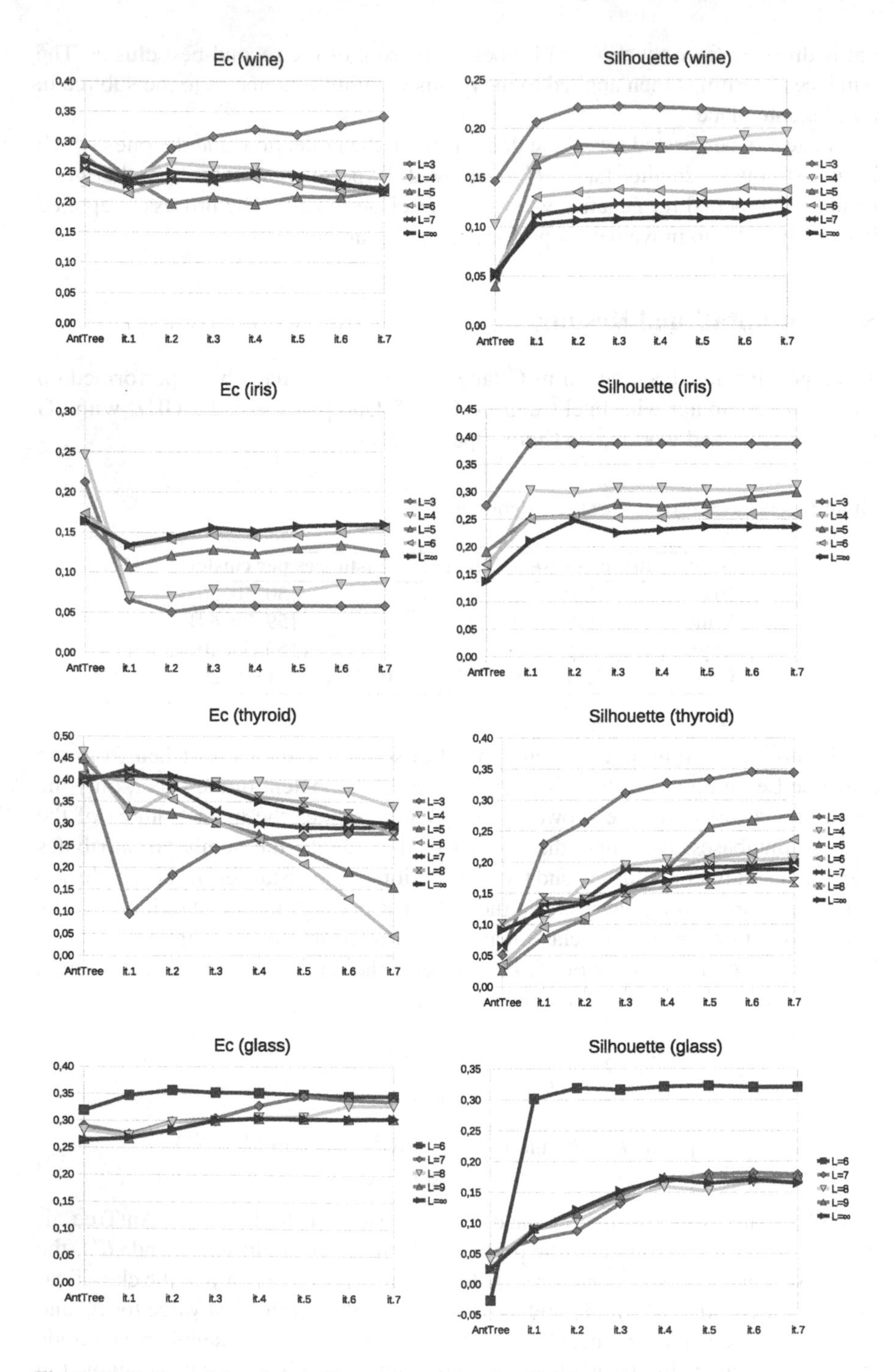

Fig. 1 Evolution of the global silhouette and the classification error

Table 4 Results for the basic and the improved AntTree

		AntTree				Improved AntTree			
Name	L	K'	T	S_U	E_c	K'	T	S_U	E_c
Wine	3	3	1	**0.146091**	0.273345	3	1	**0.222381**	0.308005
	4	4	1	0.102126	0.268330	4	1	0.195545	0.238621
	5	5	1	0.039861	0.296832	5	1	0.183337	**0.197042**
	6	6	1	0.052143	**0.233098**	6	1	0.151622	0.209801
	7	7	0	0.048773	0.255189	7	0	0.157140	0.220402
	∞	8	0	0.053283	0.267632	8	1	0.166062	0.218435
Iris	3	3	1	**0.274874**	0.212349	3	1	**0.387896**	**0.050470**
	4	4	1	0.149381	0.245996	4	1	0.318552	0.092349
	5	5	0	0.191039	0.164385	5	0	0.299148	0.107114
	6	6	1	0.167594	0.173154	6	1	0.272146	0.150962
	∞	7	0	0.137192	**0.164206**	7	0	0.247835	0.143982
Thyroid	3	3	2	0.050880	0.444860	3	2	**0.344911**	0.278374
	4	4	1	0.029641	0.463030	4	2	0.205301	0.338535
	5	5	1	0.026235	0.448381	5	1	0.275213	0.155705
	6	6	1	0.036060	0.409302	6	1	0.272143	**0.098500**
	7	7	1	0.065364	**0.393871**	7	1	0.251899	0.216822
	8	8	0	**0.100119**	0.406433	7	0	0.242150	0.254510
	∞	9	0	0.090696	0.405173	9	0	0.188188	0.299413
Glass	6	6	1	-0.027030	0.318898	6	1	**0.322598**	0.348515
	7	7	1	**0.050794**	0.290509	7	1	0.181868	0.335659
	8	8	1	0.041350	0.282699	8	1	0.169773	0.331447
	9	9	1	0.026103	0.265587	8	1	0.174215	**0.303892**
	∞	10	1	0.024735	**0.263350**	9	1	0.182384	0.310912

can identify. In all cases a test has been performed with no limit in the number of daughters a node of the tree can have, $L = \infty$. Additional tests have been performed for values of L smaller than the value reached by K' when $L = \infty$.

The global silhouette improves when the new version of the algorithm is applied. Fig. 1 shows the global silhouette and the classification error for the basic AntTree algoritm, and for each one of 7 improvement iterations. We observe that the improved AntTree algorithm generates better results than the basic AntTree: the value E_c decreases and the silouette increases when the first iteration of the improving method is applied.

The best results are obtained for iris dataset. The figura shows that when the silhouette is greater, the classification error is lower.

6 Conclusion

This paper presents a method to improve the results obtained when the AntTree algorithm is applied to cluster documents. Such method uses the silhouette value of the documents in the collection to improve the initial clustering provided by the basic AntTree algorithm. We have proven that the new method improves the solution reached by the ants. In this paper we consider the AntTree version with no threshold and no parameters, [2], but the proposed improvement method can also be applied to other AntTree variants.

Acknowledgements. This work has been financed by the Spanish Ministry of Science and Innovation (Project reference: BIA-2007-67323) and the Council of Castile and Leon (Project reference: SA130A08).

References

1. Azzag, N., Monmarche, H., Slimane, M., Venturini, G., Guinot, C.: AntTree: a new model for clustering with artificial ants. In: Proc. 2003 Congress on Evolutionary Computation, Canberra, Australia, 8-12, pp. 2642–2647. IEEE Press, Los Alamitos (2003)
2. Azzag, N., Venturini, G., Oliver, A., Guinot, C.: A hyerarchical ant based clustering algorithm and its use in three real-world applications. Eur. J. Oper. Res. 179, 906–922 (2007)
3. Bezdek, J.C., Pal, N.R.: Some new indexes of cluster validity. IEEE Trans. Systems, Man, Cybern. 28, Part B, 301–315 (1998)
4. Blake, C.L., Merz, C.J.: UCI Repository of Machine Learning Databases. University of California, Irvine, Dept. of Information and Computer Sciences (1998), `http://www.ics.uci.edu/~mlearn/MLRepository.html`
5. Chenling, L., Wenhua, Z., Jiahe, Z.: Image segmentation of MRI based on improved AntTree clustering algorithm. In: Proc. 3rd Int. Conf. on Intelligent System and Knowledge Engineering, pp. 1208–1213 (2008)
6. Halkidi, M., Batistakis, Y., Vazirgiannis, M.: On clustering validation techniques. J. Intell. Inf. Syst. 17, 107–145 (2001)
7. Hussain, A., Mahmood, N., Mahmood, K.: Texture segmentation using Ant Tree clustering. In: Proc. IEEE Int. Conf. on Engineering of Intelligent Systems (2006)
8. Jain, A.K., Dubes, R.C.: Algorithms for clustering data. Prentice Hall, Englewood Cliffs (1988)
9. Lioni, A., Sauwens, C., Theraulaz, G., Deneubourg, J.L.: The dynamics of chain formation in oecophylla longinoda. J. Insect Behav. 14, 679–696 (2001)
10. Piatrik, T., Izquierdo, E.: Hierarchical summarisation of video using Ant-Tree strategy. In: Seventh International Workshop on Content-Based Multimedia Indexing, pp. 107–112 (2009)
11. Rousseeuw, P.J.: Silhouettes: A graphical aid to the interpretation and validation of cluster analysis. J. Comp. App. Math. 20, 53–65 (1987)
12. Theraulaz, G., Bonabeau, E., Sauwens, C., Deneubourg, J.L., Lioni, A., Libert, F., Passera, L., Solé, R.V.: Model of droplet formation and dynamics in the argentine ant. Bulletin of Math. Biology (2001)
13. Yang, X., Zhao, W., Pan, L.: Graphical symbol recognition in architectural plans with an improved Ant-Tree based clustering algorithm. In: Proc. Int. Joint Conf. on Neural Networks, pp. 390–397 (2008)
14. Yang, X., Zhao, W., Chen, Y., Fang, X.: Image segmentation with a fuzzy clustering algorithm based on Ant-Tree. Signal Process 88, 2453–2462 (2008)

Solving the Parameter Setting in Multi-Objective Evolutionary Algorithms Using Grid::Cluster

Eduardo Segredo, Casiano Rodríguez, and Coromoto León

Abstract. The parameter values of a Multi-objective Evolutionary Algorithm greatly determine the behavior of the algorithm to find good solutions within a reasonable time for a particular problem. In general, static strategies consume lots of computational resources and time. In this work, a tool is used to develop a static strategy to solve the parameter setting problem, applied to the particular case of the Multi-objective 0/1 Knapsack Problem. GRID::Cluster makes feasible a dynamic on-the-fly setup of a secure and fault-tolerant virtual heterogeneous parallel machine without having administrator privileges. In the present work is used to speed-up the process of finding the best configuration, through optimal use of available resources. It allows the construction of a driver that launches, in a systematically way, different algorithm instances. Computational results show that, for a particular problem instance, the best behavior can be obtained with the same parameter values regardless of the applied algorithm. However, for different problem instances, the algorithms have to be tuned with other parameter values and this is a tedious process, since all experiments have to be repeated, for each new set of parameter values to be studied.

Keywords: Parameter Setting, Multi-objective Optimization, Multi-objective Evolutionary Algorithms, GRID::Cluster, METCO.

1 Introduction

Finding the appropriate parameter values of a Multi-objective Evolutionary Algorithm (MOEA) is a source of research emerged several decades ago [4], and during all that time, researchers have been tried to find answers to questions like: Is there

Eduardo Segredo, Casiano Rodríguez, and Coromoto León
Dpto. Estadística, I. O. y Computación. Universidad de La Laguna
La Laguna, 38271, Santa Cruz de Tenerife, Spain
e-mail: esegredo@ull.es, casiano@ull.es, cleon@ull.es

A.P. de Leon F. de Carvalho et al. (Eds.): Distrib. Computing & Artif. Intell., AISC 79, pp. 489–496.
springerlink.com

a generic set of parameter values applicable to all problems?, Is there a generic set of parameter values applicable to a particular set of problems?, Is it better preset the parameter values before the execution or modify the parameter values during the execution of the algorithm?, What parameters subset really affects the behavior of an algorithm?. These, and other questions are, nowadays, opened research topics.

The parameter values of a MOEA greatly determine the behavior of the algorithm to find good solutions within a reasonable time for a particular problem. Generally, there are two ways of making the parameter setting of a MOEA. The first one is using static strategies (or *parameter tuning*), and the second one is using dynamic strategies (or *parameter control*) [5]. Static strategies are based on finding the best combination of parameter values before the algorithm execution, and then use those values during all the algorithm run. By contrast, dynamic strategies start the execution with a set of initial values, and modify those values during the execution, depending on the algorithm behavior in each moment.

Commonly, the application of a static strategy is done by experimenting with different values and selecting the ones that produces the best results (the nearest solutions to the optimal ones). The static strategy most frequently used is a multiple execution in which several parameter values are tried to find the appropriate combination of them for a particular problem. This process can be automated by the use of a top-level driver that chooses different parameter values in a systematically way. However, the number of existing parameters in a MOEA, combined with their possible values, implies this activity consumes lots of computational resources and time. Another possibility is to use a two-level strategy or *nested* Evolution Strategy (ES), in which the top-level ES evolves the parameter values of the second one [12].

In the present work, a static strategy is used to find the best combination of parameter values for different MOEAs. The particular problem, for which the static strategy is applied, is the Multi-objective 0/1 Knapsack Problem. It is a well-known NP-*complete* problem. METCO [10] is a plugin-based framework for the resolution of Multi-objective Optimization Problems (MOPs) based on MOEAs. The problem implementation using this tool is simple and straightforward. However, due to the large amount of computational resources and time that requires the applied strategy, tools that accelerate the process of finding the best combination of parameter values are needed. GRID::Cluster is a Perl module that is used, in the present work, to build a driver which allows the parallel execution of multiple tasks, so through the optimal utilization of available resources, invested time is significantly reduced. These tasks are calls to different instances of MOEAs implemented by METCO.

The remaining work is structured as follows: in Sect. 2, a description of the implemented infrastructure to apply the aforementioned static strategy is given. The particular problem, for which parameter values are found, is presented in Sect. 3. In Sect. 4 different experiments that have been done and obtained computational results are exposed. Finally, some conclusions and some lines of future works are shown in Sect. 5.

2 Building the Environment

Several specific frameworks based on Evolutionary Algorithms have been proposed for the solution of (MOPs) [2, 6, 7, 10, 11]. Main drawbacks of using them are, on the one hand, the hard customization of the tool itself, and on the other hand, the difficult to incorporate new MOEAs and MOPs to the framework. In this work the METCO framework is used. It provides the following MOEAs: SPEA, SPEA2, NSGA-II and IBEA. With this tool, users have to perform two main tasks: an execution model has to be specified, and if neccesary, corresponding plugins have to be developed. After implementing required plugins, the tool workflow combines them, using different solvers to find the solution of a particular problem.

MOEAs have many configuration parameters, so finding the best combination of values for a given problem is a challenging task. Moreover, the complexity is exponentially increased with each new parameter to be studied. This, combined with the time invested in performing executions of a MOEA over the particular problem for which the best configuration is being found, can consume a large number of computational resources and time. It is therefore a need for tools to accelerate the process of finding the best MOEA configuration. One of these tools is GRID::Cluster [1]. The philosophy of this tool is *Zero-Administration Parallel Computing* (ZAPC). It addresses a particular subclass of users. This is a rough description of the target ones:

1. Users are scientist or programmers, familiar with some programming language like C++, Java, etc. who want to improve their application performance
2. They have access via SSH to some number - may be large - of Unix machines in the institutions they belong, and these machines have a Perl interpreter installed
3. Users are not necessarily familiar with MPI, OpenMP or other parallel computing tools, but they are willing to learn and to use parallelism to speed up the set of computationally demanding applications they have
4. It is difficult for users to convince administrators of these machines to install any additional software, and these administrators may belong to different institutions and are not willing to collaborate with users and even less among them
5. Users have a laptop or some personal computer where they have *root* permissions
6. Security and Fault Tolerance are important issues

GRID::Cluster is a Perl module which allows the writing of secure master-worker applications (or work task queueing) that fulfil the requirements of ZAPC. Communications among different machines are carried out by the use of SSH, and a public – private key based infrastructure is used to authenticate users. GRID::Cluster provides a parallel version of the classic shell backticks operator. Backticks are exceptionally useful and they are often used in scripting. The purpose of a backticks is to be able to run a command, and capture the output of that command. The backticks operator is also referred in Perl as `qx`. The `qx` method of GRID::Cluster receives a (usually very large) list of strings representing commands, and executes them in the machines of the GRID::Cluster using the master-worker/task-farming paradigm. This paradigm balances the processing load on both homogeneous and heterogeneous

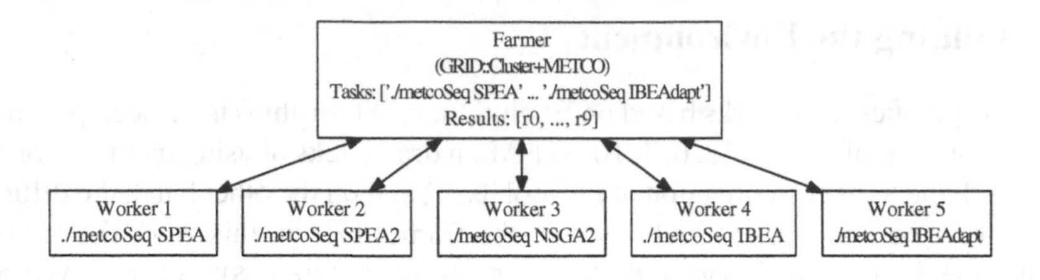

Fig. 1 GRID::Cluster uses the farm paradigm to implement the parallel version of the qx operator

machines, makes feasible the introduction of fault-tolerance hooks [13] and gives opportunities for performance tuning [3].

An execution instant of a driver based on GRID::Cluster is depicted in Fig. 1, with five involved workers and a number of tasks (calls to METCO sequential solvers) equal to the number of all possible combinations of values taken by the parameters of different MOEAs. This is the scheme which has been followed in this work.

3 The Multi-objective 0/1 Knapsack Problem

The particular problem used in the present work to find the best MOEA configuration has been chosen for two reasons. The first one, it is an academic problem, understandable and easy to formulate, and the second one, it is a NP-complete problem with real world applications, for example, in the economic and industrial domains. A 0/1 Knapsack Problem [8, 14] consists of a set of items, each one of them with a weight and a profit associated. Moreover, the knapsack is limited with an upper bound for the capacity that is capable of supporting. The main objective is to find a subset of items which maximizes the total profit, taking into account the knapsack maximum capacity constraint.

This single-objective problem can be easily extended to the multi-objective domain, using a fixed number of knapsacks. A Multiobjective 0/1 Knapsack Problem can be defined as follows:

Given a set of m items and n knapsacks, with $p_{i,j}$ the profit of item j according to knapsack i, $w_{i,j}$ the weight of item j according to knapsack i, and c_i the capacity of knapsack i. The problem consist in find a vector $x = (x_1, x_2, \cdots, x_m) \in \{0,1\}^m$, such that

$$\forall i \in \{1,2,\cdots,n\} : \sum_{j=1}^{m} w_{i,j} \cdot x_j \leq c_i \qquad (1)$$

and for which $f(\boldsymbol{x}) = (f_1(\boldsymbol{x}), f_2(\boldsymbol{x}), \cdots, f_n(\boldsymbol{x}))$ is maximum, where

$$f_i(\boldsymbol{x}) = \sum_{j=1}^{m} p_{i,j} \cdot x_j \qquad (2)$$

and $x_j = 1$ if item j is selected.

An implementation of this problem can be found on METCO [9]. In this implementation, a solution is represented by a binary vector, and infeasible solutions obtained after a mutation or a crossover are repaired by a greedy method which removes items from the solution until all knapsack capacity constraints are satisfied.

4 Computational Results

Experiments with different instances of the Multi-objective Knapsack Problem [15] have been performed. Specifically, the instances with 100 and 500 items, both of them with 2 knapsacks, have been used due to their optimal Pareto fronts are available. In the set of experiments, METCO sequential solvers have been used to execute different configurations of the implemented algorithms. These algorithms are

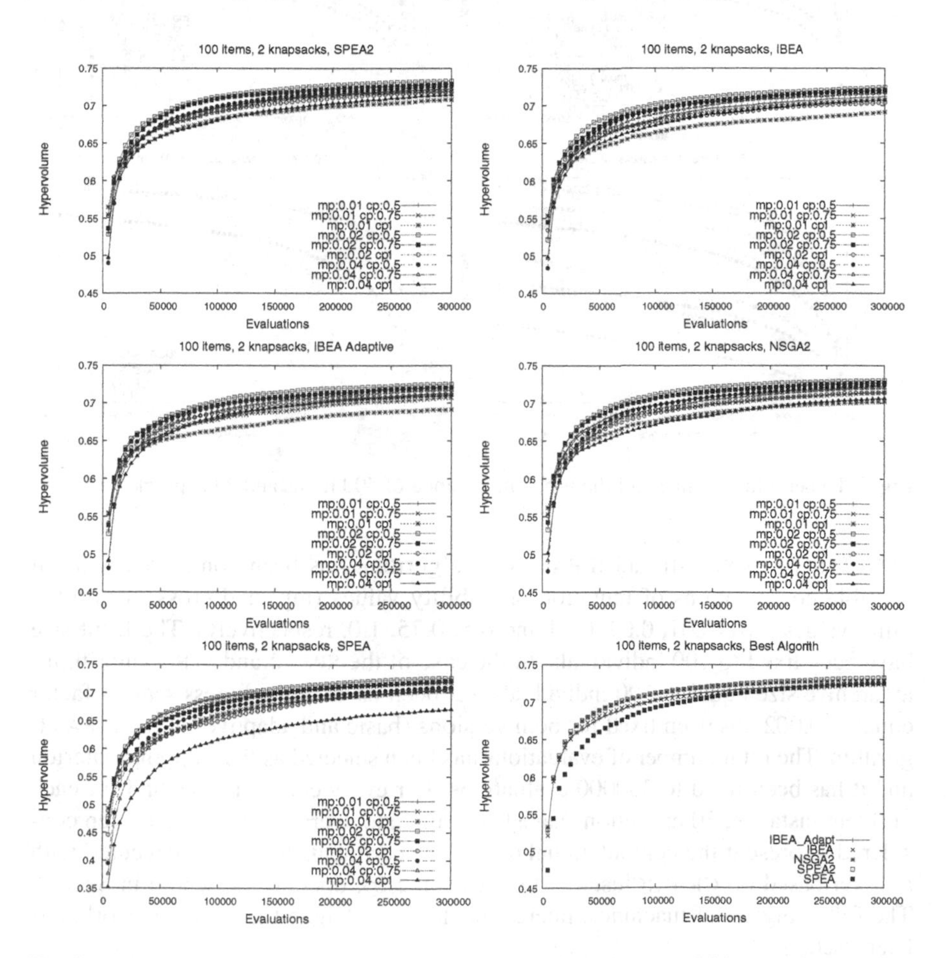

Fig. 2 Hypervolume values for the problem instance of 100 items and 2 knapsacks

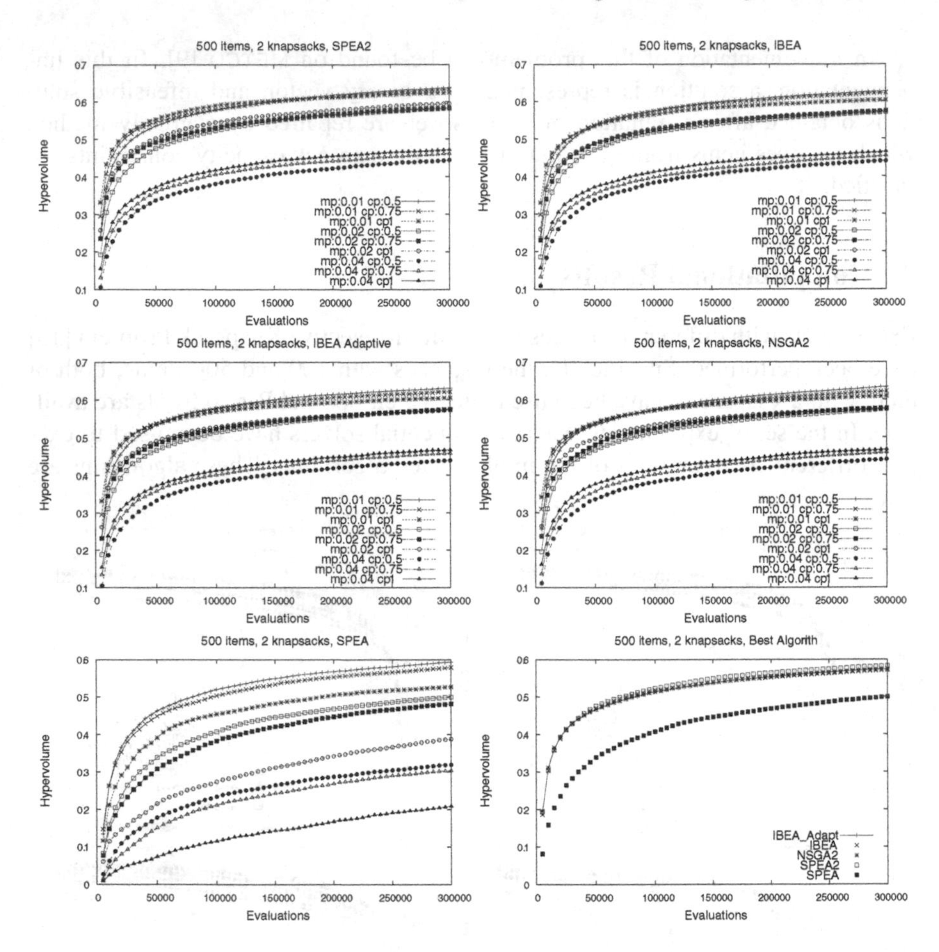

Fig. 3 Hypervolume values for the problem instance of 500 items and 2 knapsacks

SPEA2, SPEA, NSGA-II and IBEA. Every algorithm has been configured with all possible combinations of mutation probability values (mp) and crossover probability values (cp) (0.01, 0.02, 0.04 and 0.5, 0.75, 1.0, respectively). The front size have been fixed to 100 individuals. In the case of the SPEA2 and SPEA algorithms, an archive size equal to 100 individuals has been used, and a fitness scaling factor equal to 0.002 has been fixed for both versions (basic and adaptive) of the IBEA algorithm. The total number of evaluations has been selected as the stopping criterion and it has been fixed to 300000 evaluations. For every configuration, and for each problem instance, 30 executions have been run, and average values have been considered to present the computational results. Experiments have been executed with a driver based on GRID::Cluster, which uses the qx operator presented in Sect. 2. The following set of machines, interconnected by a Gigabit Ethernet network, has been used:

- Linux server with 4 Intel® Xeon™ processors at 3.00 GHz and 1 Gb of memory
- Linux server with 2 Intel® Pentium™ 4 at 3.40 GHz and 2 Gb of memory
- Linux PC with an Intel® Pentium™ 4 at 3.20 GHz and 1 Gb of memory

With obtained results, a study of the solutions quality has been performed, using a hypervolume-based metric [14]. The reference points used to calculate the hypervolume values match with the maximum and minimum values of the optimal Pareto front objective variables.

Achieved hypervolume values versus the number of performed evaluations are shown in Fig. 2 and Fig. 3, for different configurations. In the case of the instance with 100 items and 2 knapsacks, and for all algorithms, the configuration with a mutation rate equal to 0.02 and a crossover rate equal to 0.5, achieves the highest hypervolume value. For the instance with 500 items and 2 knapsacks, and also for all algorithms, the configuration with values 0.01 and 0.5, for the mutation and crossover rates, respectively, presents the best behavior. In this case, the higher the mutation rate, the lower the hypervolume value. The SPEA2 algorithm obtains the highest hypervolume values for both instances of the problem, meanwhile the SPEA algorithm has the worst behavior, in both cases.

5 Conclusions and Future Work

In the present work, a first approximation based on a static strategy, has been used to solve the Parameter Setting Problem applied to the particular case of the Multi-objective 0/1 Knapsack Problem. Sequential solvers of different MOEAs provided by METCO have been used, and to accelerate the process of finding the best configuration, a driver based on GRID::Cluster has been built.

Computational results show that, for a particular problem instance, the best behavior can be obtained with the same parameter values regardless of the applied MOEA. However, for different problem instances, MOEAs have to be tuned with other parameter values (although SPEA2 is the algorithm which have obtained the best results for both instances), and this is a tedious process, since all experiments have to be repeated, for each new set of parameter values to be studied.

The future work will be focused on the design of dynamic strategies that allow the modification of parameter values during the algorithm execution. This could remove, or at least reduce, the existing dependency between MOEAs and the particular problem (or instance) for which they are applied, and generate more generic problem solving techniques. A first approximation could be based on a multi-level strategy. The top-level would evolve parameter values of the MOEA located in the second-level, taking into account the quality of the solutions (a quality measure like the hypervolume could be used here) obtained by the last one.

Acknowledgements. This work has been partially supported by the EC (FEDER) and the Spanish Ministry of Science and Innovation inside the 'Plan Nacional de I+D+i' with the contract number TIN2008-06491-C04-02 and by the Canary Government project number PI2007/015.

References

1. GRID:Cluster Module Documentation at CPAN Website, http://search.cpan.org/dist/GRID-Cluster/
2. Bleuler, S., Laumanns, M., Thiele, L., Zitzler, E.: PISA – a platform and programming language independent interface for search algorithms. In: Fonseca, C.M., Fleming, P.J., Zitzler, E., Deb, K., Thiele, L. (eds.) EMO 2003. LNCS, vol. 2632, pp. 494–508. Springer, Heidelberg (2003)
3. Cesar, E., Moreno, A., Sorribes, J., Luque, E.: Modeling master/worker applications for automatic performance tuning. Parallel Computing Journal 3(7), 568–589 (2006)
4. De Jong, K.: Parameter setting in eas: a 30 year perspective. In: Lobo, F.G., Lima, C.F., Michalewicz, Z. (eds.) Parameter Setting in Evolutionary Algorithms, pp. 1–18. Springer, Heidelberg (2007)
5. Eiben, A.E., Michalewicz, Z., Schoenauer, M., Smith, J.E.: Parameter control in evolutionary algorithms. In: Lobo, F., Lima, C., Michalewicz, Z. (eds.) Parameter Setting in Evolutionary Algorithms, pp. 19–46. Springer, Heidelberg (2007)
6. Emmerich, M., Hosenberg, R.: TEA – A Toolbox for the Design of Parallel Evolutionary Algorithms in C++. Tech. Rep. CI-106/01, SFB 531, University of Dortmund, Germany (2001)
7. Gagné, C., Parizeau, M.: Genericity in Evolutionary Computation Software Tools: Principles and Case Study. International Journal on Artificial Intelligence Tools 15(2), 173–194 (2006)
8. Jaszkiewicz, A.: On the computational efficiency of multiple objective metaheuristics. the knapsack problem case study. European Journal of Operational Research 158, 418–433 (2004)
9. León, C., Miranda, G., Segredo, E., Segura, C.: Parallel Library of Multi-objective Evolutionary Algorithms. In: 17th Euromicro International Conference on Parallel, Distributed and Network-based Processing, pp. 28–35 (2009)
10. León, C., Miranda, G., Segura, C.: METCO: A Parallel Plugin-Based Framework for Multi-Objective Optimization. International Journal on Artificial Intelligence Tools 18(4) (2009)
11. Liefooghe, A., Basseur, M., Jourdan, L., Talbi, E.G.: ParadisEO-MOEO: A Framework for Evolutionary Multi-objective Optimization. In: Obayashi, S., Deb, K., Poloni, C., Hiroyasu, T., Murata, T. (eds.) EMO 2007. LNCS, vol. 4403, pp. 386–400. Springer, Heidelberg (2007)
12. Rechenberg, I.: Evolution strategy. Zuarda et. al. pp. 147–159 (1994)
13. Rodrigues de Souza, J., Argollo, E., Duarte, A., Rexachs, D., Luque, E.: Fault Tolerant Master-Worker over a Multi-Cluster Architecture. In: International Conference on Parallel Computing (ParCo), pp. 465–472 (2005)
14. Zitzler, E., Thiele, L.: Multiobjective Optimization Using Evolutionary Algorithms - A Comparative Case Study. In: Eiben, A.E., Bäck, T., Schoenauer, M., Schwefel, H.-P. (eds.) PPSN 1998. LNCS, vol. 1498, pp. 292–301. Springer, Heidelberg (1998), citeseer.ist.psu.edu/zitzler98multiobjective.html
15. Zitzler, E., Thiele, L.: Multiobjective Evolutionary Algorithms: A Comparative Case Study and the Strength Pareto Approach. IEEE Transactions on Evolutionary Computation 3(4), 257–271 (1999)

Constrained Trajectory Planning for Cooperative Work with Behavior Based Genetic Algorithm

Mustafa Çakır and Erhan Bütün

Abstract. In this study, subjected to the trajectory generation for cooperative work, a genetic algorithm with cultural constructs is used to search for valid and optimal solutions in task space. We develop that algorithm by reflecting the behavior of social communities with a decision maker is used to evaluate cultural adaptation level by how well phenotypes, based on quaternion representation, are fitted in goal function. Algorithm uses cognition strategy to obtain smooth trajectory considering physical restrictive structure and actuator limits by using dynamic constrains in decision engine and eliminating unexpected derivation, also avoiding local minima problem.

1 Introduction

The ultimate goal of robotics is to develop autonomous machines in interaction with environmental objects and capable of performing tasks in minimum time in order to increase the productivity and execute them without further human intervention. Thus, one of the objectives of the robotics research community is to develop algorithms that enable the manipulators to perform the tasks as fast as possible, taking into account the limits imposed by their physical characteristics and environment. To perform the task with optimum savings motion planning has the biggest role. Most of the algorithmic techniques that solve the problem of motion planning find a trajectory, without considering practical issues like temporal cost of the trajectory, regularity, etc. The optimal traveling time and the minimum mechanical energy of the actuators are considered together and this optimization problem is subject to physical constraints which include torque constraints.

Since last 20 years, the problem of motion planning for robots has been addressed with many techniques, both analytic (Zhihua et al., 2004) and evolutionary

---

Mustafa Çakır
Electronics and Communication Engineering Dep., Kocaeli University, Turkey
e-mail: mcakir@kocaeli.edu.tr

Erhan Bütün
Civil Aviation College, Kocaeli University, İzmit, Turkey
e-mail: ebutun@kocaeli.edu.tr

A.P. de Leon F. de Carvalho et al. (Eds.): Distrib. Computing & Artif. Intell., AISC 79, pp. 497–508.
springerlink.com

(Ahuactzin et al., 1992), (Rendas and Tetenoire, 1997), and (Pack et al., 1996). Most of these approaches solve the problem in the robot configuration space. In this case, the solution is presented as a continuous sequence of space configurations that does not collide with obstacles and connects the initial and final points. Many algorithms to solve this problem can be found in the literature.

The use of ant colonies was first applied to the traveling salesman problem and the quadratic assignment problem and has since been applied to other problems such as the routing in (Bell and McMullen, 2004), and space planning problem in (Bland, 1999), There has been a lot of interest in cooperative robotics in the last few years. Motion planning for a multi-robotic system refers to find trajectories for each robot to perform a common task. An analytic method is presented in (Ming Yi Ju et al., 2002) for an environment where the geometric paths of robots are pre-planned and the preprogrammed velocities are static but adjustable. The master robot always moves at a constant speed. The slave robot moves at the given velocity, selected by a tradeoff between collision trend index and velocity reduction in one collision checking time. In (Paul et al., 2004) dynamic role assignment strategy is used for mobile robots. In that strategy potential fields approach is used for path planning. In the conclusion of this study as a recommendation it is found more effective if strategy is combined with other techniques as fuzzy logic and some learning techniques. The job with n-number of precedence constraints is assigned minimizing mean tardiness on m-number of parallel robot Genetic algorithms and simulated annealing methods were used to find the solutions, which minimizes the mean tardiness in (Cakar et al., 2008). (Chen et al., 2009) proposed a novel multi-crossover genetic algorithm (GA) to identify the system parameters of a two-link robot. The resulted system model by the proposed GA is then applied to the feedback linearization control such that the two-link robot system can be transferred to a linear model with a nonlinear bounded time-varying uncertainty.

2 Algorithm

Objective of that study is generating an algorithm with a general methodology, capable of searching optimum solution trough dynamic conditions. Proposed algorithm processes the previous iterations to modify the parameters in the next iteration and accelerate search mission. Methodology provides systematic approach to wide range of problems. Algorithm was tested for the given task and scene as illustrated in Fig. 1. Proposed method is based on genetic algorithms because first of all they are powerful tools for searching in high dimensional spaces like in this case and then without dealing the structural constrains wide range of behaviors such as obstacle avoidance, energy consumption, cooperation can be taken into consider. Other advantages of GA, especially relevant that paper are flexibility and their harmony with supplementary algorithm such as fuzzy logic.

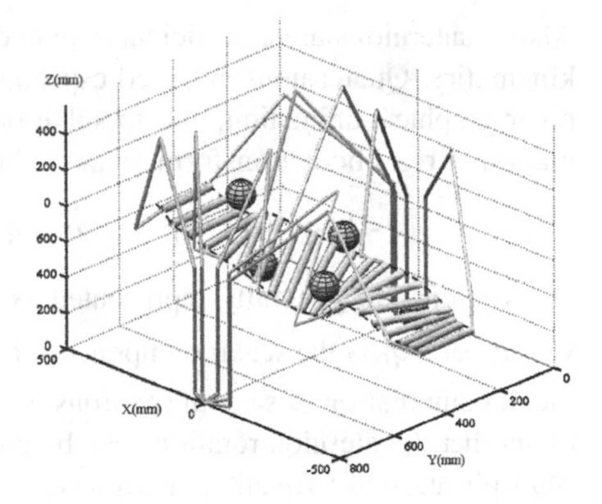

Fig. 1 Task is defined as carrying an object that is grasped by the end effectors of two articulated manipulator

The key component of the proposed algorithm is cultural constructs. A recent model of GA's are derived from models of biological adaptation based on natural selection. In this approach the genomic system is directed by inheritance of genetic material via parent and child links and the behavioral system is directed by fitness. In the place of natural selection as a process to which the individual is subjected according to one's environmental context and genomic makeup, cultural selection can be introduced as representing an internal process of decision making based on assessing consequences of actions.

3 6-Dof Robot Model

In question of that paper, a rigid body, which is not constrained to any kinematical condition, has six degrees of kinematical freedom in space. The first step to kinematical modeling is the proper assignment of coordinate frames to each link. Each coordinate system here is orthogonal, and obey right-hand rule.

We decoupled the kinematics problem into two simpler problems, as position kinematics and orientation kinematics. For a six DOF manipulator with a spherical wrist we should find first position of the intersection of the wrist axes and then the orientation of the wrist.

3.1 Quaternion

There are different kinds of representations for rotations as orthonormal matrices, Euler angles and unit quaternion in the Euclidean space. The theory of quaternions, introduced by William R. Hamilton, was expanded to include applications such as rotations in the early 20th century (Hamilton, 1969). Using the unit quaternions is a more effective, natural, and elegant way to perceive rotations compared to other methods. A comparison of these methods can be found in (Dam et al, 1998), (Schmidt and Nieman, 2001). The practical use of quaternions had been minimal in comparison with other methods but, currently, this situation has changed due to progress in robotics, animation and computer graphics technology.

Also, quaternions are an efficient way understanding many aspects of physics and kinematics. Quaternions are used especially in the area of computer vision, computer graphics, animation, and to solve optimization problems involving the estimation of rigid body transformations (Schmidt and Nieman, 2001).

$$q=(q_0,\quad w)=(q_0\quad q_1\quad q_2\quad q_3)=q_0+iq_1+jq_2+kq_3 \tag{1}$$

Quaternions are generally represented as (1), where $w=(q_1\quad q_2\quad q_3)$ form the vector part, q_0 is the scalar component of q. A single quaternion can represent either a combination of several rotations or a single rotation from a given orientation to another. Quaternion rotations can be combined simply by multiplying together. Multiplication of two unit quaternions results in another unit quaternion.

$$q_A * q_B=(q_{A0}q_{B0}-\vec{q}_A\cdot\vec{q}_B,\quad q_{A0}\cdot\vec{q}_B+q_{B0}\cdot\vec{q}_A+\vec{q}_A\times\vec{q}_B) \tag{2}$$

Quaternion multiplication is defined in (2). A quaternion in (3) represents the rotation of the 3D vector $\vec{p}$ by an angle θ about the axis N. The rotated vector, represented in Fig.16 can be obtained as fallows.

$$q=\left(\cos\left(\theta/2\right),\quad \sin\left(\theta/2\right)\cdot N\right) \tag{3}$$

$$P'=R_q(P)=qPq^*=\left(q_0^2-\vec{q}\cdot\vec{q}\right)\vec{p}+2q_0(\vec{q}\times\vec{p})+2\vec{q}(\vec{q}\cdot\vec{p}) \tag{4}$$

Fig. 2 Quaternions are hyper-complex numbers of rank 4, constituting a four dimensional vector space over the field of real numbers.

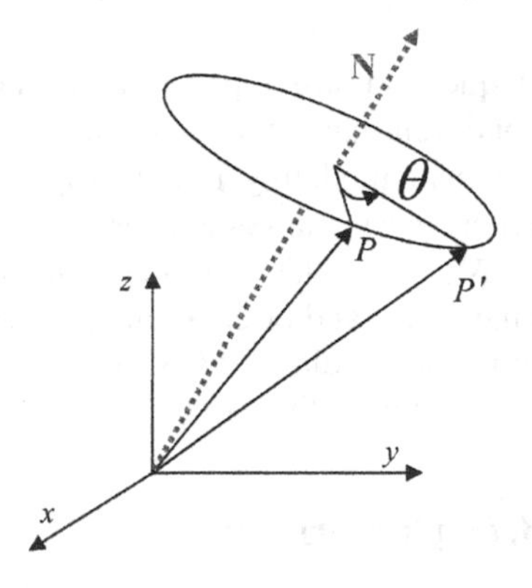

Obviously, a rotation does not affect the length of a vector, and hence $\|p'\|=\|p\|$. Thus for a proper rotation q in (2), N must be a unit vector. Also note that when q is a unit quaternion, we have the property $q^{-1}=q^*$. Considering (2), (3), and (4), the position of point P_3 in figure 3. relative to base frame is calculated with (5), where $q_{12\ldots n}=q_1 * q_2 * \cdots q_n$.

$${}^0P_3=q_{123}D_3q_{123}{}^*+q_{12}D_2q_{12}{}^*+q_1D_1q_1{}^* \tag{5}$$

Given two orientations in the 3-space, researchers conventionally used linear interpolation between the corresponding Euler angles to obtain the transition frames. However, such an interpolation algorithm has many limitations. The interpolation does not give a natural transition between the key frames, since Euler angle parameterization combines the interpolated rotations along the coordinate axes to form the resulting orientation. There is no easy way to interpolate between rotation matrices. If M_1 and M_2 are rotation matrices, the matrix $(1-t)M_1 + tM_2$ will not give a smooth movement from M_1 to M_2. In fact, the interpolated matrix is not in general a rotation matrix, which means that the rotating object will be distorted as it moves. Combination of rotations requires matrix multiplication. Combining a long sequence of matrix multiplications creates inaccuracies and sometimes even numerical instability. Quaternions have a number of advantages over matrices as a means of representing rotations. First, only four numbers are needed to represent a rotation. Composite rotations by multiplying quaternions are fast and accurate. It is easy to renormalize quaternions after many calculations and interpolate between quaternions to smoothly animate rotating object. Euler angles and the Rodrigues vector are two examples of 3-number orientation representations.

In this paper, a novel technique based on genetic algorithms is described and it is adapted to the trajectory planning problem in robotics. It can be applied to any general serial manipulator. IRB 140 shown in Fig. 4 is selected as target platform for simulations. Instead of the homogeneous transformation matrix and the parameters of Denavit-Hartenberg, quaternions are used to write the kinematic model. We obtained the dynamic model by writing Euler-Lagrange's equation using Lagrange's energy function.

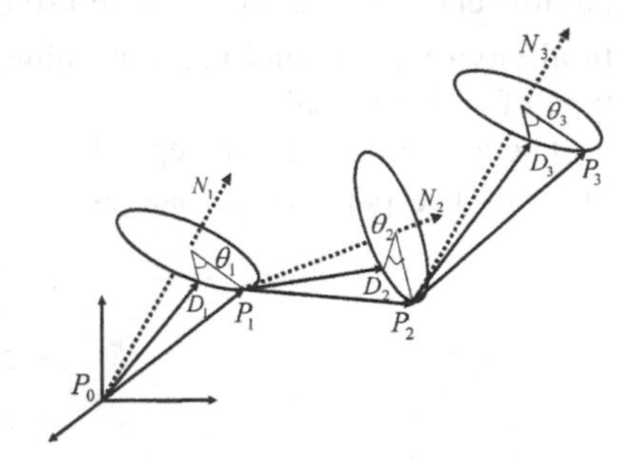

Fig. 3 Any series of rotations with respect to the universal base frame axes can be described mathematically by a series of multiplications of quaternion

The phenotype of an individual is specified by a linear combination of genotypic, cultural, and environmental contributions transmitted from its mother and father by some rule intended to approximate known rules of genetic transmission via genetic operators such as mutation and crossover, its cultural value (measured on the phenotypic scale) which may be transmitted directly or indirectly from parents, and a non-transmitted environmental value (again measured on the phenotypic scale) whose rules of transmission do not involve the parents. The effect of

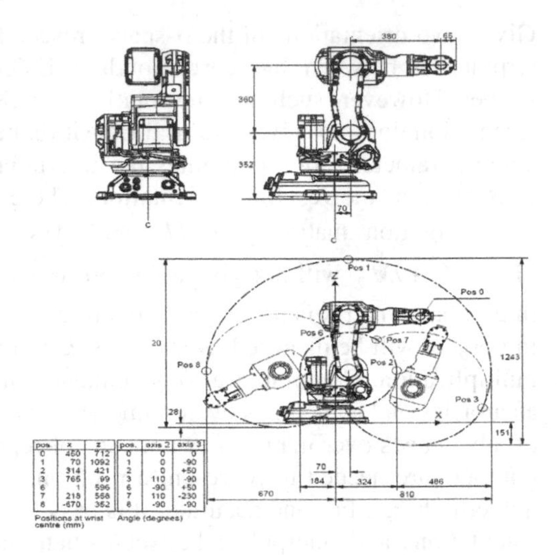

Fig. 4 A dimensional model of ABB's IRB 140 is used to reveal the capabilities of the proposed method (ABB, Product Specification 2000)

the genotype of the individual on its phenotype is denoted by the vector $\vec{G}$. The environmental contribution to an individual's phenotype will be divided into two parts; an effect influenced by cultural transmission $\vec{C}$, and a non-transmitted environmental effect $\vec{E}$, dependent only on the particular environmental experiences of the individual. The phenotype of an individual denoted by the vector $\vec{P}$ is then specified as a linear combination of the normalized genetic and environmental factors in (Otto et al., 1994).

At the starting point all variables are assumed to be normally distributed. The parameters g, c, e are used to describe the strength of the influence of genes, cultural environment, and non-transmitted environment, respectively, on the phenotype of an individual.

The above description depends on the state of the population at a specific point in time. The population changes over time and changes in variance components

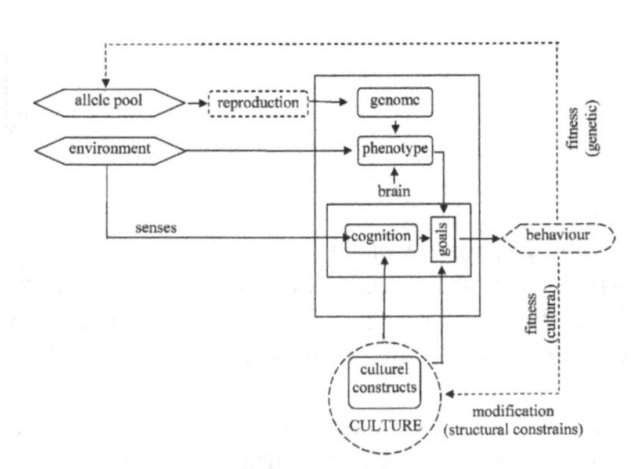

Fig. 5 Cultural model for genetic process of genetic inheritance

must be specified in a dynamically consistent manner to obtain a completely rigorous model. The determination of phenotype is given in Fig.4 as a linear function of latent variables that describe the genotypic, the cultural and the specific environmental contributions to the phenotype.

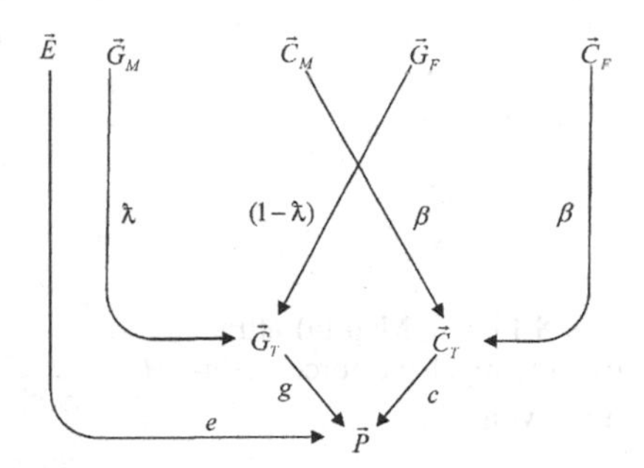

Fig. 6 Contributory effects on Phnotype $\vec{P}=g\cdot\vec{G}+c\cdot\vec{C}+e\cdot\vec{E}$

The genotype $\vec{G}$ is transmitted autonomously between generations; the latent cultural variable $\vec{C}$ is transmitted in a manner analogous to the transmission of the genotypic effects $\vec{G}$, except that $\vec{C}$ is transmitted from parents to offspring by non-genetic means, (6). The coefficients describing the transmission are independent of the genetic composition of the population and of the mating structure in the population.

$$\begin{aligned} G_{transmit}(\vec{G}_M,\vec{G}_F) &= ƛ\vec{G}_M+(1-ƛ)\vec{G}_F+\sigma\cdot S_G \\ C_{transmit}(\vec{C}_M,\vec{C}_F) &= \beta(\vec{C}_M+\vec{C}_F)+\delta\cdot S_C \end{aligned} \tag{6}$$

If S is a segregation variable caused by partition defect (mutation) and structural constrains then phenotype of individual is,

$$\vec{P}=g\cdot(ƛ\cdot\vec{G}_M+(1-ƛ)\vec{G}_F+\sigma\cdot S_G)+c\cdot(\beta\cdot(\vec{C}_M+\vec{C}_F)+\delta\cdot S_C)+e\cdot\vec{E} \tag{7}$$

3.2 Path Planning

Problem is to find a path between start-goal points that doesn't intersect with obstacles $E1, E2,\cdots,E8$ in configuration space illustrated in figure 5. Obstacles are located at $\{0,180,0,300\}$,$\{300,400,0,500\}$, $\{150,300,400,500\}$, $\{400,500,600,800\}$, $\{500,600,900,1000\}$,$\{450,490,200,300\}$,$\{100,160,800,860\}$,$\{200,600, 50\}$ coordinates where rectangles in $\{x_0,x_1,y_0,y_1\}$ representation, (x_0,y_0) indicates left bottom, (x_0,y_1) left top, (x_1,y_0) right bottom and (x_1,y_1) right top vertexes and circle in $\{x_0,y_0,r\}$ representation, (x_0,y_0) is center and r is radius. In dynamic environment simulations it is assumed that sensory information is refreshed in

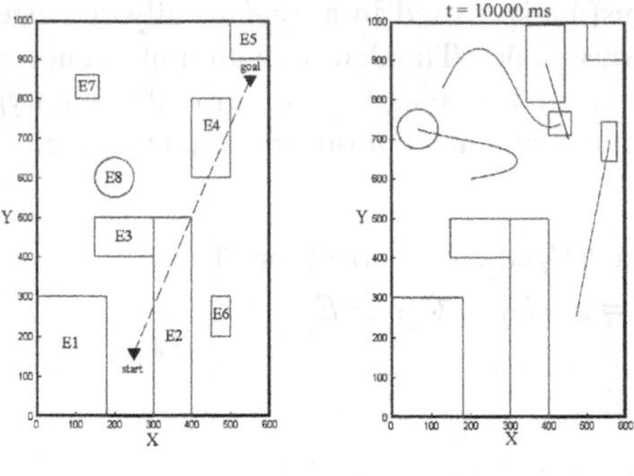

Fig. 7 Configuration space

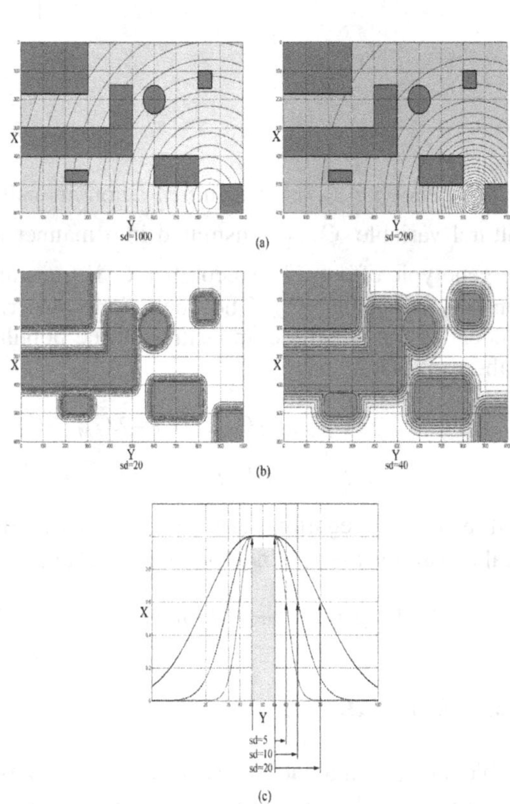

Fig. 8 Fitness Map (a) attractive (b) repulsive force variations with sd.

50 ms. periods. E4, E6, E7 and E8 are moving with linear velocity sequentially 20cm/sn, 45cm/sn, 45cm/sn, 40cm/sn. Their velocity vector is as follows.

$$\begin{aligned}
\vec{V}_{E4} &= -3x + 10y \\
\vec{V}_{E6} &= \quad 2x + 10y \\
\vec{V}_{E7} &= \qquad x + 2 \cdot \cos(t/2000)\, y \\
\vec{V}_{E8} &= (-1 + 4 \cdot \cos(t/2500))x + (1 - 0.5 \cdot \cos(t/5000)\, y
\end{aligned} \tag{8}$$

When potential fields are driven with GA different kinds of expectations can be put into optimization function. Landscape can be dynamically changed during execution. In figure 8. path planning with standard GA based potential field method is given. Local regions are defined as obstacles and filled by GA.

In proposed algorithm phenotypes are created with genetic, cultural and environmental components. Coefficients g, c, e are decided according to actual expects and with equation (7) phenotypes are produced.

By cognition process pointed in Figure 5. forces on landscape can be arranged. For example if sensory information determines target then without further intervention, region is acquitted. Thus shortest path and minimum calculation charge is achieved. Figure 11 illustrates this situation and time needed for evaluation of whole configuration space.

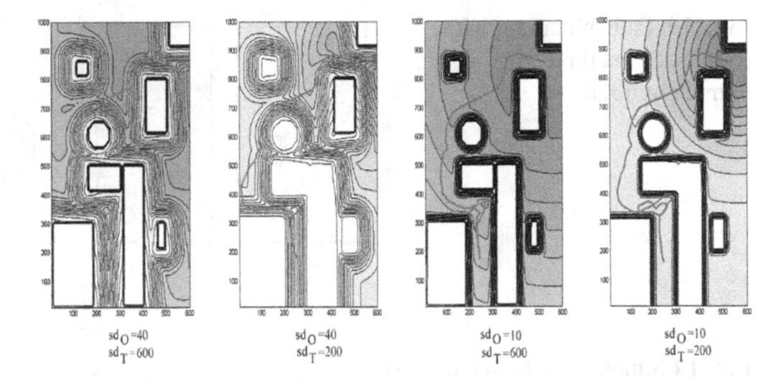

Fig. 9 Path planning with standard GA based potential field method in static environment

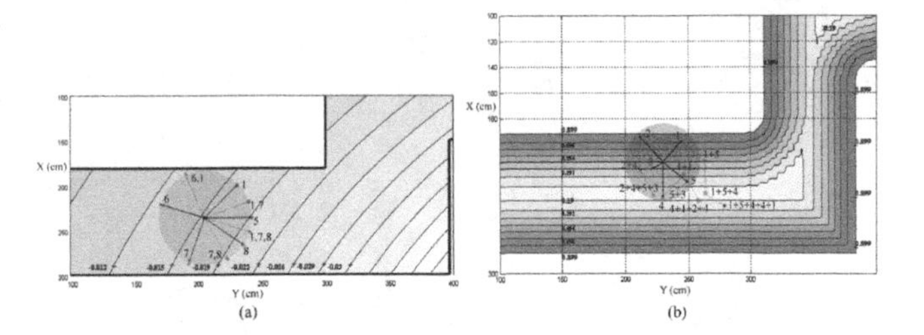

Fig. 10 Reproduction of phenotype (a) genetic (b) cultural component

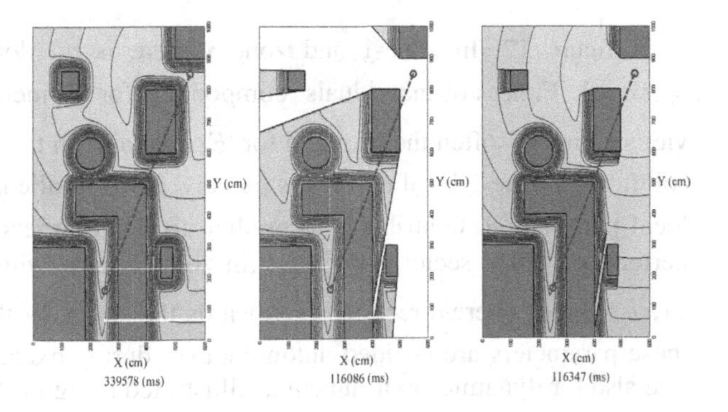

Fig. 11 Environment effects on fitness function

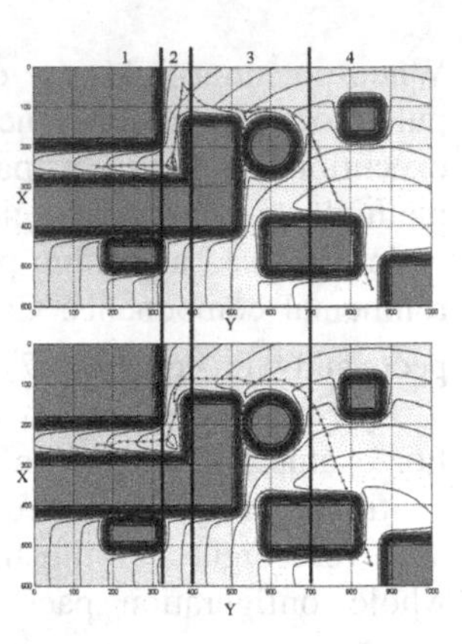

Fig. 12 Path planning with proposed method in static environment

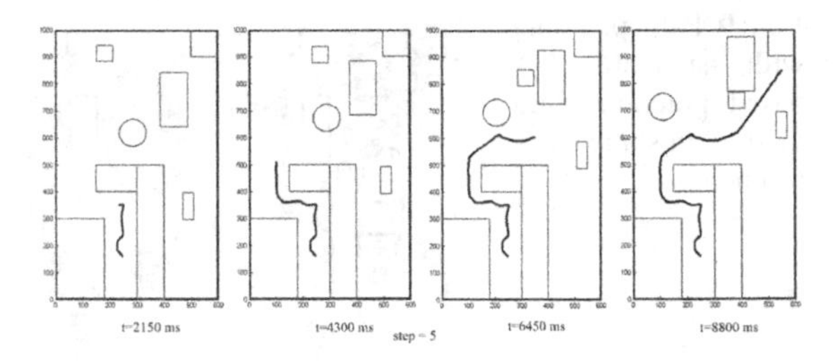

Fig. 13 Path planning with proposed method in dynamic environment

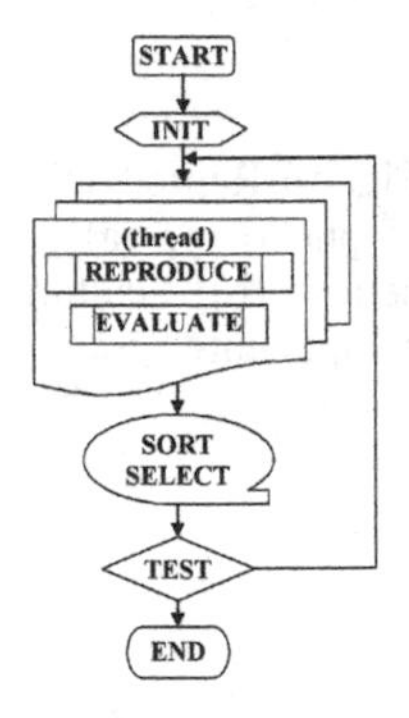

Fig. 14 Simple paralleling schema

In figure 12. In zone-1 and zone-3 there is no dominant trait on path so $g \geq c \approx e$. Fitness of individuals' components for $\vec{C}$ increases in direction of previous points to soften the path and for $\vec{E}$ increases in target direction to accelerate solution. In zone-2 local minima is perceived and $\vec{E}$ alienates the individuals from local point and $\vec{C}$ contributes to predict artificial targets that guides through experiences. In that sector $c > e > g$. ;In zone 4 same with blank regions in figure 11, $e > g > c$ where direction $\vec{E}$ is towards the target. With the proposed algorithm these parameters are decided autonomously during execution and they are effective also for dynamic environment as illustrated in figure 13.

In dynamic environments because method uses local approach and both cultural and environmental components accelerates, computing time is reasonable in real time applications. GA's routines such as reproduction and evaluation don't need previous results. In this respect overlapped operations can easily be applied. The simple paralleling schema given in figure 14 is used.

4 Conclusion

The proposed method is an effective optimization method for the trajectory planning of robot manipulators. It works well under different conditions and environments. As it can be seen from the experiments, it is able to generate optimal trajectories for the robots even in a complicated surroundings and requirements. With the proposed algorithm which carries out the real-time restrictions, generation of collision-free trajectories for the cooperating robots has been established. A novel technique based on genetic algorithms is described and adapted to the trajectory planning problem in robotics. This approach brings a methodology for influencing different kinds of requests to the genetic algorithms. Methodology gives more autonomous to the algorithm. The trajectory optimization of robotic manipulator is difficult due to the highly nonlinearity and the high search dimension of trajectory space. Most of the generated trajectories may exceed the physical constraints

The proposed GA method is an effective optimization method for the trajectory planning of robot manipulators. It works well under different conditions and environments. As it can be seen from the above simulations, it is able to generate optimal trajectories for the robots even in a complicated surroundings and requirements. We develop that algorithm by reflecting the behavior of social communities. This approach brings a methodology for influencing different kinds of requests and new operators to the simple genetic algorithms. Methodology gives more autonomous to the algorithm. If requests and restrictions are well defined at initial level, without intervention satisfactory results are obtained. We recommend adding more progressive traits for agents besides inheritances in future works through that study.

References

1. Qu, Z., Wang, J., Clinton, E.: A new Analytical Solution to Mobile Robot Trajectory Generation in the Presence of Moving Obstacles. IEEE Trans. Robotics 20, 978–993 (2004)
2. Ahuactzin, J.M., Talbi, E.G., Bessiere, P., Mazer, E.: Using genetic algorithms for robot motion planning. In: European Conference on Artificial Intelligence (ECAI 1992), pp. 671–675 (1992)
3. Rendas, M.J., Tetenoire, W.: Definition of exploratory trajectories in robotics using genetic algorithms. In: Tenth International Conference in Industrial and Engineering Applications of Artificial Intelligence and Expert Systems (IEAAIE), Atlanta, USA, pp. 211–216 (1997)

4. Pack, D., Toussaint, G., Haupt, R.: Robot trajectory planning using a genetic algorithm. In: SPIE, pp. 171–182 (1996)
5. Bell, J.E., McMullen, P.R.: Ant colony optimization techniques for the vehicle routing problem. Advanced Engineering Informatics, 41–48 (Fall, 2004)
6. Bland, J.A.: Space planning by ant colony optimization. Int. J. Comput. Appl. Technol. 12, 320–328 (1999)
7. Ju, M.Y., Liu, J.S., Hwang, K.S.: Real Time Velocity Alteration Strategy for Collision-free Trajectory Planning of two Articulated Robot Manipulator. Journal of Intelligent and Robotic Systems 33, 167–186 (2002)
8. Vallejos, P.A., Ruiz-del-Solar, J., Duvost, A.: Cooperative Strategy Using Dynamic Role Assignment and Potential Fields Path Planning. In: Proceedings of the 1st IEEE Latin American Robotics Symposium - LARS 2004 Mexico city, Mexico, pp. 48–53 (2004)
9. Cakar, T., Koker, R., Demir, H.I.: Parallel robot scheduling to minimize mean tardiness with precedence constraints using a genetic algorithm. Advances in Engineering Software 39, 47–54 (2008)
10. Chen, J.L., Chang, W.-D.: Feedback linearization control of a two-link robot using a multi-crossover genetic algorithm. Expert Systems with Applications 36, 4154–4159 (2009)
11. Hamilton, W.R.: Elements of Quaternions. Chelsea Publishing, New York (1969)
12. Dam, E.B., Koch, M., Lillholm, M.: Quaternions, Interpolation and Animation, Tech. Rep., DIKUTR-98/5, University of Copenhagen (1998)
13. Schmidt, J., Nieman, H.: Using Quaternions for Parametrizing 3-D Rotations in Unconstrained Nonlinear Optimization, Vision Modeling, and Visualization, Stuttgart, Germany, vol. 23, pp. 399–406 (2001)
14. Otto, P.S., Christiansen, B.F., Feldman, W.M.: Genetic and Cultural Inheritance of Continuous Traits. Working Paper series #0065 (July 1994)

A Hybrid Multiobjective Evolutionary Algorithm for Anomaly Intrusion Detection

Uğur Akyazı and Şima Uyar

Abstract. Intrusion detection systems (IDS) are network security tools that process local audit data or monitor network traffic to search for specific patterns or certain deviations from expected behavior. We use a multiobjective evolutionary algorithm which is hybridized with an Artificial Immune System as a method of anomaly-based IDS because of the similarity between the intrusion detection system architecture and the biological immune systems. In this study, we tested the improvements we made to jREMISA, a multiobjective evolutionary algorithm inspired artificial immune system, on the DARPA 1999 dataset and compared our results with others in literature. The almost 100% true positive rate and 0% false positive rate of our approach, under the given parameter settings and experimental conditions, shows that the improvements are successful as an anomaly-based IDS when compared with related studies.

Keywords: Anomaly-based Intrusion Detection, DARPA 1999 Dataset, Artificial Immune System, Multiobjective Evolutionary Algorithm.

1 Introduction

An intrusion detection system (IDS) is used to detect intrusions, which are actions that attempt to compromise the integrity, confidentiality or availability of a resource. IDS are classified into two groups as misuse detection and anomaly detection. In the misuse detection approach, network and system resources are examined in order to find known wrong usages by pattern matching techniques [1]. In anomaly-based IDS, the detectors construct profiles of users, servers and network connections using their normal behaviors. After the profile construction, they monitor new event data, compare them with obtained profiles and try to detect deviations. These deviations from normal behaviors are flagged as attacks.

Uğur Akyazı
Computer Eng. Department, Turkish Air Force Academy, Yesilyurt, Istanbul, Turkey
e-mail: u.akyazi@hho.edu.tr

A. Şima Uyar
Computer Eng. Department, Istanbul Technical University, Maslak, Istanbul, Turkey
e-mail: etaner@itu.edu.tr

A.P. de Leon F. de Carvalho et al. (Eds.): Distrib. Computing & Artif. Intell., AISC 79, pp. 509–516.
springerlink.com

Most real-world problems involve simultaneous optimization of several objectives. A suitable solution should have acceptable performance over all objectives. MOEAs can effectively provide several pareto optimal solutions and give the researchers the option to assess the trade-offs between different alternatives [2].

We used a multiobjective evolutionary algorithm which is hybridized with an Artificial Immune System (AIS) as a method of anomaly-based IDS because of the similarity between the IDS architecture and the biological immune system (BIS), which is a parallel and distributed adaptive system for detecting antigens.

In a previous study [3], we explored the effects of some improvements we made on jREMISA [4] using the 2000 version of the DARPA dataset for DDoS attacks. However, there are not many studies in literature with results using this version of the dataset to detect DDoS attacks. In order to compare the performance of our improved-jREMISA with other related studies, we made two types of experiments using different compositions of 1999 DARPA IDS dataset days. We observed good results with almost 100% True Positive and almost 0% False Positive rates in all conditions.

2 Related Studies

In order to be able to assess the performance of our improved-jREMISA algorithm in comparison with others, in this study, we are interested only in the IDS studies which tested their solutions on the 1999 DARPA IDS dataset.

PHAD [5] is an anomaly detection algorithm which models protocols rather than the user behavior because the majority of the attacks exploit protocol implementation bugs. It uses a time-based model and models 33 attributes which correspond to packet header fields with 1–4 bytes.

In the hybrid-IDS of [6], PHAD and NETAD [7] are experimentally improved and added as a preprocessor to Snort [8]. It operates in two phases: the filtering phase and the modeling phase.

In [9], two open-source network intrusion detection systems (Snort and Pakemon [10]) are combined with Cisco IOS Firewall [11] intrusion detection features to increase detection of attacks. IDS are used with Firewalls in order to double-check the mis-configured firewalls and to catch the insider attacks.

In [12], the results of eight different studies, [13, 14, 15, 16, 17, 18] which participated in the DARPA off-line intrusion detection evaluation in 1999, are compared.

K-Means+ID3 [19] is developed by combining two machine learning algorithms: the k-Means clustering and the ID3 decision tree learning. In the first stage, k-Means clustering is performed on training instances to obtain k disjoint clusters. In the second stage, the ID3 decision tree learning is used in each cluster.

PAYL [20] is a payload-based anomaly detector for intrusion detection. They first compute a profile byte frequency distribution and its standard deviation. Later, they use the Mahalanobis distance during the detection phase to calculate the similarity of new data against the profile. POSEIDON [21] (Payl Over Som for Intrusion DetectiON) is also payload-based, and has a two-stage architecture: the first stage consists of a Self-Organizing Map (SOM) [22], and is used to classify payload data, while the second one is a modified PAYL system.

In jREMISA [4], an Artificial Immune System [23] is used together with a Multiobjective Evolutionary Algorithm [24] in order to get good detectors with the best classifying fitness degree and multiobjective hypervolume size. Network traffic is classified as self and non-self with the help of antigen detectors which are trained using a dataset.

3 The Improved - jREMISA

In a previous study [3], we improved jREMISA [4]; a multiobjective evolutionary algorithm inspired artificial immune system, to get better true and false positive rates. We added the r-continuous evaluation method, changed the Negative Selection and the Clonal Selection structure, and redefined the objectives while keeping the general concepts the same. In the following paragraphs, the jREMISA algorithm including our improvements is explained. For details on the implementation of the original jREMISA and the improvements we made, refer to [3] and [4] respectively.

Antigen (Ag) and Antibody (Ab) chromosomes are binary arrays. Antigens are represented differently for three most common IP protocols of TCP, UDP and ICMP traffic. Ab chromosomes are composed of three parts as DNA (binary), RNA (binary), and seven state properties (integer).

Algorithm 1. The pseudocode of the improved-jREMISA algorithm

```
Procedure improved-jREMISA
Begin
repeat
  Creation of Primary TCP, UDP and ICMP Population (Pop_p)
  Empty Initialization of Secondary Population (Pop_s)
  Negative_selection(Pop_p,data_set_clean,threshold)
until (end of data_set_clean)
repeat
  FitnessFunction (ag,threshold)
  MutationCauchy(Pop_p)
  P_optimality()
  ClonalSelection(0.05)
  MutationUniform(Pop_s)
  Pop_p ← Pop_s //Copy the best Pop_s to the Pop_p of next generation
until (end of data_set_attack)
End
```

Pseudocode of the improved-jREMISA is given in Algorithm 1. Crossover is not applied since mutation is considered to be sufficient to make Ab's move in the objective search space. The primary population is separated into three groups according to the IP protocols, so that a non-TCP Ab is not compared with a TCP Ab. Every Ag in the evaluation window represents a new generation and all operations are applied to all of the Ab's in the population.

For fitness evaluations, the Hamming distance (H), which is defined as the similarity of the Ab and Ag DNA genes, is calculated. The first objective is penalized in true positives/negatives and the second objective is penalized in false positives/negatives. H should be zero in an ideal true negative and it should be equal to the length of the Ag in an ideal true positive. Cauchy Mutation is applied on the penalized Ab bits. P*-Test is applied to each Ab in order to calculate how many of the other Ab's dominate it. All the Ab's are sorted according to these domination values.

In Clonal Selection, 5% of the non-dominated Ab's of the primary population are selected with an elitist selection and copied to the secondary population. Copied Ab's are cloned three times in order to have a large population. Mutation occures during the cloning stage. Size of the secondary population is trimmed to the size of the primary population. The Ab's from the secondary population with the highest fitness values are copied to the primary population in place of the discarded Ab's.

Hamming distance evaluations in the Negative Selection, Fitness Function and Clonal Selection steps are enhanced using r-continuous bit evaluations. Two compared chromosomes need to have at least r-continuous bits the same to be considered alike.

4 Experiments

The main objective of IDS is detecting wrong, unauthorized and malicious usage of computer systems by inside and outside intruders. The key is to maximize accurate alerts (true-positive) while minimizing the occurrence of non-justified alerts (false-positive). The metrics most commonly used in the evaluation of IDS are:

- True positive (TP): a real attack correctly categorized as an attack,
- False positive (FP): a false alert erroneously raised for normal data,
- True negative (TN): normal data which correctly does not generate an alert,
- False negative (FN): a missed attack erroneously categorized as normal.

4.1 1999 DARPA IDS Dataset

Lincoln Laboratory, under the sponsorship of Defense Advanced Research Projects Agency (DARPA), created the Intrusion Detection Evaluation Dataset (IDEVAL) which is a benchmark in literature [25]. In 1998, 1999 and 2000, they gathered tcpdump, Sun BSM, process and file system information after the background activities were produced with scripts, and attacks were created [26]. The 1999 evaluation had two phases separated by about three months. During the first phase, participants were provided with three weeks of data. The first and third weeks contained no attacks, and could be used to train anomaly detection systems. During the second phase, participants were provided with two weeks of test data (weeks 4 and 5) containing 201 instances of 58 attacks [27].

4.2 Experiments with the Improved-jREMISA

We implemented the experiments on Pentium Core 2 Duo 2.4 GHz computers which run the Windows XP SP3 operating system. We performed two types of experiments using different compositions of 1999 DARPA IDS dataset days in training and testing phases of the improved-jREMISA in order to see the success of the system in different experimental environments. In the first type, the system is trained and tested with the same day of week-1 and week-4. But in the second type, different days are used for testing to measure the system performance in more realistic conditions. Each experiment has 20 runs to get the mean values of the evaluation metrics of true positive and false positive rates.

We used 20 as the r-continuous value, 50% for the affinity threshold and 300, 100, 100 for TCP, UDP and ICMP population sizes respectively. These were found to be the best parameter settings in our previous study [3] where we obtained very successful results on the DARPA LLDOS 1.0 dataset. Five days of attack-free week-1 dataset is used for Negative Selection of training and four days of week-4 with Tuesday of week-5 (Tuesday of week-4 was not given) is used for MOEA of training and also for testing in our system. We created the truthset files including the attack packets with the help of the Ethereal software [28] and identification lists given on the Lincoln Laboratory website. Secondary population Ab chromosomes which are obtained as the best results at the end of the MOEA phase are used in the Testing phase on the dataset which includes the attack traffic.

4.3 Results

In the first set of experiments, same days of week-1 and week-4 are used in the training and testing phases of our study as seen in Table 1. For example, if Wednesday of week-1 is used for Negative Selection and MOEA phases of training, Wednesday of week-4 is used for testing. These two datasets are not only different because week-1 is attack-free and week-4 is with-attack, but they are also different in the way their traffic flows and in the amount of protocols used. For example, there is 92,67 % TCP, 7,26 % UDP and 0,07 % ICMP traffic on Monday of week-1; but there is 79,08 % TCP, 17,72 % UDP and 3,20 % ICMP traffic on

Table 1 Results of the first set of experiments

Training		Testing (week-4)	True Positive (%)	False Positive (%)
Negative Selection (week-1)	MOEA (week-4)			
Monday	Monday	Monday	99,71	0
Tuesday	Tuesday	Tuesday	99,27	0,00005
Wednesday	Wednesday	Wednesday	99,45	0
Thursday	Thursday	Thursday	99,98	0
Friday	Friday	Friday	99,97	0
Average			99,68	0,00001

Monday of week-4. True Positive rates are almost 100 % and False Positive rates are almost 0 % for each day averaged over 5 days.

The second set of experiments differs from the previous set by using different days of week-4 for training and testing phases. False Positive and True Positive rates don't deviate a lot from 0% and 100% respectively. Our system also gives good results in these more realistic test conditions as seen in Table 2.

Table 2 Results of second-type experiments

Training		Testing (week-4)	True Positive (%)	False Positive (%)
Negative Selection (week-1)	MOEA (week-4)			
Composition of five days	Monday	Tuesday*	96,15	0,0003
	Monday	Wednesday	97,60	0,0038
	Monday	Thursday	95,87	0,0006
	Monday	Friday	99,99	0,0005
	Thursday	Monday	100	0,0045
	Thursday	Friday	100	0,004
Average			98,27	0,002

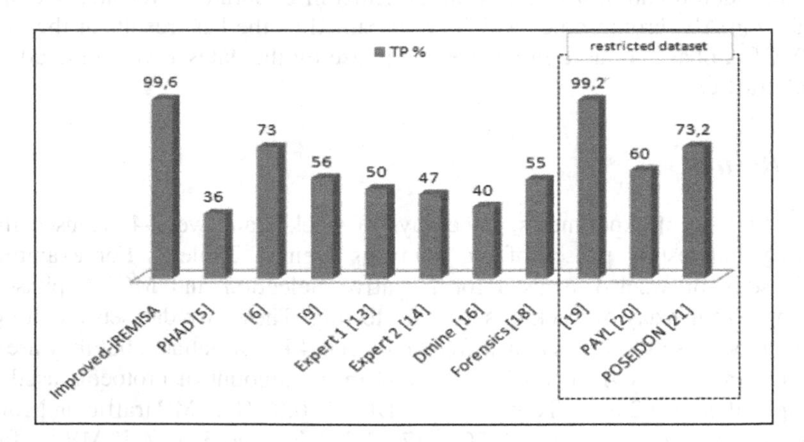

Fig. 1 TP comparison of improved-jREMISA with other studies

Table 3 FP values of the systems

Systems	improved-JREMISA	PHAD [5]	[6]	[9]	Expert 1 [13]	Expert 2 [14]	Dmine [16]	Forensics [18]	[19]	PAYL [20]	POSEIDON [21]
FP values	0,00005%	10 FA/day	N/A	N/A	10 FA/day	10 FA/day	10 FA/day	10 FA/day	4%	1%	1%

When we compare the performance of the improved-jREMISA on the first-type of experiments and other related studies over the DARPA 1999 dataset, we see that our study has better TP and FP values than the others. Last three studies of Figure 1 and Table 3 used restricted dataset as stated before.

5 Conclusion

We used a multiobjective evolutionary algorithm which is hybridized with an Artificial Immune System (AIS) as a method of anomaly-based IDS because of the similarity between the IDS architecture and the biological immune system (BIS). In a previous study [3], we improved jREMISA [4]; a multiobjective evolutionary algorithm inspired artificial immune system, to get better truc and false positive rates. We added the r-continuous evaluation method, changed the Negative Selection and the Clonal Selection structure, and redefined the objectives while keeping the general concepts the same. In this study, we performed two types of experiments using different compositions of 1999 DARPA IDS dataset days in Negative Selection, MOEA and Testing phases of the improved-jREMISA.

In the second type of experiments, different days of week-1 and week-4 are used in the training and testing phases to mimic a realistic test environment. We achieved True Positive rates close to 100% and False Positive rates close to 0% for each day as averaged over 5 days in both types of experiments. As a result, we can say that our study has better TP and FP values than other related studies over the DARPA 1999 dataset, which is a very noteworthy success as an anomaly intrusion detection system. Other common intrusion detection datasets may also be tested on this system in order to strengthen its structure.

References

1. Abraham, A., Grosan, C., Chen, Y.: Cyber Security and the Evolution of Intrusion Detection Systems. Journal of Educational Technology, Special Issue in Knowledge Management (2005), ISSN 0973-0559
2. Sbalzarini, I.S., Muller, S., Koumoutsakos, P.: Multiobjective optimization using evolutionary algorithms, Center for Turbulunce Research. In: Proc. of Summer Program (2000)
3. Akyazı, U., Uyar, A. .: Detection of DDoS Attacks via an Artificial Immune System-Inspired Multiobjective Evolutionary Algorithm. In: 7th European Event on the Application of Nature-inspired Techniques for Telecommunication Networks and other Parallel and Distributed Systems, EvoCOMNET (2010)
4. Haag, C.R., Lamont, G.B., Williams, P.D., Peterson, G.L.: An artificial immune system-inspired multiobjective evolutionary algorithm with application to the detection of distributed computer network intrusions. In: GECCO 2007: Genetic and evolutionary computation Conference, London, UK (2007)
5. Mahoney, M.V., Chan, P.K.: PHAD, Packet Header Anomaly Detection for Identifying Hostile Network Traffic. Florida Tech., technical report, 2001-04 (2001)
6. Aydin, M.A., Zaim, A.H., Ceylan, K.G.: A hybrid intrusion detection system design for computer network security. Computers and Electrical Eng. Journal 35(3), 517–526 (2009)
7. Mahoney, M.V.: Network traffic anomaly detection based on packet bytes. In: ACM Symposium on Applied Computing, SAC (2003)
8. Snort – the de facto standard for intrusion detection/prevention, http://www.snort.org (Cited March 18, 2010)
9. Kayacik, G.H., Zincir-Heywood, A.N.: Using Intrusion Detection Systems with a Firewall: Evaluation on DARPA 99 Dataset. NIMS Technical Report, #062003 (2003)

10. Takeda, K., Takefuji, Y.: Pakemon – A Rule Based Network Intrusion Detection System. Int. Journal of Knowledge-Based Intelligent Engineering Systems 5(4), 240–246 (2001)
11. Cisco IOS Firewall Intrusion Detection System, `http://www.cisco.com/en/US/docs/ios/12_0t/12_0t5/feature/guide/ios_ids.html` (Cited April 10, 2010)
12. Lippmann, R.P., Haines, J.W., Fried, D.J., Korba, J., Das, K.: Analysis and Results of the 1999 DARPA Off-Line Intrusion Detection Evaluation. In: Third International Workshop on Recent Advances in Intrusion Detection (RAID 2000), Toulouse, France (2000)
13. Neumann, P., Porras, P.: Experience with EMERALD to DATE. In: 1st USENIX Workshop on Intrusion Detection and Network Monitoring, California, pp. 73–80 (1999)
14. Vigna, G., Eckmann, S.T., Kemmerer, R.A.: The STAT Tool Suite. In: DARPA Information Survivability Conference and Exposition, DISCEX (2000)
15. Sekar, R., Uppuluri, P.: Synthesizing Fast Intrusion Prevention/Detection Systems from High-Level Specifications. In: 8th Usenix Security Symposium, Washington DC (1999)
16. Barbara, D., Wu, N., Couto, J., Jajodia, S.: ADAM: Detecting Intrusions by Data Mining. In: IEEE SMC Information Assurance Workshop, West Point, NY (2001)
17. Ghosh, A.K., Schwartzbard, A.: A Study in Using Neural Networks for Anomaly and Misuse Detection. In: USENIX Security Symposium, August 23-26, Washington D.C. (1999)
18. Tyson, W.M.: DERBI: Diagnosis, Explanation and Recovery from Computer Break-ins. Final Report, Artificial Intelligence Center, SRI Int., DARPA Project F30602-96-C-0295 (2001)
19. Gaddam, S.R., Phoha, V.V., Balagani, K.S.: A Novel Method for Supervised Anomaly Detection by Cascading K-Means Clustering and ID3 Decision Tree Learning Methods. IEEE Trans. on Knowledge and Data Engineering 19(3), 345–354 (2007)
20. Ke, W., Stolfo Salvatore, J.: Anomalous Payload-based Network Intrusion Detection. In: 7th International Symposium on Recent Advances in Intrusion Detection (RAID), pp. 203–222 (2004)
21. Bolzoni, D., Etalle, S., Hartel, P., Zambon, E.: POSEIDON: a 2-tier Anomaly-based Network Intrusion Detection System. In: Fourth IEEE International Workshop on Information Assurance IWIA 2006, pp. 144-156 (2006)
22. Kohonen, T.: Self-Organizing Maps, 3rd Extended edn. Springer Series in Information Sciences, vol. 30. Springer, Heidelberg (2001)
23. Aickelin, U., Dasgupta, D.: Artificial Immune Systems Tutorial. In: Burke, E., Kendall, G. (eds.) Search Methodologies: Introductory Tutorials in Optimization and Decision Support Methodologies, ch. 13. Springer, Heidelberg (2005)
24. Coello, C.A., Lamont, G.B., Van Veldhuizen, D.A.: Evolutionary Algorithms for Solving Multi-Objective Problems. In: Genetic and Evolutionary Computation, Springer, Heidelberg (2007)
25. Mahoney, M.V., Chan, P.K.: An Analysis of the 1999 DARPA/Lincoln Laboratory Evaluation Data for Network Anomaly Detection. In: Vigna, G., Krügel, C., Jonsson, E. (eds.) RAID 2003. LNCS, vol. 2820, pp. 220–239. Springer, Heidelberg (2003)
26. Brugger, S.T., Chow, J.: An Assessment of the DARPA IDS Evaluation Dataset Using Snort, UC Davis Technical Report CSE-2007-1, Davis, CA (2007)
27. Haines, J.W., Lippman, R., Fried, D.J., Zissman, M.A., Tran, E., Boswell, S.B.: 1999 DARPA intrusion detection evaluation: design and procedures. MIT Lincoln Laboratory Technical Report, TR-1062, Massachusetts (2001)
28. Ethereal: Open-source network protocol analyzer, `http://www.ethereal.com` (Cited March 21, 2010)

JXTA-Sim: Simulating the JXTA Search Algorithm

Sandra Garcia Esparza and René Meier

Abstract. JXTA is a set of platform-independent, open source peer-to-peer protocols that has become popular for building services and applications based on peer-to-peer overlays. Existing work towards evaluating one of JXTA's core protocols,the JXTA search algorithm, has mainly focused on testbed-based experiments. Although such evaluation configurations offer accurate results based on real deployments, scaling experiments to a large number of distributed hosts is difficult and often prohibitively expensive. Furthermore, repeating experiments using different configuration parameters might yield distorted results due to the uncontrolled nature of testbeds. Simulators offer an alternative to testbeds for evaluating large-scale applications in a controlled environment.This paper presents JXTA-Sim, a simulator for studying and evaluating the JXTA search algorithm. JXTA-Sim enables researchers to study the behavior of JXTA's search algorithm using different configuration parameters and ultimately, to test JXTA-based peer-to-peer applications.

Keywords: Peer-to-peer computing, search algorithm, simulating JXTA applications.

1 Introduction

Over the last years, features of peer-to-peer (P2P) networks, such as, self- organization, scalability and robustness, have captured the interest of researchers. Project JXTA[1] is an open-source platform originally conceived by Sun Microsystems Inc. in 2001 with the goal to standardize a set of protocols for building P2P applications. JXTA consists of a set of 6 language and platform independent protocols that allow developers to build interoperable and scalable P2P services and applications.

Sandra Garcia Esparza
CLARITY: Centre for Sensor Web Technologies, University College Dublin, Belfield, Dublin 4, Ireland
e-mail: sandra.garcia-esparza@ucd.ie

René Meier
Distributed Systems Group, School of Computer Science and Statistics, Trinity College Dublin, Dublin 2, Ireland
e-mail: rmeier@cs.tcd.ie

A.P. de Leon F. de Carvalho et al. (Eds.): Distrib. Computing & Artif. Intell., AISC 79, pp. 517–524.
springerlink.com

P2P networks are often used in applications where peers need to share information with other peers across a possibly highly-distributed network topology. Such peers are said to publish information, also called resources, and to discover resources respectively. This mechanism is key to any P2P application, needs to be efficient and scalable, and is commonly referred to as a search algorithm for P2P networks. There are different algorithms for discovering resources depending on the structure of the underlying network. In structured networks, the graph of the network follows a particular structure and resources are located using the information about this structure. On the other hand, unstructured networks do not follow a particular structure and resources are randomly placed in the topology. The JXTA search algorithm supports a hybrid of these two approaches to locate resources. JXTA's search algorithm follows a structured approach when the network is stable and an unstructured approach when the network is unstable, i.e., is exposed to frequent topology changes due to peers joining and leaving. Although JXTA has a widespread use in research and industry, its performance and scalability capabilities still remain largely unclear. Hence, we need to investigate ways to evaluate the JXTA platform and its protocols. JXTA consists of a big and complex architecture and, as a result, an evaluation may assess individual JXTA components or protocols separately.

Past evaluations [4, 3] have studied different aspects of the platform by using testbeds. However, these evaluations consider only a small number of peers due to the high cost of deploying testbeds for large scenarios. Simulations offer a good alternative to testbeds for evaluating an algorithm in such scenarios. Simulators can model an algorithm in networks with a large number of nodes without the costs testbeds introduce. Furthermore, they offer a level of control and easy modification of the simulated algorithm since they are typically deployed on relatively few (often a single) nodes.

This paper presents JXTA-Sim, a simulator for studying and evaluating the JXTA search algorithm. JXTA-Sim has two main goals. The first goal is to allow researchers to understand, study and evaluate the JXTA search algorithm. The second is to allow P2P researchers to test their applications on top of JXTA's overlay network. This will enable them to evaluate the performance of the algorithm (and ultimately their application) and compare it with other overlay algorithms. JXTA-Sim is based on PlanetSim[8], a P2P framework for implementing P2P overlay networks. This paper is organized as follows. Section 2 introduces JXTA's search algorithm. Section 3 discusses related work in the field of simulation and JXTA evaluation. Section 4 presents the design and the architecture of JXTA-Sim. Section 5 presents the evaluation experiments we have performed with JXTA-Sim. Finally, Section 6 concludes this paper and discuses issues that remain for future research.

2 The JXTA Search Algorithm

The JXTA project does not specify how to search for advertisements (i.e. XML documents describing resources) but provides a generic resolver protocol framework with a default policy that can be overwritten [1]. JXTA combines the advantages of structured and unstructured networks [5] by using a hybrid approach that combines the use of a *loosely-consistent Distributed Hash Table (DHT)* with a *limited-range rendezvous walker* [6]. The DHT is used for networks with low peer churn rates while the rendezvous walker is used for networks that change frequently.

There are two type of peers (or nodes) involved in the JXTA search algorithm. *Edge peers* are the basic peers and are responsible for publishing and discovering advertisements. *Rendezvous peers*, on the other hand, have the ability to forward discovery requests and in order to do so, they store the indexes of the advertisements. Edge peers publish and discover advertisements by pushing their indexes to their rendezvous peers. They also store advertisements in their local cache. Rendezvous peers form a rendezvous network where they propagate the queries and responses generated by peers. To keep the connections with other rendezvous peers, each of them maintains a Rendezvous Peer View (RPV), which is an ordered list of IDs of the known rendezvous peers in the group. This list does not need to be globally consistent. A peer may not be aware of the existence of other rendezvous peers and RPVs may be different on different peers.

After storing an index locally, rendezvous peers compute a hash function to identify the rendezvous peer from the RPV that has a replica of the advertisement's index. The *replication distance* is a parameter that specifies the number of peers in each direction of the rendezvous peer (up and down) that holds a copy. When an edge peer needs to retrieve an advertisement, it forwards the request to its rendezvous peer. This will execute the same hash function as the one used for the publication, to identify the rendezvous peer from its RPV that stores the index (and therefore knows the edge peer that stores the advertisement). Because the same function is executed, if the network is stable and all rendezvous peers share the same RPV, the chosen rendezvous will be the one storing the index of the advertisement. On the other hand, if it is unstable and rendezvous peers have not been able to keep their tables updated, two things can happen. If the network has not experienced many changes, we may find a replica of the index in a rendezvous peer near the original one and continue the retrieval from there. However, if the index is not found in any of the replica peers, then a walker needs to be started in both up and down directions in the vicinity of the rendezvous peer chosen from the RPV. This walking continues until the advertisement is found or until a maximum number of hops are reached.

3 Related Work

Past evaluations [4, 3] have studied different aspects of the JXTA platform by using testbeds. However, these evaluations consider only a small number of peers (up to 32). In [2] an experimental evaluation of the scalability of the JXTA search is presented based on using a testbed consisting of 5000 CPUs. This work focuses on measuring the time it takes for the rendezvous protocol to make the peerviews stable (all rendezvous have the same view) and the time it takes to retrieve an advertisement. These experiments concluded that with more than 45 rendezvous peers, the rendezvous protocol fails to make the peer views consistent. In these experiments, the rendezvous algorithm has been studied by installing the whole JXTA platform on every node. In contrast, our focus is on studying different of aspects of the search algorithm in a controlled manner. We aim to provide researchers with a tool for assessing the algorithm in application-specific scenarios while allowing for changes to the set of configuration parameters. By using a simulator, we can facilitate changes to the algorithm and re-evaluate scenarios that may scale to a considerable number of nodes. [7] discusses desirable characteristics for P2P overlay simulators. PlanetSim [8] is a discrete, event-based overlay simulator that supports implementation and validation of overlay algorithms and which provides most of those characteristics by being object-oriented, extensible and customizable. As a result, PlanetSim has been chosen as the platform for building JXTA-Sim.

4 JXTA-Sim Design and Architecture

JXTA-Sim builds on PlanetSim extending the three architecture layers: application, overlay and network. In fig. 1 depicts the architecture of JXTA-Sim. Dark colored components represent JXTA-Sim extensions that build upon the light colored PlanetSim components.

Application Layer. This layer allows applications to be assessed over different overlay schemes. JXTA-Sim defines two classes in this layer. The *EdgeNodesApplication* component represents an application that runs on an edge node. This class contains the application to be tested. JXTA-Sim currently defines a basic application that simulates edge nodes joining and leaving the network as well as publishing and retrieving advertisements. However, the *Application* class can be extended to

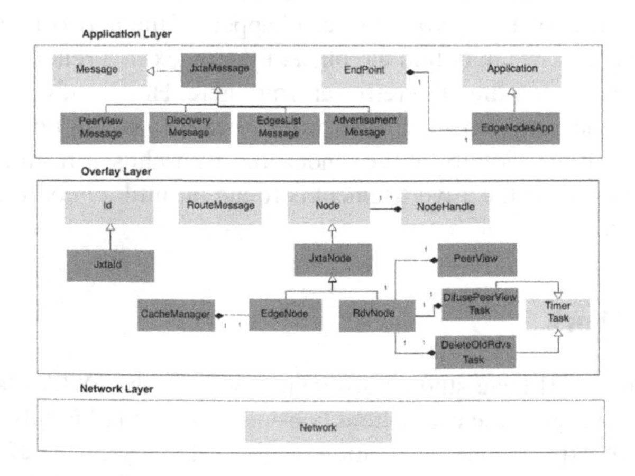

Fig. 1 The architecture of JXTA-Sim

support multiple applications concurrently. The *JxtaMessage* class has been extended from *Message* to support the different types of messages sent by JXTA peers.

Overlay Layer. This layer represents the virtual layer on top of the physical layer. As PlanetSim allows for only *Node* type, JXTA-Sim defines an intermediate *JXTA Node* class that can be of two types: *EdgNode* or *RendezvousNode*. Each node contains the methods according to their behavior. While *EdgeNode* contains methods, including, "connect to rdv", "publish adv" and "find adv", *RendezvousNode* has methods such as "store adv", "retrieve adv, "update peer view" and "diffuse peer view". In addition, *EdgeNode* has a structure to store the published advertisements, which is the *Cache Manager*. *RendezvousNode* supports a structure to maintain the connections to other rendezvous peers, called the *Rendezvous Peer View (RPV)*. JXTA-Sim periodically executes two tasks: *DifusePeerView*, which sends a subset of the entries of the peer view to a random set of rendezvous peers, and *DeleteOldRendezvous*, responsible for removing expired entries.

Network Layer. This layer represents the physical network where nodes communicate with each other by sending *Route Messages*. Currently it does not support some networking aspects, such as, latencies or node mobility. However, future versions of the network layer may simulate such characteristics. For example, in [8], the authors of PlanetSim propose an extension that supports latencies.

4.1 Configuration Parameters and Gathering Results

Currently, JXTA does not support parameters, for example, for changing the replication distance or for adjusting the maximum number of hops. We believe that allowing users to modify these parameters may improve the effectiveness of the search algorithm for certain application scenarios; for instance, it might be beneficial to have a higher replication distance in networks with high churn compared to those with a low churn. The main parameters that can be configured in JXTA-Sim are the following: number of simulation steps, max. number of edge peers connected to a rendezvous, max. number of walker hops, size of the RPV, expiration time for a RPV entry, size of edge peers cache, replication distance, number of published advertisements during the simulation, number of advertisements to be discovered and number of rendezvous that will join and leave during simulation.

4.2 Using JXTA-Sim

JXTA-Sim uses a configuration file for defining both the simulation and the search algorithm parameters. Furthermore, an event file contains a list of time-event pairs representing the application events that occur during the simulation. As an alternative, users can write a program that calls the desired events at appropriate times.

JXTA-Sim provides alternative means for capturing the results of a simulation. Using PlanetSim, JXTA-Sim allows creating a graph of the network at any time during a simulation. It also allows to see the different types of messages sent at every step of the simulation and introduces new aspects to show the number of hops of searches. Finally, JXTA-Sim captures information about the connections between peers. This enables users to observe how the edge peers connect to certain rendezvous peers and to explore a peer's RPV.

5 Evaluation

5.1 Assessing JXTA-Sim

In order to assess the simulation of the search algorithm, we perform a set of experiments that test the features of the algorithm, including adding peers, disconnecting peers, publishing advertisements and locating advertisements. These experiments generate traces and ultimately graphs for studying and verifying the behavior of the algorithm. We capture the sequence of events, such as peers joining and leaving as well as the messages sent between peers as a result of these events. This visualizes the various JXTA messages and provides a basis for verifying that they match the JXTA algorithm as described above in section 2. Due to space limitations, we only show one diagram of the results from these experiments. Fig. 2 depicts the messages sent in an experiment with 60 simulation steps in a network with 5 peers. In this experiment, a peer leaves at simulation step 28, while another peer publishes an advertisement at step 34 that is discovered by a third peer at step 53. Moreover, the messages captured in these experiments also verify that the DHT table is used (and not the walker) when the network is stable. As a result, these experiments show that our simulation of the search algorithm behaves according to the JXTA specification.

5.2 Assessing the JXTA Search Algorithm in JXTA-Sim

After assessing the simulator we now focus on evaluating the JXTA search algorithm. The performance of the algorithm under different configuration parameters is

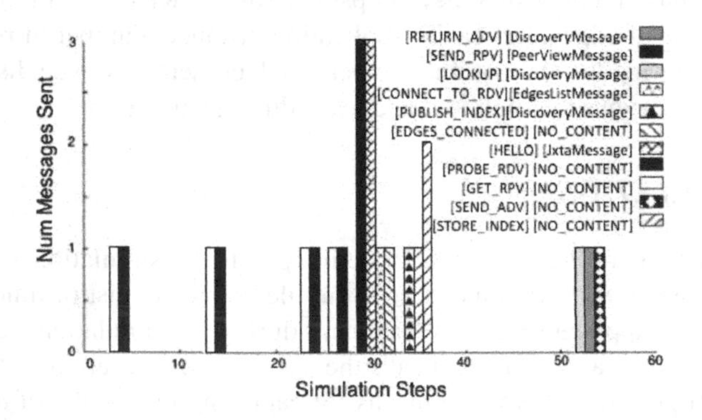

Fig. 2 Messages sent in a network with 5 rendezvous peers.

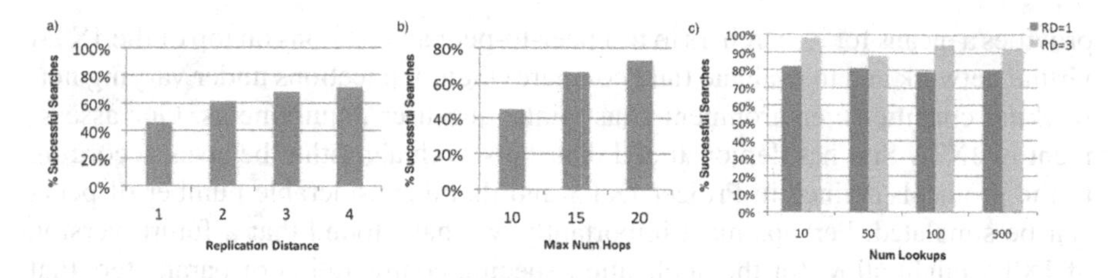

Fig. 3 Percentage of successful searches under different Replication Distance (a,c) and Max Num Hops (b) values.

evaluated by means of the percentage of successful searches and the number of hops required to locate an advertisement. The initial scenario studies the performance on a static network, where peers do not join or leave the group, and uses the following parameters: 50 rendezvous peers, 150 edge peers, replication distance (RD) = 1, and maximum number of hops = 5. This scenario was repeated for lookup searches ranging from 10 to 500 along 1000 simulation steps and found that around 95% of the advertisements were located in each setting. The refined scenario studies how churn affects the performance of the search and, most importantly, how to alleviate the effects of churn by modifying configuration parameters. Fig. 3c, shows that churn reduces the number of successful searches and that increasing RD to 3 alleviates this effect. A churn of 20% (20% of the rendezvous peers leave and 20% join during the simulation) reduces successful searches to about 80% while increasing RD increases them to above 90%.

A second scenario assesses the impact of RD and the number of hops (NH) in a network with 4000 nodes (1000 rendezvous peers and 3000 edge peers), a churn of 10%, RD= 1 and maximum number of hops = 10. Fig. 3a shows that increasing RD can improve successful searches from 45% to 73%. This again demonstrates that increasing RD may compensate the effect of churn. Fig. 3b shows that increasing NH can also improve successful searches. It can be observed that for RD=1, NH needs to be doubled to achieve the same performance increase as for a RD=4. Doubling the number of hops implies that the walker performs up to 20 hops, which increases the network traffic considerably. Hence, increasing the replication distance is likely the preferable option for compensating for churn.

These experiments demonstrate that JXTA-Sim can support a significant number of peers and that being able to modify configuration parameters can help the JXTA search algorithm to adapt to the different requirements of application scenarios. While these parameters are currently fixed to a given default value, we argue that the JXTA platform would benefit from such an extension as it will likely improve overall application performance for varying network sizes and overlay churns.

6 Conclusions and Future Work

This paper has presented JXTA-Sim, a simulator for understanding, studying and evaluating the behavior and performance of the JXTA search algorithm. JXTA-Sim provides a means for researchers to test peer-to-peer applications on top of the JXTA virtual network and to evaluate (and compare) their applications under varying networking conditions, environment constraints, and user requirements. Our assessment of JXTA-Sim has demonstrated that the search algorithm behaves according to the protocols defined in Project JXTA and that a considerable number of peers can be simulated. Perhaps most importantly, we have found that a future version of JXTA might allow for the application-specific configuration of parameters that currently support default values only. This can improve the performance of certain JXTA applications by increasing the availability of network resources and reducing the traffic in the overlay network.

JXTA-Sim currently focuses on simulating the JXTA search algorithm and, as a result, the JXTA discovery and routing protocols are presently supported only. Future work will extend JXTA-Sim to feature further JXTA protocols. The Pipe Binding Protocol, which implements virtual communication channels, and until now has only been evaluated using testbeds, might be supported, with simulated latencies. The Peer Endpoint Protocol might be supported to evaluate JXTA's concept of relay peers used for indirect routing. Finally, supporting peergroups would allow for experiments with multiple search scopes where peers belong to one or more peer-to-peer groups.

Acknowledgements. We would like to thank the PlanetSim team for their help and support during the development of JXTA-Sim.

References

1. Ahkil, B.T., Traversat, B., Arora, A., Abdelaziz, M., Duigou, M., Haywood, C., Hugly, J., Pouyoul, E., Yeager, B.: Project jxta 2.0 super-peer virtual network. Technical report, Sun Microsystems, Inc. (May 2003)
2. Antoniu, G., Cudennec, L., Jan, M., Duigou, M.: Performance scalability of the jxta p2p framework. In: IEEE International Parallel and Distributed Processing Symposium, IPDPS 2007 (March 2007)
3. Halepovic, B.T.E., Deters, R.: Performance evaluation of jxta rendezvous. In: International Symposium on Distributed Objects and Applications (DOA 2004), Agia Napa, Cyprus, October 2004. Springer, Heidelberg (2004)
4. Halepovic, E., Deters, R.: Jxta performance study. In: IEEE Pacific Rim Conference on Communications, Computers and signal Processing, PACRIM (August 2003)
5. Lua, E., Crowcroft, J., Pias, M., Sharma, R., Lim, S.: A survey and comparison of peer-to-peer overlay network schemes. IEEE Communications Surveys and Tutorials 7, 72–93 (2005)
6. Mohamed, B.T., Abdelaziz, M., Pouyoul, E.: Project jxta: A loosely-consistent dht rendezvous walker (2003)
7. Naicken, S., Livingston, B., Basu, A., Rodhetbhai, S., Wakeman, I., Chalmers, D.: The state of peer-to-peer simulators and simulations. SIGCOMM Comput. Commun. Rev. (April 2007)
8. Pujol-Ahulló, J., García-López, P., Sànchez-Artigas, M., Arrufat-Arias, M.: An extensible simulation tool for overlay networks and services. In: SAC 2009: Proceedings of the, ACM symposium on Applied Computing, ACM, New York (2009)

Scalability of Enhanced Parallel Batch Pattern BP Training Algorithm on General-Purpose Supercomputers

Volodymyr Turchenko and Lucio Grandinetti

Abstract. The development of an enhanced parallel algorithm for batch pattern training of a multilayer perceptron with the back propagation training algorithm and the research of its efficiency on general-purpose parallel computers are presented in this paper. An algorithmic description of the parallel version of the batch pattern training method is described. Several technical solutions which lead to enhancement of the parallelization efficiency of the algorithm are discussed. The efficiency of parallelization of the developed algorithm is investigated by progressively increasing the dimension of the parallelized problem on two general-purpose parallel computers. The results of the experimental researches show that (i) the enhanced version of the parallel algorithm is scalable and provides better parallelization efficiency than the old implementation; (ii) the parallelization efficiency of the algorithm is high enough for an efficient use of this algorithm on general-purpose parallel computers available within modern computational grids.

Keywords: Parallel batch pattern training, multilayer perceptron, parallelization efficiency.

1 Introduction

Artificial neural networks (NNs) have excellent abilities to model difficult nonlinear systems. They represent a very good alternative to traditional methods for solving complex problems in many fields, including image processing, predictions, pattern recognition, robotics, optimization, etc [1]. However, most NN models require high computational load in the training phase (on a range from several hours to several days). This is, indeed, the main obstacle to face for an efficient use of NNs in real-world applications. The use of general-purpose high performance computers, clusters and computational grids to speed up the training phase of NNs is one of the ways to outperform this obstacle.

Volodymyr Turchenko and Lucio Grandinetti
Department of Electronics, Informatics and Systems, University of Calabria,
via P. Bucci, 41C, 87036, Rende (CS), Italy
e-mail: turchenko@deis.unical.it, lugran@unical.it

A.P. de Leon F. de Carvalho et al. (Eds.): Distrib. Computing & Artif. Intell., AISC 79, pp. 525–532.
springerlink.com

Taking into account the parallel nature of NNs, many researchers have already focused their attention on NNs parallelization on specialized computing hardware and transputers [2-5], but these solutions require an availability of the mentioned devices for the use by wide scientific community. Instead general-purpose high performance computers and computational grids are widely used now for scientific experiments and modeling in a remote mode. There are developed several grid-based frameworks for NNs parallelization [6-7], however they do not deal with parallelization efficiency issues. The authors of [8] investigate parallel training of multi-layer perceptron (MLP) on SMP computer, cluster and computational grid using MPI parallelization which process huge number of the training patterns (around 20000) . However their implementation of relatively small MLP architecture 16-10-10-1 (16 neurons in the input layer, two hidden layers with 10 neurons in each layer and one output neuron) with 270 internal connections (number of weights of neurons and their thresholds) does not provide positive parallelization speedup due to large communication overhead, i.e. the speedup is less than 1. Small NNs models with the number of connections less than 270 are widely used for solving practical tasks due to their better generalization abilities [1]. Therefore the parallelization of small NNs models which use not very huge amount of the input data for their training is still very important research issue.

Our previous implementation of the parallel batch pattern back propagation (BP) training algorithm of MLP showed positive parallelization speedup on SMP computer [9-10]. However, the efficiency is decreasing with increasing the number of parallel processors. Therefore this algorithm will show lower efficiency on computational clusters and grids due to larger communication overhead in comparison with an SMP computer. The goal of this paper is to discuss several technical improvements of this algorithm concerning the decrease of a communication overhead and present its parallelization efficiency and scalability on general-purpose parallel computers. The rest of the paper is ordered as follows: Section 2 describes the parallel implementation of batch pattern BP training algorithm and several technical improvements of the old version, Section 3 presents the obtained experimental results and concluding remarks in Section 4 finishes this paper.

2 Parallel Batch Pattern BP Training Algorithm of Multilayer Perceptron

We have presented the MLP model and the usual sequential batch pattern training algorithm used for parallelization in [9-10]. We have used a *Master – Worker* approach for the development of the parallel version of the algorithm. The algorithms for *Master* and *Worker* processors are depicted in Fig. 1. The *Master* starts with definition (i) the number of patterns *PT* in the training data set and (ii) the number of processors *p* used for the parallel executing of the training algorithm. The *Master* divides all patterns in equal parts corresponding to the number of the *Workers* and assigns one part of patterns to himself. Then the *Master* sends to the *Workers* the numbers of the appropriate patterns to train. Each *Worker* calculates the errors

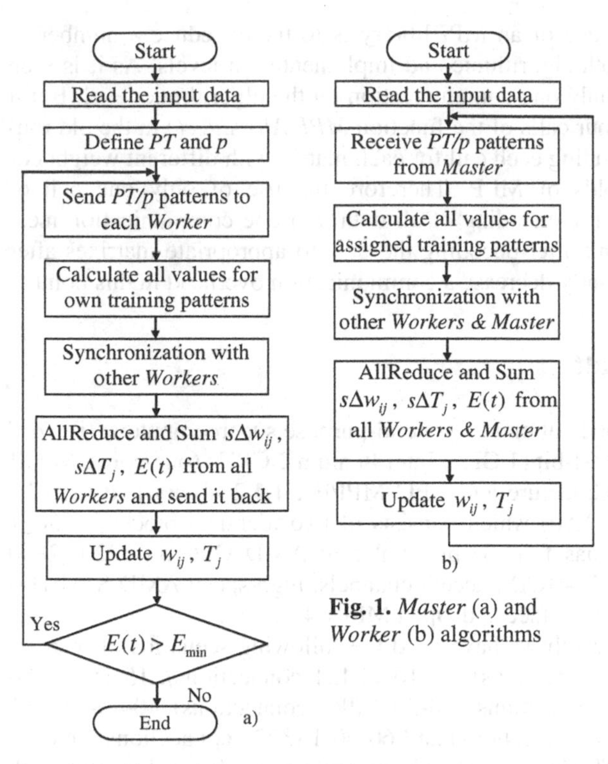

Fig. 1. *Master* (a) and *Worker* (b) algorithms

of value of MLP output and hidden layers, the partial sums of delta weights $s\Delta w_{ij}$ and delta thresholds $s\Delta T_j$ and the partial SSE value for each pattern *pt* among the *PT/p* patterns assigned to him. After processing of all assigned patterns, the global operation of reduction and summation is executed. Then the summarized values of $s\Delta w_{ij}$ and $s\Delta T_j$ are sent back to all processors working in parallel. Using a global reducing operation and simultaneously returning the reduced values back to the *Workers* allows decreasing a communication overhead in this point. Each processor uses these values $s\Delta w_{ij}$ and $s\Delta T_j$ for updating the weights and thresholds of own copy of MLP. As the summarized value of $E(t)$ is also received as a result of the reducing operation, the *Master* decides whether to continue the training or not.

The software code is developed using C programming language with the standard MPI functions. We have used MPI parallelization because now it is a de-facto standard for the development of parallel and distributed applications. The parallel part of the algorithm starts with the call of *MPI_Init()* function. The parallel processors use the synchronization point *MPI_Barrier()*. An *MPI_Allreduce()* function reduces the deltas of weights $s\Delta w_{ij}$ and thresholds $s\Delta T_j$, summarizes them and sends them back to all processors in the group. Function *MPI_Finalize()* finishes the parallel part of the algorithm.

During the experimental research [9-10] and after the discussions with MPI group of the ICL from the University of Tennessee [11] devoted to the development of the collectives of Open MPI project [12], we have added the following improvements which may lead to a reducing of a communication overhead:

- We have noticed that internal implementation of the *MPI_Allreduce()* function includes an internal synchronization point in order to provide correct internal operation of summation and sending back correct results to all processors. Therefore we have removed the function *MPI_Barrier()* from our code and received the same parallelization results;

- A general policy of usage of an MPI library is to try to reduce a number of communications on both algorithmic and implementation levels. As it is seen from Fig. 1, we have only one communication on the algorithmic level. But it was implemented by four calls of the function *MPI_Allreduce()* in the old routine, conveniently providing each call for each matrix with different weight coefficients and thresholds of MLP. Therefore the use of only one call of *MPI_Allreduce()* with pre-encoding of all data into one communication message before sending and after-decoding the data to appropriate matrixes after receiving may significantly decrease a communication overhead in this point.

3 Experimental Results

Our experiments were carried out on two general-purpose supercomputers: (i) *Crati* (NEC TX7) with thirty two 64-bit (1 GHz) Intel Itanium 2 CPUs (one-core), 64 GB of total RAM, ccNuma architecture and NEC MPI/EX 1.5.2 library and (ii) *Flamingo* (TYAN Transport VX50) which consists of two identical blocks VX50_1 and VX50_2. Each block has four 64-bit dual-core AMD Opteron 8220 (2800 MHz), 16 GB of local RAM, 4 RAM access channels, high-speed AMD-8131 Hyper Transport PCI-X tunnel interface and Open MPI 1.4.

For the experimental research we have used the following scenarios of increasing MLP sizes: 5-5-1 (36 connections), 10-10-1 (121 connections), 15-15-1 (256 connections), 20-20-1 (441 connections), 30-30-1 (961 connections), 40-40-1 (1681 connections), 50-50-1 (2601 connections) and 60-60-1 (3721 connections) and 70-70-1 (5041 connections). The number of training patterns is changed as 100, 200, 400, 600, 800, 1000, 5000 and 10000. We have chosen such MLP architectures and number of training patterns to cover both small and large parallelization problems. During the research the neurons of the hidden and output layers have logistic activation functions. The number of training epochs is fixed to 10^4. The expressions $S=Ts/Tp$ and $E=S/p\times100\%$ are used to calculate a speedup and efficiency of parallelization, where Ts is the time of sequential executing of the routine, Tp is the time of executing of the parallel version of the same routine on p processors of parallel computer.

We have run an additional experiment comparing both old and new enhanced implementations of the routine on *Crati* supercomputer in order to assess the performance of the improvements mentioned in the end of previous section. For the comparison we have chosen several scenarios of training patterns and MLP's connections (Table 1) which were researched within the old implementation in the past [9-10]. In the Table 1 we have listed the parallelization efficiencies of the both implementations on 2, 4 and 8 processors. The record "n/a" means that speedup is less than 1, i.e. it is no parallelization efficiency for this scenario. We have simply calculated the differences of parallelization efficiencies corresponding to the same scenarios. As it is seen from Table 1, the parallelization efficiency of the enhanced algorithm is better on 17.56% in average for the 20 researched scenarios. This improvement is due to a significant decrease of the communication overhead of enhanced implementation in comparison with the old implementation (Fig. 2).

Table 1 A comparison of parallelization efficiency of old and improved implementations

Pat-ns	MLP connections	Old algorithm			Enhanced algorithm			Difference		
		2	4	8	2	4	8	2	4	8
100	36 (5-5-1)	n/a	n/a	n/a	55.09	n/a	n/a	55.09	0	0
	121 (10-10-1)	57.12	n/a	n/a	75.26	46.17	14.16	18.14	46.17	14.16
	256 (15-15-1)	78.10	42.89	13.13	86.43	66.60	33.38	08.33	23.71	20.25
	441 (20-20-1)	80.38	48.51	17.91	86.81	67.45	36.41	06.43	18.94	18.50
200	36 (5-5-1)	n/a	n/a	n/a	69.64	38.24	n/a	69.64	38.24	0
	121 (10-10-1)	72.16	38.80	n/a	84.83	61.99	28.47	12.67	23.19	28.47
	256 (15-15-1)	87.77	61.37	26.73	92.65	79.00	48.81	04.88	17.63	22.08
	441 (20-20-1)	88.37	65.82	33.59	93.00	80.50	53.41	04.63	14.68	19.82
400	36 (5-5-1)	65.55	26.79	n/a	81.38	54.41	20.77	15.83	27.62	20.77
	121 (10-10-1)	83.61	51.48	20.34	91.05	77.15	45.21	07.44	25.67	24.87
	256 (15-15-1)	93.30	74.82	39.38	96.01	88.26	65.70	02.71	13.44	26.32
	441 (20-20-1)	94.24	78.95	46.34	96.33	89.20	70.49	02.09	10.25	24.15
600	36 (5-5-1)	73.93	39.35	n/a	86.12	64.27	30.04	12.19	24.92	30.04
	121 (10-10-1)	88.36	58.56	25.76	94.23	83.11	54.60	05.87	24.55	28.84
	256 (15-15-1)	95.45	81.93	45.50	97.33	91.81	74.82	01.88	09.88	29.32
	441 (20-20-1)	96.03	84.94	56.24	97.42	92.37	77.49	01.39	07.43	21.25
800	36 (5-5-1)	79.06	41.84	14.25	89.63	71.30	35.31	10.57	29.46	21.06
	121 (10-10-1)	91.09	67.80	34.15	95.26	86.65	60.79	04.17	18.85	26.64
	256 (15-15-1)	96.47	84.32	58.40	97.95	93.86	78.68	01.48	09.54	20.28
	441 (20-20-1)	96.72	86.93	62.91	98.07	94.23	82.02	01.35	07.30	19.11
Average:								**17.56**		

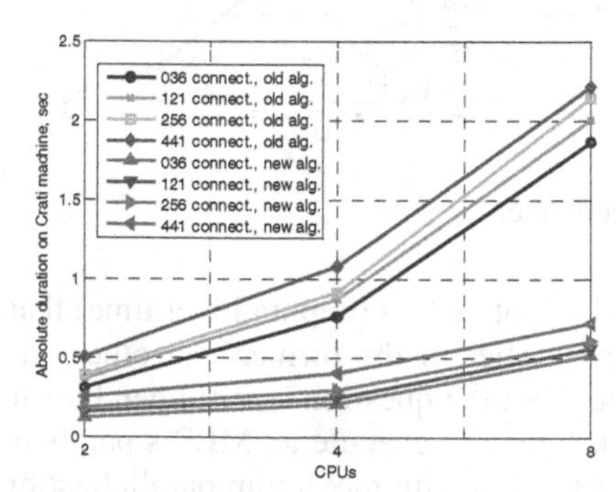

Fig. 2 Reducing a communication overhead of enhanced implementation of the routine

The parallelization efficiencies of enhanced batch pattern BP training algorithm are depicted in Fig. 3 and Fig. 4 for several parallelization scenarios on the *Crati* and *Flamingo* supercomputers respectively. As it is seen, the parallelization efficiency is increasing at increasing the number of connections and increasing the number of the training patterns. However, the parallelization efficiency is decreasing for the same scenario at increasing the number of parallel processors. The speedup is less than 1 only for one smallest scenario of 36 connections, 100 patterns and 16 processors on *Flamingo* computer and for several smallest scenarios of 36 connections, 100-1000 patterns, all processors on *Crati* computer. Therefore *Flamingo* machine

Table 2 A scalability of the enhanced parallel batch pattern BP training algorithm

Connections	Pat-s	CPUs	Efficiency on Crati, %	Efficien. on Flamingo, %
121 (10-10-1)	200	2	85	95
256 (15-15-1)	400	4	88	92
961 (30-30-1)	1000	8	83	80
2601 (50-50-1)	5000	16	92	78
3721 (60-60-1)	5000	24	85	Not available
5041 (70-70-1)	10000	28	88	Not available

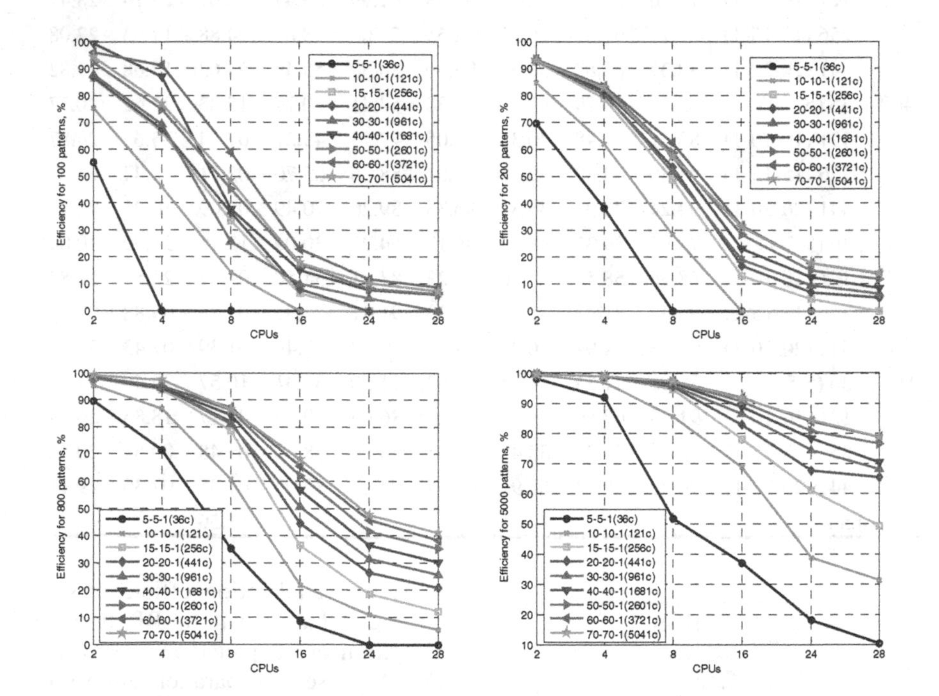

Fig. 3 Parallelization efficiency on *Crati* supercomputer

provides better parallelization efficiencies (but not faster computational time) than *Crati* machine due to better architectural properties of the former. All other scenarios provide good parallelization efficiency, but the question is still open how to choose better configuration of the parallel machine to execute an MLP's parallelization scenario within minimum execution time and with maximum parallelization efficiency.

We analyzed a scalability of the enhanced algorithm in Table 2. We consider scalability as an ability of a parallel algorithm to maintain the same parallelization efficiency when we progressively increase both the dimension of the parallelization problem and the number of processors of parallel machine. As it is seen from the

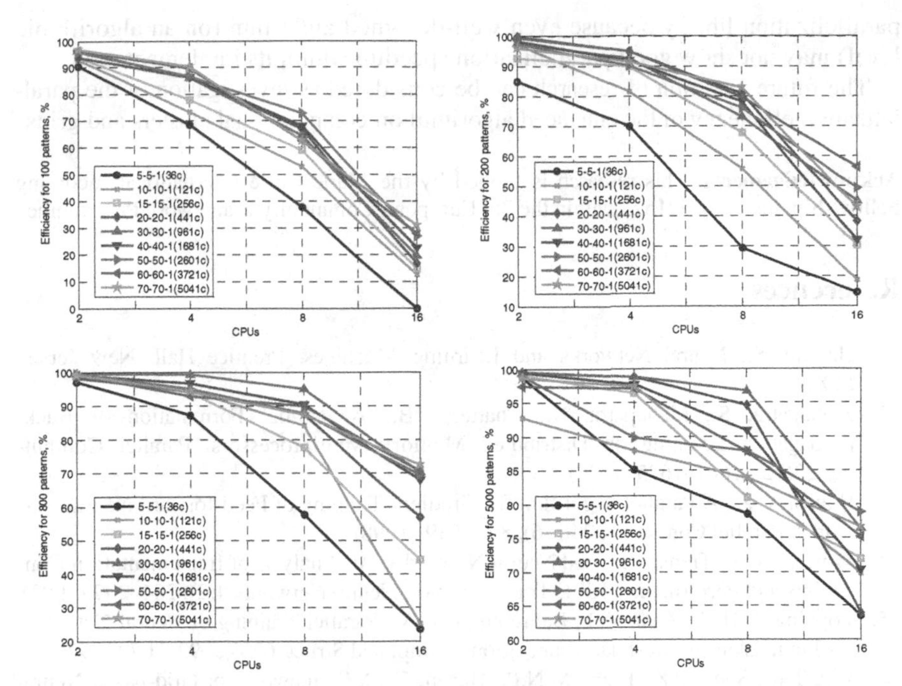

Fig. 4 Parallelization efficiency on *Flamingo* supercomputer

results, the algorithm has a good scalability, for example, it keeps the efficiency on the level of 85 percents on the *Crati* machine. At the same time we have received a slight decrease of scalability on *Flamingo* machine on 8 and 16 processors due to slower communication between two nodes (with 8 CPUs located on each node).

Also it is possible to use of the results in Figs. 3 and 4 from the practical point of view as a general reference of possible parallelization efficiencies of the batch pattern back propagation training algorithm of a MLP on general-purpose supercomputers since the presented experimental results are covered several possible parallelization scenarios.

4 Conclusions

The development of the enhanced parallel algorithm for batch pattern training of a multilayer perceptron with the back propagation training algorithm and the research of its efficiency on general-purpose parallel computers are presented in this paper. Our results show that (i) the enhanced version of the parallel algorithm is scalable and provides better parallelization efficiency than the old implementation

(ii) the parallelization efficiency of the developed algorithm is high enough for its efficient use on general-purpose parallel computers available within modern computational grid systems, (iii) it is necessary to pay attention on the implementation details of the parallel algorithm in the framework of concrete parallelization library because even well-designed algorithm (on an algorithmic level) may not show good parallelization speedup within its implementation.

The future direction of research can be considered as investigation of the parallelization efficiency of the enhanced algorithm on computational clusters and grids.

Acknowledgements. This research is funded by the Marie Curie International Incoming Fellowship grant No. 221524 within the 7th European Community Framework Programme.

References

1. Haykin, S.: Neural Networks and Learning Machines. Prentice Hall, New Jersey (2008)
2. Mahapatra, S., Mahapatra, R., Chatterji, B.: A Parallel Formulation of Back-propagation Learning on Distributed Memory Multiprocessors. Parallel Computing 22(12), 1661–1675 (1997)
3. Hanzálek, Z.: A Parallel Algorithm for Gradient Training of Feed-forward Neural Networks. Parallel Computing 24(5-6), 823–839 (1998)
4. Murre, J.M.J.: Transputers and Neural Networks: An Analysis of Implementation Constraints and Performance. IEEE Transaction on Neural Networks 4(2), 284–292 (1993)
5. Topping, B.H.V., Khan, A.I., Bahreininejad, A.: Parallel Training of Neural Networks for Finite Element Mesh Decomposition. Comp. and Struct. 63(4), 693–707 (1997)
6. Vin, T.K., Seng, P.Z., Kuan, M.N.P., Haron, F.: A Framework for Grid-based Neural Networks. In: Proc. First Intern. Conf. on Distrib. Framew. for Multim. Appl., pp. 246–250 (2005)
7. Krammer, L., Schikuta, E., Wanek, H.: A Grid-based Neural Network Execution Service Source. In: Proc. 24th IASTED Intern. Conf. on Paral and Distrib. Comp. and Netw., pp. 35–40 (2006)
8. De Llano, R.M., Bosque, J.L.: Study of Neural Net Training Methods in Parallel and Distributed Architectures. Fut. Gen. Comp. Sys. 26(2), 183–190 (2010)
9. Turchenko, V., Grandinetti, L.: Efficiency Analysis of Parallel Batch Pattern NN Training Algorithm on General-Purpose Supercomputer. In: Omatu, S., Rocha, M.P., Bravo, J., Fernández, F., Corchado, E., Bustillo, A., Corchado, J.M. (eds.) IWANN 2009. LNCS, vol. 5518, pp. 223–226. Springer, Heidelberg (2009)
10. Turchenko, V., Grandinetti, L.: Minimal Architecture and Training Parameters of Multilayer Perceptron for its Efficient Parallelization. In: Proc. 5th Intern. Work Artif. Neur. Netw. and Intel. Inform. Proces., pp. 79–87 (2009)
11. Fagg, G.E., Pjesivac-Grbovic, J., Bosilca, G., Angskun, T., Dongarra, J., Jeannot, E.: Flexible collective communication tuning architecture applied to Open MPI. In: Euro PVM/MPI (2006)
12. `http://www.open-mpi.org/`

Performance Improvement in Multipopulation Particle Swarm Algorithm

Miguel Cárdenas-Montes, Miguel A. Vega-Rodríguez, and Antonio Gómez-Iglesias

Abstract. Particle Swarm Algorithm has demonstrated to be a powerful optimizer in multitude of optimization problems. The use of multipopulation technique with periodic interchange of individuals has proved to increase the convergence toward good solutions in many other Evolutionary Algorithms. However, the policy of interchange of individuals ought to be careful studied and selected, otherwise, pernicious effects could be introduced in the optimization process. The main focus of this study is on when, how and what individuals should be exchanged between populations in order to improve the convergence. In this paper, a deep study of diverse interchange policies for multipopulation applied to Particle Swarm Optimizer is presented.

Keywords: Performance Analysis; Particle Swarm Algorithm; Multipopulation.

1 Introduction

Particle Swarm Optimisation (PSO) is an evolutionary computation technique introduced by Kennedy and Eberhart in 1995 [3] [5]. Initial simulations were modified to incorporate nearest-neighbour velocity matching, multidimensional search and acceleration by distance [5].

Miguel Cárdenas-Montes
CIEMAT, Centro de Investigaciones Energéticas Medioambientales y Tecnológicas,
Avda. Complutense 22, 28040, Madrid, Spain
e-mail: miguel.cardenas@ciemat.es

Miguel A. Vega-Rodriguez
ARCO Research Group, Dept. Technologies of Computers and Communications,
University of Extremadura, Escuela Politécnica,
Campus Universitario s/n, 10003, Cáceres, Spain
e-mail: mavega@unex.es

Antonio Gómez-Iglesias
CIEMAT, Centro de Investigaciones Energéticas Medioambientales y Tecnológicas,
Avda. Complutense 22, 28040, Madrid, Spain
e-mail: antonio.gomez@ciemat.es

A.P. de Leon F. de Carvalho et al. (Eds.): Distrib. Computing & Artif. Intell., AISC 79, pp. 533–540.
springerlink.com

The Standard Particle Swarm Optimization (SPSO) has been demonstrated to be an efficient and fast optimizer, with a wide applicability to very diverse scientific and technical problems. In spite of the efficiency demonstrated by the algorithm, also some disadvantages have appeared, mainly the premature convergence, as well as, the stagnation of the fitness improvement.

The use of more than one population, with periodic interchange of the best individuals, is a technique widely employed in diverse Evolutionary Algorithms. The multipopulation technique has proved to be excellent to accelerate the convergence and to reduce the stagnation. However, in order to maximize the advantage of this technique, the individuals interchanged and the moment of the action have to be carefully selected. Otherwise, if an incorrect moment is selected, not major advantage will be given to the algorithm.

Two factors seem to be desirable for the individuals interchanged. First at all, they should represent as good solutions as the best individuals of the population where are introduced. Second at all, they ought to increase the genetic diversity of the population.

Spite of the excellent performance of the original PSO, multitude of variations, improvements or alternatives have been proposed [1]. However, a complete characterization of the possible improvements on PSO requires a study of the behaviour of PSO in relation to multipopulation technique.

Distributed Computing is nowadays a set of computational techniques that allows scientists to approach problems with a high computational cost. Different computational models have been used in these circumstances: from Grid Computing to Desktop Computing. The Grid Computing allows computers to be connected via a special software called *Middleware*. The Middleware exports and handles all the computer resources with the goal of providing a standard layer where scientists can run their simulations and analysis, as well as, store huge volume of data [6] [4].

This paper is organized as follows: in Section 2, a resume of the Particle Swarm Optimization algorithms is introduced, as well as, some considerations about the weaknesses of the original algorithm, and the modifications necessary to tackle the multipopulation technique. In Section 3, the details of the implementation and the production setup are shown. The results are displayed and analysed in Section 4. And finally, the conclusion and the future work are presented in Section 5.

2 Particle Swarm Algorithm

In the first step of this algorithm, a set of particles are created randomly, each one represented as a point inside of a N-dimensional space. During the search process, each particle keeps track of its coordinates in the problem space that are associated with the best solution it has achieved so far. This value is termed *localbest*. Not only the best historical position of each particle is kept, also the associated fitness is stored. Another "best" value that is tracked and stored by the global version of PSO is the overall best value, and its location, obtained so far by any particle in the population. This location is called *globalbest*.

The process for implementing the global version of PSO is as follows:

1. Creation of a random initial population of particles. Each particle has a position vector and a velocity vector on N dimensions in the problem space.
2. Evaluation of the desired (benchmark function) fitness in N variables for each particle.
3. Comparison of the each particle fitness function with its *localbest*. If the current value is better than the recorded *localbest*, it is replaced. Additionally, if replacement occurs, the current position is recorded as *localbest position*.
4. For each particle, comparison of the present fitness with the global best fitness, *globalbest*. If the current fitness improves the *globalbest* fitness, it is replaced, and the current position is recorded as *globalbest position*.
5. Updating the velocity and the position of the particle according to Eqs. 1 and 2:

$$v_{id}(t+\delta t) \leftarrow v_{id}(t) + c_1 \cdot Rand() \cdot (x_{id}^{localbest} - x_{id}) + c_2 \cdot Rand() \cdot (x_{id}^{globalbest} - x_{id}) \quad (1)$$

$$x_{id}(t+\delta t) \leftarrow x_{id}(t) + v_{id} \quad (2)$$

6. If an end execution criterion – fitness threshold or number of generations– is not met, back to the step 2.

Diverse authors ([7], [2]) have demonstrated that the particles in PSO oscillate in damped sinusoidal waves until they converge to new positions. These new positions are between the *globalbest position* and their previous *localbest*. During this oscillation, a position visited can have better fitness than its previous *localbest position*, reactivating the oscillation. This movement is continuously repeated by all particles until the convergence is reached or any end execution criterion is met.

However, in some cases, where the global optimum has not a direct path between current position and the local optimum already reached, the convergence is prevented. In this case, the efficiency of the algorithm diminishes. From the computational point of view, a lot of CPU-time is wasted exploring the area of suboptimal solutions already discovered.

2.1 Multipopulation Modifications in PSO

In order to characterize intimately the capacity of multipopulation technique to improve the performance, a deep study has been performed using a catalogue of configurations and fitness functions. In all cases, three populations were implemented. This includes:

- Which individual is interchanged?
  - In the first configuration, a individual randomly selected is copied to the neighbour population (Table 1, figure A). This configuration should not produce better solutions that the equivalent –swarm size– configuration for only one population. A priory, it should represent the worst result of the three configurations.

Table 1 Representation of the three models of individuals exchanged.

A) An individual randomly selected is moved to the neighbour population.	B) The best individual of a population is moved to the neighbour population.	C) The best individual of any population is copied to the other populations.
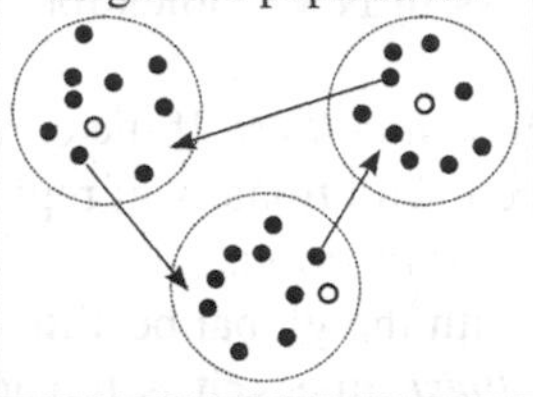	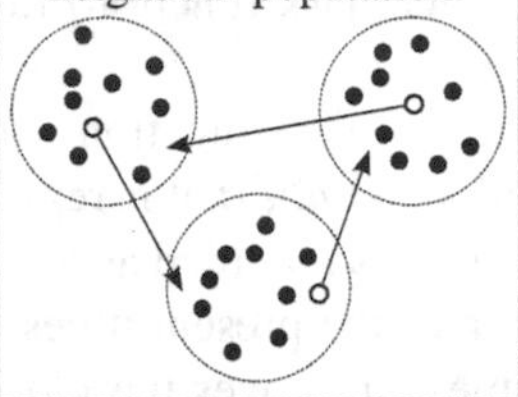	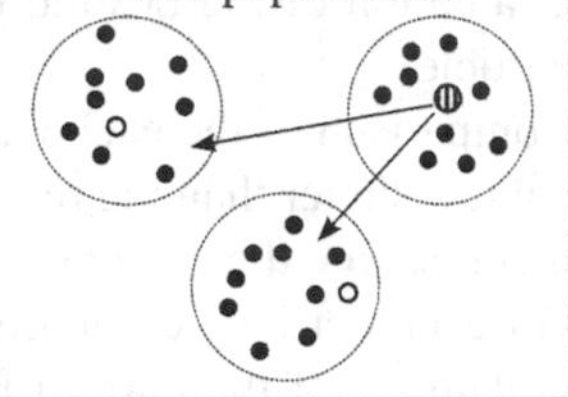

- In the second case, the best individual of one population is copied to the next population establishing a circular topology (Table 1, figure B). A net improvement, in relation with the previous configuration, is expected.
- In the third case, the best individual of any population is copied to the other populations (Table 1, figure C). For this configuration, a net improvement is also expected, however in relation with the previous configuration it is not obvious if the minor genetic diversity introduces a premature convergence.

- And, when is it more profitable to copy individuals from its original population to other population? For this characteristic three patterns have been established –when these percentages of cycles are reached, the interchange is activated.
  - 33% - 66%
  - 25% - 50% - 75%
  - 20% - 40% - 60% - 80%

3 Production Setup

The empirical study was conducted using a set of benchmarks, where diverse fitness functions widely used in these cases were employed. These functions were selected in order that the set had a mixture of multimodal (functions: f_1, f_2, f_6 and f_8) and monomodal functions (functions: f_3, f_4, f_5, f_7, f_9, f_{10} and f_{11}). For each benchmark function, a set of identical configurations was executed. These configurations represent the most characteristic values of dimensionality, population size and number of generations. Moreover, three different configurations for the individual interchanged and three more for the moment when the interchange is activated, were used. In Table 2 the benchmark functions selected are shown.

In order to avoid statistical fluctuations, a total of 350 tries of each configuration and benchmark function have been executed. In these tries, the powerful machinery of the grid was used to support the computational activity.

Each job is composed by a shellscript that handles the execution, and a tarball containing the source code of the program and the configuration files. When the job arrives to the Worker Node, it executes the instructions of the shellcript: rolling out

Table 2 Benchmark functions used in the paper.

Expression	Optimum
$f_1 = \sum_{i=1}^{D}[sin(x_i) + sin(\frac{2 \cdot x_i}{3})]$	$\approx -1.21598 \cdot D$
$f_2 = \sum_{i=1}^{D-1}[sin(x_i \cdot x_{i+1}) + sin(\frac{2 \cdot x_i \cdot x_{i+1}}{3})]$	-2D + 2
$f_3 = \sum_{i=1}^{D}[(x_i + 0.5)^2]$	0
$f_4 = \sum_{i=1}^{D}[(x_i)^2 - 10 \cdot cos(2\pi x_i) + 10]$	0
$f_5 = \sum_{i=1}^{D}[(x_i)^2]$	0
$f_6 = \sum_{i=1}^{D}[x_i \cdot sin(10 \cdot \pi \cdot x_i)]$	$\approx -1.95 \cdot D$
$f_7 = 20 + 20 \cdot \exp(-20 \cdot \exp(-0.2\sqrt{\frac{\sum_{i=1}^{D} x_i^2}{D}})) - \exp(\sum_{i=1}^{D} \frac{cos(2\pi x_i)}{D})$	0
$f_8 = 418.9828 \cdot D - \sum_{i=1}^{D}[x_i \cdot sin(\sqrt{\lvert x_i \rvert})]$	0
$f_9 = \sum_{i=1}^{D-1}[100 \cdot (x_{i+1} - x_i^2)^2 + (x_i - 1)^2]$	0
$f_{10} = \sum_{i=1}^{D}[i \cdot (x_i)^2]$	0
$f_{11} = \sum_{i=1}^{D}[(x_i)^2] + [\sum_{i=1}^{D}(\frac{i}{2} \cdot x_i)]^2 + [\sum_{i=1}^{D}(\frac{i}{2} \cdot x_i)]^4$	0

the tarball, compiling the source code and executing the 50 tries of each configuration for a benchmark function, and finally resuming the result files in a tarball. When the job finishes, it recuperates the results tarball. As pseudorandom number generator, a subroutine based on Mersenne Twister has been used.

All cases share some common parameters, such as, $c_1 = c_2 = 1$ in Eq. 1, and the maximum velocity, $V_{max} = 2$. Taking into account the configuration for the individual interchanged and the moment of activation, a total of 72 configurations for the 11 fitness functions are established, being the total of the cases 792.

The whole production takes a total of 4,330.2 hours, being 618.6 hours by run (7 runs). The number of jobs executed to complete the production was 693, and the number of tries was 831,600.

4 Analysis and Results

The results of the empirical study are presented in Table 3. In this table, the configuration which obtains the best result for each fitness function is presented. If more than one configuration produces the same best result, none is represented.

The analysis of Table 3 shows a dominance of interchange patterns where best individuals are involved. Furthermore, a tendency toward more number of interchanges (pattern: 20%-40%-60%-80%) is also remarked. This tendency will be more clearly drawn in Table 4. Spite of this trend, a certain dispersion of best results can be observed. Even the worst a priory configuration (individuals randomly selected) obtains several best results.

Configurations, where only individuals randomly selected are interchanged, have the capacity to produce best results. This fact underlines the relevance of the genetic diversity of the individuals in the swarm. Stagnation in the convergence process are frequently due to a lack of genetically different individuals, able to explore areas far away of the local minima. Therefore, the individuals randomly selected provide genetic richness to the target swarm avoiding stagnation.

Table 3 Results of the benchmark for diverse interchange patterns. For each configuration (dimensionality, population size and generations) the best pattern of interchange for each function is presented.

D.	P.	G.	Random			Circular Best			Global Best		
			33%	25%	20%	33%	25%	20%	33%	25%	20%
100	10	10^2	f_8		f_7	f_6	f_1 f_9	f_3 f_4 f_5 f_{10} f_{11}			f_2
		10^3	f_{11}	f_7	f_8	f_1	f_2	f_3 f_4 f_6 f_{10}		f_9	f_5
		10^4		f_5 f_9			f_8	f_3	f_{11}	f_4 f_{10}	f_2 f_6
	100	10^2				f_8 f_9		f_1 f_5	f_4 f_{11}	f_{10}	f_2 f_3 f_6
		10^3	f_8		f_{11}	f_4 f_7	f_2	f_3 f_{10}	f_9	f_5	f_6
		10^4		f_{11}		f_4	f_9	f_2 f_3		f_5	f_6 f_{10}
50	10	10^2		f_8	f_5		f_6 f_{10}	f_7 f_9	f_{11}	f_1 f_3 f_4	f_2
		10^3	f_7 f_8			f_6	f_2 f_{11}	f_1 f_3 f_{10}	f_9	f_4	f_5
		10^4	f_{11}			f_4			f_2 f_9	f_5 f_6	f_3 f_{10}
	100	10^2		f_9		f_4 f_{11}	f_8	f_2		f_1 f_6	f_3 f_5 f_{10}
		10^3						f_2 f_{10}		f_4 f_5 f_{11}	f_3 f_6 f_9
		10^4		f_{11}			f_6	f_3 f_9 f_{10}	f_4	f_2 f_5	
20	10	10^2		f_9	f_{11}		f_5	f_3 f_4	f_6	f_2	f_1 f_8 f_{10}
		10^3		f_2			f_5 f_9	f_3 f_{10}		f_6	f_4 f_{11}
		10^4		f_{11}		f_3	f_2	f_4		f_5 f_6 f_8 f_9 f_{10}	
	100	10^2			f_4	f_8	f_{10}			f_3 f_6 f_9 f_{11}	f_1 f_2 f_5
		10^3			f_8	f_3 f_9 f_{11}	f_2 f_5	f_4 f_{10}			f_6
		10^4	f_{10}			f_5 f_8 f_9 f_{11}				f_3	f_4
10	10	10^2				f_5 f_{11}	f_8	f_9		f_2 f_3 f_4 f_6	f_1 f_7 f_{10}
		10^3	f_4		f_3	f_5	f_8	f_9 f_{11}		f_2 f_6	f_{10}
		10^4		f_5	f_4			f_2 f_8 f_9		f_3	f_6 f_{10} f_{11}
	100	10^2		f_9	f_3	f_4	f_{10}	f_8	f_5 f_{11}	f_1 f_6	f_2
		10^3			f_2 f_8	f_4 f_5	f_9	f_{11}			f_3 f_6 f_{10}
		10^4		f_9 f_{10}		f_4		f_2	f_3 f_{11}	f_5	f_8

In Table 4, a digest of the data of Table 3 is shown. The analysis of data exposes that the performance of any interchange pattern involving best individuals has better performance than patterns involving only randomly selected individuals. This consideration is obvious, however, the number of best results for randomly selected individuals are not negligible in relation to the other two selection modes. As consequence, the importance of the genetic diversity is again underlined. This case clearly shows the importance of keeping a genetic diversity in order to avoid the stagnation.

Taking into consideration this argument, it is foreseeable that an algorithm interchanging best individuals and other randomly selected ought to reach higher level of convergence toward good solutions in relation to other interchanging only best solutions.

Regarding the number of interchanges, the Table 4 shows an augmentation of the best results as much as the number of interchanges grows. However, this trend

Table 4 Number of best results obtained for each configuration.

	Random	Circular Best	Global Best	
33%	8	27	14	**49**
25%	14	23	41	**78**
20%	12	43	42	**97**
	34	**93**	**97**	**Totals**

Table 5 Number of best results obtained for each configuration in function of the character of fitness function.

	Multimodal functions				Monomodal functions			
	Random	Circular Best	Global Best		Random	Circular Best	Global Best	
11	3	6	2	33%	5	21	12	**38**
30	2	12	16	25%	12	11	25	**48**
33	4	10	19	20%	8	33	23	**64**
Totals	**9**	**28**	**37**		**25**	**65**	**60**	**Totals**

seems to have an upper and asymptotic limit. The number of best results obtained augments from 25% to 20% more smoothly than from 33% to 25%.

In Table 5 an alternative digest of data of Table 3 is presented. In this case the analysis is performed in function of the character of fitness function: multimodal or monomodal. The same considerations expressed in the analysis of Table 4 can be applied to the analysis of this table.

5 Conclusion and Future Work

Particle Swarm Optimiser has proved sufficiently to be an efficient optimiser through a great number of use-cases. In this research paper, a complete set of configurations and fitness functions has been used in order to characterize the behaviour of the PSO in relation to multipopulations.

Taking into account only the results of the two configurations exchanging best individuals, we can not infer which one produces better results. As future work, the use of statistical inference to decide if the performance are equal, it is proposed.

Furthermore, regarding the results attained, we can conclude that as much as higher is the number of interchanges between the subpopulations, better are the results obtained by the algorithm. However, this improvement seems to be an asymptotic limit.

On the other hand, it is relevant to underline the capacity of the interchange of randomly selected individuals to produce best results. This capacity induces to introduce for future tests a measure of genetic diversity of the individuals interchanged. This parameter will allow to select individuals which keep an equilibrium between fitness and genetic diversity. A possibility is to interchange the individual that maximize expression $\frac{Fitness(P_j)-Fitness(X_i)}{P_{jd}-X_{id}}$ in relation to the best individual in the target subpopulation.

Acknowledgements. The research leading to these results has received funding by the Spanish Ministry of Science and Innovation and ERDF (the European Regional Development Fund), under the contract TIN2008-06491-C04-04 (the MSTAR project) and from the European Communitys Seventh Framework Programme (FP7/2007-2013) under grant agreement number 211804 (EUFORIA). Numerical simulations have been carried out in the resources of the National Grid Initiative of Spain. The author thanks the Spanish Network for e-Science (CAC-2007-52) for their support.

References

1. Cárdenas-Montes, M., Vega-Rodríguez, M.A., Gómez-Iglesias, A., Morales-Ramos, E.: Empirical Study of Performance of Particle Swarm Optimization Algorithms Using Grid Computing. In: International Workshop on Nature Inspired Cooperative Strategies for Optimization, Granada, Spain (2010)
2. Clerc, M., Kennedy, J.: The Particle Swarm: Explosion, Stability and Convergence in a Multi-dimensional Complex Space. IEEE Transaction on Evolutionary Computation 6, 58–73 (2002)
3. Eberhart, R.C., Morgan, Y.S.: Computational Intelligence: Concepts to Implementations. Kaufmann Publishers, San Francisco (2007)
4. Foster, I., Kesselman, C. (eds.): The Grid: Blueprint for a New Computing Infrastructure, 1st edn. Morgan Kaufmann Publishers, San Francisco (1998) ISBN: 1558604758
5. Kennedy, J., Eberhart, R.C.: Particle swarm optimization. In: Proceedings of the IEEE International Conference on Neural Networks, vol. IV, pp. 1942–1948. IEEE Service Center, Perth (1995)
6. Li, B., Baker, M.: The Grid Core Technologies. John Wiley & Sons Ltd., Chichester (2005)
7. Ozcan, E., Mohan, C.K.: Particle Swarm Optimization: Surfing the waves. In: Congress on Evolutionary Computation, pp. 1939–1944. Washington (July 1999)

A New Memetic Algorithm for the Two-Dimensional Bin-Packing Problem with Rotations

A. Fernández, C. Gil, A.L. Márquez, R. Baños, M.G. Montoya, and A. Alcayde

Abstract. The two-dimensional bin-packing problem (2D-BPP) with rotations is an important optimization problem which has a large number of practical applications. It consists of the non-overlapping placement of a set of rectangular pieces in the lowest number of bins of a homogenous size, with the edges of these pieces always parallel to the sides of bins, and with free 90 degrees rotation. A large number of methods have been proposed to solve this problem, including heuristic and meta-heuristic approaches. This paper presents a new memetic algorithm to solve the 2D-BPP that incorporates some operators specially designed for this problem. The performance of this memetic algorithm is compared with two other heuristics previously proposed by other authors in ten classes of frequently used benchmark problems. It is observed that, in some cases, the method here proposed is able to equal or even outperform to the results of the other two heuristics in most test problems.

Keywords: two-dimensional bin-packing problem, memetic algorithms, heuristics.

1 Introduction

The bin-packing problem (BPP) and its multi-dimensional variants, have a large number of practical applications, including production planning, project selection, multiprocessor scheduling, data storage in computers, packing objects in boxes, assigning advertisements to newspaper columns, etc. This family of problems are included in the category of NP-hard [5] problems, and there is therefore no known method that can solve the problem in a polynomial time.

Given a collection of objects (pieces) characterized by having different heights and weights, the two-dimensional version of the BPP (2D-BPP) consists of packing all the objects in the minimum number of bins (containers). In contrast with

A. Fernández, C. Gil, A.L. Márquez, R. Baños, M.G. Montoya, and A. Alcayde
Dpt. Computer Architecture and Electronics, University of Almería,
Carretera de Sacramento s/n, Cañada San Urbano, 04120 Almería (Spain)
Tel./Fax: (+34) 950015710 / 950015486
email: {afmolina,cgilm,almarquez,rbanos,dgil,aalcayde}@ual.es

A.P. de Leon F. de Carvalho et al. (Eds.): Distrib. Computing & Artif. Intell., AISC 79, pp. 541–548.
springerlink.com

other packing problems, such as the strip-packing problem [6], the bins in the 2D-BPP are size-limited. The classical approaches to solve the 1D-BPP include the heuristic placement routines Next-Fit, First-Fit, and Best-Fit, which have also been adapted to the two-dimensional case [14]. However, their poor performance when applied to large and complex problems has led some researchers to propose more sophisticated methods, including meta-heuristic approaches. This paper presents a new memetic algorithm that incorporates evolutionary and local search operators specifically designed to store objects successfully in different bins. Section 2 gives a general formulation of the two-dimensional bin packing problem with rotations. Section 3 presents the general operation of the memetic algorithm and its specific operators. Section 4 describes the set of 500 test instances used in the experimental execution and presents the main results obtained by memetic algorithm in comparison with other two heuristic approaches recently proposed by other authors. Section 5 summarizes the conclusions and the future work.

2 The 2D Bin-Packing Problem with Rotations

The 2D-BPP with rotations can be defined as follows: Given a set of n rectangular pieces (objects) where h_i and w_i are the height and weigth of object i, respectively (i=1,2,...,n), and given an unlimited number of bins, all of which have a height H and width W, the goal is to insert all the pieces without overlap in the minimum number of bins (nBIN). The pieces can be rotated 90 degrees, which increases the search space and therefore the difficulty to reach the optimal solution, but also provides the advantage of increasing the possibility of reducing the number of bins required to store the pieces. The 2D-BPP has often been analyzed with fixed orientation of pieces [7]. In the last decade, some papers have dealt with the version of 2D-BPP in which the pieces can be rotated by 90 degrees [4,10]. As the problem is NP-hard, some authors have proposed using heuristic approaches, including evolutionary algorithms [13] and tabu search [8]. Ben-Mohamed-Ahemed and Yassine [10] have recently developed a hybrid algorithm (SACO) which is based on ant colony optimization and IMA heuristic, which was proposed by El-Hayek et al. [4] and extends the best-fit heuristic using a list of available areas in the bins.

3 A New Memetic Algorithm for the 2D-BPP

Memetic Algorithms (MA) [11] are heuristic methods inspired by Darwinian principles of natural evolution and Dawkins notion of meme [2], defined as a unit of cultural evolution that can exhibit local refinement. In practice, MAs apply similar operators to those used in evolutionary algorithms (EAs) while also applying a local search process to refine agents [12]. The present work puts forward a new memetic algorithm for the 2D-BPP named MA2dbpp, which consists of a population of individuals (agents) that are optimized using the typical evolutionary operators (mutation, crossover, and selection) and a local-search optimizer. Further, MA2dbpp also uses of a list of available spaces in the bins, in order to improve the performance of the operators. Figure 1 shows the flow diagram of the algorithm,

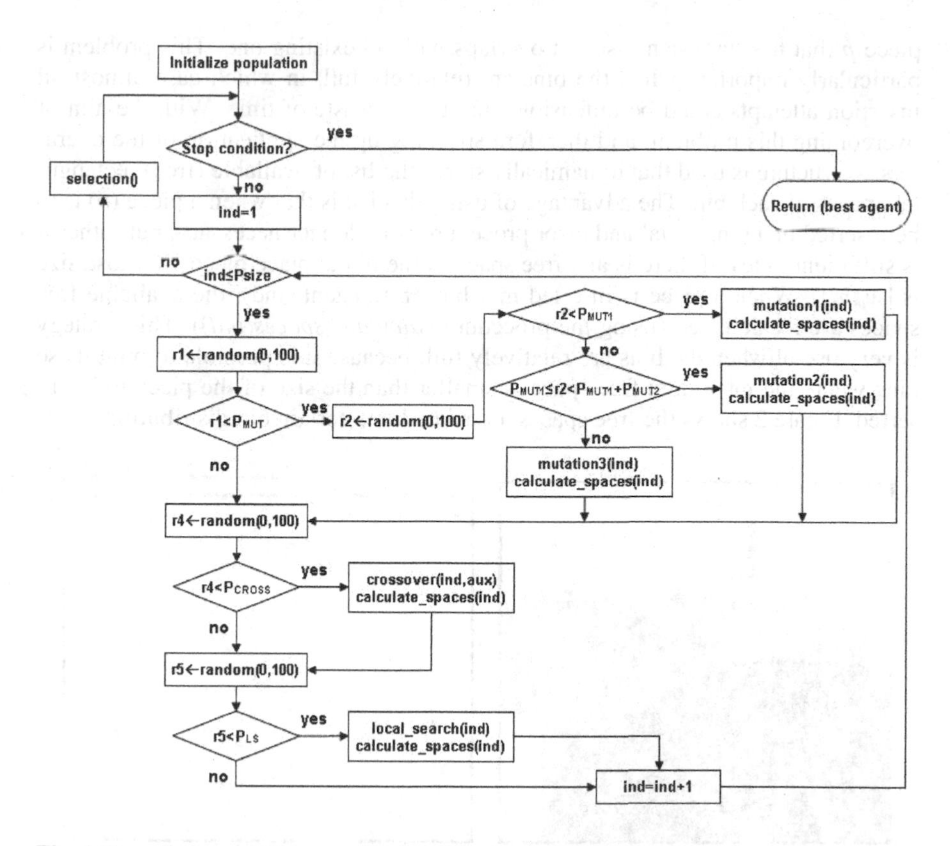

Fig. 1 Flow diagram of the memetic algorithm for the 2D-BPP.

which starts by initializing the population of agents, each of which is considered a solution of the problem, i.e. a set of bins where all the items are stored. This initialization consists of including the pieces in a random order in the bins, such that if a piece cannot be located in a previously opened bin (due to the overlaps), it is then inserted in a new one. While the stop condition is not fulfilled, all the agents apply mutation, crossover and local search with a given probability. Each time an operator is applied, and therefore the distribution of a bin changes, the list of available areas (free space) is updated. The selection consists of passing a percentage of solutions to the next iteration according to the roulette mechanism [3] (here, only the better 50% of agents pass to the next iteration, while the remaining 50% are substituted by the better ones using the roulette mechanism). When the stop condition is fulfilled (30,000 evaluations of the fitness function without reducing the number of bins), MA2dbpp finishes then returning the best agent.

3.1 Dynamic Storage of Free Spaces in the Bins

The main difficulty of applying operators to reduce the number of bins in the 2D-BPP is to determine the feasible and unfeasible solutions, i.e. whether or not a piece p that has just been inserted overlaps with an existing one. This problem is particularly important when the bins are relatively full, in which case almost all insertion attempts could be unfeasible, and thus a waste of time. With the aim of overcoming this problem, and therefore speeding up the application of the operators, a structure is used that dynamically stores the list of available (free) rectangular spaces of each bin. The advantage of using this list is that when a piece (p) is to be inserted in a bin, a trial and error procedure is no longer necessary, but rather it is sufficient to test if there is any free space in the bin to place piece p (whose size is $[w_p, h_p]$). When a piece is inserted in a bin of an agent (ind), the available free spaces are recalculated (using the procedure *calculate_spaces(ind)*). This strategy is very useful when the bins are relatively full, because it is possible to omit these bins where the maximum free space is smaller than the size of the piece to be inserted. Figure 2 shows the free spaces associated to a certain bin distribution.

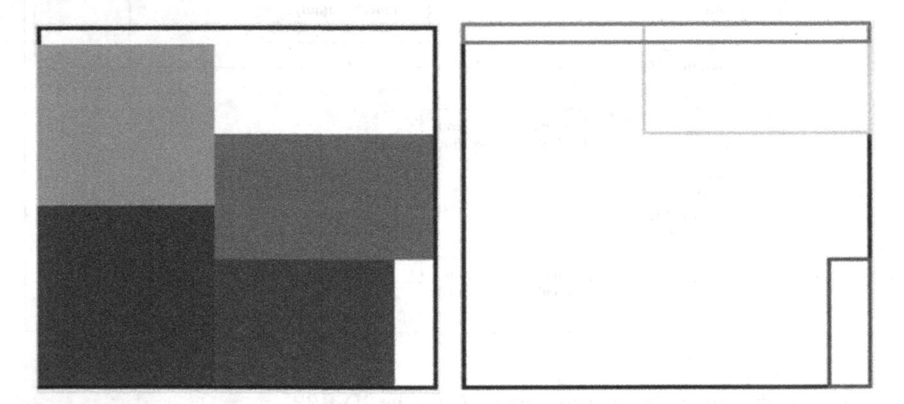

Fig. 2 Current storage in a certain bin (left), and the available spaces (right)

3.2 Description of the Operators Used in MA2dbpp

As shown in Figure 1, the memetic algorithm here presented (MA2dbpp) applies mutation, crossover, selection and local search operators. In contrast with [4] where the authors consider the pieces in a given order to apply best-fit, MA2dbpp uses mutation and crossover operators that do not need to sort the pieces. The three mutation operators, the crossover and the local optimizer are now described:

- *Mutation1*: a piece is randomly taken from one bin and it is stored in another randomly chosen one only if the available space is large enough.
- *Mutation2*: a piece is chosen from the bin with most available space, and it is stored in another randomly chosen bin only if there is free space.
- *Mutation3*: a piece is chosen from the bin with most available space, and it is stored in the empties remaining bin only if there is free space. If the storage is not possible, the piece is inserted in a new bin in the lower left corner.

- *Crossover*: The child (CH) agent is formed by considering bins of two random parents (A1, A2): takes the fullest bin of A1, plus the bins of A2, but discarding the pieces already taken from A1 in order not to duplicate pieces.
- *Local optimizer*: the least occupied bin is taken and each available space is tested to see whether a piece from the remaining bins can fit.

4 Results and Discussion

The algorithms are evaluated on two well-known sets of benchmarks. The first set includes six classes (I, II, III, IV, V, VI) [1], for which all the pieces have been generated in the same range and are defined as follows:

Class I: w_i and h_i are randomly selected within [1,10], W=H=10
Class II: w_i and h_i are randomly selected within [1,10], W=H=30
Class III: w_i and h_i are randomly selected within [1,35], W=H=40
Class IV: w_i and h_i are randomly selected within [1,35], W=H=100
Class V: w_i and h_i are randomly selected within [1,100], W=H=100
Class VI: w_i and h_i are randomly selected within [1,100], W=H=300

The second set [9] includes four classes, where the bin dimensions are W=H=100 for all classes, and the pieces are generated as follows:

Class VII: Type A with 70% probability, Type B,C,D, each with 10% probability.
Class VIII: Type B with 70% probability, Type A,C,D, each with 10% probability.
Class IX: Type C with 70% probability, Type A, B, D, each with 10% probability.
Class X: Type D with 70% probability, Type A, B, C, each with 10% probability.
Type A: w_i random within [2W/3,W], h_i random in [1,H/2].
Type B: w_i random within [1,W/2], h_i random in [2H/3,H].
Type C: w_i random within [W/2,W], h_i random in [H/2,H].
Type D: w_i random within [1,W/2], h_i random in [1,H/2].

Each one of the ten classes includes a different number of objects: n={20,40,60,80,100}. Finally, for each class and value of n ten different instances are generated. Therefore, a total of 500 benchmark instances are used in this study.

The performance of the memetic algorithm presented in Section 3 is compared with two other methods: IMA [10], and SACO [8]. When applying heuristic methods to optimization problems it is advisable to perform a sensitivity analysis. The MA implementation here presented has a large number of parameters, as commented in Section 3, which is why the effect of modifying some of these parameters has been analyzed, while another subset has been fixed to certain values. Thus, fixed values have been established for the following parameters: $P_{mutation}$ (15%), $P_{crossover}$ (50%). The probability of applying the three mutation operators are: ($P_{mutation1}$=30%, $P_{mutation2}$=50%, $P_{mutation3}$=20%). On the other hand, different population sizes (agents of the population, P_{size}) are used: P_{size}={50, 100, 200}.

Tables 1 and 2 show the results obtained by the MA-2dbpp, IMA and SACO. The first three columns show the characteristics of the test problems, the fourth

Table 1 Comparing MA, IMA and SACO in Berkey and Wang's benchmarks.

Class	WxH	N	nBin (IMA)	nBin (SACO)	nBin (MA2dbpp)		
					P_{size}=50	P_{size}=100	P_{size}=200
I	10x10	20	6.6	6.6	6.6	6.6	6.6
		40	12.9	12.9	**12.8**	**12.8**	**12.8**
		60	19.5	19.5	19.5	19.5	19.5
		80	27.0	27.0	27.0	27.0	27.0
		100	31.3	31.1	31.4	31.3	31.3
II	30x30	20	1.0	1.0	1.0	1.0	1.0
		40	1.9	1.9	2.0	1.9	2.0
		60	2.5	2.5	2.6	2.6	2.6
		80	3.1	3.1	3.3	3.2	3.3
		100	3.9	3.9	4.0	4.0	4.0
III	40x40	20	4.7	4.7	4.7	4.7	4.7
		40	9.4	9.4	**9.3**	**9.2**	**9.3**
		60	13.5	13.5	13.5	**13.4**	13.5
		80	18.4	18.4	18.6	18.4	18.4
		100	22.2	21.9	22.3	22.1	22.1
IV	100x100	20	1.0	1.0	1.0	1.0	1.0
		40	1.9	1.9	1.9	1.9	1.9
		60	2.5	2.5	2.5	2.5	2.5
		80	3.1	3.0	3.3	3.3	3.3
		100	3.7	3.7	4.0	3.9	3.9
V	100x100	20	5.9	5.9	5.9	5.9	5.9
		40	11.4	11.9	11.4	11.4	11.4
		60	17.4	17.4	17.4	17.4	17.4
		80	23.9	23.9	24	23.9	24.0
		100	27.9	27.9	28.2	27.9	28.0
VI	300x300	20	1.0	1.0	1.0	1.0	1.0
		40	1.7	1.7	1.9	1.9	1.9
		60	2.1	2.1	2.2	2.2	2.2
		80	3.0	3.0	3.0	3.0	3.0
		100	3.2	3.1	3.5	3.5	3.5

and fifth columns show the average number of bins of ten runs for each problem instance obtained by IMA and SACO, while the last three columns show the average number of bins obtained by MA2dbpp using three population sizes. It is also seen that MA2dbpp equals and even outperforms the other methods in most cases (bold numbers), specially with an intermediate number of pieces (n={40,60,80}).

The effect of using different population sizes considering the average results of the ten runs for each of the 50 test problems is now analyzed. Table 3 summarizes the cases where each one of the three P_{size} configurations is able to outperform to the other cases: MA2dbpp-50agents is not able to improve on MA2dbpp-100, and it is only able to improve on MA2dbpp-200 in one case. Therefore, it can be concluded that using medium and large populations is better than using smaller populations.

Table 2 Comparing MA, IMA and SACO in Martello and Vigo's benchmarks.

Class	WxH	N	nBin (IMA)	nBin (SACO)	nBin (MA2dbpp)		
					P_{size}=50	P_{size}=100	P_{size}=200
VII	100x100	20	5.2	5.2	5.2	5.2	5.2
		40	10.4	10.4	**10.3**	**10.3**	**10.2**
		60	14.7	14.7	14.7	14.7	**14.6**
		80	21.2	21.2	21.2	21.2	**21.1**
		100	25.3	25.3	25.8	25.7	25.3
VIII	100x100	20	5.3	5.3	5.3	5.3	5.3
		40	10.4	10.4	10.4	10.4	10.4
		60	15.0	15.0	15.0	15.0	**14.8**
		80	20.8	20.7	21.1	20.9	20.8
		100	25.7	25.6	26.2	25.9	25.7
IX	100x100	20	14.3	14.3	14.3	14.3	14.3
		40	27.5	27.5	27.5	27.5	27.5
		60	43.5	43.5	43.5	43.5	43.5
		80	57.3	57.3	57.3	57.3	57.3
		100	69.3	69.2	69.3	69.3	69.3
X	100x100	20	4.1	4.1	4.1	4.1	4.1
		40	7.3	7.3	7.3	7.3	7.3
		60	10.1	10.0	**9.9**	**9.9**	**9.9**
		80	12.8	12.7	12.9	12.8	12.9
		100	15.8	15.7	16.0	16.0	16.1

Table 3 Effect of using different population sizes in the MA2dbpp.

	MA2dbpp(P_{size}50)	MA2dbpp(P_{size}100)	MA2dbpp(P_{size}200)
MA2dbpp (P_{size}50)		0	1
MA2dbpp (P_{size}100)	14		8
MA2dbpp (P_{size}200)	12	7	

5 Conclusions

This paper has proposed a new memetic algorithm that incorporates specific operators to solve the 2D-BPP with rotations. It uses a dynamic structure that stores, in real-time, the maximum spaces generated by applying different operators (three different mutation operators, and a crossover one). The performance of this memetic algorithm is compared with those obtained by another heuristic technique (IMA) and a hybrid meta-heuristic one (SACO) from the literature. The results show that all the methods obtain similar results, but in some test instances the MA2dbpp is able to improve on some of the results obtained by the other two methods. The results obtained by the memetic algorithm in this problem reinforce the previous conclusions of other authors about the good performance of this meta-heuristic to solve NP-hard optimization problems. Future research should be focused on extending the memetic algorithm with the aim of solving multi-objective formulations of this problem that consider other objectives in addition to minimizing the number of bins.

Acknowledgements. This work has been financed by the Spanish Ministry of Science and Innovation (TIN2008-01117), and the Excellence Project of Junta de Andalucía (P07-TIC02988), financed by the European Regional Development Fund (ERDF).

References

[1] Berkey, J.O., Wang, P.Y.: Two dimensional finite bin-packing algorithms. J. Oper. Res. Soc. 38, 423–429 (1987), http://www.jstor.org/pss/2582731
[2] Dawkins, R.: The selfish gene. Oxford University Press, New York (1976)
[3] Kad, D.-J.: An analysis of the behavior of a class of genetic adaptive systems. Diss. Abstr. Int. 36(10), 5140B (1975)
[4] El Hayek, J., Moukrim, A., Negre, E.S.: New resolution algorithm and pretreatments for the two-dimensional bin-packing problem. Comput. Oper. Res. 35, 3184–3201 (2008)
[5] Garey, M.R., Johnson, D.S.: Computers and intractability: a guide to the theory of NP-completeness. W.H. Freeman & Company, San Francisco (1979)
[6] Hopper, E., Turton, B.C.H.: A review of the application of meta-heuristic algorithms to 2D Strip Packing Problems. Artif. Intell. Rev. 16, 257–300 (2001)
[7] Lodi, A., Martello, S., Vigo, D.: Recent advances on two-dimensional bin-packing problems. Discret. Appl. Math. 123, 379–396 (2002)
[8] Lodi, A., Martello, S., Vigo, D.: TSpack: A unified tabu search code for multi-dimensional bin packing problems. Ann. Oper. Res. 131, 203–213 (2004)
[9] Martello, S., Vigo, D.: Exact solution of the two-dimensional finite bin packing problem. Manag. Sci. 44, 388–399 (1998)
[10] Mohamed-Ahemed, B., Yassine, A.: Optimization by ant colony hybrid for the bin-packing problem. World Acad. of Sci. Eng. and Technol. 49, 354–357 (2009)
[11] Moscato, P.: On evolution, search, optimization, genetic algorithms and martial arts: towards memetic algorithms. California Institute of Technology Technical Report C3P 826, Pasadena, CA (1989)
[12] Moscato, P.: Memetic algorithms: A short introduction, New Ideas in Optimization, pp. 219–234. McGraw-Hill, New York (1999)
[13] Stawowy, A.: Evolutionary based heuristic for bin packing problem. Comput. Ind. Eng. 55(2), 465–474 (2008)
[14] Wong, L., Lee, L.S.: Heuristic placement routines for two-dimensional bin packing problem. J. Math Stat. 5(4), 334–341 (2009)

A Meta Heuristic Solution for Closest String Problem Using Ant Colony System

Faranak Bahredar, Hossein Erfani, H.Haj Seyed Javadi, and Nafiseh Masaeli

Abstract. Suppose $\sum$ is the alphabet set and S is the set of strings with equal length over alphabet $\sum$. The closest string problem seeks for a string over $\sum$ that minimizes the maximum hamming distance with other strings in *S*. The closest string problem is NP-complete. This problem has particular importance in computational biology and coding theory. In this paper we present an algorithm based on ant colony system. The proposed algorithm can solve closest string problem with reasonable time complexity. Experimental results have shown the correctness of algorithm. At the end, a comparison with one Meta heuristic algorithm is also given.

Keywords: Closest string, Ant colony system, Meta heuristic.

1 Introduction

Closest string problem (CSP) has been extensively studied in computational biology [5, 6, 12, 7, 15] and coding theory [10]. This problem finds application in PCR prime design [8, 3, 16], genetic probe design [8], motif finding [8] and antisense drug design [1, 16]. In all these applications, a common task is to design a new DNA or protein sequence that is very similar to each given sequence.

The closest string problem and other related problems are NP-complete [10, 13]. Researchers have developed approximation algorithms and fixed-parameter algorithms for solving this problem. Despite of its hardness, polynomial-time approximation scheme (PTAS) presents a natural and efficient way [11]. But this algorithm and other related algorithms do not give an exact and efficient solution.

Faranak Bahredar and Nafiseh Masaeli
Department of Information Technology, Payam Noor University
e-mail: bahredar_f@yahoo.com, Nafiseh_masaeli@yahoo.com

Hossein Erfani
Computer Engineering Department, Science and Research Branch,
Islamic Azad University
e-mail: hossein.erfani@gmail.com

H. Haj Seyed Javadi
Department of Mathematics and Computer Science, Shahed University, Tehran, Iran
e-mail: h.s.javadi@shahed.ac.ir

A.P. de Leon F. de Carvalho et al. (Eds.): Distrib. Computing & Artif. Intell., AISC 79, pp. 549–557.
springerlink.com

So [6] offers an algorithm in which it solves the problem with $O(nL + nd\,d^d)$ time complexity, which is linear in respect of both L and n to be fixed parameter. Here the algorithm growth is restricted by $O(d^d)$. The CSP is fixed-parameter tractable when some parameters are fixed. An upper bound is presented in [1]. An improved and optimized lower bound with $f(1/\varepsilon)|\Sigma|^{O(\log(1/\varepsilon))}$ time complexity is studied in [7], in which ε is number of mismatches and f is an arbitrary function.

As sequential algorithms have been developed, some parallel algorithms have been presented. The algorithm [17], uses genetic algorithm idea and presents a suitable solution for this problem. In all these methods, the lengths of input sequences are average. For string with different length, a parallel algorithm is presented in [4].

In [14] the ACS-CSP algorithm is presented which is a Meta heuristic algorithm. This algorithm just solves the inputs with little number of strings and bounded alphabet.

In this paper we present an algorithm that finds closest string using ant colony system. In proposed algorithm, every ant indicates a solution. This algorithm finds the closest string based on the remaining pheromone in each path. The proposed algorithm solves the closest string problem in optimal time.

The paper is organized as follows. Section 2 gives a mathematic description of the closest string problem and has a short look on ant colony system. The algorithm is introduced in section 3. Section 4 provides an overview of results on a set of standard test problems and comparisons of proposed algorithm with ACS-CSP algorithm. At last Section 5 is dedicated to the discussion of the main characteristics of our algorithm.

2 Preliminaries

2.1 Formulation of CSP

Suppose Σ is a finite symbol set and $|\Sigma|$ is its cardinality.

Definition: Let $x = x_1, x_2, ..., x_n$ and $y = y_1, y_2, ..., y_n$ be two strings with length n over Σ. The hamming distance $d_H(x, y)$ between x and y is defined as :

$$d_H(x,y) = \sum_{i=1}^{n} \varepsilon(x_i, y_i), \varepsilon(x,y) = \begin{cases} 1 & x \neq y \\ 0 & x = y \end{cases} \quad (1)$$

Given: A set $S = \{s_1, s_2, ..., s_n\}$ of strings over a finite symbol set Σ, such that $i = \{1, 2, ..., n\}$, $|s_i| \leq L$ and $0 \leq d \leq L$.

Objective: s is the closest string, iff there is no string s' with: $\max_{i=1,\dots,n} d_H(s,s') < \max_{i=1,\dots,n} d_H(s,s_i)$.

2.2 Ant Colony System

Ant colony system (ACS) is inspired by the forgoing behavior of ant colonies concerning in particular how they can find shortest paths between food source and their colony. In real world, ants (initially) wander randomly and return to their colony after finding food while leaving down pheromone trails. If other ants find such a path, they are likely not to keep traveling at random, but instead they follow the trail, returning and reinforcing it if they eventually find food. Over time, however, the pheromone starts to evaporate, then a shortest path gets marched over faster and thus the pheromone density remains high.

As the amount of pheromone trail is increasing, more ants will traverse that path. Thus the pheromone evaporates in long paths and more ants will be absorbed to the shortest path.

Ant colony system was initially introduced by Marco Dorigo in collaboration with Alberto Colorni and Vittrio Maniezzo [9]. They used ACS in order to solve the traveling salesman problem (TSP).

The ant colony system works as follows: At first the equal amount of pheromone is associated to each path. The virtual ant k which is located in node r, selects the node j by the following probability function:

$$S = \begin{cases} \arg\max_{u \in j_k(r)} \{[\tau(r,u)].[\eta(r,u)]^\beta\} & q \le q_0 \\ \rho & otherwise \end{cases} \tag{2}$$

Where q is a random number uniformly distinguished in [0..1], q_0 is a parameter in which $0 \le q_0 \le 1$, $\tau(r,u)$ is the amount of pheromone in the path between nodes r and u, and $\eta(r,u)$ is an inverse function of distance between r and u. β is the parameter which determines the relative importance of pheromone in the path and the length of the path. S is a random variable which is selected according to the probability distribution that mentioned bellow:

$$p_k = \begin{cases} \dfrac{[\tau(r,s)].[\eta(r,s)]^\beta}{\sum_{u \in j_k(r)}[\tau(r,u)].[\eta(r,u)]^\beta} & s \in j_k(r) \\ 0 & otherwise \end{cases} \tag{3}$$

Where $j_k(r)$ is the set of nodes that remain to be visited by ant K positioned in city r. $p_k(r,s)$ gives the probability with which ant K in city r chooses to move to node s. In a few number of choices, action criterion is equal to the probability function P. In other situations, transition takes place between nodes connected by short edges and with a large amount of pheromone. The parameter q_0 determines the relative importance of exploitation versus exploration.

While building a solution, ants visits edges and decay their pheromone level by applying the local update rule:

$$\tau(r,s) = (1-\alpha).\tau(r,s) + \rho.\Delta\tau(r,s) \tag{4}$$

Where $\tau(r,s)$ denotes the pheromone level of the edge between nodes r and s. ρ is a parameter related to the evaporation time for pheromone.

After one tour is finished, the global update is performed. In ACS only the globally best ant is allowed to deposit pheromone. The amount of pheromone that the best ant deposited on each path is the inverse function of tour length. Shortest path will have most pheromone:

$$\tau(r,s) = (1-\alpha).\tau(r,s) + \alpha.\Delta\tau(r,s) \tag{5}$$

Where α is the pheromone decay parameter $(0 \le \alpha \le 1)$ and $\Delta\tau(r,s)$ is the inverse of length of the globally best tour from the beginning of trail.

3 Proposed Algorithm

The genetic algorithm presented in [19] for solving the closest string problem has some limitations, which consists of first generation number and strings length. By increasing these factors, the quality of answer decreases. There by, final result is closely dependent on choosing the first generation.

The algorithm that we propose is also based on ant colony system. In this algorithm every ant produces a solution according to formula 2. As mentioned in fig 2, at first a number of ants are initialized to find the solution. Then a matrix H will be initialized, in which the number of rows indicate the length of input sequences L. In matrix the h_{ij} is equal to the frequency of letter which is placed in the *j-th* index of *i-th* position of all input sequences (fig 1). *J* is indicative of the index of alphabet letters in Σ.

$\Sigma = \{a, b, c, d\}$, $|\Sigma| = 4$

S= {abcd, aacc, bdca, cbdd}

$$\begin{array}{c} \\ 1 \\ 2 \\ 3 \\ 4 \end{array} \begin{array}{c} \begin{array}{cccc} a & b & c & d \end{array} \\ \begin{pmatrix} 2 & 1 & 1 & 0 \\ 1 & 2 & 0 & 1 \\ 0 & 0 & 3 & 1 \\ 1 & 0 & 1 & 2 \end{pmatrix} \end{array}$$

Fig. 1 Example of matrix H

In fact, the heuristic function finds the frequency of every letter in every position of input strings. Each ant starts from the first node and chooses one of the alphabet letters as a next node, randomly. Then according to formula 3, the probability of each path will be computed. Then based on the probability associated to each path, one of the alphabet letters will be selected. Later on, after one ant passed one complete tour, local updating is started. In this stage, $\eta(r,s)$ is computed using $1/1\text{-}d_{r,s}$, in which $d_{r,s}$ is the sum of hamming distances. After every ant passes the entire path, the global update operation will be done. Otherwise the algorithm starts to find the closest string, again from the beginning.

α indicates degree of the ant's willingness to choose a path. β indicates the willingness to heuristically choose a path. q_0 shows the probability of choosing the best letter. Evaporation on the edges is controlled by ρ . path is $\tau(r,s)$ which is indicative of *i-th* position's in the solution for *j-th* letter of Σ. Ant's population is shown by AntNo.

Fig. 2 The proposed Algorithm

The proposed algorithm
1: initialize Ant system parameter α, β, q_0, t_0, ρ, path, AntNo.
2: calculate Heuristic
3: while (satisfy termination criterion)
4: for each Ant in Ants
5: \\ Construct solution according to the Ant system formula
6: for each i, where $0 \le i \le L$, do
7: for each letter, where $0 \le$ letter $\le$ Alphabet Number, do
8: $p_i = \tau(i, letter)^{\alpha} * \tau(i, letter)^{\beta}$
9: Choose one Alphabet according to the allocated p_i, the letter which has more probability than others will be randomly chosen.
10: end for
11: end for
12: \\ Local update
13: $\tau(i, letter) = (1-\rho).(i, letter) + \rho t_0$
14: end for
15: Calculate Hamming Distance for solution
16: find the best answer
17: end While
18: end.

4 Experimental Results

In this section we study the proposed algorithm on set of data. At first we find the optimal parameters by changing the parameters and then we compare our algorithm results with ACS-CSP algorithm. Codes are written in C# programming language and have been executed on Pentium IV computer, Intel 2.5GH processor and 3GB main memory.

At first a set of data containing 10000 strings with similar length 15, is generated. These sequences are generated randomly. After running 500 epochs, the hamming distance of closest string with others is achieved. This operation is done

20 times for each variation in parameters. In all figures the vertical axis indicates the hamming distance of closest string with other strings in the set.

Fig 3 shows the result of changing in evaporation parameter ρ. By increasing the evaporation amount, the hamming distance of the closest string with other strings increases. As the ρ decreases, the probability that one ant chooses a path with more pheromone will increase.

Fig. 3 Variation of ρ in proposed algorithm

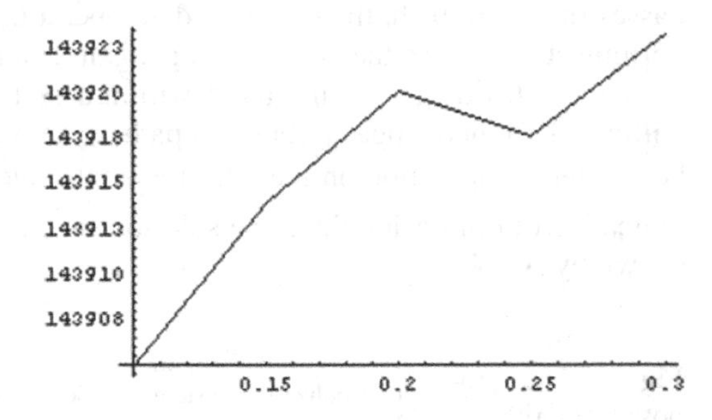

By increasing the β, the hamming distance with other sequences is decreased (fig 4). This reduction is due to correctness of heuristic function and it means that explorative information has positive effect on ant's decision for choosing the next symbol.

Fig. 4 Variation of β in proposed algorithm

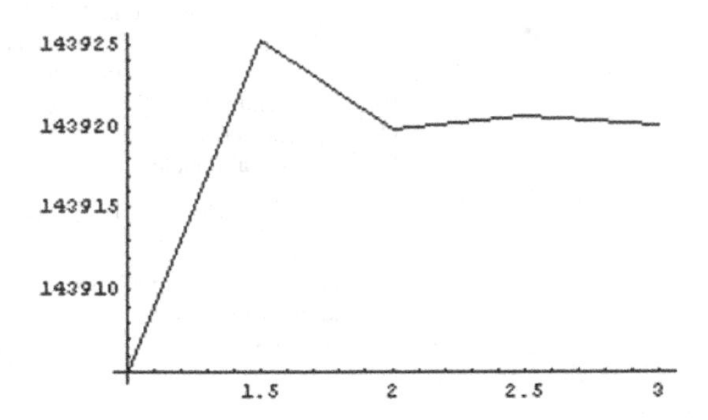

After acquiring the optimal parameters, we compare our algorithm with ACS-CSP one. Both algorithms are executed 20 times on input set. In closest string problem the size of problem is dependent on some parameters such as alphabet numbers, input sequences numbers and the length of input sequences.

By increasing the alphabet size, our algorithm is more efficient than ACS-CSP algorithm. This phenomenon is because of first generation limitation in ACS-CSP algorithm. But in our algorithm the number of ants does not have any effects on choosing letters. This trait is shown in fig 5. In fig 5, *S* is constant, but the number of alphabet is different. Here the numbers are in this group{5, 10, 15, 20, 25}. By

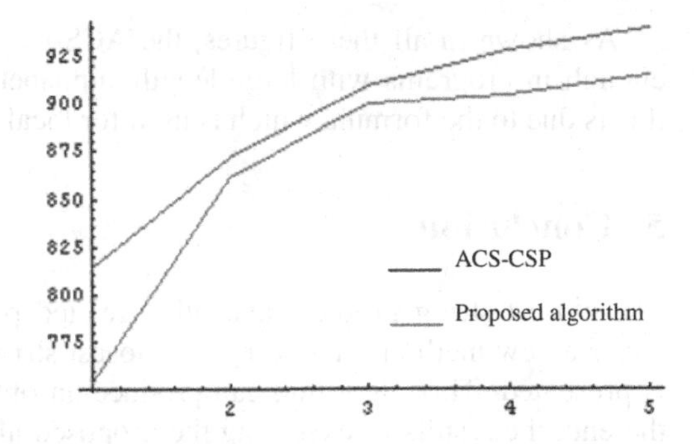

Fig. 5 Comparison of ACS-CSP and proposed algorithm with variation in alphabet size

increasing the number of alphabets, proposed algorithm gives better results than ACS-CSP algorithm. The final string has less hamming distance with other strings.

As shown in fig 6, by increasing the length of input strings, proposed algorithm works better. We have tested both algorithms with *L*= 50, 100, 500, 1000, 1500, where *n* is 100.

The result of proposed algorithm execution and ACS-CSP algorithm is compared in table 1. Here the number of input strings is different but the length of input strings as constant. As shown, the average hamming distance in proposed algorithm is less than the one in ACS-CSP algorithm.

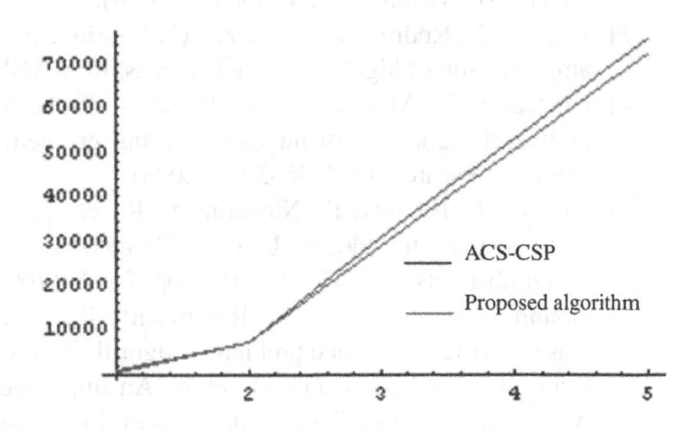

Fig. 6 Comparison of ACS-CSP and proposed algorithm based on different length

Table 1 Comparison of ACS-CSP and proposed algorithm based on different sequences number

Proposed algorithm	ACS-CSP	String #
344	353	n= 50
695	702	n= 100
1064	1074	n= 150
1436	1452	n= 200
1805	1820	n= 250

As shown in all these figures, the ACS-CSP algorithm does not operate well enough in programs with large length, alphabet size and number of strings. And this is due to the formula which is used for local and global updating.

5 Conclusion

The closest string problem and other related problems are NP-complete. In this paper a new method for finding the closest string based on the ant colony system is presented. This algorithm can produce an optimal solution in suitable time. In the end, the results of executing the proposed algorithm are compared with ACS-CSP algorithm. The experimental results indicate that the proposed algorithm has better performance than the ACS-CSP. Also it shows that by increasing the number of n and alphabet, our algorithm is considerably better results.

References

[1] Meneses, C.N., Lu, Z., Oliveria, A.S., Pardolos, P.M., et al.: Optimal Solutions for the Closest String Problem via Integer Programing. INFORMS J. Computing 16(4), 419–429 (2004)

[2] Deng, X., Li, G., Li, Z., Ma, B., Wang, L., et al.: Genetic design of drugs without side effects. SIAM Journal on Computing 32(4), 1073–1090 (2003)

[3] Dopazo, J., Rodríguez, A., Sáiz, J.C., Sobrino, F., et al.: Design of primers for PCR amplification of highly variable genomes. In: CABIOS, vol. 9, pp. 123–125 (1993)

[4] Gomes, F.C., Meneses, C.N., Pardalos, P.M., Viana, G.V.R., et al.: A parallel multistart algorithm for the closest string problem. SienceDirect. Computers & Operations Research 35, 3636–3643 (2008)

[5] Gramm, J., Hüffner, F., Niedermeier, R., et al.: Closest strings, primer design, and motif search. In: Florea, L. (ed.) Currents in Computational Molecular Biology, poster abstracts of RECOMB 2002, pp. 74–75 (2002)

[6] Gramm, J., Niedermeier, R., Rossmanith, P., et al.: Fixed-parameter algorithms for closest string and related problems. Algorithmica 37, 25–42 (2003)

[7] Wang, J., Chen, J., Huang, M., et al.: An Improved Lower Bound on Approximation Alogorithms for the Closest Substring Problem. Science Direct Information Processing Letters 107, 24–28 (2008)

[8] Lanctot, K., Li, M., Ma, B., Wang, S., Zhang, L., et al.: Distinguishing string search problems. In: Proceedings of the 10th Annual ACM-SIAM Symposium on Discrete Algorithms (SODA), pp. 633–642 (1999)

[9] Dorigo, M., Maniezzo, V., Colorni, A., et al.: The Ant System Optimization by a colony of cooperating agents. IEEE Transactions on Systems, Man, and Cybernetics-Part B 26(1), 29–41 (1996)

[10] Frances, M., Litman, A.: On Covering Problems of Codes. Theoretical Computer System 30, 113–119 (1997)

[11] Li, M., Ma, B., Wang, L., et al.: On the Closest String and Substring Problems. J. ACM 49(2), 151–171 (2002)

[12] Meneses, C.N., Lu, Z., Oliveira, C.A.S., Pardalos, P.M., et al.: Optimal solutions for the closest-string problem via integer programming. INFORMS Journal on Computing, 419–429 (2004)
[13] Evans, P.A., Smith, A.D., Li, M., Ma, B., Wang, L.: Complexity of Finding Similar Regions in Approximating Closest Substring Problems. Fundamentals of Computation Theory, 210–221 (2003)
[14] Faro, S., Pappalardo, E.: ANT-CSP: An Ant Colony Optimization for the Closest String Problem. In: SOFSEM 2010. LNCS, vol. 5901, pp. 370–381. Springer, Heidelberg (2010)
[15] Wang, L., Dong, L.: Randomized algorithms for motif detection. Journal of Bioinformatics and Computational Biology 3(5), 1039–1052 (2005)
[16] Wang, Y., Chen, W., Li, X., Cheng, B., et al.: Degenerated primer design to amplify the heavy chain variable region from immunoglobulin cDNA. BMC Bioinformatics 7(suppl. 4), S9, 123–130 (2006)
[17] Liu, X., He, H., Sykora, O., et al.: Parallel Genetic Algorithm and Parallel Annealing Algorithm for the Closest String Problem, pp. 591–597 (2005)

A New Parallel Cooperative Model for Trajectory Based Metaheuristics

Gabriel Luque, Francisco Luna, and Enrique Alba

Abstract. This paper proposes and studies the behavior of a new parallel cooperative model for trajectory based metaheuristics. Algorithms based on the exploration of the neighborhood of a single solution like simulated annealing (SA) have offered very accurate results for a large number of real-world problems. Although this kind of algorithms are quite efficient, more improvements are needed to address the large temporal complexity of industrial problems. One possible way to improve the performance is the utilization of parallel methods. The field of parallel models for trajectory methods has not been deeply studied. The new proposed parallel cooperative model allows both to reduce the global execution time and to improve the efficacy. We have evaluated this model in two very different techniques (SA and PALS) solving a real-world problem (the DNA Fragment Assembly).

1 Introduction

Metaheuristics are general heuristics that provide sub-optimal solutions in a reasonable time for various optimization problems. According to the number of solutions they manage during optimization process, they fall into two categories [3]: trajectory based methods and population based techniques. A population based metaheuristic makes use of a randomly generated population of solutions. The initial population is enhanced iteratively so that, at each generation of the process, either the whole population or just a part is replaced by newly generated individuals (often the best ones). On the other hand, a trajectory based algorithm starts with a single initial solution which, at each step of the search, is replaced by another (often better) solution found in its neighborhood.

Although metaheuristics allow to significantly reduce computational time of the search process, the exploration remains time-consuming for many industrial and engineering problems. In this context, parallelism emerges as a useful strategy to reduce this computational times down to affordable values [1]. The point is that the

Gabriel Luque, Francisco Luna, and Enrique Alba
ETSI Informática, University of Málaga, 29071 Málaga, Spain
e-mail: {gabriel, flv, eat}@lcc.uma.es

A.P. de Leon F. de Carvalho et al. (Eds.): Distrib. Computing & Artif. Intell., AISC 79, pp. 559–567.
springerlink.com

parallel versions of metaheuristics allows not only to speed up the computations, but also to improve the quality of the provided solutions. For both trajectory-based and population-based metaheuristics, different parallel models have been proposed in the literature. In general, these parallel models are mostly oriented to study parallel population-based algorithms, but it actually exists a gap on parallel models for single solution methods from which something could be gained for other researchers.

The focus of this paper is on parallel trajectory-based metaheuristics. Usually, three major parallel models exist for this kind of algorithms [5]: the parallel exploration of the neighborhood, the parallel evaluation of each solution, and the multi-start model. The two first models speed up the execution of the method without changing the semantics of the algorithm in comparison with a sequential exploration. The last one is maybe more interesting from the algorithmic point of view since it can change the behavior of the method with respect to its serial counterpart. The multi-start model lies in launching in parallel several independent or cooperative homo/heterogeneous algorithms. Usually, in its cooperative mode, subalgorithms of the parallel multi-start model exchange information (solutions) during execution and when the target subalgorithm receives a solution, it continues the search using the previous one or the newly received one according to a selection scheme. The problem of this classical cooperative model is that some interesting information is lost since either the new solution is discarded (it is not chosen by the selection scheme) and no new information is incorporated, or it is accepted and the previous historical information of the subalgorithm is lost.

This work proposes a new parallel yet simple model that extends the cooperative multi-start model to avoid the aforementioned flaw. Several approaches of the literature have also dealt with this issue. For example, Cadenas et al. [4] proposed a centralized method in which a coordinator gathers solutions from subalgorithms and then, by using a knowledge extraction mechanism (with fuzzy rules), the behavior of the subalgorithms is updated. A similar approach has been used in [10]. Ribeiro and Rosseti [11] have evaluated a parallel GRASP where cooperation between the subalgorithms is implemented via path-relinking and a centralized pool of elite solutions. The main drawback of these approaches is precisely their centralized strategy, which limits the scalability of the algorithms when the number of subalgorithms is increased. On the contrary, our proposal is a fully decentralized approach in which the information contained in the incoming solution is combined with the local solution of the target subalgorithm by using a simple yet effective method (a crossover operator). This parallel model has been evaluated in two very different methods: simulated annealing [7], a classical and general purpose algorithm, and the problem aware local search method (PALS) [2], a problem specific technique. To ensure the impact of our model, we use a real-world problem: the DNA fragment assembly.

This paper is organized as follows. Section 2 presents our proposed parallel model. The experimental design, the problem addressed, the algorithms, and the discussion of the results are given in Section 3. The last section summarizes the conclusions and provides some hints on the future work.

2 Proposed Parallel Model

Our goal is to design a new parallel model for trajectory based metaheuristics which allows to reduce the global execution time but, at the same time, it also improves the efficacy of the exploration of the search space. A number of papers has been devoted to this topic for parallel approaches involving population based methods (some of them also involving trajectory-based ones) but it is not a very studied field for pure parallel trajectory based metaheuristics.

Since we want to improve the efficacy of the resulting parallel algorithm, we focus on the multi-start cooperative paradigm. As discussed in the introduction, classical approaches of multi-start models for trajectory-based metaheuristics loses search information. Indeed, when a subalgorithm receives a solution from other subalgorithm, it has to choose whether it continues the search either with the current one or the newly received one, losing the stored information from the discarded solution.

We propose a new model in which we do not have to choose between the two solutions, but generate a new solution with the main characteristics of both solutions. This is performed by using an operator (similar to the recombination operator of population based method) which combines both solutions.

In the previous existing multi-start models, the features of the incoming solution were not very important rather than it fitness value, but now, this issue can provoke an important impact in the search behaviour. Different possible mechanisms are analyzed here:

- **Predefined:** in this case, each subalgorithm receives a single solution (the sending island is defined by the topology). Therefore, any subalgorithm only receives a single solution which is combined with the local one.
- **Depending of the fitness value:** in this case, each subalgorithm receives a solution from each subalgorithm which composes the global method. Now, the subalgorithm has to select one solution from this set of candidate solutions, that will be combined with the current one. In this strategy, the selection mechanism is based on the fitness value of the incoming solutions. In this study, we analyze two different techniques: select the solution with best fitness, and select the solution with the worst one.
- **Depending of the features of the solution:** as in the previous one, each subalgorithm receives several solutions (one per subalgorithm) and it has to select one. In this case, the selection will be performed by using a genotypic distance (a diversity measure) among the solutions. Again, two different approaches are analyzed: select the closest solution, and select the farthest one. The distance depends on the representation used.

3 Experimentation

This section includes the experimental study performed to assess the suitability of our proposal. We first describe the optimization problem used as testbed, while the algorithms involved are shown next. The final part is devoted to present the experimental design used and discuss the results obtained.

3.1 Testbed: The DNA Fragment Assembly Problem

In order to determine the function of a specific genes, scientists have learned to read the sequence of nucleotides comprising a DNA sequence in a process called DNA sequencing. Currently, it is impossible to read a complete DNA sequence due to technological limitations. The procedure therefore proceeds by making multiple exact copies of the original DNA sequence, which are then cut into short fragments at random positions. These steps take place in the laboratory. After the fragment set is obtained, a traditional assemble approach is followed in this order: overlap, layout, and then consensus. To ensure that enough fragments overlap, the reading of fragments continues until a coverage is satisfied. The quality of a consensus is then measured by looking at the distribution of the coverage. Coverage at a base position is defined as the number of fragments at that position. It is a measure of the redundancy of the fragment data, and it denotes the number of fragments, on average, in which a given nucleotide in the target DNA is expected to appear. The higher the coverage, the fewer number of the gaps, and the better the result.

Table 1 Information of datasets.

Parameters	**Instance**				
	M15421		**J02459**	**BX842596**	
Coverage	5	7	7	4	7
Fragment length	398	383	405	708	703
Nb. of fragments	127	177	352	442	773

To test and analyze the performance of our algorithm we have generated several problem instances with GenFrag [6]. We have chosen three sequences from the NCBI web site[1]: a human apolopoprotein HUMAPOBF, with accession number M15421, which is 10,089 bases long; the complete genome of bacteriophage lambda, with accession number J02459, which is 20k bases long; and the Neurospora crassa (common bread mold) BAC, with accession number BX842596, which is 77,292 bases long.

3.2 Algorithms

In this section, we describe the algorithms used to test our parallel model. In order to achieve a more relevant contribution, we have selected two very different algorithms: a well-known and general purpose algorithm, SA, and a specific and problem dependant technique, PALS.

3.2.1 Simulated Annealing

Simulated Annealing (SA) [7] is a generalization of the Metropolis heuristic. Indeed, SA consists of a sequence of executions of Metropolis with a progressive

[1] `http://www.ncbi.nlm.nih.gov/`

decrement of the temperature starting from a high temperature, where almost any move is accepted, to a low temperature, where the search resembles hill climbing. In fact, it can be seen as a hill-climber with an internal mechanism to escape local optima. In SA, the solution s' is accepted as the new current solution if $\delta \leq 0$ holds, where $\delta = f(s') - f(s)$. To allow escaping from a local optimum, moves that increase the energy function are accepted with a decreasing probability $exp(-\delta/T)$ if $\delta > 0$, where T is a parameter called the "temperature". The decreasing values of T are controlled by a cooling schedule, which specifies the temperature values at each stage of the algorithm, what represents an important decision for its application.

For the DNA fragment assembly, the algorithm uses a permutation representation of integer values. A permutation of integers represents a sequence of fragment numbers, where successive fragments overlap. The solution in this representation requires a list of fragments assigned with a unique integer ID.

As fitness function, we use the one proposed by Parsons, Forrest, and Burks [9]. This fitness function sums the overlap score for adjacent fragments in a given solution. When this fitness function is used, the objective is to maximize such a score. It means that the best individual will have the highest score.

Finally, to explore the neighborhood of a solution, we use a movement operator which given a solution s, and two positions i and j, reverses the subpermutation between the positions i and j.

3.2.2 Problem Aware Local Search

PALS [2] is a method specifically designed for the DNA fragment assembly problem, although it can be extended to other problems. It is a variation of the Lin's 2-opt [8] for the DNA field, which does not only use the overlap among the fragments, but it also takes into account the number of contigs[2] in an intelligent manner that have been created or destroyed.

This algorithm works on a single solution which is iteratively modified by applying movements in a structured manner. A movement here is the same as the perturbation operator used in SA. The key step in PALS is the calculation of the variation in the overlap (Δ_f) and in the number of contigs (Δ_c) between the current and the perturbed solutions. This calculation is computationally light since we do not compute neither the fitness function nor the number of contigs, but an estimation of the variation of these values. To do this, we only need to analyze the affected fragments by the tentative movement.

In each iteration, PALS makes these calculations for all possible movements, storing the candidates in a list L. Our proposed method only considers the candidate the movements that do not increase the number of contigs ($\Delta_c \leq 0$). Once it has completed the previous calculations, the method selects and applies a movement of the list L. The algorithm stops when no more candidate movements are generated.

[2] The number of final sequences generated by the solution, being a single one the optimal value.

3.3 Experimental Design

This section provides the reader with the details of the experiments performed to evaluate the new parallel model proposed for trajectory-based metaheuristics. Different versions have been used according to the criterion for selecting the incoming solution that will be combined with the current one (Section 2). We use the terminology ALG_SEL, where ALG represent the algorithms (**SA** or **PALS**) and SEL the selection method (**pre** for predefined one, **bs** for the best solution, **ws** for the worse solution, **cs** for the closest solution, and **fs** for the farthest one). Since the algorithms work with permutations, the classical order-based operator (OX) is used for combining both the incoming and the current solutions. The order-based operator first copies the fragment ID between two random positions in one solution into the new solution corresponding positions. It then copies the rest of the fragments from the other solution into the new solution in the relative order presented in that solution. If the fragment ID is already present in the new solution, then we skip that fragment. The method preserves the feasibility of solution.

For a complete definition of the method, we need to specify the stop condition and the exchange period (number of evaluations between two successive interchanges) for both methods and, in the case of SA, we have also to define the cooling schedule for the temperature. To perform a fair comparison, the stop criterion will be to find the optimal solution, and the exchange period will be 1000 evaluations for all the versions. In SA, we update the temperature each 500 evaluations using a decay factor α of 0.99 ($next\_T = \alpha \cdot current\_T$). All the parallel algorithms uses eight processors (one per subalgorithm).

We will also compare our proposed model with the serial version (**sSA** and **sPALS**), a parallel version using the multi-start no-cooperative model, also known as independent run model (**iSA** and **iPALS**), and a parallel version using the classical multi-start cooperative model (**cSA** and **cPALS**), in which incoming solutions just replace the current one.

The experiments have been executed on a Intel Pentium IV 2.8GHz with 512MB running SuSE Linux 8.1. Because of the stochastic nature of the algorithms, we perform 30 independent runs of each test to gather meaningful experimental data and apply statistical confidence metrics to validate our results.

3.4 Results

In order to compare the algorithms, we have measured the average number of evaluations (in thousands) and the global wall-clock time (in seconds) needed by the algorithms to find the optimal solution (recall that this is the stopping condition). We also want to note that when we refer to number of evaluations, we are really indicating partial function evaluations since both SA and PALS only perform a complete evaluation for the initial solution.

Table 2 Normalized number of variables for all the methods (smaller values are better). The algorithms are ordered by this score.

Alg.	Score	Alg.	Score
SA_fs	0.82	**PALS_fs**	0.80
SA_ws	0.83	**PALS_ws**	0.82
SA_bs	0.86	**PALS_bs**	0.83
SA_pre	0.87	**PALS_pre**	0.86
SA_cs	0.91	**PALS_cs**	0.89
cSA	0.93	**cPALS**	0.90
iSA	0.97	**iPALS**	0.96
sSA	1	**sPALS**	1

Let's first compare the numerical performance of the different algorithms. Since there are many different problem instances and analyzing them thoroughly would hinder us from drawing clear conclusions, we have summarized the results in Table 2 as follows: we have normalized the number of evaluations needed to find the optimum for each problem instance with respect to the worst (maximum) value obtained by any of the proposed algorithms (we have considered separately PALS and SA), so we can easily compare them without scaling problems. As so, the values in Table 2 are the average values over all the DNA fragment assembly instances, and the algorithms are ranked according to this value.

Quite interesting conclusions can be drawn from this table. First, we can observe that the classical serial versions of the algorithms are the worse ones according to this ranking, indicating that the use of the parallelism is beneficial for this problem. In fact, the parallel methods have reduced the computational cost (measured as number of evaluations) between a 3% (independent runs model) to a 20% (one of our proposed models).

Table 3 Normalized speedups for all the methods (larger values are better). The algorithms are ordered by this score.

Alg.	Score	Alg.	Score
iSA	7.78	**iPALS**	7.8
cSA	7.51	**cPALS**	7.49
SA_pre	7.47	**PALS_pre**	7.49
SA_ws	7.21	**PALS_fs**	7.19
SA_fs	7.16	**PALS_cs**	7.19
SA_cs	7.15	**PALS_ws**	7.15
SA_bs	7.09	**PALS_bs**	7.11

A second important conclusion is that all the variants of our model outperform the traditional parallel models for trajectory based methods. These results show that the exploration scheme induced by our model is more efficient than the other parallel algorithms in the context of this problem. Analyzing the different variants of the proposed model, it can be seen that the ones that make use of some information from the incoming solutions (either fitness or distance) converge faster towards the optimal solution (with the exception of the **_cs** strategy). Our hypothesis is that for the problem at hand, the diversity is very important, so the methods having an enhanced diversification capability have achieved better results. Indeed, the strategy that requires the lowest number of evaluation to find the optimum solution is the one that incorporates more diversity, i.e., the **_fs** scheme (it combines the current solution with the more different solution among the candidates). Finally, we can notice that the results for each variant of the proposed parallel model are very robust with respect to the algorithms, obtaining almost the same profit in both methods.

The second relevant issue to be analyzed when working with parallel algorithms is the execution time. In this work, we compare the wall-clock time of the parallel algorithm that runs on one single processor to the wall-clock time required by the same algorithm on m processors (*weak speedup*). Due to space constraints, a similar analysis is performed for this performance measure so that the average behavior over all the instance is presented in Table 3.

The obtained speedup values are near to 8 (quasi-linear) and they show that all the parallel algorithms are able to profit quite well from the parallel computing platform. Three different categories according to the speedup value can be distinguished: the first one composed by the parallel methods following the independent runs model; in the second category, we can find **cXXX** and **XXX_pre** (being **XXX** either **SA** or **PALS**); and the third one contains the remainder settings. These categories correspond with the communication overhead of the methods: the methods in the first category do not interchange any information; in the second one, each subpopulation only communicates with a single subpopulation; finally, the third one, all the subalgorithm send information to all the other ones. Therefore, the small loss of parallel performance is due to the amount of information transmitted.

4 Conclusions

In this paper, we have developed a new parallel model for trajectory based methods. We have tested this model using two very different algorithms, SA and PALS, to solve a real-world problem from bioinformatics. We have also studied different versions of that models.

The results have shown that our proposed method is more efficient than the classical one with respect to the execution time and the number of evaluation performed. Our study has also indicated that the utilization of solutions which incorporate more diversity (using the solution with the farthest genotypic distance to the current one) is beneficial for the global search.

As future work, we plan to use more sophisticated methods for combining the solutions such as path relinking, and to evaluate other different selection strategies for choosing the solution to be combined. We also want to extend this study to other problems for generalizing the conclusions of this paper.

Acknowledgements. This work has been partially funded by the "Consejería de Innovación, Ciencia y Empresa", Junta de Andalucía under contract P07-TIC-03044, and the Spanish Ministry of Science and Innovation and FEDER under contract TIN2008-06491-C04-01. Francisco Luna acknowledges support from the grant BES-2006-13075 funded by the Spanish government.

References

1. Alba, E. (ed.): Parallel Metaheuristics: A New Class of Algorithms. Wiley, Chichester (2005)
2. Alba, E., Luque, G.: A new local search algorithm for the DNA fragment assembly problem. In: Cotta, C., van Hemert, J. (eds.) EvoCOP 2007. LNCS, vol. 4446, pp. 1–12. Springer, Heidelberg (2007)

3. Blum, C., Roli, A.: Metaheuristics in combinatorial optimization: Overview and conceptual comparison. ACM Computing Surveys 35(3), 268–308 (2003)
4. Cadenas, J., Garrido, M., Muñoz, E.: Using machine learning in a cooperative hybrid parallel strategy of metaheuristics. Information Sciences 179(19), 3255–3267 (2009)
5. Crainic, T.G., Toulouse, M.: Parallel Strategies for Metaheuristics. In: Glover, F.W., Kochenberger, G.A. (eds.) Handbook of Metaheuristics, Norwell, MA, USA. Kluwer Academic Publishers, Dordrecht (2003)
6. Engle, M.L., Burks, C.: Artificially generated data sets for testing DNA fragment assembly algorithms. Genomics 16 (1993)
7. Kirkpatrick, S., Gellatt, C., Vecchi, M.: Optimization by Simulated Annealing. Science 220(4598), 671–680 (1983)
8. Lin, S., Kernighan, B.: An Effective Heuristic Algorithm for TSP. Operations Research 21, 498–516 (1973)
9. Parsons, R., Forrest, S., Burks, C.: Genetic algorithms, operators, and DNA fragment assembly. Machine Learning 21, 11–33 (1995)
10. Pelta, D., Sancho-Royo, A., Cruz, C., Verdegay, J.L.: Using memory and fuzzy rules in a co-operative multi-thread strategy for optimization. Information Sciences 176(13), 1849–1868 (2006)
11. Ribeiro, C.C., Rosseti, I.: Efficient parallel cooperative implementations of grasp heuristics. Parallel Computing 33(1), 21–35 (2007)

Using a Parallel Team of Multiobjective Evolutionary Algorithms to Solve the Motif Discovery Problem

David L. González–Álvarez, Miguel A. Vega–Rodríguez, Juan A. Gómez–Pulido, and Juan M. Sánchez–Pérez

Abstract. This paper proposes the use of a parallel multiobjective evolutionary technique to predict patterns, motifs, in real deoxyribonucleic acid (DNA) sequences. DNA analysis is a very important branch within bioinformatics, resulting in a large number of NP-hard optimization problems such as multiple alignment, motif finding, or protein folding. In this work we study the use of a multiobjective evolutionary algorithms team to solve the Motif Discovery Problem. According to this, we have designed a parallel heuristic that allows the collaborative work of four algorithms, two population-based algorithms: Differential Evolution with Pareto Tournaments and Nondominated Sorting Genetic Algorithm II, and two trajectory-based algorithms: Multiobjective Variable Neighborhood Search and Multiobjective Skewed Variable Neighborhood Search. In this way, we take advantage of the properties of different algorithms, getting to expand the search space covered in our problem. As we will see, the results obtained by our team significantly improve the results published in previous research.

Keywords: Bioinformatics, motif discovery, parallel team, evolutionary algorithms, multiobjective optimization.

1 Introduction

Bioinformatics arises from the need to work specifically with a large amount of deoxyribonucleic acid (DNA) and protein sequences stored in databases. This information is currently used in many research works [1], ranging from multiple sequence alignment, DNA fragments assembly, or genomic mapping; to the prediction of DNA motifs, the search of these motifs in sequences of other species, or protein folding. In this paper we predict motifs using evolutionary techniques, solving the Motif Discovery Problem (MDP). The MDP aims to maximize three conflicting

David L. González–Álvarez, Miguel A. Vega–Rodríguez, Juan A. Gómez–Pulido, and Juan M. Sánchez–Pérez
University of Extremadura, Polytechnic School, Dept. TC2, 10.003, Cáceres, Spain
e-mail: {dlga,mavega,jangomez,sanperez}@unex.es

A.P. de Leon F. de Carvalho et al. (Eds.): Distrib. Computing & Artif. Intell., AISC 79, pp. 569–576.
springerlink.com

objectives: support, motif length, and similarity. So we apply multiobjective optimization (MOO) to obtain motifs in the most efficient way. To solve the MDP we have designed and configured a parallel strategy composed by four evolutionary algorithms with different properties. The algorithms chosen to be part of our parallel team are based on Differential Evolution, Nondominated Sorting Genetic Algorithm II, and Variable Neighborhood Search. From the latter we have also included a skewed variant. We have selected two population-based algorithms and two trajectory-based algorithms. Including different types of algorithms, we enrich the parallel team with the characteristics of each one of them. Recent trends are intended for using parallel strategies and we can find in the literature many parallel teams of evolutionary algorithms used successfully to solve real-world problems. The main objective of our team is exploiting the benefits of each algorithm, causing a more robust motif discovery. We also study the performance of our parallel team using different number of cores. The results obtained by our parallel heuristic improve other state-of-the-art methods of finding motifs such as AlignACE, MEME, and Weeder. Furthermore, our technique achieves better results than approaches from other researchers in the field.

This paper is organized as follows. In Section 2 we describe the motif discovery problem. Section 3 details the adjustments and modifications made on the evolutionary algorithms used in our parallel team. Section 4 shows the results obtained by our team, including the experiments made with it. Furthermore, we compare the team results with those achieved by the algorithms independently and with other techniques and algorithms for discovering DNA motifs. Finally, we explain the conclusions and future lines in Section 5.

2 Motif Discovery Problem

Sequence motifs are becoming increasingly important in the analysis of gene regulation. Sequence motifs are short DNA recurring patterns that are assumed to have a biological function. Motifs often indicate sequence specific binding sites for proteins and transcription factors [2]. Others are involved in important processes at the RNA level, including ribosome binding, mRNA processing, and transcription termination. The increase of bioinformatics works that investigate the DNA sequence motifs makes this an interesting problem in the DNA analysis field. A recent publication have proposed a new method for discovering motifs using a multiobjective approach [3]. This method maximizes three objectives: support, motif length, and similarity. Support indicates the number of sequences used to form the motif. If we do not find some candidate motif in any of the data set sequences, this sequence will not be taken into account in the motif creation. Motif length is the number of nucleotides that compose the motif. Finally, similarity is the objective that maximizes the resemblance of subsequences that make the resulting motif. To calculate it we must first generate a position weight matrix from the motif found and we calculate the dominance value (*dv*) of every nucleotide at each motif position. Then we select the highest value of each motif position $dv_{max}(i)$. With all these data, we can obtain the similarity value of a motif averaging all the dominance value for every weight matrix column. As it is indicated in equation (1). A more detailed description of this process is included in [4].

$$Similarity(Motif) = \frac{\sum_{i=1}^{l} dv_{max}(i)}{length} \quad (1)$$

3 Methodology

The MDP is an NP-hard optimization problem to predict patterns, motifs, in real DNA sequences. According to this, the use of evolutionary techniques can be a great way to get quality solutions in a reasonable time, but additionally if we combine properties of different algorithms in a parallel team, we will obtain better solutions more quickly. In this section we describe the four selected algorithms and we explain the operation of the parallel heuristic created. We have chosen four algorithms to build our parallel team, two population-based algorithms and two trajectory-based algorithms. The first population-based algorithm is an adaptation of the Differential Evolution, incorporating the concept of Pareto Tournaments (DEPT) in order to address the multiobjective optimization of this problem. The fundamental idea of DEPT is to define a schema to generate trial individuals, taking advantage of differences between the population members. Algorithm 1 shows an outline of the algorithm. The most important point of this code is in line 8, the *paretoTournaments* function performs a tournament between the solution we are processing and the trial individual. The winner will evolve to the next generation. The description of the parameters and a more detailed description of the code can be found in [4]. The second population-based algorithm is the Nondominated Sorting Genetic Algorithm II (NSGA-II) created by Deb et al. [5]. This algorithm is a standard in multiobjective optimization and provides reliability to the results produced by other algorithms. For more information, including pseudocode see [5]. The two trajectory-based algorithms implemented for our parallel team are adaptations of the Variable Neighborhood Search (VNS) algorithm, named Multiobjective Variable Neighborhood Search (MO–VNS) and Multiobjective Skewed Variable Neighborhood Search (MO–SVNS). Their essence lies in a systematic change of neighborhoods through mutation and local search functions. The first algorithm is a multiobjective

Algorithm 1. Pseudocode of DEPT

```
1: g ← 0 {initial generation}
2: population P[g] ← createRandomPopulation(NP) {NP = population size}
3: evaluatePopulation (P[g])
4: while not time limit do
5:     for i = 0 to NP do
6:         xtrial ← createTrialIndividual(P[g](i))
7:         evaluateIndividual(xtrial)
8:         P[g](i) ← paretoTournaments(P[g](j),xtrial)
9:     end for
10:    g ← g+1
11: end while
```

Algorithm 2. Pseudocode of MO-VNS

```
1: paretoFrontSolutions ← 0
2: while not time limit do
3:     S ← generateInitialRandomSolution()
4:     if isNotDominatedByAnyParetoFrontSolution(S) then
5:         /* We add the solution and clean the Pareto Front */
6:     end if
7:     k ← 1 //We initialize the neighborhood environment
8:     while k < k_max do
9:         S2 ← mutationAndLocalSearchFunctions(S,k)
10:        if isNotDominatedByAnyParetoFrontSolution(S2) then
11:            /* We add the solution and clean the Pareto Front */
12:        end if
13:        if S2 dominates S then
14:            S ← S2
15:            k ← 1
16:        else
17:            k ← k+1 //We increase the neighborhood environment
18:        end if
19:    end while
20: end while
```

Algorithm 3. Pseudocode of MO-SVNS

```
1: if S2 dominates ( S + α * distance (S2,S)) then
2:     S ← S2
3:     k ← 1
4: else
5:     k ← k+1
6: end if
```

version of the VNS algorithm, an outline of this new algorithm is shown in Algorithm 2. The most important function is the mutation and local search function (line 9). This function examines the neighborhoods by applying different levels of mutation and intensity of local search. The MO–SVNS algorithm is a skewed variant of the VNS algorithm. MO–SVNS does not always take the best solution as a reference to evolve. This algorithm allows the use of solutions slightly worse to avoid falling into local maxima. The outline of MO–SVNS is similar to the MO–VNS algorithm except the code between lines 13 and 18. The new code lines are shown in Algorithm 3. In [6] we describe in more detail these two algorithms. To combine the properties of these four algorithms we have designed a parallel heuristic, which allows the collaborative work of all the algorithms in a parallel team. This team has been implemented using Message Passing Interface (MPI). The master process pseudocode is shown in Algorithm 4. Firstly, the master process initializes the population randomly. Then it executes the algorithms (initial tests have been conducted with a parallel team of four cores, one instance of each algorithm) using the instruction *MPI_Comm_spawn_multiple*. At this point the master process stops, waiting for the algorithm results. The algorithms start their execution after receiving the population from the master process. When the synchronization time finishes, the algorithms send their Pareto fronts to the master process. In this process we store all the solutions and then we obtain the final Pareto front, deleting the solutions that are dominated by some other one. The resulting population is sent to the algorithms in the next synchronization. In this way, the algorithms work together to discover

Algorithm 4. Pseudocode of Team Master Process

```
1: population ← generateInitialPopulation()
2: /* We execute the corresponding algorithms */
3: for i = 0 to NumberOfSynchronizations do
4:     for j = 0 to NumberOfProcesses do
5:         sendPopulation(j,population)
6:     end for
7:     /* The master process waits the Pareto fronts */
8:     for j = 0 to NumberOfProcesses do
9:         solutions ← receiveParetoFrontSolutions(i)
10:        paretoFront ← addAlgorithmSolutions(solutions)
11:    end for
12:    paretoFront ← cleanParetoFrontSolutions(paretoFront)
13:    population ← obtainNewPopulation(paretoFront)
14: end for
```

quality motifs. If an algorithm is not able to evolve, the results obtained by any other algorithm ca help it to overcome this impasse.

4 Experiments and Comparisons

In this section we include the experiments performed with the algorithms and the parallel team. The experimental results are expressed in hypervolume averages, by performing 30 independent executions. The reference volume is calculated using the maximum values of each objective in each data set. All executions have been performed at one minute, fixing the synchronization time of the parallel team to twelve seconds (a total of five synchonizations). In our experiments, we used twelve real data sets [7] selected from the TRANSFAC database [8], corresponding to different species: three from the fly ('dm' instances), three from human ('hm'), three from mouse ('mus'), and three from yeast ('yst'). In addition, we selected data sets of different sizes (number of nucleotides) and with different number of sequences to ensure the good performance of our heuristics. We obtain the best configuration of DEPT from [4] with *population size = 200*, *crossover factor = 0.25*, *mutation factor = 0.02*, and *rand/1/binomial* selection schema. The best configuration of NSGA-II has been obtained following the same procedure as in [4] with *population size = 200*, *crossover probability = 0.75*, and *mutation rate = 0.25*. Finally, we obtain the best configuration of MO–VNS and MO–SVNS algorithms from [6] with *local search depth = 300*, *mutation shift = 0.20*, and $\alpha = 1$ in the MO–SVNS algorithm. The individuals of the population include the necessary information to build a motif, for further information see references [4] and [6]. The first experiment compares the results obtained by each algorithm with those obtained by the parallel team (the initial parallel team executes one instance of each algorithm). In Table 1 we see how our parallel team obtains higher hypervolume values than the individual algorithms, achieving minimal standard deviations in the data. So we conclude that the collaborative work of the four algorithms improves the quality of the solutions discovered in executions with the same time. Once tested the parallel team with 4 cores, we have performed experiments with different numbers of cores: 8, 16, 32, 64, and 128 cores (2, 4, 8, 16, and 32 instances of each algorithm, respectively). Table 2 shows

Table 1 Comparison between the algorithms and the parallel team: A_B where A is the hypervolume average and B is the standard deviation (in bold we show the best results).

Data set	DEPT	MOVNS	NSGAII	MOSVNS	4 Cores
dm01r	$0{,}405_{0{,}3089}$	$0{,}724_{0{,}0354}$	$0{,}749_{0{,}0159}$	$0{,}723_{0{,}0480}$	$\boldsymbol{0{,}779_{0.0050}}$
dm04r	$0{,}216_{0{,}2288}$	$0{,}728_{0{,}0260}$	$0{,}758_{0{,}0181}$	$0{,}737_{0{,}0176}$	$\boldsymbol{0{,}791_{0{,}0064}}$
dm05r	$0{,}166_{0{,}1798}$	$0{,}773_{0{,}0167}$	$0{,}786_{0{,}0112}$	$0{,}760_{0{,}0754}$	$\boldsymbol{0{,}813_{0{,}0036}}$
hm03r	$\boldsymbol{0{,}690_{0{,}0184}}$	$0{,}576_{0{,}0317}$	$0{,}657_{0{,}0111}$	$0{,}565_{0{,}0521}$	$0{,}685_{0{,}0109}$
hm04r	$0{,}478_{0{,}0169}$	$0{,}458_{0{,}0283}$	$0{,}543_{0{,}0223}$	$0{,}453_{0{,}0317}$	$\boldsymbol{0{,}571_{0{,}0220}}$
hm16r	$0{,}730_{0{,}0251}$	$0{,}676_{0{,}0331}$	$0{,}702_{0{,}0285}$	$0{,}673_{0{,}0266}$	$\boldsymbol{0{,}768_{0{,}0206}}$
mus02r	$\boldsymbol{0{,}674_{0{,}0036}}$	$0{,}582_{0{,}0347}$	$0{,}645_{0{,}0153}$	$0{,}584_{0{,}0386}$	$\boldsymbol{0{,}674_{0{,}0074}}$
mus07r	$0{,}625_{0{,}2376}$	$0{,}725_{0{,}0167}$	$0{,}730_{0{,}0178}$	$0{,}723_{0{,}0240}$	$\boldsymbol{0{,}773_{0{,}0038}}$
mus11r	$\boldsymbol{0{,}638_{0{,}0075}}$	$0{,}524_{0{,}0390}$	$0{,}598_{0{,}0128}$	$0{,}515_{0{,}0464}$	$\boldsymbol{0{,}638_{0{,}0102}}$
yst03r	$0{,}687_{0{,}0043}$	$0{,}637_{0{,}0241}$	$0{,}674_{0{,}0098}$	$0{,}636_{0{,}0330}$	$\boldsymbol{0{,}698_{0{,}0057}}$
yst04r	$0{,}737_{0{,}0144}$	$0{,}681_{0{,}0313}$	$0{,}726_{0{,}0150}$	$0{,}682_{0{,}0314}$	$\boldsymbol{0{,}755_{0{,}0063}}$
yst08r	$0{,}718_{0{,}0236}$	$0{,}628_{0{,}0340}$	$0{,}707_{0{,}0124}$	$0{,}635_{0{,}0306}$	$\boldsymbol{0{,}734_{0{,}0043}}$

Table 2 Results of the parallel team using different numbers of cores: A_B where A is the hypervolume average and B is the standard deviation (in bold we show the best results)

Data set	4 Cores	8 Cores	16 Cores	32 Cores	64 Cores	128 Cores
dm01r	$0{,}779_{0{,}0050}$	$0{,}784_{0{,}0047}$	$0{,}789_{0{,}0026}$	$0{,}790_{0{,}0027}$	$0{,}792_{0{,}0025}$	$\boldsymbol{0{,}794_{0{,}0022}}$
dm04r	$0{,}791_{0{,}0064}$	$0{,}798_{0{,}0031}$	$0{,}802_{0{,}0024}$	$0{,}803_{0{,}0021}$	$0{,}804_{0{,}0017}$	$\boldsymbol{0{,}806_{0{,}0018}}$
dm05r	$0{,}813_{0{,}0036}$	$0{,}818_{0{,}0039}$	$0{,}823_{0{,}0026}$	$0{,}824_{0{,}0021}$	$0{,}825_{0{,}0017}$	$\boldsymbol{0{,}828_{0{,}0010}}$
hm03r	$0{,}685_{0{,}0109}$	$0{,}692_{0{,}0088}$	$0{,}695_{0{,}0093}$	$\boldsymbol{0{,}697_{0{,}0089}}$	$0{,}695_{0{,}0080}$	$0{,}692_{0{,}0080}$
hm04r	$0{,}571_{0{,}0220}$	$0{,}597_{0{,}0159}$	$0{,}610_{0{,}0125}$	$0{,}614_{0{,}0115}$	$0{,}626_{0{,}0086}$	$\boldsymbol{0{,}630_{0{,}0101}}$
hm16r	$0{,}768_{0{,}0206}$	$0{,}781_{0{,}0144}$	$0{,}794_{0{,}0170}$	$0{,}796_{0{,}0139}$	$0{,}801_{0{,}0110}$	$\boldsymbol{0{,}806_{0{,}0115}}$
mus02r	$0{,}674_{0{,}0074}$	$0{,}681_{0{,}0066}$	$\boldsymbol{0{,}685_{0{,}0063}}$	$0{,}683_{0{,}0067}$	$0{,}682_{0{,}0062}$	$0{,}684_{0{,}0079}$
mus07r	$0{,}773_{0{,}0038}$	$0{,}775_{0{,}0035}$	$0{,}780_{0{,}0024}$	$0{,}781_{0{,}0022}$	$0{,}782_{0{,}0023}$	$0{,}784_{\boldsymbol{0{,}0028}}$
mus11r	$0{,}638_{0{,}0102}$	$0{,}648_{0{,}0073}$	$0{,}651_{0{,}0067}$	$0{,}652_{0{,}0064}$	$\boldsymbol{0{,}655_{0{,}0092}}$	$0{,}654_{0{,}0078}$
yst03r	$0{,}698_{0{,}0057}$	$0{,}701_{0{,}0062}$	$0{,}700_{0{,}0064}$	$0{,}700_{0{,}0071}$	$0{,}700_{0{,}0066}$	$\boldsymbol{0{,}706_{0{,}0074}}$
yst04r	$0{,}755_{0{,}0063}$	$0{,}760_{0{,}0044}$	$0{,}763_{0{,}0047}$	$0{,}765_{0{,}0050}$	$\boldsymbol{0{,}767_{0{,}0051}}$	$\boldsymbol{0{,}767_{0{,}0065}}$
yst08r	$0{,}734_{0{,}0043}$	$\boldsymbol{0{,}736_{0{,}0051}}$	$\boldsymbol{0{,}736_{0{,}0103}}$	$0{,}728_{0{,}0123}$	$0{,}726_{0{,}0094}$	$0{,}728_{0{,}0089}$

these results. We can see how increasing the number of instances of each algorithm, we get to improve the hypervolume results. However, in some data sets we fail to improve the results significantly. This is because we are repeating algorithms without providing new ways to evolve. Bioinformaticians can choose to invest funds in cores to achieve more robust motifs, or they may choose to discover good motifs cheaply, using a 4-core parallel team. We opted for the latter option.

To demonstrate the effectiveness and efficiency of our parallel team we have compared it with other multiobjective algorithm: MOGAMOD algorithm [3], and with other well-known motif discovery methods as AlignACE [9], MEME [10], and Weeder [11]. We could not perform this comparison using hypervolume because unfortunately, we have not this information. In order to compare with [3], we concentrate our comparisons on yst04r, yst08r, and hm03r data sets. Firstly, in Table 3 we compare our parallel team with the best configuration of MOGAMOD algorithm. We have done similarity and length comparisons for the higher value of support of each data set. We can see how our parallel team gets higher similarity motifs, keeping constant the other two objectives: support and motif length. We also can see how the parallel team discovers longer motifs than the other methods with the same support and similarity. In both comparisons our heuristic achieves higher values for the corresponding objective. Besides comparing the results with MOGAMOD algorithm, we have compared our discovered motifs with other well-known methods in the bioinformatics field as AlignACE, MEME, and Weeder. Table 5 gives the results

Table 3 Motif similarity and length comparison (in bold we show our results).

			Similarity Comparison		Length Comparison		
			Team	MOGAMOD	Team	MOGAMOD	
Data set	Support	Length	Similarity	Similarity	Length	Length	Similarity
yst04r	7	9	**0.952**	0.80	**23**	9	0.80
		8	**0.964**	0.84	**17**	8	0.84
yst08r	11	11	**0.901**	0.77	**24**	11	0.77
		10	**0.918**	0.80	**19**	10	0.80
hm03r	10	11	**0.864**	0.74	**23**	11	0.74
		10	**0.870**	0.79	**16**	10	0.79
		9	**0.889**	0.81	**14**	9	0.81

Table 4 Comparison of the predicted motifs by five methods (in bold we show our results).

Data set	Method	Support	Length	Similarity	Predicted motif
yst04r	AlignACE	N/A	10	N/A	CGGGATTCCA
	MEME	N/A	11	N/A	CGGGATTCCCC
	Weeder	N/A	10	N/A	TTTTCTGGCA
	MOGAMOD	5	14	0.84	CGAGCTTCCACTAA
		6	14	0.77	CGGGATTCCTCTAT
	Team	**7**	**16**	**0.848**	**TTTTTTTTTCTTTTCT**
		7	**14**	**0.877**	**TTTATTTTTCTTTT**
yst08r	AlignACE	N/A	12	N/A	TGATTGCACTGA
	MEME	N/A	11	N/A	CACCCAGACAC
	Weeder	N/A	10	N/A	ACACCCAGAC
	MOGAMOD	7	15	0.84	TCTGGCATCCAGTTT
		7	15	0.87	GCGACTGGGTGCCTG
		8	14	0.83	GCCAGAAAAAGGCG
		8	13	0.85	ACACCCAGACATC
	Team	**11**	**20**	**0.791**	**TTTTTTTTTTTTTATTTTTTT**
		11	**16**	**0.812**	**TTTTTTTTTTTTTATTT**
hm03r	AlignACE	N/A	13	N/A	TGTGGATAAAAAA
	MEME	N/A	20	N/A	AGTGTAGATAAAAGAAAAAC
	Weeder	N/A	10	N/A	TGATCACTGG
	MOGAMOD	7	22	0.74	TATCATCCCTGCCTAGACACAA
		7	18	0.82	TGACTCTGTCCCTAGTCT
		10	11	0.74	TTTTTTCACCA
		10	10	0.79	CCCAGCTTAG
		10	9	0.81	AGTGGGTCC
	Team	**10**	**25**	**0.728**	**AAAAAAAAAAACAGTGAAACAAAAA**
		10	**22**	**0.750**	**AAAAAAAACAGTGAAACAATAA**

of this comparison. We can see how our parallel team achieves longer solutions than the other methods, maintaining high values in each objective. Our heuristic finds longer motifs that have high similarity values with maximum values of support. As we can see, the solutions always maintain a balance among the values of the three objectives. We see how as the support and the motif length values increase, the similarity value decreases. However, with the same value of support, as the motif length decreases, the similarity value raises.

5 Conclusions and Future Lines

In this work we propose the use of a parallel team of multiobjective evolutionary algorithms to discover motifs. The objective of our approach has been to try to exploit the different skills of each algorithm, getting better solutions. We have demonstrated that the use of a parallel strategy achieves better results than those obtained by individual algorithms. Moreover, we analyze the performance of our parallel team using different number of cores and we conclude that, although we can get better solutions using many cores, the cost that this entails may not be profitable. As future work we will implement new parallel heuristics, testing new algorithms distribution functions. Furthermore, we will incorporate new algorithms to the parallel team, trying to improve the results obtained so far. Another possible future work is to study new ways to parallelize our algorithms, a possible way could be using OpenMP [12].

Acknowledgements. This work was partially funded by the Spanish Ministry of Science and Innovation and ERDF (the European Regional Development Fund), under the contract TIN2008-06491-C04-04 (the M* project). Thanks also to the Fundación Valhondo, for the economic support offered to David L. González–Álvarez to make this research.

References

1. Dopazo, J., Zanders, E., Dragoni, I., Amphlett, G., Falciani, F.: Methods and approaches in the analysis of gene expression data. Journal of immunological methods 250(1-2), 93–112 (2001)
2. D'haeseleer, P.: What are DNA sequence motifs? Nature Biotechnology 24(4), 423–425 (2006)
3. Kaya, M.: MOGAMOD: Multi–objective genetic algorithm for motif discovery. Expert Systems with Applications: An International Journal 36(2), 1039–1047 (2009)
4. González-Álvarez, D.L., Vega-Rodríguez, M.A., Gómez-Pulido, J.A., Sánchez-Pérez, J.M.: Solving the Motif Discovery Problem by Using Differential Evolution with Pareto Tournaments. In: Proceedings of the 2010 IEEE Congress on Evolutionary Computation (CEC 2010). IEEE Computer Society, Los Alamitos (2010)
5. Deb, K., Pratap, A., Agarwal, S., Meyarivan, T.: A fast and elitist multi–objective genetic algorithm: NSGA II. IEEE Transactions on Evolutionary Computation 6, 182–197 (2002)
6. González-Álvarez, D.L., Vega-Rodríguez, M.A., Gómez-Pulido, J.A., Sánchez-Pérez, J.M.: A Multiobjective Variable Neighborhood Search for Solving the Motif Discovery Problem. In: Advances in Intelligent and Soft Computing. Springer, Heidelberg (2010)
7. Tompa, M., et al.: Assessing computational tools for the discovery of transcription factor binding sites. Nature Biotechnology 23(1), 137–144 (2005)
8. Wingender, E., Dietze, P., Karas, H., Knüppel, R.: TRANSFAC: a database on transcription factors and their DNA binding sites. Nucleic Acids Research 24(1), 238–241 (1996)
9. Roth, F.P., Hughes, J.D., Estep, P.W., Church, G.M.: Finding DNA regulatory motifs within unaligned noncoding sequences clustered by whole genome mRNA quantitation. Nature Biotechnology 16(10), 939–945 (1998)
10. Bailey, T.L., Elkan, C.: Fitting a mixture model by expectation maximization to discover motifs in biopolymers. In: Proceedings of the Second International Conference on Intelligent Systems for Molecular Biology, pp. 28–36. AAAI Press, Menlo Park (1994)
11. Pavesi, G., Mereghetti, P., Mauri, G., Pesolev, G.: Weeder Web: discovery of transcription factor binding sites in a set of sequences from co–regulared genes. Nucleic Acids Research 32, 199–203 (2004)
12. OpenMP, `http://openmp.org/`

Rule-Based System to Improve Performance on Mash-up Web Applications

Carlos Guerrero, Carlos Juiz, and Ramon Puigjaner

Abstract. Web cache performance has been reduced in Web 2.0 applications due to the increase of the update rate of the contents and of the personalization of the web pages. This problem must be minimized by the caching of content fragments instead of the complete web page. We propose a rule-based optimization algorithm to define the fragments design that experiment a best performance. This algorithm uses characterization parameters of the fragment contents to find the optimized solution.

Keywords: Web caching, Performance engineering, Web 2.0, Mashup applications.

1 Introduction

The way that the users use modern web applications has change completely during the last years. Nowadays, web applications are more personalized and content sources are completely decentralized [11]. This affects the applicability of technical solutions for the system architecture. Particularly, it affects performance and solutions as web caching. Web caching is based in re-usability of request results between different users and between requests separated by a short period of time. Modern web applications, as mash-ups applications and Web 2.0-based systems, have a high personalization grade and they are built, at most of the times, using different content sources [9]. Those facts produce that invalidation periods are too shorts -more than the time period between two requests of the same user on the same web page- and that the probability of two users requesting the same page is very small due to high personalization grade of web contents. One solution for the problem of request re-usability is to reduce the minimum cacheable unit. Instead of using web pages, web cache systems could manage parts of these documents (fragments). ESI (Edge Side Includes) is a "de facto" standard which is used to define web pages as an aggregation of fragments.

Carlos Guerrero, Carlos Juiz, and Ramon Puigjaner
Universitat de les Illes Balears, Crta. Valldemossa km 7.5, Illes Balears (Spain)
e-mail: carlos.guerrero, cjuiz, putxi@uib.es

A.P. de Leon F. de Carvalho et al. (Eds.): Distrib. Computing & Artif. Intell., AISC 79, pp. 577–584.
springerlink.com

The number of research studies about how fragment-based systems influence the performance is very large, but these studies are not applied to the use of a web caching system of fragments for mash up applications. The main contributions of the research presented in this paper are: (a) The study of which characterization parameters, of the web page fragments, influence in the performance of the cache architecture. (b) The applicability of these characterization parameters in optimization techniques to determine an optimal fragments design of web pages. (c) The study of the benefits of a rule-based system which uses content fragment characterization parameters to determine a good performance solution for the problem of how to fragment the page

2 Web Caching for Web 2.0 Content Aggregation Applications

Content aggregation technologies are systems that combine content from different sources to create new content elements. I.e., web pages are created by the aggregation of independent contents. Therefore, we could distinguish between two different content element types: pure content elements (which are created for a system or an user) and aggregated content elements (which are created from the combination of others content elements). Authors of [2] define a representation for fragment-based web applications using a graph (object dependence graphs, ODGs). We have adapted that representation for our propose. We use a Directed Acyclic Graph (DAG) where the edges of the graph represents the aggregation of contents. We use labelled edges in the DAG to represent when two elements are joined in the cache or in the server. The vertices represent the pure and aggregated content fragments. Sink vertices correspond to pure elements and source vertices to user web pages. The rest of the vertices correspond to aggregated elements.

If the web pages are created in the server (by the joining of pure and aggregated contents or fragments), the web cache layer is only able to cache the final and indivisible web page. In the other hand, the cache is able to manage and to store the fragments independently if the process of joining the fragments is done in it . The fragments of this second solution have smaller update rates and bigger request rates than the final web pages of the first solution [8, 6, 7]. This benefits the hit ratio of the web cache. But this second solution has the problem of the overhead of joining fragments [5, 2]. This damages the user observed response time. From a performance point of view, we consider that the optimal solution is in the middle of the two before. A solution where some fragments are joined in the server (which less benefit the hit ratio and/or more damage the response time) and other fragments are delivered to the web cache independently.

The problem we need to solve is an optimization problem. We want to find a solution that would produce the best performance in the web cache. This solution (or output) is a coloured graph that represent the state of the different fragments in the web page design (*joined* or *split*). The number of possible design solutions is the total number of variations of repetitions of 2 states choose p (where p is the number of edges) and is calculated as $V'(2,p) = 2^p$.

The input parameters of the algorithm must be easily measured. Performance parameters are not considered because they show transient behaviour after a change in the content. Another limitation is the level in which our system works (the application level). So the number of possible motorization points is quite reduced. In one hand, we could use web mining techniques (web structure mining) to extract data from our system. We will consider the graph structure of the different fragments as inputs parameters. In the other hand, web cache systems have traditionally used parameters as size, update ratios, request ratios and response times to improve caching algorithms [3, 10]. We will also take them into account.

3 Performance Rule-Based Optimization Algorithm Using Content Fragments Characterization Parameters

The first step of our research work is to determine if the parameters commented in the previous section can be used to predict the improvement of the performance of the web cache system. If these parameters show some kind of relation with the web cache performance, we will be able to use them as inputs of an optimization algorithm. In order to accomplish this goal we will study the correlation between the cache performance metrics and these characterization parameters. The parameters we have taken into account and that we have analysed are the next: (i) number of aggregations that a fragment has (represented as the number of child of one node); (ii) number of fragments that aggregates a given fragment (represented as the number of fathers of one node); (iii) size of the content of a fragment (size of all the HTML code corresponding to a given fragment); (iv) service time of a fragment (time required in the server to generate the content of a given fragment); (v) user request rate; and (vi) update content rate

Figure 1 shows the architecture of the emulation environment we have created. In this emulation environment content fragments can be joined in the server (edges labeled as *joined*) or in the web cache (edges labelled as *split*). To create the content model and the user emulation model we have randomly created values for the parameters using an uniform distribution in order to cover a wide range of values. The maximum and minimum values and the gap between samples are presented in Table 1. To study the correlation between performance improvement and characterization parameters we have created two scenarios (one where all the edges are labelled as *joined* and other where they are labelled as *split*) and we have study the web cache response time speedup between both scenarios. Finally we have study the correlation between these values and the characterization parameters of the content fragments.

After running the emulation during the enough time to obtain means with the enough confidential interval (95%) we analysed the correlation between the response time of the web cache and the characterization parameters. We use the Pearson Correlation Factor to study the correlation between the samples(values close to 0 indicates no correlation and values close to 1 or -1 indicates correlation). Template size (-0.01) and service time (-0.03) show correlation values close to 0. Father

Table 1 Characterization parameters values for the fragments of the emulated model

Parameter	Minimum	Maximum	Gap
Fathers number	1	20	1
Children number	1	20	1
Template Size (KB)	5	300	1
Size (KB)	1	50	1
Service Time (ms)	1	100	5
Request rate (s^{-1})	1/10	1/300	1/10
Update rate (s^{-1})	1/60	1/3600	1/30

number (0.45), child number (0.31), and fragment size (0.41) show enough correlation to be taken into account. Finally, request and update rates do not show a clear correlation (0.07 and -0.11 respectively). But if we analysed the relation between both parameters instead their absolutes values, we found a considerable correlation (-0.34). We present the samples most representative in Figure 2.

We interpretate these results as the performance improvement of having split fragments is bigger when: (a) The higher the size of the fragments is. (b) The higher the number of fathers a fragment has. (c) The higher the number of children a fragment has. (d) The bigger the difference between request rate and update rate is. Thus, we conclude that the fragment size, the difference between update rate and request rate, and the number of fathers and children of a fragment are suitable to be the inputs variables of our optimization algorithm.

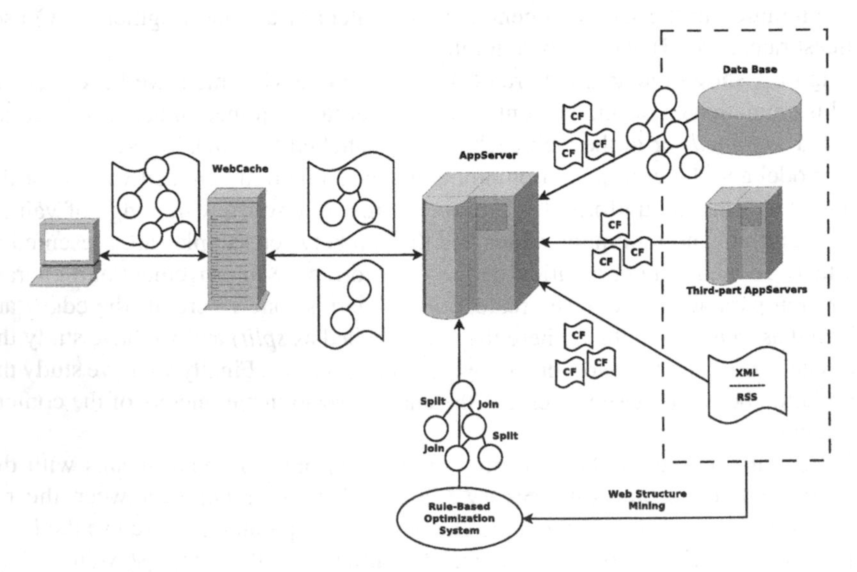

Fig. 1 Architecture of the emulation environment.

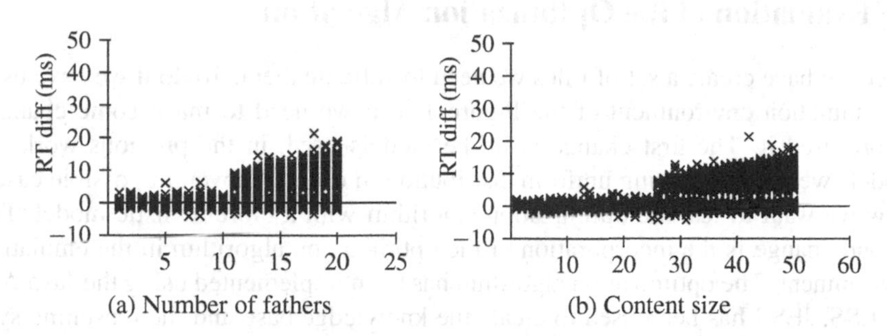

Fig. 2 Response time difference between samples of the joined and split model related with the number of fathers of the samples (a) and with the size of the fragments (b).

Once we have study the correlation between characterization parameters and performance improvement we are able to create a set of rules to be used in a optimization algorithm to find the fragments design that offer the best performance. The inputs of the rules are the characterization parameters and the output is the values of the coloured edges of the DAG that represents the web page design. I.e., the left-hand side (LHS) of our rules uses the characterization parameters of our fragments, and the right-hand side (RHS) conclude about the state or color of and edge.

We used the clustering techniques that implements WEKA to create classes of our samples in order to create the rules. The attributes used to create our classes are the characterization parameters and the performance speedud between the *joined* and *split* scenario. After trying different clustering algorithms we decide to establish 6 different classes. Our interpretation of these six classes is explained below. The samples of pairs of fragments (a template fragment or father with a content fragment or child) experiment a better performance when they both are: (a) split if only one fragment has a a higher update rate than request rate. (b) joined if both fragments have a higher update rate than request rate. (c) split if both fragments have a smaller update rate than request rate and the child fragment has more than one father (it is included in more than one template). (d) joined if fragments have a smaller update rate than request rate and the child fragment has only one father and the number of children of the father fragment (template) has only one child. (e) joined if both fragments have a smaller update rate than request rate and the child fragment has only one father and the number of children of the father fragment has more than one child and the size of the content fragment (child) is smaller than 25 KB. (f) split if both fragments have a smaller update rate than request rate and the child fragment has only one father and the number of children of the father fragment has more than one child and the size of the content fragment (child) is bigger than 25 KB.

We translate the previous conclusions to a decision tree (Figure 3). The proposed knowledge base (group of rules or decision tree) is very simple. This is a necessary feature in our system. In web systems, the characterization parameters, that we have used to deduce the fragment design, change continually. This determines the need of continually updating the fragments design after some period of time. So we need a simple rule system to reduce the computation requirements of the algorithm.

4 Evaluation of the Optimization Algorithm

Once we have create a set of rules we need to validate them. To do it we have used the emulation environment of the Figure 1. But we need to make some changes before used it. The first change is in the models used, in the previous work the models were created using uniform distribution in order to cover all possible cases. Now we want to test the optimization algorithm with a more realistic model. The second change is the incorporation of the optimization algorithm in the emulation environment. The optimization algorithm has been implemented using the Java API of JESS. JESS has been used to create the knowledge base and the reasoning system to optimize the fragment design. JESS is an rule engine environment to give to Java projects the capacity to "reason" using knowledge supplied in the form of declarative rules.

The date from real systems is not always available. From the structure point of view, we have analysed the structure and size of fragment of the New York Times web page. Using a web structure mining process we could determine the content fragments (each one of the news on a front page), their structure and the size of these fragments. From the user point of view, we need to know which pages are more requested for read and for update the content (popularity) and which is the grade of load of the user activity. This data is not available in most of web sites, neither in the case of the New York Times. The last choice we have is to use statistical data. Research works about Web 2.0 ([4, 1]) have concluded that the popularity of web objects follows a power law statistical distribution with parameter $\alpha = 0.83$ (for read requests) and $\alpha = 0.54$ (for write requests) both with $R^2 = 0.99$. The workload of the system depends on the web site we analyse and on the period of time analysed. We

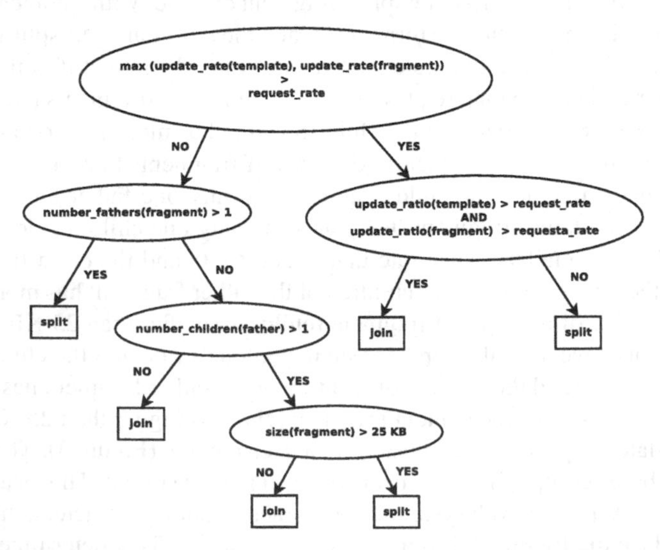

Fig. 3 Decision tree for the optimization algorithm of web fragment designs. The rules for the JESS environment are created directly from this decision tree.

Table 2 Speedup between the experiments

Experiment	Speedup Complete	Fragmented	Optimized
Complete	*	0.8800	0.8354
Fragmented	1.1364	*	0.9542
Optimized	1.1970	1.048	*

decided to test our system over a low utilization scenario due to the scaling problems of the clients simulator. But we consider that, over high utilization scenarios, our solution would show better results.

We have tested the design optimization of our algorithm over a group of 200 pages created using the parameters explained above. To determine if our fragment designs experiments a better performance (Optimized experiment) we have compare the results with the two extreme scenarios: (i) Complete experiment: where all the fragments are joined in the server and the web cache is not able to manage content fragments; (ii) Fragment experiment: where all the content fragments are split and the server does not aggregate any content.

We executed the three experiments measuring the user observed response time over 200.000 requests. Analysing the means of each single request we have calculated the speedup between the three experiments. The mean values of these speedups are shown in Table 2. The speedup of the optimized experiment over the fragmented experiment is 1.0488 and over the complete experiment is 1.1970. This values could be considered not very high but we have to notice that the speedup of the complete experiment over the fragmented one is only 1.1364. So we have a small gap between the extreme solution results. We explain those results because the low load of the server resources.

5 Conclusions

We have presented an rule-based optimization algorithm to improve the web cache performance on content aggregation systems. The output of the algorithm is the content fragments structure that provides the best cache performance. This output is defined as the state between a pair of fragments (*joined* or *split*). The inputs of the algorithm (the conditional part of the rules) are a set of characterization parameters of the content fragments and their structure.

We have implemented a core system which use our optimization algorithm and we have tested it in an emulation scenario. The model used for the emulation has been mined from the web of the New York Times. Some parameters of the model were not available and they have been modelled by statistical studies. The system has been tested over a scenario with a low resource utilization. The obtained results shows that our optimization algorithm offers a better user response time than the two extreme scenarios.

As future work we will repeat the same experiment but in a high resource utilization scenario and we will analyse how the load of the system could influence the improvement offered by our optimized fragment structures. Another future work is to re-apply data mining techniques over the results of the experiments. For some of the samples of the results, the complete or fragmented experiments present a best response time than for the optimized one. It is a small number of samples, but we need to understand under which conditions it takes place to add them in our rules.

References

1. Cha, M., Kwak, H., Rodriguez, P., Ahn, Y.-Y., Moon, S.: I tube, you tube, everybody tubes: analyzing the world's largest user generated content video system. In: IMC 2007: Proceedings of the 7th ACM SIGCOMM conference on Internet measurement, San Diego, California, USA, pp. 1–14. ACM, New York (2007)
2. Challenger, J., Dantzig, P., Iyengar, A., Witting, K.: A fragment-based approach for efficiently creating dynamic web content. ACM Trans. Internet Technol. 5(2), 359–389 (2005)
3. Cherkasova, L.: Improving www proxies performance with greedy-dual-size-frequency caching policy. Technical report
4. Duarte, F., Mattos, B., Bestavros, A., Almeida, V., Almeida, J.: Traffic characteristics and communication patterns in blogosphere. In: Proceedings of the International AAAI Conference on Weblogs and Social Media (2007)
5. Guerrero, C., Juiz, C., Puigjaner, R.: The applicability of balanced esi for web caching. In: Proceedings of the 3rd International Conference on Web Information Systems and Technologies (March 2007)
6. Guerrero, C., Juiz, C., Puigjaner, R.: Web performance engineering based on ontological languages and semantic web. Int. J. Comput. Appl. Technol. 33(4), 300–311 (2009)
7. Hassan, O.A.-H., Ramaswamy, L., Miller, J.A.: Mace: A dynamic caching framework for mashups. In: ICWS 2000: Proceedings of the IEEE 7th International Conference on Web Services, IEEE, Los Alamitos (2009)
8. Leighton, T.: Improving performance on the internet. Queue 6(6), 20–29 (2008)
9. Mislove, A., Marcon, M., Gummadi, K.P., Druschel, P., Bhattacharjee, B.: Measurement and analysis of online social networks. In: IMC 2007: Proceedings of the 7th ACM SIGCOMM conference on Internet measurement, pp. 29–42. ACM, New York (2007)
10. Nagaraj, S.: Web Caching and Its Applications. Kluwer Academic Publishers, Dordrecht (2004)
11. Schneider, F., Agarwal, S., Alpcan, T., Feldmann, A.: The new web: Characterizing ajax traffic. In: Claypool, M., Uhlig, S. (eds.) PAM 2008. LNCS, vol. 4979, pp. 31–40. Springer, Heidelberg (2008)

Information Extraction from Heterogeneous Web Sites Using Clue Complement Process Based on a User's Instantiated Example

Junya Shimada, Hironori Oka, Masanori Akiyoshi, and Norihisa Komoda

Abstract. Since the growth of the Internet, World Wide Web has become significant infrastructure in various fields such as business, commerce, education and so on. Accordingly, a user has gathered information by using the Internet. However due to increasing Web pages, it becomes difficult for a user to collect desirable information. Advanced Web search engines may provide solution to some extent, it is still up to a user to summarize or extract meaningful information from such retrieval results. Based on this viewpoints, this paper addresses a generation method of table-style data from heterogeneous Web pages that reflects a user's intention. To achieve it, the method utilize a user's instantiated example in a table in addition to column labels as the table. Based on a user's instantiated example, meaningful information are extracted using pattern matching and N-gram method. We apply this method to 57 pages with 27 travel agencies whether the proposed method is effective or not. As the result, 88% was precision rate and 68% was recall rate.

Keywords: User's instantiated example, Information extraction, Clue complement process, Table-style data.

1 Introduction

In recent years, the Internet provide various information to users and advanced Web search engines that have specific techniques have been developed, for instance, Page Ranking technique [1][2][3], List by snippet [4][5][6], Summarization [7][8] and so forth. However, in case that a user extracts a part of necessary information for the problem-solving through entirely reading the related pages, it still takes lots of time

Junya Shimada, Masanori Akiyoshi, and Norihisa Komoda
Osaka University, 2-1 Yamadaoka Suita, Osaka, Japan
e-mail: shimada.junya@ist.osaka-u.ac.jp,
akiyoshi@ist.osaka-u.ac.jp, komoda@ist.osaka-u.ac.jp

Hironori Oka
Codetoys, 2-6-8 Nishitenma Kita, Osaka, Japan
e-mail: oka@codetoys.co.jp

A.P. de Leon F. de Carvalho et al. (Eds.): Distrib. Computing & Artif. Intell., AISC 79, pp. 585–592.
springerlink.com

to achieve it. Also a user sometimes have to compare other sites to obtain necessary information, which is usually attained by using a table format. When making a table, "Copy and Paste" action on the Web page is manually done. Portal sites showing comparison table like "kakaku.com (http://kakaku.com/)" enable users to compare various products. However, their sites do not reflect user's intention.

Therefore there is a strong need to support making up such a table simply from an inputted example as a user's intention as to extracting necessary information. Google starts to provide experimental service towards such a user's need as "Google Squared (http://www.google.com/squared)" It extracts the information as a table when a user inputs a search keyword. This seems to be slightly biased to full-automatic processing, because the table labels on columns are configured without a user's intention. This paper addresses a method of being mostly full-automatic processing, which means the table labels on columns are provided by a user and corresponding data are extracted by using a user's instantiated example as a reference. Through preliminary investigation on this viewpoint, we propose a method to design the system and show reasonable efficacy.

2 Extraction Method Using Clue Complement Process Based on a User'S Instantiated Example

Fig. 1 shows the outline of system to extract desirable information from retrieved Web pages and user's instantiated examples. Then the system output is a filled table with necessary information. Assume that retrieved Web pages are collected by using "keyword search" along with a user's intention. Then collected Web pages are stored in a local database with HTML format data. As an indicated example in Fig. 1, a user inputs "schedule", "recommendation point" and so on to compare which tour is more attractive to investigate in further detail manner. The top row is a user's provided example when he look into the Web page, then the rest of information is obtained by the specific feature of appearing words, length of text portions and position in pages. Then we describe the recognition of extraction pattern and extraction method based on a user's instantiated example.

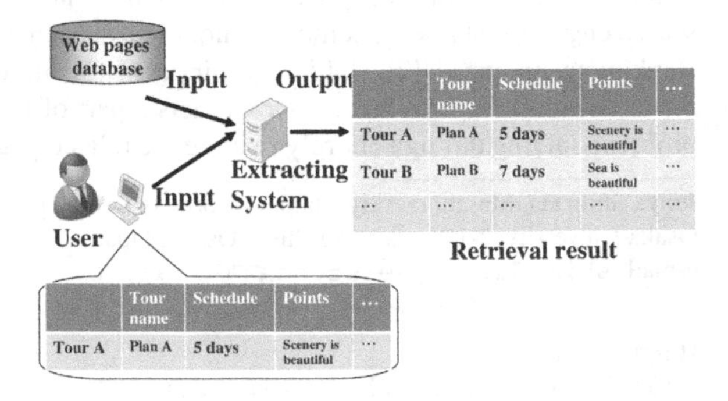

Fig. 1 Outline of the system

Based on relations among the label and data about the label which a user inputs, our system recognizes the extraction pattern. We focus on "Exist to be near" which means possible data related to column label are placed across "tag" and "delimiter". This time, we focus on "delimiter". So based on the pattern, data is extracted from other Web sites.

Extraction method based on a user's instantiated example

Based on the extraction pattern, our system carries on the following 4 steps and extracts the data. Fig. 2 shows the process flow of extraction method.

2.1 Extraction by Label Matching

Fig. 3 shows the extraction method by label matching. This time, we focus on delimiters. Delimiters we used here are ":"(colon), ";"(semicolon) and "white space". Then we extract the data about the label.

When user's instantiated example is "Air Lines", Japan Air Lines, Korean Air Lines, and Air France Air Lines are extracted from Web sites stored in the local database as necessary information.

2.2 Elimination of Unnecessary Words

In the extracted data, there are unnecessary words which do not have any relevance to the label. So, we need to eliminate the unnecessary words using the user's

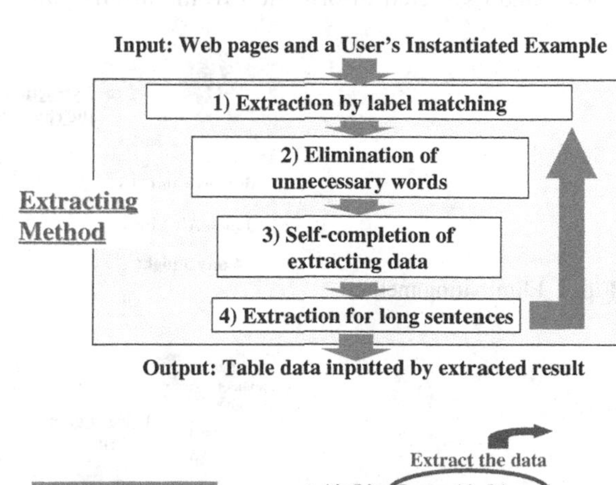

Fig. 2 Extracting method

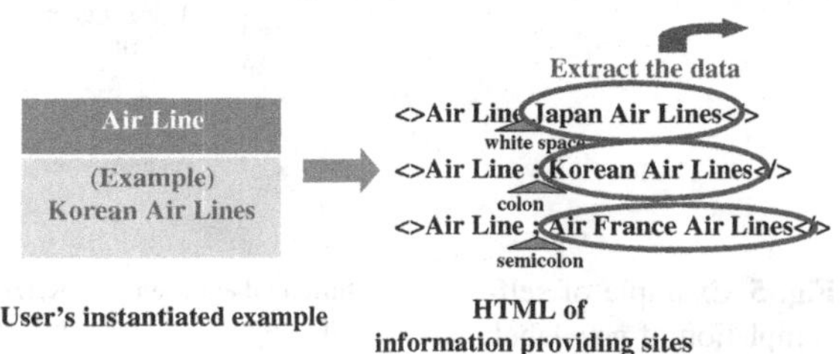

Fig. 3 Example of extraction by label matching

instantiated example. Fig. 4 shows the eliminating method. We distinguish the extracting data from following 3 relations.

- Proper name
- Synonym between label and data
- Numeral representation

After extracted data, they are extracted the characteristics based on the user's instantiated example. Based on the user's example, column's words are distinguished from proper name or synonym between label and data or number. Words which do not have same characteristics among the column's words are eliminated as unnecessary words. A user's example and the extracted data have common characteristics, when the label and a user's example is Fig. 4. They include "xx air lines", so 4-day 3-night is regarded as an unnecessary words. Then the word is eliminated. In this way, unnecessary words are eliminated and only words what seem to be necessary are remained.

2.3 *Self-completion of Extracting Data*

2.3.1 Self-completion of New Label

After extracting some information using "label" depicted in a user's provided table, the mechanism intends to complete possible labels by analyzing such extracted information.

Fig. 5 shows overall flow. Initially the column label is "minimum num. of persons", and extracted information from various pages with "Exist to be near" rule

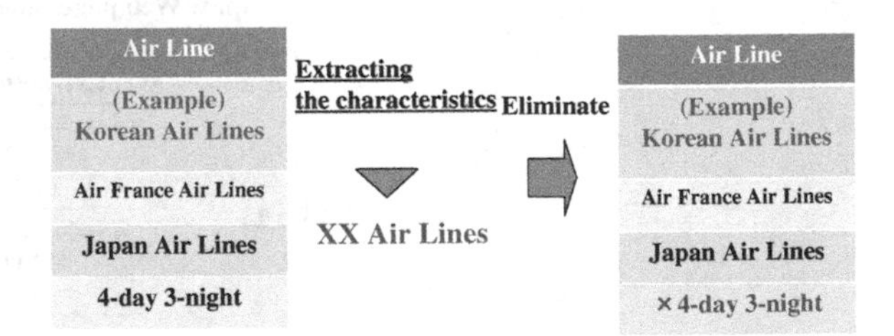

Fig. 4 Eliminating method

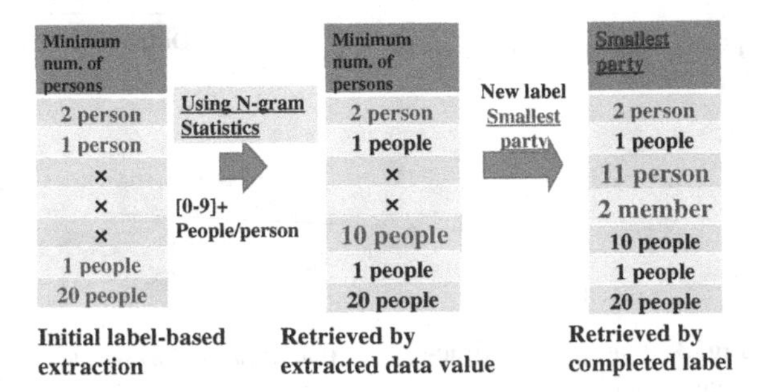

Fig. 5 Example of self-completion of new label

are "xx people" in addition to a user's instantiated "2 person". And about extracted data, they are grasped characteristics of frequently-appearing expression using N-gram method. Then regular expression, "[0-9]+[person, people]" are generated and used again for extracting in target pages. Then the red-colored "10 people" without the label "Minimum num. of persons" is extracted as a new one. Then the "label" from "Exist to be near" rule on this "10 people" is found as a new label. Such newly identified label "Smallest party" is used again for extracting and finally the red-colored "2 member" is collected. As explained above, the initial "column label in a user's table" is self-completed in "bootstrap" manner.

2.3.2 Self-completion of Extracting Data which Has No Label

In some web pages, there are no label because the data about the label is easily guessed. In this case, using N-gram method, data about the label are extracted. A user's example is compared with the other web pages using N-gram method. Then most common words are regarded as the extracting pattern, and based on the pattern, extraction is done, in which the search name is the words in the other web pages.

2.4 Extraction for Long Sentences

Fig. 6 shows extraction method for long sentences. If data about a user's instantiated example are multiline, this method is used. Long sentences which seem what a user wants are extracted using the number of words. And the long sentences are compared with data of a user's instantiated example. If the long sentences and data of a user's instantiated example have the similar type of words, the long sentences are extracted as a user wants. On the contrary, if the long sentences and data of a user's instantiated example do not have, the process is repeated.

Then, Fig. 7 shows extracting method on portion of sentences to be extracted. We extract data using the words and appearance position. And changing the windows width and threshold amount about words, appropriate sentences are extracted.

Fig. 8 shows process of extracting long sentences. Using 2 loop about the threshold and the windows width, long sentences are extracted.

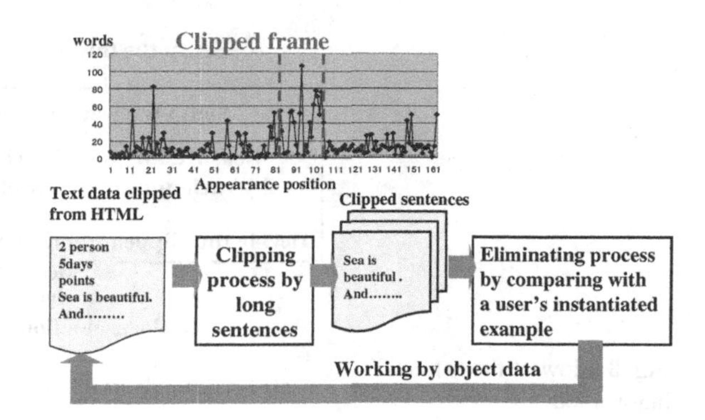

Fig. 6 Example of extracting method for long sentences

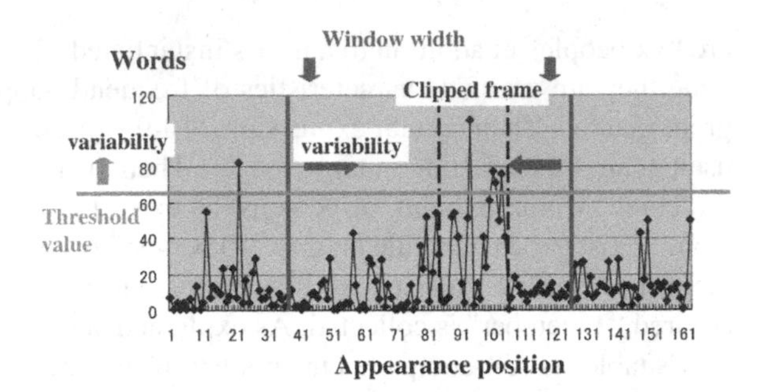

Fig. 7 Extracting method of long sentences

At first, the threshold value is set up as average amount of the whole words because the words of long sentences seem to be larger number than the average amount. Second, the window width is set up as the wide of appearance position. Then, displaying the window width, the words' variation are measured. The group of long sentences are described in many line. So, the number of words are changed by appearance position. If the words' variation are large, the appearance position of the group is extracted as a user wants. After extracted data, the window width is changed more shortly. Then, the appearance position of the group which seem to be data what a user wants is extracted until the window width is 0.

If the window width is 0, the threshold amount is changed. The number of the threshold amount is larger when the threshold amount is changed. Then, after changed the threshold amount, the same method is repeated. And the group clipped frame is stored.

After these processes, appearance position of the group clipped frame and the wide of window width at that time are compared. Then, if appearance position of the group and the wide of window width are nearly distance, the wide of window width is extracted as a user really wants. In this way, long sentences are extraced.

3 Experiment

We make evaluation between proposed method and extracted data by hands calculating precision rate and recall.

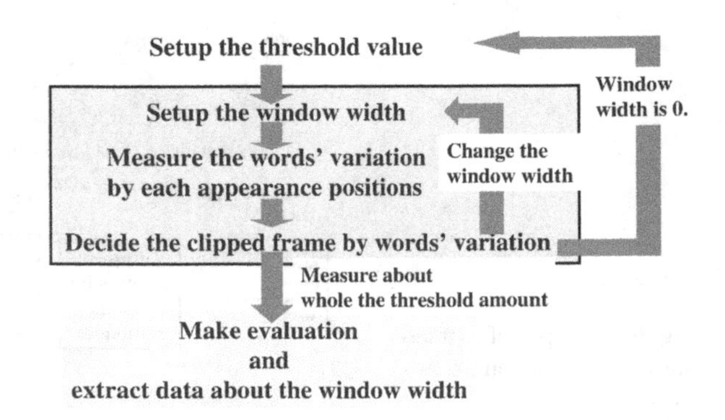

Fig. 8 Flow of the extracting method

Fig. 9 User's instantiated example

Pay	Fuel surcharge	Days	Period of time	Airline	Tour escort	Minimu m num. of persons	Points
¥79800 ~ ¥90000	¥12000	5 days	4th Jan~2 4th Mar	Korean Air Lines	None	2 person	Sea is beautiful.

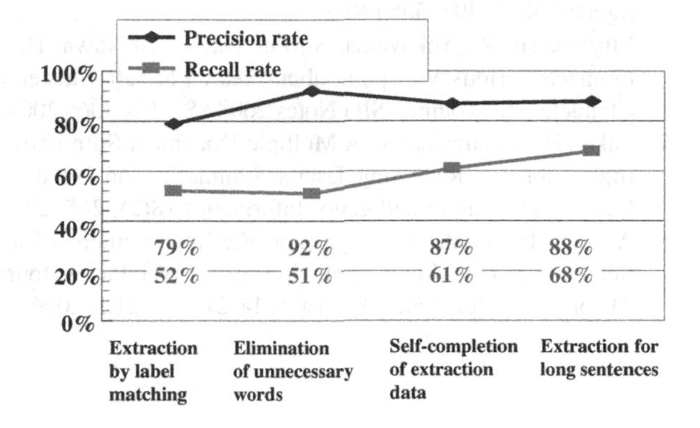

Fig. 10 Experimental result

- As for paris tour sites in the travel sites. 57 pages with 27 travel agencies
- Fig. 9 shows 8 labels and data about the labels as a user's instantiated example.

We calculate the precision rate and recall rate. Fig. 10 shows the comparison result with the extracted data using our proposed method and extracted data by hands.

As the result, the precision rate is 88% and the recall rate is 68%.

4 Conclusion

This paper describes a generation method of table-style data by extracting appropriate portions in Web pages. Instead of using "pattern" or "template" with dictionary, a user's instantiated example in a table is used as clues in extracting process. In this process, extracting mechanisms are proposed to enhance extraction functionality. We conducted an experiment whether our proposed method could extract more efficiently. As the result, 88% was precision rate and 68% was recall rate.

References

1. Okamura, H., Miyauchi, S., Dohi, T.: A Web Page Ranking Algorithm Based on a Markov Decision Process. The IEICE transitions on information and systems(Japanese edition) J89-D(2), 210–219 (2006)
2. Aratani, H., Fujita, S., Sugawara, K.: Extremely Precise Finding Methodology on the Mutual Evolution Method Among Web Pages, IEICE technical report, Artificial Intelligence and Knowledge-based Processing 105(105), 1–6 (May 2005)

3. Kawamae, N., Aoki, T., Yasuda, H.: Page Ranking Method of Search System Considering Difference of Access to the Pages. In: Proc. of the IEICE General Conference, vol. (1), p. 47 (March 2000)
4. Watanabe, N., Okamoto, M., Kikuchi, M., Iida, T., Hattori, M.: Influence of Presentation Style in Web-Search Result Recommenndation, IPSJ SIG technical reports 2009 (28), 61-68 (March 2009)
5. Toda, H., Yasuda, N., Okumura, M., Matsuura, Y., Kataoka, R.: Snippet Generation for Geographic Information Retrieval. Transition of the Japanese Society for Artificial Intelligence 24(6), 494–506 (2009)
6. Muramatsu, R., Yokoyama, S., Fukuta, N., Ishikawa, H.: Architect Snippets with Harmonized Various Viewpoint about Search Result Cluster with Consideration of Word's Characteristic Volume. SIG Notes 2008 88, 301–306 (2008)
7. Sakai, H., Masuyama, S.: A Multiple-Document Summarization System Introducing User Interaction for Reflecting User's Summarization Needs. Journal of Japan Society for Fuzzy Theory and Intelligence Informatics 18(2), 265–279 (2006)
8. Aratani, H., Fujita, S., Sugawara, K.: Improvement of a Re-ranking Method for Web Search Based on Mutual Evaliation among Web Pages. Journal of Japan Society for Fuzzy Theory and Intelligent Informatics 18(2), 196–212 (2006)

Information Extraction from Heterogenous Web Sites Using Additional Search of Related Contents Based on a User's Instantiated Example

Yuki Mitsui, Hironori Oka, Masanori Akiyoshi, and Norihisa Komoda

Abstract. Recently, since the growth of the Internet, WorldWide Web has become significant infrastructure in various fields such as business, commerce, education and so on. Accordingly, a user has gathered information by using the Internet. However due to the flood of Web pages, it becomes difficult for a user to collect desirable information. Advanced Web search engines may provide solution to some extent, it is still up to a user to summarize or extract meaningful information from such retrieval results. Based on this viewpoints, we addressed a generation method of table-style data from heterogeneous Webpages that reflects a user's intention. However if original pages have less information, our system may not extract sufficient information. To improve this problem, we address a method that searches related page contents automatically. We apply this method to shopping sites and the experimental result shows it improves recall rate.

Keywords: Information Extraction, Additional Search, User's Instantiated Example.

1 Introduction

Recently the Internet provides users to gather various information and advanced Web search engines that have specific techniques have been developed, for instance, "Page Ranking technique"[1, 2, 3], "List by snippet"[4, 5, 6], "Summarization and

Yuki Mitsui, Masanori Akiyoshi, and Norihisa Komoda
Osaka University, 2-1 Yamadaoka Suita, Osaka, Japan
e-mail: mitsui.yuuki@ist.osaka-u.ac.jp,
akiyoshi@ist.osaka-u.ac.jp, komoda@ist.osaka-u.ac.jp

Hironori Oka
Codetoys Codetoys, 2-6-8 Nishitenma Kita, Osaka, Japan
e-mail: oka@codetoys.co.jp

A.P. de Leon F. de Carvalho et al. (Eds.): Distrib. Computing & Artif. Intell., AISC 79, pp. 593–600.
springerlink.com

so forth"[7, 8]. However, in case that a user extracts a part of necessary information for the problem-solving by entirely reading the related pages by the retrieval, it still takes lots of time to achieve it. Also a user sometimes have to compare other sites to obtain necessary information, which is usually attained by using a table format. When making a table, "Copy and Paste" action on the Web page is manually done.

Therefore there is a strong need to support making up such a table simply from an inputted example as a user's intention as to extract necessary information. Google starts to provide experimental service towards such a user's need as "Google Squared (http://www.google.com/squared)". It extracts the information as a table when a user inputs a search keyword. This seems to be slightly biased to full-automatic processing, because the table labels on columns are configured without a user's intention. Based on this viewpoints, we addressed a method of being mostly full-automatic processing, which means the table labels on columns and instantiated data examples as a reference are provided by a user and corresponding data are extracted [9]. However if original pages have less information, our system may not extract sufficient information. In this case, a user has to extract from other pages manually. This paper addresses a method that searches related page contents automatically to improve this problem.

2 Generation Method of Table-Style Data Based on a User's Instantiated Example

Fig. 1 shows the outline of system to extract desirable information from retrieved Web pages with a user's table format and instantiated examples. Then the system output is a filled table with necessary information. Assume that retrieved Web pages are collected by using "keyword search" along with a user's intention. Then collected Web pages are stored in a local database with HTML format data. As an indicated example in Fig. 1, a user inputs "Tour name", "schedule" and so on to compare which tour is more attractive to investigate in further detail manner. The top row is a user's provided example when he looks into the Web page, then the rest of information is obtained by the specific feature of appearing words, length of text portions and position in pages by using thesaurus.

Web sites mainly consists of HTML tag and text. Even if it extracts from only text, it is difficult to extract using existing Web Wrapper[10, 11, 12, 13] that is based on pattern because we want to extract various Web sites that have diversified structure. The columns are assumed targeted Web sites and the rows are assumed direction words (user labels) in a user's instantiated example, so we use this user labels as a clue. Because a user label is made from only one example of the Web site, it is difficult to extract information from other Web sites using a only user indicated label. We tried to improve by "Elimination of Unnecessary Words and Adding Extraction". Fig. 2 shows the outline of extraction method based on this viewpoint.

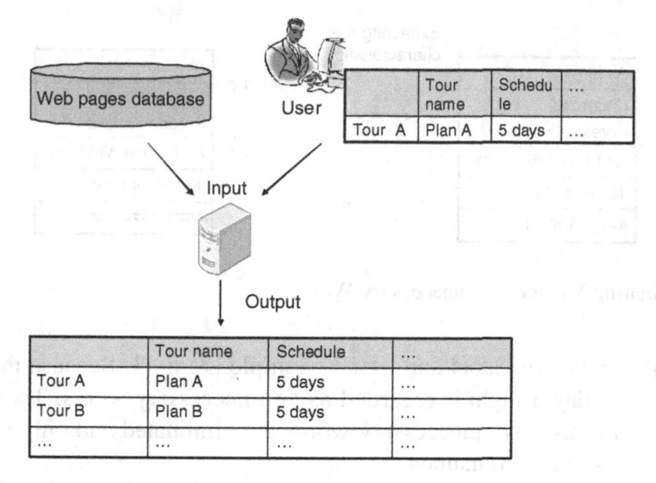

Fig. 1 Outline of Extraction System based on a User's Instantiated Example

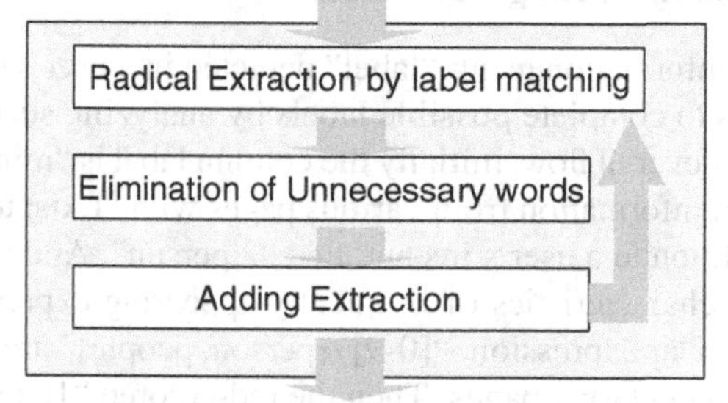

Fig. 2 Overview of Extracting Method

2.1 Extraction by Label Matching

At first, we focus on "Exist to be near" which means possible data related to column label are placed across "tag" and "delimiter". Delimiters we used here are "colon" and "space". Then we extract the data about the label.

2.2 Elimination of Unnecessary Words

In the extracted data, there are unnecessary words which do not have no relation to the label. So, we need to eliminate the unnecessary words using the user's instantiated example. Fig. 3 shows the eliminating method.

After extracted data, they are extracted the characteristics based on a user's instantiated example. A user's example and the extracted data have common

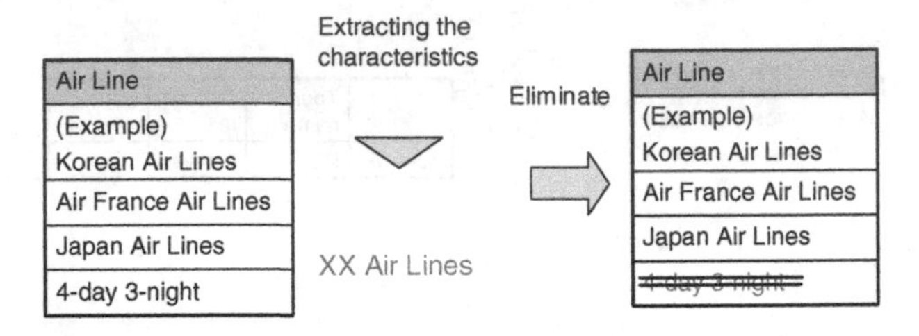

Fig. 3 Eliminating Method of Unnecessary Words

characteristics, when the label and user's example is Fig. 3. Because they include "xx air lines", 4-day 3-night is regarded as an unnecessary words. Then the word is eliminated. In this way, unnecessary words are eliminated and only words what seem to be necessary are remained.

2.3 Adding Extraction

2.3.1 Self-completion Extracting New Label

After extracting some information using "label" depicted in a user's provided table, the mechanism intends to complete possible labels by analyzing such extracted information. Fig. 4 shows overall flow. Initially the column label is "minimum num. of persons", and extracted information from various pages with "Exist to be near" rule are "xx people" in addition to a user's instantiated "2 person". And about extracted data, they are grasped characteristics of frequently appearing expression using N-gram method. Then regular expression, "[0-9]+[person, people]" are generated and used again for extracting in target pages. Then the red-colored "10 people" without the label "Minimum num. of persons" is extracted as a new one. Then the "label" from "Exist to be near" rule on this "10 people" is found as a new label. Such newly identified label "Smallest party" is used again for extracting and finally the red-colored "2 member" is collected. As explained above, the initial "column label in a user's table" is self-completed in "bootstrap" manner.

2.3.2 Self-completion Extracting Data Which Has No Label

In some of the web pages, there are no label. In this case, using N-gram method, data about the label are extracted. A user's example is compared with the other web pages using N-gram method. Then most common words are regarded as the extracting pattern, and based on the pattern, extraction is executed again in which the search name is the words in the other web pages.

2.4 Problem of a User's Instantiated Example-Based Extraction

If original pages have less information, our system may not extract sufficient information in the above-mentioned method, because targeted pages of this method are

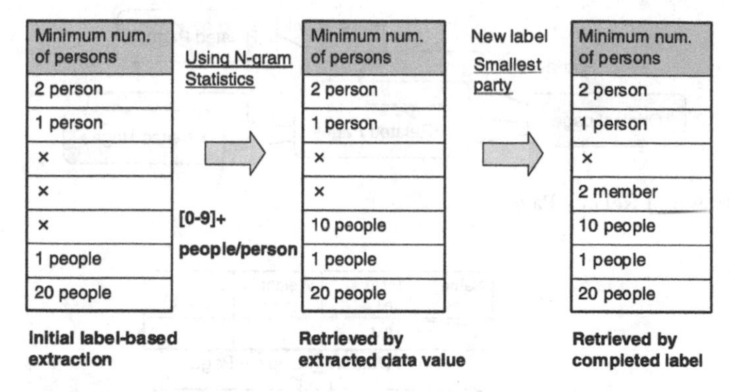

Fig. 4 Extracting Method of New Label

limited to only ones initially collected before the extraction process. In that case, users have to get information other sites, and fill the table manually. We propose a new method below that searches related contents and fills the table automatically to reduce a user's work.

3 Method of Using Additional Search of Related Contents

As mentioned in section 2, our system may not satisfy to extract necessary information. We address a method of additional search related page contents to improve this problem.

3.1 Retrival of Related Pages

We use two types of links. Fig. 5 shows that first we extract links from the original page that are other domains as related contents pages, and also extract the links from these related contents pages as related one. These two type of links are sufficient because there are few cases that more than three hopped linked pages provide necessary information as to the originally intended information.

3.2 Extraction from Related Pages

Acquired related pages may contain unrelated contents as to original page's contents, we should select the best matching page. As Fig. 6 shows, we choose one from all related pages by extracting from each related page and each result is compared with the extraction result of original pages. The related page that is best corresponding original one is selected, and by using this result of extraction from related page we can fill the table cell that cannot fill by our previous method.

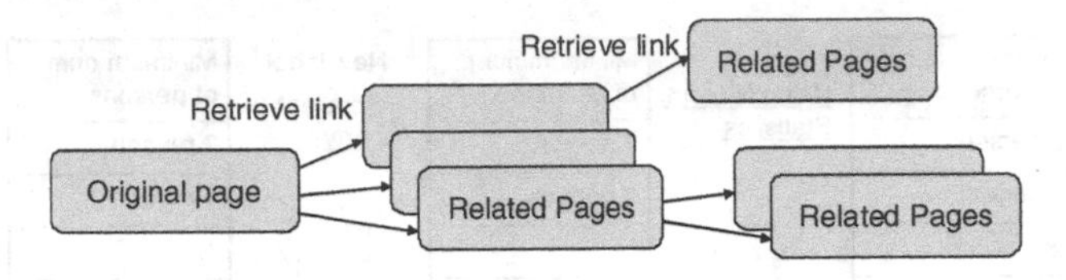

Fig. 5 Retrieval of Related Pages

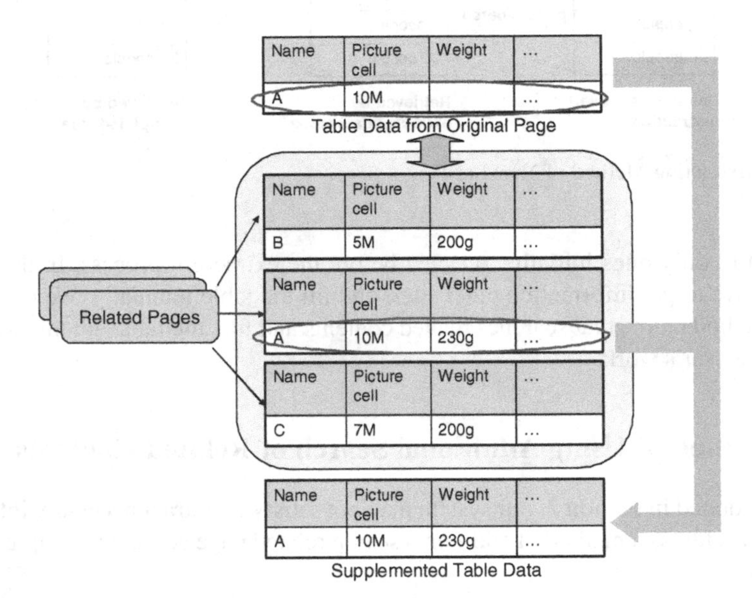

Fig. 6 Extraction from Related Pages

4 Evaluation Experiment

We apply our extraction method in section 3 to shopping sites to verify its availability.

There is a shopping site that introduces a digital camera as a user's instantiated example and 10 pages from other shopping sites as input. We should extract 11 pages that is provided by 11 company. The targeted table that has 12 labels. There is 111 cells that we should extract. Labels are limited text that can copy and paste from target pages. Table 1 shows the result.

Table 1 Result of Applying Our Proposed Method Experiment

	Previous Method	Proposed Method
Precision	86%	78%
Recall	38%	46%

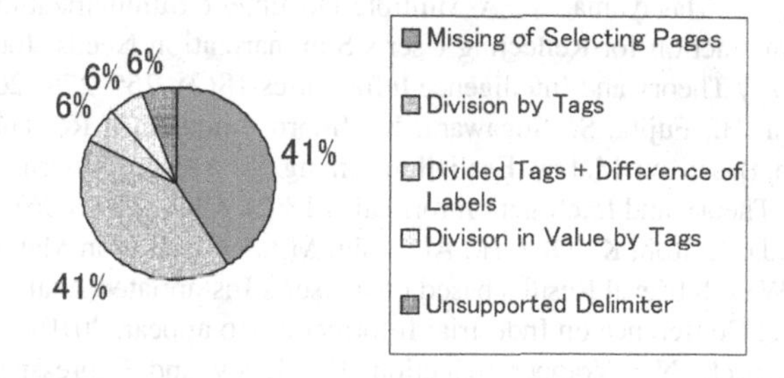

Fig. 7 Reason of Miss Extraction at Additional Search

Recall rate of our previous method is 38%, that of our proposed method is improved up to 46%. However precision rate of proposed method slightly decreased. It is considered that this is caused by extracting wrong data from related pages. More specifically, it may be caused that targeted data become widespread.

5 Conclusion

This paper describes a generation method of table-style data that uses additional Search related page contents. Recall rate is improved about 8% than our previous method. We should address a new method that find related content pages that are not linked a original page, a method that remove unrelated pages and so on in the future.

References

1. Okamura, H., Miyauchi, S., Dohi, T.: A Web Page Ranking Algorithm Based on a Markov Decision Process. The IEICE transitions on information and systems(Japanese edition) J89-D(2), 210–219 (2006)
2. Aratani, H., Fujita, S., Sugawara, K.: Extremely Precise Finding Methodology on the Mutual Evolution Method Among Web Pages, IEICE technical report, Artificial Intelligence and Knowledge-based Processing 105(105), 1–6 (May 2005)
3. Kawamae, N., Aoki, T., Yasuda, H.: Page Ranking Method of Search System Considering Difference of Access to the Pages. In: Proc. of the IEICE General Conference, vol. (1), p. 47 (2000)
4. Watanabe, N., Okamoto, M., Kikuchi, M., Iida, T., Hattori, M.: Influence of Presentation Style in Web-Search Result Recommenndation, IPSJ SIG technical reports 2009 (28), 61–68 (2009)
5. Toda, H., Yasuda, N., Okumura, M., Matsuura, Y., Kataoka, R.: Snippet Generation for Geographic Information Retrieval. Transition of the Japanese Society for Artificial Intelligence 24(6), 494–506 (2009)
6. Muramatsu, R., Yokoyama, S., Fukuta, N., Ishikawa, H.: Architect Snippets with Harmonized Various Viewpoint about Search Result Cluster with Consideration of Word's Characteristic Volume. SIG Notes 2008, 301–306 (2008)

7. Sakai, H., Masuyama, S.: A Multiple-Document Summarization System Introducing User Interaction for Reflecting User's Summarization Needs. Journal of Japan Society for Fuzzy Theory and Intelligence Informatics 18(2), 265–279 (2006)
8. Aratani, H., Fujita, S., Sugawara, K.: Improvement of a Re-ranking Method for Web Search Based on Mutual Evaliation among Web Pages. Journal of Japan Society for Fuzzy Theory and Intelligent Informatics 18(2), 196–212 (2006)
9. Shimada, J., Itoh, K., Oka, H., Akiyoshi, M.: A Generation Method of Table-style Data from Web Retrieval Results based on a User's Instantiated Example. In: 8th IEEE International Conference on Industrial Informatics (to appear, 2010)
10. Kushmerick, N.: Wrapper Induction. Efficiency and Expressiveness, Articial Intelligence 118, 15–68 (2000)
11. Muslea, I., Minton, S., Knoblock, C.: Active Learning for Hierarchical Wrapper Induction. In: Proceedings of the 16th national conference on Artificial Intelligence (AAAI 1999) (1999)
12. Uematsu, Y., Uchiyama, T., Kataoka, R., Matsui, T., Ohwada, H.: Information Extraction Using Specie Rule Wrapper Array. SIG Notes 2007 (6), 117–123 (2007)
13. Chang, C.-H., Lui, S.-C.: IEPAD: Information Extraction Based on Pattern Discovery. In: Proceedings of the 10th international conference on World Wide Web, pp. 681–688 (2001)

Bridging together Semantic Web and Model-Driven Engineering

Manuel Álvarez Álvarez, B. Cristina Pelayo G-Bustelo, Oscar Sanjuán-Martínez, and Juan Manuel Cueva Lovelle

Abstract. Ontologies are part of Semantic Web as models are part of Model-Driven Engineering, they can be seen as abstract, simplified views of the world. The possibility of transforming ontologies into software models, and vice versa, will bring both spaces together helping to achieve knowledge reuse. Both ontologies and models can assist in the domain analysis for the development of Domain-Specific Languages, so new transformations can be built to derivate DSLs from ontologies or models. This paper shows the current work in progress to build all these transformations and the concepts involved.

Keywords: Semantic Web, Model-Driven Engineering, Ontology, Model, Domain-Specific Language, OWL, UML.

1 Introduction

The purpose behind ontologies and models is not the shame but they share one key characteristic, which is abstraction. They are reduced renderings of the domain they represent, so the real world can be described either by ontologies or by models. There are several languages for ontology representation and OWL is currently one of the most popular. In the context of Model-Driven Engineering there is the UML, which is described by the OMG standard MOF, MOF is a metamodeling architecture used to define metamodels. The Odontology Definition Metamodel (ODM) is a recently adopted standard from the Object Management Group playing a central role for bridging Model-Driven Engineering and Semantic Web. It is a family of MOF metamodels that reflect the abstract syntax of several knowledge representation languages like OWL.

Manuel Álvarez Álvarez, B. Cristina Pelayo G-Bustelo, Oscar Sanjuán-Martínez, and Juan Manuel Cueva Lovelle
University of Oviedo, Department of Computer Science, Sciences Building, C/ Calvo Sotelo s/n 33007, Oviedo, Asturias, Spain
e-mail: UO156949@uniovi.es, crispelayo@uniovi.es, osanjuan@uniovi.es, cueva@uniovi.es

A.P. de Leon F. de Carvalho et al. (Eds.): Distrib. Computing & Artif. Intell., AISC 79, pp. 601–604.
springerlink.com

2 Proposal

The first objective of this paper is to be able to take existing OWL ontologies and transform them to their equivalent ODM models, or take ODM models and transform them to OWL ontologies. Thanks to this mechanism, Semantic Web experts can focus on ontologies while Model-Driven experts focus on models, and they will be able to share their work achieving knowledge reuse.

The second objective is to automatically build DSLs from previous step models. Using the models as the abstract syntax of the DSL, the only need is to generate the concrete syntax that can be either textual or graphical. According to [2] and [3] ontologies assist in the initial phases of domain understanding during the development of a domain-specific language. Ontologies represent the elements of a domain through a vocabulary and relationships between these elements.

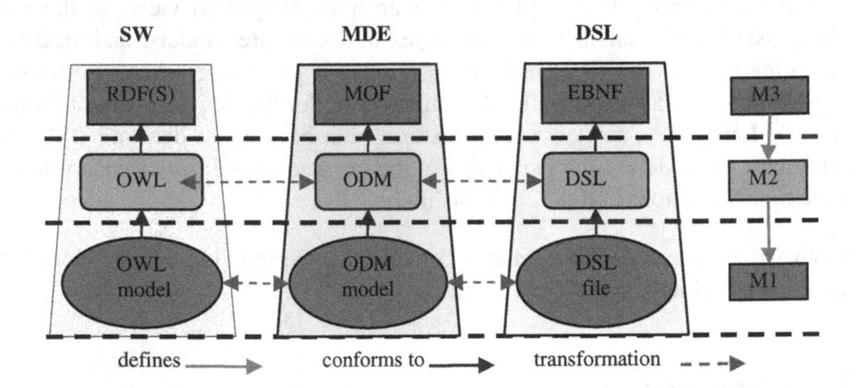

Fig. 1 Modelling spaces involved

Figure 1 [1], shows the three modelling spaces involved in this paper, a modelling space is a modelling architecture based on a particular super-metamodel.

3 Current Work

The first step is to build an implementation of the ODM specification (defined by OMG with MOF) based on Ecore, since Ecore is an implementation of EMOF it's quite a straightforward process and won't be described here.

Next step is trying to construct the transformations between OWL and ODM. In [4] Guillaume Hillairet builds them with ATL, sadly he uses ATL functions not working with current versions (for the extraction of XML), so we propose an alternative solution based on XML Schema.

A XSD can be seen as a metamodel for XML documents, taking this into account, the EMF is capable of automatically building an equivalent Ecore metamodel. One of OWL syntaxes is based on XML and has a valid XSD, so we can use it to generate an ECORE metamodel and import OWL files. The following table shows an example of the results of importing a model from an OWL file:

Source OWL file	<owlx:Ontology owlx:name="http://www.example.org/wine" xmlns:owlx="http://www.w3.org/2003/05/owl-xml" />
Imported ECORE model	<_xml:DocumentRoot xmi:version="2.0" xmlns:xmi="http://www.omg.org/XMI" xmlns:_xml="http://www.w3.org/2003/05/owl-xml"> <ontology name="http://www.example.org/wine"/> </_xml:DocumentRoot>

Next we can build the ATL transformations between the ODM metamodel and the XSD generated metamodel. Figures 2 and 3 show how the two metamodels represent the ontology element.

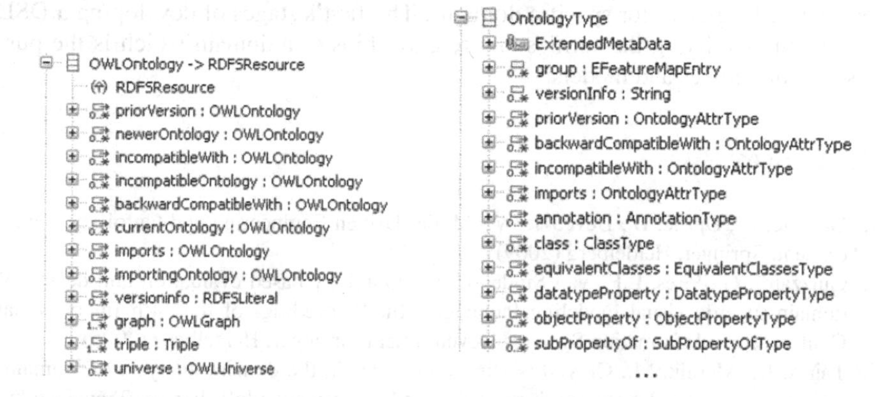

Fig. 2 ODM ontology element

Fig. 3 XSD ontology element

Last step is to derivate the DSL from the models from previous steps. In [3], the development of a DSL assisted by ontologies is described through a case study, manually building a DSL for air traffic communications. We believe their approach is correct, but we are trying to go one step further automatically building the DSL.

For this purpose we will be using Xtex which is a tool for building DSLs. We already have the abstract syntax thanks to the ODM model, so the only thing left is the concrete syntax of the language. According to Richard Gronback "The default nature of Xtext is to begin with a grammar and produce an Ecore model, rather than begin with an Ecore model and derive the grammar" [5], we'll try to overcome the problem of being a concrete syntax first problem by defining XSL transformations between the ODM metamodel and the Xtext grammar expressed in EBNF. Xtext allows importing the models instead of automatically building them from the concrete syntax; we will use this feature in the XSL transformation so the concrete syntax of the DSL will point to the ODM metamodel for its abstract syntax. Next we show an example of a fragment from a possible EBNF generated grammar:

```
grammar es.uniovi.miw.tfm.xtext.OWLText with org.eclipse.xtext.common.Terminals
import "platform:/resource/es.uniovi.miw.tfm.odm/model/ODM.ecore"
RDFSLiteral : 'text:' lexicalForm=STRING;
OWLOntology : 'Ontology' '{'
        'version' ':' (versioninfo=RDFSLiteral) ';'
'}';
```

4 Conclusions

Ontologies and models have something in common; both of them are tools for representing reality through abstractions. There is a great knowledge base developed with ontologies and also with models, so this work shows a possible way to interconnect and reuse all this knowledge.

This paper shows how ontologies and also models can be used to develop representative languages for specific domains. The first's stages of developing a DSL consist on identifying the entities and relationships of a domain which is the purposes of ontologies and models.

References

1. Gasevic, D., Djuric, D., Devedzic, V.: Model Driven Engineering and Ontology Development. Springer, Heidelberg (2009)
2. Guizzardi, G., Pires, L.F., van Sinderen, M.: Ontology-based evaluation and design of domain-specific visual modeling languages. In: Proceedings of the 14th International Conference on Information Systems Development. Springer, Heidelberg (2005)
3. Tairas, R., Mernik, M., Gray, J.: Using ontologies in the domain analysis of Domain-Specific languages. In: Proceedings of the 1st International Workshop on Transforming and Weaving Ontologies in Model Driven Engineering 2008. CEUR Workshop Proceedings, vol. 395 (2008) CEUR-WS.org
4. Hillairet, G.: ATL Use Case - ODM Implementation (Bridging UML and OWL) (2007), `http://www.eclipse.org/m2m/atl/usecases/ODMImplementation/`
5. Gronback, R.C.: Eclipse Modeling Project: A Domain-Specific Language (DSL) Toolkit (2009), ISBN: 978-0-321-53407-1

Novel Chatterbot System Utilizing Web Information

Miki Ueno, Naoki Mori, and Keinosuke Matsumoto

Abstract. Recently, the use of various chatterbots has been proposed to simulate conversation with human users. Several chatterbots can talk with users very well without a high-level contextual understanding. However, it may be difficult for chatterbots to reply to specific and interesting sentences because chatterbots lack intelligence. To solve this problem, we propose a novel chatterbot that can directly use Web information. We carried out computational experiments by applying the proposed chatterbot to "2channel" (2ch) and "Twitter".

Keywords: Chatterbot, Web Information, Estimating User's Interests, Twitter.

1 Introduction

Attempting to create an intelligent conversation system is one of the most important and interesting themes in computer engineering. However, approaches that use natural language processing and artificial intelligence techniques have not yet been able to create an enjoyable experience for users. On the other hand, chatterbots[1] have become a well-known method for simulating human conversation. The main objective of chatterbots is to produce interesting conversation, and many chatterbots do not care to simulate the actual human thought. Eliza[2], one of the very first chatterbots, could simulate a Rogerian psychotherapist. The Eliza concept was simple, utilizing a pattern matching algorithm and sentence reconstruction without processing the natural language. Many chatterbots have been proposed since Eliza[3, 4, 5]; however, the conversational level of the proposed chatterbots has not been sufficient to satisfy users.

To solve this problem, we propose a novel chatterbot that directly uses Web information. Computational experiments were carried out by applying the proposed

Miki Ueno, Naoki Mori, and Keinosuke Matsumoto
Department of Computer and Systems Sciences, College of Engineering,
Osaka Prefecture University, 1-1 Gakuencho, Nakaku, Sakai, Osaka 599-8531, Japan
e-mail: ueno@ss.cs.osakafu-u.ac.jp, mori@cs.osakafu-u.ac.jp,
matsu@cs.osakafu-u.ac.jp

A.P. de Leon F. de Carvalho et al. (Eds.): Distrib. Computing & Artif. Intell., AISC 79, pp. 605–612.
springerlink.com

chatterbot, which has a database of sentences from "2channel" (2ch) message boards, to Twitter.

In this paper, we first present the constitution of the proposed chatterbot in Section 2. We propose a novel chatterbot in Section 3. Computer experiments are described in Section 4, while Section 5 introduces possible applications of the proposed chatterbot. Finally, in Section 6, we present the conclusions of this study.

2 Constitution of the Proposed Chatterbot

In this study, the proposed chatterbot is constituted in the following manner.

Interpretation. The chatterbot receives an input sentence and formats this input for the following processes. The first step of interpretation is morphological analysis of the input. Here, we utilize Sen[6], which is one of the leading pure Java morphological analysis libraries for Japanese.

Replying. The chatterbot replies to the user with an appropriate expression. We adopt short-term memory for reasonable conversation and a user logging system for retrieving information on the specific user. The proposed chatterbot also checks Wikipedia[7] to formulate an appropriate response to unknown keywords.

Learning. The chatterbot can memorize new statements and revise its memory. If the chatterbot fails to supply an appropriate response using its memory, it will ask the user about the user's input sentence or any unknown keywords. The user can teach the chatterbot the meaning of new words, or force the chatterbot to forget a specific part of its memory.

Personality. Since the chatterbot matures through conversation with users, the personality of the chatterbot is crucial for forging an emotional bond between the users and the chatterbot. The proposed chatterbot maintains a self-portrait and introduces unique topics of conversation.

Fig. 1 shows the outline of proposed chatterbot.

3 Proposed Method

3.1 Basic Concept

We propose a novel chatterbot that uses Web information. A simple approach to utilize Web information is to extract the sentences from Wikipedia[7] or from search engine results. In this study, we propose a novel method for finding appropriate sentences from a Bulletin Board System (BBS) that has been categorized into typical fields in order to answer the specific and deep topics[10].

3.2 Target BBS

We selected 2ch[8] as the target BBS because 2ch is the most comprehensive forum in Japan and covers diverse fields of interest. The top level unit is called "category".

Several boards belong to a "category". It has more than 600 active boards including "Social News", "Computers", and "Cooking". Each board usually has many active threads that have main topics for discussion.

3.3 Utilizing BBS Information

The following methods are adopted to use the information from the BBS.

1. If the chatterbot knows the user's interests before the session, the chatterbot will try to use words that appear frequently on the board related to the user's interests.
2. The chatterbot determines the user's interests automatically by using statistical information from the user's conversation log.

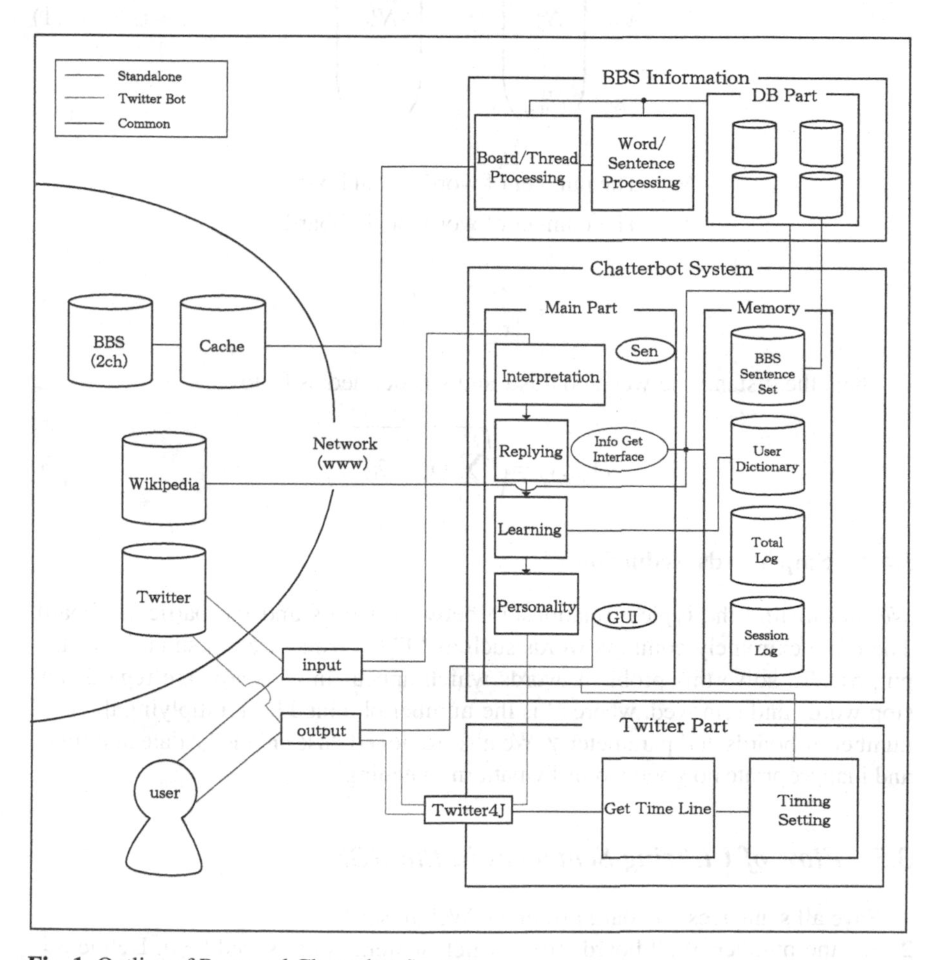

Fig. 1 Outline of Proposed Chatterbot System

3.4 Distance between Boards

3.4.1 Definition of Distance

In this study, we define the distance between two boards in the 2ch BBS as a simple Euclidean distance. If the similarity of two boards is high, the distance between those 2 boards will be low.

The words set throughout the entire BBS is defined as W, and the i-th word is denoted as w_i. We set $|W| = M$. Then the feature vectors of board x and y, $\hat{x}$ and $\hat{y}$ respectively, are defined as follows.

$$x = \begin{pmatrix} N^x_{w_1} \\ \vdots \\ N^x_{w_i} \\ \vdots \\ N^x_{w_M} \end{pmatrix}, \ y = \begin{pmatrix} N^y_{w_1} \\ \vdots \\ N^y_{w_i} \\ \vdots \\ N^y_{w_M} \end{pmatrix} \tag{1}$$

$N^x_{w_i}$: The number of words w_i in board x

$N^y_{w_i}$: The number of words w_i in board y

$$\hat{x} = \frac{x}{|x|}, \ \hat{y} = \frac{y}{|y|} \tag{2}$$

Then, the distance between boards x and y is defined as follows.

$$D(x,y) = \sqrt{\sum_{i=1}^{M} (\hat{x}_i - \hat{y}_i)^2} \tag{3}$$

3.4.2 Stop Words Reduction

We aim to find the typical relationship between words and the particular board. Therefore extremely common words such as “I” or “you” are not suitable for this purpose. To solve this problem, words which appear in N boards are regarded as stop words and removed, where N is the number obtained by multiplying the total number of boards and parameter γ. We also removed “user name”, “date and time” and inappropriate no good words by pattern matching.

3.5 Flow of Utilizing Sentences in the BBS

1. Save all sentences on board i from the Web to set S_i.
2. Let the number of all boards from which sentences are saved be n. Define $S = \bigcup_{i=1}^{n} S_i (\neq \phi)$ as the total set for n boards.
3. Set the probability of using the positive set p.

4. Let the first noun in the sentence be k. Set the positive set as $S^{\text{true}} = \{x | x \in S, x \text{ is the sentence which contains word } k\}$.
   Let $S^{\text{false}} = S \setminus S^{\text{true}}$.
5. Select a sentence randomly from S^{true} with probability p, or from S^{false} with probability of $1-p$, and output this sentence after the formatting procedure. If $S^{\text{true}} = \phi$ or $S^{\text{false}} = \phi$, a sentence is selected from S.

4 Computer Experiments

We evaluate whether the board's distance can be practically used to detect user interests.

4.1 Experimental Setup and Results

In this study, we selected 2ch[8] for the computer experiments because it is one of the largest BBSs in Japan and its BBS topics are categorized well (Top → Category → Board → Thread). First, we selected five boards: Math, Physics, Literature, Beauty (topics about cosmetics), and Jobs (topics about employment). We only focus on nouns when analyzing target words.

Table 1 shows the top 10 most frequently used words on each board. Table 2 shows the distance between each pair of boards.

To investigate the effects of applying BBS information to determine user interests, we conducted the following questionnaire survey of users. Specifically, we evaluated the extent to which the conversational quality of the proposed chatterbot improved. The procedure for the experiment is presented below. We set the number of user inputs of each trial to 10 times.

Table 1 Top 10 words on each board ($\gamma = 0.6$)

Mathematics	Physics	Literature	Beauty	Jobs
mathematics	time	smile	face	informal decision
proof	existence	novel	male	company
book	physics	work of art	female	enterprise
definition	light	book	eye	work
1	universe	novelist	hair	interview
number	understanding	literature	skin	university
understanding	explanation	love	love	age
case	human	human	nose	activity
existence	earth	Japan	beautiful	recruit
you	case	age	effect	day

Note: Japanese results translated into English.

Table 2 Distance between boards in 2ch BBS ($\gamma = 0.6$)

	Math	Physics	Literature	Beauty	Jobs	Norm
Math	0	0.87	1.04	1.18	1.12	26975.91
Physics	0.87	0	1.02	1.17	1.13	28850.04
Literature	1.04	1.02	0	1.03	1.06	37687.74
Beauty	1.18	1.17	1.03	0	1.10	34642.16
Jobs	1.12	1.13	1.06	1.10	0	31950.71

rounded to two decimal places

Table 3 Evaluation of chatterbots before and after utilization of BBS information

	No BBS Info	Use BBS Info
User1(Math)	1/5	5/5
User2(Fashion)	2/5	4/5
User3(TV Game)	2/5	4/5

BBS Experiment

1. The target user converses with the original chatterbot freely.
2. The target user then selects one board of interest.
3. We adjust the chatterbot settings to use sentences from the board selected in Step 2.
4. The target user converses with the chatterbot that is using board information.
5. Repeat Step 1 $\sim$ Step 4 five times and count the number of trials that user could be satisfied the chatterbot outputs from the view point of their interests in conversations of Step 1 and Step 4.

The results of the questionnaire survey are shown in Table 3, where the words after user name (Math, Fashion, TV Game) were user's interests. $n/5$ represents that user could be satisfied chatterbot outputs n times in 5 trials of BBS Experiment.

4.2 Discussion

The results shown in Table 2 indicate that the distance between Math and Physics is smaller than the distance between Math and Beauty. Since Math and Physics have several top 10 words in common in Table 1, it is natural that the distance between Math and Physics is small. On the other hand, since Math and Beauty have no words in common in Table 1 the distance between Math and Beauty is large. The results in Table 1 and Table 2 are reasonable, and appropriate values were obtained by checking the similarity of the different boards.

Table 3 shows that utilizing board information is effective for satisfying the user's interests.

5 Application

We developed two types of applications for the proposed chatterbot. Since our chatterbot engine is independent part, we can easily apply the proposed chatterbot to other applications.

5.1 Standalone Application

This is a local application with a GUI. Although it is possible to obtain Web information from Wikipedia or a BBS simultaneously via a network, this application also can be executed without a network by using a local dictionary. The GUI handles displaying characters, obtaining user input and printing responses of the chatterbot.

5.2 Twitter Bot

Twitter is a social network-based communication service that enables its users to send, receive and view short messages known as tweets. Tweets are text-based posts of up to 140 characters that are displayed on the author's profile page and delivered to the author's subscribers, who are known as followers. We applied the proposed chatterbot as a Twitter bot using the Java-based Twitter API called Twitter4J[11]. We have asked several testers to try our chatterbot in Twitter, and obtained positive opinions.

6 Conclusion

In this paper, we proposed a novel chatterbot that directly uses Web information. Through the results of several computer experiments we confirmed that the proposed chatterbot is effective in generating replies that can satisfy users who have a deep knowledge about particular fields.

The following objectives will be studied in future research.

1. Constructing a large-scale database for the chatterbot's memory.
2. Developing a filtering mechanism for reply sentence selection.
3. Including a personality in the chatterbot.
4. Developing an evolving chatterbot by means of evolutionary computation.

Acknowledgements. This research was supported in part by a Grant-in-Aid for Scientific Research (C), 22500208, 2010-2014 from the Ministry of Education, Culture, Sports, Science and Technology.

References

1. Chatterbot Is Thinking,
   http://www.ycf.nanet.co.jp/~skato/muno/index1.shtml
2. Weizenbaum, J.: Eliza - a computer program for the study of natural language communication between man and machine. Communications of the ACM (9) (1966)

3. Chamberlain, W.: The Policeman's Beard is Half Constructed, Warner Books (1984)
4. Guzeldere, G., Franchi, S.: Dialogues with colorful personalities of early AI. Stanford Humanities Review 4(2) (1995)
5. Vrajitoru, D., Ratkiewicz, J.: Evolutionary Sentence Combination for Chatterbots. In: International Conference on Artificial Intelligence and Applications (AIA 2004), pp. 287–292. ACTA Press (2004)
6. Sen, https://sen.dev.java.net/
7. Wikipedia, http://wikipedia.org/
8. 2 channel, http://www.2ch.net/
9. Twitter, http://twitter.com/
10. Yutaka, M., Ohsawa, Y., Ishizuka, M.: Mining and Summarizing Conversational Data on Electrical Message Boards. In: The 16th Annual Conference of Japanese Society for Artificial Intelligence (2002)
11. Twitter4J, http://yusuke.homeip.net/twitter4j/

Exploring the Advances in Semantic Search Engines

Walter Renteria-Agualimpia, Francisco J. López-Pellicer,
Pedro R. Muro-Medrano, Javier Nogueras-Iso, and F.Javier Zarazaga-Soria

Abstract. With the vertiginous volume information growing, the amount of answers provided by traditional search engines and satisfying syntactically the user queries has enlarged directly. In order to reduce this problem the race to develop Semantic Search Engines (SSE) is increasingly popular. Currently, there are multiple proposals for Semantic Search Engines, and they are using a wide range of methods for matching the semantics behind user queries and the indexed collection of resources. In this work we survey the semantic search engines domain, and present a miscellaneous of perspectives about the different classification of approaches. We have created a comparative scheme and identified the prevalent research directions in SSE.

Keywords: Semantic Search Engines, Semantic Web, Information Retrieval.

1 Introduction

With the vertiginous increase of volume information on the Web, the results provided by traditional search engines in response to user queries do no longer satisfy the needs of specific communities of users. There is an increasing amount of answers that satisfies the terms contained in user queries. However, these answers are not precise enough for some users demanding a more refined list of results according to the semantics of their queries. This open problem has motivated a new era of search systems that have received the name of Semantic Search Engines.

A Semantic Search Engine (SSE) can be understood as a semantic Web application that can answer questions based on the meaning of users query specification, resources in the repositories and in many cases it is based on predefined domain semantics or a knowledge model. SSE can return relevant results on your topics that do not necessarily mention the word you searched for explicitly.

The goal of this work is to study and discuss various widespread research directions in semantic search engines, as well as identifying common features and main

Walter Renteria-Agualimpia, Francisco J. López-Pellicer, Pedro R. Muro-Medrano, Javier Nogueras-Iso, and F.Javier Zarazaga-Soria
Computer Science and Systems Engineering Department, University of Zaragoza. Zaragoza, Spain
e-mail: {walterra,fjlopez,prmuro,jnog,javy}@unizar.es

A.P. de Leon F. de Carvalho et al. (Eds.): Distrib. Computing & Artif. Intell., AISC 79, pp. 613–620.
springerlink.com

approaches used in them. In this work we can get an overview of current approaches to semantic search and its state of development, but not an exhaustive review of all implemented systems.

The rest of the paper is structured as follows. The next section shows several schemes of classification approaches, in order to identify what kind of semantic search approach is the base for each SSE. And then we analyze the trend and prevalent research directions in the SSE domain. Section 3 provides a comprehensive analysis based on a survey of more than 30 SSE. We discuss and compare the classifications of approaches. All results of this analysis are available online as linked data (see sect. 3). Finally we summarize our main conclusions in section 4.

2 Different Schemes for Comparison and Classification of SSE

Currently, there are multiple proposals for Semantic Search Engines, and they are using a wide range of methods for matching the semantics behind user queries and the indexed collection of resources. Several authors have studied the current status of semantic search engines from different viewpoints. We present a review of approaches from a research perspective.

A first review was presented by Miller et al. [5]. They present a classification based on the intention of users. That is, if the users want to navigate to a particular intended document, this approach is called: *Navigational Searches.* On the other hand, there maybe users trying to locate a number of documents, which together will give them the information they are trying to find. This is *Research Searches.*

Mangold [2] presents a categorization scheme that he uses to classify different approaches for semantic search along several dimensions. His classification is based on the next criteria: architecture, coupling, transparency, user context, query modification, ontology structure and technology. The analyzed approaches were implemented by the next technologies: SHOE, Inquirus2, TAP, Hybrid spreading Activation, ISRA, Librarian agent, SCORE, TRUST, Audio, and Ontogator.

The next author viewpoints are more focused in the semantic processing method to resolve queries. Mäkelä identifies five distinct research directions emerged and prevalent research directions in semantic search, based on similarity of research goals[1]. He observed that sometimes the categories do not differ much in methodology, but they seem sufficiently separate. His classification is:

- *Augmenting Traditional Keyword Search with Semantic Techniques:* more specific ontological techniques are used. i.e, Terms are expanded to their synonym and meronym sets [9]. Direct ontological Browning is supported. The intention is to find related concepts as the writer of the document [15].
- *Basic Concept Location*: The main goal is to locate instances of the core semantic web formed by concepts, instances and relationships. Users can choose the class of instances by means of ontological navigation [7].
- *Complex Constraint Queries*: Many SSE with this approach are based on navigating the ontology as the last approach [13]. One way is based on a global intersection of distinct selectors, constraining do not need to be ontological.
- *Problem Solving*: The SSE use ontological knowledge to solve a problem; searching for solutions by inference and other reasoning techniques [14, 16].

- *Connecting Path Discovery:* The SSE are based on the ideas of a vast amount of varied semantic data will be available to be mined for semantic connections. The major technical problems are the locating complex and hidden relations.

Hildebrand et al. [3] systematically scanned proceedings about Web Semantics to compile a list of end-user applications described or referred to. For each system they collected basic characteristics such as the intended purpose, intended users, the scope, the triple store and the technique or software used for literal indexing, giving a total of 35 systems. Based on the data resulting from the survey they perform a more thorough analysis of the three individual phases in the search process: *query construction in section, search algorithms, presentation of the results.*

Now, in the search algorithms stage, we can find the semantic component, that is, the main interest in this work. The cores of SS approaches identified are the:

- *Graph Traversal*: Takes only the structure of the graph into account. It uses weighted graph search algorithm. Weights reflect the importance of relations.
- *Query Expansion*: Thesaurus relations are used for query expansion. Semantic matching with hierarchical broader, narrower and the associative related term.
- *Spread Activation*: It uses weights as well as the number of incoming links.
- *RDFS/OWL Reasoning*: Has the ability to influence the search results. RDFS. Some SSE support OWL reasoning based on logic programming or rules [12].

Dietze and Schroeder [6] suggest a new classification approaches. They developed an interesting study about 27 SSE and use a classification based on 9 criteria: structured/unstructured file, ontologies, text mining type, number of documents, type of documents, clustering, result type, highlighting, scientifically evaluated.

Dong et al. [4] present a extended classification: Semantic Search (SS) Algorithm based on the Graph, SS Methodology on Distributed Hash Tables (DHT), Logics (DL)-based Information retrieval (IR) Thesaurus-DL form Knowledge Base (TK), DAML+OIL-based Semantic Search, Keyword-based Search Engines combined with Semantic Techniques, SSE based on Ontology Annotations, Agent-based SSE, SS Engine and XML Objects, Semantic Multi-media SE.

Finally, Grimes [11] presents an extensive classification of approaches:

- *Related searches/queries:* The SSE recommends searches that are in some "sense" similar to the user search.
- *Reference results*: SSE is responding with resources that define the search terms, via a dictionary look-up, or elaborately, pulling Wikipedia pages.
- *Semantically annotated results*: SSE returns pages or documents with highlighting of text features, especially named or pattern-defined entities.
- *Full-text similarity search:* SSE use a block of text ranging submitted from a phrase to a full document, rather than a few keywords.
- *Search on semantic/syntactic annotations*. Users define the semantic of search by means of indicate the syntactic role the term play.
- *Concept search*: The SSE identifies specific concept to seek the original and their equivalent concepts semantically.
- *Ontology-based search:* SSE can understand hierarchical relationships of entities and concepts as in taxonomy, and more complex inter-entity relations.

- *Semantic Web Search*: SSE capture data relationships and make the resulting "Web of data" queryable.
- *Faceted search:* It provides a means of exploring results according to a set of predefined, high-level categories called facets.
- *Clustered search:* It is like faceted search, but without the predefined categories. Meaning is inferred from topics extracted from the search results.
- *Natural language search:* The SSE understands the semantic behind the questions, and present answers in natural language.

A summary about the classifications is presented in Table 1. The 5th column shows the final classification based on Grimes, because this is more extensive than other perspectives. The goal is identify the main active areas in SSE domain.

Table 1 Comparison of semantic search Approaches

Mäkelä	Hildebrand	Dong et al.	Grimes	Proposed final classification
Connecting Path Discovery	Graph Traversal	SS Algorithm based on the Graph	-	SS based on Graphs
Augmenting Traditional Keyword Search with Semantic Techniques	Query Expansion	Keyword-based SE with Semantic Techniques	Related Searches/Queries	Related Searches/Queries
	Spread Activation	SSE based on Ontology Annotations	Search on Semantic/Syntactic Annot.	Search on Semantic/Syntactic Annotations
			Semant. Annot. Results	Semant. Annot. R.
Problem Solving Complex Constraint Queries	RDFS/OWL Reasoning	Agent-based SSE (DL)-based on IR TK DAML+OIL-based SS	Ontology-based Search	Ontology-based Search
-	-	SSE and XMLObjects Semantic Multimedia SE	Semantic Web Search	Semantic Web Search
-	-	SS Methodology on DHT	Reference Results	Reference Results
-	-	-	Full-Text Similarity S	Full-Text Similarity S
Basic Concept Location	-	-	Concept Search	Concept Search
			Faceted Search	Faceted Search
			Clustered Search	Clustered Search
-	-	-	Natural Language Search	Natural Language S.

3 Analysis of Current SSE

The methodology used in this work was based on 4 steps. The first step was to review the applications available in the Web, publications and projects in the state of art. Then, we evaluated a series of parameters (see below for the list of parameters and their description) for each semantic search engine. The third step was to contact some authors because the available information for some engines was uncomplete. However, in some cases we could not contact some of the authors. Then we complement the analysis with the several previous works and their operation mode and results [1, 2, 4]. Finally the complete research results are detailed in an updatable technical report and all results are published in the research group portal IAAA[1] by means of a RDF file, and linked data in order to obtain more feedback.

We have studied different scheme classifications and the several author viewpoints about main semantic search approaches. Numerous criteria and parameters have been used in this purpose. Our objective is not to reward or dismiss those proposals; but to identify the predominant or prevalent active areas or approaches by means of exploring many semantic search engines at present.

Researchers and developers are aware of the need to improve traditional engines, including features like: *user feedback; results explanation* and compressive presentation of results; and more dialogue with the users about possible problem with their request, e.g *ambiguity advertisement.*

Many of these aspects are related to human understanding, but it is important to study the *interoperability*, that is, to analyze what kind of interoperability do SSE present? Is the SSE a machine or informatic agent queryable? We have summarised the SSE exploration in Table 2, which show the following 8 parameters:

- *Main approach(es):* This field identifies the type of approach used by each SSE. The type of approach was presented in Table 1; it was obtained by means of unifying the Grimes classifications with the other approaches unmentioned. The complete results are available in the RDF file cite above.
- *Features:* It is a description about the main SSE qualities.
- *Type of Result*: It specifies the query result: summary, link, free text or other.
- *User feedback:* This is useful when there are multiple controlled terms that match with the free text input semantically. There are two ways. The first one is called "pre-query disambiguation", allow us to select the intended term before it is processed by the search algorithm [10]. The second way is called "post-query disambiguation"; feedback is taking into account on the results.
- *Multilingual*: Multiple language support.
- *Interoperability*: It evaluates if the SSE is able to exchange machine-understable content by mean of a standard protocol.
- *Result explanation*: Here we recognize if the SSE argue the query answer, justifying by means a graph, conceptual structure or other.
- *Ambiguity alarm*: In many cases, there are results that match with the query. SSE must advert to user about the different senses that satisfy the query.

Additional we present two features available in online version, as following:

- *Geospatial component*: It allows evaluate as if the SSE takes into account additional richness aspects, such as *geospatial location information* when is required to complement or clarify the semantic or to confirm the result sense. i.e Washington state instead of Washington president (see RDF online).
- *Availability*: It examines if the Web application is available now (see RDF).

[1] http://iaaa.cps.unizar.es/openknowledge/papers/2010/dcai/sse/

It is worth noting that we had to face the problem that some systems, Web applications and publications describe their approaches from a very abstract viewpoint. For this reason we relied on the given information without knowing the deep details, but assigning and classifying the SSE according to their external description and comparing with similar semantic search engines.ïï

Table 2 Comparison of semantic search Engines

Engine	Main Approach(es)	Features	Type of Result	M[a]	Interoperability	RE[b]	AA[c]
SenseBot	Concept Search	Text mining	Summary	Yes	SOAP, REST	No	No
BotPowerseet	Natural Language Processing (NLP)	Free text input, disambiguate.	Summary	Yes	-	Yes	Yes
DeepDyve	Semantic/Syntactic Annot., Reference results	Analysis across large amounts of data	Summary	Yes	-	No	No
Cognition	NLP	Business, APIs	Link	Yes	API	Yes	Yes
Hakia	Related searches, NLP	Excellent resumes	Link & Free text	Yes	Yes	Yes	No
TrueKnowledge	Ontology-based search, Semantically annot. results	Questions – answering	Summary and classification	No	Direct Answer API, Query API	Yes	Yes
Open Mind Common Sense	NLP, concepts search	Learn general knowledge	Free text	No	-	No	No
Swoogle	Semantic Web search	Semantic Web documents.	OWL, RDF	No	REST web service	No	No
TrueVert	Concept search, NLP and Clustered results	model of word relations in context	Free text	Yes	-	No	No
Wolfram Alpha	Reference results, Ontology-based search, Clustered search	Web, parallel computing, mathematical, grid knowledge	Taxonomy, graph	Yes	REST API	Yes	No
Duck Duck Go	Clustered search, NLP	Zero-click Info above links, Disambiguation	Summary	Yes	XML-based API	-	Yes

[a] Multilingual, [b] Result explanation, [c] Ambiguity alarm

In table, the symbol "-" represents unknown information. The main parameter of comparison in the table is the second column "*Main approach(es)*". It allows us to identify the research areas with the more intense activity in the semantic search.

Which is the prevalent Semantic Search Approach? Have the Semantic Search Engines analized a unique approach? Taking into account these questions and the information provided in Tables 2 and 3, we can see five main groups with major activity, i.e. a significant number of SSE using that approaches. Those groups are: in first place, Concept Search, Faceted Search, Clustered Search; then Search

Table 3 Summary of prevalent research directions in SSEï

Group	Approach(es)	Number of SSE using this approach
Group 1	SS based on Graphs	0
Group 2	Related Searches/Queries, Search on Semantic/Syntactic Annotations, Semantically Annotated Results	9
Group 3	Ontology-based Search	9
Group 4	Semantic Web Search	8
Group 5	Reference Results	4
Group 6	Full-Text Similarity Search	0
Group 7	Concept Search, Faceted Search, Clustered Search	11
Group 8	Natural Language Search	10

Engines based on NLP; in third place, SE based on Related Searches/Queries, Search on Semantic/Syntactic Annotations and Semantically Annotated; then Ontology-based Search; and finally Semantic Web Search. We have analized several SSE implementing different approaches, and based on combinations of the last groups mentioned. Probably these research directions will be the dominant approaches.

4 Conclusions

There is one common idea in the majority of approaches, that is, the machines must understand the meaning behind the Query and Data sources in order to return answers based on the meaning. Maybe this is the main requirement for a SSE. Intuitively we can say that many SSE will be based on a similar core, including conceptual structures such as ontologies, and founded on main components to process queries in form of natural language. A direct consequence is the need to develop SSE allowing the users to play a part of the answers, before and after the query, that is, pre-query disambiguation, advertisement of ambiguity presence, and feedback to improve futures answers.

Another aspect related to the semantic legibility intrinsically is the system ability to explain results, that is, what was the form to generate one or other result? In this aspect, many systems are working to improve the visualization and interpretation form strongly, e.g. some SSE such as *Wolfram*[8], *Google*[2], and Kolline [6] provide visualization of results by means of concept connection graph or surfable graph.

Acknowledgments. This work has been partially funded by the Spanish government through the projects "España Virtual" (ref. CENIT 2008-1030) and TIN2009-10971, and the Government of Aragon through the project PI075/08.

[2] At the time of writing this paper, one could reach the Google tool by selecting "Show options" in the main page of Google.

References

1. Mäkelä, E.: Survey of semantic search research. In: Proc. of the Seminar on Knowledge Management on the Semantic Web (2005)
2. Mangold, C.: A survey and classification of semantic search approaches. International Journal of Metadata, Semantics and Ontologies 2(1), 23–34 (2007)
3. Hildebrand, M., Ossenbruggen, J., Van Hardman, L.: An analysis of search-based user interaction on the semantic web. Report, CWI, Amsterdam, Holland (2007)
4. Dong, H., Hussain, F.K., Chang, E.: A survey in semantic search technologies. In: Second IEEE International Conference on Digital Ecosystems and Technologies (2008)
5. Guha, R., McCool, R., Miller, E.: Semantic Search. In: Proceedings of the WWW 2003, Budapest (2003)
6. Figueira, F., Porto de Albuquerque, J., Resende, A., Geus, de Geus, P.L., Olso, G.: A visualization interface for interactive search refinement. In: Proc. 3rd Annual Workshop on Human-Computer Interaction and IR, Washington DC, pp. 46–49 (2009)
7. Mäkelä, E., Hyvönen, E., Saarela, S., Viljanen, K.: OntoViews - A Tool for Creating Semantic Web Portals. In: Proc. of the 3rd International Semantic Web Conf. (2004)
8. Wolfram alpha (March 2010), system available at, http://www.wolframalpha.com/
9. Buscaldi, D., Rosso, P., Arnal, E.S.: A wordnet-based query expansion method for geographical information retrieval. In: Working Notes for the CLEF Workshop (2005)
10. Hildebrand, M., van Ossenbruggen, J., Hardman, L.: /facet: A Browser for Heterogeneous Semantic Web Repositories. In: Cruz, I., Decker, S., Allemang, D., Preist, C., Schwabe, D., Mika, P., Uschold, M., Aroyo, L.M. (eds.) ISWC 2006. LNCS, vol. 4273, pp. 272–285. Springer, Heidelberg (2006)
11. Informationweek (March-2010), http://intelligent-nterprise.informationweek.com/channels/information_management/showArticle.jhtml;jsessionid=QTPB04LAEZJMHQE1GHOSKHWATMY32JVN?articleID=222400100
12. UMBC: F-OWL: An OWL Inference Engine in Flora-2 (April 2010), http://fowl.sourceforge.net/
13. Zhang, L., Yu, Y., Yang, Y., Zhou, J., Lin, C.: An Enhanced Model for Searching in Semantic Portals. In: WWW 2005: Proceedings of the 14th international conference on World Wide Web, pp. 453–462. ACM Press, New York (2005)
14. Duke, A., Glover, T., Davies, J.: Squirrel: An Advanced Semantic Search and Browse Facility. In: Franconi, E., Kifer, M., May, W. (eds.) ESWC 2007. LNCS, vol. 4519, pp. 341–355. Springer, Heidelberg (2007)
15. Rocha, C., Schwabe, D., de Aragao, M.P.: A hybrid approach for searching in the semantic web. In: Proc. of the 13th international conf. on World Wide Web, pp. 374–383 (2004)
16. Wine Agent 1.0. (March 2010), http://onto.stanford.edu:8080/wino/index.jsp
17. Guha, R., McCool, R., Miller, E.: Semantic search. In: WWW 2003: Proc. of the 12th international conference on World Wide Web, pp. 700–709. ACM Press, New York (2003)

Control Performance Assessment: A General Survey

Daniel Gómez, Eduardo J. Moya, and Enrique Baeyens

Abstract. This paper reviews the different indexes and benchmarks used in the control performance assessment field of industrial processes. They are usually implemented to detect and diagnose malfunctions and disturbances in industrial controllers. This survey is just an overview of the methods and tools used in the control performance assessment/monitoring (CPA/CPM) technology which has been deeply studied over the last two decades.

Keywords: Control Performance Assessment, Monitoring, Minimum Variance, MPC, LQG, Fault Detection, Fault Diagnosis, Oscillations, Static Friction, OPC, Industrial Processes.

1 Introduction

Current control engineers must face complex technical processes in the modern industry. However, the fact that those processes run in companies which operate in a highly global and competitive market cannot be forgotten. A process will be valid as long as its throughput is high enough from the economic and financial health of the company point of view. Control performance assessment and monitoring is an important technology to evaluate the efficiency of the automation systems which usually run in the factory plants of manufacturing companies. Control loops malfunctions, including sensors and actuators, are usual and their effects introduce enormous variations in the process, decreasing machines effectiveness, increasing costs and altering the final quality of the product. The main goal of the CPA technology is to implement an online automated process which gives information to determine if the specified objectives and the defined output are being met according to the value of the controlled variables in the process and to assess the control system performance.

Daniel Gómez
Fundación CARTIF, Parque Tecnológico de Boecillo, Parcela 205, 47151 Valladolid, Spain
e-mail: dangom@cartif.es

Eduardo J. Moya and Enrique Baeyens
Departamento Ingeniería de Sistemas y Automática, Universidad de Valladolid,
Paseo del Cauce s/n, 47011, Valladolid, Spain
e-mail: {edumoy,enrbae}@eis.uva.es

A.P. de Leon F. de Carvalho et al. (Eds.): Distrib. Computing & Artif. Intell., AISC 79, pp. 621–628.
springerlink.com

This paper is divided as follows: next section deals with the OPC technology which is the most used to access the data of an actual industrial process. Third section is about the main benchmarking methods. Section four deals with an overview of the different methods to detect and diagnose process disturbances and final section ends with the conclusions. Excellent basic references about previous reviews can be found in [12](CPA indexes), [20] (disturbances detection and diagnosis) and [16].

2 OPC

OPC (OLE for Process Control) is the standard of communication in the control and supervision field in the manufacturing and process companies. OPC establishes an open interface in order to access data in the automation industry. These open interfaces are well defined and maintained by the OPC Foundation. There has been different standards based on the DCOM/COM technology like OPC Data Access (OPC DA), OPC Alarms and Events (OPC A&E) and OPC Historical Data Access (OPC HDA). OPC DA is used to read/write control variables. OPC A&E is about the reception of alarms and events notifications. Finally, OPC HDA access archived data. These specifications have been widely approved and implemented by almost every system focused in the world of automation technology.

The OPC Unified Architecture (OPC UA) specification has been born to replace the former specifications. It is platform independent with a data model capable of dealing with very complex systems and a broad range of applications, from embedded systems, PLC (programmable logic controllers), DCS (distributed control systems) to MES (management execution systems) and ERP (enterprise resource planning) systems. OPC UA is built on different layers supported by two fundamental pillars, [14]. The first pillar is the transportation mechanism. OPC UA defines two optimized methods based on different cases. First one is based on an optimized binary protocol based on TCP (UA TCP) of high efficiency oriented to Intranet communications. The second mechanism is based on Web services (SOAP/HTTPs) oriented to less constrained communications or the Internet. The second pillar is the address space model (meta model) which defines the rules and base types to model the data. OPC UA services are above those two pillars and define the corresponding methods that OPC UA clients must use to access the OPC UA servers data. The definition of those services are defined in an abstract way to make them independent of the transport mechanisms (*codification*: UA Binary, UA XML, *security*: UA-SecureConversation, WS-SecureConversation, *transport*: UA TCP, SOAP/HTTP(s)) and the used programming language in order to be widen in the future. The Base OPC UA Information Model uses the concepts of the address space model to define its own types and rules. In turn, other models of information are built over the base one, from both OPC (DA, AE, HDA) and other organizations (IEC, EDDL, FDT, PLCopen) or companies. The information models are used for the creation of specific data that OPC UA servers provide access.

3 Control Performance Assessment

Control performance assessment is related to the deviation of the controlled variable from its set-point (SP) of a manufacturing control system. These deviations can be computed just by a single index, the Control Performance Index (CPI). Its value ranges from zero to infinity, where 1 means that the current operational status is the optimum and if it is above 1 indicates that thc current operational status is better than that one taken as reference (values close to 0 mean that the current status is poorly controlled). The numerator of the index is the optimum value and the denominator is the actual current value of the control loop or process which is being evaluated. Traditionally, the most used performance indexes have been the rise time, overshoot, settling-time, steady-state error, etc., specially on SISO systems. However, the most used one is the variance on automated control loops, [9].

3.1 Historical Data Benchmarks

These types of indexes compare the current control system with a historical data set which where recorded when the system was working fine and tuned from the control/maintenance engineers point of view, [6]. Ideal moments to collect and gather that kind of data is just right after the commissioning and start up of the control system, where poorly controlled effects are fixed and the system is well tuned.

3.2 MVC- Based Benchmarks

Harris, [9], presented a normalized index comparing the actual controller performance with the minimum variance controller. This index is based on the fact that the system output depends on two terms: the variance of the minimum variance controller (MVC, which is independent of the control law) and the variance of the suboptimum controller (SOC, which depends on the used control law). In order to calculate the Harris index for a stable feedback and minimum phase SISO system, the use of an auto-regressive of finite length model was proposed. Their parameters are calculated using a linear regression of the process data with the delay time being known. This concept was also proposed for feedforward systems.

For MIMO systems, the concept of interaction matrix was introduced as a generalization of the scalar delay time term of SISO systems. The same methods applied to SISO systems can be implemented in the MIMO case, but, in general, the interaction matrix calculation is very complex although it can be obtained from data of the closed loop and known delay times. The minimum variance is calculated for each system output using the concept of delay time input/output (similar to the SISO case). [10] proposes a similar method but only the order of the interaction matrix is required to be known.

[8] introduced the generalized minimum variance index which minimizes the variance of a "generalized output" defined as a weighted combination of the true

output and the control law. If it is compared to the Harris index, its main difference is that it uses a penalization over an excessive control action. Choosing the right weights, the system shows high gain in low frequencies, low gain in high frequencies and robustness.

These indexes minimize the variance in stationary state over a finite length of time. Extensions to these methods have been proposed in order to assess the transitory state also. In order to avoid the need of knowing the delay time, the concept of extended horizon performance index was introduced.

3.3 MPC- Based Benchmarks

These methods compare the current performance of the control system with generalized predictive control (GPC) algorithms. If the controller is a GPC, the method shows the performance of the control system but, in any other case, it shows information about the improvement over its replacement with a GPC controller. In case of a SISO system, its estimation can be made from closed loop data and information about the delay time. In case of MIMO systems, the model of the system is required. These indexes evaluate both, the performance of the control system in transitory and stationary states.

3.4 LQG-Based Benchmarks

Huang and Shah, [11], presented an index taking the LQG controller as reference. This controller penalizes the control actions but it requires the knowledge of the system model. The solution to this problem is very complex and it requires too much computational cost compared to those methods based on the minimum variance calculation. [11] recommended the use of a predictive model control approximation using a finite generalized predictive control solution (GPC). [3] proposed a simpler method based on subspace identification approximation.

3.5 Restrictive Structure

In spite of the MVC concept and design, most of the real industrial controllers are PID (almost 90%). This way, their actions, structure and order are restricted. Therefore, more realistic performance indexes are required. Approximations which calculate a low limit of the variance restricted to a PID (Optimal PID Benchmarking) have been developed by [13]. There are also OPID approximations based on IMC models. [7] shows a restricted LQG control (RS-LQG) applied to PID controllers and feedback/feedforward compensators.

4 Disturbances

One of the reasons a control loop has a poor performance can be due to external disturbances. A classification of disturbances based on time can be: Slow developing as the soiling of a heat exchanger, abrupt, as, for example, when a compressor gets stuck and dynamic disturbances which last hours or days. This section refers to dynamic disturbances. First, methods to detect them are explained along with the different methods to be diagnosed in the following section.

4.1 Disturbances Detection

Disturbances are divided into two groups: oscillatory and non-oscillatory. Besides, there are non stationary dynamic disturbances. For example, an oscillation which appears and disappears or its magnitude changes. Oscillatory disturbances detection can be made in both, time and frequency domains. A non-oscillatory disturbance is more difficult to characterize using its temporal evolution so frequency domain based methods are preferred. Usually, non stationary disturbances are detected using wavelet methods, based on a combination of time and frequency domains.

4.1.1 Oscillatory Disturbances

Oscillatory disturbances detection methods are classified based on the time domain, the autocovariance function or the frequency domain. In case of time domain, a noise filter is necessary. A proposed online method considers the integrated absolute error (IAE) between two zero crossings of the signal. If the IAE is high enough, a counter is incremented. If that value reaches a specified threshold, an oscillation has been detected. [18] explains a method based on the ARMA model poles. Noise problem is eliminated from the control signal error thanks to the autocovariance function. [21] uses the regular zero crossing that the autocovariance function usually shows. Methods based on the frequency domain are more difficult to solve and they are based on the detection of two wide peaks which indicate the presence of multiple non-sinusoidal oscillations.

4.1.2 Non-oscillatory Disturbances

This type of steady disturbances are characterized by their spectrum, which could be a wide bandwidth or the presence of multiple peaks. The detection problem requires a measure from a certain distance in order to detect similarities and determine several measurement groups with similar spectrum. [22] uses spectral methods based on the principal components analysis (PCA). There are methods based on the independent components analysis (ICA). There are also those ones based on the non-negative factorization matrix (NFM) and the use of the autocorrelation to detect non-oscillatory disturbances.

4.1.3 Non-stationary Oscillations

To detect this type of oscillations, a combination of time and frequency domains is used and control loop data is treated with wavelet analysis, [15]. However, wavelet graphics are very difficult to evaluate in an automatic way. Then, this is a tool only suitable to be assessed by human experts.

4.2 Disturbances Diagnosis

Once a disturbance has been detected, it has to be diagnosed. There is a distinction between non linear and linear sources. Among non-linear sources, we can find: Controlled valves with excessive static friction, start-up and stop control, non linearities which lead the system to limit cycles and unstable hydrodynamics like slow flows. The diagnosis problem is divided into two separate objectives. First, the main disturbance has to be distinguished between the secondary disturbances due to the fact that these last ones will disappear when the main one is fixed. Second step consists of testing the main disturbance in order to diagnose its cause.

4.2.1 Non Linear Sources

Non linear time series analysis, limit cycles and valve diagnosis are found among these methods. There has to be taken into account that non linear disturbances are stronger the nearer to the source. This is due to the fact that the system acts as a filter. A non linear time series is a time series generated by the output of a non linear system where the presence of phase couplings between different frequency bands is usual, [19].Permanent limit cycles are common in non linear systems. The limit cycle wave shape is periodic but not sinusoidal. Therefore, it has harmonics which can be detected to diagnose the non linearity, [21]. This section also deals with the valves diagnostic techniques. The main problems that can be found within this field are dead zones and static friction. A valve diagnosis is simple if the controller output (OP), or the flow throughout the valve (MV) or the valve position are measured. In this case, an OP-MV graph with a 45° straight line indicates a good behavior and any deviation, like dead zones, can be visually diagnosed. Unfortunately, the flow throughout the valve is not usually measured unless it is regulated by a control loop. In this case, it is necessary to determine a possible fault from the process variable (PV) and OP data. The Horch method is based on the cross correlations between the controller output and the process variable. This method separates the static friction from other oscillatory sources. The objective of this method is to detect abrupt changes in the process variable. This is achieved using the probability distribution of the derived signal. In case of dry friction, there is a Gaussian distribution whereas in any other case, the distribution shows two peaks. [2] calculates non-linear indexes (NLI) and no-gaussian indexes (NGI). Both indexes are calculated from the bicoherence signal. [19] shows a simple algorithm which detects if the signal period is symmetric. Other algorithm is also based on the relation between the controller output and the valve position. [17] suggests a method based on the change of the three first order parameters plus the process delay time.

4.2.2 Linear Sources

Linear sources for poorly control loops are due to bad tuning, interaction among controllers and erroneous structure selection. To detect this type of faults, different algorithms can be found in [1].

5 Conclusions

CPM/CPA technology has been, in the past 20 years, of extensive research in both, academic and industrial field. Its introduction in industry has been made from large industrial processes such as refineries or paper industry where cost reduction is a huge amount of money. This paper has reviewed this technology, explaining different performance indexes and methodologies for detection of disturbances in industrial processes. Also, there has been a review of the OPC standard technology, which is very important for all of these methods since the CPM technology relies on real process data analysis. Over the next few years, this technology will continue to be of great interest both in theory and industrial. Examples are the existence of commercial products from companies such as AspenTech, ABB or Honeywell.

Acknowledgments. The authors would like to acknowledge the financial assistance of the Ministry of Science and Innovation of Spain under grants BES-2006-12849 and DPI2008-05795 which have partially funded this work.

References

1. Bauer, M., Thornhill, N.F., Meaburn, A.: Specifying the directionality of fault propagation paths using transfer entropy. In: DYCOPS, Boston, USA, vol. 7 (2004)
2. Choudhury, M.A.A., Kariwala, V., Shah, S.L., Douke, H., Takada, H., Thornhill, N.F.: A simple test to confirm control valve stiction. In: IFAC World Congr, Praha (2005)
3. Dai, C., Yang, S.H.: Controller performance assessment with a LQG benchmark obtained by using the subspace method. In: Proc. Control 2004. University of Bath, UK (2004)
4. Gao, J., Patwardhan, R.S., Akamatsu, K., Hashimoto, Y., Emoto, G., Shah, S.L.: Performance evaluation of two industrial MPC controllers. Control Eng. Pract. 11, 1371–1387 (2003)
5. Grimble, M.J.: Restricted structure LQG optimal control for continuous-time systems. Control Theory and Appl. 147, 185–195 (2000)
6. Grimble, M.J.: Controller performance benchmarking and tuning using generalised minimum variance control. Autom. 38, 2111–2119 (2002)
7. Harris, T.: Assessment of closed loop performance. Can J. of Chem. Eng. 67, 856–861 (1989)
8. Huang, B., Ding, S.X., Qin, J.: Closed-loop subspace identification: An orthogonal projection approach. J. of Process Control 15, 53–66 (2005)
9. Huang, B., Shah, S.L.: Performance assessment of control loops. Advances in Industrial Control. Springer, Heidelberg (1999)

10. Jelali, M.: An overview of control performance assessment technology and industrial applications. Control Eng. Pract. 14, 441–466 (2006)
11. Ko, B.S., Edgar, T.F.: Assessment of achievable PI control performance for linear processes with dead time. In: Proc. Am Control Conf., Philadelphia, USA (1998)
12. Mahnke, W., Leitner, S.H., Damm, M.: OPC Unified Architecture. Springer, Heidelberg (2009)
13. Matsuo, T., Sasaoka, H., Yamashita, Y.: Detection and diagnosis of oscillations in process plants. In: Palade, V., Howlett, R.J., Jain, L. (eds.) KES 2003. LNCS, vol. 2773, pp. 1258–1264. Springer, Heidelberg (2003)
14. Ordys, A.W., Uduehi, D., Johnson, M.A.: Process control performance assessment: from theory to implementation. Advances in Industrial Control. Springer, Heidelberg (2007)
15. Rossi, M., Scali, C.: A comparison of techniques for automatic detection of stiction: simulation and application to industrial data. J. of Process Control 15, 514–550 (2005)
16. Salsbury, T.I., Singhal, A.: A new approach for ARMA pole estimation using higher-order crossings. In: Proc. ACC 2005, Portland, USA (2005)
17. Thornhill, N.F., Cox, J.W., Paulonis, M.: Diagnosis of plant-wide oscillation through data-driven analysis and process understanding. Control Eng. Pract. 11, 1481–1490 (2003)
18. Thornhill, N.F., Horch, A.: Advances and new directions in plant-wide controller performance assessment. In: Proc. ADCHEM 2006, Gramado, Brazil (2006)
19. Thornhill, N.F., Huang, B., Zhang, H.: Detection of multiple oscillations in control loops. J of Process Control 13, 91–100 (2003)
20. Thornhill, N.F., Shah, S.L., Huang, B., Vishnubhotla, A.: Spectral principal component analysis of dynamic process data. Control Eng. Pract. 10, 833–846 (2002)

Improving Optical WDM Networks by Using a Multi-core Version of Differential Evolution with Pareto Tournaments

Álvaro Rubio-Largo, Miguel A. Vega-Rodríguez, Juan A. Gómez-Pulido, and Juan M. Sánchez-Pérez

Abstract. Wavelength Division Multiplexing (WDM) in optical networks is the most favorable technology to exploit the huge bandwidth of this kind of networks. A problem occurs when it is necessary to establish a set of demands. This problem is called in the literature as Routing and Wavelength Assignment problem (RWA problem). In this paper we have used multiobjective evolutionary computing for solving the Static-RWA problem (demands are given in advance). We have implemented a population-based algorithm, Differential Evolution but incorporating the Pareto Tournament concept (DEPT). By using OpenMP, we have exploited the use of different multi-core systems (2, 4 and 8 cores), obtaining an average efficiency of 93.46% with our approach. To ensure that our heuristic obtains relevant results we have compared it with a parallel version of the standard algorithm NSGA-II. Furthermore we have compared the obtained results with other approaches and we can conclude that the DEPT algorithm has obtained better results.

Keywords: Multi-core, Optical networks, Multiobjective Evolutionary Algorithm.

1 Introduction

The most favorable technology to exploit the enormous bandwidth of optical networks is based on Wavelength Division Multiplexing (WDM). This technique multiplies the available capacity of an optical fiber by adding new channels, each channel on a new wavelength of light. The main aim of this technology is to ensure fluent communications between several devices [1]. A problem occurs when it is necessary to establish a set of demands, this problem is called in the literature as Routing and Wavelength Assignment problem (RWA problem). There are two types of RWA problem, depending on the demands, we could refer to a static problem whether the demands are given in advance (Static-RWA), and if the demands are given in real time we refer to a dynamic problem (Dynamic-RWA).

Álvaro Rubio-Largo, Miguel A. Vega-Rodríguez, Juan A. Gómez-Pulido, and Juan M. Sánchez-Pérez
University of Extremadura, Polytechnic School, Dept. TC2, 10003, Cáceres, Spain
e-mail: {arl,mavega,jangomez,sanperez}@unex.es

A.P. de Leon F. de Carvalho et al. (Eds.): Distrib. Computing & Artif. Intell., AISC 79, pp. 629–636.
springerlink.com

In this paper we have decided to develop a Multiobjective Evolutionary Algorithm (MOEA) for solving the Static-RWA problem (the most usual one). The Differential Evolution (DE) [2] is the selected algorithm, but adding the Pareto Tournament concept (DEPT). Additionally, using OpenMP we have demonstrated that this approach is highly suitable to be parallelized. After accomplishing several experiments with different multi-core systems (2, 4 and 8 cores) and with different instances of a real-world topology (NTT Japan), we obtained an average efficiency of 93.46 per cent. We have also implemented a multi-core version of a well-known MOEA, Non-dominated Sorting Genetic Algorithm (NSGA-II) [3], with the aim of making comparisons between them. As we can see later, our approach obtains better efficiency and greater results than the NSGA-II algorithm. Finally, we have compared the DEPT algorithm with other approaches of the literature for this specific problem and we conclude that this heuristic achieves better results than any other for this problem.

The rest of this paper is organized as follows. The RWA problem in a formal way is presented in Section 2. A description of the multi-core versions of the DEPT and NSGA-II algorithms appears in Section 3. In Section 4 we present several experiments with different multi-core systems, making a comparison between DEPT and NSGA-II. A comparison with other approaches appears in Section 5. Finally, the conclusions and future work are left for Section 6.

2 RWA Problem

An optical network is normally modeled as a direct graph $G = (V, E, C)$, where V is the set of nodes, E is the set of links between nodes and C is the set of available wavelengths for each optical link in E.

- $(i, j) \in E$: Optical link from node i to node j.
- $c_{ij} \in C$: Number of channels or different wavelengths at link (i, j).
- $u = (s, d)$: Unicast request u with source node s and a destination node d, where $s, d \in V$.
- U : Set of unicast request, where $U = \{\ u \mid u \text{ is an unicast request}\}$.
- $u^{\lambda}_{i,j}$: Wavelength λ assigned to the unicast request u at link (i, j).
- l_u : Lightpath or set of links between a source node s_u and destination node d_u; with the corresponding wavelength assignment in each link (i, j).
- L_u : Solution of the RWA problem considering the set of U requests.

Notice that $L_u = \{l_u | l_u \text{ is the set of links with their corresponding wavelength assignment }\}$. Using the above definitions, the RWA problem may be stated as a Multiobjective Optimization Problem (MOOP) [4], searching the best solution L_u that simultaneously minimizes the Number of hops (y_1) and the Number of wavelength conversions (y_2):

$$y1 = \sum_{u \in U} \sum_{(i,j) \in l_u} \Phi_j \ where \begin{Bmatrix} \Phi_j = 1 \ \text{if} \ (i,j) \in l_u \\ \Phi_j = 0 \ \text{if} \ otherwise \end{Bmatrix} \tag{1}$$

$$y2 = \sum_{u \in U} \sum_{j \in V} \varphi_j \; where \begin{Bmatrix} \varphi_j = 1 \text{ if } i \in V switches \; \lambda \\ \varphi_j = 0 \text{ if } \quad otherwise \end{Bmatrix} \quad (2)$$

Furthermore, we have to fulfill *the wavelength conflict constraint* [5]: Two different unicast transmissions must be allocated with different wavelengths when they are transmitted through the same optical link (i, j).

3 Multi-core Versions of DEPT and NSGA-II

As the Static-RWA problem is a MOOP, it is necessary to compare candidate solutions, the most common way to compare them is known as Pareto dominance. For further details about the Pareto dominance, please refer [6].

The individual in this paper was designed as it is shown in Fig. 1. We have developed an adapted version of Yen´s algorithm [7]. We modified it by introducing an heuristic to assign the wavelengths. Whenever possible, we try to assign always the same wavelength that was previously assigned (previous hop), but if it is not possible or it is the first assignation, we assign the first free wavelength. Using this modified version, we create a list of possible routes (including wavelengths) and we select one, and we store it in a vector. We repeat this process for every lightpath.

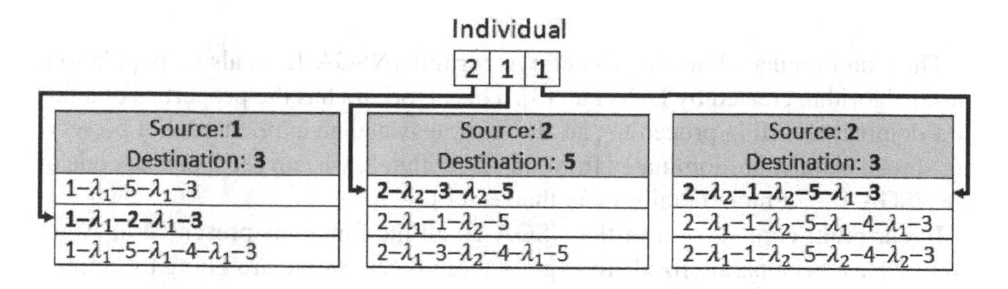

Fig. 1 Structure of an Individual

The Differential Evolution is a population-based algorithm created by Rainer Storn and Ken Price [2]. This algorithm generates new individuals taking advantage of the differences among various randomly selected individuals of the population. In Algorithm 1 we can see the pseudocode of the Differential Evolution adapted to a multiobjective context, by introducing the Pareto Tournament concept (DEPT). For further information about this algorithm, please refer [8].

Analyzing the Algorithm 1, we notice that the two *for* loops (lines 3 and 4) do not present any data dependency so we could parallelized any of them, but we have selected the inner loop (line 4) because in this way, we can also execute the DEPT algorithm by time instead of by generations. Our multi-core version of the DEPT consists in dividing this loop between different threads in every generation. In a more formal way, if we dispose a $N - core$ system and a population size of PS, then each thread is going to execute PS/N iterations of this *for* loop.

Algorithm 1. DEPT Pseudocode

```
1: P ← generateRandomPopulation(PopSize)
2: EvaluatePopulation (P)
3: for i = 0 to MAX_GENERATIONS do
4:     for i = 0 to PopSize do
5:         X_target ← P[i]
6:         X_trial ← GenerateTrialIndividual(X_target)
7:         EvaluateIndividual(X_trial)
8:         P[i] ← ParetoTournament(X_target, X_trial)
9:     end for
10: end for
```

Algorithm 2. NSGA-II Pseudocode

```
1: P ← GenerateFatherPopulation(PopSize)
2: P ← FastNonDominatedSort(P, PopSize)
3: for i = 0 to MAX_GENERATIONS do
4:     Q ← GenerateChildPopulation(PopSize)
5:     R ← P ∪ Q
6:     R ← FastNonDominatedSort(R, 2 * PopSize)
7:     P ← GenerateNewFatherPopulation(R, PopSize)
8:     P ← FastNonDominatedSort(P, PopSize)
9: end for
```

The Non-dominated Sorting Genetic Algorithm (NSGA-II) is also a population-based algorithm created by Deb et al. [3]. This algorithm has the properties of a fast non-dominated sorting procedure, an elitist strategy and an explicit goal of preserving spread on the non-dominated front. In Algorithm 2, we can see a pseudocode of the NSGA-II. For more details about that, refer [3].

In our multi-core version of the NSGA-II, all the functions presented in Algorithm 2 have been parallelized. To explain these functions we are going to suppose that we have a N-core system and a population size of PS for populations P and Q, and $2*PS$ for population R. In *GenerateFatherPopulation*, each thread creates randomly PS/N individuals. In function *FastNonDominatedSort*, we classify (in a parallel way) all individuals of the population into sets of pareto front, subsequently we sort the population by sets. On function *GenerateChildPopulation* each thread creates PS/N new individuals by using binary tournaments, crossover and mutation. In *GenerateNewFatherPopulation* we have parallelized the calculus of the crowding distance of all individuals, after that, we sort the population R by sets of pareto fronts and in case of tie by crowding distance, finally we generate the new father population for the next generation.

4 Experimental Results

In this section we present several experiments with different multi-core systems (2, 4 and 8 cores). For each multi-core system we have calculated the average speedup

Table 1 Performance of the DEPT and the NSGA-II algorithms using different multi-core systems. ($\overline{X}$ represents the average time (in seconds). *Std* represents the standard deviation. S_p represents the speedup for p processors and E_p represents the efficiency for p processors)

		NTT $c_{ij}=8$						NTT $c_{ij}=10$					
		$\|U\|=10$		$\|U\|=20$		$\|U\|=30$		$\|U\|=10$		$\|U\|=20$		$\|U\|=40$	
		DEPT	NSGA-II	DEPT	NSGA-II	DEPT	NSGA-II	DEPT	NSGA-II	DEPT	NSGA-II	DEPT	NSGA-II
Seq.	$\overline{X}$	169.69	172.57	320.02	322.04	422.35	775.76	207.13	210.47	398.57	398.32	693.55	683.59
	Std	0.09	0.17	0.32	0.14	0.75	4.03	0.12	0.36	0.27	0.15	1.19	0.59
2-Core	$\overline{X}$	86.45	92.22	161.70	168.52	214.34	437.86	104.75	110.62	201.43	208.03	353.65	355.30
	Std	0.06	0.11	0.14	0.10	0.29	1.77	0.04	0.10	0.14	0.10	0.81	0.69
	S_2	**1.96**	1.87	**1.98**	1.91	**1.97**	1.77	**1.98**	1.90	**1.98**	1.91	**1.96**	1.92
	E_2	**98.15%**	93.56%	**98.95%**	95.55%	**98.52%**	88.59%	**98.87%**	95.13%	**98.93%**	95.73%	**98.06%**	96.20%
4-Core	$\overline{X}$	44.96	51.86	84.70	92.76	112.29	262.84	54.74	62.18	105.24	113.44	184.38	192.28
	Std	0.02	0.12	0.10	0.09	0.27	1.32	0.02	0.10	0.07	0.08	0.31	0.48
	S_4	**3.77**	3.33	**3.78**	3.47	**3.76**	2.95	**3.78**	3.38	**3.79**	3.51	**3.76**	3.56
	E_4	**94.35%**	83.20%	**94.45%**	86.80%	**94.03%**	73.78%	**94.60%**	84.62%	**94.68%**	87.78%	**94.04%**	88.88%
8-Core	$\overline{X}$	24.31	31.79	45.43	54.05	61.06	174.12	29.60	37.39	56.62	65.88	99.18	111.47
	Std	0.01	0.08	0.04	0.08	0.32	1.13	0.01	0.14	0.04	0.09	0.11	0.94
	S_8	**6.98**	5.43	**7.04**	5.96	**6.92**	4.46	**7.00**	5.63	**7.04**	6.05	**6.99**	6.13
	E_8	**87.26%**	67.87%	**88.05%**	74.48%	**86.46%**	55.69%	**87.48%**	70.37%	**88.00%**	75.58%	**87.41%**	76.66%

and efficiency for 30 independent runs, with the purpose of ensuring a statistical relevance. This experiment was carried out using Nippon Telegraph and Telephone topology (NTT network), and six different sets of demands taken from [9].

The best configuration for the DEPT algorithm was obtained from [8], number of generations = 700, k-shortest-paths=25, population size= 50, crossover factor=0.05, mutation factor=0.5 and the DEPT schema=RandToBest/1/Binomial.

The best configuration for the NSGA-II was carried out by using the same metodology as [8], the best configuration is: number of generations = 700, k-shortest-path=25, population size = 50, crossover probability=0.8, crossover schema=1-point and mutation rate = 0.35.

As we can see in Table 1, our multiobjective Differential Evolution achieves better speedups and efficiencies in all sets of demands for the NTT topology. In Table 2, we can notice that the DEPT algorithm obtains an upper average efficiency than the NSGA-II. With 2 cores, the difference is about 4.5%, however with 8 cores the difference grows to 17.33%, it means that if we use other systems with more cores the difference will grow exponentially. We can conclude that our approach is very suitable to be parallelized.

Table 2 Summary of the average efficiency by both algorithms per multi-core system and also a global average efficiency.

		2-Core	4-Core	8-Core	*Global* $\overline{X}_{Efficiency}$
DEPT	$\overline{X}_{Speedup}$	1.972	3.774	6.995	
	$\overline{X}_{Efficiency}$	98.58%	94.36%	87.44%	**93.46%**
NSGA-II	$\overline{X}_{Speedup}$	1.883	3.367	5.608	
	$\overline{X}_{Efficiency}$	94.13%	84.18%	70.11%	**82.80%**

Table 3 Comparison of results between DEPT and NSGA-II (Hypervolume)

	$NTT\ c_{ij}=8$			$NTT\ c_{ij}=10$		
	$\lvert U\rvert=10$	$\lvert U\rvert=20$	$\lvert U\rvert=30$	$\lvert U\rvert=10$	$\lvert U\rvert=20$	$\lvert U\rvert=40$
DEPT	0.933	0.841	0.813	0.933	0.845	0.697
NSGA-II	0.933	0.797	0.713	0.933	0.809	0.567

Furthermore, we have also compared the results obtained by both algorithms using the hypervolume concept [6]. Taking the results generated in the sequencial experiment (remember that the multi-core versions only accelerate the excutions, obtaining results of identical quality), we have composed a final Pareto front for each algorithm. These optimal pareto fronts are obtained by selecting the Pareto front of all points earned in each of the 30 runs of the sequential experiment. We have used as reference points to calculate the hypervolume, $r_{min}=(0,0)$ and $r_{max}=(999,99)$, where, inside the parentheses, the first value refers to y_1 and the second refers to y_2. In Table 3 we can see that the DEPT obtains better hypervolume than the NSGA-II in all instances.

5 Comparisons with Other Approaches

In this section we compare the DEPT algorithms with approaches of other authors, and also with the NSGA-II algorithm. To be fair with the other authors, we use the same metrics and test instances as them. These metrics determinate the performance of the algorithms in quality ($M1$), distribution ($M2$) and extension ($M3$) of the Pareto front, they were taken from [10]. From [4] we took the fourth metric ($M4$), it calculates the percentage of generated solutions which not belong to the Pareto front. For further information about these metrics, please, refer to [10] and [4]. Also we present the average of these four metrics, and we denote it as R. In order to calculate this average, the metrics $M1$ and $M4$ were slightly modified, thus, $M1^* = 1 - M1$ and $M4^* = 1 - M4$, where the best value is now 1 and the worst 0.

For each set of unicast demands U and capacities c_{ij} of wavelengths in the network, a set of optimal solutions approximated to the Pareto front are calculated using the following procedure: the algorithm was executed 10 times; a set including all the obtained solutions was generated; the dominated solutions were deleted and a new approximated set to the Pareto front was generated. It is denoted as Y_{known}.

In [11], the number of blocked unicast requests (NB) was also calculated in average for the 10 runs. They developed the next algorithms [11]:BIANT (Bicriterion Ant), COMP (COMPETants), MOAQ (Multiple Objective Ant Q Algorithm), MOACS (Multi-Objective Ant Colony System), M3AS (Multiobjective Max-Min Ant System), MAS (Multiobjective Ant System), PACO (Pareto Ant Colony Optimization), MOA (Multiobjective Omicron ACO), 3SPFF (3-Shortest Path routing, First-Fit wavelength assignment), 3SPLU (3-Shortest Path routing, Least-Used wavelength assignment), 3SPMU (3-Shortest Path routing, Most-Used wavelength assignment), 3SPRR (3-Shortest Path routing, Random wavelength assignment), SPFF (Shortest Path Dijkstra routing, First-Fit wavelength assignment), SPLU (Shortest Path Dijkstra routing, Least-Used wavelength assignment), SPMU (Shortest Path Dijkstra routing, Most-Used wavelength assignment), SPRR (Shortest Path Dijkstra routing, Random wavelength assignment) . Considering the results

Table 4 Comparison with other approaches [11]

	NTT ($c_{ij}=8.$ $\lvert U\rvert=10$)						NTT ($c_{ij}=8.$ $\lvert U\rvert=20$)						NTT ($c_{ij}=10.$ $\lvert U\rvert=20$)						NTT ($c_{ij}=10.$ $\lvert U\rvert=40$)					
	NB	M1*	M2	M3	M4*	R	NB	M1*	M2	M3	M4*	R	NB	M1*	M2	M3	M4*	R	NB	M1*	M2	M3	M4*	R
DEPT	0	1	1	1	1	**1**	0	0.95	1	1	0.53	**0.87**	0	0.88	0.98	1	0.27	**0.78**	0	0.92	1	1	0.17	**0.77**
NSGA-II	0	0.89	0.90	0.91	0.24	**0.73**	0	0.96	0.88	0.89	0.15	**0.72**	0	0.96	0.91	0.96	0.15	**0.74**	0	0.97	0.88	0.88	0.18	**0.73**
BIANT	0	0.99	0.1	0.05	0.9	0.51	0	0.82	0.95	0.71	0.02	0.62	0	0.76	0.98	0.62	0	0.59	0	0.64	1	1	0	0.66
COMP	0	0.85	0.8	0.54	0.3	0.62	0	0.74	0.98	0.76	0	0.62	0	0.7	0.87	1	0	0.64	0	0.46	0.82	0.91	0	0.55
MOAQ	0	0.86	1	1	0.3	0.79	0	0.69	0.9	1	0	0.65	0	0.57	0.73	0.89	0	0.55	0	0.37	0.7	0.92	0	0.5
MOACS	0	0.98	0.2	0.11	0.8	0.52	0	0.78	0.81	0.8	0	0.6	0	0.76	0.91	0.85	0	0.63	0	0.61	0.93	0.89	0	0.61
M3AS	0	1	0	0	1	0.5	0	0.82	0.86	0.77	0	0.61	0	0.75	0.8	0.6	0	0.54	0	0.52	0.88	0.88	0	0.57
MAS	0	0.98	0.2	0.1	0.7	0.49	0	0.79	1	0.86	0	0.66	0	0.6	0.81	0.84	0.02	0.57	0	0.6	0.93	0.97	0	0.62
PACO	0	0.99	0.1	0.05	0.9	0.51	0	0.77	0.8	0.77	0	0.59	0	0.77	0.95	0.82	0	0.63	0	0.55	0.82	0.84	0	0.55
MOA	0	1	0	0	1	0.5	0	0.82	0.94	0.88	0	0.66	0	0.79	1	0.84	0	0.66	0	0.6	0.87	0.88	0	0.59
3SPFF	0	0	0	0	0	0	0	0	0.03	0.06	0	0.02	0	0	0.1	0.15	0	0.06	0	0.3	0.41	0.3	0	0.25
3SPLU	0	1	0	0	1	0.5	0	0.35	0.08	0.07	0.65	0.29	0	0.43	0.07	0.17	0.2	0.22	0	0.86	0.38	0.26	0.13	0.41
3SPMU	0	0.03	0	0	0	0.01	0	0.08	0.11	0.09	0	0.07	0	0.05	0.13	0.15	0	0.08	0	0	0.32	0.24	0	0.14
3SPRR	0	1	0	0	1	0.5	0	0.36	0.14	0.1	0.1	0.17	0	0.39	0.12	0.2	0	0.18	0	0.81	0.31	0.24	0.08	0.36
SPFF	0	0	0	0	0	0	2	N/A	N/A	N/A	N/A	N/A	2	N/A	N/A	N/A	N/A	N/A	4	N/A	N/A	N/A	N/A	N/A
SPLU	0	1	0	0	1	0.5	2	N/A	N/A	N/A	N/A	N/A	2	N/A	N/A	N/A	N/A	N/A	4	N/A	N/A	N/A	N/A	N/A
SPMU	0	0.03	0	0	0	0.01	2	N/A	N/A	N/A	N/A	N/A	2	N/A	N/A	N/A	N/A	N/A	4	N/A	N/A	N/A	N/A	N/A
SPRR	0	1	0	0	1	0.5	2	N/A	N/A	N/A	N/A	N/A	2	N/A	N/A	N/A	N/A	N/A	4	N/A	N/A	N/A	N/A	N/A

Table 5 Comparison with other approaches (Metric M4* in %)

	NTT $c_{ij}=8$			NTT $c_{ij}=10$			
	$\lvert U\rvert=10$	$\lvert U\rvert=20$	$\lvert U\rvert=30$	$\lvert U\rvert=10$	$\lvert U\rvert=20$	$\lvert U\rvert=40$	$\overline{X}$
DEPT	100%	53.3%	18.3%	100%	26.6%	16.6%	**52.47%**
NSGA-II	24.2%	14.7%	16.7%	16.7%	14.9%	17.8%	17.5%
MOACS [12]	5%	7.5%	3%	3.3%	5.1%	4.2%	4.68%
M3AS [12]	0%	2.5%	7%	6.6%	5.1%	8.4%	4.93%
TA-MOEA [13]	N/A	N/A	N/A	N/A	N/A	N/A	6.1%

shown in Table 4, we notice that the DEPT algorithm overcomes all algorithms in all set of demands.

In Table 5 we present a comparison with [12] and [13] (using M4* in %). Two Ant Colony Optimization algorithms are presented in [12], MOACS and M3AS. In [13] a team that combines seven MOEAs is presented (TA-MOEA). Analyzing these results, we conclude that the DEPT achieves better results than other approaches.

6 Conclusions and Future Work

In this paper we have purposed a multi-core version of the Differential Evolution with Pareto Tournaments (DEPT) for solving the Static-RWA problem. As we have seen previously, our approach is very suitable to be parallelized. We have obtained an average efficiency about 93.46% in several experiments using different multi-core systems and instances. We have developed a multi-core version of a well-known MOEA (NSGA-II) and we notice that our approach obtains better efficiency and greater results than NSGA-II, also greater results than other approaches.

As future work, we intend to develop more multi-core MOEAs with the aim of make comparisons with the DEPT. Another idea would be to apply our multi-core approach to the Dynamic-RWA problem.

Acknowledgements. This work has been partially funded by the Spanish Ministry of Education and Science and ERDF (the European Regional Development Fund), under contract TIN2008-06491-C04-04 (the M* project). Álvaro Rubio-Largo is supported by the research grant PRE09010 from Junta de Extremadura (Spain).

References

1. Hamad, A.M., Kamal, A.E.: A survey of multicasting protocols for broadcast-and-select single-hop networks. IEEE Network 16, 36–48 (2002)
2. Storn, R., Price, K.: Differential Evolution - A Simple Evolution Strategy for Fast Optimization. Dr. Dobb 22(4), 18–24 (1997)
3. Deb, K., Pratap, A., Agarwal, S., Meyarivan, T.: A Fast Elitist Multi-Objective Genetic Algorithm: NSGA-II. IEEE Transactions on Evolutionary Computation 6, 182–197 (2000)
4. Deb, K.: Multi-Objective Optimization Using Evolutionary Algorithms. John Wiley & Sons, Inc., New York (2001)
5. Gagnaire, M., Koubaa, M., Puech, N.: Network Dimensioning under Scheduled and Random Lightpath Demands in All-Optical WDM Networks. IEEE Journal on Selected Areas in Communications 25(S-9), 58–67 (2007)
6. Zitzler, E., Thiele, L.: Multiobjective Optimization Using Evolutionary Algorithms - A Comparative Case Study, pp. 292–301. Springer, Heidelberg (1998)
7. Yen, J.Y.: Finding the K Shortest loopless paths in a Network. Manage Sci. 17(11), 712–716 (2003)
8. Rubio-Largo, A., Vega-Rodríguez, M.A., Gómez-Pulido, J.A., Sánchez-Pérez, J.M.: A Differential Evolution with Pareto Tournaments for solving the Routing and Wavelength Assignment Problem in WDM Networks. In: Proceedings of the 2010 IEEE Congress on Evolutionary Computation, CEC 2010 (2010)
9. Schaerer, M., Barán, B.: A Multiobjective Ant Colony System for Vehicle Routing Problem with Time Windows. In: IASTED International Conference on Applied Informatics, pp. 97–102 (2003)
10. Zitzler, E., Deb, K., Thiele, L.: Comparison of Multiobjective Evolutionary Algorithms: Empirical Results. Evolutionary Computation 8, 173–195 (2000)
11. Arteta, A., Barán, B., Pinto, D.: Routing and Wavelength Assignment over WDM Optical Networks: a comparison between MOACOs and classical approaches. In: LANC 2007: Proceedings of the 4th international IFIP/ACM Latin American conference on Networking, pp. 53–63. ACM, New York (2007)
12. Insfrán, C., Pinto, D., Barán, B.: Diseño de Topologías Virtuales en Redes Ópticas. Un enfoque basado en Colonia de Hormigas. In: XXXII Latin-American Conference on Informatics 2006 - CLEI 2006, vol. 8, pp. 173–195 (2006)
13. Fernandez, J.M., Vila, P., Calle, E., Marzo, J.L.: Design of Virtual Topologies using the Elitist Team of Multiobjective Evolutionary Algorithms. In: Proceedings of International Symposium on Performance Evaluation of Computer and Telecommunication Systems - SPECTS 2007, pp. 266–271 (2007)

Solving the General Routing Problem by Artificial Ants

María-Luisa Pérez-Delgado

Abstract. Routing Problems arise in several areas of distribution management and logistics and their practical significance is widely known. These problems are usually difficult to solve. Therefore, heuristic methods are applied to try to solve them. This paper describes the application of artificial ant colonies to solve the General Routing Problem. For this, the problem is first transformed into a node-routing problem. The transformed problem is solved by applying an ant-based algorithm which has been widely applied to node-routing problems, obtaining good results.

Keywords: artificial-ants, routing problems.

1 Introduction

The General Routing Problem (GRP) is the problem of finding a minimum cost route for a single vehicle, subject to the condition that the vehicle visits certain nodes and edges of a network, [16], [21]. The GRP has many practical applications, [1], [9], [28]; and it includes some well-known routing problems as special cases: the Rural Postman Problem, [11]; the Chinese Postman Problem, [10]; the Road Travelling Salesman Problem, [12]; and the Graphical Travelling Salesman Problem, [5].

This problem was introduced by Orloff, [20]. It is an NP-hard problem, [16]. Therefore, heuristic solutions are usually studied to solve it.

Despite the practical importance of this problem, relatively few studies have been published on the GRP. Heuristic solutions were proposed in [13] and [22]. An exact algorithm was proposed by Corberán et al. to solve to optimality undirected problems of moderate size (with no more than 200 nodes), [3]. Another interesting papers related to this problem are [4], [17], [18], and the recent work of Reinelt and Theis, [23].

María-Luisa Pérez-Delgado
Escuela Politécnica Superior de Zamora. Universidad de Salamanca,
Av. Requejo, 33, C.P. 49022, Zamora, Spain
e-mail: mlperez@usal.es

A.P. de Leon F. de Carvalho et al. (Eds.): Distrib. Computing & Artif. Intell., AISC 79, pp. 637–644.
springerlink.com

Blais and Laporte proposed a graph transformation to convert the GRP into a node-routing problem or an arc-routing problem, [2]. If the GRP is directed, it can be transformed into an asymmetric Traveling Salesman Problem (TSP). If the problem is mixed or undirect, it can be transformed into an asymmetric Generalized Traveling Salesman Problem (GTSP), which can be transformed into an asymmetric TSP.

The use of graph transformations allows to apply to a problem a solution method developed for other problems.

In this paper the transformation proposed by Blais and Laporte is considered and the TSP is solved by an heuristic method: ant-based algorithms. These algorithms have been successfully applied to several NP-hard combinatorial optimization problems. The algorithm selected is the one called MAX-MIN Ant System, [24], one of the best performing ant-based algorithms.

The inspiring source of ants-algorithms is the foraging behaviour of real ants that consents to find shortest paths between food sources and the nest. While walking from food sources to the nest and vice versa, ants release a chemical substance, called pheromone, on the ground, and the direction chosen by the following ants is the path marked with a stronger pheromone concentrations.

The document is structured as follows. First a formulation of the GRP is presented, to allow a better understanding of the transformation described in Sec. 3. Next, the MAX-MIN Ant System algorithm is described. We describe the method of solution of the problem, based on the refered algorithm. Computational results are presented in Sec. 5 and, finally, the conclusions of the paper are presented.

2 Problem Definition and Formulation

Given a graph $G = (V, E \bigcup A)$ with vertex set V, edge set E, arcs set A, a cost function c for each link $(i, j) \in E \bigcup A$, a set $E_R \subseteq E$ of required edges, a set $A_R \subseteq A$ of required arcs and a set $V_R \subseteq R$ of required nodes, the GRP is the problem of finding a minimum cost closed walk (a tour) passing through each required connection $e \in E_R \bigcup A_R$ and through each required node $i \in V_R$ at least once, [20].

3 The Transformation

The first step consists of replacing each edge $(i, j) \in E$ by two opposite arcs (i, j) and (j, i), with the same cost as the edge. All such arcs are included in A to yield an extended arc set A'. If (i, j) is a required edge, then the arcs (i, j) and (j, i) are included in A_R, to yield A'_R. At the end of the first step, the transformed graph $G' = (V, A')$ is directed.

The next step consists of transforming the problem on G' into an equivalent node-routing problem on a complete graph, $H = (W, B)$. In this graph, W consists of one vertex for each arc of A'_R and each node of V_R, and B is the set of all arcs linking two vertices of W.

Each required vertex, $i \in V_R$, generates a required node v_{ii}. Each required arc, $(i,j) \in A'_R$, generates a required node v_{ij}. We can observe that each required edge, $(i,j) \in E_R$, is replaced by two nodes, v_{ij} and v_{ji}, only one of which is required.

Each vertex pair, (v_{ij}, v_{kl}), in the transformed problem defines an arc of the new graph having cost $c'_{jl} = dist(j,k) + c_{kl}$, where $c'_{ii} = 0$. $dist(j,k)$ is the cost of a shortest path in G' from node i to node k, calculated by the Floyd algorithm.

At the end of this step, the original routing problem on G has been transformed into an equivalent directed Generalized Traveling Salesman Problem (GTSP) on H. The GTSP consists of determining a least cost Hamiltonian circuit passing exactly once through each of m sets or clusters of nodes, [15]. Any required node, arc and edge of G defines such a set. A node or an arc defines a set with one element, whereas an edge defines a set with two elements. Solving the GTSP on H ensured that each required connection will be traversed once and each required node will also be traversed once.

The last step consists of transforming the GTSP into a Traveling Salesman Problem (TSP) using the rules described in Noon and Bean, [19]. The TSP consists of finding the shortest closed path by which every city out of a set of cities is visited once and only once, [6], [14]. The transformation generates a TSP with the same number of nodes than the GTSP. Weights are modified in such a way that guarantees that an optimal tour visits all nodes that belong to the same cluster in the original problem before moving on to the next cluster.

When a cluster contains two nodes, v_{ij} and v_{ji}, the two intra-cluster arcs are assigned a cost equal to an arbitrarily large negative constant; given any vertex v_{kl} outside the cluster, the cost of arc (v_{ij}, v_{kl}) is substituted by that or arc (v_{ji}, v_{kl}) and the cost of arc (v_{ji}, v_{kl}) is substituted by that or arc (v_{ij}, v_{kl}). Because of the low cost of intra-cluster arcs, any TSP solution on the transformed graph will use the two v_{ij} and v_{ji} in succession. If v_{ji} precedes in the solution, this means edge (v_i, v_j) is entered through v_i in the GRP defined on G.

4 The MAX-MIN Ant System Algorithm

The MAX-MIN Ant System algorithm (MMAS) was proposed by Stützle and Hoos, [25], [24]. It is an improvement over the first ant-based algorithm, called Ant-System, [7], [8], and it is one of the best performing ant-based algorithms.

It was first applied to solve the TSP, [26]. Let $G = (V,E)$ be the graph associated to the TSP, where V is the set of n cities or nodes in the problem, E is the set of connections among the nodes, and d is a function that associates a cost to each element in E. To solve the problem, we consider a set of m ants cooperating to find a solution to the TSP (a tour). To each connection, (i,j), of the TSP graph, a value τ_{ij} (called pheromone) is associated with $\tau_{ij} \in (0,1]$. The pheromone allows ants to communicate among themselves, contributing in this way to the solution of the problem. At each iteration of the algorithm, each one of the m ants looks for a solution to the problem. Once all the ants have found a solution, the pheromone of the connections is updated, thus contributing to make the connections pertaining to the best solution more desirable for the ants in the next iteration. The process is repeated until the solution converges or until completing the maximum number of iterations allowed for the algorithm to be performed.

To allow each ant to build a valid solution to the problem, visiting each city once and only once, each ant has an associated data structure called the tabu list, which stores the cities that have already been visited by the ant. When the ant begins the search for a new solution, its tabu list is empty. Each time an ant visits a city, it is added to its tabu list. When it has completed the path, all the cities will be in such a list.

Each ant generates a complete tour starting at a randomly selected city and choosing the next city of its path as a function of the probabilistic state transition rule (1), which defines the probability with which ant k chooses to move from city i to city j at iteration t.

$$p_{ij}^{k}(t) = \frac{[\tau_{ij}(t)]^{\alpha}\,[\eta_{ij}(t)]^{\beta}}{\sum_{l \in N_i^k} [\tau_{lj}(t)]^{\alpha}\,[\eta_{lj}(t)]^{\beta}} \text{ if } j \in N_i^k \tag{1}$$

where $\tau_{ij}(t)$ is the pheromone associated to the connection (i,j) at time t, $\eta_{ij}(t)$ is called visibility of the connection (i,j), N_i^k is the feasible neighborhood for ant k, whereas the parameters α and β determine the relative influence of the pheromone and the visibility, respectively. For the TSP, the visibility of a connection is a fixed value equal to the inverse of the distance associated to such a connection: $\eta_{ij} = 1/d_{ij}$. The feasible neighborhood for ant k, now placed on city i, N_i^k, is the set of cities not yet visited by ant k and accessible from city i.

If $j \notin N_i^k$ we have $p_{ij}^{k}(t) = 0$.

The state transition rule (1) indicates that the ants prefer to move to cities near the present one and connected to it with arcs or edges having a high amount of pheromone.

To update the pheromone, the expression (2) is applied to all the connections in the graph.

$$\tau_{ij}(t+1) = (1-\rho)\tau_{ij}(t) + \Delta\tau_{ij}^{best} \tag{2}$$

where ρ is a parameter called evaporation rate of the pheromone, $0 < \rho \leq 1$, which determines the fraction of pheromone eliminated from each connection. It represents the effect of pheromone evaporation over time observed in natural ants. $\Delta\tau_{ij}^{best}$ is the amount of pheromone deposited on the connections belonging to the path of the best ant, whose value is given by expression (3).

$$\Delta\tau_{ij}^{best} = \begin{cases} \frac{1}{L_{best}} & \text{if } (i,j) \text{ is part of the tour of the best ant} \\ 0 & \text{otherwise} \end{cases} \tag{3}$$

with L_{best} being the length of the tour of the best ant.

Therefore, first a fraction of the pheromone associated with each connection is evaporated, which avoids an unlimited buildup of it; moreover, it represents the phenomenon observed in natural ant colonies. Next, the best ant deposits an amount of pheromone on the connections of its tour, proportional to the length of that tour.

To select the best ant we can take the iteration-best ant (the one that generated the best solution at the present iteration) or the global-best ant (the one that generated the best solution so far). Although in many problems it is sufficient to consider iteration-best, in other cases better results have been obtained by gradually increasing the frequency of the selection of global-best as iterations of the algorithm proceed, [24].

To avoid search stagnation, the MAX-MIN algorithm limits the pheromone trails to the interval $[\tau_{min}, \tau_{max}]$, been $\tau_{min} > 0$. Both values must be determined for each particular problem, the first one being more critical. In [27] it is shown why these values can be calculated in a heuristic way.

Before starting the search for a solution, the pheromone of all connections is set to the value τ_{max}, which permits a greater exploration of the search space at the beginning of the algorithm. Moreover, when we apply the update rule for the trial, the pheromone remains on the connections of the better solutions with high values, and it is reduced on the bad ones. When pheromone is updated, the values are forced to the interval indicated, so that each pheromone trail greater than τ_{max} is set to the maximum, and those lower than τ_{min} equal the minimum.

While iterations proceed, if search stagnation is detected, a re-initialization is applied, setting all the pheromone trails to τ_{max} again. Stagnation occurs when all the ants follow the same path and construct the same tour.

The algorithm is usually combined with some improvement heuristic, such as 2-opt or 3-opt, which usually improves the results. This technique is commonly used in all ant-based algorithm, as in other metaheuristics.

5 Computational Results

The solution has been coded using C language. The tests have been performed on a personal computer with Intel Centrino Core 2 Duo processor, 2.2 GHz, with 2G RAM memory and working on Linux Operating System.

Table 1 First group of test problems

PROBLEM	$\lvert E\rvert$	$\lvert V_R\rvert$	$\lvert E_R\rvert$	$\lvert E_{NR}\rvert$	OPT
GRP1	116	22	61	113	9659
GRP2	116	22	64	110	10277
GRP3	116	21	61	113	9536
GRP4	116	31	88	86	11420
GRP5	116	24	72	102	10490
GRP6	116	6	126	48	12335
GRP7	116	38	52	122	9985
GRP8	116	27	81	93	9674
GRP9	116	0	59	115	12108
GRP10	116	31	87	87	11260

Table 2 Second group of test problems

PROBLEM	$\|V\|$	$\|V_R\|$	$\|E_R\|$	$\|E_{NR}\|$	OPT
ALBA_3_1	116	44	51	123	9830
ALBA_3_2	116	48	46	128	9812
ALBA_3_3	116	57	44	130	10100
ALBA_3_4	116	46	49	125	9558
ALBA_3_5	116	43	57	117	10033
ALBA_5_1	116	15	88	86	11581
ALBA_5_2	116	16	92	82	11581
ALBA_5_3	116	17	92	82	10497
ALBA_5_4	116	25	88	86	10734
ALBA_5_5	116	14	91	83	10911

The solution has been tested on 20 GRP instances described in [3], defined on the graph of a Spanish city called Albaida. The first group of problems, labeled GRP1 to GRP10, includes 10 GRP instances generated from the graph by visually selecting the required edges. The last group includes 10 instances randomly generated from the graph of the city. Tables 1 and 2 show information about the problems. For each problem it is shown its name (first column), the number of nodes of the graph, $|V|$;

Table 3 Solution obtained by applying ants

PROBLEM	$\|W\|$	Best	%OPT	$\bar{T}(seg.)$
GRP1	144	9753	0.97	14.95
GRP2	150	10410	1.29	16.90
GRP3	143	9661	1.31	14.85
GRP4	207	11704	2.48	41.95
GRP5	168	10543	0.50	25.35
GRP6	258	12542	1.67	49.22
GRP7	142	10078	0.93	14.30
GRP8	189	9864	1.96	29.25
GRP9	118	12252	1.18	26.9
GRP10	205	11596	2.98	39.9
ALBA_3_1	146	9924	0.95	17.40
ALBA_3_2	140	9893	0.82	11.95
ALBA_3_3	130	10355	2.52	14.70
ALBA_3_4	144	9730	1.79	13.40
ALBA_3_5	157	10225	1.91	21.05
ALBA_5_1	191	11865	2.45	34.8
ALBA_5_2	200	11885	2.62	39.29
ALBA_5_3	201	10937	4.19	34.2
ALBA_5_4	201	11316	5.42	33.7
ALBA_5_5	196	11175	2.41	31.8

the number of required nodes, $|V_R|$; the number of required edges, $|E_R|$; the number of not required edges, $|E_{NR}|$; and the optimum, OPT.

Twenty independent runs were performed for each problem. The values considered for the parameters of the MMAS algorithm are: $\alpha = 1$, $\beta = 2$, $\tau_{min} = 0.01$, $\tau_{max} = 1$, $\rho = 0.1$. The Table 3 shows the results: the best solution reached for the problem, (Best); the percentaje over the optimum of the best solution, (%OPT); and the average time in seconds to reach a solution, ($\bar{T}(seg.)$). The second column shows the number of nodes of the transformed problem, ($|W|$).

We observe that the solution reached by the ants is close to the optimum.

The number of nodes of the transformed problem is proportional to the number of required edges and required nodes of the initial problem. The problem labeled GRP9 generates the transformed problem with less nodes: from an initial problem with 116 nodes, a transformed problem with 118 nodes is obtained. In this problem all the nodes in the graph are end-points of the required edges. The transformed problem corresponding to GRP6 includes more nodes than the others, because of the original problem includes more required edges.

The average time required to reach a feasible solution is less than 50 seconds for all problems.

6 Conclusions

The GRP is a NP-hard combinatorial optimization problem, and exact solution methods are available only for small problems. Therefore, the application of heuristic solutions is necessary when big problems are considered.

This paper presents the application of artificial ants to the GRP. The use of a graph transformation allows as to solve a GRP as a node-routing problem. The results obtained in the computational test are close to the best known solution for the problems considered.

References

1. Assard, A.A., Golden, B.L.: Arc routing methods and applications. In: Ball, M.G., Magnanti, T.L., Monma, C.L., Nemhauser, G.L. (eds.) Network Routing. Handbooks in Operations Research and Management Science, vol. 8, pp. 375–483. Elsevier, Amsterdam (1995)
2. Blais, M., Laporte, G.: Exact solution of the generalized routing problem through graph transformations. The J. of the Oper. Res. Soc. 54, 906–910 (2003)
3. Corberán, A., Letchford, N., Sanchís, J.M.: A cutting plane algorithm for the general routing problem. Math. Program. Ser. A 90, 291–316 (2001)
4. Corberán, A., Sanchís, J.M.: The general routing problem polyhedron: Facets from the RPP and GTSP polyhedra. Eur. J. Oper. Res. 108(3), 538–550 (1998)
5. Cornuejols, G., Fonlupt, J., Naddef, D.: The traveling salesman on a graph and some related integer polyhedra. Math. Program. 33, 1–27 (1985)

6. Lawler, E.L., Lenstra, J.K., Rinnooy Kan, A.H.G., Shmoys, D.B. (eds.): The Traveling Salesman Problem: A Guided Tour of Combinatorial Optimization. John Wiley & Sons, New York (1985)
7. Dorigo, M, Gambardella, L.M.: Ant colony system: A cooperative learning approach to the traveling salesman problem. IRIDIA. 96-05. Université Libre de Bruxelles (1996)
8. Dorigo, M., Maniezzo, V., Colorni, A.: Ant System: an atocatalytic optimizing process. Dipartamento di Electtronica e Informazione - Politecnico di Milano, Italia (1991)
9. Dror, M.: Arc Routing: Theory, Solutions and Applications. Kluwer Academic Publishers, Boston (2000)
10. Eiselt, H.A., Gendreau, M., Laporte, G.: Arc routing problems, Part I: The chinese postman problem. Oper. Res. 43(3), 231–244 (1995)
11. Eiselt, H.A., Gendreau, M., Laporte, G.: Arc routing problems, Part II: The rural postman problem. Oper. Res. 43(3), 399–414 (1995)
12. Fleischmann, B.: A cutting-plane procedure for the travelling salesman on a road network. Eur. J. Oper. Res. 21, 307–317 (1985)
13. Jansen, K.: An approximation algorithm for the general routing problem. Inf. Proc. Letts 41, 333–339 (1992)
14. Laporte, G.: The traveling salesman problem: An overview of exact and approximate algorithms. Eur. J. Oper. Res. 59(2), 231–247 (1992)
15. Laporte, G., Nobert, Y.: Generalized traveling salesman through n sets of nodes: an integer programming approach. INFOR 21, 61–75 (1983)
16. Lenstra, J.K., Rinnooy Kan, A.H.G.: On general routing problems. Networks 6, 273–380 (1976)
17. Letchford, A.N.: New inequalities for the general routing problem. Eur. J. Oper. Res. 96(2), 317–322 (1997)
18. Letchford, A.N.: The general routing polyhedron: A unifying framework. Eur. J. Oper. Res. 112(1), 122–133 (1999)
19. Noon, C.E., Bean, J.C.: An efficient transformation of the generalized traveling salesman problem. INFOR 31, 39–43 (1993)
20. Orloff, C.S.: A Fundamental Problem in Vehicle Routing. Networks 4, 35–64 (1974)
21. Orloff, C.S.: On general routing problems. Networks 6, 281–284 (1976)
22. Orloff, C.S., Caprera, D.: Reduction and solution of large scale vehicle routing problems. Transp. Sci. 19, 361–373 (1976)
23. Reinelt, G., Theis, D.O.: On the general routing polytope. Discrete Applied Mathematics 156, 368–384 (2008)
24. Stützle, T., Hoos, H.: MAX-MIN ant system. Futur. Gener. Comput. Syst. 16(8), 889–914 (2000)
25. Stützle, T., Hoos, H.: Improving the ant-system: A detailed report on the MAX-MIN ant system. AIDA-96-12. FG Intellektik, TH Darmstadt (1996)
26. Stützle, T., Hoos, H.: The MAX-MIN ant system and local search for the traveling salesman problem. In: Baeck, T., Michalewicz, Z., Yao, X. (eds.) Proc. of the IEEE Int. Conf. on Evol. Comput., pp. 309–314 (1997)
27. Stützle, T.: A short convergence proof for a class of ant colony optimization algorithms. IEEE Trans. Evol. Comput. 6(4), 358–365 (2002)
28. Toth, P., Vigo, D. (eds.): The Vehicle Routing Problem. Society for Industrial and Applied Mathematics, Philadelphia (2002)

Eliminating Datacenter Idle Power with Dynamic and Intelligent VM Relocation

Takahiro Hirofuchi, Hidemoto Nakada, Hirotaka Ogawa, Satoshi Itoh, and Satoshi Sekiguchi

Abstract. We are developing an advanced IaaS (Infrastructure-as-a-Service) datacenter management system that dynamically minimizes running physical servers depending on resource utilization. The management system periodically monitors the loading of a datacenter, and dynamically repacks virtual machines (VMs) into optimal physical servers. Live migration of VMs and the standby mode of physical servers are automatically orchestrated by a genetic algorithm (GA) engine. A preliminary experiment showed that our first prototype system correctly worked for a proof-of-concept datacenter.

1 Introduction

IaaS is an emerging cloud service that provides virtualized hardware resources for customers over the Internet. A service provider runs thousands of physical machines in a 24/7 manner; the key to success in datacenter business is to reduce running cost as much as possible, thereby achieving more competitive pricing than other providers.

Live migration of VMs is considered promising to reduce idle power of IaaS datacenters. It is possible to dynamically reduce running physical servers in response to the loading of a datacenter; in off-peak hours like early mornings, most VMs are operating at lower utilization levels, which are relocated to fewer physical servers to power off unused servers and facilities.

To the best of our knowledge, however, commercial IaaS providers do not employ live migration for dynamic repacking of VMs. They assign a fixed number of VMs to one physical server, and never change VM locations while the VMs are running. Although dynamic VM packing has been discussed in research papers, there still lacks practice and experience bridging the gap between academia and industry.

Takahiro Hirofuchi, Hidemoto Nakada, Hirotaka Ogawa, Satoshi Itoh, and Satoshi Sekiguchi
National Institute of Advanced Industrial Science and Technology (AIST),
Central 2, Umezono 1-1-1, Tsukuba, Japan 305-8568
e-mail: t.hirofuchi_at_aist.go.jp

A.P. de Leon F. de Carvalho et al. (Eds.): Distrib. Computing & Artif. Intell., AISC 79, pp. 645–648.
springerlink.com

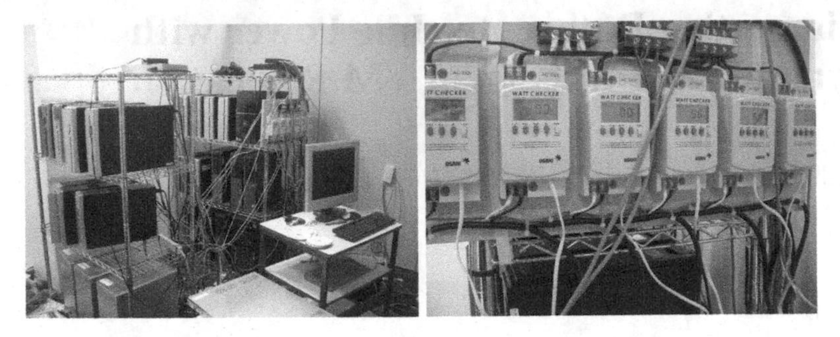

Fig. 1 Prototype Datacenter Room (left) and Power Dataloggers (right)

In our project, we present an IaaS datacenter management system that dynamically reduces/increases running physical servers depending on resource utilization. Live migration of VMs is exploited to optimize VM locations for reducing the number of running physical servers, and ACPI S3 (the standby mode) of physical hardware is utilized to eliminating idle power of unused servers. Our GA engine [2] quickly determines near-optimal VM locations, solving a multi-dimensional bin packing problem with a heuristic method; VMs, consuming different amounts of CPU, memory, and I/O resources, are packed into the smallest number of physical servers. A prototype datacenter room is built to evaluate how our management system reduces power consumption. Through this project, we aim to clarify the feasible system design and its implementation meeting reliability and scalability in production-level datacenter environments.

This paper is the work-in-progress report of our ongoing project. The overall design of our system is introduced and then a preliminary experiment is reported.

2 System Overview and Preliminary Experiment

We have developed a prototype datacenter room in which power consumption of physical servers is recorded with power dataloggers (Figure 1). All physical servers are capable of the ACPI S3 suspend/resume; suspended nodes are resumed by means of an out-of-band hardware management mechanism, Intel AMT (Active Management Technology), which is more reliable and powerful than Wake-On-LAN. Power consumption in the suspend mode is approximately only 8W (i.e., 10% of active server power).

The overview of our system is illustrated in Figure 2. The management node monitors VM's resource usage and periodically relocates VMs to optimal locations. In every 5 seconds, our GA engine calculates optimal VM locations that meet both criteria of VM performance and server power as much as possible. Then, the management node conducts relocations; which wakes up suspended server nodes if needed, starts live migration of VMs, and sends unused server nodes to sleep.

In our preliminary experiments, we launched 64 VMs on 7 physical servers. A video-on-demand streaming server was installed into each VM. When many client

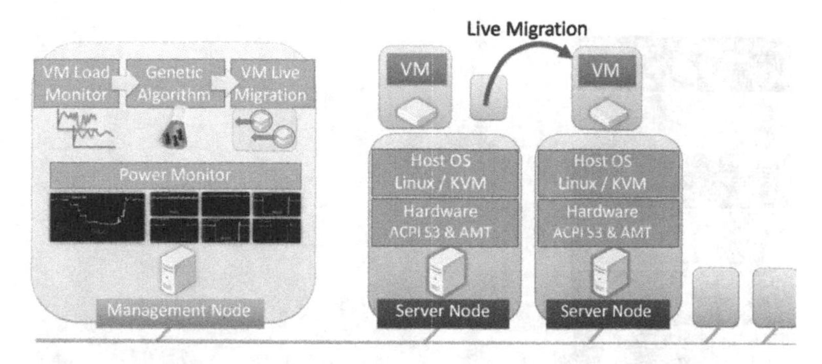

Fig. 2 System Overview

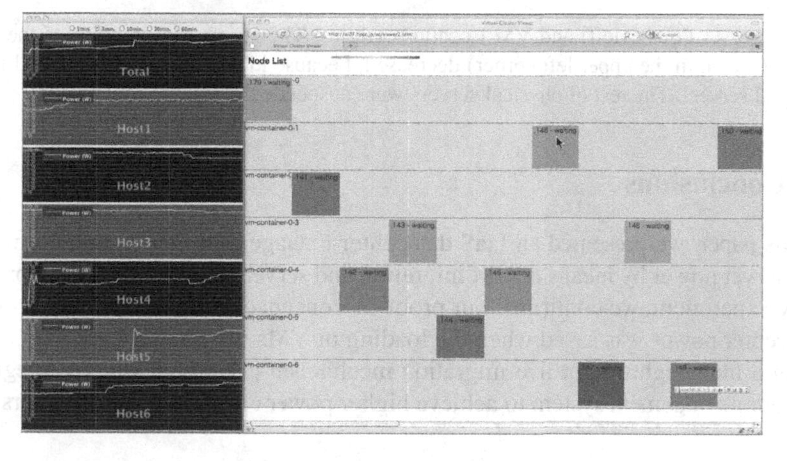

Fig. 3 Server Power (left) and VM Locations (right). The power consumption of the datacenter (shown in the upper left corner) was high, because heavily-loaded VMs were distributed to all physical servers.

users started watching videos over the Internet, VMs were heavily loaded and then the management node distributed VMs to all physical servers. Figure 3 shows our power monitors and VM location viewer at this moment. All physical servers were active, consuming 80-100W of power, respectively. The total power consumption of this datacenter was approximately 500W. As shown in Figure 4, when all client users left this streaming service, all VMs were relocated into 3 physical servers. 4 physical servers in the suspend mode were consuming 8W. The total power consumption went down to approximately 250W, only half the value of the maximum loading. During this experiment, all video streamings were smoothly played without any visible frame drops. VMs were continuously transmitting video frames while they were being migrated.

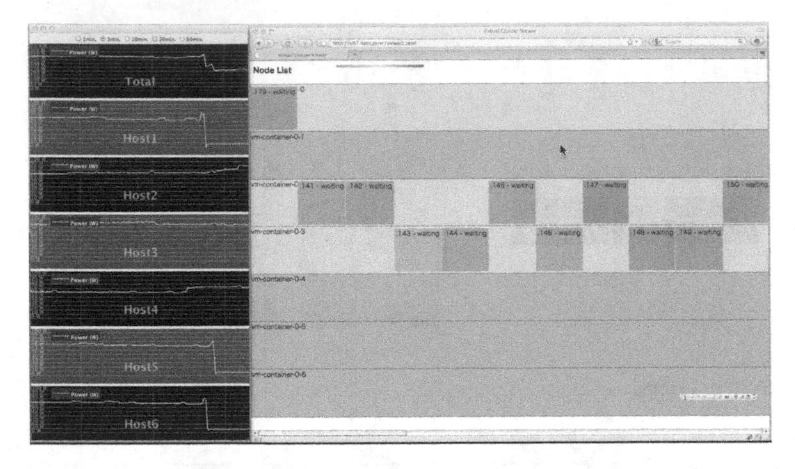

Fig. 4 Server Power (left) and VM Locations (right). The power consumption of the datacenter (shown in the upper left corner) decreased, because idle VMs were aggregated into 3 physical servers. The rest of physical servers were suspended.

3 Conclusions

In this paper, we presented an IaaS datacenter management system that eliminates idle server power by means of VM migration and server standby. Through a preliminary experiment, we confirmed our proof-of-concept datacenter correctly worked; datacenter power was saved when the loading of VMs was low. We are now developing a more lightweight live migration mechanism [1], which will be integrated into our management system to achieve higher power efficiency of datacenters.

Acknowledgements. This work was partially supported by KAKENHI (20700038) and JST/CREST ULP.

References

1. Hirofuchi, T., Nakada, H., Itoh, S., Sekiguchi, S.: Enabling instantaneous relocation of virtual machines with a lightweight vmm extension. In: Proceedings of the 10th IEEE/ACM International Symposium on Cluster, Cloud and Grid Computing (CCGrid 2010), IEEE Computer Society, Los Alamitos (2010)
2. Nakada, H., Hirofuchi, T., Ogawa, H., Itoh, S.: Toward virtual machine packing optimization based on genetic algorithm. In: Omatu, S., Rocha, M.P., Bravo, J., Fernández, F., Corchado, E., Bustillo, A., Corchado, J.M. (eds.) IWANN 2009. LNCS, vol. 5518, pp. 651–654. Springer, Heidelberg (2009)

Bee Colony System: Preciseness and Speed in Discrete Optimization

Sadegh Nourossana, H. Haj Seyyed Javadi, Hossein Erfani, and Amir Masoud Rahmani

Abstract. One of the useful patterns to create algorithms capable of solving complex problems is the foraging behavior of bees in finding food sources. In this article, a method has been presented for solving the complex problems in discrete spaces by simulation of this behavior of bees and also considering a memory for these bees. The proposed method has been successfully applied to solve the traveling salesman problem. The simulation results show the high ability of this algorithm in compare with the similar ones.

Keywords: Artificial life; Bee colony system; Discrete Optimization; Traveling salesman problem.

1 Introduction

Reaching higher quality production with less charge, to make more benefits is a basic need in the daily increasing competitions in today's world of business and industry, which implies the increasing importance of optimization. On the other hand, with the rapid expansion of the global economy, to remain in the competition field, companies have to equip their production lines and produce more up-to-date items to attract more customers. It has led to the increase of the size of optimization problems which again shows the importance of this subject. Optimization problems like scheduling problems are mostly of NP-Complete or NP-Hard class.

This work aims to explore an improvement heuristic method based on foraging behavior of bees' colony, which is used to solve the traveling salesman problem. In this method, which is called Bee Colony System (BCS), we consider a limited memory for the bees' colony, gained through the overall experience of the bees. This research is inspired by the works done by D.KARABOGA [1] who analyzed

Sadegh Nourossana, Hossein Erfani, and Amir Masoud Rahmani
Computer Engineering Department, Science and Research Branch,
Islamic Azad University, Tehran, Iran

H. Haj Seyyed Javadi
Department of Mathematics and Computer Science, Shahed University, Tehran, Iran

A.P. de Leon F. de Carvalho et al. (Eds.): Distrib. Computing & Artif. Intell., AISC 79, pp. 649–655.
springerlink.com

the foraging behavior of honey bee swarm and proposed a new algorithm simulating this behavior for solving multi-dimensional and multi-modal optimization problems, called Artificial Bee Colony (ABC).

the presented algorithm (BCS) is mentioned in section 2. The solution of the traveling salesman problem using the BCS algorithm and the results are consequently presented in sections 3 and 4. Finally, this paper ends up with conclusions.

2 Bee Colony System Algorithm

Different algorithms has already been presented inspired by the bees behavior, some of which are based on the foraging behavior of bees. These algorithms has been successfully applied to solve different problems [3-6]. This behavior was firstly biologically simulated by Yonezawa and Kikuchi [7]. Their simulation results proved the efficiency of this foraging behavior and became a stepping stone for the presentation of different algorithms based on this behavior of bees. The so called "Bee System" algorithm, which was the first one based on this behavior, was presented by Lucic and Teodorovic [8], the result of which showed the capability of this type of algorithm in solving complex problems.

In this section we propose a new algorithm based on this foraging behavior of bees. In the presented algorithm, we consider a limitation for the dance area and the number of dancing bees, in such a way that only those bees which have found better paths to the food source can enter the dance area and make dance. Each member of the colony observes all the dances and selects one as the preferred path, but memorizes all of them. This memorization, which is the distinction of the BCS algorithm, will later help the bee to improve the preferred path, in such a way that each bee tries to optimize the preferred path using its own deduction and other bees' experiences which has previously memorized while observing the dances. In case that any bee can find a path better than its own preferred path, this bee will be substituted for the dancing bee which has proposed this preferred path in the dance area, this way the preferred paths will be highly optimized. If a dancing bee has not been replaced for a specific period of time, it will leave the dance area automatically. Whenever there is free space in the dancing area, some of the scout bees will try to find new paths, and the one which finds the best path will take this free space. There are three types of bees engaged in this algorithm:

(a) *Employed Bees.* Bees which have found better results in compare with their counterparts are called employed bees. They are the only bees which dance in the dancing area. So the number of employed bees is exactly equal to the capacity of the dancing area. Besides proposing their path in the dancing area, they try to find better path near their own path too.

(b) *Recruit Bees.*Bees which try to optimize the selected path using their own deduction and other bees' experiences are called recruit bees. They do it by choosing one of the proposed paths in the dancing area as the preferred path, and try to improve it by neighborhood search.

(c) *Scout Bees.* Bees which search for new paths in search space, using only their own deduction, are called scout bees.

The BCS algorithm consists of three main steps: Attraction, Foraging, Data update.

Attraction: In this step, employed bees would be evaluated based on the quality of their individual solution. Those whose solutions are of higher quality will attract more recruit bees. At this time, each recruit bee will be attracted to one path (preferred path) and also memorizes the other paths proposed by the employed bees in the dance area. The employed bees will also observe and memorize the other proposed paths, but they will take their own path as the preferred path.

Foraging: In this step, each employed or recruit bee will try to find a better path by moving around in the neighborhood of its preferred path. The neighborhood paths, which are in accordance to the bee's deduction and memorization of other paths, are more attractive to it to move around.

Data update: In this step, the new paths of the employed and recruit bees will be compared with their preferred paths, In case that any bee has found a path better than its own preferred path, this bee will be substituted for the dancing bee which has proposed this preferred path in the dance area. Also if a preferred path has not been improved for a specific period of time, it will be omitted by replacement of the bee which has proposed this path in the dance area. Whenever a free space is produced in the dancing area by replacement of these bees, scout bees will try to find new paths, and the one which finds the best path will take this free space.

3 Applying BCS to Solve Traveling Salesman Problem

3.1 Attraction

- To solve the TSP using BCS algorithm, at first scout bees start searching to find initial solutions. It is done by locating each scout bee in one of the cities randomly. Then it will choose one city to move to, using the probability function (2), and travels to it. This action will be repeated by each scout bee to make a complete tour containing all the cities.

$$P_k(i,j) = \begin{cases} 0 & j \in l \\ \dfrac{1/\eta(i,j)}{\sum_{s=1, s \notin l}^{n} 1/\eta(i,s)} & j \notin l \end{cases} \tag{1}$$

Where $P_k(i, j)$ is the probability with which the kth scout bee moves from city i to j, $\eta(i,j)$ the distance between i and j cities, n the number of cities, and l a list of all the visited cities so far.

To evaluate the quality of each solution we consider a fitness value for each one, which is calculated as bellows:

$$Fit = \frac{1}{Tl} \tag{2}$$

Where *Fit* is the fitness value of the solution, and *Tl* the tour length of the solution.

Then the best scout bees are selected for the attraction section, whose number is equal to the capacity of the dance area. Each dancing bee will attract some recruit bees. The probability of attraction of each recruit bee by the employed bees is calculated by the Eq. (3).

$$R_k = \frac{Fit_k}{\sum_{e=1}^{n} Fit_e} \tag{3}$$

Where R_k is the probability of a recruit bee being attracted by the k-th employed bee, and Fit_k is the fitness value of k-th employed bee.

Each bee remembers the helpful information of all the proposed tours before starting the foraging stage in each step. Since all the bees keep the same information, we can consider only one memory for the whole colony. It will reduce the information redundancy which decreases the capacity of the needed memory, and consequently decreases the time of processing.

Each bee keeps the information of all tours in its mind, and considers a specific value in its memory for each path which links two cities. At the beginning of each foraging stage, the value of each path in its mind is 1. This value will be increased by the number of tours which include this path. For example if a path is included in two proposed tours, this number will be added to its initial value and the final value will become 3, 1+2=3.

To increase the speed of the program, this evaluation is only used in the first step of foraging. But in this step and all the other steps, when the bees with weak solutions are substituted by the new ones in the data update, for each bee that leaves the dancing area, the value of all paths in its tour will be decreased by 1. For example if the value of a path between two cities is 9 and one of the bees whose tour includes this path leaves the dancing area, the value of the path will be decreased once and becomes 8, 9-1=8. And for each bee that enters the dancing area, the value of all the paths included in its tour will be added by 1. For example, if the value of a path is 5 and the tours of 2 newly entered bees include this path, the value will be added twice and becomes 7, 5+1+1=7.

After the specification of the preferred path and memory of all the bees, they leave the hive and start neighborhood search to find new paths.

3.2 Foraging (Neighborhood Search)

Each preferred path which a bee takes is a complete tour which passes through all the cities. So each city is connected to two other cities which are called nearby neighborhoods of this city.

In a complete tour, each city has only two nearby neighborhoods. The decision of each bee for changing these nearby neighborhoods results in the invention of new tours which are considered as neighborhood tours of the preferred path.

In the suggestive model of the neighborhood search, each bee tries to follow its own preferred path with the probability ω, and with the probability $(1-\omega)$ tries to make better paths by changing the nearby neighborhoods of its preferred path cities. The value of ω is calculated by Eq. (4).

$$\omega = \frac{Problem\ Size - Search\ range}{Problem\ Size} \quad (4)$$

Where *Problem size* is the number of all cities of the problem, and *Search range* is a positive parameter which identifies the extension of the neighborhood searching area.

This way, each bee begins to make a new tour. It will be randomly located in a city and selects the next city by following the below rules:

(a) When a bee has decided to follow its preferred path, and none of the two nearby neighborhoods have been visited. In this case it will choose one of them randomly and moves to it.
(b) When a bee has decided to follow its preferred path, but there is only one nearby neighborhood unvisited. So it will move to this unvisited city.
(c) When a bee has decided to follow its preferred path, but both of the nearby neighborhoods have been already visited. In this case the bee will select the next city based on the probability function (5).

$$I(i,j) = \begin{cases} 0 & j \in l \\ \dfrac{[m(i,j)][1/\eta(i,j)]^{\beta}}{\sum\limits_{s=1, s\notin l}^{n} [m(i,j)][1/\eta(i,s)]^{\beta}} & j \notin l \end{cases} \quad (5)$$

Where $I(i,j)$ is Where $I(i,j)$ is the the probability with which the bee moves from city i to j, $\eta(i,j)$ the distance between i and j cities, β positive parameter, whose values determine the relative importance of memory versus heuristic information, n the number of cities, and l a list of all the visited cities so far.

(d) When a bee has decided not to follow its preferred path and choose a new nearby neighborhood, in this case it will do the same as in rule c.

4 Experimental Results

The problems EIL51, Berlin52, St70, Pr76, KroA100, Eil101, D198, Tsp225, A280, Lin318, Pcb442, Rat783 and PR1002 has been selected from TSPLIB for testing the algorithm.

To study and evaluate some of the specifications of the algorithm such as cooperation between employed bees, tradeoff between time and precision among employed bees, and the effects of the number of the recruit bees we have appointed EIL51, KroA100 among the above mentioned problems. The program is written in C# language and all the tests have been done on an Intel Core 2 Duo 2.13GHZ, 4MB Cache, 1GB Ram.

We present the results gained from the algorithm execution on some of the benchmark problems The results after 20 trials with stop condition of $n^2log_2(n)$ iteration are discussed in the below table. In all the tests, the number of recruit bees is zero, the search range 8, $\beta = 4$, Pl=20, and n is the problem size.

We compare the results gained by the BCS algorithm with that of some o ther algorithms in solving the same problems to evaluate the capability of this algorithm.

Ant colony system (ACS) [9], and bee system (BS) [6] algorithms are selected to be compared with the presented algorithm BCS. Both of ACS and BS algorithms are in swarm intelligence category and have gained satisfying results in solving traveling salesman problem. The ACS is based on the ants' behavior and their communication through pheromone trail, and the BS algorithm is based on the foraging behavior of honey bees. The results of ACS and BS are directly taken from their researchers' paper.

Figure 1 gives a clear image of the best results gained by the above mentioned 2 algorithms and BCS algorithm in solving the traveling salesman problems.

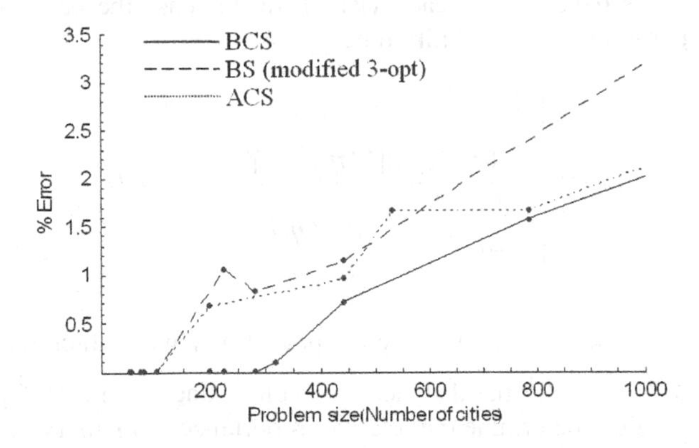

Fig. 1 Comparison between BS, ACS, and BCS best results

5 Conclusions

A novel algorithm, inspired from foraging behavior of the bee colony, has been presented in this paper. It was successfully applied to solve the traveling salesman problem. The results were perfectly satisfying, which clearly showed that the presented algorithm is far too capable than the other algorithms such as ant colony system and bee system. It implies the high ability of this algorithm in solving complex problems in discrete spaces. Easiness in implementation and quickness in getting the satisfying results are of the benefits of this algorithm.

In future, the evaluation of this algorithm in solving other problems by applying it in different subjects is one of the most important fields which can contribute to the improve-ment of the idea. Also it is suggested that by studying more carefully on the bees' behavior in nature, researcher try to optimize this algorithm or propose new ones. I hope the pre-sented algorithm can bc successfully merged with other algorithms to make more efficient hybrid algorithms too.

References

1. Basturk, B., Karaboga, D.: An Artificial Bee Colony (ABC) Algorithm for Numeric Function Optimization. In: IEEE Swarm Intelligence Symposium 2006, indianapolis, Indiana, USA (2006)
2. Chan, F.T.S., Tiwari, M.K. (eds.): Swarm Intelligence: Focus on Ant and Particle Swarm Optimization, p. 532. Itech Education and Publishing, Vienna (December 2007), ISBN 978-3-902613-09-7
3. Teodorovic, D., Dell'Orco, M.: Bee Colony Optimization - A Cooperative Learning Approach to Complex Transportation Problems. Advanced OR and AI Methods in Transportation, 51–60 (2005)
4. Lucic, P.: Modeling Transportation Problems Using Concepts of Swarm In-telligence and Soft Computing, PhD Thesis, Civil Engineering, Faculty of the Virginia Polytechnic Institute and State University (2002)
5. Luckic, P., Teodorovic, D.: Transportation Modeling: An Artificial Life Approach. In: ICTAI 2002 14th IEEE International Conference on Tools with Artificial Intelligence, pp. 216–223 (2002)
6. Lucic, P., Teodorovic, D.: Computing with Bees: Attacking Complex Trans-portation Engineering Problems. International Journal on Artificial Intelligence Tools 12(3), 375–394 (2003)
7. Yonezawa, Y., Kikuchi, T.: Ecological Algorithm for Optimal Ordering Used by Collective Honey Bee Behavior. In: 7th International Symposium on Micro Machine and Human Science, pp. 249–256 (1996)
8. Lucic, P., Teodorovic, D.: Bee system: Modeling Combinatorial Optimization Transportation Engineering Problems by Swarm Intelligence. In: Preprints of the TRISTAN IV Triennial Symposium on Transportation Analysis, Sao Miguel, Azores Islands, pp. 441–445 (2001)
9. Dorigo, M., Gambardella, L.M.: Ant colonies for the travelling salesman problem. Biosystems 43, 73–81 (1997)

An Iconic Notation for Describing the Composition between Relations of a Qualitative Model Based on Trajectories

F.J. González-Cabrera, M. Serrano-Montero, J.C. Peris-Broch,
M.T. Escrig-Monferrer, and J.V. Álvarez-Bravo

Abstract. In this paper an iconic notation for describing the composition between the relations of a new qualitative representation model based on trajectories in two dimensions is presented. This qualitative representation model represents a new intuitive approach for describing the spatiotemporal features of two mobile entities through the relations between its trajectories. In order to describe the composition between relations in terms of a transitivity operation, an iconic notation derived from the Conceptual Neighborhood Graph is provided. Finally, in order to illustrate this iconic notation, some examples of composition between relations are presented.

Keywords: Qualitative Representation model, Qualitative Spatio-temporal Reasoning, Qualitative Trajectories, Composition Table.

1 Introduction

Once a representation model about a specific system is defined, the implementation of an inference process for reasoning about the properties of this former one is a fundamental task. If the model is based on the relations between two entities a composition operation based on the transitive property can be implemented. This task can become quiet difficult depending on the complexity of the system and the granularity of the model. This last factor, that considers the level of detail of the model, determines the number of relations. If this number is large (high granularity) a suitable solution for collecting properly all the compositions between relations is needed. The most common way to deal with it is through a composition

---

F.J. González-Cabrera, M. Serrano-Montero, and J.V. Álvarez-Bravo
Dpto. de Informática, Universidad de Valladolid, Escuela Universitaria de Informática (Campus de Segovia), Pza. de Santa Eulalia 9-11, Segovia, Spain
e-mail: {fjgonzalez,mserrano,jvalvarez}@infor.uva.es

J.C. Peris-Broch and M.T. Escrig-Monferrer*
Dpto. de Lenguajes y Sistemas Informáticos, * Dpto. de Ingeniería y Ciencia de los Computadores Universidad Jaume I, Avda. Vicent Sos Baynat s/n, Castellón, Spain
e-mail: {jperis,escrigm}@.uji.es

A.P. de Leon F. de Carvalho et al. (Eds.): Distrib. Computing & Artif. Intell., AISC 79, pp. 657–664.
springerlink.com

table (Randell and Cohn 1989, Egenhofer 1991, Freksa 1992a, Freksa 1992b, Freksa & Zimmermann 1992, Zimmermann and Freksa 1993, Hemandez 1994, Zimmerman & Freksa 1996, Cohn et al. 1997, Van de Weghe 2005). A Composition table contains all the compositions between two relations, given all the relations of a specific representation model and it provides any piece of information through a simple table look-up operation. In this paper an iconic notation for implementing the composition table is presented. This iconic notation, derived from the Conceptual Neighbourhood Graph, has been proposed for representing the composition between the 48 relations of the model.

The paper is organized as follows: in section 2 the qualitative representation model is described. In this regard, the complete table of qualitative relations and its corresponding Conceptual Neighborhood Graph are presented. In section 3, the iconic notation for representing properly the composition operation under this approach is described. Finally in section 4 conclusions and some prospects are shown.

2 The Qualitative Representation Model

If anyone would have to explain how objects are moving around, may be he or she would use sentences such as: "the object is currently to my left moving toward the right" or "at this instant of time it is crossing just in front of me, moving toward the right". All these qualitative descriptions contain spatiotemporal information in terms of geometrical and temporal statements and can be used for distinguishing behaviors and making reasoning processes. In the model presented in this work the spatiotemporal features are described through the relations between the trajectories of the two moving objects. In this context:

- The moving objects are oriented points presenting two qualitative spatial dichotomies: the front-back and left-right dichotomies (Moratz 2006) (figure 1a).
- The motion of a single moving object is described in terms of a trajectory. Under this approach a trajectory is a spatiotemporal entity characterized, given an instant of time, through a rectilinear projection of its current object-face direction (figure 1b).
- The relations between two moving object trajectories arise as a result of describing qualitatively some spatial properties.
- Generalization when systems consist of more than two objects is achieved relating the objects to each others.

2.1 *The Qualitative Spatial Features*

In order to characterize a dynamical system consisting of several moving objects a set of features, expressed in term of relations between some spatial properties of the system, are proposed. These relations represent relative positions between some important entities of the system given an instant of time: the trajectories (T),

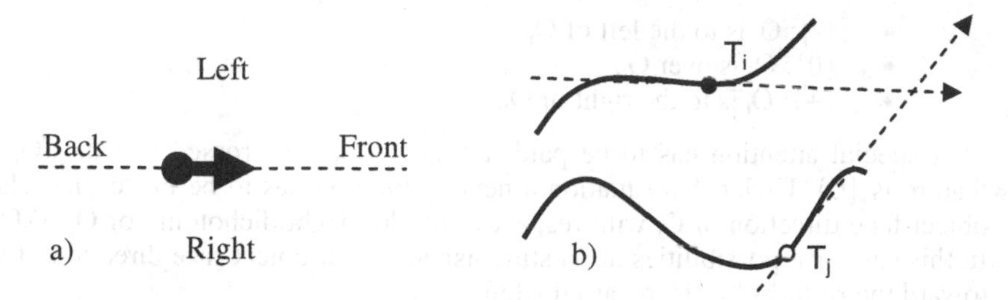

Fig. 1 a) An oriented object and its two qualitative spatial dichotomies. b) Relation between two moving objects trajectories through the rectilinear projections of its corresponding object-face directions given an instant of time.

the crossing point between two trajectories (C) and the moving objects (O). These last ones are described through its object-face directions (D), so any relation involving objects is made in terms of its front/back (FB) and left/right (LR) dichotomies. According to that, the following features are considered:

- **(T_iT_j)**. The relative position between the trajectories of two moving objects.
- **($O_iO_j^{LR}$)**. The relative position of O_i with respect to the left-right dichotomy of O_j.
- **(CO_i^{FB}) (CO_j^{FB})**. The relative position of the C with respect to the front-back dichotomy of the two moving objects.
- **($O_iO_j^{FB}$)**. The relative position of O_i with respect to the front-back dichotomy of O_j when the trajectories are superimposed.

In this context "i" and "j" are two labels denoting the moving objects (for instance. i,j∈ (1,2,3,4,…n), where "n" is the number of moving objects).

2.1.1 The Relative Position between the Trajectories of Two Moving Objects

The relative position between two trajectories of two moving objects (T_iT_j) can be depicted in terms of the angle between its rectilinear projections and the minimum distance between them. In this regard five distinctions are considered (fig.2).

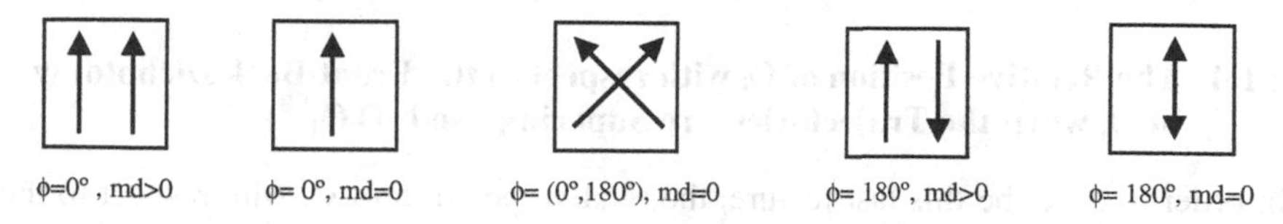

Fig. 2 The iconic notation for representing the relative position between trajectories (T_iT_j).

2.1.2 The Relative Position of O_i with respect to the Left/Right Dichotomy of O_j, ($O_iO_j^{LR}$)

The second spatial feature makes reference to the relative position of O_i with respect to the left/right dichotomy of O_j, given an instant of time. Three situations are distinguished:

- {-}: O_i is to the left of O_j.
- {0}: O_i is over O_j.
- {+}: O_i is to the right of O_j.

A special attention has to be paid if trajectories are crossed and the $(O_iO_j^{LR})$ feature is {0}. Under this situation a new distinction has to be made: the relative object-face direction of O_i with respect to the left/right dichotomy of O_j $(D_iO_j^{LR})$. In this case two possibilities are distinguished: when object-face direction of O_i is toward the right {+} or is toward the left {-}.

2.1.3 The Relative Position of C with respect to the Front/Back Dichotomy of the Two Moving Objects $\{(CO_i^{FB})\ (CO_j^{FB})\}$

The third feature describes, given an instant of time, the relative position of the crossing point between the trajectories of two moving objects with respect to the front-back dichotomy of these ones. Three possible spatial situations are distinguished between the crossing point and one moving object:

- {+}: The crossing point is in front of the moving object.
- {0}: The moving object is over the crossing point.
- {-}: The crossing point is behind the moving object.

In accordance with that, eight different configurations are identified if the relative position of the crossing point with respect to the object-face directions of the two moving objects $\{(CO_i^{FB})\ (CO_j^{FB})\}$ is considered [{+,+),{0,+},{+,-},{0,-},{-,+}, {-,0}, {-,-}, {+,0}]. For not crossed projection relations (labeled as {↑↑,↑,↑↓,↔}), $\{(CO_i^{FB})\ (CO_j^{FB})\}$ presents an ambivalence since the position of the crossing point can be located simultaneously in two different positions. So, for relations labeled as {↑↑} or {↑↓}, the crossing point is located at (+∞) or (-∞). Hence, if relation is labeled as {↑↑} its corresponding $\{(CO_i^{FB})\ (CO_j^{FB})\}$ is {+,+} or {-,-} and if it is labeled as {↑↓} then its corresponding $\{(CO_i^{FB})\ (CO_j^{FB})\}$ is {+,-} or {-,+}. For relations labeled as {↑,↔} the ambivalence is because of the crossing point is located over the two moving objects. Finally, the {0,0} relation (the two moving objects are simultaneously over the crossing point) is not considered since both objects can not be simultaneously at the same position.

2.1.4 The Relative Position of O_i with respect to the Front/Back Dichotomy of O_j when the Trajectories Are Superimposed $(O_iO_j^{FB})$

In order to describe this last feature, the relative position of O_i with respect to the front/back dichotomy of O_j a comparison is performed by superimposing the two trajectories through a rotation from O_i toward O_j with respect to the crossing point. According to that, three relative positions of are distinguished:

- {+}: O_i is in front of O_j.
- {0}: O_i is at the same position of O_j.
- {-}: O_i is behind O_j.

In connection with this, the $\{(CO_i^{FB})\ (CO_j^{FB})\}$ parameter determines how the superposition is performed. So, if $\{(CO_i^{FB})\ (CO_j^{FB})\}$ is {+,+} or {-,-}, the rotation has to be performed in such a way that the objects are finally facing in the same direction. On the other hand, if $\{(CO_i^{FB})\ (CO_j^{FB})\}$ is {+,-} or {-,+}, the rotation has to be performed in such a way that the objects are finally facing in different directions. This criterion is adopted in order to maintain the representation consistency. Finally, for crossed trajectories when $\{(CO_i^{FB})\ (CO_j^{FB})\}$ is {-,0}or {+,0}, the $(O_iO_j^{FB})$ parameter presents an ambivalence since in these situations the criterion of rotation is not well defined. Therefore, under this situation both values have to be considered.

2.2 *The Qualitative Relations*

Under this approach a qualitative relation, noted as R_{ij}, describes the relation of a moving object noted as "i" with respect to another one noted as "j". The notation for representing these qualitative relations has to combine all the considered spatial aspects in a compact way. In the fig. 3 the complete description is presented.

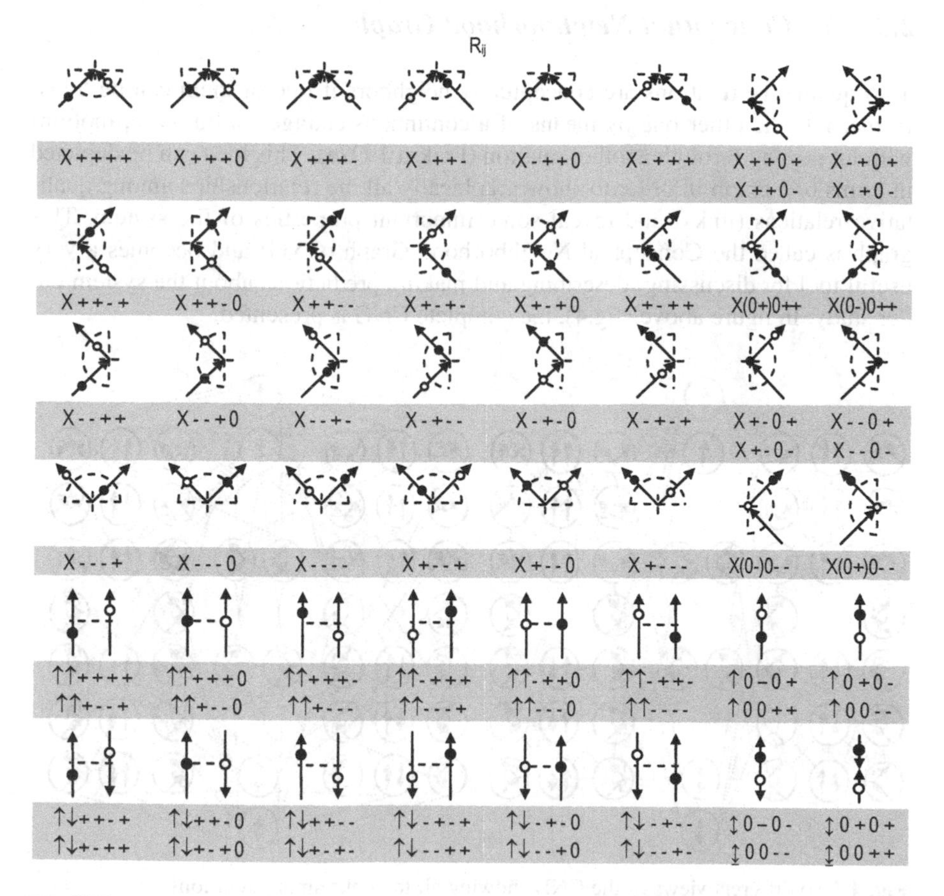

Fig. 3 The complete description and its corresponding iconic representation.

The general structure of the adopted notation is presented as follows:

$$R_{ij}=[(T_iT_j)\ (O_iO_j^{LR})\ (CO_i^{FB})\ (CO_j^{FB})\ (O_iO_j^{FB})],\ \text{if}\ (O_iO_j^{LR})\neq\{0\}$$

or

$$R_{ij}=[(T_iT_j)\ ((O_iO_j^{LR})\ (D_iO_j^{LR}))\ (CO_i^{FB})\ (CO_j^{FB})\ (O_iO_j^{FB})],\ \text{if}\ (O_iO_j^{LR})=\{0\}$$

The qualitative representation model is finally described through 48 relations. This set is achieved making all the possible combinations among the values of the chosen spatial parameters and rejecting all the inconsistencies once the physical restrictions of the system are imposed. In order to improve the comprehension of this description, a meaningful iconic representation is also included. In this graphical representation the arrows depict the projections of the object-face direction of each moving object, the hollow-dot represents the object "i" and the filled-dot represents the object "j".

2.3 The Conceptual Neighborhood Graph

Two qualitative relations are conceptually neighbors if one of them can be transformer into the other one by means of a continuous change (in our case, motion) without passing through another relation (Freksa 1992a). This idea can be depicted in terms of a graph in order to show graphically all the relationships among qualitative relations (links) and reveal some important properties of the system. This graph is called the Conceptual Neighborhood Graph (CNG) and becomes a very useful tool for discussing, describing and making predictions about the system under study. In figure above (fig.4), the complete CNG is presented.

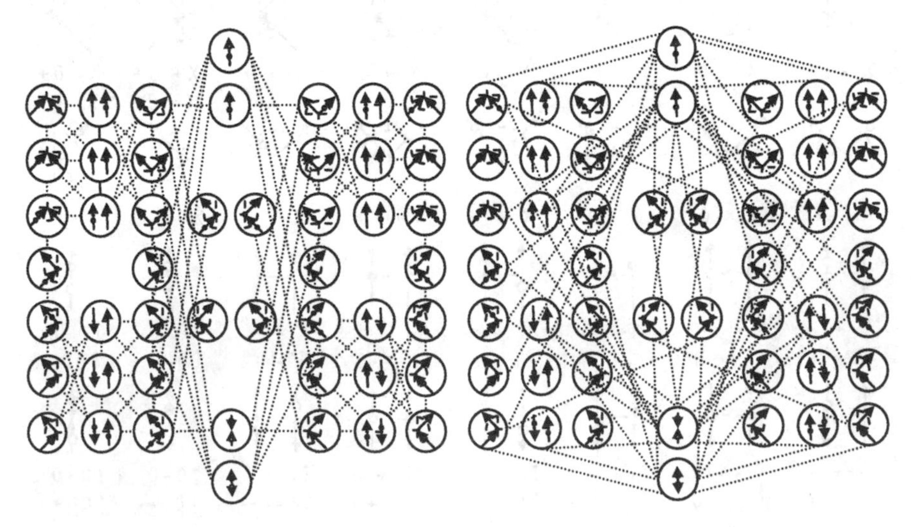

Fig. 4 Two different views of the CND showing all the links among relations

3 The Iconic Notation for Representing the Composition

In order to illustrate the importance of finding a suitable notation for describing the composition operation between relations for our model and some related examples are presented. The composition operation, also known as Basic Step Inference Process (BSIP), stands for the basic operation to obtain all the elements of the composition table given a representation model. In our case, the composition operation consists in obtaining the relation (or set of relations), noted as R_{13}, through a transitive operation between two relations, noted as R_{12} and R_{23}. In other words, if R_{12} represents how the object "1" is moving with respect to the object "2" and R_{23} represents how the object "2" is moving with respect to the object "3" then R_{13} represents all the relation of the model that, describing how the object "1" is moving with respect to "3", satisfy the transitivity properties between the relations R_{12} and R_{13}. The most common nomenclature for denoting this composition operation is:

$$R_{13}=R_{12}\otimes R_{23}.$$

Given that our model is defined through a set of 48 relations, a composition table of 48x48 elements has to be implemented. Some elements of this table are shown in figure 5a. As it can be observed, the number of relations yielded as a result of the composition operation may be quite large. For that reason a suitable notation for describing the composition table is needed. In this regard, an iconic notation derived from the CNG is proposed. This iconic notation uses the relative position of the qualitative relations inside the CNG in order to make references to any of them. In figure below (figure 5b), this iconic notation is used for representing the same composition operations presented at figure 5a.

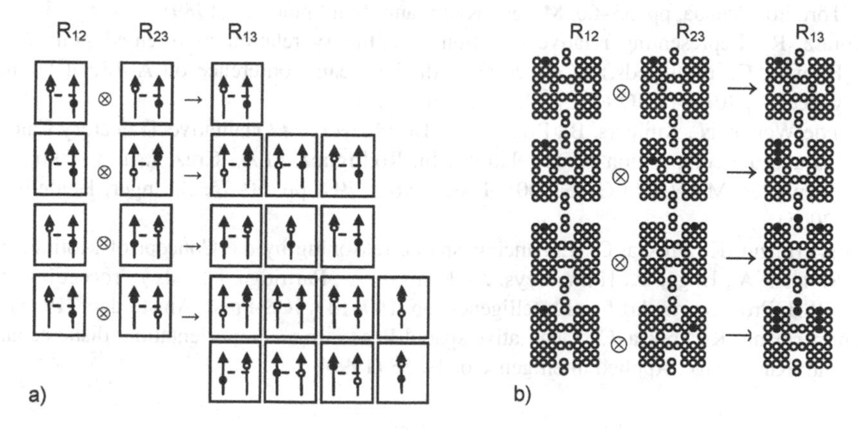

Fig. 5 a) Some examples of BSIP and b) its corresponding iconic notation

4 Conclusions and Prospects

In this work an iconic notation for representing the composition operation between the relations of a new qualitative representation model based on trajectories is presented. This notation has been derived from the Conceptual Neighborhood Graph of this model. In this regard this iconic notation is able to represent, in a compact way, any composition operation yielding a high number of relations.

As current work, the complete composition table is being generated in order to be used in cognitive processes.

References

Cohn, A.G., Bennett, B., Gooday, J., Gotts, N.M.: Representing and Reasoning with Qualitative Spatial Relations about Regions. In: Stock, O. (ed.) Spatial and Temporal Reasoning, pp. 97–134. Kluwer Academic Publishers, Dordrecht (1997)

Egenhofer, M.: Reasoning about Binary Topological Relations. In: Günther, O., Schek, H.-J. (eds.) SSD 1991. LNCS, vol. 525, pp. 143–160. Springer, Heidelberg (1991)

Freksa, C.: Temporal Reasoning Based on Semi-Intervals. Artificial Intelligence 54, 199–227 (1992a)

Freksa, C.: Using orientation information for qualitative spatial reasoning. In: Frank, M.A., Campari, I., Formentini, U. (eds.) Theories and Methods of Spatio-temporal Reasoning in Geographyc Space, pp. 162–178. Springer, Berlin (1992b)

Freksa, C., Zimmermann, K.: On the utilizacion of spatial structures for cognitively plausible and efficient reasoning. In: IEEE international Conference on Systems, Man and Cybernetics, pp. 261–266. IEEE press, Los Alamitos (1992)

Hernández, D.: Qualitative Representation of Spatial Knowledge. LNCS, vol. 804. Springer, Heidelberg (1994)

Randell, D.A., Cohn, A.G.: Modelling Topological and Metrical Properties of Physical Processes. In: Brachman, R., Levesque, H., Reiter, R. (eds.) Proceedings of the 1st International Conference on Principles of Knowledge Representation and Reasoning (KR), Toronto, Canada, pp. 55–66. Morgan Kaufmann, San Francisco (1989)

Moratz, R.: Representing relative direction as a binary relation of oriented points. In: Brewka, G., et al. (eds.) ECAI 2006: 17th European Conference on Artificial Intelligence, pp. 407–411. IOS Press, Amsterdam (2006)

Van de Weghe, N., Kuijpers, B., Bogaert, P., De Maeyer, P.: Qualitative Trajectory Calculus and the composition of its relations. In: Rodríguez, M.A., Cruz, I., Levashkin, S., Egenhofer, M.J. (eds.) GeoS 2005. LNCS, vol. 3799, pp. 60–76. Springer, Heidelberg (2005)

Zimmermann, K., Freksa, C.: Enhancing Spatial reasoning by the Concept of Motion. In: Sloman, A., Hogg, D., Humphreys, A., Ramsay, A., Partridge, D. (eds.) Proceedings of AISB. Prospects in Artificial Intelligence, pp. 140–147. IOS-Press, Amsterdam (1993)

Zimmermann, K., Freksa, C.: Qualitative spatial Reasoning using orientation, distance and path Knowledge. Applied intelligence 6, 49–58 (1996)

Speaker Adaptation and Speech-Spectral Deformation

Yoshinao Shiraki

Abstract. We study the relation between a spectral deformation in speech processing and a geometrical deformation theory. We show that topological field theory yields the systematic treatment of these two methods. Some of the examples and the application to speech-spectra of classical mathematical ideas are discussed.

Keywords: Speech-Spectrum, Speaker Adaptation, Geometric Deformation.

1 Introduction

In this note, we study the relation between a speaker adaptation method[5, 7] and a geometrical deformation theory[2] including a sound-morphing scheme[3]. Here, we consider the speaker adaptation method to be VFS (Vector Field Smoothing, refer to Ohkura [5]). Given a spectral-codebook to be adapted to a new speaker and an input-speech of the speaker, VFS consists of a 2-step procedure as follows.

Step1-Local procedure : Calculate "local correspondence" between the codebook and the input-speech in each phoneme.
Step2-Global procedure : Modify "local correspondence" properly based on "global structure" which is assumed to lie common in the codebook and the input-speech.

The procedure of VFS for codebook adaptation is considered equivalent to the piece-wise-linear adaptation method [7] originated in Shikano's work[6]. In an actual procedure of VFS, getting the difference-spectrum in each cluster corresponds to calculating "local correspondence" and smoothness of the codebook corresponds to common "global structure" respectively. The adaptation method is hereafter referred to as VFS if it has the 2-step procedure mentioned above. One of the most important problems in VFS is how to get a nice combination between "local correspondence" and "global structure". We consider the problem above to be the study of differentiable manifolds and differentiable maps. Then two manifolds are considered equivalent if they are diffeomorphic: there exists a differentiable map from one to the other with a differentiable inverse.

Yoshinao Shiraki
Toho University
e-mail: shiraki@is.sci.toho-u.ac.jp

A.P. de Leon F. de Carvalho et al. (Eds.): Distrib. Computing & Artif. Intell., AISC 79, pp. 665–672.
springerlink.com

In Sect.2, we briefly review parts of topological field theory. In Sect.3, we summarize the relation between algebraic structure and the local operation in geometry. In Sect.4, some applications to deformations of speech-spectra are referenced and some constructive algorithms for the deformation are also proposed.

2 Topological Field Theory: TFT

Let us briefly review the basic notions of topological field theory (hereafter we write TFT) and Morse theory according to Fukaya[2]. These theories relate the data derived from the local properties of a manifold and the global data defined by considering of a whole manifold. We shall begin with a given n-dimensional manifold M, and the following is a rough sketch of procedure of TFT.

Step1-Choice of structure : Choose a proper structure on M and so forth. For example, a Riemannian metric, a Morse function.
Step2-Giving partial differential equation : Using the structure above, give a partial differential equation (hereafter we write PDE) on M.
Step3-Definition of numerical number : Define the numerical number by means of counting the number of the elements of the solution space of the PDE above.
Step4-Topological invariance check : Show that the numerical number is independent of the structure chosen in *Step1*. That is, check its topological invariance.

Where a Morse function f on M is a smooth function such that the Hessian of f is nonsingular at every critical point, and it's known that every smooth manifold M possesses a Morse function (see e.g. Theorem 2.5 in [4]).

Now we refer to useful memoranda for the procedure of TFT as follows.

Memo : In many cases, the number given in *Step3* is not a topological invariant. In such a case, we should define a group algebraically using the numerical number and then show its topological invariance.

From the view point of TFT mentioned above, we can regard Morse theory as an example of TFT. In Morse theory[4], we use a space of maps from 1-dimension to M as the space of PDE in *Step2*. Therefore, the differential equation considered in Morse theory is an ordinary differential equation (ODE).

Let M be an n-dimensional closed orientable manifold and consider a Morse function f on M. A Morse function and a Riemannian metric are the structure chosen in *Step1*.

In Morse theory, we consider the following ODE

$$\frac{d}{dt}\phi(t) = -\nabla_{\phi(t)} f, \tag{1}$$

where ϕ is a curve of $\phi : \mathbb{R} \to M$. $\mathbb{R}$ is not compact, so it seems natural that we consider the boundary condition to construct a solution space of equation (1). For this purpose, we use critical points of the Morse function. A set of critical points is given by the following equation:

$$Cr(M,f) := \{p \in M | df(p) = 0\}. \tag{2}$$

Because the function f is a Morse function, the set $Cr(M,f)$ is finite. For $p \in Cr(M,f)$, let $\mu(p)$ be the Morse index of p. Namely, $\mu(p)$ is the number of negative-eigenvalue of Hessian. Using this as the boundary condition, we define $\mathcal{M}(M,f:p,q)$ as follows for two critical points $p,q \in Cr(M,f)$

$$\mathcal{M}(M,f:p,q) := \left\{ l : \mathbb{R} \to M \;\middle|\; \begin{array}{l} \frac{dl}{dt} = -\nabla f \\ \lim_{t \to -\infty} \phi(t) = p \\ \lim_{t \to \infty} \phi(t) = q \end{array} \right\}. \tag{3}$$

According to *Step3* in **2.1** we count the cardinality of the set $\mathcal{M}(M,f:p,q)$. In general, to count the cardinality of set, the dimension of the set must be 0-dimension. For almost every Morse function f, $\mathcal{M}(M,f:p,q)$ is a manifold and we can obtain its dimension as follows:

$$\dim \mathcal{M}(M,f:p,q) = \mu(p) - \mu(q). \tag{4}$$

$\mathbb{R}$ has an action on $\mathcal{M}(M,f:p,q)$ as $\phi(t+s) = \phi(t) \circ \phi(s)$, so we consider the quotient space $\bar{\mathcal{M}}(M,f:p,q)$ with respect to this action. We can count the number of elements in the quotient space if its dimension is zero. The dimension is calculated as follows

$$\dim \bar{\mathcal{M}}(M,f:p,q) = \mu(p) - \mu(q) - 1. \tag{5}$$

In this case, however, because the set of critical points depends on the choice of a Morse function, the number of the solution space is not a topological invariant.

So we do algebraic construction according to *Memo* in **2.1**. Let define the graded Abelian group $C_\bullet(M;f)$ as follows:

$$C_k(M;f) := \bigoplus_{\substack{p \in Cr(M,f) \\ \mu(p)=k}} \mathbb{Z} \cdot [p], \tag{6}$$

where the right-hand side is the free module on $\mathbb{Z}$ in which the number of generating elements is equal to the number of critical points of index k. We define a map $\partial_k : C_k(M;f) \to C_{k-1}(M;f)$ on the group as follows:

$$\partial_k([p]) := \sum_{\substack{q \in Cr(M,f) \\ \mu(q)=k-1}} \sharp \bar{\mathcal{M}}(M,f:p,q)[q], \tag{7}$$

where $\sharp$ means counting the number; more precisely speaking, consider $\bar{\mathcal{M}}(M,f:p,q)[p]$ to be a 0-dimensional oriented manifold and $\sharp$ means counting the number of signed number of the manifold.

The formulation above of chain complex is due to Witten[9]. In this construction, the following theorem holds.

Theorem 1. *$(C_\bullet(M;f),\partial)$ is a chain complex. Namely, $\partial_{k-1}\circ\partial_k=0$. Furthermore, homology group of $(C_\bullet(M;f),\partial)$ is isomorphic to homology group of M: $H_*((C_\bullet(M;f),\partial))\cong H_*(M)$.*

(With respect to chain complex and homology group, and see the proof e.g. in [1]). Smale used *Theorem* 1 in handlebody approach to Morse theory in order to solve the generalized Poincarè's conjecture in higher dimension [8]. Floer used this construction in order to define the infinite dimensional homology in Morse theory[1]. The complex of this construction type is called Witten's complex. In this note, we construct the complex of deformation of speech-spectra in Sect.4 following Witten.

3 Generalized Poincarè's Conjecture

In this section, we describe briefly the solution of the generalized Poincarè's conjecture in higher dimension by Smale [8]. Hereafter we write the generalized Poincarè's conjecture as GPC. This solution has a significant meaning when we try to combine actually the local and the global data of a manifold. Here we state Whitney's cancellation theorem of intersections. This theorem plays a central role in differential topology in higher dimensions.

Whitney's cancellation theorem of intersections. *Let M be an n-dimensional closed simply-connected manifold ($n\geq 5$), X_1,X_2 be orientable sub-manifold of M such that* $\dim M=\dim X_1+\dim X_2$ *and $s=[X_1]\cdot[X_2]$ be their intersection number of them ($[z]$ is a homology class of cycle z). Then, by means of isotopic-deformation we can make the intersection number of X_1,X_2 to be $|s|$.*

Before we describe the cancellation of critical points we refer to GPC itself and give a rough sketch of the solution of GPC as follows. The sketch gives us clearly an actual process of the deformations. From this sketch, we can consider the solution of GPC in higher dimensions to be a certain kind of TFT.

GPC. *Every closed n-manifold which has the homotopy type of n-sphere S^n (called homotopy sphere) is homeomorphic to the n-sphere S^n.*

Sketch of solution of GPC in higher dimensions

Step1 : Put a Morse function f on homotopy sphere M.
Step2 : Put the gradient flow $\phi(t)$ of $-\nabla f$ between critical points of the Morse function f defined above.
Step3 : Define the numerical number of the orbits based on the information of homology $H_*(M)$.
Step4 : Using Whitney's cancellation theorem of intersections, convert the algebraic number defined in *Step3* into the corresponding geometrical number.
Step5 : Using the number given in *Step4*, cancel a pair of critical points gradually until only two critical points remain on M. (This manifold is homeomorphic to S^n.)

In this subsection, we describe how to cancel a pair of critical points from the standpoint of TFT while Smale used cancellation of handles in his handlebody approach

[8]. Note that defining the number in *Step3* of **3.1** is analogous to Witten's complex. In fact, $\sharp\bar{\mathscr{M}}(M, f: p, q)$ defined in **2.2** is the algebraic number of the orbit of the gradient flow $\phi(t)$ connecting critical points p, q. We can deform a Morse function f isotopically so that the deformed Morse function $\tilde{f}$ has critical points just the two critical points having been canceled, if the following two conditions are satisfied for two critical points p, q of the Morse function f.

Cond1 :$\mu(p) = \mu(q) + 1$
Cond2 :There exists just one flow $\phi(t)$ connecting p and q.

These two conditions are not necessary and sufficient for cancellation of critical points p, q. So let us define a stable manifold S_p and an unstable manifold U_p for a critical point p of an n-dimensional manifold M as follows:

$$S_p = \left\{ x \in M \mid \lim_{t \longrightarrow \infty} \phi(t) = p \right\}, \tag{8}$$

$$U_p = \left\{ x \in M \mid \lim_{t \longrightarrow -\infty} \phi(t) = p \right\}. \tag{9}$$

The dimensions of S_p and U_p are $n - k$ and k, respectively. In fact, TFT formulation uses these submanifolds in the form of equation (3), and the following equation holds:

$$\mathscr{M}(M, f: p, q) = S_q \cap U_p. \tag{10}$$

Hereafter we assume Cond1: $\mu(p) = \mu(q) + 1$ for two critical points p, q. Let $c \in \mathbb{R}$ be such that $f(q) < c < f(p)$, and denote a neighborhood $N := f^{-1}(c)$. Then S_q and U_p intersect N transversally [8]. We define $S_q(c) := N \cap S_q, U_p(c) := N \cap U_p$ and $\dim U_p(c) = k - 1, \dim S_q(c) = n - k$. Therefore, if $S_q(c)$ and $U_p(c)$ intersect transversally, their intersection number is finite. If the condition *Cond2* holds, $S_q(c)$ and $U_p(c)$ intersect at only one point. This condition is geometrical. Now we refer to algebraic condition corresponding to the geometrical one as follows.

Intersection number of S_q and U_p. With respect to the intersection number in a neighborhood N of homology class defined by $[U_p(c)]$ and $[S_q(c)]$, the following equation holds:

$$H_0(M) \ni [U_p(c)] \cdot [S_q(c)] = \pm 1. \tag{11}$$

Cancellation theorem of critical points. *Let p, q be a pair of critical points satisfying the* **Cond1** *and the equation (11) and M be an n-dimensional simply-connected manifold ($n \geq 6$). Then, we can deform a Morse function isotopically to cancel two critical points p, q.*

4 TFT on Speech-Spectra

In this section, we try to construct TFT on speech-spectra following some examples of TFT: Morse theory and Witten's complex. We use Morse theory as the basis of the construction and also use a set of critical points of Morse function as the boundary conditions. Hereafter let a manifold M be orientable.

4.1 Witten's Complex Based on Orientation

We set a Morse function f on a manifold M so that $S_q(c)$ and $U_p(c)$ intersect transversally for any critical points. The gradient flow $\phi(t)$ is said to be of *Morse-Smale type* if for any two critical points p,q the stable and unstable manifolds S_q and U_p intersect transversally. We define Cr_k as follows:

$$Cr_k := \{p \in Cr(M,f) | \mu(p) = k\}, \tag{12}$$

and let the linear space C_k whose basis are Cr_k be defined as follows:

$$C_k := \{ \sum_{p \in Cr_k} a_p[p] | a_p \in \mathbb{R}\}. \tag{13}$$

We define the boundary operator ∂ based on the two orientations defined on an image of the gradient flow $\phi(t)$ as follows:

Procedeure of Witten's complex based on orientations

Step1 : Determine the orientation of a manifold M.
Step2 : For each critical point $p \in Cr$ determine arbitrarily the orientation of the unstable manifold U_p.
Step3 : Determine the orientation of the stable manifold S_p so that the orientation, which was determined from the intersection of the orientations of S_p and U_p, at p is $+1$, because S_p intersects U_p transversally.
Step4 : For $p,q \in Cr$ determine the orientation of $S_q \cap U_p$ so that this orientation is equal to that of the intersection of the orientations of S_q and U_p.
Step5 : When $\mu(p) - \mu(q) = 1$ holds, the following equation holds:

$$S_q \cap U_p = \cup \gamma_i(p,q), \tag{14}$$

where $\gamma_i(p,q)$ is an image of integral curve of $\phi(t)$ and is homeomorphic to $\mathbb{R}$. Therefore, the orientation determined on $S_q \cap U_p$ determines the orientation of each element $\gamma_i(p,q)$. Note that $\gamma_i(p,q)$ has its own orientation preserving the gradient flow $\phi(t)$.
Step6 : Assign a number $+1$ or -1 to $\nu(\gamma_i(p,q))$ according to whether the two orientations of $\gamma_i(p,q)$ coincide: one orientation of $\gamma_i(p,q)$ is determined by *Step4* and another is determined by $\gamma_i(p,q)$ itself.

When $\mu(p) - \mu(q) = 1$, using $\nu(\gamma_i(p,q))$ defined in *Step6* let us define

$$n(p,q) := \sum_i \nu(\gamma_i(p,q)). \tag{15}$$

We define the boundary operation $\partial : C_k \to C_{k-1}$ as follows:

$$\partial(\sum_{p\in Cr_k} a_p[p]) := \sum_{p\in Cr_k} a_p\partial[p], \tag{16}$$

$$\partial[p] := \sum_{q\in Cr_{k-1}} n(p,q)[q]. \tag{17}$$

In this construction, the following theorem holds.

Theorem 2. (C_*, ∂) defined above is a chain complex. Namely, $\partial_{k-1}\circ\partial_k = 0$. (See the proof e.g. [1])

4.2 *Examples*

The followings are some examples of Witten's complex for 1- and 2-manifold. We omit the case of a manifold with boundary.

4.2.1 1-Dimensional Case

Let M^1 be a 1-dimensional speech-spectrum envelope in log-scale. After normalizing M^1 properly, let f be a height function of M^1. Then, f is a Morse function. In fact, the Hessians at peak and bottom points are positive or negative definite, so all critical points are nondegenerate and the Morse index is 0 or 1.

In this case, there is a 1-dimensional homology group $H_1(M^1)$ corresponding to peaks, and each element of the homology is not independent. That is, the homology group is not changed if the number of peaks (or bottom) is increased or decreased by deformations of spectrum. This fact means that the homology group is homotopy equivalent and also holds in the case of 0 dimensional homology group corresponding to bottoms. We consider the case of M^1 having two peaks. We denote two bottoms and peaks $p_1^0, p_2^0, p_1^1, p_2^1$. In this case each Morse index is $0,0,1,1$ and the calculation of Witten's complex is as follows:

$$\partial p_1^1 = (+1)p_1^0 + (-1)p_2^0 = p_1^0 - p_2^0, \tag{18}$$

$$\partial p_2^1 = (+1)p_2^0 + (-1)p_1^0 = p_2^0 - p_1^0. \tag{19}$$

4.2.2 2-Dimensional Case

Let M^2 be a 2-dimensional power spectrum in log-scale. After normalizing M^2 properly, let f be a height function of M^2. Then f is a Morse function. In fact, the Hessians at peak and bottom points are positive or negative definite, so all critical points are nondegenerate and the Morse index is 0 or 1 or 2.

Let us consider the case of M^2 having two peaks. We denote two bottoms and peaks p^0, p^1, p_1^2, p_2^2. In this case each Morse index is $0,1,2,2$ and the calculation of Witten's complex is as follows:

$$\partial p_1^2 = (-1)p^1 = -p^1,\ \ \partial p_2^2 = (+1)p^1 = p^1, \tag{20}$$

$$\partial : C_1 \to C_0,\ \ \partial p^1 = (+1+(-1))p^0 = 0. \tag{21}$$

5 Conclusion

We described the relation between a speaker adaptation method such as VFS and a geometrical deromation. We showed that TFT yields the systematic treatment of these two methods. A kind of complex algebraically combines the local and the global information of a speech-spectrum. Consequently, it gave us numerical and algebraic invariants of speech-spectra. Some of the examples and the application to speech-spectra were also given.

References

1. Floer, A.: Witten's complex and infinite dimensional Morse theory. J. Diff. Geom. 30, 207–221 (1989)
2. Fukaya, K.: Floer homology of connected sum of homology 3-spheres. Topology 35(1), 89–136 (1996)
3. Ho, C.H., Rentzos, D., Vaseghi, S.: Formant Model estimation and transformation for Voice Morphing. In: Proc. of ICSLP 2002, pp. 2149–2152 (2002)
4. Milnor, J.W.: Lectures on the h-cobordism theorem. Math Notes 1. Princeton University Press, Princeton (1965)
5. Ohkura, K., Sugiyama, M., Sagayama, S.: Speaker adaptation based on transfer vector field smoothing method with continuous mixture density HMMs. IEICE Trans. J76-D-II(12), 2469–2476 (1993)
6. Shikano, K., Lee, K.-F., Reddy, R.: Speaker adaptation through vector quantization. In: Proc. of ICASSP 1986, vol. 49(5), pp. 2643–2646 (1986)
7. Shiraki, Y., Honda, M.: Speaker adaptation algorithms based on piece-wise moving adaptive segment quantization method. In: Proc. of ICASSP 1990, vol. S12(5), pp. 657–660 (1990)
8. Smale, S.: Generalized Poincarè's conjecture in dimensions greater than four. Ann. Math. 74, 391–406 (1961)
9. Witten, E.: Supersymmetry and Morse theory. J. Diff. Geom. 17, 661–692 (1982)

A Hybrid Parallel Approach for 3D Reconstruction of Cellular Specimens

M. Laura da Silva, Javier Roca-Piera, and José Jesús Fernández

Abstract. Electron tomography combines the acquisition of projection images through electronic microscope and techniques of tomographic reconstruction to allow structure determination of complex biological specimens. This kind of applications requires an extensive use of computational resources and considerable processing time because high resolution 3D reconstructions are demanded. The new tendency of high performance computing heads for hierarchical computational systems, where several shared memory nodes with multi-core CPUs are connected. In this work, we propose a hybrid parallel implementation for tomographic reconstruction of cellular specimens. Our results show that the balanced and adaptative algorithm allows an ideal speedup factor when large datasets are used.

Keywords: parallel programming, heterogeneous sytems, load balancing.

1 Introduction

The study of 3D structure of cellular specimens is essential for understanding the cellular role played by the specimen in the environment where it is located [1]. The electron microscope allows us to tilt the specimen around one or more axes and to take views from different directions collecting the projection images in digital format. Electron Tomography (ET) makes possible to determine the 3D structure of biological samples from 2D projection images obtained by electron microscope[2]. Furthermore, because of the resolution needs, ET of complex biological specimens requires large projection images. So, ET requires an extensive use of computational resources and considerable processing time to allow the 3D structure of cellular specimens [3]. High performace computing (HPC) has been widely investigated for

M. Laura da Silva and Javier Roca-Piera
Dpt. Computer Architecture, University of Almería, 04120 Almería, Spain
e-mail: mlauradsh@ual.es, jroca@ual.es

José Jesús Fernández
National Center for Biotechnology (CSIC), Cantoblanco, 28049 Madrid, Spain
e-mail: jjfdez@ual.es

A.P. de Leon F. de Carvalho et al. (Eds.): Distrib. Computing & Artif. Intell., AISC 79, pp. 673–680.
springerlink.com

many years in the field of ET, and [3, 4, 5] show that HPC allows determination of 3D structure of large volumes in reasonable computation time.

On the other hand, Moore's law estimated that the number of transistors that can be placed on an integrated circuit would double approximately every two years [6, 7]. Given the actual physical limitation of this prediction, new types of architectures begin to appear getting less computing time. Nowadays, supercomputers are based on hierarchical computer systems which consist on several shared memory multicore nodes interconnected. Therefore the parallelism of this new generation of computing systems must be exploited at two levels: one level of parallelism distributed among interconnected nodes and a second level of parallelism shared within the node itself [8, 9, 10].

In this paper, we propose a hybrid parallel implementation, BAHPTomo, that exploits the parallelism of the heterogeneous architectures for 3D reconstruction of cellular specimens. This program has a testing step to obtain the executing times of each node. We show a method to obtain a balanced static distribution of workload in Sec. 3.1.1 and a method to do a optimal choice of processors in Sec. 3.1.2. Both methods use the times of the testing step as benchmark. In order to evaluate the eficiency of our approach, we have implemented another hybrid program, HPTomo, that does not carry out the testing step and it considers the same number of processors in each node. The comparison of these algorithms leads us to affirm that the balanced distribution of workload and the optimal choice of processors determine the goodness in the execution times obtained. In fact, BAHPTomo achieves a speedup nearly at 30 when large datasets are used.

2 Electron Tomography

Tomography refers to the cross-sectional imaging of an object viewed from different angles. The electron tomography (ET) consists on the three dimensional (3D) reconstruction of a object from the projected two-dimensional slices which were obtained through the electron microscope. The biological specimen is placed inside the electron microscope, it is tilted over a limited range and electron beams will cross the specimen resulting a projection image with the same object area (see Fig. 1). The specimen is tilted typically from -70° to $+70^\circ$, at small tilt increments (1°–2°). These projection images will be acquired using the so-called single-axis tilt geometry and they will be recorded for each tilt angle via usually in CCD cameras.

In the field of ET these projection images are known as sinograms. Under the assumption that projection images represent the amount of mass density encountered by imaging rays, the reconstruction method known as Weighted BackProjection (WBP) simply distributes the known specimen mass present in projection images evenly over the reconstruction volume. When this process is repeated for a series of projection images recorded at different tilt angles, backprojection rays from the different images intersect and reinforce each other at the points where mass is found in the original structure.

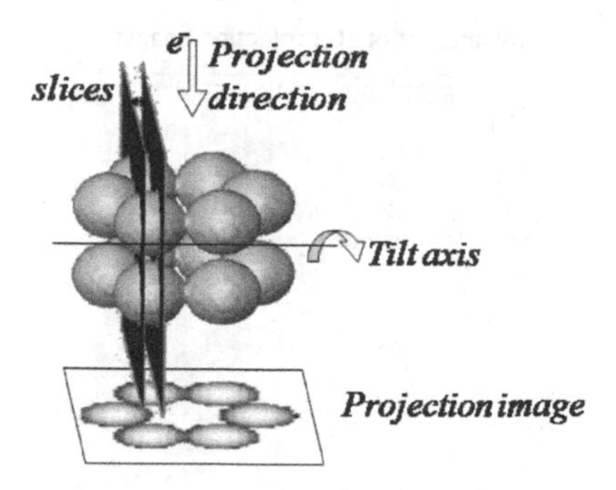

Fig. 1 Acquisition of a 2D projection image while the object is tilted around an axis.

3 New Tendencies of High Performance Computing and ET

Parallel computing has been widely investigated for many years as a means to provide high-performance computational facilities for large-scale and grand-challenge applications [3, 11, 12]. HPC addresses the computational requirements of different applications by means of the use of parallel computing on supercomputers or networks of workstations, sophisticated code optimization techniques, intelligent use of the hierarchical systems in the computers and awareness of communication latencies. In ET, the reconstruction files are usually large and, as a consequence, the processing time needed is considerable. Parallelization strategies with data decomposition provide solutions to this kind of problem. The single-axis tilt geometry in ET involves the application of a computational model widely used in parallel computing known as SPMD (single-program multiple-data). In this model, all nodes of the parallel system run the same program for different data subdomains. In ET, the SPMD model consists on the decomposition of the volume in subsets of 2D slices which will be distributed among different nodes. This computational model led us to implement different strategies based on MPI parallel master-slave paradigm to study the tradeoff between distributed load and the number of nodes that perform the processing in distributed systems [11].

On the other hand, given the actual physical limitation of this Moore's prediction [7], new types of architectures begin to appear getting less computing time. The new architectures are based on a hierarchical computer system consisting of a distributed memory system where each node is a shared memory system with several cores and different architectural features. The SPMD parallel computation model and the new architectures lead us now to study the data parallelism at two levels: one level of parallelism distributed among the interconnected nodes and a second level of parallelism shared within the node itself (see Fig. 2). Therefore in this case, the SPMD model assumes that the different data subdomains will be distributed among the nodes and they will be again distributed within each node assigning the same workload to each core (each thread) of the shared memory system. We can observe this fact in Fig. 2.

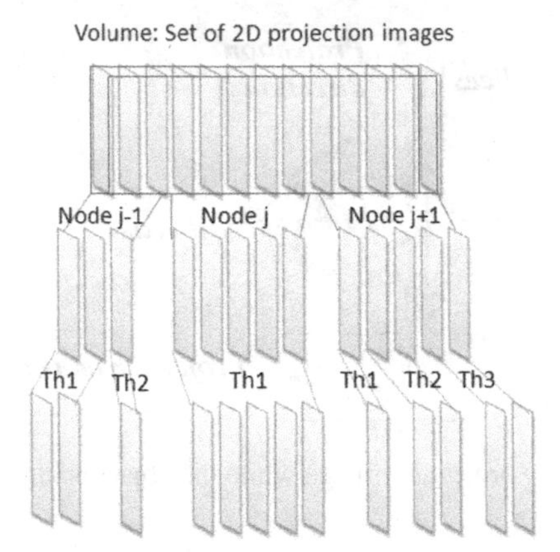

Fig. 2 Decomposition of the global 3D problem into multiple, independent reconstruction problems of slabs (i.e. subsets) of slices that are assigned to different nodes in the parallel computer. Each column represent a 2D slice orthogonal to the tilt axis of the volume to be reconstructed.

3.1 Hybrid Approaches for 3D Reconstruction of Cellular Specimens

In this paper, we propose a parallel algorithm with centralized load balancing for the 3D reconstruction of cellular specimens, BAHPTomo (Balancing Adaptative algorithm for Heterogeneous Parallel systems in Tomography). Our approach has two steps. During the first step, the program evaluates the performance of each node in the distributed system. To this end, node 0 sends the same sinogram to each node using MPI and each node creates a thread to perform the processing of the sinogram. Finally, node 0 gathers the time spent by each node in the processing of the sinogram. It can be observed in Fig. 3, in the first diagram of step 0.

In the second step, node 0 does a final distribution of workload among nodes. Node 0 decides what is the best choice of cores for each node and what is the optimal workload distribution among nodes and cores. The data subsets are sent from node 0 to each node. Each one receives the new workload and it creates as threads as cores for processing of the different data subset, as Fig. 3 shows in the first diagram of step 1. Thus, each node will have a different subset of reconstructed images (see the last diagram of Fig. 3). These slices or reconstructed images will form the 3D reconstruction of the biological specimen.

In order to evaluate the algorithm eficiency described above, other reconstruction algorithm was implemented. This algorithm has been called HPTomo (Hybrid Parallel algorithm for Tomography). It does not perform the testing step and it takes an equal number of processors in each node. The criterion for the workload distribution is according to the number of processors on each node.

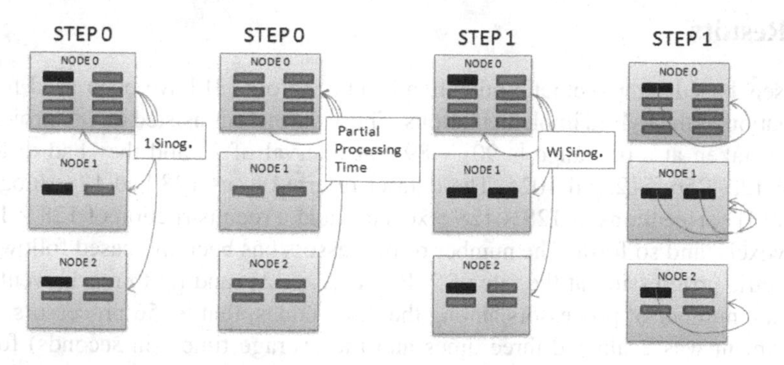

Fig. 3 Diagrams of the different steps that BAHPTomo follows.

3.1.1 Static Load Balancing

The different features of each node of the heterogeneous system should be considered to do a balanced distribution of workload [13, 14, 15]. BAHPTomo takes into account the one hand the time spent in processing a sinogram on each node and on the other hand, the number of processors available on each node. This load balancing strategy will be formalized mathematically below.

We consider a heterogeneous computing system composed of N nodes, where each node is a shared memory system consisting on p_j processors with $j = 0, \ldots, N-1$. TS is the total number of sinograms to be distributed between the different nodes and W_j is partial work assigned to each node. The apportionment of workload W_j proposed in this article can be mathematically expressed as follows:

$$W_j = \frac{k_j * TS}{\sum_{j=0}^{N-1} k_j}, \tag{1}$$

where $k_j = p_j * t_{min}/t_j$ and t_{min} is the time spent by the fastest node for the processing of a sinogram, and t_j is the time spent by each node for processing of the same sinogram.

3.1.2 Optimal Choice of Processors

The nodes of a heterogeneous system can be composed of the same or different number of processors with different computational performances. We will have to decide how many processors choose in each node for obtaining the best performance. Different processor combinations have been tested and we have concluded that to choose the larger number of processors in the fastest node is the optimal choice. This conclussion has been taken into account in our algorithm (BAHPTomo) so that not only the application adapts to the architecture, also the architectures adapts to the application.

4 Results

Datasets based on a synthetic mitochondrion phantom [3] have been used for the evaluation of the hybrid implementations. These datasets consisted of 180 projection images taken at a tilt range $[-90°, +89°]$ at interval of $1°$ and they had different sizes: 128, 256, 512 and 1024. The dataset referred to as 128 had 128 sinograms of 180 1D projections of 128×128 pixels to yield a reconstruction of $128 \times 128 \times 128$ voxels, and so forth. The number of processors has been increased following a geometric progression at the rate of 2, $P \in \{2, \ldots, 32\}$, and up to reach eventually the total number of processors among the three nodes, that is, 56 processors. Each experiment was evaluated three times and the average times (in seconds) for the reading, processing, writing, comunications, testing, balancing and total time were computed.

4.1 Preliminary Study of Our Heterogeneous Cluster

Our models were evaluated in a heterogeneous cluster, which has three nodes with different architectural features. The first node consists on 8 processors Opteron Quad Core, it has 64 GB of RAM and the memory access is NUMA. The second node has 2 processors Intel Xeon Quad Core and it has 16 GB of RAM. Finally, the third node consists on 8 processors Intel Itanium Dual Core, it has 64 GB of RAM and the memory access is NUMA. We can notice in Table 1 that N2 is 1,56 times faster than N1, N2 is 4 times faster than N3 and N1 is 2,6 faster than N3. Several tests have been done concluding that the best distribution of processors is obtained when more processors of the fastest node are chosen. The BAHPTomo algorithms take into account this event.

Table 1 Processing time of 1 sinogram in each node.

Node	Sin. 128x180	Sin. 256x180	Sin. 512x180	Sin. 1024x180
N0: Opteron	0,0745709	0,287073	1,12996	4,56124
N1: Xeon	0,0477531	0,179863	0,700783	2,80184
N2: Itanium	0,193778	0,738971	2,88673	11,5836

4.2 Speed-up for Heterogeneous Clusters

New indicators for the measurement of the performance must be used in heterogeneous environments. We will consider the heterogeneous speed-up suggested in [14], that is, $HS = T_1/T_N$, where T_1 is the execution time in the fastest node, T_N is the execution time using N nodes. Following the same notation that Eq. 1, the ideal value of HS will be HSIdeal $= \sum_{j=0}^{P-1} t_{min}/t_j = \sum_{j=0}^{N-1} t_{min} * p_j/t_j$, where P is the total number of processors.

We can observe in Fig. 4 that BAHPTomo achieves the ideal speedup when the number of processors and the size of datasets are increased. However, we can see

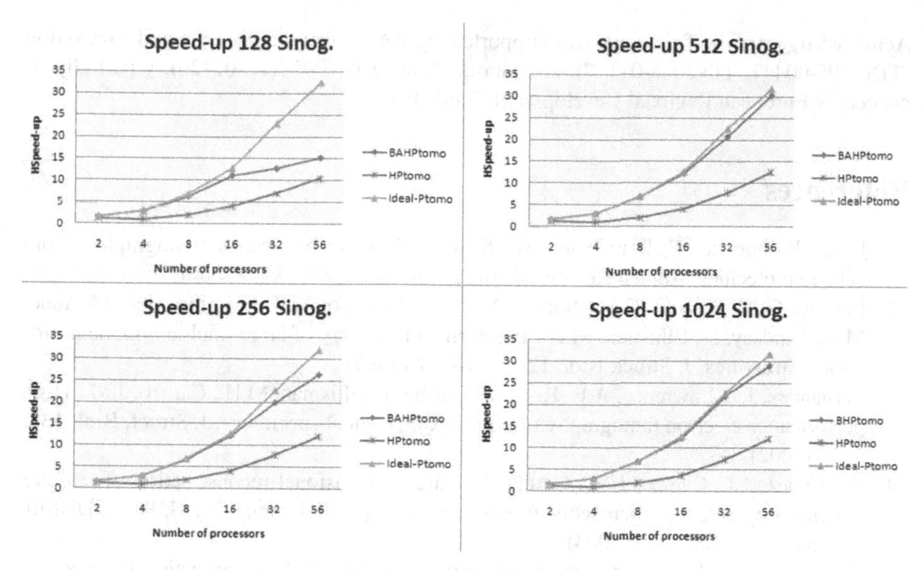

Fig. 4 Speedup for datasets of 128, 256, 512 and 1024 sinograms.

too, if we works with a dataset of 128 and the algorithm is runned in 32 or 56 processors, the speed-up of BAHPtomo decreases. This inflection point occurs because each node has few sinograms to process and then the load balancing does not mean a great advantage between the algorithms. In fact, we can sense that the performance of BAHPtomo converges from 56 processors on.The best speedup obtained is when BAHPtomo algorithm is used. We can obtain a speedup almost of 30 as it can be seen in Fig. 4.

5 Conclusions

In this work, the computational requirements to allow the 3D reconstruction of cellular specimens through ET have been shown. The new tendencies of high performance computing lead us to implement hybrid algorithms in order to exploit the parallelism at node and core levels. So, a hybrid C algorithm have been implemented using MPI and POSIXThread libraries. Static centralized load balancing and optimal choice of processors are taken into account in this algorithm (BAHPtomo) and a balancing method has been proposed. BAHPTomo has been evaluated in a heterogeneous cluster and it was compared with another algorithm (HPTomo) which does not take into account the different features. The results have shown that BAHPtomo algorithm gets an ideal speedup when large datasets are used. In fact, the penalties for testing time are offset by optimal load distribution. Our results demonstrate that to use our suggested balancing method has been crucial to achieve a speedup nearly at 30.

Acknowledgements. This work was supported by the Ministry of Science and Innovation (TIN2005-00447, TIN2008-01117), la Junta de Andalucía (P06-TIC-01426), y partially financed by European Regional Development Fund (ERDF).

References

1. Lucic, V., Foerster, F., Baumeister, W.: Structural studies by electron tomography: From cells to molecules. Annual Review of Biochemistry 74, 833–865 (2005)
2. Perkins, G.A., Renken, C.W., Song, J.Y., Frey, T.G., Young, S.J., Lamont, S., Martone, M.E., Lindsey, S., Ellisman, M.H.: Electron tomography of large, multicomponent biological structures. J. Struct. Biol. 120, 219–227 (1997)
3. Fernández, J.J., Lawrence, A.F., Roca, J., García, I., Ellisman, M.H., Carazo, J.M.: High performance electron tomography of complex biological specimens. J. Struct. Biol. 138, 6–20 (2002)
4. Fernández, J.J., Carazo, J.M., García, I.: Three-dimensional reconstruction of cellular structures by electron microscope tomography and parallel computing. J. Paral. Distrib. Computing 64, 285–300 (2004)
5. Fernández, J.J.: High performance computing in structural determination by electron cryomicroscopy. J. Struct. Biol. 165, 1–6 (2008)
6. Moore, G.E.: Cramming more components onto integrated circuits. Electronics 38(8) (1965)
7. Strohmaier, E., Dongarra, J.J., Meuer, H.W., Simon, H.D.: Recent trends in the marketplace of high-performance computing. Parallel Computing 31, 261–273 (2005)
8. Cappello, F., Etiemble, D.: MPI versus MPI+OpenMP on the IBM SP for the NAS Benchmarks. In: Proc. of the Supercomputing, SC (2000)
9. Rabenseifner, R.: Hybrid Parallel Programming: Performance Problems and Chances. In: Proceedings of the 45th Cray User Group Conference, Ohio, May 12-16 (2003)
10. Rabenseifner, R., Hager, G., Jost, G.: Hybrid MPI/OpenMP parallel programming on clusters of multi-core SMP nodes. In: Proc. of 17th Euromicro Int. Conference on Parallel, Distributed, and Network-Based Processing (PDP 2009), pp. 427–236 (2009)
11. da Silva, M.L., Roca-Piera, J., Fernández, J.J.: Evaluation of Master-Slave Approaches for 3D Reconstruction in Electron Tomography. In: Omatu, S., Rocha, M.P., Bravo, J., Fernández, F., Corchado, E., Bustillo, A., Corchado, J.M. (eds.) IWANN 2009. LNCS, vol. 5518, pp. 227–231. Springer, Heidelberg (2009)
12. Álvarez, J.A., Roca-Piera, J., Fernández, J.J.: From structured to object oriented programming in parallel algorithms for 3D image reconstruction. In: Proceedings of the 8th Workshop on Parallel/High-Performance Object-Oriented Scientific Computing, POOSC (2009), doi:10.1145/1595655.1595656
13. Xiao, L., Zhang, X., Qu, Y.: Effective Load Sharing on Heterogeneous Networks of Workstations. In: Parallel and Distributed Processing Symposium (IPDPS 2000), Mexico, pp. 1–5 (2000), doi:10.1109/IPDPS.2000.846016
14. Martínez, J.A., Garzón, E.M., Plaza, A., García, I.: ADITHE: An approach to optimise iterative computation on heterogeneous multiprocessors. J. Supercomput. (2009), doi:10.1007/s11227-009-0350-1
15. Álvarez, J.A., Roca-Piera, J., Fernández, J.J.: A load balancing framework in multithreaded tomographic reconstruction. Parallel Computing: Architectures, Algorithms and Applications (NIC)

Data Fusion for Face Recognition

Jamal Ahmad Dargham, Ali Chekima, Ervin Moung, and S. Omatu

Abstract. Face recognition is an important biometric because of its potential applications in many fields, such as access control, surveillance, and human-computer interface. In this paper, we propose a rule-based face recognition system that fuses the output of two face recognition systems based on principal component analysis (PCA). One system uses the face image while the other use the Radon transform of the same face image. In addition, both systems use the Euclidean distance is the matching criteria. Both systems are trained using the same training images database, and fed with the same test input image at same time and the recognition result of each system is serving as input for the fusion decision stage. The proposed system is found to be better (97% recognition rate for recall and 93% for reject) than either system alone

1 Introduction

Over the past two decades face recognition has received considerable interest as it is a widely accepted biometric because of its potential applications in many fields, such as access control, surveillance, human-computer interface, and so on. Several methods are available to extract and represent the facial features. One of the widely used representations of the face region is Eigenfaces [3], which are based on principal component analysis (PCA). The goal of the Eigenface method is to linearly project the image space to a feature space of a lower dimension.

Karsili and Acan [4] face recognition system is a hybrid of radon transform and PCA. Radon transform is used to produce a rotational invariant features from images. The rotational invariant features are used as input for kernel PCA. The experiment were evaluated using FERET database and a 75% (rank 1) recognition rate was reported using frontal facial images. Jadhao and Holambe [5] described a

Jamal Ahmad Dargham, Ali Chekima, and Ervin Moung
School of Engineering and Information Technology, University Malaysia Sabah,
Locked Bag 2073, Teluk Likas, 88999 Kota Kinabalu, Sabah, Malaysia
email: {jamalad,chekima}@ums.edu.my, menirva.com@gmail.com

S. Omatu
Faculty of Engineering, Department of Electronics, Information and Communication Engineering, Osaka Institute of Technology, 5-16-1, Omiya, Asahi-ku, Osaka, 535-8585, Japan
email: omatu@rsh.oit.ac.jp

A.P. de Leon F. de Carvalho et al. (Eds.): Distrib. Computing & Artif. Intell., AISC 79, pp. 681–688.
springerlink.com

new technique for face recognition which uses Radon transform and Fourier transform to derive the directional spatial frequency features. The Radon transform is used for the line integral of features and the DFT is used to enhance the features. They used FERET and ORL databases and achieved recognition rates 99% and 97%, respectively.

Jamikaliza et al. [6] face recognition method used PCA as feature extractor and Fuzzy ArtMap (FAM) neural network as classifier. The advantage of their method is whenever a new data is added to the database, training stage is not needed. Recognition rate of 97% is reported using local dataset and 98% using ORL dataset. Chunming et al. [7] proposed a face recognition method called Statistical PCA. Their method is based on 2D matrices rather than 1D vectors image and have some advantages over normal PCA; simple computation and has a high recognition rate (95%) on CVL face database. In this paper, a data fusion of two face recognition system is discussed. The system is discussed in more detail in section 3.

1.1 Radon Transform

The Radon Transform (RT) is one of the techniques used to detect features within an image. Let f(x,y) be an image function, the Radon transform is defined as:

$$R(\theta,\rho) = \int_{-\infty}^{\infty} (\rho\cos\theta - S\sin\theta, \rho\sin\theta + S\cos\theta)dS \tag{1}$$

where S is defined as integral along a line through the image, θ is the angle and ρ is the distance of the line from the origin of the coordinate system as shown in Fig. 1.

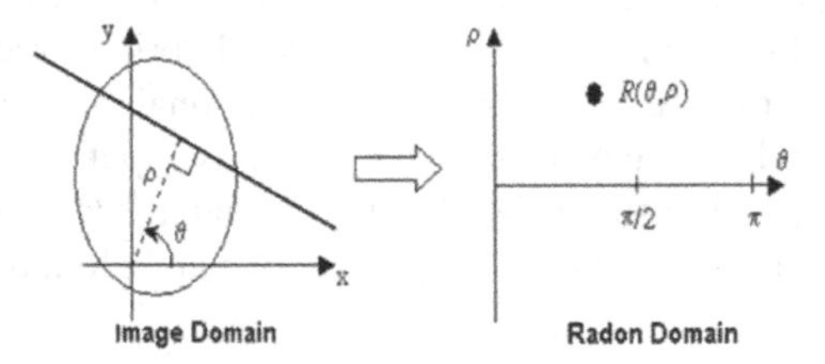

Fig. 1 Image domain and Radon domain [8].

1.2 Principal Component Analysis

Principal Component Analysis (PCA) is a statistical linear transform method [1], often used for data analysis and identifying patterns in data. PCA can be used for data compression by reducing the number dimensions without much loss of information. One of the most widely used representations of the face region is eigenface . Eigenface method uses PCA to linearly project the image space to a low dimensional feature space.

Consider a set of M sample data, $\mathbf{X} = [\, \mathbf{x}_1\, \mathbf{x}_2 \ldots \mathbf{x}_M \,]$, where $\mathbf{x}_i$ is a 1-D vector image and each data with n-dimensional space. The total scatter matrix $\mathbf{S}_t$ of $\mathbf{X}$ defined as,

$$S_t = \sum_{i=0}^{M} (x_i - \mu)(x_i - \mu)^T \tag{2}$$

where $\mu \in \Re^n$ is the mean image of all samples. Finding a set of n-dimensional eigenvectors of $\mathbf{S}_t$: $\mathbf{S}_t\mathbf{V}_j = \lambda_j \mathbf{V}_j$ where j = 1, 2, …, k and k < M. $\mathbf{V} = [\, \mathbf{V}_1\, \mathbf{V}_2 \ldots \mathbf{V}_k \,]$ is a set of n-dimensional eigenvectors of $\mathbf{S}_t$ corresponding to first k largest eigenvalues $[\, \lambda_1\, \lambda_2 \ldots \lambda_k \,]$.

$\mathbf{V}$ were then normalized so that $\|\mathbf{V}_j\| = 1$. $\mathbf{V}_j$ has the same dimension as the original image. Thus they are referred as eigenfaces.

The linear transformation mapping the original n-dimensional image to k-dimensional feature space defined as $\mathbf{Y}_i = \mathbf{V}^T \times \mathbf{X}_i$. $\mathbf{Y}_i \in \Re^k$, $\mathbf{X}_i \in \Re^n$, where k < n and $\mathbf{V} \in \Re^{nxk}$.

2 Face Databases

A total of 500 frontal face images were selected from the FERET database. They represent 200 different individuals. 100 individuals are used for training and recall testing, and the other 100 different individuals are used for reject testing only. All the 500 images were cropped to get only the desired face part of each person (from forehead to the chin). All images have been adjusted so that both eyes of an individual are aligned in the same horizontal line. All images are resized to 60x60 pixels.

2.1 Training Database

Each of the 100 persons in the training and testing database (Known Database) has four images. Three images will be used for training and one image will be used for testing. Fig. 2 shows an example of training and testing database. The four images are of the same person. The three images on the left, Fig. 2(a), will be used for training while the image on the right, Fig. 2(b), will be used for testing.

2.2 Testing Databases

Two testing databases were created. The first database, known test database, has 100 images of the 100 persons in the training database. This database will be used

(a) (b)

Fig. 2 An example of the known database.

to test the recall capability of the face recognition system. The second database, unknown test database, has also 100 images of 100 different persons. This database will be used to test the rejection capability of the system.

3 Results and Discussion

3.1 PCA (Without Radon Transform) Based System

Fig. 3(a) shows the block diagram of eigenface based face recognition method. A 2-D face images are concatenated to form 1-D image vectors. A zero mean 1-D training images set are computed. PCA is then applied on the collection of 1-D zero-mean images set vector to produce a low-dimensional features vector.

3.2 PCA (With Radon Transform) Based System

Fig. 3(b) shows the block diagram of Radon Transform (RT)+PCA based face recognition method. The Radon transform will be applied on the image to compute its 2-D projection image along angles varying from 0 to 180°. The result of the projection is the sum of the intensities of the pixels in each direction. All the projections of the image are concatenated to form 1-D radon transform's vector.

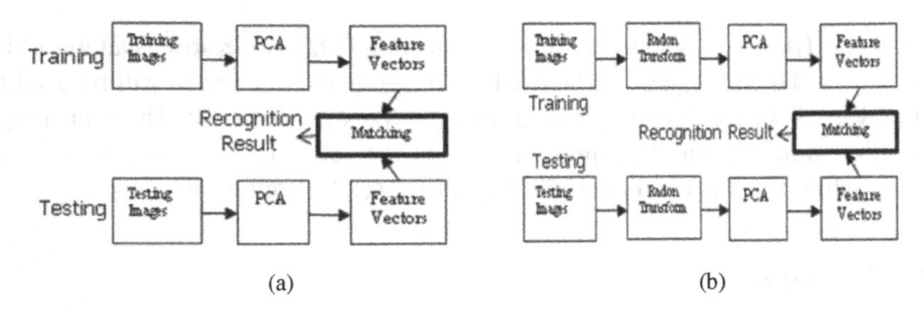

(a) (b)

Fig. 3 (a) Block diagram of PCA (without Radon Transform) face recognition system (b) Block diagram of PCA (with Radon Transform) face recognition system

The 1-D radon transform's vectors for all training images are computed. PCA is then applied on the collection of 1-D radon transform's vector to produce a low-dimensional features vector.

3.3 Matching Task

For the matching task of both systems, the Euclidean distance is used. If the Euclidean distance between test image y and image x in the training database, d(x,y) is smaller than a given threshold t, then images y and x are assumed to be of the same person. The threshold t is the largest Euclidean distance between any two face images in the training database, divided by a threshold tuning value (Tcpara) as given in equation (3).

$$t = \frac{\left[\max\left\{\left\|\Omega_j - \Omega_k\right\|\right\}\right]}{Tcpara} \tag{3}$$

where j, k = 1, 2, ..., M. M is the total number of training images, and Ω is the reduced dimension images.

To measure the performance of both systems, several performance metrics are used. These are:

i. For Recall
   - ***Correct Classification.*** If a test image $\mathbf{y}_i$ is correctly matched to an image $\mathbf{x}_i$ of the same person in the training database.
   - ***False Acceptance.*** If test image $\mathbf{y}_i$ is incorrectly matched with image $\mathbf{x}_j$, where ***i*** and ***j*** are not the same person
   - ***False Rejection.*** If image $\mathbf{y}_i$ is of a person ***i*** in the training database but rejected by the system.

ii. For Reject
   - ***Correct Classification.*** If $\mathbf{y}_i$, from the unknown test database is rejected by the program
   - ***False Acceptance.*** If image $\mathbf{y}_i$ is accepted by the program.

3.4 Setting the Threshold Tuning Parameter

The value of the threshold tuning parameter can be used to tune the performance of the system to have either high correct recall with high false acceptance rate for application such as boarder monitoring or high correct rejection rate for unknown persons for application such as access control. For this work, the threshold tuning parameter was set so that the system has equal correct recall rate and correct rejection rate. The Tcpara value mentioned in equation (3) that was chosen for each system is shown in Fig. 4(a) and Fig. 4(b).

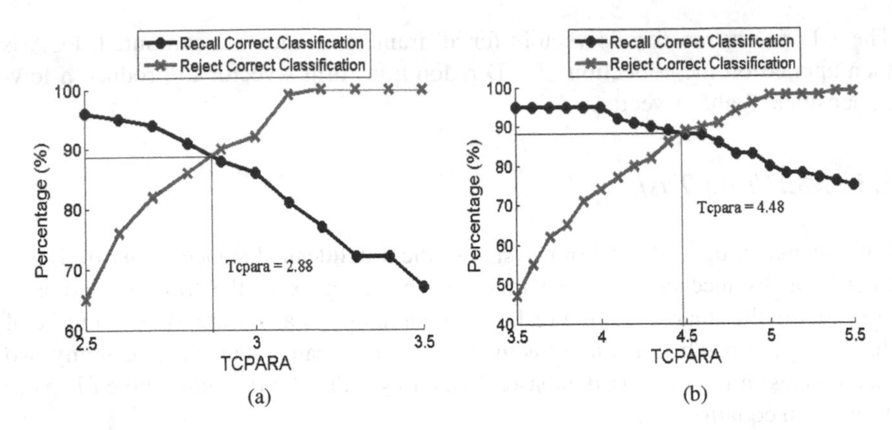

Fig. 4 Correct Classification result for (a) PCA (without Radon Transform) system and (b) PCA (with Radon Transform) system

3.5 Data Fusion Stage

Fig. 5 shows the block diagram of the fusion of both face recognition method mentioned in 3.1 and 3.2. The output of each system will serve as input for the fusion decision stage. The fusion decision stage is a module that consists of several rules. The rules are:

- If both systems report a match is found, then the fusion system reports a match is found.
- If both systems report a match not found, then the fusion system reports a match not found.
- If either system reports a match is found while the other system reports a no match is found, then the fusion system reports a match not found

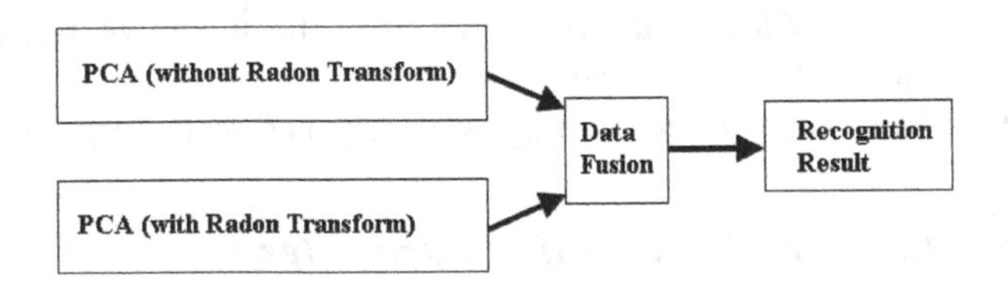

Fig. 5 Block diagram of the whole system.

The fusion decision rules can be summarize as an OR operator as shown in Table 1. A 0 in Table 1 indicates a no match found while a 1 indicates a match is found.

Experiments are performed using the fusion decision rule mentioned above and the results are shown in Fig.6. As can be seen from Fig. 6, the fusion of both

Table 1 Fusion Decision Rules

PCA	PCA with RT	Fusion System
0	0	0
1	0	1
0	1	1
1	1	1

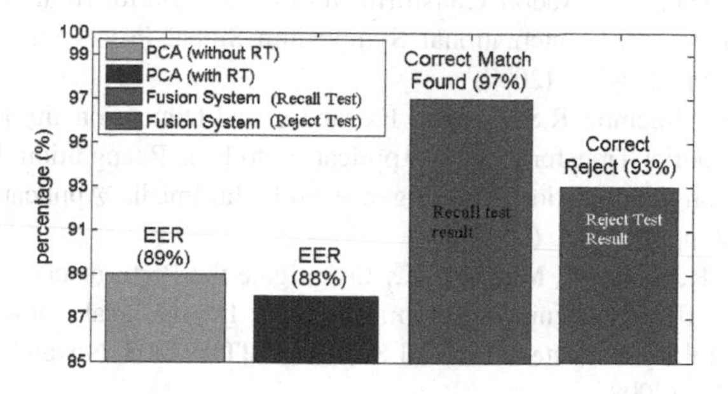

Fig. 6 Comparison of recognition results.

Table 2 Comparison of several PCA-based face recognition methods on FERET database.

Methods	Recognition Rate
(Radon+Kernel PCA) – Karsili and Acan [4]	75%
(Gabor-based kernel PCA) - Chengjun Liu [2]	99.5%
Proposed method	97% (recall), 93% (reject)

systems increases the recognition rate to 97% for recall and 93% for reject. Although, it is difficult to make meaningful comparisons between any two systems unless tested on the same test bed, we have compared our system with two similar systems reported in the literature as shown in Table 2.

4 Conclusion

In this paper a face recognition system based on principal component analysis (PCA) that fuses the eigenvectors from the face and the Radon transform of the same face is proposed. The proposed system fuses both systems using a set of rules. Experimental results on the FERET database have shown that the proposed system outperforms each of the individual system when used separately.

References

[1] Jolliffe, I.T.: Principal Component Analysis, 2nd edn. Series: Springer Series in Statistics, vol. XXIX, 487, p. 28. Springer, NY (2002)

[2] Liu, C.: Gabor-Based Kernel PCA with Fractional Power Polynomial Models for Face Recognition. IEEE Transactions on Pattern Analysis and Machine Intelligence 26(5), 572–581 (2004)

[3] Turk, M., Pentland, A.: Eigenfaces for Recognition. Journal of Cognitive Neuroscience 3(1), 71–86 (1991)

[4] Karsili, L., Acan, A.: A Radon Transform and PCA Hybrid for High Performance Face Recognition. In: IEEE International Symposium Signal Processing and Information Technology, pp. 246–251 (2007)

[5] Jadhao, D.V., Holambe, R.S.: Feature Extraction and Dimensionality Reduction Using Radon and Fourier Transforms with Application to Face Recognition. In: International Conference on Computational Intelligence and Multimedia Applications, December 13-15, vol. 2, pp. 254–260 (2007)

[6] Abdul, K.J., Rubiyah, Y., Marzuki, K.: Investigate the Performance of Fuzzy Artmap Classifier for Face Recognition System. In: IEEE International Conference on Signal Image Technology and Internet Based Systems, SITIS 2008, November 30-December 3, pp. 254–259 (2008)

[7] Chunming, L., Yanhua, D., Hongtao, M., Yushan, L.: A Statistical PCA Method for Face Recognition. In: Second International Symposium on Intelligent Information Technology Application, IITA 2008, December 20-22, pp. 376–380 (2008)

[8] http://idlastro.gsfc.nasa.gov/idl_html_help/RADON.html

Object Signature Features Selection for Handwritten Jawi Recognition

Mohammad Faidzul Nasrudin, Khairuddin Omar, Choong-Yeun Liong, and Mohamad Shanudin Zakaria

Abstract. The trace transform allows one to construct an unlimited number of image features that are invariant to a chosen group of image transformations. Object signature that is in the form of string of numbers is one kind of the transform features. In this paper, we demonstrate a wrapper method along with several ranking evaluation measurements to select useful features for the recognition of handwritten Jawi images. We compare the result of the recognition with those obtained by using methods where features are randomly selected or no feature selection at all. The proposed methods seem to be most promising.

Keywords: feature selection, object signature, handwritten Jawi recognition, trace transform.

1 Introduction

The object signature feature developed by [1][2] has shown its usefulness for various applications such as face recognition [3][4], Korean character recognition [5] and image database retrieval [2]. The object signature feature is invariant to affine distortion. It is based on the trace transform that theoretically allows us to use an unlimited number of features, which are mathematically appropriate, but perceptually indescribable. Although most of them will not be useful; nevertheless, one can investigate the features and make the appropriate choice for the specific task with the help of experimentation

Feature selection methods in classification can be divided into three categories [6]. The first category, referred to as filter, defined as a preprocessing step and can be independent from learning. A filter method assesses the relevance of features

Mohammad Faidzul Nasrudin, Khairuddin Omar, and Mohamad Shanudin Zakaria
Centre for Artificial Intelligence Technology (CAIT), Faculty of Information Science and Technology, Universiti Kebangsaan Malaysia, 43600 UKM Bangi, Selangor D.E., Malaysia
e-mail: {mfn,ko}@ftsm.ukm.my, msz@ftsm.ukm.my

Choong-Yeun Liong
Centre for Modelling and Data Analysis (DELTA), School of Mathematical Sciences, Faculty of Science and Technology, Universiti Kebangsaan Malaysia, 43600 UKM Bangi, Selangor D.E., Malaysia
e-mail: lg@ukm.my

A.P. de Leon F. de Carvalho et al. (Eds.): Distrib. Computing & Artif. Intell., AISC 79, pp. 689–698.
springerlink.com

by looking at the intrinsic properties of the data [7]. This method calculates a score for each feature and then selects features according to the scores [8]. Information gain and chi-square, according to [9][10], are among the most effective filter methods of feature selection for classification. In the second category, which is named wrapper [11], utilizes the learning system as a black box to score subsets of features. In wrapper, a search procedure in the space of possible feature subsets is defined, and various subsets of features are generated and evaluated by a specific classification algorithm. The third category called the embedded method [12] performs the feature selection within the process of training.

Concerning the selection of the object signature, to the best of our knowledge, there have not been any work dedicated to it. The challenge in this study is that the object signature descriptor is a string of numbers, which is like a signature of the object. The task of identifying an object is by comparing two strings of numbers (one from the test image and the other from the reference image), that are circularly shifted and possibly scaled versions of each other. The only reported method for object signatures comparison is by computing their correlation coefficient for all possible shifts [1]. To express this as a distance, the inverse cosine of the maximum value of the correlation coefficient is taken. The distances are ranked and the smallest number indicates the two most similar signatures. Since the shifting value varies from image to image, existing filter feature selection methods for classification are not suitable for the object signature. Obviously, wrapper method that "wrapped" around the classification model is a preferred choice. In such case, the classification model refers to similarity measurement between two "signatures".

Jawi is a cursive script that was derived from the Arabic alphabets and was adopted for the use of Malay language writing. Jawi can largely be found in old Malay manuscripts that have not been fully digitized yet. Features based on the trace transform has shown its effectiveness for printed Jawi character recognition [13]. In the study, features were selected based on manual selection with trial approach. Instead in this study, we would like to propose a wrapper that evaluates the usefulness of feature subsets based on recognition performance determined by ranking evaluation measurements, the mean average precision (MAP) [14] and normalized discounted cumulative gain (NDCG) [15]. The recognition performance of the feature subset from the wrapper method is compared to the recognition performance from other selection methods. In terms of data, we use handwritten Jawi images instead of the printed ones.

In Section 2 we present the background to object signature from the trace transform. In Sections 3 and 4, we elaborate on the experiments and the results and discussion, respectively. We conclude in Section 5.

2 Object Signature from the Trace Transform

Let us imagine an image $f(x, y)$ criss-crossed by all possible lines $l(r,\theta,t)$ that one can draw on it (refers Fig.1). Let $L(r,\theta)$ denote the set of all lines, the trace transform is a function $g(T,f,r,\theta)$ defined on $L(r,\theta)$ with the help of trace functional T (some functional of the image function $f(x, y)$ when it is considered along the line $l(r,\theta)$, as a function of parameter t), then

$$g(T, f, r, \theta) = T\left[f(r, \theta, t)\right] \tag{1}$$

One then calculates another functional, P, along the columns of the transform, i.e. over parameter r, and finally a functional Φ over the string of numbers created this way, i.e. over parameter θ [16][17][18]. The result is a single number called triple feature Π that is defined as:

$$\Pi(f) = \Phi\left[P\left[T\left[f(r, \theta, t)\right]\right]\right] \tag{2}$$

where Π represents the extracted triple feature of image $f(x, y)$.

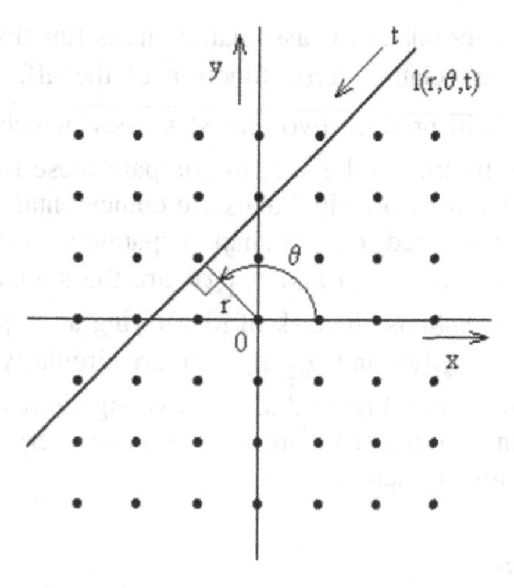

Fig. 1 Definition of the parameters of an image $f(x, y)$ and tracing line $l(r, \theta, t)$

The extracted triple feature is influenced strongly by the properties of the chosen functionals T, P and Φ. For practical applications, such as feature extraction, these functionals may be chosen so that the triple feature has intended properties such as invariant to affine distortion. Using an appropriate combination of the functionals T, P and Φ selected by a suitable feature selection method, thousands of triple features can be generated. For example, [11] and [19] proposed the functionals one should use in order to produce features invariant to rotation, translation and scaling for fish image database retrieval and insect footprint recognition respectively.

In this paper we are not going to attempt to characterize an object by a single number produced by the cascaded application of the three carefully chosen functional T, P and Φ, but instead we are going to use only the first two functionals. Using only functionals T and P will allow us to characterize an object by a string of numbers. The object signature is a function, called the associated circus, $h_a(\phi)$, defined in terms of the function $h(\phi)$, which is produced by applying functionals T and P:

$$h_a(\phi) \equiv |h(\phi)|^{-1/(\lambda_P K_T - K_P)} \tag{3}$$

Parameters λ_P, K_T and K_P are some real valued numbers which characterize functionals T and P (for details refer to [2]). If $\lambda_P K_T - K_P = 0$, the associated circus is defined as:

$$h_a(\phi) \equiv \sqrt{\left|\frac{dh(\phi)}{d\phi}\right|} \tag{4}$$

If we plot in polar coordinates the associated circus function of the original image, $h_{a1}(\phi)$, and the associated circus function of the affinely distorted image, $h_{a2}(\phi)$, the functions will produce two closed shapes, which are connected by a linear transformation. In order to be able to compare these two shapes, they have to be normalized so that their principal axes are coincidental. This can be done by a linear transformation applied to each shape separately as described in detail in [1]. The normalized shapes $h_{n1}(\phi)$ and $h_{n2}(\phi)$ are the signatures of the two images. For practical applications, the task of identifying an object is just comparing two strings of numbers, $h_{n1}(\phi)$ and $h_{n2}(\phi)$, that are circularly shifted and possibly scaled versions of each other. Figure 2 shows two signatures in polar coordinates of a Jawi character in two different font types that are very similar in shape and differ mainly by rotation and scaling.

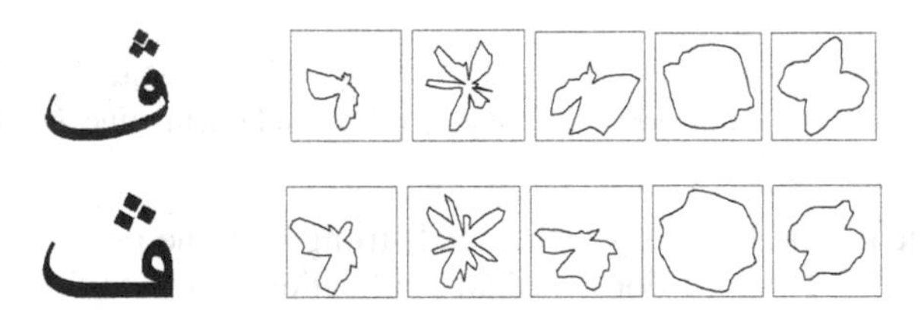

Fig. 2 The signatures of Jawi character "Pa" in two different font types

3 Experiments on Handwritten Jawi Recognition

Generally, we divided all the experiments into three steps. Firstly, we performed the feature selection based on the proposed ranking evaluation measurements. Then, based on those selected features subset, we run another experiment to compare their recognition performance with other four methods of choosing feature subsets. Lastly, we run an experiment to recognize all collected handwritten images using the best feature subset.

3.1 Data

We collected nine sets of scanned articles written by nine different writers. Each article was designed such that all possible combinations of 36 Jawi characters exist at least once in the text. This was to ensure that all kinds of character combinations were tested since character shape changes depending on the character position in a word. All the scanned pages were decomposed into a set of 4835 sub word images using Connected Component segmentation algorithm. Each article contains 213 distinct sub words. In order to reduce the time and magnitude of the computation, for feature selection, we randomly selected only 2 sets of images. Each set contains 213 of those distinct sub word images. One as the test set and another as the reference set. For the comparison experiments, we generated another 16 sets of images randomly. Eight of the image sets were used for testing and the rest as the reference. For the final experiment, all the 4835 sub word images were randomly divided into 9 sets of images for cross validation procedure.

3.2 Feature Extraction

For the trace transform method we computed the object signature, h_a, by applying the functionals T and P. We tested seven different T functionals and eleven P functionals. The T functionals were:

- T_1: Integral of $f(t)$, where $f(t)$ is the value of the image function along the tracing line;
- T_2: Max of $|f(t)|$;
- T_3: Integral of $|f'(t)|$;
- T_4: Integral of $|f''(t)|$;
- T_5: Lp quasi-norm ($p = 0.5$) = q^2, where q = integral of $\sqrt{|f(t)|}$;
- T_6: Median R+: $f(t-c)$ where c is the median abscissa;
- T_7: Weighted Median R+: $f(t-c)$ where c is the median abscissa and the weights are $|f(t)|(t-c)$.

The first seven P functionals are the same as T_1 to T_7, called P_1 to P_7, respectively. In addition, the following four P functionals were used:

- P_8: t – (median index dividing the integral of $|f(t)|$);
- P_9: (Average of t) – (index max of $|f(t)|$);
- P_{10}: t – (gravity center of $|f(t)|$);
- P_{11}: t – (median index dividing integral of $\sqrt{|f(t)|}$);

where $f(t)$ is the value of the image at sample point t along the tracing line. In the definitions of T functionals, R+ means that the integration is over the positive values of the variable of integration. The explanation on the weighted median can be found in [1]. We then generated all possible pairs using these seven T functionals and eleven P functionals. In total, there were $7 \times 11 = 77$ candidate circus functions or signatures to characterize an image.

Each image was traced by lines one pixel apart, i.e. the value of parameter p for two successive parallel lines used differed by 1. For each value of p, 48 different orientations were used. This means that the orientations of the lines with the same p differed by 7.5 degrees. Each line was sampled with points one pixel apart, that is to say parameter t took discrete values with steps equal to 1 inter-pixel distance.

For the comparison of two signature values, we computed a novel distance measure called normalized circular cross-distance, a modified version of the normalized circular cross correlation function used in [20]. The multiplicative expression in the original normalized circular cross correlation function is substituted by a distance measure. The normalized circular cross-distance, *NCXD*, is defined as:

$$NCXD(d) = \sum_{i=1}^{N} \left| \frac{h_{at}(i)}{\sqrt{\sum_{i=1}^{N} h_{at}(i)^2}} - \frac{h_{al}(i-d)}{\sqrt{\sum_{i=1}^{N} h_{al}(i)^2}} \right| \quad (5)$$

where h_{at} and h_{al} are the signature values of the test image and the reference image respectively, N is the length of the signatures and d is the shift. Two signatures are most similar when the *NCXD* is minimum. We chose the minimum value over 48 shifts (equals to 48 different orientations used). Then, we used the sum of these *NCXD* numbers across all signatures as measure of similarity of two images. We then ranked the numbers. The smallest number indicates the two most similar images.

3.3 Feature Selection

Not all signatures are useful. Features that have all zeros or all in one fixed value will be discarded. For the rest, we applied the hill climbing search on a test set for selecting feature subsets. Given a set of features $S_1 = \{f_1, \hbar\ , f_n\}$ the algorithm works as follows:

1. Start with the feature f_s that individually performs best on the test set and put it into the set of best features B_1; then set $S_2 = S_1 \setminus \{f_s\}$.
2. For $k = 2, \ldots, n$ do:
   2.1 Evaluate the ranking performance of $B_{k-1} \cup \{f_k\}$ on the test set.
   2.2 If the set produces lower MAP or NDCG value, then the $\{f_k\}$ is discarded. Otherwise, $\{f_k\}$ is added to the set of best features $B_k = B_{k-1} \cup \{f_k\}$ and set $S_{k+1} = S_k \setminus \{f_k\}$.

When all object signatures are tested, it terminates. The feature subset B_k finally selected is the one that performs best among all subsets considered by the algorithm.

Each run generated results in ranks. For that we adopted two widely-used measures in evaluation of ranking methods for information retrieval, which are MAP and NDCG. MAP is a measure on precision of ranking while assuming that there are two types of item, which are positive (relevant) and negative (irrelevant). Precision at n measures the accuracy of the top n sorted items and is defined as:

$$P(n) = \frac{I_n}{n} \quad (6)$$

where I_n is the number of positive items within the top n. Average precision of a query, AP, is defined as:

$$AP = \sum_{n=1}^{N} \frac{P(n) \times pos(n)}{I} \quad (7)$$

where n represents position, N is the number of results, $pos(n)$ is a binary function indicating whether the item at position n is positive, and I is the number of positive items. MAP is defined as AP averaged over all sorted items.

NDCG is designed for measuring ranking accuracies when there are multiple levels of relevant judgment. NDCG at position n in sorted items is defined as:

$$NDCG(n) = Z_n \sum_{j=1}^{n} \frac{2^{R(j)} - 1}{\log(1+j)} \quad (8)$$

where n denotes position, $R(j)$ denotes score for rank j, and Z_n is a normalization factor to guarantee that NDCG is equal to 1 in a perfect ranking. In evaluation, NDCG is further averaged over all queries.

After we got the selected feature subset from the MAP and NDCG, we then compared them with four other feature subsets. The first subset called All-77, consist of all possible signatures which is 77 in total. The signatures of the other three subsets, Random-1, Random-2 and Random-3, are randomly generated. The number of feature chosen is relatively equal to the number of feature selected by the MAP and NDCG methods. Therefore, the number of features chosen for the random methods are 25, 23 and 23 respectively.

4 Results and Discussion

Each test image is correlated against each of the 213 reference images. The correlation result is put in a rank that is measured by the MAP and NDCG. Feature that increases the MAP or NDCG value will be selected. The feature selection experiment showed that 20 and 23 out of the 77 features were selected by the MAP and NDCG methods respectively. Based on those selected feature subsets, we then run comparison experiments with the other four feature subsets. Each method

Table 1 Average percentage of correct sub words recognition and the number of feature used based on the MAP, NDCG, All-77, Random-1, Random-2 and Random-3 feature subsets

Method	Average percentage of correct recognition					Number of features
	Top-1	1 – 5	6 – 10	11 – 15	16 –	
MAP	24.46	41.58	8.92	5.89	43.61	20
NDCG	24.99	42.15	8.08	6.10	43.66	23
All-77	21.13	36.36	9.34	7.67	46.64	77
Random-1	13.04	22.43	7.67	7.67	62.23	25
Random-2	14.82	27.07	7.25	7.62	58.06	23
Random-3	12.99	22.80	6.88	6.73	63.59	23

Table 2 Average percentage of correct sub words recognition based on MAP and NDCG using 9-fold cross-validation on all sub word images

Method	Average percentage of correct recognition				
	Top-1	1 – 5	6 – 10	11 – 15	16 –
MAP	60.19	78.34	6.18	3.66	11.82
NDCG	60.16	78.20	6.02	3.64	12.14

will produce 9 results and the average computed. The results of the comparison experiments on sub words recognition based on the feature subsets selected by the MAP, NDCG and the four other methods are presented in Table 1. The result of the final experiment using 9-fold cross-validation on all the sub word images set is presented in Table 2.

From the results presented in Table 1, we can see that the best recognition method is the one based on the features selected by MAP and NDCG. Using all possible features (All-77) from the trace transform functionals combination for the recognition had produced a lower result. Using a random approach is not the best way either. From Table 2, the recognizer based on features selected by MAP and NDCG has shown decent results for the recognition of the handwritten Jawi images in the test set. The methods had recognized up to 78.34% of all the sub words images for the top-5 recognition.

5 Conclusion

Feature selection of object signature based on the trace transform for the recognition of handwritten Jawi images has been demonstrated. Object signature theoretically allows us to use an unlimited number of features. Using a proper feature selection method, thousands of relevant features can easily be selected. We proved that using a wrapper method along with several ranking evaluation measurements (MAP and NDCG) to select useful features is better than random selections or no feature selection at all. These feature selection schemes have greatly enhanced the applicability of the trace transform based method.

Acknowledgments. The authors would like to thank the University for the Research Grants No. UKM-GUP-TMK-07-02-034 and UKM-AP-ICT-17-2009.

References

1. Kadyrov, A., Petrou, M.: Object Signatures Invariant to Affine Distortions Derived from the Trace Transform. Image and Vision Computing 21(13-14), 1135–1143 (2003)
2. Kadyrov, A., Petrou, M.: Object Descriptors Invariant to Affine Distortions. In: Proceedings BMVC 2001, Manchester, UK, vol. 2, pp. 391–400 (2001)
3. Srisuk, S., Petrou, M., Kurutach, W., Kadyrov, A.: Face Authentication using the Trace Transform. In: Proceedings of the 2003 IEEE Computer Society Conference on Computer Vision and Pattern Recognition, vol. 1, pp. 305–312 (2003)
4. Srisuk, S., Petrou, M., Kurutach, W., Kadyrov, A.: A Face Authentication System using the Trace Transform. Pattern Analysis and Applications 8(1-2), 50–61 (2005)
5. Kadyrov, A., Petrou, M., Park, J.: Korean Character Recognition with the Trace Transform. In: Proceedings of the International Conference on Integration of Multimedia Contents, ICIM 2001, Chosun University, Gwangju, South Korea, November 15, pp. 7–12 (2001)
6. Guyon, I., Elisseeff, A.: An Introduction to Variable and Feature Selection. Journal of Machine Learning Research 3, 1157–1182 (2003)
7. Saeys, Y., Inza, I., Larranaga, P.: A Review of Feature Selection Techniques in Bioinformatics. Bioinformatics 23(19), 2507–2517 (2007)
8. Mladenic, D., Grobelnik, M.: Feature Selection for Unbalanced Class Distribution and Naïve Bayes. In: Proceedings of the Sixteenth International Conference on Machine Learning (ICML), pp. 258–267 (1999)
9. Yang, Y., Pedersen, J.O.: A Comparative Study on Feature Selection in Text Categorization. In: Proceedings of 14th International Conference on Machine Learning, pp. 412–420 (1997)
10. Forman, G.: An Extensive Empirical Study of Feature Selection Metrics for Text Classification. Journal of Machine Learning Research 3, 1289–1305 (2003)
11. Kohavi, R., John, G.H.: Wrappers for Feature Selection. Artificial Intelligence 97, 273–324 (1997)
12. Breiman, L., Friedman, J.H., Olshen, R.A., Stone, C.J.: Classification and Regression Trees. Wadsworth and Brooks, Pacific Grove (1984)
13. Nasrudin, M.F., Omar, K., Liong, C.-Y., Zakaria, M.S.: Invariant Features from the Trace Transform for Jawi Character Recognition. In: Omatu, S., Rocha, M.P., Bravo, J., Fernández, F., Corchado, E., Bustillo, A., Corchado, J.M. (eds.) IWANN 2009. LNCS, vol. 5518, pp. 256–263. Springer, Heidelberg (2009)
14. Yates, R.B., Neto, B.R.: Modern Information Retrieval. Addison Wesley, Redwood City (1999)
15. Jarvelin, K., Kekalainen, J.: Cumulated Gain-based Evaluation of IR Techniques. ACM Transactions on Information Systems 20(4), 422–446 (2002)
16. Kadyrov, A., Petrou, M.: The Trace Transform and Its Applications. IEEE Transactions on Pattern Analysis and Machine Intelligence, PAMI 23, 811–828 (2001)

17. Kadyrov, A., Petrou, M.: The Trace Transform as a Tool to Invariant Feature Construction. In: Proceedings ICPR 1998, Brisbane, Australia, pp. 1037–1039 (1998)
18. Kadyrov, A., Fedotov, N.: Triple Features. Pattern Recognition and Image Analysis: Advances in Mathematical Theory and Applications 5(4), 546–556 (1995)
19. Shin, B.S., Cha, E.Y., Cho, K.W., Klette, R., Woo, Y.W.: Effective Feature Extraction by Trace Transform for Insect Footprint Recognition, MI-tech Report Series, Computer Science Department, The University of Auckland, New Zealand, Multimedia Imaging Report 12 (2008)
20. Azarnasab, E.: Robot-in-the-loop Simulation to Support Multi-Robot System Development: A Dynamic Team Formation Example, M.Sc. Thesis, College of Arts and Sciences, Georgia State University (2007)

Author Index